Introductory Algebra

SEVENTH EDITION

GO TO STUDENT SERVICES FINANCIAL AID

Introductory Algebra

SEVENTH EDITION

Marvin L. Bittinger
Indiana University—Purdue University at Indianapolis

Mervin L. Keedy
Purdue University

ADDISON-WESLEY PUBLISHING COMPANY

Reading, Massachusetts • Menlo Park, California • New York
Don Mills, Ontario • Wokingham, England • Amsterdam
Bonn • Sydney • Singapore • Tokyo • Madrid
San Juan • Milan • Paris

Sponsoring Editor	Jason A. Jordan
Special Projects Editor	Susan Gleason
Managing Editor	Karen Guardino
Production Supervisor	Jack Casteel
Marketing Manager	Kate Derrick
Text Design	Bruce Kortebein, The Design Office
Art, Design, Editorial, and Production Services	Geri Davis and Martha Morong, Quadrata, Inc.
Prepress Buying Manager	Sarah McCracken
Art Buyer	Joseph Vetere
Electronic Illustration	Scientific Illustrators, Precision Graphics, and Monotype, Inc.
Manufacturing Manager	Roy Logan
Cover Design Director	Peter Blaiwas
Cover Design	Marshall Henrichs
Composition	Beacon Graphics Corporation
Printer	Banta Company

Photo Credits

1, © The Telegraph Colour Library, FPG International **41,** Comstock **44,** NASA **67,** Thia Konig, © Tony Stone Worldwide **111,** J. Zuckerman, Westlight **152,** Mackson, FPG International **166,** Tom Tracy, FPG International **177,** Phototake **191,** NASA **197,** The Image Bank **255,** © James Watt, WaterHouse Stock Photography **305,** Bill Hickey, The Image Bank **319,** Tony Stone Worldwide **365,** Jon Riley, Tony Stone Worldwide **368,** © Jeffrey Sylvester, FPG International **391,** Larry Keenan Assoc., The Image Bank **451,** C. G. Randall, FPG International **475,** Denny's **484,** Terry Vine, Tony Stone Worldwide **485,** The Image Bank **493,** Tony Stone Worldwide **501,** William Warren, Westlight **516,** The Image Bank **545 (top),** Uniphoto New York **545 (bottom),** The Image Bank **553,** D. Phillipe, FPG International **580,** Clyde Smith, FPG International **586,** Comstock **593,** CEZUS, FPG International

Reprinted with corrections, April 1995.

Library of Congress Cataloging-in-Publication Data

Bittinger, Marvin L.
Introductory Algebra/Marvin L. Bittinger, Mervin L. Keedy.—7th ed.
p. cm.
Keedy's name appears first on the previous ed.
ISBN 0-201-59561-3
1. Algebra. I. Keedy, Mervin Laverne. II. Title.
QA152.2.K43 1994
512'.9—dc20 94-17324
CIP

2 3 4 5 6 7 8 9 10—BAM—979695

Contents

3 *Polynomials: Operations* 177

4 *Polynomials: Factoring* 255

Preface

Intended for students who have not studied algebra but have a firm background in basic mathematics, this text is appropriate for a one-term course in introductory algebra. It is the second in a series of texts that includes the following:

Bittinger/Keedy: *Basic Mathematics,* Seventh Edition,

Bittinger/Keedy: *Introductory Algebra,* Seventh Edition,

Bittinger/Keedy: *Intermediate Algebra,* Seventh Edition.

What's New in the Seventh Edition?

Introductory Algebra, Seventh Edition, is a significant revision of the Sixth Edition, especially with respect to design, an all-new art program, pedagogy, and an enhanced supplements package. Its unique approach, which has been developed and refined over many years, is designed to help students both learn *and* retain mathematical skills. It is our belief that the Seventh Edition will *continue* to help today's students through pedagogical use of full color, updated applications, and thorough applications of geometry. As part of *MathMax: The Bittinger/Keedy System of Instruction,* it is accompanied by an extremely comprehensive and well-integrated supplements package to provide maximum support for both instructor and student.

The style, format, and approach of the Sixth Edition have been strengthened in this new edition in a number of ways.

USE OF COLOR The text is now printed in an extremely functional use of full color, evident in striking new design elements and artwork on nearly every page of the text. The use of color has been carried out in a methodical and precise manner so that its use carries a consistent meaning, which enhances the readability of the text for both student and instructor.

For example, the use of both red and blue in mathematical art increases understanding of the concepts. When two lines are graphed using the

This number decreases by 1 each time.

$$
\begin{aligned}
4 \cdot (-5) &= -20 \\
3 \cdot (-5) &= -15 \\
2 \cdot (-5) &= -10 \\
1 \cdot (-5) &= -5 \\
0 \cdot (-5) &= 0 \\
-1 \cdot (-5) &= 5 \\
-2 \cdot (-5) &= 10 \\
-3 \cdot (-5) &= 15
\end{aligned}
$$

This number increases by 5 each time.

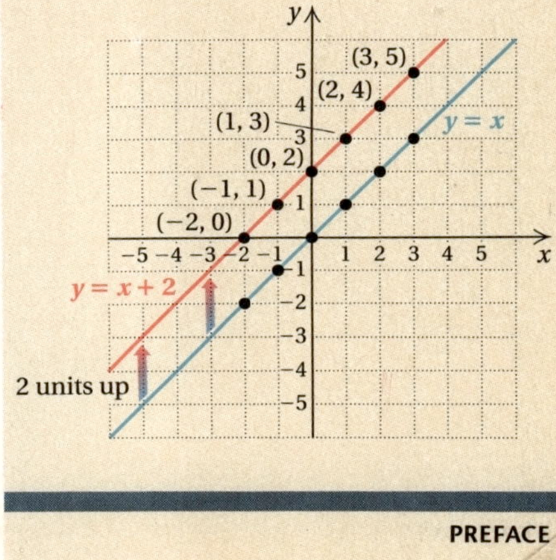

same set of axes, one is usually red and the other blue. Note that equation labels are the same color as the line to assist student understanding.

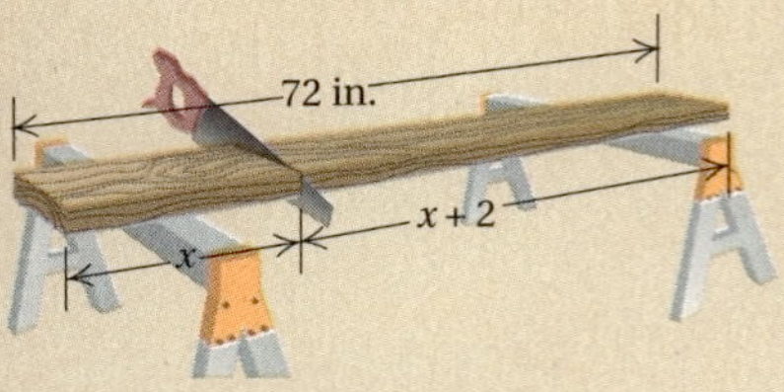

New Art Program All art in both exposition and the answer section is new. The use of full color in the art program greatly enhances the learning process.

Updated Applications Extensive research has been done to make the Seventh Edition's applications even more up-to-date and realistic. Not only are 20% of the exercises new to this edition, but many are drawn from the fields of business and economics, life and physical sciences, social sciences, and areas of general interest such as sports and daily life. To encourage students to "see mathematics" around them every day, we use graphs and drawings similar to those found in today's newspapers and magazines.

Critical Thinking Each chapter now ends with a set of Critical Thinking exercises, which includes Calculator Connections, Extended Synthesis Exercises, and Exercises for Thinking and Writing (all described more fully below).

Scientific and Graphing Calculators Instruction and exercises for the scientific and graphing calculators are now covered in many different locations. A calculator icon 🖩 highlights exercises that lend themselves to practice with a scientific calculator. Calculator Connection exercises occur in the Critical Thinking sections at the end of each chapter. (These exercises occasionally cover calculator procedures, but for the most part provide critical thinking exercises using a calculator.) Last, there is a special new appendix on calculator keystroke instruction (including the graphing calculator), prepared with the assistance of Rheta Beaver of Valencia Community College.

The Bittinger/Keedy System of Instruction

Following are distinctive features of the Bittinger/Keedy System of Instruction that work to ensure learning success for developmental math students.

Careful Development of Concepts We have divided each section into discrete and manageable learning objectives. Within the presentation of each objective, there is a careful buildup of understanding through a series of developmental examples. These examples enable students to thoroughly understand the mathematical concepts involved at each step.

Focus on Understanding Throughout the text, we present the appropriate mathematical rationale for a topic, rather than simply listing rules and procedures. For example, when manipulating rational expressions, we remove factors of 1 rather than cancel (although cancellation is mentioned with appropriate cautions). This method helps prevent student errors in simplification. The notion of multiplying by 1 is a theme that is carried out throughout the book and the series, providing a rationale for many other procedures.

Problem Solving We include real-life applications and problem-solving techniques to motivate students and encourage them to think about how mathematics can be used in their everyday life. The basis for problem solving is a five-step process (*Familiarize, Translate, Solve, Check,* and *State*) established early in the text and used henceforth.

Learning Aids

Interactive Worktext Approach The pedagogy of this text is designed as an interaction between the student and the exposition, annotated examples, margin exercises, and exercise sets. This approach provides students with a clear set of learning objectives, involves them with the development of the material, and provides immediate and continual reinforcement.

Section objectives are keyed by letter not only to appropriate objectives of the section, but also to exercises in the exercise sets and answers to review exercises and test questions, so that students can easily find appropriate review material if they are unable to work a particular exercise.

Numerous *margin exercises* throughout the text provide immediate reinforcement of the concepts covered in each section.

For Extra Help Many valuable study aids accompany this text. Below each list of section objectives are references to appropriate videotape, audiotape, and tutorial software programs, to make it easy for the student to find the correct support materials. The text exercises that appear on the videotapes are listed in an index at the back of the text as well as in the Instructor's Resource Guide.

Exercise Sets The exercises are paired, meaning that each even-numbered exercise is very much like the odd-numbered one that precedes it. Answers to the odd-numbered exercises are given at the back of the book, whereas those for the even-numbered exercises are not. This provides the instructor with many options. If an instructor wants the student to have answers, the odds are assigned. If an instructor wants the student to be able to practice (as on a test) with no answers, the evens are assigned. Thus each exercise set serves as two exercise sets. If an instructor wants the student to have all the answers, a complete answer book is available.

OPPORTUNITIES FOR CRITICAL THINKING In response to the recommendations of both instructors and educational organizations, we provide many opportunities for students to synthesize concepts, verbalize mathematics, and think critically.

Synthesis Exercises at the end of most exercise sets require students to synthesize learning objectives from the section being studied and preceding sections in the book.

Critical Thinking exercise sets occur at the end of each chapter. These exercise sets provide further opportunity for critical thinking by providing three types of exercises:

- *Calculator Connections* review keystrokes and provide exercises for a scientific calculator.
- *Extended Synthesis Exercises* call for students to further synthesize objectives from the chapter being studied and preceding chapters, thereby building critical-thinking skills.
- *Exercises for Thinking and Writing* encourage students to both think and write about key mathematical ideas in the chapter.

SKILL MAINTENANCE A well-received feature of preceding editions, the Skill Maintenance exercises have been enhanced by the inclusion of 50% more exercises in this edition. They occur at the end of most exercise sets. Although these exercises can review any objective of preceding chapters, they tend to focus on four specific objectives, called *Objectives for Review.* These objectives are listed at the beginning of each chapter and are covered in each *Summary and Review* and *Chapter Test* at the end of each chapter. The Objectives for Review are also included in a consistent manner in the Printed Test Bank that the instructor uses for testing.

The *Summary and Review* at the end of each chapter provides an extensive set of review exercises along with a list of important formulas and properties covered in that chapter.

We also include a *Cumulative Review* at the end of each chapter (except Chapters R and 1), which reviews material from all preceding chapters.

At the back of the text are answers to all end-of-chapter review exercises, together with section and objective references, so that students know exactly what material to restudy if they miss a review exercise.

TESTING The following assessment opportunities exist in the text.

The *Diagnostic Pretest,* provided at the beginning of the text, can place students in the appropriate chapter for their skill level by identifying familiar material and specific trouble areas.

Chapter Pretests can then be used to place students in a specific section of the chapter, allowing them to concentrate on topics with which they have particular difficulty.

Chapter Tests allow students to review and test comprehension of chapter skills, as well as the four Objectives for Review from earlier chapters. Answers to all Chapter Test questions are found at the back of the book, along with appropriate section and objective references.

Supplements for the Instructor

Teacher's Edition

The Teacher's Edition is a specially bound version of the student text with answers to all exercises in the margins, the exercise sets, and the chapter tests printed in a special color. It also includes answers to all the Critical Thinking exercises at the back of the text.

Instructor's Solutions Manual

The Instructor's Solutions Manual by Judith A. Penna contains brief worked-out solutions to all even-numbered exercises in the exercise sets.

Instructor's Resource Guide

The Instructor's Resource Guide contains the following:

- Conversion Guide.
- Extra practice exercises (with answers) for some of the most difficult topics in the text.
- Answers to the Critical Thinking exercises.
- Number lines and grids that can be used as transparency masters for teaching aids and for test preparation.
- Indexes to the videotapes, audiotapes, and tutorial software that accompany the text.
- Instructions for using the Math Hotline.
- Essays on setting up learning labs and testing centers, together with a directory of learning lab coordinators who are available to answer questions.

Printed Test Bank

Prepared by Donna DeSpain, the Printed Test Bank is an extensive collection of alternate chapter test forms, including the following:

- 4 alternate test forms for each chapter, with questions in the same topic order as the objectives presented in the chapter.
- 4 alternate test forms for each chapter, modeled after the Chapter Tests in the text.
- 3 alternate test forms for each chapter, designed for a 50-minute class period.
- 2 multiple-choice test forms for each chapter.
- 2 cumulative review tests for each chapter (with the exception of Chapter 1).
- 8 alternate forms of the final examination, 3 with questions organized by chapter, 3 with questions scrambled, as in the cumulative reviews, and 2 with multiple-choice questions.

Answer Book

The Answer Book contains answers to all exercises in the exercise sets in the text. Instructors may make quick reference to all answers or have quantities of these booklets made available for sale if they want students to have all the answers.

Computerized Testing: Omnitest³

Addison-Wesley's algorithm-driven computerized testing system for Macintosh and DOS computers features a brand-new graphical user interface for the DOS version and a substantial increase in the number of test items available for each chapter of the text.

The new graphical user interface for DOS is a Windows look-alike. It allows users to choose items by test item number or by reviewing all the test items available for a specific text objective. Users can choose the exact iteration of the test item they wish to have on their test or allow the computer to generate iterations for them. Users can also preview all the items for a test on screen and make changes to them during the preview process. They can control the format of the test, including the appearance of the test header, the spacing between items, and the layout of the test and the answer sheet. In addition, users can now save the exact form of the test they have created so that they can modify it for later use. Users can also enter their own items using Omnitest³'s WYSIWYG editor, and have access as well to a library of preloaded graphics.

Both the DOS and Macintosh versions of Omnitest³ for *Introductory Algebra* contain over 1500 items; 1000 of these are algorithm-driven—capable of generating hundreds of alternate versions. Omnitest³ for *Introductory Algebra* features at least one algorithm-driven multiple-choice and free-response item for *each* text objective, as well as a selection of static items. Many objectives are covered by *several* multiple-choice and free-response algorithm-based items—the coverage is comparable to the exercise coverage in the text's Summary and Review sections. Each chapter also includes a selection of Thinking and Writing questions.

Omnitest³ also includes preloaded chapter tests, cumulative tests, and tests designed to parallel state competency examinations to make building your own tests easier than ever before!

Course Management and Testing System

InterAct Math Plus for Windows or Macintosh (available from Addison-Wesley) combines course management and on-line testing with the features of the basic tutorial software (see "Supplements for the Student") to create an invaluable teaching resource. Consult your Addison-Wesley representative for details.

Supplements for the Student

Student's Solutions Manual

The Student's Solutions Manual by Judith A. Penna contains completely worked-out solutions with step-by-step annotations for all the odd-numbered exercises in the exercise sets in the text. It may be purchased by your students from Addison-Wesley Publishing Company.

"Math Makes a Difference" Videotapes

"Math Makes a Difference" is new to this edition of *Introductory Algebra.* It is a complete revision of the existing series of videotapes, based on extensive input from both students and instructors. "Math Makes a

Difference" features a team of mathematics teachers who present comprehensive coverage of each section of the text:

Marvin Bittinger, *Indiana University—Purdue University at Indianapolis*

Carilynn Bouie, *Chattanooga State Technical Community College*

Michael Butler, *College of the Redwoods*

Patricia Cleary, *University of Delaware*

Bettyann Daley, *University of Delaware*

Barbara Johnson, *Indiana University—Purdue University at Indianapolis*

Joanne Peeples, *El Paso Community College*

Anita Polk-Conley, *Chattanooga State Technical Community College*

Since the format is a lecture to a group of students, each videotape is interactive and engaging. Lecturers use odd-numbered exercises from the text as examples—these are listed in the videotape indexes in the Instructor's Resource Guide and at the back of the text. Icons at the beginning of each text section reference the appropriate videotape number.

A complete set of "Math Makes a Difference" videotapes is free to qualifying adopters.

Audiotapes

The audiotapes are designed to lead students through the material in each text section. Bill Saler, the narrator, explains solution steps to examples, cautions students about common errors, and instructs them at certain points to stop the tape and do exercises in the margin. He then reviews the margin exercise solutions, pointing out potential errors. Icons at the beginning of each section reference the appropriate audiotape number.

The audiotapes are free to qualifying adopters.

The Math Hotline

Prepared by Larry A. Bittinger, the Math Hotline is open 24 hours a day at 1-800-333-4227 so that students can obtain detailed hints for exercises through a voice menu system. This system is broken down by text, chapter, section, and exercise number. Exercises covered include all the odd-numbered exercises in the exercise sets, with the exception of the Skill Maintenance and Synthesis exercises.

InterAct Math Tutorial Software

InterAct Math Tutorial Software, new to this edition of *Introductory Algebra,* has been developed and designed by professional software engineers working closely with a team of experienced developmental math teachers.

InterAct Math Tutorial Software includes exercises that are linked one-to-one with the odd-numbered exercises in the textbook and require the same computational and problem-solving skills as their companion exercises in the text. Each exercise has an example and an interactive guided solution that are designed to involve students in the solution process and to help them identify precisely where they are having trouble. In addition, the software recognizes common student errors and provides students with appropriate customized feedback.

With its sophisticated answer recognition capabilities, InterAct Math Tutorial Software recognizes appropriate forms of the same answer for any kind of input. It also tracks student activity and scores for each section, which can then be printed out. Icons at the beginning of each text section reference the appropriate disk number.

Available for both DOS-based and Macintosh computers, the software is free to qualifying adopters.

We, your authors, have committed ourselves to writing a usable, understandable, accomplishable, error-free book that will extend the student's knowledge and enjoyment of mathematics. Students and instructors will undoubtedly have many general impressions and attitudes that form during their semester or two in a mathematics course. To help us to continually improve the text and to support the instructor's goals, we invite correspondence from both students and instructors to:

Marv Bittinger and Mike Keedy
c/o Marv Bittinger
3011 Whispering Trail
Carmel, IN 46033

Acknowledgments

Many have helped to mold the Seventh Edition by reviewing, answering surveys, participating in focus groups, filling out questionnaires, and spending time with us on their campuses. We owe a special debt of gratitude to the InterAct Math writers and reviewers for their persistence and pursuit of quality. Our deepest appreciation to all of you and in particular to the following:

Rheta Beaver, *Valencia Community College*
Carole Bergen, *Mercy College*
Mary-Jean Brod, *The University of Montana*
Laurence Chernoff, *Miami-Dade Community College—Kendall Campus*
Karen Clark, *Tacoma Community College*
Camille Cochrane, *Shelton State Community College*
Elaine Craft, *Chesterfield–Marlboro Technical College*
Nancy C. Davis, *Brunswick Community College*
Gudryn Doherty, *Community College of Denver*
Janice Eckmier, *California State University—Northridge*
Janice Gahan-Rech, *University of Nebraska at Omaha*
Roberta Hinkle Gansman, *Guilford Technical Community College*
Mary Lou Hart, *Brevard Community College—Melborne*
Phyllis A. Jore, *Valencia Community College—East Campus*
Richard Langlie, *North Hennepin Community College*
William Livingston, *Missouri Southern State College*
Annette Magyar, *Southwestern Michigan College*
Gary W. Martin, *DeVry Institute of Technology*
Gael T. Mericle, *Mankato State University*
Carol Metz, *Westchester Community College*
Jane Pinnow, *University of Wisconsin—Parkside*
Michael R. Schultz, *Central Maine Technical College*
Randolph J. Taylor, *Las Positas College*
Linda Verceles, *Edison State Community College*
Lynden Weberg, *University of Wisconsin—River Falls*
Lorena Wolff, *Westark Community College*
Deborah Woods, *University of Cincinnati—Raymond Walters College*

We also wish to thank the following people who reviewed the videotapes:

Kathleen Bavelas, *Manchester Community—Technical College*
Julane B. Crabtree, *Johnson County Community College*
Helen Hancock, *Shoreline Community College*
Marti Hidden, *Sacramento City College*
Louis Levy, *Northland Pioneer College*
Aimee Martin, *Amarillo College*
Letty Ann Macdonald, *Piedmont Virginia Community College*
Bill Thieman, *Ventura College*
Cindie Wade, *St. Clair County Community College*

We also wish to thank many people on the Bittinger/Keedy team at Addison-Wesley for the endless hours of hard work and unwavering support. The editorial, design, and production coordination of Quadrata, Inc., was exceptional as always—we especially appreciate their wholehearted dedication and hard work over the many years of our association.

In particular, we thank Judy Beecher for her editorial assistance, without which these books would not exist. Judy Penna has always provided steadfast, quality leadership in the preparation of the solutions manuals and the supervision of all printed supplements. We also gratefully acknowledge a strong supporting cast: Donna DeSpain, for the printed test banks; Bill Saler, for the audiotapes; Larry A. Bittinger, for the Math Hotline; and Larry Bittinger, Laurie Hurley, and Barbara Johnson, for their usual fine quality in the proofreading and pursuit of errors.

M.L.B.
M.L.K.

The Steps to Success

The following six pages show you how to use *Introductory Algebra* to maximize understanding while making studying easier.

1 **Use the chapter opener to begin your work with the chapter.**

The chapter introduction provides an overall view of the chapter's content.

2 Solving Equations and Inequalities

INTRODUCTION
In this chapter, we use the manipulations discussed in Chapter 1 to solve equations and inequalities. We then use equations and inequalities to solve problems.

AN APPLICATION
You see a flash of lightning. Your distance from the storm is M miles. You determine that distance by counting the number of seconds n that it takes the sound of the thunder to reach you and then multiplying by $\frac{1}{5}$. A formula relating M and n is $M = \frac{1}{5}n$.
Suppose we are 2 mi from the storm. How many seconds does it take for the sound of the thunder to reach us?

THE MATHEMATICS
We substitute 2 for M and solve for n:
$$2 = \frac{1}{5}n.$$
To find n, we solve this equation.

An application drawn from a chapter example or exercise immediately shows how a key chapter concept applies to the real world.

2 **Read the Objectives for Review, which list four objectives from preceding chapters that will be reinforced for skill maintenance in this chapter and its test.**

OBJECTIVES FOR REVIEW
The review objectives to be tested in addition to t…
[3.2a, b] Use the power rule to raise powers to p…
[3.4a, c] Add and subtract polynomials.
[4.6a] Factor polynomials completely.
[…a] Solve problems involving quadratic eq…

Pretest: Chapter 5

1. Find the LCM of $x^2 + 5x + 6$ and x^2 …

Perform the indicated operations and sim…

2. $\frac{b-1}{2-b} + \frac{b^2-3}{b^2-4}$
3. $\frac{4y-4}{y^2-y-2} - \frac{3y-5}{y^2-y-2}$
4. $\frac{4}{a+2} + \frac{3}{a}$
5. $\frac{x}{x+1} - \frac{x}{x-1} + \frac{2x^2}{x^2-1}$
6. $\frac{4x+8}{x+1} \cdot \frac{x^2-2x-3}{2x^2-8}$
7. $\frac{x+3}{x^2-9} \div \frac{x+3}{x^2-6x+9}$
8. Simplify: $\dfrac{\frac{1}{x}+\frac{1}{y}}{\frac{1}{x}-\frac{1}{y}}$.

Solve.

9. $\frac{1}{x+4} = \frac{5}{x}$
10. $\frac{3}{x-2} + \frac{x}{2} = \frac{6}{2x-4}$
11. Solve $R = \frac{1}{3}M(a-b)$ for M.
12. It takes 6 hr for a paper carrier to deliver 200 papers. At this rate, how long would it… to deliver 350 papers?
13. One data-entry clerk can key in a report in 6 hr. Another can key in the same report in 5 hr. How long would it take them to key in the same report working together?
14. One car travels 20 mph faster than another. While one car travels 300 mi, the other travels 400 mi. Find their speeds.

3 **Take the Chapter Pretest to assess your own strengths and weaknesses in the upcoming material.**

Each section of the chapter is designed as an interaction between you and the written explanations, the annotated examples, the margin exercises, and the exercise set.

4

Read the objectives listed in the margin and keyed to the text.

Objectives provide an instant outline of the section.

Important definitions, rules, and procedures are highlighted in boxes.

As you study the examples, note the detailed annotations and color highlights that help you on your way.

You are encouraged to do the margin exercises as you work through the material.

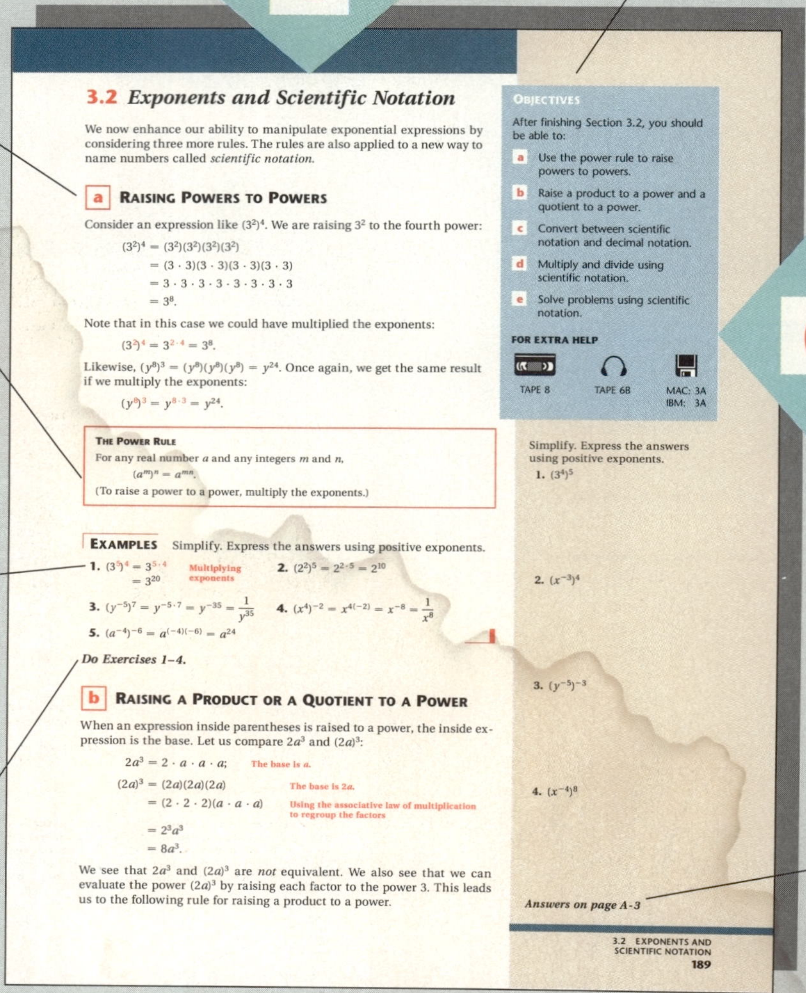

3.2 *Exponents and Scientific Notation*

We now enhance our ability to manipulate exponential expressions by considering three more rules. The rules are also applied to a new way to name numbers called *scientific notation.*

a Raising Powers to Powers

Consider an expression like $(3^2)^4$. We are raising 3^2 to the fourth power:

$$(3^2)^4 = (3^2)(3^2)(3^2)(3^2) = (3\cdot 3)(3\cdot 3)(3\cdot 3)(3\cdot 3) = 3\cdot3\cdot3\cdot3\cdot3\cdot3\cdot3\cdot3 = 3^8.$$

Note that in this case we could have multiplied the exponents:

$$(3^2)^4 = 3^{2\cdot 4} = 3^8.$$

Likewise, $(y^8)^3 = (y^8)(y^8)(y^8) = y^{24}$. Once again, we get the same result if we multiply the exponents:

$$(y^8)^3 = y^{8\cdot 3} = y^{24}.$$

The Power Rule
For any real number a and any integers m and n,
$$(a^m)^n = a^{mn}.$$
(To raise a power to a power, multiply the exponents.)

Examples Simplify. Express the answers using positive exponents.

1. $(3^5)^4 = 3^{5\cdot 4} = 3^{20}$ Multiplying exponents
2. $(2^2)^5 = 2^{2\cdot 5} = 2^{10}$
3. $(y^{-5})^7 = y^{-5\cdot 7} = y^{-35} = \frac{1}{y^{35}}$
4. $(x^4)^{-2} = x^{4(-2)} = x^{-8} = \frac{1}{x^8}$
5. $(a^{-4})^{-6} = a^{(-4)(-6)} = a^{24}$

Do Exercises 1–4.

b Raising a Product or a Quotient to a Power

When an expression inside parentheses is raised to a power, the inside expression is the base. Let us compare $2a^3$ and $(2a)^3$:

$2a^3 = 2\cdot a\cdot a\cdot a;$ The base is a.

$(2a)^3 = (2a)(2a)(2a)$ The base is $2a$.

$= (2\cdot 2\cdot 2)(a\cdot a\cdot a)$ Using the associative law of multiplication to regroup the factors

$= 2^3a^3$

$= 8a^3.$

We see that $2a^3$ and $(2a)^3$ are *not* equivalent. We also see that we can evaluate the power $(2a)^3$ by raising each factor to the power 3. This leads us to the following rule for raising a product to a power.

Objectives

After finishing Section 3.2, you should be able to:

- a Use the power rule to raise powers to powers.
- b Raise a product to a power and a quotient to a power.
- c Convert between scientific notation and decimal notation.
- d Multiply and divide using scientific notation.
- e Solve problems using scientific notation.

For Extra Help

TAPE 8 TAPE 6B MAC: 3A IBM: 3A

Simplify. Express the answers using positive exponents.

1. $(3^4)^5$
2. $(x^{-3})^4$
3. $(y^{-5})^{-3}$
4. $(x^{-4})^8$

Answers on page A-3

3.2 EXPONENTS AND SCIENTIFIC NOTATION 189

6

The videotape, audiotape, and software references provide extra help for each section.

Answers to the margin exercises are given at the back of the book.

Exercise Sets provide for a wealth of practice with chapter concepts.

5

Exercises are keyed to objectives in the text.

The Student's Solutions Manual and the Math Hotline are available for immediate help with the exercises.

NAME SECTION DATE

Exercise Set 5.3

a Find the LCM.

1. 12, 27 **2.** 10, 15 **3.** 8, 9 **4.** 12, 18

5. 6, 9, 21 **6.** 8, 36, 40 **7.** 24, 36, 40 **8.** 4, 5, 20

9. 10, 100, 500 **10.** 28, 42, 60

b Add, first finding the LCD. Simplify, if possible.

11. $\frac{7}{24}+\frac{11}{18}$ **12.** $\frac{7}{60}+\frac{2}{25}$ **13.** $\frac{1}{6}+\frac{3}{40}$

14. $\frac{5}{24}+\frac{3}{20}$ **15.** $\frac{1}{20}+\frac{1}{30}+\frac{2}{45}$

c Find the LCM.

17. $6x^2$, $12x^3$ **18.** 2*a*

19. $2x^2$, $6xy$, $18y^2$ **20.** p^3

21. $2(y-3)$, $6(y-3)$ **22.** 5(

23. t, $t+2$, $t-2$ **24.** *y*,

25. x^2-4, x^2+5x+6 **26.** x^2

ANSWERS

1. 2. 3. 4. 5. 6. 7. 8. 9. 10. 11.

ANSWERS

21. 22. 23. 24. 25. 26. 27. 28. 29. 30. 31. 32. 33.

21. The largest regulation soccer field is 100 yd wide and 130 yd long. Find the length of a diagonal of such a field.

22. How long is a guy wire reaching from the top of a 12-ft pole to a point 8 ft from the pole?

23. An airplane is flying at an altitude of 4100 ft. The slanted distance directly to the airport is 15,100 ft. How far is the airplane horizontally from the airport?

24. A surveyor had poles located at points *P*, *Q*, and *R*. The distances that the surveyor was able to measure are marked on the drawing. What is the approximate distance from *P* to *R*?

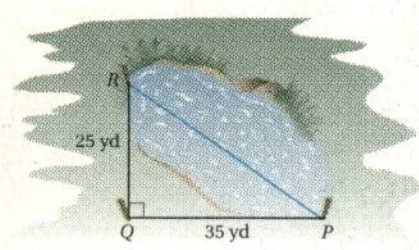

SKILL MAINTENANCE

Solve.

25. $5x+7=8y,$ $3x=8y-4$

26. $5x+y=17,$ $-5x+2y=10$

27. $3x-4y=-11,$ $5x+6y=12$

28. $x+y=-9,$ $x-y=-11$

SYNTHESIS

29. Two cars leave a service station at the same time. One car travels east at a speed of 50 mph, and the other travels south at a speed of 60 mph. After one-half hour, how far apart are they?

30. The length and the width of a rectangle are given by consecutive integers. The area of the rectangle is 90 cm². Find the length of a diagonal of the rectangle.

Find *x*.

31.

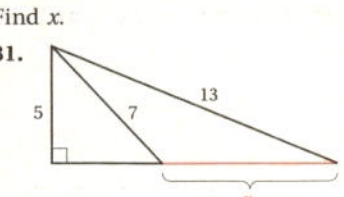

32.

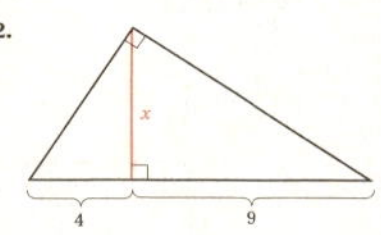

33.

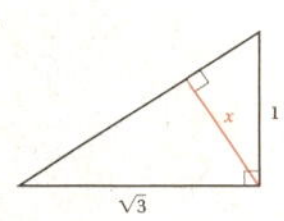

CHAPTER 8 RADICAL EXPRESSIONS
544

Skill Maintenance exercises at the end of most exercise sets review objectives from earlier chapters, especially the Objectives for Review that will be tested on the chapter test.

Synthesis exercises help you to synthesize objectives and provide insight into the material.

Critical Thinking exercises at the end of each chapter (optional) provide further opportunity to synthesize concepts and think critically.

7

Calculator Connections review important keystrokes and provide exercises for the scientific calculator and the graphing calculator. (See the calculator appendix at the end of the book for basic instruction.)

Extended Synthesis exercises call for you to synthesize many objectives.

Exercises for Thinking and Writing encourage you to think and write about key ideas.

CRITICAL THINKING

CALCULATOR CONNECTION

1. a) Approximate each of the following and look for patterns:

$\sqrt{49}$, $\sqrt{490}$, $\sqrt{4900}$, $\sqrt{49{,}000}$, $\sqrt{490{,}000}$.

b) On the basis of the patterns you found in part (a), approximate $\sqrt{4{,}900{,}000}$ and $\sqrt{49{,}000{,}000}$ without using your calculator.

Find the missing lengths.

2.
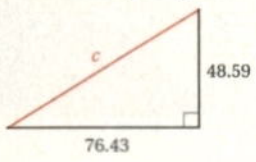

3.
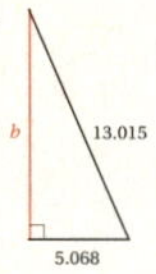

4.
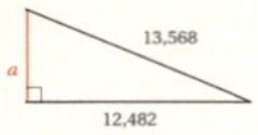

5.
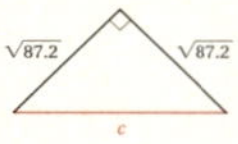

Assume that x is a positive number. Place one of $<$, $=$, or $>$ in each blank to make a true sentence.

6. $14 \,\blacksquare\, \sqrt{195}$
7. $\sqrt{450} \,\blacksquare\, 15\sqrt{2}$
8. $\sqrt{15}\sqrt{17} \,\blacksquare\, 16$
9. $25 \,\blacksquare\, \sqrt{625}$
10. $7\sqrt{2} \,\blacksquare\, 3\sqrt{11}$
11. $5\sqrt{0.64x} \,\blacksquare\, 4\sqrt{x}$
12. $100\sqrt{90x} \,\blacksquare\, 90\sqrt{100x}$
13. $\sqrt{12x} + 4\sqrt{2x} \,\blacksquare\, 4\sqrt{5x}$
14. $5\sqrt{7} \,\blacksquare\, 4\sqrt{11}$
15. $\sqrt{12x} - 4\sqrt{2x} \,\blacksquare\, 4\sqrt{3x}$

Use a graphing calculator to find the points of intersection of each pair of graphs.

16. $y = \sqrt{x},\ y = x^2 - 2x$
17. $y = \sqrt{4 - x},\ y = x^2 - 5$
18. $y = \sqrt{x + 7},\ y = x - 5$
Relate this to Example 3 in Section 8.5.
19. $y = \sqrt{x} - 1,\ y = \sqrt{x - 5}$
Relate this to Example 5 in Section 8.5.

Wind Chill Temperature We can use square roots to consider an application involving the effect of wind on the feeling of cold in the winter. Because wind speed enhances the loss of heat from the skin, we feel colder when there is wind than when there is not. The *wind chill temperature* is what the temperature would have to be with no wind in order to give the same chilling effect. A formula for finding the wind chill temperature, T_w, is

$$T_w = 91.4 - \frac{(10.45 + 6.68\sqrt{v} - 0.447v)(457 - 5T)}{110},$$

where T is the actual temperature given by a thermometer, in degrees Fahrenheit, and v is the wind speed, in miles per hour.

Use a calculator to find the wind chill temperature in each case. Round to the nearest degree.

20. $T = 30°F$, $v = 25$ mph
21. $T = 10°F$, $v = 25$ mph
22. $T = 20°F$, $v = 20$ mph
23. $T = 20°F$, $v = 40$ mph
24. $T = -10°F$, $v = 30$ mph
25. $T = -30°F$, $v = 30$ mph

(continued)

CRITICAL THINKING: CHAPTER 8

545

CRITICAL THINKING

EXTENDED SYNTHESIS EXERCISES

1. Find a real number between 5 and 6.
2. Find a real number between $\sqrt{9}$ and $\sqrt{25}$.
3. Find a real number between $\sqrt{9}$ and $\sqrt{16}$.
4. Find a real number between $\sqrt{10}$ and $\sqrt{11}$.

An *equilateral triangle* is shown below.

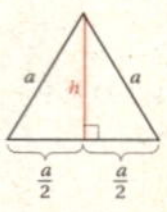

5. Find an expression for its height h in terms of a.
6. Find an expression for its area A in terms of a.

Simplify.

7. $\sqrt{x^{8n}}$
8. $\sqrt{0.04x^{4n}}$
9. Determine whether it is true that

$\sqrt{A} - \sqrt{B} = \sqrt{A - B}$.

10. In baseball, a third-baseman fields a line drive on the left-field line about 20 ft behind the bag. How far is the throw to first base?

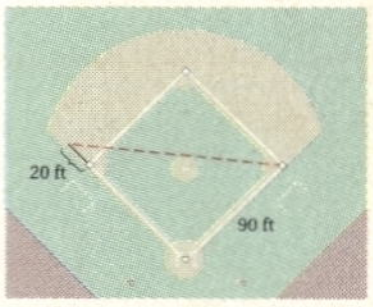

11. A tent has an opening in the shape of an isosceles triangle. The base of the triangle is 10 ft and the two congruent sides are each 8 ft. Find the height of the tent.

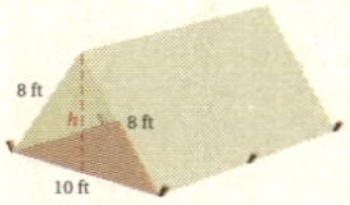

One important ... use organized ... to make disco... part of the *Fa*... problem-solvi... inadvertently ... step. Use an o... Exercises 12 a...

12. The sides ... of the sides are prime numbers that differ by 50. Find a set of numbers a, b, and c that satisfies this condition.
13. When asked his age, Augustus DeMorgan, a famous nineteenth-century English mathematician, said, "I was x years old in the year x^2." Find the year in which DeMorgan was born.

EXERCISES FOR THINKING AND WRITING

1. Explain why the following is incorrect:

$$\sqrt{\frac{9 + 100}{25}} = \frac{3 + 10}{5}.$$

Determine whether each of the sentences in Exercises 2–4 is true or false for all real numbers and whether the statement in Exercise 5 is true or false. Explain your answers.

2. $\sqrt{5x^2} = x\sqrt{5}$
3. $\sqrt{b^2 - 4} = b - 2$
4. $\sqrt{x^2 + 16} = x + 4$
5. The solution of $\sqrt{11 - 2x} = -3$ is 1.

CHAPTER 8 RADICAL EXPRESSIONS

546

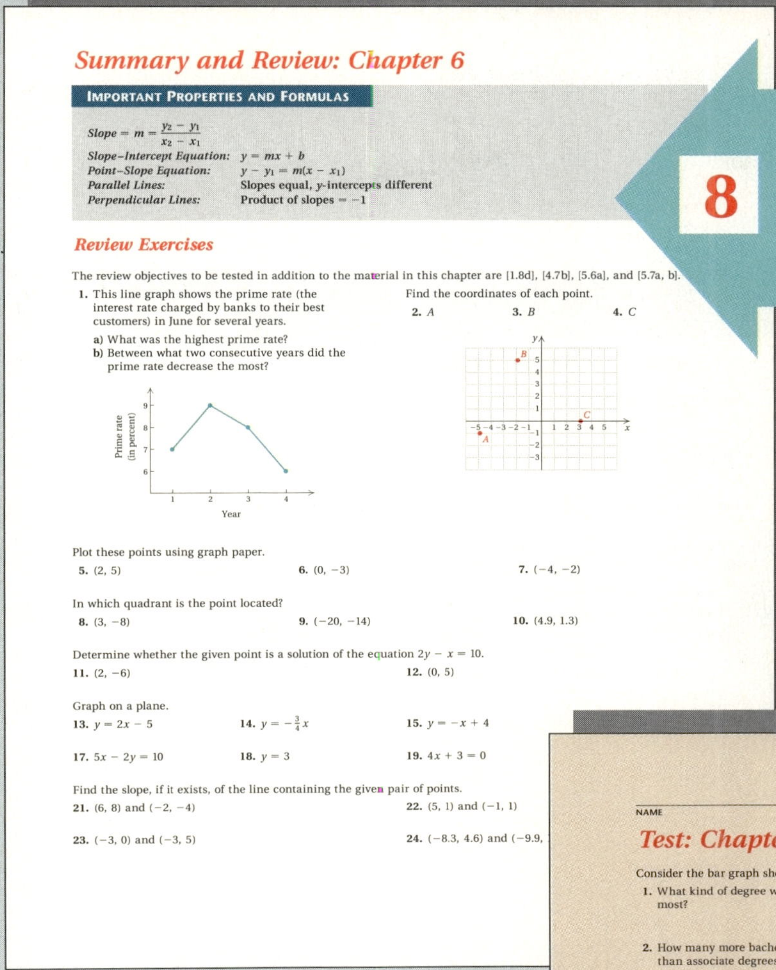

Summary and Review: Chapter 6

IMPORTANT PROPERTIES AND FORMULAS

Slope $= m = \frac{y_2 - y_1}{x_2 - x_1}$
Slope–Intercept Equation: $y = mx + b$
Point–Slope Equation: $y - y_1 = m(x - x_1)$
Parallel Lines: Slopes equal, y-intercepts different
Perpendicular Lines: Product of slopes $= -1$

Review Exercises

The review objectives to be tested in addition to the material in this chapter are [1.8d], [4.7b], [5.6a], and [5.7a, b].

1. This line graph shows the prime rate (the interest rate charged by banks to their best customers) in June for several years.
a) What was the highest prime rate?
b) Between what two consecutive years did the prime rate decrease the most?

Find the coordinates of each point.
2. A **3.** B **4.** C

Plot these points using graph paper.
5. (2, 5) **6.** (0, −3) **7.** (−4, −2)

In which quadrant is the point located?
8. (3, −8) **9.** (−20, −14) **10.** (4.9, 1.3)

Determine whether the given point is a solution of the equation $2y - x = 10$.
11. (2, −6) **12.** (0, 5)

Graph on a plane.
13. $y = 2x - 5$ **14.** $y = -\frac{3}{4}x$ **15.** $y = -x + 4$
17. $5x - 2y = 10$ **18.** $y = 3$ **19.** $4x + 3 = 0$

Find the slope, if it exists, of the line containing the given pair of points.
21. (6, 8) and (−2, −4) **22.** (5, 1) and (−1, 1)
23. (−3, 0) and (−3, 5) **24.** (−8.3, 4.6) and (−9.9,

8 The Summary and Review at the end of each chapter provides an extensive set of review exercises.

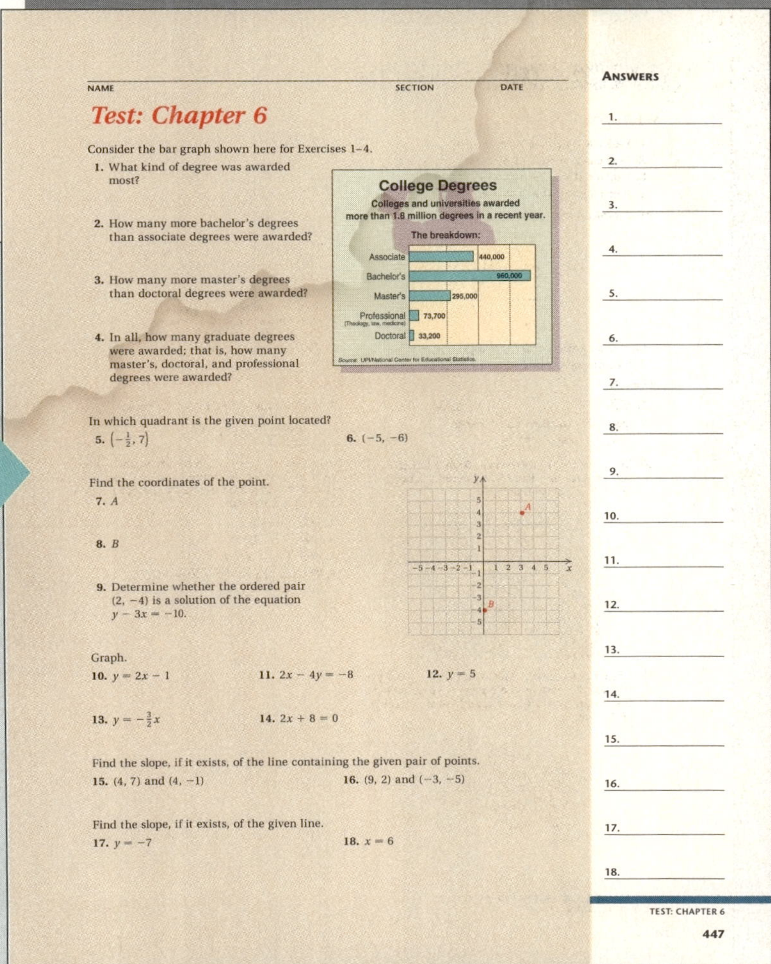

NAME SECTION DATE

Test: Chapter 6

Consider the bar graph shown here for Exercises 1–4.

1. What kind of degree was awarded most?
2. How many more bachelor's degrees than associate degrees were awarded?
3. How many more master's degrees than doctoral degrees were awarded?
4. In all, how many graduate degrees were awarded; that is, how many master's, doctoral, and professional degrees were awarded?

In which quadrant is the given point located?
5. $\left(-\frac{1}{2}, 7\right)$ **6.** (−5, −6)

Find the coordinates of the point.
7. A
8. B

9. Determine whether the ordered pair (2, −4) is a solution of the equation $y - 3x = -10$.

Graph.
10. $y = 2x - 1$ **11.** $2x - 4y = -8$ **12.** $y = 5$
13. $y = -\frac{3}{2}x$ **14.** $2x + 8 = 0$

Find the slope, if it exists, of the line containing the given pair of points.
15. (4, 7) and (4, −1) **16.** (9, 2) and (−3, −5)

Find the slope, if it exists, of the given line.
17. $y = -7$ **18.** $x = 6$

ANSWERS 1. 2. 3. 4. 5. 6. 7. 8. 9. 10. 11. 12. 13. 14. 15. 16. 17. 18.

TEST: CHAPTER 6 447

9 The Chapter Test at the end of each chapter allows you to review and test your comprehension of chapter skills as well as the four Objectives for Review from preceding chapters.

The Cumulative Review at the end of each chapter provides you with a review of material from all preceding chapters. This is an excellent tool for skill maintenance that provides continual review for the Final Examination.

10

Cumulative Review: Chapters 1–7

Compute and simplify.

1. $-2[1.4 - (-0.8 - 1.2)]$

2. $(1.3 \times 10^8)(2.4 \times 10^{-10})$

3. $\left(-\frac{1}{6}\right) \div \left(\frac{2}{9}\right)$

4. $\frac{2^{12}2^{-7}}{2^8}$

Simplify.

5. $\frac{x^2 - 9}{2x^2 - 7x + 3}$

6. $\frac{t^2 - 16}{(t + 4)^2}$

7. $\frac{x - \frac{x}{x + 2}}{\frac{2}{x} - \frac{1}{x + 2}}$

Perform the indicated operations and simplify.

8. $(1 - 3x^2)(2 - 4x^2)$

9. $(2a^2b - 5ab^2)^2$

10. $(3x^2 + 4y)(3x^2 - 4y)$

11. $-2x^2(x - 2x^2 + 3x^3)$

12. $(1 + 2x)(4x^2 - 2x + 1)$

13. $\left(8 - \frac{1}{3}x\right)\left(8 + \frac{1}{3}x\right)$

14. $(-8y^2 - y + 2) - (y^3 - 6y^2 + y - 5)$

15. $(2x^3 - 3x^2 - x - 1) \div (2x - 1)$

16. $\frac{7}{5x - 25} + \frac{x + 7}{5 - x}$

17. $\frac{2x - 1}{x - 2} - \frac{2x}{2 - x}$

18. $\frac{y^2 + y}{y^2 + y - 2} \cdot \frac{y + 2}{y^2 - 1}$

19. $\frac{7x + 7}{x^2 - 2x} \div \frac{14}{3x - 6}$

Factor completely.

20. $6x^5 - 36x^3 + 9x^2$

21. $16y^4 - 81$

22. $3x^2 + 10x - 8$

23. $4x^4 - 12x^2y + 9y^2$

24. $3m^3 + 6m^2 - 45m$

25. $x^3 + x^2 - x - 1$

Solve.

26. $3x - 4(x + 1) = 5$

27. $x(2x - 5) = 0$

28. $5x + 3 \geq 6(x - 4) + 7$

29. $1.5x - 2.3x = 0.4(x - 0.9)$

30. $2x^2 = 338$

31. $3x^2 + 15 = 14x$

32. $\frac{2}{x} - \frac{3}{x - 2} = \frac{1}{x}$

33. $1 + \frac{3}{x} + \frac{x}{x + 1} = \frac{1}{x^2 + x}$

34. $y = 2x - 9$,
$2x + 3y = -3$

35. $6x + 3y = -6$,
$-2x + 5y = 14$

36. $2x = y - 2$,
$3y - 6x = 6$

37. $\frac{1}{x} - \frac{1}{y} = \frac{1}{xy}$, for x

CUMULATIVE REVIEW: CHAPTERS 1–7
499

For more detailed information on how to use MathMax: The Bittinger/Keedy System of Instruction, please see the preface or contact your local Addison-Wesley representative.

Introductory Algebra

SEVENTH EDITION

Diagnostic Pretest

Chapter R

Perform the indicated operations and simplify if possible.

1. $\frac{8}{9} \cdot \frac{3}{5}$

2. $\frac{1}{4} + \frac{2}{3}$

3. $4.94 \div 0.19$

4. $12.04 - 1.057$

Chapter 1

Compute and simplify.

5. $3.8 + (-4.62) - (-2)$

6. $-9(1.3)$

7. A small business made a profit of $135.97 on Tuesday. The next day, it had a loss of $145.90. Find the total profit or loss.

8. Remove parentheses and simplify:
$3[11(a - 2) - 2(3 - a)]$.

Chapter 2

Solve.

9. $2(x - 1) = 4(x + 2)$

10. $4 - 13x \leq 10x - 5$

11. A 36-in. string is cut into two pieces. One piece is three times as long as the other. How long are the pieces?

12. A family spent $270 one month on clothing. This was 18% of its income. What was the family's income?

Chapter 3

Simplify.

13. $\frac{x^2y^2}{x^{-2}y^3}$

14. $(-2x)^3(2x^4)^2$

15. Subtract: $(x^2 + 3x - 1) - (2x^2 - 5)$.

16. Multiply: $(2x^2 + 3)(2x^2 - 3)$.

Chapter 4

Factor completely.

17. $2x^2 - 162$

18. $5x^2 - 14x - 3$

Solve.

19. $x^2 + 3x = 10$

20. The width of a rectangle is 9 m less than the length. The area is 136 m^2. Find the width and the length.

CHAPTER 5

21. Divide and simplify:

$$\frac{2x^3 + 6x^2}{x^2 + 10x + 25} \div \frac{4x^3 - 36x}{x^2 + x - 20}.$$

22. Add and simplify:

$$\frac{1 - x}{x^2 + x} + \frac{x}{x^2 + 3x + 2}.$$

Solve.

23. $\frac{2}{x + 4} = \frac{1}{x}$

24. One car travels 15 mph faster than another. While one car travels 165 mi, the other travels 120 mi. How fast is each car?

CHAPTER 6

25. Graph: $y = -2x + 1$.

26. Find the slope and the y-intercept of $2x + 3y = 8$.

27. Find an equation of the line containing the pair of points (3, 2) and (4, −1).

28. Graph: $x - 2y \leq 6$.

CHAPTER 7

Solve.

29. $x + y = 5,$
$2x + 3y = 7$

30. $2x + 4y = 5,$
$3x - 2y = 9$

31. Solution A is 20% alcohol and solution B is 50% alcohol. How much of each should be used to make 50 L of a solution that is 35% alcohol?

32. Two cars leave campus at the same time going in the same direction. One travels 56 mph and the other travels 62 mph. In how many hours will they be 60 mi apart?

CHAPTER 8

33. Multiply and simplify:

$$\sqrt{2x^2y} \cdot \sqrt{6xy^3}.$$

34. Divide and simplify:

$$\frac{\sqrt{5x^3}}{\sqrt{45xy^2}}.$$

35. Rationalize the denominator:

$$\frac{3}{2 - \sqrt{3}}.$$

36. Solve: $\sqrt{2x + 4} - 1 = 8$.

CHAPTER 9

Solve.

37. $3x^2 + 2x = 1$

38. $2x^2 + 10 = x$

39. The hypotenuse of a right triangle is 34 m. One leg is 14 m longer than the other. Find the lengths of the legs.

40. Graph: $y = x^2 - 4x + 1$.

R
Prealgebra Review

INTRODUCTION

This chapter is a review of skills basic to a study of algebra. If you do well on the diagnostic pretest on p. xxvii or the chapter pretest on the next page, you may be able to skip this chapter. Consult with your instructor if you need advice.

AN APPLICATION

It takes Jupiter 12 years and Saturn 30 years to revolve around the sun. How often will Jupiter and Saturn appear in the same position?

THE MATHEMATICS

We can use least common multiples to solve problems related to orbiting planets. The least common multiple of 12, or $2 \cdot 2 \cdot 3$, and 30, or $2 \cdot 3 \cdot 5$, is $2 \cdot 2 \cdot 3 \cdot 5$, or 60. Thus the planets will be in the same position once every 60 years.

Pretest: Chapter R

1. Find the prime factorization of 248.

2. Find the LCM: 12, 24, 42.

3. Write an expression equivalent to $\frac{2}{3}$ by multiplying by 1 using $\frac{5}{5}$.

4. Write an expression equivalent to $\frac{11}{12}$ with a denominator of 48.

Simplify.

5. $\frac{46}{128}$

6. $\frac{28}{42}$

Compute and simplify.

7. $\frac{3}{5} \div \frac{6}{11}$

8. $\frac{3}{7} - \frac{1}{3}$

9. $\frac{3}{10} + \frac{1}{5}$

10. $\frac{4}{7} \cdot \frac{5}{12}$

11. Convert to fractional notation (do not simplify): 32.17.

12. Convert to decimal notation: $\frac{789}{10{,}000}$.

13. Add: $8.25 + 91 + 34.7862$.

14. Subtract: $230 - 17.95$.

15. Multiply: 34.78×10.08.

16. Divide: $78.12 \div 6.3$.

17. Convert to decimal notation: $\frac{13}{9}$.

18. Round to the nearest hundredth: 345.8395.

19. Round to the nearest tenth: 345.8395.

20. Convert to decimal notation: 11.6%.

21. Convert to fractional notation: 87%.

22. Convert to percent notation: $\frac{7}{8}$.

23. Write exponential notation: $5 \cdot 5 \cdot 5 \cdot 5$.

24. Evaluate: 2^3.

25. Evaluate: $(1.1)^2$.

26. Calculate: $9 \cdot 3 + 24 \div 4 - 5^2 + 10$.

R.1 *Factoring and LCMs*

a FACTORS AND PRIME FACTORIZATIONS

We begin our review with *factoring*, which is a necessary skill for addition and subtraction with fractional notation. Factoring is also an important skill in algebra. You will eventually learn to factor algebraic expressions.

The numbers we will be factoring are from the set of **natural numbers:**

1, 2, 3, 4, 5, and so on.

Consider the product $12 = 3 \cdot 4$. We say that 3 and 4 are **factors** of 12 and that $3 \cdot 4$ is a **factorization** of 12. Since $12 = 12 \cdot 1$, we also know that 12 and 1 are factors of 12 and that $12 \cdot 1$ is a factorization of 12.

> To *factor* a number N means to express N as a product.
>
> A *factor* of a number N is a number that can be used to express N as a product.
>
> A *factorization* of a number N is an expression that names the number as a product of natural numbers.

EXAMPLE 1 Find all the factors of 8.

We first find some factorizations:

$$8 = 2 \cdot 4, \qquad 8 = 1 \cdot 8, \qquad 8 = 2 \cdot 2 \cdot 2.$$

The factors of 8 are 1, 2, 4, and 8.

Note that the word "factor" is used both as a noun and as a verb. You **factor** when you express a number as a product. The numbers you multiply together to get the product are **factors**.

EXAMPLE 2 Find all the factors of 12.

We first find some factorizations:

$$12 = 1 \cdot 12, \qquad 12 = 2 \cdot 6, \qquad 12 = 3 \cdot 4, \qquad 12 = 2 \cdot 2 \cdot 3.$$

The factors of 12 are 1, 2, 3, 4, 6, and 12.

Do Exercises 1–4 (in the margin at the right).

> A natural number that has *exactly two different* factors, itself and 1, is called a *prime number.*

EXAMPLE 3 Which of these numbers are prime? 7, 4, 11, 18, 1

7 is prime. It has exactly two different factors, 7 and 1.
4 is not prime. It has three different factors, 1, 2, and 4.
11 is prime. It has exactly two different factors, 11 and 1.
18 is not prime. It has factors 1, 2, 3, 6, 9, and 18.
1 is not prime. It does not have two *different* factors.

OBJECTIVES

After finishing Section R.1, you should be able to:

a Find all the factors of numbers and find prime factorizations of numbers.

b Find the LCM of two or more numbers using prime factorizations.

FOR EXTRA HELP

TAPE 1 | TAPE 1A | MAC: R IBM: R

Find all the factors of the number.

1. 9

2. 16

3. 18

4. 24

Answers on page A-1

5. Which of these numbers are prime?

8, 6, 13, 14, 1

Find the prime factorization.

6. 48

7. 50

8. 770

Answers on page A-1

The following is a table of the prime numbers from 2 to 157. There are more extensive tables, but these prime numbers will be the most helpful to you in this text.

A TABLE OF PRIMES

2, 3, 5, 7, 11, 13, 17, 19, 23, 29, 31, 37, 41, 43, 47, 53, 59, 61, 67, 71, 73, 79, 83, 89, 97, 101, 103, 107, 109, 113, 127, 131, 137, 139, 149, 151, 157

Do Exercise 5.

If a natural number, other than 1, is not prime, we call it **composite**. Every composite number can be factored into a product of prime numbers. Such a factorization is called a **prime factorization.**

EXAMPLE 4 Find the prime factorization of 36.

We begin by factoring 36 any way we can. One way is like this:

$$36 = 4 \cdot 9.$$

The factors 4 and 9 are not prime, so we factor them:

$$\begin{aligned} 36 &= \;\;4 \quad\cdot\quad 9 \\ &\quad\;\;\downarrow \qquad\;\; \downarrow \\ &= 2 \cdot 2 \cdot 3 \cdot 3 \end{aligned}$$

The factors in the last factorization are all prime, so we now have the *prime factorization* of 36. Note that 1 is *not* part of this factorization because it is not prime.

Another way to find the prime factorization of 36 is like this:

$$36 = 2 \cdot 18 = 2 \cdot 3 \cdot 6 = 2 \cdot 3 \cdot 2 \cdot 3.$$

In effect, we begin factoring any way we can think of and keep factoring until all factors are prime. Using a **factor tree** might also be helpful.

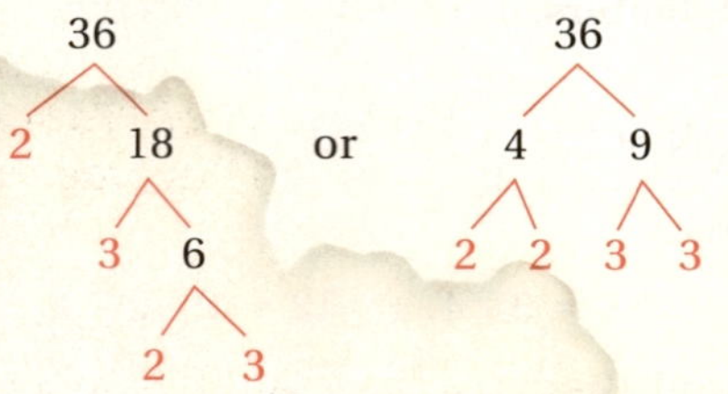

EXAMPLE 5 Find the prime factorization of 60.

This time, we use the list of primes from the table. We go through the table until we find a prime that is a factor of 60. The first such prime is 2.

$$60 = 2 \cdot 30$$

We keep dividing by 2 until it is not possible to do so.

$$60 = 2 \cdot 2 \cdot 15$$

Now we go to the next prime in the table that is a factor of 60. It is 3.

$$60 = 2 \cdot 2 \cdot 3 \cdot 5$$

Each factor in $2 \cdot 2 \cdot 3 \cdot 5$ is a prime. Thus this is the prime factorization.

Do Exercises 6–8.

b Least Common Multiples

Least common multiples are used to add and subtract with fractional notation.

The **multiples** of a number all have that number as a factor. For example, the multiples of 2 are

$$2, \quad 4, \quad 6, \quad 8, \quad 10, \quad 12, \quad 14, \quad 16, \ldots.$$

We could name each of them in such a way as to show 2 as a factor. For example, $14 = 2 \cdot 7$.

The multiples of 3 all have 3 as a factor:

$$3, \quad 6, \quad 9, \quad 12, \quad 15, \quad 18, \ldots.$$

Two or more numbers always have many multiples in common. From lists of multiples, we can find common multiples.

EXAMPLE 6 Find the common multiples of 2 and 3.

We make lists of their multiples and circle the multiples that appear in both lists.

2, 4, (6), 8, 10, (12), 14, 16, (18), 20, 22, (24), 26, 28, (30), 32, 34, (36), . . . ;
3, (6), 9, (12), 15, (18), 21, (24), 27, (30), 33, (36), . . .

The common multiples of 2 and 3 are

$$6, \quad 12, \quad 18, \quad 24, \quad 30, \quad 36, \ldots.$$

Do Exercise 9.

In Example 6, we found common multiples of 2 and 3. The *least,* or smallest, of those common multiples is 6. We abbreviate **least common multiple** as **LCM**.

There are several methods that work well for finding the LCM of several numbers. Some of these do not work well in algebra, especially when we consider expressions with variables such as $4ab$ and $12abc$. We now review a method that will work in arithmetic *and in algebra as well.* To see how it works, let us look at the prime factorizations of 9 and 15 in order to find the LCM:

$$9 = 3 \cdot 3, \qquad 15 = 3 \cdot 5.$$

Any multiple of 9 must have *two* 3's as factors. Any multiple of 15 must have *one* 3 and *one* 5 as factors. The smallest number satisfying all of these conditions is

Two 3's; 9 is a factor

$$3 \cdot 3 \cdot 5 = 45.$$

One 3, one 5; 15 is a factor

The LCM must have all the factors of 9 and all the factors of 15, but the factors cannot be repeated when they are common to both numbers.

To find the LCM of several numbers:

a) Write the prime factorization of each number.

b) Form the LCM by writing the product of the different factors from step (a), using each factor the greatest number of times it occurs in any one factorization.

9. Find the common multiples of 3 and 5 by making lists of multiples.

Answer on page A-1

Find the LCM by factoring.

10. 8 and 10

11. 18 and 27

12. Find the LCM of 18, 24, and 30.

Find the LCM.

13. 3, 18

14. 12, 24

Find the LCM.

15. 4, 9

16. 5, 6, 7

Answers on page A-1

EXAMPLE 7 Find the LCM of 40 and 100.

a) We find the prime factorizations:

$$40 = 2 \cdot 2 \cdot 2 \cdot 5,$$
$$100 = 2 \cdot 2 \cdot 5 \cdot 5.$$

b) We write 2 as a factor three times (the greatest number of times it occurs in any one factorization). We write 5 as a factor two times (the greatest number of times it occurs in any one factorization).

The LCM is $2 \cdot 2 \cdot 2 \cdot 5 \cdot 5$, or 200.

Do Exercises 10 and 11.

EXAMPLE 8 Find the LCM of 27, 90, and 84.

a) We factor:

$$27 = 3 \cdot 3 \cdot 3,$$
$$90 = 2 \cdot 3 \cdot 3 \cdot 5,$$
$$84 = 2 \cdot 2 \cdot 3 \cdot 7.$$

b) We write 2 as a factor two times, 3 three times, 5 one time, and 7 one time.

The LCM is $2 \cdot 2 \cdot 3 \cdot 3 \cdot 3 \cdot 5 \cdot 7$, or 3780.

Do Exercise 12.

EXAMPLE 9 Find the LCM of 7 and 21.

Since 7 is prime, it has no prime factorization. We still need it as a factor, however:

$$7 = 7,$$
$$21 = 3 \cdot 7.$$

The LCM is $7 \cdot 3$, or 21.

> If one number is a factor of another, then the LCM is the larger of the two numbers.

Do Exercises 13 and 14.

EXAMPLE 10 Find the LCM of 8 and 9.

We have

$$8 = 2 \cdot 2 \cdot 2,$$
$$9 = 3 \cdot 3.$$

The LCM is $2 \cdot 2 \cdot 2 \cdot 3 \cdot 3$, or 72.

> If two or more numbers have no common prime factor, then the LCM is the product of the numbers.

Do Exercises 15 and 16.

NAME SECTION DATE

Exercise Set R.1

Always review the objectives before doing an exercise set. See page 3. Note how the objectives are keyed to the exercises.

a Find all the factors of the number.

1. 20 **2.** 36 **3.** 72 **4.** 81

Find the prime factorization of the number.

5. 15 **6.** 14 **7.** 22 **8.** 33

9. 9 **10.** 25 **11.** 49 **12.** 121

13. 18 **14.** 24 **15.** 40 **16.** 56

17. 90 **18.** 120 **19.** 210 **20.** 330

21. 91 **22.** 143 **23.** 119 **24.** 221

b Find the prime factorization of the numbers. Then find the LCM.

25. 4, 5 **26.** 18, 40 **27.** 24, 36 **28.** 24, 27

29. 3, 15 **30.** 20, 40 **31.** 30, 40 **32.** 50, 60

33. 13, 23 **34.** 12, 18 **35.** 18, 30 **36.** 45, 72

ANSWERS

1. _____
2. _____
3. _____
4. _____
5. _____
6. _____
7. _____
8. _____
9. _____
10. _____
11. _____
12. _____
13. _____
14. _____
15. _____
16. _____
17. _____
18. _____
19. _____
20. _____
21. _____
22. _____
23. _____
24. _____
25. _____
26. _____
27. _____
28. _____
29. _____
30. _____
31. _____
32. _____
33. _____
34. _____
35. _____
36. _____

ANSWERS

37. ______
38. ______
39. ______
40. ______
41. ______
42. ______
43. ______
44. ______
45. ______
46. ______
47. ______
48. ______
49. ______
50. ______
51. ______
52. ______
53. a) ______
b) ______
c) ______
d) ______
54. ______
55. ______
56. ______
57. ______
58. ______

37. 30, 36 **38.** 30, 50 **39.** 24, 30 **40.** 60, 70

41. 17, 29 **42.** 18, 24 **43.** 12, 28 **44.** 35, 45

45. 2, 3, 5 **46.** 3, 5, 7 **47.** 24, 36, 12 **48.** 8, 16, 22

49. 5, 12, 15 **50.** 12, 18, 40 **51.** 6, 12, 18 **52.** 24, 35, 45

SYNTHESIS

Synthesis exercises are extra and optional, and usually more challenging, requiring you to put together objectives of this section or preceding sections of the text. Any exercises marked with a 🖩 are to be worked with a calculator.

53. Consider the numbers 8 and 12. Determine whether each of the following is the LCM of 8 and 12. Tell why or why not.

a) $2 \cdot 2 \cdot 3 \cdot 3$ b) $2 \cdot 2 \cdot 3$ c) $2 \cdot 3 \cdot 3$ d) $2 \cdot 2 \cdot 2 \cdot 3$

🖩 Use a calculator to find the LCM of the numbers.

54. 288, 324 **55.** 2700, 7800

Planetary orbits and LCMs. The earth, Jupiter, Saturn, and Uranus all revolve around the sun. The earth takes 1 yr, Jupiter 12 yr, Saturn 30 yr, and Uranus 84 yr. On a certain night, you look at all the planets and wonder how many years it will be before they all have the same position again. To find out, you find the LCM of 12, 30, and 84. It will be that number of years.

56. How often will Jupiter and Saturn appear in the same position?

57. How often will Saturn and Uranus appear in the same position?

58. How often will Jupiter, Saturn, and Uranus appear in the same position?

R.2 *Fractional Notation*

We now review fractional notation and its use with addition, subtraction, multiplication, and division of arithmetic numbers.

One feature of algebra that distinguishes it from arithmetic is the use of letters, or variables, to represent numbers. We will make use of such letters in this section.

OBJECTIVES

After finishing Section R.2, you should be able to:

- **a** Find equivalent fractional expressions by multiplying by 1.
- **b** Simplify fractional notation.
- **c** Add, subtract, multiply, and divide using fractional notation.

FOR EXTRA HELP

TAPE 1 TAPE 1A MAC: R IBM: R

a EQUIVALENT EXPRESSIONS AND FRACTIONAL NOTATION

An example of **fractional notation** for a number is

$$\frac{2}{3} \begin{array}{l} \longleftarrow \text{Numerator} \\ \longleftarrow \text{Denominator} \end{array}$$

The top number is called the **numerator**, and the bottom number is called the **denominator**.

The **whole numbers** consist of the natural numbers and 0:

$$0, \quad 1, \quad 2, \quad 3, \quad 4, \quad 5, \ldots .$$

The **arithmetic numbers,** also called the **nonnegative rational numbers,** consist of the whole numbers and the fractions, such as $\frac{2}{3}$ and $\frac{9}{5}$. The arithmetic numbers can also be described as follows.

> The *arithmetic numbers* are the whole numbers and the fractions, such as $\frac{3}{4}$, $\frac{6}{5}$, or 8. All of these numbers can be named with fractional notation $\frac{a}{b}$, where a and b are whole numbers and $b \neq 0$.

Note that all whole numbers are also arithmetic numbers. We can say this because we can name a whole number like 8 with fractional notation as $\frac{8}{1}$. We call 8 and $\frac{8}{1}$ **equivalent expressions.**

Being able to find an equivalent expression is critical to a study of algebra. Some simple but powerful properties of numbers that allow us to find equivalent expressions are the identity properties of 0 and 1.

> **THE IDENTITY PROPERTY OF 0**
>
> For any number a,
>
> $$a + 0 = a.$$
>
> (Adding 0 to any number gives that same number.)

> **THE IDENTITY PROPERTY OF 1**
>
> For any number a,
>
> $$a \cdot 1 = a.$$
>
> (Multiplying any number by 1 gives that same number.)

1. Write a fractional expression equivalent to $\frac{2}{3}$ with a denominator of 12.

2. Write a fractional expression equivalent to $\frac{3}{4}$ with a denominator of 28.

3. Multiply by 1 to find three different fractional expressions for $\frac{7}{8}$.

Answers on page A-1

Here are some ways to name the number 1:

$$\frac{5}{5},\quad \frac{3}{3},\quad \text{and}\quad \frac{26}{26}.$$

The following property allows us to find equivalent fractional expressions, that is, find other names for arithmetic numbers.

EQUIVALENT EXPRESSIONS FOR 1

For any number a, $a \neq 0$,

$$\frac{a}{a} = 1.$$

We can use the identity property of 1 and the preceding result to find equivalent fractional expressions.

EXAMPLE 1 Write a fractional expression equivalent to $\frac{2}{3}$ with a denominator of 15.

Note that $15 = 3 \cdot 5$. We want fractional notation for $\frac{2}{3}$ that has a denominator of 15, but the denominator 3 is missing a factor of 5. We multiply by 1, using $\frac{5}{5}$ as an equivalent expression for 1. Recall from arithmetic that to multiply with fractional notation, we multiply numerators and denominators:

$$\frac{2}{3} = \frac{2}{3} \cdot 1 \qquad \text{Using the identity property of 1}$$

$$= \frac{2}{3} \cdot \frac{5}{5} \qquad \text{Using } \tfrac{5}{5} \text{ for 1}$$

$$= \frac{10}{15}. \qquad \text{Multiplying numerators and denominators}$$

Do Exercises 1–3.

b SIMPLIFYING EXPRESSIONS

We know that $\frac{1}{2}$, $\frac{2}{4}$, $\frac{4}{8}$, and so on, all name the same number. Any arithmetic number can be named in many ways. The **simplest fractional notation** is the notation that has the smallest numerator and denominator. We call the process of finding the simplest fractional notation **simplifying**. We reverse the process of Example 1 by first factoring the numerator and the denominator. Then we factor the fractional expression and remove a factor of 1 using the identity property of 1.

EXAMPLE 2 Simplify: $\dfrac{10}{15}$.

$$\frac{10}{15} = \frac{2 \cdot 5}{3 \cdot 5} \qquad \text{Factoring the numerator and the denominator. In this case, each is the prime factorization.}$$

$$= \frac{2}{3} \cdot \frac{5}{5} \qquad \text{Factoring the fractional expression}$$

$$= \frac{2}{3} \cdot 1$$

$$= \frac{2}{3} \qquad \text{Using the identity property of 1 (removing a factor of 1)}$$

EXAMPLE 3 Simplify: $\frac{36}{24}$.

$$\frac{36}{24} = \frac{6 \cdot 6}{4 \cdot 6} \qquad \text{Factoring the numerator and the denominator}$$

$$= \frac{3 \cdot 2 \cdot 6}{2 \cdot 2 \cdot 6} \qquad \text{Factoring further}$$

$$= \frac{3}{2} \cdot \frac{2 \cdot 6}{2 \cdot 6} \qquad \text{Factoring the fractional expression}$$

$$= \frac{3}{2} \cdot 1$$

$$= \frac{3}{2} \qquad \text{Removing a factor of 1}$$

It is always a good idea to check at the end to see if you have indeed factored out all the common factors of the numerator and the denominator.

Simplify.

4. $\frac{18}{45}$

5. $\frac{38}{18}$

6. $\frac{72}{27}$

CANCELING

Canceling is a shortcut that you may have used to remove a factor of 1 when working with fractional notation. With *great* concern, we mention it as a possibility of speeding up your work. You should use canceling only when removing common factors in numerators and denominators. Each common factor allows us to remove a factor of 1 in a product. Canceling *may not* be done in sums or when adding expressions together. Our concern is that "canceling" be done with care and understanding. Example 3 might have been done faster as follows:

$$\frac{36}{24} = \frac{3 \cdot \not{2} \cdot \not{6}}{2 \cdot \not{2} \cdot \not{6}} = \frac{3}{2}, \quad \text{or} \quad \frac{36}{24} = \frac{3 \cdot \not{12}}{2 \cdot \not{12}} = \frac{3}{2}, \quad \text{or} \quad \frac{\overset{\overset{3}{\not{18}}}{\not{36}}}{\underset{\underset{2}{\not{12}}}{\not{24}}} = \frac{3}{2}.$$

Caution! The difficulty with canceling is that it is often applied incorrectly in situations like these:

$$\frac{\not{2} + 3}{\not{2}} = 3, \qquad \frac{\not{4} + 1}{\not{4} + 2} = \frac{1}{2}, \qquad \frac{\not{1}5}{\not{5}4} = \frac{1}{4}.$$

Wrong! Wrong! Wrong!

$$\frac{2 + 3}{2} = \frac{5}{2} \qquad \frac{4 + 1}{4 + 2} = \frac{5}{6} \qquad \frac{15}{54} = \frac{5}{18}$$

In each of these situations, the expressions canceled out were *not* factors of 1. Factors are parts of products. For example, in $2 \cdot 3$, 2 and 3 are factors, but in $2 + 3$, 2 and 3 are *not* factors. **If you can't factor, you can't cancel! If in doubt, don't cancel!**

Do Exercises 4–6.

Answers on page A-1

Simplify.

7. $\frac{27}{54}$

8. $\frac{48}{12}$

Multiply and simplify.

9. $\frac{6}{5} \cdot \frac{25}{12}$

10. $\frac{3}{8} \cdot \frac{5}{3} \cdot \frac{7}{2}$

Answers on page A-1

The number of factors in the numerator and the denominator may not always be the same. If not, we can always insert the number 1 as a factor. The identity property of 1 allows us to do that.

EXAMPLE 4 Simplify: $\frac{18}{72}$.

$$\frac{18}{72} = \frac{2 \cdot 9}{4 \cdot 2 \cdot 9} = \frac{1 \cdot \not{2} \cdot \not{9}}{4 \cdot \not{2} \cdot \not{9}}$$ Using the identity property of 1 to rename $2 \cdot 9$ as $1 \cdot 2 \cdot 9$

$$= \frac{1}{4}$$ Removing a factor of 1: $\frac{2 \cdot 9}{2 \cdot 9} = 1$

EXAMPLE 5 Simplify: $\frac{72}{9}$.

$$\frac{72}{9} = \frac{8 \cdot 9}{1 \cdot 9}$$ Factoring and inserting a factor of 1 in the denominator

$$= \frac{8 \cdot \not{9}}{1 \cdot \not{9}}$$

$$= \frac{8}{1}$$ Removing a factor of 1: $\frac{9}{9} = 1$

$$= 8$$ Simplifying

Do Exercises 7 and 8.

c MULTIPLICATION, ADDITION, SUBTRACTION, AND DIVISION

After we have performed an operation of multiplication, addition, subtraction, or division, the answer may or may not be simplified. We simplify, if at all possible. Let us continue as we have in the preceding examples.

MULTIPLICATION

EXAMPLE 6 Multiply and simplify: $\frac{5}{6} \cdot \frac{9}{25}$.

$$\frac{5}{6} \cdot \frac{9}{25} = \frac{5 \cdot 9}{6 \cdot 25}$$ Multiplying numerators and denominators

$$= \frac{1 \cdot 5 \cdot 3 \cdot 3}{2 \cdot 3 \cdot 5 \cdot 5}$$ Factoring the numerator and the denominator

$$= \frac{\not{3} \cdot \not{5} \cdot 1 \cdot 3}{\not{3} \cdot \not{5} \cdot 2 \cdot 5}$$ Removing a factor of 1: $\frac{3 \cdot 5}{3 \cdot 5} = 1$

$$= \frac{3}{10}$$ Simplifying

Do Exercises 9 and 10.

ADDITION

When denominators are the same, we can add by adding the numerators and keeping the same denominator.

EXAMPLE 7 Add and simplify: $\frac{4}{8} + \frac{5}{8}$.

The common denominator is 8. We add the numerators and keep the common denominator:

$$\frac{4}{8} + \frac{5}{8} = \frac{4+5}{8} = \frac{9}{8}.$$

In arithmetic, we generally write $\frac{9}{8}$ as $1\frac{1}{8}$. In algebra, you will find that *improper* symbols such as $\frac{9}{8}$ are more useful and are quite *proper* for our purposes.

When denominators are different, we use the identity property of 1 and multiply to find a common denominator. The smallest such denominator is called the lowest or **least common denominator.** That number is the least common multiple of the original denominators. The least common denominator is often abbreviated **LCD**.

EXAMPLE 8 Add and simplify: $\frac{3}{8} + \frac{5}{12}$.

The LCM of the denominators, 8 and 12, is 24. Thus the LCD is 24. We multiply each fraction by 1 to obtain the LCD:

$$\frac{3}{8} + \frac{5}{12} = \frac{3}{8} \cdot \frac{3}{3} + \frac{5}{12} \cdot \frac{2}{2}$$

Multiplying by 1. Since $3 \cdot 8 = 24$, we multiply the first number by $\frac{3}{3}$. Since $2 \cdot 12 = 24$, we multiply the second number by $\frac{2}{2}$.

$$= \frac{9}{24} + \frac{10}{24}$$

$$= \frac{19}{24}.$$

Do Exercises 11–14.

SUBTRACTION

When subtracting, we also multiply by 1 to obtain the LCD.

EXAMPLE 9 Subtract and simplify: $\frac{9}{8} - \frac{4}{5}$.

$$\frac{9}{8} - \frac{4}{5} = \frac{9}{8} \cdot \frac{5}{5} - \frac{4}{5} \cdot \frac{8}{8}$$

The LCD is 40.

$$= \frac{45}{40} - \frac{32}{40}$$

$$= \frac{13}{40}$$

Add and simplify.

11. $\frac{4}{5} + \frac{3}{5}$

12. $\frac{5}{6} + \frac{7}{6}$

13. $\frac{5}{6} + \frac{7}{10}$

14. $\frac{1}{4} + \frac{1}{2}$

Answers on page A-1

Subtract and simplify.

15. $\frac{7}{8} - \frac{2}{5}$

16. $\frac{5}{12} - \frac{2}{9}$

Find the reciprocal.

17. $\frac{4}{11}$

18. $\frac{15}{7}$

19. 5

20. $\frac{1}{3}$

21. Divide by multiplying by 1:

$$\frac{\frac{3}{5}}{\frac{4}{7}}.$$

Answers on page A-1

EXAMPLE 10 Subtract and simplify: $\frac{7}{10} - \frac{1}{5}$.

$$\frac{7}{10} - \frac{1}{5} = \frac{7}{10} - \frac{1}{5} \cdot \frac{2}{2} \qquad \text{The LCD is 10.}$$

$$= \frac{7}{10} - \frac{2}{10}$$

$$= \frac{5}{10} = \frac{1 \cdot \cancel{5}}{2 \cdot \cancel{5}} = \frac{1}{2} \qquad \text{Removing a factor of 1: } \frac{5}{5} = 1$$

Do Exercises 15 and 16.

Reciprocals

Two numbers whose product is 1 are called **reciprocals**, or **multiplicative inverses,** of each other. All the arithmetic numbers, except zero, have reciprocals.

EXAMPLES

11. The reciprocal of $\frac{2}{3}$ is $\frac{3}{2}$ because $\frac{2}{3} \cdot \frac{3}{2} = \frac{6}{6} = 1$.

12. The reciprocal of 9 is $\frac{1}{9}$ because $9 \cdot \frac{1}{9} = \frac{9}{9} = 1$.

13. The reciprocal of $\frac{1}{4}$ is 4 because $\frac{1}{4} \cdot 4 = 1$.

Do Exercises 17–20.

Reciprocals and Division

The number 1 and reciprocals can be used to justify a fast way to divide arithmetic numbers. We multiply by 1, carefully choosing the expression for 1.

EXAMPLE 14 Divide $\frac{2}{3}$ by $\frac{7}{5}$.

This is a symbol for 1.

$$\frac{2}{3} \div \frac{7}{5} = \frac{\frac{2}{3}}{\frac{7}{5}} = \frac{\frac{2}{3}}{\frac{7}{5}} \cdot \frac{\frac{5}{7}}{\frac{5}{7}} \qquad \text{Multiplying by } \frac{\frac{5}{7}}{\frac{5}{7}}\text{. We use } \frac{5}{7} \text{ because it is the reciprocal of } \frac{7}{5}.$$

$$= \frac{\frac{2}{3} \cdot \frac{5}{7}}{\frac{7}{5} \cdot \frac{5}{7}} \qquad \text{Multiplying numerators and denominators}$$

$$= \frac{\frac{10}{21}}{\frac{35}{35}}$$

$$= \frac{\frac{10}{21}}{1} = \frac{10}{21} \qquad \text{Simplifying}$$

After multiplying, we had a denominator of 1. That was because we used $\frac{5}{7}$, the reciprocal of the divisor, for both the numerator and the denominator of the symbol for 1.

Do Exercise 21.

When multiplying by 1 to divide, we get a denominator of 1. What do we get in the numerator? In Example 14, we got $\frac{2}{3} \times \frac{5}{7}$. This is the product of $\frac{2}{3}$, the dividend, and $\frac{5}{7}$, the reciprocal of the divisor.

To divide, multiply by the reciprocal of the divisor:

$$\frac{a}{b} \div \frac{c}{d} = \frac{a}{b} \cdot \frac{d}{c}.$$

EXAMPLE 15 Divide by multiplying by the reciprocal of the divisor: $\frac{1}{2} \div \frac{3}{5}$.

$$\frac{1}{2} \div \frac{3}{5} = \frac{1}{2} \cdot \frac{5}{3} \qquad \frac{5}{3} \text{ is the reciprocal of } \frac{3}{5}$$

$$= \frac{5}{6} \qquad \text{Multiplying}$$

After dividing, simplification is often possible and should be done.

EXAMPLE 16 Divide and simplify: $\frac{2}{3} \div \frac{4}{9}$.

$$\frac{2}{3} \div \frac{4}{9} = \frac{2}{3} \cdot \frac{9}{4} \qquad \frac{9}{4} \text{ is the reciprocal of } \frac{4}{9}$$

$$= \frac{\cancel{2} \cdot \cancel{3} \cdot 3}{\cancel{3} \cdot \cancel{2} \cdot 2} \qquad \text{Removing a factor of 1: } \frac{2 \cdot 3}{2 \cdot 3} = 1$$

$$= \frac{3}{2}$$

Do Exercises 22–24.

EXAMPLE 17 Divide and simplify: $\frac{5}{6} \div 30$.

$$\frac{5}{6} \div 30 = \frac{5}{6} \div \frac{30}{1} = \frac{5}{6} \cdot \frac{1}{30} = \underbrace{\frac{\cancel{5} \cdot 1}{6 \cdot \cancel{5} \cdot 6} = \frac{1}{6 \cdot 6}} = \frac{1}{36}$$

Removing a factor of 1: $\frac{5}{5} = 1$

EXAMPLE 18 Divide and simplify: $24 \div \frac{3}{8}$.

$$24 \div \frac{3}{8} = \frac{24}{1} \div \frac{3}{8} = \frac{24}{1} \cdot \frac{8}{3} = \frac{24 \cdot 8}{1 \cdot 3} = \underbrace{\frac{\cancel{3} \cdot 8 \cdot 8}{1 \cdot \cancel{3}} = \frac{8 \cdot 8}{1}} = 64$$

Removing a factor of 1: $\frac{3}{3} = 1$

Do Exercises 25 and 26.

Divide by multiplying by the reciprocal of the divisor. Then simplify.

22. $\frac{4}{3} \div \frac{7}{2}$

23. $\frac{5}{4} \div \frac{3}{2}$

24. $\frac{\frac{2}{9}}{\frac{5}{12}}$

Divide and simplify.

25. $\frac{7}{8} \div 56$

26. $36 \div \frac{4}{9}$

Answers on page A-1

From time to time, you will find some *Sidelights* like the one below. Although they are optional, you may find them helpful and of interest. They will include study tips, career opportunities involving mathematics, applications, computer–calculator exercises, and many other mathematical topics.

SIDELIGHTS

CALCULATOR CORNER: NUMBER PATTERNS

There are many interesting number patterns in mathematics. Look for a pattern in the following. You can use a calculator for the computations.

$$6^2 = 36$$
$$66^2 = 4356$$
$$666^2 = 443556$$
$$6666^2 = 44435556$$

Do you see a pattern? If so, find 66666^2 without using a calculator.

EXERCISES

In each of the following, do the first four calculations using a calculator. Look for a pattern. Use the pattern to do the last calculation without a calculator.

1. 3^2, 33^2, 333^2, 3333^2, 33333^2
2. $9 \cdot 6$, $99 \cdot 66$, $999 \cdot 666$, $9999 \cdot 6666$, $99999 \cdot 66666$
3. $37 \cdot 3$, $37 \cdot 33$, $37 \cdot 333$, $37 \cdot 3333$, $37 \cdot 33333$
4. $37 \cdot 3$, $37 \cdot 6$, $37 \cdot 9$, $37 \cdot 12$, $37 \cdot 15$
5. $(9 \cdot 9) + 7$, $(98 \cdot 9) + 6$, $(987 \cdot 9) + 5$, $(9876 \cdot 9) + 4$, $(98765 \cdot 9) + 3$
6. $(1 \cdot 8) + 1$, $(12 \cdot 8) + 2$, $(123 \cdot 8) + 3$, $(1234 \cdot 8) + 4$, $(12345 \cdot 8) + 5$
7. $8 \cdot 6$, $68 \cdot 6$, $668 \cdot 6$, $6668 \cdot 6$, $66668 \cdot 6$
8. $6 \cdot 4$, $66 \cdot 44$, $666 \cdot 444$, $6666 \cdot 4444$, $66666 \cdot 44444$
9. 1^2, 11^2, 111^2, 1111^2, 11111^2
10. 9^2, 99^2, 999^2, 9999^2, 99999^2
11. $77 \cdot 78$, $777 \cdot 78$, $7777 \cdot 78$, $77777 \cdot 78$, $777777 \cdot 78$
12. $1 \cdot 13 \cdot 76{,}923$, $2 \cdot 13 \cdot 76{,}923$, $3 \cdot 13 \cdot 76{,}923$, $4 \cdot 13 \cdot 76{,}923$, $5 \cdot 13 \cdot 76{,}923$

NAME SECTION DATE

Exercise Set R.2

a Write an equivalent expression for each of the following. Use the indicated name for 1.

1. $\frac{3}{4}$ $\left(\text{Use } \frac{3}{3} \text{ for } 1.\right)$

2. $\frac{5}{6}$ $\left(\text{Use } \frac{10}{10} \text{ for } 1.\right)$

3. $\frac{3}{5}$ $\left(\text{Use } \frac{20}{20} \text{ for } 1.\right)$

4. $\frac{8}{9}$ $\left(\text{Use } \frac{4}{4} \text{ for } 1.\right)$

5. $\frac{13}{20}$ $\left(\text{Use } \frac{8}{8} \text{ for } 1.\right)$

6. $\frac{13}{32}$ $\left(\text{Use } \frac{40}{40} \text{ for } 1.\right)$

Write an equivalent expression with the given denominator.

7. $\frac{7}{8}$ (Denominator: 24)

8. $\frac{5}{6}$ (Denominator: 48)

9. $\frac{5}{4}$ (Denominator: 16)

10. $\frac{2}{9}$ (Denominator: 54)

b Simplify.

11. $\frac{18}{27}$

12. $\frac{49}{56}$

13. $\frac{56}{14}$

14. $\frac{48}{27}$

15. $\frac{6}{42}$

16. $\frac{13}{104}$

17. $\frac{56}{7}$

18. $\frac{132}{11}$

19. $\frac{19}{76}$

20. $\frac{17}{51}$

21. $\frac{100}{20}$

22. $\frac{150}{25}$

23. $\frac{425}{525}$

24. $\frac{625}{325}$

25. $\frac{2600}{1400}$

26. $\frac{4800}{1600}$

27. $\frac{8 \cdot x}{6 \cdot x}$

28. $\frac{13 \cdot v}{39 \cdot v}$

ANSWERS

1. _____
2. _____
3. _____
4. _____
5. _____
6. _____
7. _____
8. _____
9. _____
10. _____
11. _____
12. _____
13. _____
14. _____
15. _____
16. _____
17. _____
18. _____
19. _____
20. _____
21. _____
22. _____
23. _____
24. _____
25. _____
26. _____
27. _____
28. _____

ANSWERS

29. ______
30. ______
31. ______
32. ______
33. ______
34. ______
35. ______
36. ______
37. ______
38. ______
39. ______
40. ______
41. ______
42. ______
43. ______
44. ______
45. ______
46. ______
47. ______
48. ______
49. ______
50. ______
51. ______
52. ______
53. ______
54. ______
55. ______
56. ______
57. ______
58. ______
59. ______
60. ______
61. ______
62. ______
63. ______
64. ______
65. ______
66. ______
67. ______

c Compute and simplify.

29. $\frac{1}{3} \cdot \frac{1}{4}$ **30.** $\frac{15}{16} \cdot \frac{8}{5}$ **31.** $\frac{15}{4} \cdot \frac{3}{4}$ **32.** $\frac{10}{11} \cdot \frac{11}{10}$

33. $\frac{1}{3} + \frac{1}{3}$ **34.** $\frac{1}{4} + \frac{1}{3}$ **35.** $\frac{4}{9} + \frac{13}{18}$ **36.** $\frac{4}{5} + \frac{8}{15}$

37. $\frac{3}{10} + \frac{8}{15}$ **38.** $\frac{9}{8} + \frac{7}{12}$ **39.** $\frac{5}{4} - \frac{3}{4}$ **40.** $\frac{12}{5} - \frac{2}{5}$

41. $\frac{11}{12} - \frac{3}{8}$ **42.** $\frac{15}{16} - \frac{5}{12}$ **43.** $\frac{11}{12} - \frac{2}{5}$ **44.** $\frac{15}{16} - \frac{2}{3}$

45. $\frac{7}{6} \div \frac{3}{5}$ **46.** $\frac{7}{5} \div \frac{3}{4}$ **47.** $\frac{8}{9} \div \frac{4}{15}$ **48.** $\frac{3}{4} \div \frac{3}{7}$

49. $\frac{1}{8} \div \frac{1}{4}$ **50.** $\frac{1}{20} \div \frac{1}{5}$ **51.** $\frac{\frac{13}{12}}{\frac{39}{5}}$ **52.** $\frac{\frac{17}{6}}{\frac{3}{8}}$

53. $100 \div \frac{1}{5}$ **54.** $78 \div \frac{1}{6}$ **55.** $\frac{3}{4} \div 10$ **56.** $\frac{5}{6} \div 15$

SKILL MAINTENANCE

This heading indicates that the exercises that follow are *skill maintenance exercises,* which review any skill previously studied in the text. You can expect such exercises in almost every exercise set.

Find the prime factorization.

57. 28 **58.** 56 **59.** 1000 **60.** 192

Find the LCM.

61. 16, 24 **62.** 28, 49, 56

SYNTHESIS

Simplify.

63. $\frac{192}{256}$ **64.** $\frac{p \cdot q}{r \cdot q}$ **65.** $\frac{64 \cdot a \cdot b}{16 \cdot a \cdot b}$

66. $\frac{4 \cdot 9 \cdot 24}{2 \cdot 8 \cdot 15}$ **67.** $\frac{36 \cdot (2 \cdot h)}{8 \cdot (9 \cdot h)}$

R.3 *Decimal Notation*

OBJECTIVES

After finishing Section R.3, you should be able to:

a Convert between decimal notation and fractional notation.

b Add, subtract, multiply, and divide using decimal notation.

c Round numbers to a specified decimal place.

FOR EXTRA HELP

TAPE 1 | TAPE 1B | MAC: R IBM: R

Let us say that the cost of a sound system is

$1768.95.

This is given in **decimal notation.** The following place-value chart shows the place value of each digit in 1768.95.

PLACE-VALUE CHART								
Ten Thousands	Thousands	Hundreds	Tens	Ones	Ten*ths*	Hundred*ths*	Thousand*ths*	Ten-Thousand*ths*
10,000	1000	100	10	1	$\frac{1}{10}$	$\frac{1}{100}$	$\frac{1}{1000}$	$\frac{1}{10,000}$
	1	7	6	8 .	9	5		

a Converting Between Certain Decimal Notation and Fractional Notation

Decimals are defined in terms of fractions—for example,

$$0.1 = \frac{1}{10}, \qquad 0.6875 = \frac{6875}{10,000}, \qquad 53.47 = \frac{5347}{100}.$$

We see that the number of zeros in the denominator is the same as the number of decimal places in the number. From these examples, we obtain the following procedure for converting from this type of decimal notation to fractional notation.

> To convert from decimal notation to fractional notation:
>
> **a)** Count the number of decimal places. — 4.98 (2 places)
>
> **b)** Move the decimal point that many places to the right. — 4.98. (Move 2 places.)
>
> **c)** Write the result over a denominator with that number of zeros. — $\frac{498}{100}$ (2 zeros)

EXAMPLE 1 Convert 0.876 to fractional notation. Do not simplify.

0.876 (3 places) 0.876. (3 places) $0.876 = \frac{876}{1000}$ (3 zeros)

Convert to fractional notation. Do not simplify.

1. 0.568

2. 2.3

3. 89.04

Answers on page A-1

Convert to decimal notation.

4. $\frac{4131}{1000}$

5. $\frac{4131}{10,000}$

6. $\frac{573}{100}$

Add.

7. 69 + 1.785 + 213.67

8. 17.95 + 14.68 + 236

Answers on page A-1

EXAMPLE 2 Convert 1.5018 to fractional notation. Do not simplify.

1.5018 1.5018. (4 places) $1.5018 = \frac{15,018}{10,000}$ (4 zeros)

Do Exercises 1–3 on the preceding page.

> To convert from fractional notation to decimal notation when the denominator is a number like 10, 100, or 1000:
>
> a) Count the number of zeros. $\frac{8679}{1000}$ 3 zeros
>
> b) Move the decimal point that number of places to the left. Leave off the denominator. 8.679. Move 3 places.

EXAMPLE 3 Convert to decimal notation: $\frac{123,067}{10,000}$.

$\frac{123,067}{10,000}$ (4 zeros) 12.3067. (4 places) $\frac{123,067}{10,000} = 12.3067$

Do Exercises 4–6.

b ADDITION, SUBTRACTION, MULTIPLICATION, AND DIVISION

ADDITION

> Adding with decimal notation is similar to adding whole numbers. First we line up the decimal points. Then we add the thousandths, then the hundredths, and so on, carrying if necessary.

EXAMPLE 4 Add: 74 + 26.46 + 0.998.

```
  1 1 1
  7 4.
  2 6.4 6
+   0.9 9 8
-----------
1 0 1.4 5 8
```

You may put extra zeros to the right of any decimal point so there are the same number of decimal places, but this is not necessary. If you did, the preceding problem would look like this:

```
  1 1 1
  7 4.0 0 0
  2 6.4 6 0
+   0.9 9 8
-----------
1 0 1.4 5 8
```

Do Exercises 7 and 8.

SUBTRACTION

> Subtracting with decimal notation is similar to subtracting whole numbers. First we line up the decimal points. Then we subtract the thousandths, then the hundredths, the tenths, and so on, borrowing if necessary. Extra zeros can be added if needed.

EXAMPLES

5. Subtract: 76.14 − 18.953.

$$\begin{array}{r} \overset{6}{\cancel{7}}\,\overset{15}{\overset{5}{\cancel{6}}}.\overset{10}{\overset{0}{\cancel{1}}}\,\overset{13}{\overset{3}{\cancel{4}}}\,\overset{10}{\cancel{0}} \\ -\;1\;8.9\;5\;3 \\ \hline 5\;7.1\;8\;7 \end{array}$$

6. Subtract: 200 − 0.68.

$$\begin{array}{r} \overset{1}{\cancel{2}}\,\overset{9}{\cancel{0}}\,\overset{9}{\cancel{0}}.\overset{9}{\cancel{0}}\,\overset{10}{\cancel{0}} \\ -\quad 0.6\;8 \\ \hline 1\;9\;9.3\;2 \end{array}$$

Do Exercises 9–12.

Subtract.

9. 29.35 − 1.674

10. 92.375 − 27.692

11. 100 − 0.41

12. 240 − 0.117

MULTIPLICATION

Look at this product.

$$\underset{\text{2 places}}{5.14} \times \underset{\text{1 place}}{0.8} = \frac{514}{100} \times \frac{8}{10} = \frac{514 \times 8}{100 \times 10} = \frac{4112}{1000} = \underset{\text{3 places}}{4.112}$$

We can also do this more quickly by multiplying the whole numbers 8 and 514 and determining the position of the decimal point.

> To multiply with decimal notation:
>
> a) Ignore the decimal points and multiply as whole numbers.
>
> b) Place the decimal point in the result of step (a) by adding the number of decimal places in the original factors.

EXAMPLE 7 Multiply: 5.14 × 0.8.

a) Ignore the decimal points and multiply as whole numbers.

$$\begin{array}{r} \overset{1}{5}.\overset{3}{1}\;4 \\ \times\quad 0.8 \\ \hline 4\;1\;1\;2 \end{array}$$

b) Place the decimal point in the result of step (a) by adding the number of decimal places in the original factors.

$$\begin{array}{rl} 5.1\;4 & \leftarrow \text{2 decimal places} \\ \times\quad 0.8 & \leftarrow \text{1 decimal place} \\ \hline 4.1\;1\;2 & \uparrow \text{3 decimal places} \end{array}$$

Do Exercises 13–15.

Multiply.

13. $\begin{array}{r} 6.5\;2 \\ \times\quad 0.9 \\ \hline \end{array}$

14. $\begin{array}{r} 6.5\;2 \\ \times\,0.0\;9 \\ \hline \end{array}$

15. $\begin{array}{r} 5\;6.7\;6 \\ \times\,0.9\;0\;8 \\ \hline \end{array}$

Answers on page A-1

Divide.

16. $7\overline{)342.3}$

17. $16\overline{)253.12}$

Divide.

18. $25\overline{)32}$

19. $38\overline{)682.1}$

Answers on page A-1

DIVISION

Note that 37.6 ÷ 8 = 4.7 because 8 × 4.7 = 37.6. If we write this as

```
      4.7
  8 ) 3 7.6
      3 2
        5 6
        5 6
          0
```

we see how the following method can be used to divide by a whole number.

> To divide by a whole number:
>
> a) Place the decimal point in the quotient directly above the decimal point in the dividend.
>
> b) Divide as whole numbers.

EXAMPLE 8 Divide: 216.75 ÷ 25.

a)

```
              .
  2 5 ) 2 1 6.7 5
              ↑
```

Place the decimal point.

b)

```
            8.6 7
  2 5 ) 2 1 6.7 5
        2 0 0
          1 6 7
          1 5 0
            1 7 5
            1 7 5
                0
```

Divide as though dividing whole numbers.

Do Exercises 16 and 17.

It is sometimes helpful to write extra zeros to the right of the decimal point. The answer is not changed. Remember that the decimal point for a whole number, though not normally written, is to the right of the number.

EXAMPLE 9 Divide: 54 ÷ 8.

a)

```
          .
  8 ) 5 4.
```

b)

```
        6.7 5
  8 ) 5 4.0 0
      4 8
        6 0
        5 6
          4 0
          4 0
            0
```

Extra zeros are written to the right of the decimal point as needed.

Do Exercises 18 and 19.

> To divide when the divisor is not a whole number:
>
> a) Move the decimal point in the divisor as many places to the right as it takes to make it a whole number. Move the decimal point in the dividend the same number of places to the right and place the decimal point in the quotient.
>
> b) Divide as whole numbers, adding zeros if necessary.

EXAMPLE 10 Divide: 83.79 ÷ 0.098.

a)

```
                  .
0.0 9 8. ) 8 3.7 9 0.
```

b)

```
                 8 5 5.
0.0 9 8^ ) 8 3.7 9 0^
           7 8 4
             5 3 9
             4 9 0
               4 9 0
               4 9 0
                   0
```

Do Exercises 20 and 21.

Divide.

20. $0.024\overline{)20.544}$

21. $4.6\overline{)3.91}$

Converting from Fractional Notation to Decimal Notation

To convert from fractional notation to decimal notation when the denominator is not a number like 10, 100, or 1000, we divide the numerator by the denominator.

EXAMPLE 11 Convert to decimal notation: $\frac{5}{16}$.

```
      0.3 1 2 5
1 6 ) 5.0 0 0 0
      4 8
        2 0
        1 6
          4 0
          3 2
            8 0
            8 0
              0
```

If we get a remainder of 0, the decimal terminates. Decimal notation is 0.3125.

EXAMPLE 12 Convert to decimal notation: $\frac{7}{12}$.

```
      0.5 8 3 3
1 2 ) 7.0 0 0 0
      6 0
      1 0 0
        9 6
          4 0
          3 6
            4 0
            3 6
              4
```

The number 4 repeats as a remainder, so the digits will repeat in the quotient. Therefore,

$$\frac{7}{12} = 0.583333\ldots.$$

Convert to decimal notation.

22. $\frac{5}{8}$

23. $\frac{2}{3}$

24. $\frac{84}{11}$

Answers on page A-1

Round to the nearest tenth.

25. 2.76

26. 13.85

27. 7.009

Round to the nearest hundredth.

28. 7.834

29. 34.675

30. 0.025

Round to the nearest thousandth.

31. 0.9434

32. 8.0038

33. 43.1119

34. 37.4005

Round 7459.3549 to the nearest:

35. thousandth.

36. hundredth.

37. tenth.

38. one.

39. ten.

Answers on page A-1

Instead of dots, we often put a bar over the repeating part—in this case, only the 3. Thus,

$$\frac{7}{12} = 0.58\overline{3}.$$

Do Exercises 22–24 on the preceding page.

c Rounding

When working with decimal notation, we often shorten notation by **rounding**. Although there are many ways to round, we will use the following.

> To round to a certain place:
> a) Locate the digit in that place.
> b) Then consider the digit to its right.
> c) If the digit to the right is 5 or higher, round up; if the digit to the right is less than 5, round down.

Example 13 Round 3872.2459 to the nearest tenth.

a) We locate the digit in the tenths place.

3 8 7 2.2 4 5 9 (↑ under the 2)

b) Then we consider the next digit to the right.

3 8 7 2.2 4 5 9 (↑ under the 4)

c) Since that digit is less than 5, we round down.

3 8 7 2.2 ← This is the answer.

Note that 3872.3 is *not* a correct answer to Example 13. It is incorrect to round from the ten-thousandths place over, as follows:

3872.246, 3872.25, 3872.3

Example 14 Round 3872.2459 to the nearest thousandth, hundredth, tenth, one, ten, hundred, and thousand.

thousandth:	3872.246
hundredth:	3872.25
tenth:	3872.2
one:	3872
ten:	3870
hundred:	3900
thousand:	4000

Do Exercises 25–39.

In rounding, we sometimes use the symbol ≈, which means "is approximately equal to." Thus,

$$46.124 \approx 46.1.$$

NAME SECTION DATE

Exercise Set R.3

a Convert to fractional notation. Do not simplify.

1. 5.3 **2.** 2.7 **3.** 0.67 **4.** 0.93

5. 2.0007 **6.** 4.0008 **7.** 7889.8 **8.** 1122.3

Convert to decimal notation.

9. $\frac{1}{10}$ **10.** $\frac{1}{100}$ **11.** $\frac{1}{10{,}000}$ **12.** $\frac{1}{1000}$

13. $\frac{9999}{1000}$ **14.** $\frac{39}{10{,}000}$ **15.** $\frac{4578}{10{,}000}$ **16.** $\frac{94}{100{,}000}$

b Add.

17. $\begin{array}{r} 415.78 \\ +\ 29.16 \\ \hline \end{array}$ **18.** $\begin{array}{r} 708.99 \\ +\ 75.48 \\ \hline \end{array}$ **19.** $\begin{array}{r} 234.000 \\ +156.617 \\ \hline \end{array}$ **20.** $\begin{array}{r} 1345.12 \\ +\ 566.98 \\ \hline \end{array}$

21. $85 + 67.95 + 2.774$ **22.** $119 + 43.74 + 18.876$

23. $17.95 + 16.99 + 28.85$ **24.** $14.59 + 16.79 + 19.95$

Subtract.

25. $\begin{array}{r} 78.110 \\ -45.876 \\ \hline \end{array}$ **26.** $\begin{array}{r} 14.080 \\ -\ 9.199 \\ \hline \end{array}$ **27.** $\begin{array}{r} 38.7 \\ -11.865 \\ \hline \end{array}$ **28.** $\begin{array}{r} 300. \\ -\ 24.677 \\ \hline \end{array}$

29. $57.86 - 9.95$ **30.** $2.6 - 1.08$ **31.** $3 - 1.0807$ **32.** $5 - 3.4051$

Multiply.

33. $\begin{array}{r} 7.34 \\ \times\ 1.8 \\ \hline \end{array}$ **34.** $\begin{array}{r} 6.55 \\ \times\ 3.2 \\ \hline \end{array}$ **35.** $\begin{array}{r} 0.86 \\ \times 0.93 \\ \hline \end{array}$ **36.** $\begin{array}{r} 0.028 \\ \times 7.409 \\ \hline \end{array}$

Answers

1.
2.
3.
4.
5.
6.
7.
8.
9.
10.
11.
12.
13.
14.
15.
16.
17.
18.
19.
20.
21.
22.
23.
24.
25.
26.
27.
28.
29.
30.
31.
32.
33.
34.
35.
36.

ANSWERS

37. ____
38. ____
39. ____
40. ____
41. ____
42. ____
43. ____
44. ____
45. ____
46. ____
47. ____
48. ____
49. ____
50. ____
51. ____
52. ____
53. ____
54. ____
55. ____
56. ____
57. ____
58. ____
59. ____
60. ____
61. ____
62. ____
63. ____
64. ____
65. ____
66. ____
67. ____
68. ____
69. ____
70. ____
71. ____
72. ____
73. ____
74. ____
75. ____
76. ____
77. ____
78. ____

37. 17.95×10

38. 17.95×100

39. 18.94×0.01

40. 18.94×0.1

41. 0.457×3.08

42. 0.0024×0.015

43. 3.642×0.99

44. 287.4×1.08

Divide.

45. $72\overline{)165.6}$

46. $5.2\overline{)44.2}$

47. $8.5\overline{)44.2}$

48. $7.8\overline{)72.54}$

49. $9.9\overline{)0.2277}$

50. $100\overline{)95}$

51. $0.64\overline{)12}$

52. $1.6\overline{)75}$

53. $1.05\overline{)693}$

54. $25\overline{)4}$

55. $8.6\overline{)5.848}$

56. $0.47\overline{)0.1222}$

Convert to decimal notation.

57. $\frac{11}{32}$

58. $\frac{17}{32}$

59. $\frac{13}{11}$

60. $\frac{17}{12}$

61. $\frac{5}{9}$

62. $\frac{5}{6}$

63. $\frac{19}{9}$

64. $\frac{9}{11}$

c Round to the nearest hundredth, tenth, one, ten, and hundred.

65. 745.06534

66. 317.18565

67. 6780.50568

68. 840.15493

Round to the nearest cent and to the nearest dollar (nearest one).

69. \$17.988

70. \$20.492

71. \$346.075

72. \$4.718

Round to the nearest dollar.

73. \$16.95

74. \$17.50

75. \$189.50

76. \$567.24

Divide and round to the nearest ten-thousandth, thousandth, hundredth, tenth, and one.

77. $\frac{1000}{81}$

78. $\frac{23}{17}$

R.4 Percent Notation

After finishing Section R.4, you should be able to:

- **a** Convert from percent notation to decimal notation.
- **b** Convert from percent notation to fractional notation.
- **c** Convert from decimal notation to percent notation.
- **d** Convert from fractional notation to percent notation.

FOR EXTRA HELP

TAPE 2 | TAPE 1B | MAC: R IBM: R

a CONVERTING TO DECIMAL NOTATION

The average family spends 28% of its income for food. What does this mean? It means that of every \$100 earned, \$28 is spent for food. Thus, 28% is a ratio of 28 to 100.

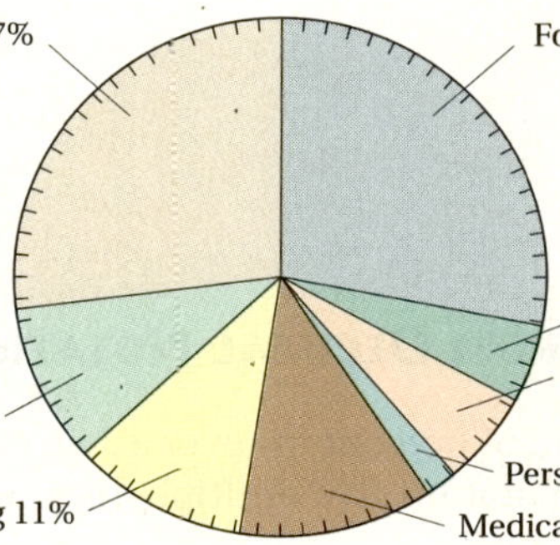

The percent symbol % means "per hundred." We can regard the percent symbol as part of a name for a number. For example,

$$28\% \text{ is defined to mean } 28 \times 0.01 \text{ or } 28 \times \frac{1}{100} \text{ or } \frac{28}{100}.$$

> $n\%$ means $n \times 0.01$, or $n \times \frac{1}{100}$, or $\frac{n}{100}$.

EXAMPLE 1 Convert 78.5% to decimal notation.

$78.5\% = 78.5 \times 0.01$ Replacing % by $\times$ 0.01

$= 0.785$

> To convert from percent notation to decimal notation, move the decimal point two places to the left and drop the percent symbol.

EXAMPLE 2 Convert 43.67% to decimal notation.

43.67% 0.43.67 43.67% = 0.4367

Move the decimal point two places to the left.

Do Exercises 1 and 2.

b CONVERTING TO FRACTIONAL NOTATION

EXAMPLE 3 Convert 88% to fractional notation.

$88\% = 88 \times \frac{1}{100}$ Replacing % by $\times \frac{1}{100}$

$= \frac{88}{100}$ You need not simplify.

Convert to decimal notation.

1. 87.3%

2. 100%

Answers on page A-1

Convert to fractional notation.

3. 53%

4. 45.6%

5. 0.23%

Convert to percent notation.

6. 6.77

7. 0.9944

Convert to percent notation.

8. $\frac{1}{4}$

9. $\frac{7}{8}$

10. $\frac{2}{3}$

Answers on page A-1

EXAMPLE 4 Convert 34.7% to fractional notation.

$$\begin{aligned} 34.7\% &= 34.7 \times \frac{1}{100} && \text{Replacing \% by } \times\frac{1}{100} \\ &= \frac{34.7}{100} \\ &= \frac{34.7}{100}\cdot\frac{10}{10} && \text{Multiplying by 1 to get a whole number in the numerator} \\ &= \frac{347}{1000} \end{aligned}$$

Do Exercises 3–5.

c CONVERTING FROM DECIMAL NOTATION

By applying the definition of percent in reverse, we can convert from decimal notation to percent notation. We multiply by 1, expressing it as 100×0.01 and replacing $\times\, 0.01$ by %.

EXAMPLE 5 Convert 0.93 to percent notation.

$$\begin{aligned} 0.93 &= 0.93 \times 1 \\ &= 0.93 \times 100\% && \text{Expressing 1 as 100\%} \\ &= 0.93 \times (100 \times 0.01) && \text{Expressing 100\% as } 100 \times 0.01 \\ &= (0.93 \times 100) \times 0.01 \\ &= 93 \times 0.01 \\ &= 93\% && \text{Replacing } \times\, 0.01 \text{ by \%} \end{aligned}$$

To convert from decimal notation to percent notation, move the decimal point two places to the right and write a percent symbol.

EXAMPLE 6 Convert 0.032 to percent notation.

0.032 0.03.2 $0.032 = 3.2\%$

Move the decimal point two places to the right.

Do Exercises 6 and 7.

d CONVERTING FROM FRACTIONAL NOTATION

We can also convert from fractional notation to percent notation by converting first to decimal notation.

EXAMPLE 7 Convert $\frac{5}{8}$ to percent notation.

$$\frac{5}{8} = 0.625 = 62.5\%$$

Do Exercises 8–10.

NAME SECTION DATE

Exercise Set R.4

a Convert to decimal notation.

1. 63% **2.** 64% **3.** 94.1% **4.** 34.6%

5. 1% **6.** 100% **7.** 0.61% **8.** 125%

9. 240% **10.** 0.73% **11.** 3.25% **12.** 2.3%

b Convert to fractional notation.

13. 60% **14.** 40% **15.** 28.9% **16.** 37.5%

17. 110% **18.** 120% **19.** 0.042% **20.** 0.68%

21. 250% **22.** 3.2% **23.** 3.47% **24.** 12.557%

c Convert to percent notation.

25. 1 **26.** 8.56 **27.** 0.996 **28.** 0.83

29. 0.0047 **30.** 2 **31.** 0.072 **32.** 1.34

Answers

1. ______
2. ______
3. ______
4. ______
5. ______
6. ______
7. ______
8. ______
9. ______
10. ______
11. ______
12. ______
13. ______
14. ______
15. ______
16. ______
17. ______
18. ______
19. ______
20. ______
21. ______
22. ______
23. ______
24. ______
25. ______
26. ______
27. ______
28. ______
29. ______
30. ______
31. ______
32. ______

ANSWERS

33. ______
34. ______
35. ______
36. ______
37. ______
38. ______
39. ______
40. ______
41. ______
42. ______
43. ______
44. ______
45. ______
46. ______
47. ______
48. ______
49. ______
50. ______
51. ______
52. ______
53. ______
54. ______
55. ______
56. ______
57. ______
58. ______
59. ______
60. ______
61. ______
62. ______
63. ______
64. ______
65. ______
66. ______
67. ______

33. 9.2 **34.** 0.013 **35.** 0.0068 **36.** 0.675

d Convert to percent notation.

37. $\frac{1}{6}$ **38.** $\frac{1}{5}$ **39.** $\frac{13}{20}$ **40.** $\frac{14}{25}$

41. $\frac{29}{100}$ **42.** $\frac{123}{100}$ **43.** $\frac{8}{10}$ **44.** $\frac{7}{10}$

45. $\frac{3}{5}$ **46.** $\frac{17}{50}$ **47.** $\frac{2}{3}$ **48.** $\frac{7}{8}$

49. $\frac{7}{4}$ **50.** $\frac{3}{8}$ **51.** $\frac{3}{4}$ **52.** $\frac{99.4}{100}$

SKILL MAINTENANCE

Convert to decimal notation.

53. $\frac{9}{4}$ **54.** $\frac{11}{8}$ **55.** $\frac{17}{12}$ **56.** $\frac{8}{9}$ **57.** $\frac{10}{11}$ **58.** $\frac{17}{11}$

SYNTHESIS

Simplify. Express the answer in percent notation.

59. 18% + 14% **60.** 84% − 12% **61.** 1 − 30%

62. 92% − 10% **63.** 27 × 100% **64.** 42% − (1 − 58%)

65. 3(1 + 15%) **66.** 7(1% + 13%) **67.** $\frac{100\%}{40}$

R.5 *Exponential Notation and Order of Operations*

a EXPONENTIAL NOTATION

Exponents provide a shorter way of writing products. A product in which the factors are the same is called a **power**. For

$10 \cdot 10 \cdot 10$, we write 10^3.

3 factors

This is read "ten to the third power." We call the number 3 an **exponent** and we say that 10 is the **base**. An exponent of 2 or greater tells how many times the base is used as a factor. For example,

$$a \cdot a \cdot a \cdot a = a^4.$$

In this case, the exponent is 4 and the base is a. An expression for a power is called **exponential notation.**

a^n — This is the exponent. (the n)

This is the base. (the a)

EXAMPLE 1 Write exponential notation for $10 \cdot 10 \cdot 10 \cdot 10 \cdot 10$.

$$10 \cdot 10 \cdot 10 \cdot 10 \cdot 10 = 10^5$$

Do Exercises 1–3.

b EVALUATING EXPONENTIAL EXPRESSIONS

EXAMPLE 2 Evaluate: 5^2.

$$5^2 = 5 \cdot 5 = 25$$

EXAMPLE 3 Evaluate: 3^4.

We have

$$3^4 = 3 \cdot 3 \cdot 3 \cdot 3 = 9 \cdot 9 = 81.$$

We could also carry out the calculation as follows:

$$3^4 = 3 \cdot 3 \cdot 3 \cdot 3 = 9 \cdot 3 \cdot 3 = 27 \cdot 3 = 81.$$

EXPONENTIAL NOTATION

For any natural number n greater than or equal to 2,

$$b^n = \underbrace{b \cdot b \cdot b \cdot b \cdots b}_{n \text{ factors}}.$$

Do Exercises 4–6.

OBJECTIVES

After finishing Section R.5, you should be able to:

- **a** Write exponential notation for a product.
- **b** Evaluate exponential expressions.
- **c** Simplify expressions using the rules for order of operations.

FOR EXTRA HELP

TAPE 2   TAPE 2A   MAC: R IBM: R

Write exponential notation.

1. $4 \cdot 4 \cdot 4$

2. $6 \cdot 6 \cdot 6 \cdot 6 \cdot 6$

3. 1.08×1.08

Evaluate.

4. 10^4

5. 8^3

6. $(1.1)^3$

Answers on page A-1

Calculate.

7. $18 - 4 \times 3 + 7$

8. $(2 \times 5)^3$

9. 2×5^3

10. $8 + 2 \times 5^3 - 4 \cdot 20$

Answers on page A-1

C ORDER OF OPERATIONS

What does $5 \times 2 + 4$ mean? If we multiply 5 by 2 and add 4, we get 14. If we add 2 and 4 and multiply by 5, we get 30. Since our results are different, we see that the order in which we carry out operations is important. To tell which operation to do first, we can use grouping symbols such as parentheses ( ), or brackets [ ], or braces { }. For example,

$$(3 \times 5) + 6 = 15 + 6 = 21,$$

but

$$3 \times (5 + 6) = 3 \times 11 = 33.$$

Grouping symbols tell us what to do first. If there are no grouping symbols, we have agreements about the order in which operations should be done.

RULES FOR ORDER OF OPERATIONS

1. Do all calculations within grouping symbols before operations outside.
2. Evaluate all exponential expressions.
3. Do all multiplications and divisions in order from left to right.
4. Do all additions and subtractions in order from left to right.

EXAMPLE 4 Calculate: $15 - 2 \times 5 + 3$.

$15 - 2 \times 5 + 3 = 15 - 10 + 3$ **Multiplying**

$= 5 + 3$ **Subtracting**

$= 8$ **Adding**

Always calculate within parentheses first. When there are exponents and no parentheses, simplify powers before multiplying or dividing.

EXAMPLE 5 Calculate: $(3 \times 4)^2$.

$(3 \times 4)^2 = (12)^2$ **Working within parentheses first**

$= 144$ **Evaluating the exponential expression**

EXAMPLE 6 Calculate: 3×4^2.

We have

$3 \times 4^2 = 3 \times 16$ **Evaluating the exponential expression**

$= 48.$ **Multiplying**

Note that $(3 \times 4)^2 \neq 3 \times 4^2$.

EXAMPLE 7 Calculate: $7 + 3 \times 4^2 - 29$.

$7 + 3 \times 4^2 - 29 = 7 + 3 \times 16 - 29$ **There are no parentheses, so we find 4^2 first.**

$= 7 + 48 - 29$ **Multiplying second**

$= 55 - 29$ **Adding**

$= 26$ **Subtracting**

Do Exercises 7–10.

EXAMPLE 8 Calculate: $2.56 \div 1.6 \div 0.4$.

$$2.56 \div 1.6 \div 0.4 = 1.6 \div 0.4 \quad \text{Doing the divisions in order from left to right}$$
$$= 4 \quad \text{Doing the second division}$$

EXAMPLE 9 Calculate: $1000 \cdot \frac{1}{10} \div \frac{4}{5}$.

$$1000 \cdot \frac{1}{10} \div \frac{4}{5} = 100 \div \frac{4}{5} \quad \text{Doing the multiplication}$$
$$= 125 \quad \text{Dividing}$$

Do Exercises 11 and 12.

Calculate.

11. $51.2 \div 0.64 \div 40$

12. $1000 \div \frac{1}{10} \cdot \frac{4}{5}$

Sometimes combinations of grouping symbols are used, as in

$$5[4 + (8 - 2)].$$

The rules still apply. We begin with the innermost grouping symbols—in this case, the parentheses—and work to the outside.

EXAMPLE 10 Calculate: $5[4 + (8 - 2)]$.

$$5[4 + (8 - 2)] = 5[4 + 6] \quad \text{Subtracting within the parentheses first}$$
$$= 5[10] \quad \text{Adding inside the brackets}$$
$$= 50 \quad \text{Multiplying}$$

A fraction bar can play the role of a grouping symbol, although such a symbol is not as evident as the others.

Calculate.

13. $4[(8 - 3) + 7]$

14. $\dfrac{13(10 - 6) + 4.9}{5^2 - 3^2}$

EXAMPLE 11 Calculate: $\dfrac{12(9 - 7) + 4 \cdot 5}{3^4 + 2^3}$.

An equivalent expression with brackets as grouping symbols is

$$[12(9 - 7) + 4 \cdot 5] \div [3^4 + 2^3].$$

What this shows, in effect, is that we do the calculations in the numerator and then in the denominator, and divide the results:

$$\frac{12(9 - 7) + 4 \cdot 5}{3^4 + 2^3} = \frac{12(2) + 4 \cdot 5}{81 + 8}$$
$$= \frac{24 + 20}{89}$$
$$= \frac{44}{89}.$$

Do Exercises 13 and 14.

Answers on page A-1

SIDELIGHTS

FACTORS AND SUMS

To *factor* a number is to express it as a product. Since $15 = 5 \cdot 3$, we say that 15 is *factored* and that 5 and 3 are *factors* of 15. In the table below, the top number has been factored in such a way that the sum of the factors is the bottom number. For example, in the first column, 56 has been factored as $7 \cdot 8$, and $7 + 8 = 15$, the bottom number. Such thinking will be important in a later chapter of this text.

PRODUCT	56	63	36	72	140	96		168	110			
FACTOR	7									9	24	3
FACTOR	8						8	8		10	18	
SUM	15	16	20	38	24	20	14		21			24

EXERCISE
Find the missing numbers in the table.

NAME SECTION DATE

Exercise Set R.5

a Write exponential notation.

1. $5 \times 5 \times 5 \times 5$

2. $3 \times 3 \times 3 \times 3 \times 3$

3. $10 \cdot 10 \cdot 10$

4. $10 \times 10 \times 10 \times 10 \times 10$

5. $1 \cdot 1 \cdot 1$

6. $18 \cdot 18$

b Evaluate.

7. 7^2

8. 4^3

9. 9^5

10. 12^4

11. 10^2

12. 1^5

13. 1^4

14. $(1.8)^2$

15. $(2.3)^2$

16. $(0.1)^3$

17. $(0.2)^3$

18. $(14.8)^2$

19. $(20.4)^2$

20. $\left(\frac{4}{5}\right)^2$

21. $\left(\frac{3}{8}\right)^2$

22. 2^4

23. 5^3

24. $(1.4)^3$

25. $1000 \times (1.02)^3$

26. $2000 \times (1.06)^2$

c Calculate.

27. $9 + 2 \times 8$

28. $14 + 6 \times 6$

29. $9 \times 8 + 7 \times 6$

30. $30 \times 5 + 2 \times 2$

31. $39 - 4 \times 2 + 2$

32. $14 - 2 \times 6 + 7$

33. $9 \div 3 + 16 \div 8$

34. $32 - 8 \div 4 - 2$

35. $7 + 10 - 10 \div 2$

36. $(5 \cdot 4)^2$

37. $(6 \cdot 3)^2$

38. $3 \cdot 2^3$

ANSWERS

1. ______
2. ______
3. ______
4. ______
5. ______
6. ______
7. ______
8. ______
9. ______
10. ______
11. ______
12. ______
13. ______
14. ______
15. ______
16. ______
17. ______
18. ______
19. ______
20. ______
21. ______
22. ______
23. ______
24. ______
25. ______
26. ______
27. ______
28. ______
29. ______
30. ______
31. ______
32. ______
33. ______
34. ______
35. ______
36. ______
37. ______
38. ______

ANSWERS

39. ____
40. ____
41. ____
42. ____
43. ____
44. ____
45. ____
46. ____
47. ____
48. ____
49. ____
50. ____
51. ____
52. ____
53. ____
54. ____
55. ____
56. ____
57. ____
58. ____
59. ____
60. ____
61. ____
62. ____
63. ____
64. ____
65. ____
66. ____
67. ____
68. ____
69. ____
70. ____
71. ____
72. ____
73. ____
74. ____
75. ____
76. ____
77. ____

39. $4 \cdot 5^2$ **40.** $(7 + 3)^2$ **41.** $(8 + 2)^3$

42. $7 + 2^2$ **43.** $6 + 4^2$ **44.** $(5 - 2)^2$

45. $(3 - 2)^2$ **46.** $10 - 3^2$ **47.** $4^3 \div 8 - 4$

48. $20 + 4^3 \div 8 - 4$ **49.** $120 - 3^3 \cdot 4 \div 6$ **50.** $7 \times 3^4 + 18$

51. $6[9 + (3 + 4)]$ **52.** $8[(13 + 6) - 11]$ **53.** $8 + (7 + 9)$

54. $(8 + 7) + 9$ **55.** $15(4 + 2)$ **56.** $15 \cdot 4 + 15 \cdot 2$

57. $12 - (8 - 4)$ **58.** $(12 - 8) - 4$ **59.** $1000 \div 100 \div 10$

60. $256 \div 32 \div 4$ **61.** $2000 \div \frac{3}{50} \cdot \frac{3}{2}$ **62.** $400 \times 0.64 \div 3.2$

63. $\dfrac{80 - 6^2}{9^2 + 3^2}$ **64.** $\dfrac{5^2 + 4^3 - 3}{9^2 - 2^2 + 1^5}$ **65.** $\dfrac{3(6 + 7) - 5 \cdot 4}{6 \cdot 7 + 8(4 - 1)}$

66. $\dfrac{20(8 - 3) - 4(10 - 3)}{10(6 + 2) + 2(5 + 2)}$

SKILL MAINTENANCE

67. Find percent notation: $\frac{5}{16}$.

68. Simplify: $\frac{125}{325}$.

69. Find the prime factorization of 48.

70. Find the LCM of 12, 24, and 56.

71. Simplify: $\frac{64}{96}$.

72. Find percent notation: $\frac{17}{32}$.

SYNTHESIS

Write each of the following with a single exponent.

73. $\dfrac{10^5}{10^3}$ **74.** $\dfrac{10^7}{10^2}$ **75.** $\dfrac{5^4}{5^2}$ **76.** $\dfrac{2^8}{8^2}$

77. *Five 5's.* We can use five 5's and any combination of grouping symbols to represent the numbers 0 through 10. For example,

$$0 = 5 \cdot 5 \cdot 5(5 - 5), \quad 1 = \frac{5 + 5}{5} - \frac{5}{5}, \quad 2 = \frac{5 \cdot 5 - 5}{5 + 5}.$$

Often more than one way to make a representation is possible. Use five 5's to represent the numbers 3 through 10.

Summary and Review: Chapter R

Important Properties and Formulas

Identity Property of 0: $a + 0 = a$

Identity Property of 1: $a \cdot 1 = a$

Equivalent Expressions for 1: $\frac{a}{a} = 1, \quad a \neq 0$

$n\% = n \times 0.01 = n \times \frac{1}{100} = \frac{n}{100}$

Exponential Notation: $a^n = \underbrace{a \cdot a \cdot a \cdots a}_{n \text{ factors}}$

Review Exercises

The review exercises that follow are for practice. Answers are at the back of the book. If you miss an exercise, restudy the section and objective indicated alongside the answer.

Find the prime factorization.

1. 92

2. 1400

Find the LCM.

3. 13, 32

4. 5, 18, 45

Write an equivalent expression using the indicated number for 1.

5. $\frac{2}{5}$ $\left(\text{Use } \frac{6}{6} \text{ for 1.}\right)$

6. $\frac{12}{23}$ $\left(\text{Use } \frac{8}{8} \text{ for 1.}\right)$

Write an equivalent expression with the given denominator.

7. $\frac{5}{8}$ (Denominator: 64)

8. $\frac{13}{12}$ (Denominator: 84)

Simplify.

9. $\frac{20}{48}$

10. $\frac{1020}{1820}$

Compute and simplify.

11. $\frac{4}{9} + \frac{5}{12}$

12. $\frac{3}{4} \div 3$

13. $\frac{2}{3} - \frac{1}{15}$

14. $\frac{9}{10} \cdot \frac{16}{5}$

15. Convert to fractional notation: 17.97.

16. Convert to decimal notation: $\frac{2337}{10{,}000}$.

Add.

17.
$$\begin{array}{r} 2344.56 \\ +\quad 98.345 \\ \hline \end{array}$$

18. $6.04 + 78 + 1.9898$

Subtract.

19. $20.4 - 11.058$

20.
$$\begin{array}{r} 789.032 \\ -\ 655.768 \\ \hline \end{array}$$

Multiply.

21.
$$\begin{array}{r} 17.95 \\ \times\quad 24 \\ \hline \end{array}$$

22.
$$\begin{array}{r} 56.95 \\ \times\quad 1.94 \\ \hline \end{array}$$

Divide.

23. $2.8\,\overline{)\,155.68}$

24. $52\,\overline{)\,23.4}$

25. Convert to decimal notation: $\frac{19}{12}$.

26. Round to the nearest tenth: 34.067.

27. Convert to decimal notation: 4.7%.

28. Convert to fractional notation: 60%.

Convert to percent notation.

29. 0.886

30. $\frac{5}{8}$

31. $\frac{29}{25}$

32. Write exponential notation: $6 \cdot 6 \cdot 6$.

33. Evaluate: $(1.06)^2$.

Calculate and compare answers to Exercises 34–36.

34. $120 - 6^2 \div 4 + 8$

35. $(120 - 6^2) \div 4 + 8$

36. $(120 - 6^2) \div (4 + 8)$

37. Calculate: $\frac{4(18 - 8) + 7 \cdot 9}{9^2 - 8^2}$.

SYNTHESIS

38. For health reasons, there must be 6 parts per billion (ppb) of chlorine in a swimming pool. What percent of total volume is this?

NAME SECTION DATE

Test: Chapter R

1. Find the prime factorization of 300.

2. Find the LCM of 15, 24, and 60.

3. Write an expression equivalent to $\frac{3}{7}$ using $\frac{7}{7}$ as a name for 1.

4. Write an equivalent expression with the given denominator:

$\frac{11}{16}$. (Denominator: 48)

Simplify.

5. $\frac{16}{24}$

6. $\frac{925}{1525}$

Compute and simplify.

7. $\frac{10}{27} \div \frac{8}{3}$

8. $\frac{9}{10} - \frac{5}{8}$

9. Convert to fractional notation (do not simplify): 6.78.

10. Convert to decimal notation: $\frac{1895}{1000}$.

11. Add: 7.14 + 89 + 2.8787.

12. Subtract: 1800 − 3.42.

ANSWERS

1. __________

2. __________

3. __________

4. __________

5. __________

6. __________

7. __________

8. __________

9. __________

10. __________

11. __________

12. __________

ANSWERS

13. ____

14. ____

15. ____

16. ____

17. ____

18. ____

19. ____

20. ____

21. ____

22. ____

23. ____

24. ____

25. ____

13. Multiply: $\begin{array}{r} 123.6 \\ \times \quad 3.52 \\ \hline \end{array}$

14. Divide: $7.2 \overline{)11.52}$.

15. Convert to decimal notation: $\frac{23}{11}$.

16. Round 234.7284 to the nearest tenth.

17. Round 234.7284 to the nearest thousandth.

18. Convert to decimal notation: 0.7%.

19. Convert to fractional notation: 91%.

20. Convert to percent notation: $\frac{11}{25}$.

21. Evaluate: 5^4.

22. Evaluate: $(1.2)^2$.

23. Calculate: $200 - 2^3 + 5 + 10$.

24. Calculate: $8000 \div 0.16 \div 2.5$.

SYNTHESIS

25. Simplify: $\frac{13,860}{42,000}$.

1

Introduction to Real Numbers and Algebraic Expressions

INTRODUCTION

In this chapter, we consider the number system used most in algebra. It is called the real-number system. We will learn to add, subtract, multiply, and divide real numbers and to manipulate certain expressions. Such manipulation will be important when we solve equations and problems in Chapter 2.

AN APPLICATION

It is three seconds before liftoff of the space shuttle. Tell what integer corresponds to this situation.

THE MATHEMATICS

Three seconds before liftoff corresponds to

$-3.$

↑

This is a negative number.

Objectives for Review

The review objectives to be tested in addition to the material in this chapter are as follows.

[R.1a, b] **Find all the factors of numbers, find prime factorizations, and find the LCM of two or more numbers.**

[R.2b, c] **Simplify fractional notation, and add, subtract, multiply, and divide using fractional notation.**

[R.4a, d] **Convert from percent notation to decimal notation, and convert from fractional notation to percent notation.**

[R.5b, c] **Evaluate exponential expressions, and simplify expressions using the rules for order of operations.**

Pretest: Chapter 1

1. Evaluate $x/2y$ when $x = 5$ and $y = 8$.

2. Write an algebraic expression: Seventy-eight percent of some number.

3. Find the area of a rectangle when the length is 22.5 ft and the width is 16 ft.

4. Find $-x$ when $x = -12$.

Use either $<$ or $>$ for ■ to write a true sentence.

5. 0 ■ -5

6. 10 ■ -5

7. -35 ■ -45

8. $-\frac{2}{3}$ ■ $\frac{4}{5}$

Find the absolute value.

9. $|-12|$

10. $|2.3|$

11. $|0|$

Find the opposite, or additive inverse.

12. 5.4

13. $-\frac{2}{3}$

Compute and simplify.

14. $-9 + (-8)$

15. $20.2 - (-18.4)$

16. $-\frac{5}{6} - \frac{3}{10}$

17. $-11.5 + 6.5$

18. $-9(-7)$

19. $\frac{5}{8}\left(-\frac{2}{3}\right)$

20. $-19.6 \div 0.2$

21. $-56 \div (-7)$

22. $12 - (-6) + 14 - 8$

23. $20 - 10 \div 5 + 2^3$

Multiply.

24. $9(z - 2)$

25. $-2(2a + b - 5c)$

Factor.

26. $4x - 12$

27. $6y - 9z - 18$

Simplify.

28. $3y - 7 - 2(2y + 3)$

29. $\{2[3(y + 1) - 4] - [5(y - 3) - 5]\}$

30. Write an inequality with the same meaning as $x > 12$.

1

Introduction to Real Numbers and Algebraic Expressions

INTRODUCTION

In this chapter, we consider the number system used most in algebra. It is called the real-number system. We will learn to add, subtract, multiply, and divide real numbers and to manipulate certain expressions. Such manipulation will be important when we solve equations and problems in Chapter 2.

AN APPLICATION

It is three seconds before liftoff of the space shuttle. Tell what integer corresponds to this situation.

THE MATHEMATICS

Three seconds before liftoff corresponds to

$-3.$

↑

This is a negative number.

Objectives for Review

The review objectives to be tested in addition to the material in this chapter are as follows.

[R.1a, b] Find all the factors of numbers, find prime factorizations, and find the LCM of two or more numbers.

[R.2b, c] Simplify fractional notation, and add, subtract, multiply, and divide using fractional notation.

[R.4a, d] Convert from percent notation to decimal notation, and convert from fractional notation to percent notation.

[R.5b, c] Evaluate exponential expressions, and simplify expressions using the rules for order of operations.

Pretest: Chapter 1

1. Evaluate $x/2y$ when $x = 5$ and $y = 8$.

2. Write an algebraic expression: Seventy-eight percent of some number.

3. Find the area of a rectangle when the length is 22.5 ft and the width is 16 ft.

4. Find $-x$ when $x = -12$.

Use either $<$ or $>$ for ▇ to write a true sentence.

5. 0 ▇ -5

6. 10 ▇ -5

7. -35 ▇ -45

8. $-\frac{2}{3}$ ▇ $\frac{4}{5}$

Find the absolute value.

9. $|-12|$

10. $|2.3|$

11. $|0|$

Find the opposite, or additive inverse.

12. 5.4

13. $-\frac{2}{3}$

Compute and simplify.

14. $-9 + (-8)$

15. $20.2 - (-18.4)$

16. $-\frac{5}{6} - \frac{3}{10}$

17. $-11.5 + 6.5$

18. $-9(-7)$

19. $\frac{5}{8}\left(-\frac{2}{3}\right)$

20. $-19.6 \div 0.2$

21. $-56 \div (-7)$

22. $12 - (-6) + 14 - 8$

23. $20 - 10 \div 5 + 2^3$

Multiply.

24. $9(z - 2)$

25. $-2(2a + b - 5c)$

Factor.

26. $4x - 12$

27. $6y - 9z - 18$

Simplify.

28. $3y - 7 - 2(2y + 3)$

29. $\{2[3(y + 1) - 4] - [5(y - 3) - 5]\}$

30. Write an inequality with the same meaning as $x > 12$.

1.1 *Introduction to Algebra*

Many types of problems require the use of equations in order to be solved effectively. The study of algebra involves the use of equations to solve problems. Equations are constructed from algebraic expressions. The purpose of this section is to introduce you to the types of expressions encountered in algebra.

OBJECTIVES

After finishing Section 1.1, you should be able to:

a Evaluate algebraic expressions by substitution.

b Translate phrases to algebraic expressions.

FOR EXTRA HELP

TAPE 2 | TAPE 2A | MAC: 1A IBM: 1A

a ALGEBRAIC EXPRESSIONS

In arithmetic, you have worked with expressions such as

$$37 + 86, \quad 7 \times 8, \quad 19 - 7, \quad \text{and} \quad \frac{3}{8}.$$

In algebra, we use certain letters for numbers and work with *algebraic expressions* such as

$$x + 86, \quad 7 \times t, \quad 19 - y, \quad \text{and} \quad \frac{a}{b}.$$

Sometimes a letter can represent various numbers. In that case, we call the letter a **variable**. Let $a =$ your age. Then a is a variable since a changes from year to year. Sometimes a letter can stand for just one number. In that case, we call the letter a **constant**. Let $b =$ your date of birth. Then b is a constant.

How do algebraic expressions occur? Most often they are found in problem-solving situations. For example, consider the bar graph shown in the margin, which you might see in a magazine. Suppose we want to know how many more flights there are on the Los Angeles–San Francisco route than on the New York–Chicago route.

In algebra, we translate the problem into an equation. It might be done as follows.

New York–Chicago route	plus	How much	is	Los Angeles–San Francisco route
↓	↓	↓	↓	↓
2710	+	x	=	2822

Note that we have an algebraic expression on the left. To find the number x, we can subtract 2710 on both sides of the equation:

$$2710 + x - 2710 = 2822 - 2710$$
$$x = 2822 - 2710.$$

Then we carry out the subtraction and obtain the answer, 112 flights.

In arithmetic, you probably would do this subtraction right away without considering an equation. In algebra, you will find most problems difficult without first solving the equation.

Do Exercise 1.

1. Translate this problem to an equation. Use the graph below.

How many more flights are there on the Dallas–Houston route than on the New York–Boston route?

Answer on page A-1

2. Evaluate $a + b$ when $a = 38$ and $b = 26$.

An **algebraic expression** consists of variables, numerals, and operation signs. When we replace a variable by a number, we say that we are **substituting** for the variable. This process is called **evaluating the expression.**

EXAMPLE 1 Evaluate $x + y$ when $x = 37$ and $y = 29$.

We substitute 37 for x and 29 for y and carry out the addition:

$$x + y = 37 + 29 = 66.$$

3. Evaluate $x - y$ when $x = 57$ and $y = 29$.

The number 66 is called the **value** of the expression.

Algebraic expressions involving multiplication can be written in several ways. For example, "8 times a" can be written as $8 \times a$, $8 \cdot a$, $8(a)$, or simply $8a$. Two letters written together without an operation symbol, such as ab, also indicates a multiplication.

4. Evaluate $4t$ when $t = 15$.

EXAMPLE 2 Evaluate $3y$ when $y = 14$.

$$3y = 3(14) = 42$$

Do Exercises 2–4.

5. Find the area of a rectangle when l is 24 ft and w is 8 ft.

EXAMPLE 3 The area A of a rectangle of length l and width w is given by the formula $A = lw$. Find the area when l is 24.5 in. and w is 16 in.

We substitute 24.5 in. for l and 16 in. for w and carry out the multiplication:

$$\begin{aligned} A = lw &= (24.5 \text{ in.})(16 \text{ in.}) \\ &= (24.5)(16)(\text{in.})(\text{in.}) \\ &= 392 \text{ in}^2\text{, or 392 square inches.} \end{aligned}$$

6. Find the orbiting time of the satellite in Example 4 when the velocity is 8000 mph.

Do Exercise 5.

Algebraic expressions involving division can also be written in several ways. For example, "8 divided by t" can be written as $8 \div t$, $\frac{8}{t}$, or $8/t$, where the fraction bar is a division symbol.

EXAMPLE 4 The time needed for a satellite to orbit the earth is determined by the height of the satellite above the earth's surface and the speed, or velocity, of the satellite. If a satellite is orbiting 300 mi above the earth's surface, it travels about 27,000 mi in one orbit. The time t, in hours, that it takes to orbit the earth one time is given by

$$t = \frac{27{,}000}{v},$$

where v is the velocity of the satellite in miles per hour. Find the orbiting time of the satellite when the velocity v is 10,000 mph.

We substitute 10,000 for v and carry out the division:

$$t = \frac{27{,}000}{v} = \frac{27{,}000}{10{,}000} = 2.7 \text{ hr.}$$

Answers on page A-1

Do Exercise 6.

EXAMPLE 5 Evaluate $\frac{a}{b}$ when $a = 63$ and $b = 9$.

We substitute 63 for a and 9 for b and carry out the division:

$$\frac{a}{b} = \frac{63}{9} = 7.$$

EXAMPLE 6 Evaluate $\frac{12m}{n}$ when $m = 8$ and $n = 16$.

$$\frac{12m}{n} = \frac{12 \cdot 8}{16} = \frac{96}{16} = 6$$

Do Exercises 7 and 8.

7. Evaluate a/b when $a = 200$ and $b = 8$.

8. Evaluate $10p/q$ when $p = 40$ and $q = 25$.

b TRANSLATING TO ALGEBRAIC EXPRESSIONS

In algebra, we translate problems to equations. The different parts of an equation are translations of word phrases to algebraic expressions. It is easier to translate if we know that certain words translate to certain operation symbols.

KEY WORDS			
ADDITION (+)	SUBTRACTION (−)	MULTIPLICATION (·)	DIVISION (÷)
add sum plus more than increased by	subtract difference minus less than decreased by take from	multiply product times twice of	divide quotient divided by

EXAMPLE 7 Translate to an algebraic expression:

Twice (or two times) some number.

Think of some number, say, 8. What number is twice 8? It is 16. How did you get 16? You multiplied by 2. Do the same thing using a variable. We can use any variable we wish, such as x, y, m, or n. Let us use y to stand for some number. If we multiply by 2, we get an expression

$$y \times 2, \quad 2 \times y, \quad 2 \cdot y, \quad \text{or} \quad 2y.$$

EXAMPLE 8 Translate to an algebraic expression:

Thirty-eight percent of some number.

The word "of" translates to a multiplication symbol, so we get the following expressions as a translation:

$$38\% \cdot n, \quad 0.38 \times n, \quad \text{or} \quad 0.38n.$$

Answers on page A-1

Translate to an algebraic expression.

9. Eight less than some number

10. Eight more than some number

11. Four less than some number

12. Half of a number

13. Six more than eight times some number

14. The difference of two numbers

15. Fifty-nine percent of some number

16. Two hundred less than the product of two numbers

17. The sum of two numbers

Answers on page A-1

EXAMPLE 9 Translate to an algebraic expression:

Seven less than some number.

We let

x represent the number.

Now if the number were 23, then the translation would be $23 - 7$. If we knew the number to be 345, then the translation would be $345 - 7$. If the number is x, then the translation is

$$x - 7.$$

Note that $7 - x$ is *not* a correct translation of the expression in Example 9. The expression $7 - x$ is a translation of "seven minus some number" or "some number less than seven."

EXAMPLE 10 Translate to an algebraic expression:

Eighteen more than a number.

We let

$t =$ the number.

Now if the number were 26, then the translation would be $26 + 18$. If we knew the number to be 174, then the translation would be $174 + 18$. If the number is t, then the translation is

$$t + 18.$$

EXAMPLE 11 Translate to an algebraic expression:

A number divided by 5.

We let

$m =$ the number.

Now if the number were 76, then the translation would be $76 \div 5$, or $76/5$, or $\frac{76}{5}$. If the number were 213, then the translation would be $213 \div 5$, or $213/5$, or $\frac{213}{5}$. If the number is m, then the translation is

$$m \div 5, \quad m/5, \quad \text{or} \quad \frac{m}{5}.$$

EXAMPLE 12 Translate each of the following phrases to an algebraic expression.

PHRASE	ALGEBRAIC EXPRESSION
Five more than some number	$n + 5$, or $5 + n$
Half of a number	$\frac{1}{2}t$, $\frac{t}{2}$, or $t/2$
Five more than three times some number	$3p + 5$, or $5 + 3p$
The difference of two numbers	$x - y$
Six less than the product of two numbers	$mn - 6$
Seventy-six percent of some number	$76\%z$, or $0.76z$

Do Exercises 9–17.

NAME SECTION DATE

Exercise Set 1.1

a Substitute to find values of the expressions.

1. Lawanda is 6 yr younger than her husband Emmett. Suppose that the variable x stands for Emmett's age. Then $x - 6$ stands for Lawanda's age. How old is Lawanda when Emmett is 29? 34? 47?

2. Felipe took five times as long to do a job as Marina did. Suppose that t stands for the time it takes Marina to do a job. Then $5t$ stands for the time it takes Felipe. How long did it take Felipe if Marina took 30 sec? 90 sec? 2 min?

3. The area A of a triangle with base b and height h is given by $A = \frac{1}{2}bh$. Find the area when $b = 45$ m (meters) and $h = 86$ m.

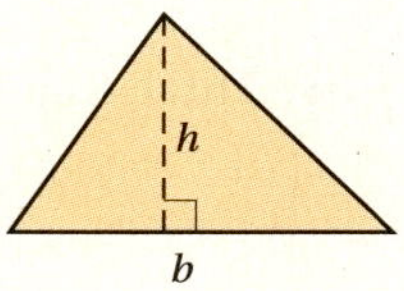

4. The area A of a parallelogram with base b and height h is given by $A = bh$. Find the area of the parallelogram when the height is 15.4 cm (centimeters) and the base is 6.5 cm.

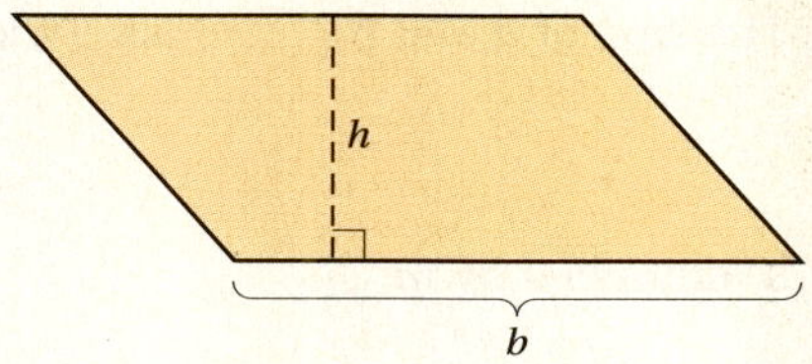

5. A driver who drives at a speed of r mph for t hr will travel a distance d mi given by $d = rt$ mi. How far will a driver travel at a speed of 65 mph for 4 hr?

6. The simple interest I on a principal of P dollars at interest rate r for time t, in years, is given by $I = Prt$. Find the simple interest on a principal of \$4800 at 9% for 2 yr. (*Hint:* 9% = 0.09.)

Evaluate.

7. $8x$ when $x = 7$

8. $6y$ when $y = 7$

9. $\frac{a}{b}$ when $a = 24$ and $b = 3$

10. $\frac{p}{q}$ when $p = 16$ and $q = 2$

11. $\frac{3p}{q}$ when $p = 2$ and $q = 6$

12. $\frac{5y}{z}$ when $y = 15$ and $z = 25$

13. $\frac{x + y}{5}$ when $x = 10$ and $y = 20$

14. $\frac{p + q}{2}$ when $p = 2$ and $q = 16$

15. $\frac{x - y}{8}$ when $x = 20$ and $y = 4$

16. $\frac{m - n}{5}$ when $m = 16$ and $n = 6$

ANSWERS

1. ______
2. ______
3. ______
4. ______
5. ______
6. ______
7. ______
8. ______
9. ______
10. ______
11. ______
12. ______
13. ______
14. ______
15. ______
16. ______

ANSWERS

17. ____
18. ____
19. ____
20. ____
21. ____
22. ____
23. ____
24. ____
25. ____
26. ____
27. ____
28. ____
29. ____
30. ____
31. ____
32. ____
33. ____
34. ____
35. ____
36. ____
37. ____
38. ____
39. ____
40. ____
41. ____
42. ____
43. ____
44. ____
45. ____
46. ____
47. ____
48. ____
49. ____
50. ____
51. ____
52. ____

b Translate to an algebraic expression.

17. 7 more than b

18. 9 more than t

19. 12 less than c

20. 14 less than d

21. 4 increased by q

22. 13 increased by z

23. b more than a

24. c more than d

25. x less than y

26. c less than h

27. x added to w

28. s added to t

29. m subtracted from n

30. p subtracted from q

31. The sum of r and s

32. The sum of a and b

33. Twice z

34. Three times q

35. 3 multiplied by m

36. The product of 8 and t

37. The product of 89% and some number

38. 67% of some number

39. A driver drove at a speed of 55 mph for t hours. How far did the driver travel?

40. An executive assistant has d dollars before going to an office supply store. The assistant bought some fax paper for $18.95. How much did he have after the purchase?

SKILL MAINTENANCE

Find the prime factorization.

41. 54

42. 32

43. 108

44. 192

Find the LCM.

45. 6, 18

46. 6, 24, 32

47. 10, 20, 30

48. 16, 24

SYNTHESIS

Translate to an algebraic expression.

49. Some number x plus three times y

50. Some number a plus 2 plus b

51. A number that is 3 less than twice x

52. Your age in 5 years, if you are a years old now

1.2 *The Real Numbers*

A **set** is a collection of objects. For our purposes, we will most often be considering sets of numbers. One way to name a set uses what is called **roster notation.** For example, roster notation for the set containing the numbers 0, 2, and 5 is {0, 2, 5}.

Sets that are parts of other sets are called **subsets**. In this section, we become acquainted with the set of *real numbers* and its various subsets.

Two important subsets of the real numbers are listed below using roster notation.

> Natural numbers = {1, 2, 3, . . .}. These are the numbers used for counting.

> Whole numbers = {0, 1, 2, 3, . . .}. This is the set of natural numbers with 0 included.

We can represent these sets on a number line. The natural numbers are those to the right of zero. The whole numbers are the natural numbers and zero.

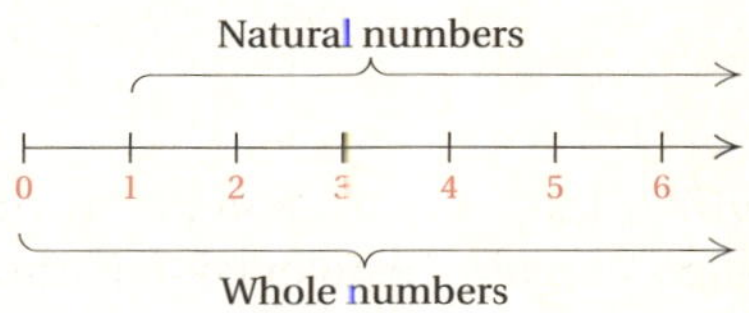

We create a new set, called the *integers,* by starting with the whole numbers, 0, 1, 2, 3, and so on. For each natural number 1, 2, 3, and so on, we obtain a new number to the left of zero on the number line:

For the number 1, there will be an *opposite* number -1 (negative 1).

For the number 2, there will be an *opposite* number -2 (negative 2).

For the number 3, there will be an *opposite* number -3 (negative 3), and so on.

The **integers** consist of the whole numbers and these new numbers. We picture them on a number line as follows.

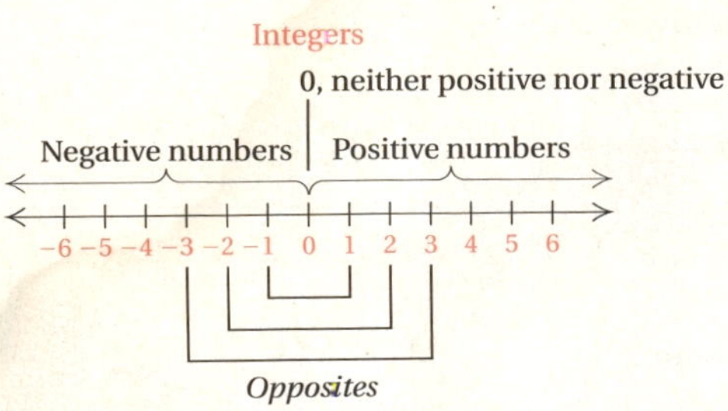

We call the newly obtained negative numbers **negative integers.** The natural numbers are called **positive integers.** Zero is neither positive nor negative. We call -1 and 1 opposites of each other. Similarly, -2 and 2 are

OBJECTIVES

After finishing Section 1.2, you should be able to:

a Name the integer that corresponds to a real-world situation.

b Graph rational numbers on a number line.

c Convert from fractional notation to decimal notation for a rational number.

d Determine which of two real numbers is greater and indicate which, using $<$ or $>$; given an inequality like $a < b$, write another inequality with the same meaning. Determine whether an inequality like $-3 \leq 5$ is true or false.

e Find the absolute value of a real number.

FOR EXTRA HELP

TAPE 3 | TAPE 2B | MAC: 1A IBM: 1A

opposites, −3 and 3 are opposites, −100 and 100 are opposites, and 0 is its own opposite. This gives us the set of integers, which extends infinitely on the number line to the left and right of zero.

The set of integers = $\{\ldots, -5, -4, -3, -2, -1, 0, 1, 2, 3, 4, 5, \ldots\}$.

a INTEGERS AND THE REAL WORLD

Integers correspond to many real-world problems and situations. The following examples will help you get ready to translate problem situations that involve integers to mathematical language.

EXAMPLE 1 Tell which integer corresponds to this situation: The temperature is 3 degrees below zero.

3° below zero is −3°

EXAMPLE 2 Tell which integer corresponds to this situation: A contestant missed a $600 question on the television game show "Jeopardy."

Missing a $600 question causes a $600 loss on the score—that is, the contestant earns −600 dollars.

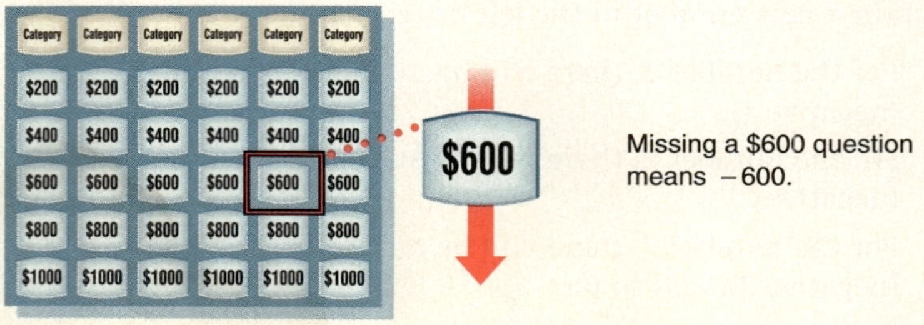

EXAMPLE 3 Tell which integer corresponds to this situation: Death Valley is 280 ft below sea level.

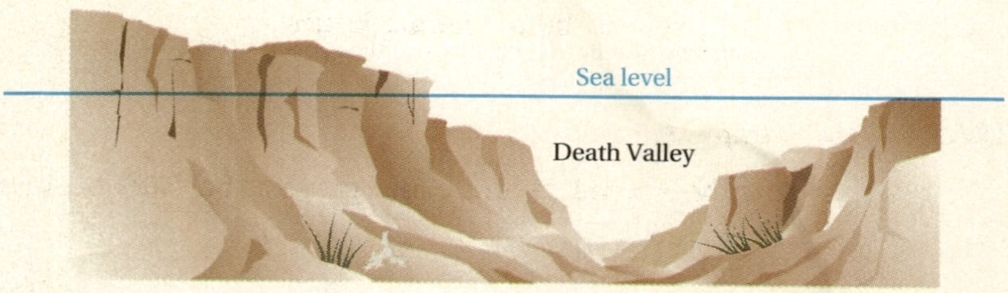

The integer −280 corresponds to the situation. The elevation is −280 ft.

EXAMPLE 4 Tell which integers correspond to this situation: A salesperson made a \$78 sale on Monday, but lost \$57 through a return of merchandise on Tuesday.

The integers 78 and -57 correspond to the situation. The integer 78 corresponds to the sale on Monday and -57 corresponds to the return on Tuesday.

Do Exercises 1–4.

b THE RATIONAL NUMBERS

We created the set of integers by obtaining a negative number for each natural number. To create a larger number system, called the set of **rational numbers,** we consider quotients of integers with nonzero divisors. The following are rational numbers:

$$\frac{2}{3},\quad -\frac{2}{3},\quad \frac{7}{1},\quad 4,\quad -3,\quad 0,\quad \frac{23}{-8},\quad 2.4,\quad -0.17,\quad 10\frac{1}{2}.$$

The number $-\frac{2}{3}$ (read "negative two-thirds") can also be named $\frac{2}{-3}$ or $\frac{-2}{3}$. The number 2.4 can be named $\frac{24}{10}$ or $\frac{12}{5}$, and -0.17 can be named $-\frac{17}{100}$.

Note that this new set of numbers, the rational numbers, contains the whole numbers, the integers, and the arithmetic numbers (also called the nonnegative rational numbers). We can describe the set of rational numbers using **set-builder notation,** as follows.

> The set of rational numbers $= \left\{\frac{a}{b} \,\middle|\, a \text{ and } b \text{ are integers and } b \neq 0\right\}$.
>
> (This is read "the set of numbers $\frac{a}{b}$, where a and b are integers and $b \neq 0$.")

We picture the rational numbers on a number line, as follows. There is a point on the line for every rational number.

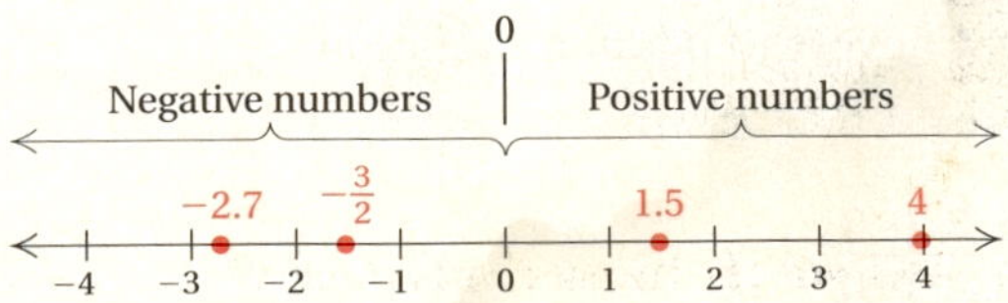

To **graph** a number means to find and mark its point on the number line. Some numbers are graphed in the preceding figure.

EXAMPLE 5 Graph: $\frac{5}{2}$.

The number $\frac{5}{2}$ can be named $2\frac{1}{2}$, or 2.5. Its graph is halfway between 2 and 3.

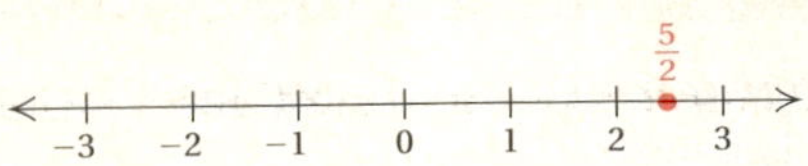

Tell which integers correspond to the given situation.

1. The halfback gained 8 yd on the first down. The quarterback was sacked for a 5-yd loss on the second down.

2. The highest temperature ever recorded in the United States was 134° in Death Valley on July 10, 1913. The coldest temperature ever recorded in the United States was 76° below zero in Tanana, Alaska, in January of 1886.

3. At 10 sec before liftoff, ignition occurs. At 156 sec after liftoff, the first stage is detached from the rocket.

4. A student owes \$147 to the bookstore. The student has \$289 in a savings account.

Answers on page A-1

Graph on a number line.

5. $-\dfrac{7}{2}$

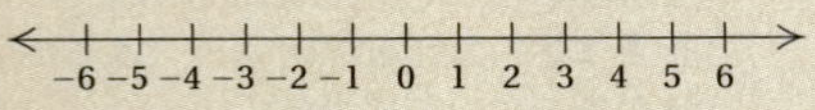

6. 1.4

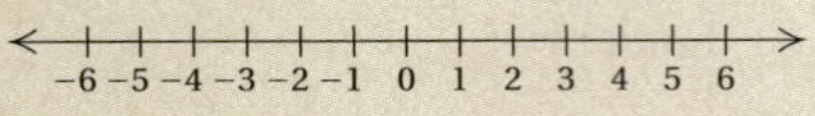

7. $-\dfrac{11}{4}$

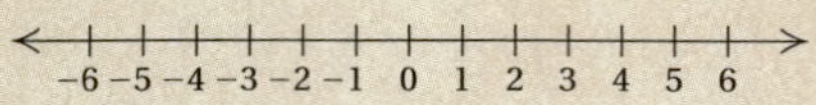

Answers on page A-1

EXAMPLE 6 Graph: -3.2.

The graph of -3.2 is $\frac{2}{10}$ of the way from -3 to -4.

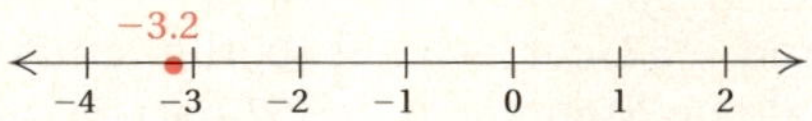

EXAMPLE 7 Graph: $\frac{13}{8}$.

The number $\frac{13}{8}$ can be named $1\frac{5}{8}$, or 1.625. The graph is about $\frac{6}{10}$ of the way from 1 to 2.

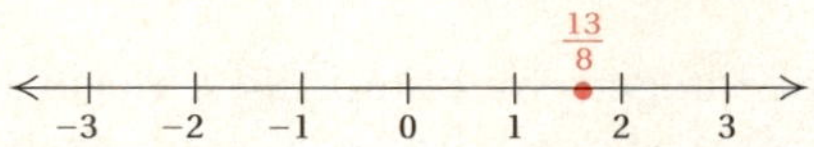

Do Exercises 5–7.

c NOTATION FOR RATIONAL NUMBERS

Each rational number can be named using fractional or decimal notation.

EXAMPLE 8 Convert to decimal notation: $-\frac{5}{8}$.

We first find decimal notation for $\frac{5}{8}$. Since $\frac{5}{8}$ means $5 \div 8$, we divide.

```
     0.6 2 5
8 ) 5.0 0 0
    4 8
      2 0
      1 6
        4 0
        4 0
          0
```

Thus, $\frac{5}{8} = 0.625$, so $-\frac{5}{8} = -0.625$.

Decimal notation for $-\frac{5}{8}$ is -0.625. We consider -0.625 to be a **terminating decimal.** Decimal notation for some numbers repeats.

EXAMPLE 9 Convert to decimal notation: $\frac{7}{11}$.

We divide.

```
        0.6 3 6 3...
1 1 ) 7.0 0 0 0
      6 6
        4 0
        3 3
          7 0
          6 6
            4 0
            3 3
              7
```

$$\frac{7}{11} = 0.\overline{63}$$

We can abbreviate repeating decimal notation by writing a bar over the repeating part, in this case, $0.\overline{63}$.

The following are other examples to show how each rational number can be named using fractional or decimal notation:

$$0 = \frac{0}{8}, \qquad \frac{27}{100} = 0.27, \qquad -8\frac{3}{4} = -8.75, \qquad \frac{-13}{6} = -2.1\overline{6}.$$

Do Exercises 8–10.

d THE REAL NUMBERS AND ORDER

Every rational number has a point on the number line. However, there are some points on the line for which there is no rational number. These points correspond to what are called **irrational numbers.**

What kinds of numbers are irrational? One example is the number π, which is used in finding the area and the circumference of a circle: $A = \pi r^2$ and $C = 2\pi r$.

Another example of an irrational number is the square root of 2, named $\sqrt{2}$.

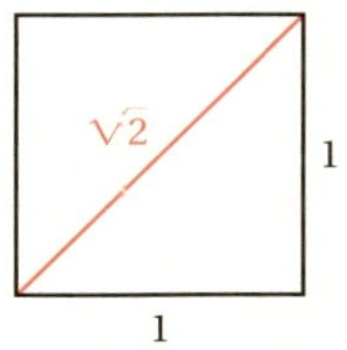

It is the length of the diagonal of a square with sides of length 1. It is also the number that when multiplied by itself gives 2. There is no rational number that can be multiplied by itself to get 2. But the following are rational *approximations*:

1.4 is an approximation of $\sqrt{2}$ because $(1.4)^2 = 1.96$;

1.41 is a better approximation because $(1.41)^2 = 1.9881$;

1.4142 is an even better approximation because $(1.4142)^2 = 1.99996164$.

We can find rational approximations for square roots using a calculator.

Decimal notation for rational numbers *either* terminates *or* repeats. Decimal notation for irrational numbers *neither* terminates *nor* repeats. Some other examples of irrational numbers are π, $\sqrt{3}$, $-\sqrt{8}$, $\sqrt{11}$, and 0.121221222122221.... Whenever we take the square root of a number that is not a perfect square, we will get an irrational number.

The rational numbers and the irrational numbers together correspond to all the points on a number line and make up what is called the **real-number system.**

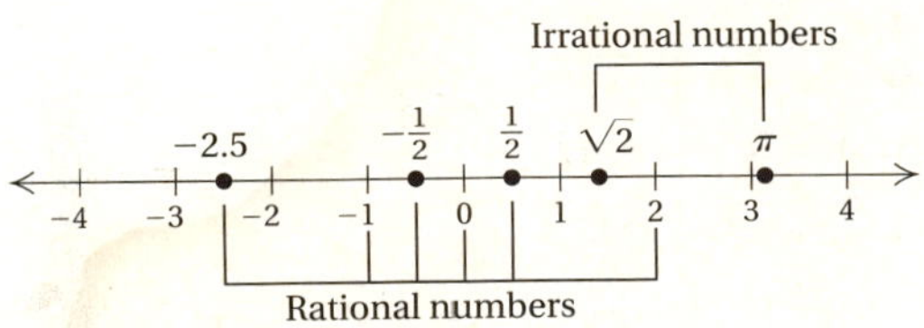

The set of real numbers	=	The set of all numbers corresponding to points on the number line.

Convert to decimal notation.

8. $-\frac{3}{8}$

9. $-\frac{6}{11}$

10. $\frac{4}{3}$

Answers on page A-1

Use either $<$ or $>$ for ■ to write a true sentence.

11. -3 ■ 7

12. -8 ■ -5

13. 7 ■ -10

14. 3.1 ■ -9.5

15. $-\frac{2}{3}$ ■ -1

16. $-\frac{11}{8}$ ■ $\frac{23}{15}$

17. $-\frac{2}{3}$ ■ $-\frac{5}{9}$

18. -4.78 ■ -5.01

Answers on page A-1

The real numbers consist of the rational numbers and the irrational numbers. The following figure shows the relationship between various kinds of numbers.

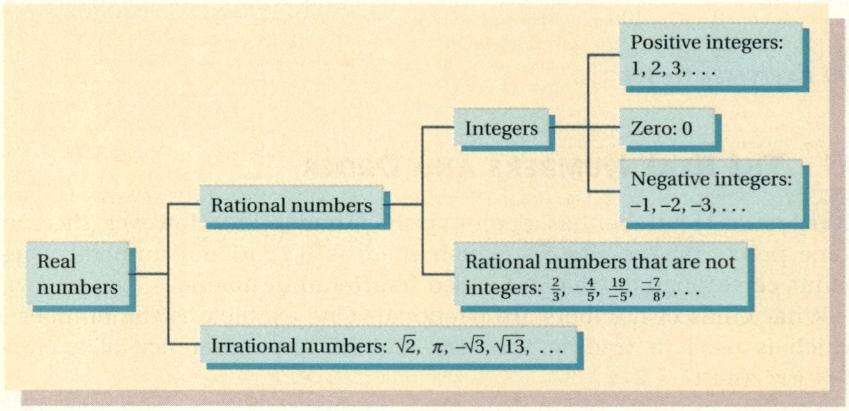

Order

Real numbers are named in order on the number line, with larger numbers named farther to the right. For any two numbers on the line, the one to the left is less than the one to the right.

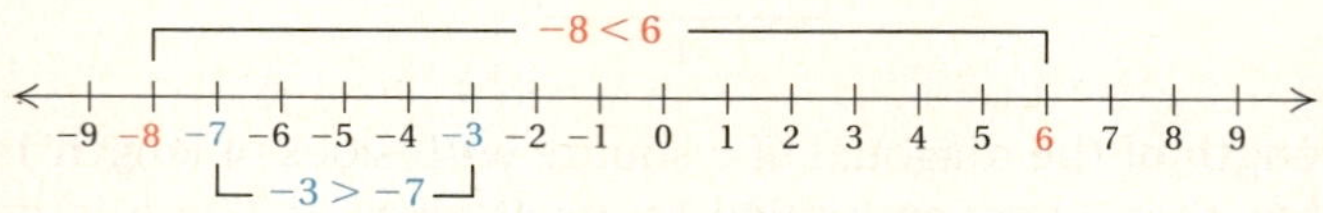

We use the symbol $<$ to mean "**is less than.**" The sentence $-8 < 6$ means "-8 is less than 6." The symbol $>$ means "**is greater than.**" The sentence $-3 > -7$ means "-3 is greater than -7."

Examples Use either $<$ or $>$ for ■ to write a true sentence.

10. 2 ■ 9 — Since 2 is to the left of 9, 2 is less than 9, so $2 < 9$.

11. -7 ■ 3 — Since -7 is to the left of 3, we have $-7 < 3$.

12. 6 ■ -12 — Since 6 is to the right of -12, then $6 > -12$.

13. -18 ■ -5 — Since -18 is to the left of -5, we have $-18 < -5$.

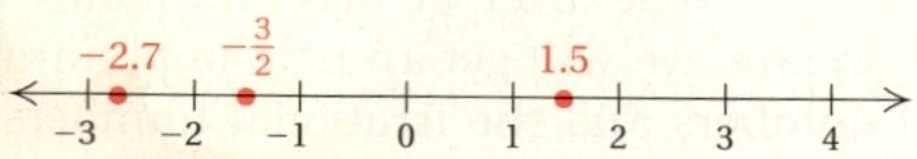

14. -2.7 ■ $-\frac{3}{2}$ — The answer is $-2.7 < -\frac{3}{2}$.

15. 1.5 ■ -2.7 — The answer is $1.5 > -2.7$.

16. 1.38 ■ 1.83 — The answer is $1.38 < 1.83$.

17. -3.45 ■ 1.32 — The answer is $-3.45 < 1.32$.

18. -4 ■ 0 — The answer is $-4 < 0$.

19. 5.8 ■ 0 — The answer is $5.8 > 0$.

20. $\frac{5}{8}$ ■ $\frac{7}{11}$ — We convert to decimal notation: $\frac{5}{8} = 0.625$ and $\frac{7}{11} = 0.6363\ldots$. Thus, $\frac{5}{8} < \frac{7}{11}$.

Do Exercises 11–18.

Note that both $-8 < 6$ and $6 > -8$ are true. These are **inequalities**. Every true inequality yields another true inequality when we interchange the numbers or variables and reverse the direction of the inequality sign.

$a < b$ also has the meaning $b > a$.

EXAMPLES Write another inequality with the same meaning.

21. $a < -5$ The inequality $-5 > a$ has the same meaning.

22. $-3 > -8$ The inequality $-8 < -3$ has the same meaning.

A helpful mental device is to think of an inequality sign as an "arrow" with the arrow pointing to the smaller number.

Do Exercises 19 and 20.

Note that all positive real numbers are greater than zero and all negative real numbers are less than zero.

If x is a positive real number, then $x > 0$.
If x is a negative real number, then $x < 0$.

Expressions like $a \leq b$ and $b \geq a$ are also inequalities. We read $a \leq b$ as "***a* is less than or equal to *b*.**" We read $a \geq b$ as "***a* is greater than or equal to *b*.**"

EXAMPLES Write true or false for each statement.

23. $-3 \leq 5$ True since $-3 < 5$ is true

24. $-3 \leq -3$ True since $-3 = -3$ is true

25. $-5 \geq 4$ False since neither $-5 > 4$ nor $-5 = 4$ is true

Do Exercises 21–23.

e ABSOLUTE VALUE

From the number line, we see that numbers like 4 and -4 are the same distance from zero. Distance is always a nonnegative number. We call the distance from zero on a number line the **absolute value** of the number.

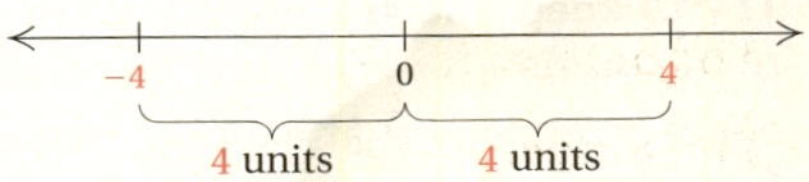

The *absolute value* of a number is its distance from zero on a number line. We use the symbol $|x|$ to represent the absolute value of a number x.

To find absolute value:

1. If a number is negative, make it positive.
2. If a number is positive or zero, leave it alone.

Write another inequality with the same meaning.

19. $-5 < 7$

20. $x > 4$

Write true or false.

21. $-4 \leq -6$

22. $7 \geq 7$

23. $-2 \leq 3$

Answers on page A-1

Find the absolute value.

24. $|8|$

25. $|0|$

26. $|-9|$

27. $\left|-\frac{2}{3}\right|$

28. $|5.6|$

Answers on page A-1

EXAMPLES Find the absolute value.

26. $|-7|$ The distance of -7 from 0 is 7, so $|-7| = 7$.

27. $|12|$ The distance of 12 from 0 is 12, so $|12| = 12$.

28. $|0|$ The distance of 0 from 0 is 0, so $|0| = 0$.

29. $\left|\frac{3}{2}\right| = \frac{3}{2}$

30. $|-2.73| = 2.73$

Do Exercises 24–28.

From time to time, you will find some *Sidelights* like the one below. Although they are optional, you may find them helpful and of interest. They will include study tips, career opportunities involving mathematics, applications, computer–calculator exercises, and many other mathematical topics.

SIDELIGHTS

STUDY TIPS

There are many ways in which you can enhance your use of this book, and they have been outlined in the chart entitled "The Steps to Success" on the inside front cover. If you have not read that outline, do so now, before you begin the exercise set on the following page.

Here we highlight a few points that we consider most helpful when learning from this text.

- **Be sure to note the special symbols [a], [b], [c], and so on, that correspond to the objectives you are to be able to perform.** They appear in many places throughout the text. The first time you see them is at the beginning of each section. The second time is in the headings of each subsection. The third time is in the exercise set, as follows. You will also find them in the answers to the summary and review exercises, the chapter tests, and the cumulative reviews. These allow you to reference back when you need to review a topic.
- **Be sure to note the symbols in the margin under the list of objectives at the beginning of each section.** These refer to the many distinctive study aids that accompany the book.
- **Be sure to read and study each step of each example.** The examples include important side comments that explain each step. These carefully chosen examples and notes prepare you for success in the exercise set.
- **Be sure to stop and do the margin exercises as you study a section.** When our students come to us troubled about how they are doing in the course, the first question we ask is "Are you doing the margin exercises when directed to do so?" This is one of the most effective ways to enhance your ability to learn mathematics from this text. Don't deprive yourself of its benefits!
- **When you study the book, don't mark points you think are important, but mark the points you do not understand!** This book includes many design features that highlight important points. Use your efforts to mark where you are having trouble. Then when you go to class, a math lab, or a tutoring session, you will be prepared to ask questions that home in on your difficulties rather than spending time going over what you already understand.
- **If you are having trouble, consider using the Student's Solutions Manual that contains worked-out solutions to the odd-numbered exercises in the exercise sets.**
- **Try to keep one section ahead of your syllabus.** If you study ahead of your lectures, you can concentrate on what is being explained in your lectures, rather than trying to write everything down. You can then take notes only of special points or of questions related to what is happening in class.

NAME SECTION DATE

Exercise Set 1.2

a Tell which integers correspond to the situation.

1. The Dead Sea, between Jordan and Israel, is 1286 ft below sea level, whereas Mt. Rainier in Washington State is 13,804 ft above sea level.

2. The temperature on Wednesday was 18° above zero. On Thursday, it was 2° below zero.

3. During a certain time period, the United States had a deficit of \$8 million in foreign trade.

4. A student deposited \$750 in a savings account. Two weeks later, she withdrew \$125.

5. It is 3 sec before liftoff of the space shuttle. 128 sec after liftoff, the shuttle loses its solid fuel rockets.

6. During a video game, a player intercepted a missile worth 30 points, lost a starship worth 140 points, and captured a base worth 300 points.

b Graph the number on the number line.

7. $\frac{10}{3}$

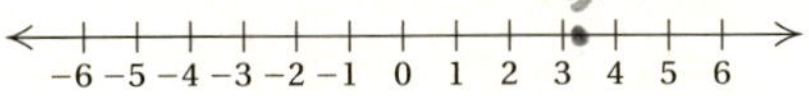

8. $-\frac{17}{4}$

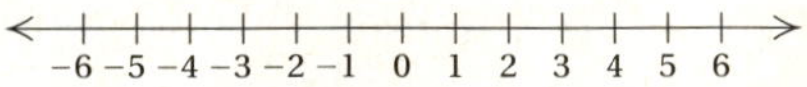

9. -5.2

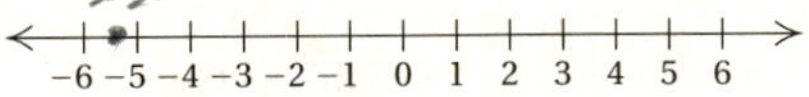

10. 4.78

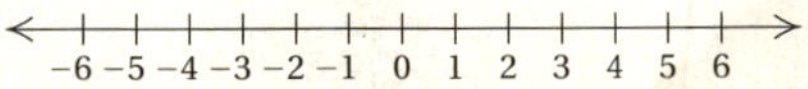

c Convert to decimal notation.

11. $-\frac{7}{8}$

12. $-\frac{1}{8}$

13. $\frac{5}{6}$

14. $\frac{5}{3}$

15. $\frac{7}{6}$

16. $\frac{5}{12}$

17. $\frac{2}{3}$

18. $\frac{1}{4}$

19. $-\frac{1}{2}$

20. $\frac{5}{8}$

21. $\frac{1}{10}$

22. $-\frac{7}{20}$

Answers

1.

2.

3.

4.

5.

6.

7. See graph.

8. See graph.

9. See graph.

10. See graph.

11.

12.

13.

14.

15.

16.

17.

18.

19.

20.

21.

22.

ANSWERS

23. ______
24. ______
25. ______
26. ______
27. ______
28. ______
29. ______
30. ______
31. ______
32. ______
33. ______
34. ______
35. ______
36. ______
37. ______
38. ______
39. ______
40. ______
41. ______
42. ______
43. ______
44. ______
45. ______
46. ______
47. ______
48. ______
49. ______
50. ______
51. ______
52. ______
53. ______
54. ______
55. ______
56. ______
57. ______
58. ______
59. ______
60. ______
61. ______
62. ______
63. ______
64. ______

d Use either < or > for ■ to write a true sentence.

23. 8 ■ 0 **24.** 3 ■ 0 **25.** −8 ■ 3 **26.** 6 ■ −6

27. −8 ■ 8 **28.** 0 ■ −9 **29.** −8 ■ −5 **30.** −4 ■ −3

31. −5 ■ −11 **32.** −3 ■ −4 **33.** −6 ■ −5 **34.** −10 ■ −14

35. 2.14 ■ 1.24 **36.** −3.3 ■ −2.2 **37.** −14.5 ■ 0.011

38. 17.2 ■ −1.67 **39.** −12.88 ■ −6.45 **40.** −14.34 ■ −17.88

41. $\frac{5}{12}$ ■ $\frac{11}{25}$ **42.** $-\frac{13}{16}$ ■ $-\frac{5}{9}$

Write true or false.

43. $-3 \geq -11$ **44.** $5 \leq -5$ **45.** $0 \geq 8$ **46.** $-5 \leq 7$

Write an inequality with the same meaning.

47. $-6 > x$ **48.** $x < 8$ **49.** $-10 \leq y$ **50.** $12 \geq t$

e Find the absolute value.

51. $|-3|$ **52.** $|-7|$ **53.** $|10|$ **54.** $|11|$

55. $|0|$ **56.** $|-4|$ **57.** $|-24|$ **58.** $|325|$

59. $\left|-\frac{2}{3}\right|$ **60.** $\left|-\frac{10}{7}\right|$ **61.** $\left|\frac{0}{4}\right|$ **62.** $|14.8|$

SYNTHESIS

List in order from the least to the greatest.

63. $-\frac{2}{3}, \frac{1}{2}, -\frac{3}{4}, -\frac{5}{6}, \frac{3}{8}, \frac{1}{6}$

64. $-8\frac{7}{8}, 7^1, -5, |-6|, 4, |3|, -8\frac{5}{8}, -100, 0, 1^7, \frac{14}{4}, \frac{-67}{8}$

1.3 *Addition of Real Numbers*

In this section, we consider addition of real numbers. First, to gain an understanding, we add using a number line. Then we consider rules for addition.

Addition of numbers can be illustrated on a number line. To do the addition $a + b$, we start at a, and then move according to b.

a) If b is positive, we move to the right.

b) If b is negative, we move to the left.

c) If b is 0, we stay at a.

EXAMPLE 1 Add: $3 + (-5)$.

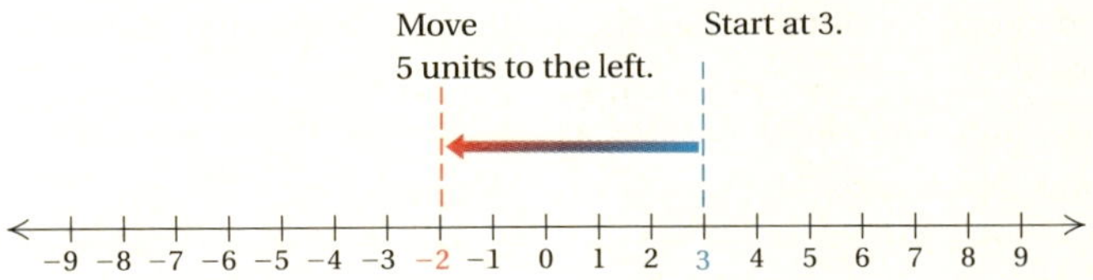

$3 + (-5) = -2$

EXAMPLE 2 Add: $-4 + (-3)$.

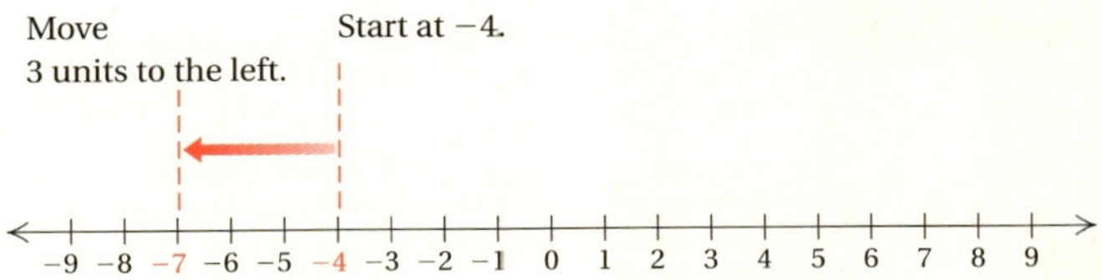

$-4 + (-3) = -7$

EXAMPLE 3 Add: $-4 + 9$.

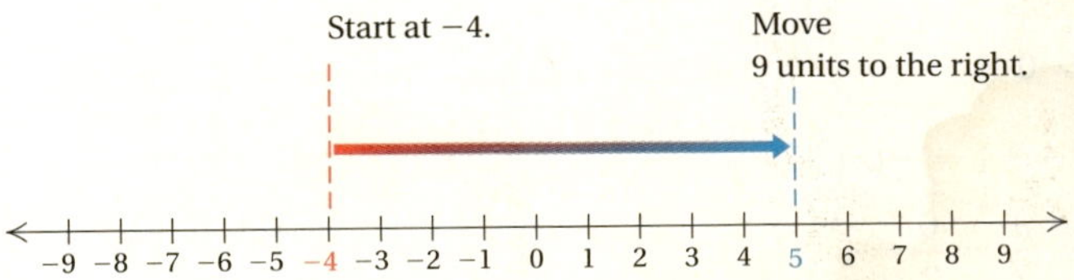

$-4 + 9 = 5$

EXAMPLE 4 Add: $-5.2 + 0$.

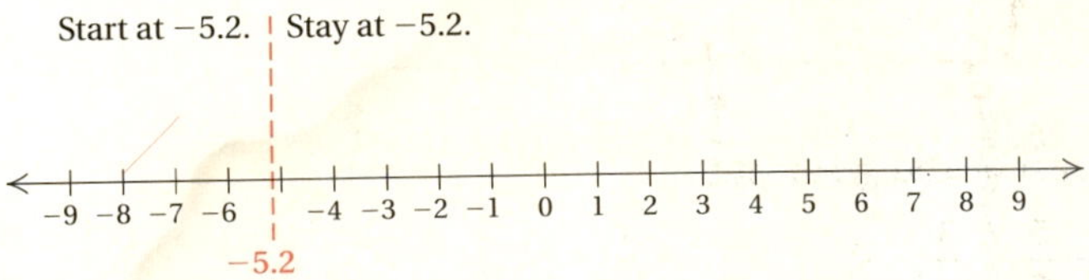

$-5.2 + 0 = -5.2$

Do Exercises 1–6.

OBJECTIVES

After finishing Section 1.3, you should be able to:

a Add real numbers without using a number line.

b Find the opposite, or additive inverse, of a real number.

FOR EXTRA HELP

TAPE 3 | TAPE 2B | MAC: 1A IBM: 1A

Add using a number line.

1. $0 + (-6)$

2. $1 + (-4)$

3. $-3 + (-5)$

4. $-3 + 7$

5. $-5.4 + 5.4$

6. $-\frac{5}{2} + \frac{1}{2}$

Answers on page A-1

Add without using a number line.

7. $-5 + (-6)$

8. $-9 + (-3)$

9. $-4 + 6$

10. $-7 + 3$

11. $5 + (-7)$

12. $-20 + 20$

13. $-11 + (-11)$

14. $10 + (-7)$

15. $-0.17 + 0.7$

16. $-6.4 + 8.7$

17. $-4.5 + (-3.2)$

18. $-8.6 + 2.4$

19. $\frac{5}{9} + \left(-\frac{7}{9}\right)$

20. $-\frac{1}{5} + \left(-\frac{3}{4}\right)$

Answers on page A-1

a ADDING WITHOUT A NUMBER LINE

You may have noticed some patterns in the preceding examples. These lead us to rules for adding without using a number line that are more efficient for adding larger or more complicated numbers.

RULES FOR ADDITION OF REAL NUMBERS

1. *Positive numbers:* Add the same as arithmetic numbers. The answer is positive.
2. *Negative numbers:* Add absolute values. The answer is negative.
3. *A positive and a negative number:* Subtract absolute values. Then:
   a) If the positive number has the greater absolute value, the answer is positive.
   b) If the negative number has the greater absolute value, the answer is negative.
   c) If the numbers have the same absolute value, the answer is 0.
4. *One number is zero:* The sum is the other number.

Rule 4 is known as the **identity property of 0.** It says that for any real number a, $a + 0 = a$.

EXAMPLES Add without using a number line.

5. $-12 + (-7) = -19$ — Two negatives. *Think:* Add the absolute values, 12 and 7, getting 19. Make the answer *negative*, −19.

6. $-1.4 + 8.5 = 7.1$ — The absolute values are 1.4 and 8.5. The difference is 7.1. The positive number has the larger absolute value, so the answer is *positive*, 7.1.

7. $-36 + 21 = -15$ — The absolute values are 36 and 21. The difference is 15. The negative number has the larger absolute value, so the answer is *negative*, −15.

8. $1.5 + (-1.5) = 0$ — The numbers have the same absolute value. The sum is 0.

9. $-\frac{7}{8} + 0 = -\frac{7}{8}$ — One number is zero. The sum is $-\frac{7}{8}$.

10. $-9.2 + 3.1 = -6.1$

11. $-\frac{3}{2} + \frac{9}{2} = \frac{6}{2} = 3$

12. $-\frac{2}{3} + \frac{5}{8} = -\frac{16}{24} + \frac{15}{24} = -\frac{1}{24}$

Do Exercises 7–20.

Suppose we want to add several numbers, some positive and some negative, as follows. How can we proceed?

$$15 + (-2) + 7 + 14 + (-5) + (-12)$$

We can change grouping and order as we please when adding. For instance, we can group the positive numbers together and the negative numbers together and add them separately. Then we add the two results.

EXAMPLE 13 Add: $15 + (-2) + 7 + 14 + (-5) + (-12)$.

a) $15 + 7 + 14 = 36$ Adding the positive numbers

b) $-2 + (-5) + (-12) = -19$ Adding the negative numbers

c) $36 + (-19) = 17$ Adding the results

We can also add the numbers in any other order we wish, say, from left to right as follows:

$$\begin{aligned} 15 + (-2) + 7 + 14 + (-5) + (-12) &= 13 + 7 + 14 + (-5) + (-12) \\ &= 20 + 14 + (-5) + (-12) \\ &= 34 + (-5) + (-12) \\ &= 29 + (-12) \\ &= 17 \end{aligned}$$

Do Exercises 21–24.

b OPPOSITES, OR ADDITIVE INVERSES

Suppose we add two numbers that are **opposites**, such as 6 and −6. The result is 0. When opposites are added, the result is always 0. Such numbers are also called **additive inverses.** Every real number has an opposite, or additive inverse.

> Two numbers whose sum is 0 are called *opposites*, or *additive inverses*, of each other.

EXAMPLES Find the opposite of each number.

14. 34 The opposite of 34 is −34 because $34 + (-34) = 0$.

15. −8 The opposite of −8 is 8 because $-8 + 8 = 0$.

16. 0 The opposite of 0 is 0 because $0 + 0 = 0$.

17. $-\frac{7}{8}$ The opposite of $-\frac{7}{8}$ is $\frac{7}{8}$ because $-\frac{7}{8} + \frac{7}{8} = 0$.

Do Exercises 25–30.

To name the opposite, we use the symbol −, as follows.

> The opposite, or additive inverse, of a number a can be named $-a$ (read "the opposite of a," or "the additive inverse of a").

Note that if we take a number, say, 8, and find its opposite, −8, and then find the opposite of the result, we will have the original number, 8, again.

Add.

21. $(-15) + (-37) + 25 + 42 + (-59) + (-14)$

22. $42 + (-81) + (-28) + 24 + 18 + (-31)$

23. $-2.5 + (-10) + 6 + (-7.5)$

24. $-35 + 17 + 14 + (-27) + 31 + (-12)$

Find the opposite.

25. −4

26. 8.7

27. −7.74

28. $-\frac{8}{9}$

29. 0

30. 12

Answers on page A-1

Find $-x$ and $-(-x)$ when x is each of the following.

31. 14

32. 1

33. -19

34. -1.6

35. $\frac{2}{3}$

36. $-\frac{9}{8}$

Find the opposite. (Change the sign.)

37. -4

38. -13.4

39. 0

40. $\frac{1}{4}$

Answers on page A-1

The opposite of the opposite of a number is the number itself. (The additive inverse of the additive inverse of a number is the number itself.) That is, for any number a,

$$-(-a) = a.$$

EXAMPLE 18 Find $-x$ and $-(-x)$ when $x = 16$.

If $x = 16$, then $-x = -16$. **The opposite of 16 is −16.**

If $x = 16$, then $-(-x) = -(-16) = 16$. **The opposite of the opposite of 16 is 16.**

EXAMPLE 19 Find $-x$ and $-(-x)$ when $x = -3$.

If $x = -3$, then $-x = -(-3) = 3$.

If $x = -3$, then $-(-x) = -(-(-3)) = -3$.

Note that in Example 19 we used an extra set of parentheses to show that we are substituting the negative number -3 for x. Symbolism like $--x$ is not considered meaningful.

Do Exercises 31–36.

A symbol such as -8 is usually read "negative 8." It could be read "the additive inverse of 8," because the additive inverse of 8 is negative 8. It could also be read "the opposite of 8," because the opposite of 8 is -8. Thus a symbol like -8 can be read in more than one way. A symbol like $-x$, which has a variable, should be read "the opposite of x" or "the additive inverse of x" and *not* "negative x," because we do not know whether x represents a positive number, a negative number, or 0. You can verify this by referring to the preceding examples.

We can use the symbolism $-a$ to restate the definition of opposite, or additive inverse.

For any real number a, the *opposite,* or *additive inverse,* of a, which is $-a$, is such that

$$a + (-a) = (-a) + a = 0.$$

Signs of Numbers

A negative number is sometimes said to have a "negative sign." A positive number is said to have a "positive sign." When we replace a number by its opposite, we can say that we have "changed its sign."

EXAMPLES Find the opposite. (Change the sign.)

20. -3 $\quad -(-3) = 3$ **The opposite of −3 is 3.**

21. -10 $\quad -(-10) = 10$

22. 0 $\quad -(0) = 0$

23. 14 $\quad -(14) = -14$

Do Exercises 37–40.

NAME SECTION DATE

Exercise Set 1.3

a Add. Do not use a number line except as a check.

1. $2 + (-9)$

2. $-5 + 2$

3. $-11 + 5$

4. $4 + (-3)$

5. $-6 + 6$

6. $8 + (-8)$

7. $-3 + (-5)$

8. $-4 + (-6)$

9. $-7 + 0$

10. $-13 + 0$

11. $0 + (-27)$

12. $0 + (-35)$

13. $17 + (-17)$

14. $-15 + 15$

15. $-17 + (-25)$

16. $-24 + (-17)$

17. $18 + (-18)$

18. $-13 + 13$

19. $-28 + 28$

20. $11 + (-11)$

21. $8 + (-5)$

22. $-7 + 8$

23. $-4 + (-5)$

24. $10 + (-12)$

25. $13 + (-6)$

26. $-3 + 14$

27. $-25 + 25$

28. $50 + (-50)$

29. $53 + (-18)$

30. $75 + (-45)$

31. $-8.5 + 4.7$

32. $-4.6 + 1.9$

33. $-2.8 + (-5.3)$

34. $-7.9 + (-6.5)$

35. $-\frac{3}{5} + \frac{2}{5}$

36. $-\frac{4}{3} + \frac{2}{3}$

ANSWERS

1. ______
2. ______
3. ______
4. ______
5. ______
6. ______
7. ______
8. ______
9. ______
10. ______
11. ______
12. ______
13. ______
14. ______
15. ______
16. ______
17. ______
18. ______
19. ______
20. ______
21. ______
22. ______
23. ______
24. ______
25. ______
26. ______
27. ______
28. ______
29. ______
30. ______
31. ______
32. ______
33. ______
34. ______
35. ______
36. ______

Answers

37. ______
38. ______
39. ______
40. ______
41. ______
42. ______
43. ______
44. ______
45. ______
46. ______
47. ______
48. ______
49. ______
50. ______
51. ______
52. ______
53. ______
54. ______
55. ______
56. ______
57. ______
58. ______
59. ______
60. ______
61. ______
62. ______
63. ______
64. ______
65. ______
66. ______
67. ______
68. ______
69. ______
70. ______
71. ______
72. ______
73. ______
74. ______
75. ______
76. ______
77. ______
78. ______

37. $-\frac{3}{9} + \left(-\frac{5}{9}\right)$ **38.** $-\frac{4}{7} + \left(-\frac{6}{7}\right)$ **39.** $-\frac{5}{8} + \frac{1}{4}$ **40.** $-\frac{5}{6} + \frac{2}{3}$

41. $-\frac{5}{8} + \left(-\frac{1}{6}\right)$ **42.** $-\frac{5}{6} + \left(-\frac{2}{9}\right)$ **43.** $-\frac{3}{8} + \frac{5}{12}$ **44.** $-\frac{7}{16} + \frac{7}{8}$

45. $76 + (-15) + (-18) + (-6)$ **46.** $29 + (-45) + 18 + 32 + (-96)$

47. $-44 + \left(-\frac{3}{8}\right) + 95 + \left(-\frac{5}{8}\right)$ **48.** $24 + 3.1 + (-44) + (-8.2) + 63$

49. $98 + (-54) + 113 + (-998) + 44 + (-612)$

50. $-458 + (-124) + 1025 + (-917) + 218$

b Find the opposite, or additive inverse.

51. 24 **52.** -64 **53.** -26.9 **54.** 48.2

Find $-x$ when x is each of the following.

55. 8 **56.** -27 **57.** $-\frac{13}{8}$ **58.** $\frac{1}{236}$

Find $-(-x)$ when x is each of the following.

59. -43 **60.** 39 **61.** $\frac{4}{3}$ **62.** -7.1

Find the opposite. (Change the sign.)

63. -24 **64.** -12.3 **65.** $-\frac{3}{8}$ **66.** 10

SKILL MAINTENANCE

Convert to decimal notation.

67. 57% **68.** 49% **69.** 52.9% **70.** 71.3%

Convert to percent notation.

71. $\frac{5}{4}$ **72.** $\frac{1}{8}$ **73.** $\frac{13}{25}$ **74.** $\frac{13}{32}$

SYNTHESIS

75. For what numbers x is $-x$ negative?

76. For what numbers x is $-x$ positive?

Tell whether the sum is positive, negative, or zero.

77. If a is positive and b is negative, $-a + b$ is ______.

78. If $a = b$ and a and b are negative, $-a + (-b)$ is ______.

1.4 *Subtraction of Real Numbers*

a SUBTRACTION

We now consider subtraction of real numbers. Subtraction is defined as follows.

> The difference $a - b$ is the number that when added to b gives a.

For example, $45 - 17 = 28$ because $28 + 17 = 45$. Let us consider an example whose answer is a negative number.

EXAMPLE 1 Subtract: $5 - 8$.

Think: $5 - 8$ is the number that when added to 8 gives 5. What number can we add to 8 to get 5? The number must be negative. The number is -3:

$$5 - 8 = -3.$$

That is, $5 - 8 = -3$ because $5 = -3 + 8$.

Do Exercises 1–3.

The definition above does *not* provide the most efficient way to do subtraction. From that definition, however, we can develop a faster way to subtract. Look for a pattern in the following examples.

SUBTRACTIONS	ADDING AN OPPOSITE
$5 - 8 = -3$	$5 + (-8) = -3$
$-6 - 4 = -10$	$-6 + (-4) = -10$
$-7 - (-10) = 3$	$-7 + 10 = 3$
$-7 - (-2) = -5$	$-7 + 2 = -5$

Do Exercises 4–7.

Perhaps you have noticed that we can subtract by adding the opposite of the number being subtracted. This can always be done.

> For any real numbers a and b,
>
> $$a - b = a + (-b).$$
>
> (To subtract, add the opposite, or additive inverse, of the number being subtracted.)

This is the method normally used for quick subtraction of real numbers.

OBJECTIVES

After finishing Section 1.4, you should be able to:

a Subtract real numbers and simplify combinations of additions and subtractions.

b Solve problems involving addition and subtraction of real numbers.

FOR EXTRA HELP

TAPE 3 TAPE 3A MAC: 1A IBM: 1A

Subtract.

1. $-6 - 4$

Think: What number can be added to 4 to get -6?

2. $-7 - (-10)$

Think: What number can be added to -10 to get -7?

3. $-7 - (-2)$

Think: What number can be added to -2 to get -7?

Complete the addition and compare with the subtraction.

4. $4 - 6 = -2$;
$4 + (-6) =$ ______

5. $-3 - 8 = -11$;
$-3 + (-8) =$ ______

6. $-5 - (-9) = 4$;
$-5 + 9 =$ ______

7. $-5 - (-3) = -2$;
$-5 + 3 =$ ______

Answers on page A-1

Subtract.

8. $2 - 8$

9. $-6 - 10$

10. $12.4 - 5.3$

11. $-8 - (-11)$

12. $-8 - (-8)$

13. $\frac{2}{3} - \left(-\frac{5}{6}\right)$

Read each of the following. Then subtract by adding the opposite of the number being subtracted.

14. $3 - 11$

15. $12 - 5$

16. $-12 - (-9)$

17. $-12.4 - 10.9$

18. $-\frac{4}{5} - \left(-\frac{4}{5}\right)$

Simplify.

19. $-6 - (-2) - (-4) - 12 + 3$

20. $9 - (-6) + 7 - 11 - 14 - (-20)$

21. $-9.6 + 7.4 - (-3.9) - (-11)$

Answers on page A-1

EXAMPLES Subtract.

2. $2 - 6 = 2 + (-6) = -4$ — The opposite of 6 is −6. We change the subtraction to addition and add the opposite.

3. $4 - (-9) = 4 + 9 = 13$ — The opposite of −9 is 9. We change the subtraction to addition and add the opposite.

4. $-4.2 - (-3.6) = -4.2 + 3.6 = -0.6$ — Adding the opposite. *Check:* $-0.6 + (-3.6) = -4.2$.

5. $-\frac{1}{2} - \left(-\frac{3}{4}\right) = -\frac{1}{2} + \frac{3}{4} = \frac{1}{4}$ — Adding the opposite. *Check:* $\frac{1}{4} + \left(-\frac{3}{4}\right) = -\frac{1}{2}$.

Do Exercises 8–13.

EXAMPLES Read each of the following. Then subtract by adding the opposite of the number being subtracted.

6. $3 - 5$;
$3 - 5 = 3 + (-5) = -2$

Read "three minus five is three plus the opposite of five"

7. $\frac{1}{8} - \frac{7}{8}$;
$\frac{1}{8} - \frac{7}{8} = \frac{1}{8} + \left(-\frac{7}{8}\right) = -\frac{6}{8}$, or $-\frac{3}{4}$

Read "one-eighth minus seven-eighths is one-eighth plus the opposite of seven-eighths"

8. $-4.6 - (-9.8)$;
$-4.6 - (-9.8) = -4.6 + 9.8 = 5.2$

Read "negative four point six minus negative nine point eight is negative four point six plus the opposite of negative nine point eight"

9. $-\frac{3}{4} - \frac{7}{5}$;
$-\frac{3}{4} - \frac{7}{5} = -\frac{3}{4} + \left(-\frac{7}{5}\right) = -\frac{15}{20} + \left(-\frac{28}{20}\right) = -\frac{43}{20}$

Read "negative three-fourths minus seven-fifths is negative three-fourths plus the opposite of seven-fifths"

Do Exercises 14–18.

When several additions and subtractions occur together, we can make them all additions.

EXAMPLES Simplify.

10. $8 - (-4) - 2 - (-4) + 2 = 8 + 4 + (-2) + 4 + 2 = 16$ — Adding the opposites where subtraction is indicated

11. $8.2 - (-6.1) + 2.3 - (-4) = 8.2 + 6.1 + 2.3 + 4 = 20.6$

Do Exercises 19–21.

b PROBLEM SOLVING

Let us see how we can use subtraction of real numbers to solve problems.

EXAMPLE 12 The lowest point in Asia is the Dead Sea, which is 400 m below sea level. The lowest point in the United States is Death Valley, which is 86 m below sea level. How much higher is Death Valley than the Dead Sea?

It is helpful to draw a picture of the situation.

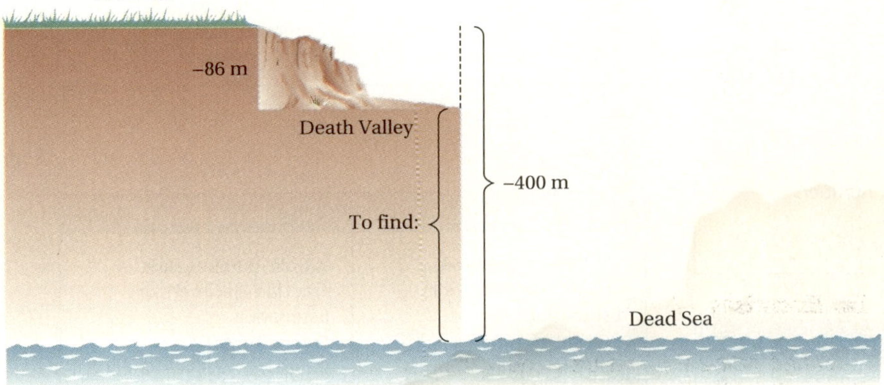

We see that -86 is the higher altitude at Death Valley and -400 is the lower altitude at the Dead Sea. To find how much higher Death Valley is, we subtract:

$$-86 - (-400) = -86 + 400 = 314.$$

Death Valley is 314 m higher than the Dead Sea.

Do Exercises 22 and 23.

Solve.

22. Bill owns a small sports card business. It made a profit of \$18 on Monday. There was a loss of \$7 on Tuesday. On Wednesday there was a loss of \$5, and on Thursday there was a profit of \$11. Find the total profit or loss.

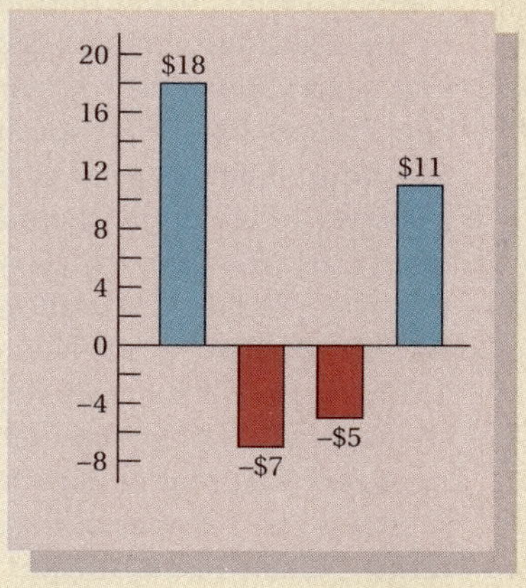

23. In Churchill, Manitoba, Canada, the average daily low temperature in January is $-31°C$. The average daily low temperature in Key West, Florida, is 19°C. How much higher is the average daily low temperature in Key West, Florida?

Answers on page A-1

SIDELIGHTS

CAREERS AND THEIR USES OF MATHEMATICS

Students typically ask the question "Why do we have to study mathematics?" One answer is that you will use this mathematics in the next course. Although it is a correct answer, it sometimes frustrates students, because this answer can be given in the next mathematics course, and the next one, and so on. Sometimes an answer can be given by applications like those you have seen or will see in this book. Another answer is that you are living in a society in which mathematics becomes more and more critical with each passing day. Evidence of this was provided recently by a nationwide symposium sponsored by the National Research Council's Mathematical Sciences Education Board. Results showed that "Other than demographic factors, the *strongest* predictor of earnings nine years after high school is the number of mathematics courses taken." This is a significant testimony to the need for you to take as many mathematics courses as possible.

We try to provide other answers to "Why do we have to study mathematics?" in what follows. We have listed several occupations that are attractive and popular to students and various kinds of mathematics that are useful in that occupation.

DOCTOR	LAWYER
Equations Percent notation Graphing Statistics Geometry Measurement Estimation Exponents Logic	Equations Percent notation Graphing Probability Statistics Ratio and proportion Area and volume Negative numbers Formulas Calculator skills

PILOT	FIREFIGHTER
Equations Percent notation Graphing Trigonometry Angles and geometry Calculator skills Computer skills Ratio and proportion Vectors	Percent notation Graphing Estimation Formulas Angles and geometry Probability Statistics Area and geometry Square roots Exponents Pythagorean theorem

ACCOUNTANT OR BUSINESS MANAGER	TRAVEL AGENT
Computer skills Calculator skills Equations Systems of equations Formulas Probability Statistics Ratio and proportion Percent notation Estimation	Whole-number skills Fraction/decimal skills Estimation Percent notation Equations Calculator skills Computer skills

LIBRARIAN	MACHINIST
Whole-number skills Fraction/decimal skills Estimation Percent notation Ratio and proportion Area and perimeter Formulas Calculator skills Computer skills	Whole-number skills Fraction/decimal skills Estimation Percent notation Length, area, volume, and perimeter Angle measures Geometry Pythagorean theorem Square roots Equations Formulas Graphing Calculator skills Computer skills Metric measures

NURSE	POLICE OFFICER
Whole-number skills Fraction/decimal skills Estimation Percent notation Ratio and proportion Estimation Equations English/Metric measurement Probability Statistics Formulas Exponents and scientific notation Calculator skills Computer skills	Whole-number skills Fraction/decimal skills Estimation Percent notation Ratio and proportion Geometry Negative numbers Probability Statistics Calculator skills

NAME ______________________ SECTION __________ DATE __________

Exercise Set 1.4

a Subtract.

1. $2 - 9$

2. $3 - 8$

3. $0 - 4$

4. $0 - 9$

5. $-8 - (-2)$

6. $-6 - (-8)$

7. $-11 - (-11)$

8. $-6 - (-6)$

9. $12 - 16$

10. $14 - 19$

11. $20 - 27$

12. $30 - 4$

13. $-9 - (-3)$

14. $-7 - (-9)$

15. $-40 - (-40)$

16. $-9 - (-9)$

17. $7 - 7$

18. $9 - 9$

19. $7 - (-7)$

20. $4 - (-4)$

21. $8 - (-3)$

22. $-7 - 4$

23. $-6 - 8$

24. $6 - (-10)$

25. $-4 - (-9)$

26. $-14 - 2$

27. $1 - 8$

28. $2 - 8$

29. $-6 - (-5)$

30. $-4 - (-3)$

31. $8 - (-10)$

32. $5 - (-6)$

33. $0 - 10$

34. $0 - 18$

35. $-5 - (-2)$

36. $-3 - (-1)$

ANSWERS

1. ______
2. ______
3. ______
4. ______
5. ______
6. ______
7. ______
8. ______
9. ______
10. ______
11. ______
12. ______
13. ______
14. ______
15. ______
16. ______
17. ______
18. ______
19. ______
20. ______
21. ______
22. ______
23. ______
24. ______
25. ______
26. ______
27. ______
28. ______
29. ______
30. ______
31. ______
32. ______
33. ______
34. ______
35. ______
36. ______

ANSWERS

37. ______
38. ______
39. ______
40. ______
41. ______
42. ______
43. ______
44. ______
45. ______
46. ______
47. ______
48. ______
49. ______
50. ______
51. ______
52. ______
53. ______
54. ______
55. ______
56. ______
57. ______
58. ______
59. ______
60. ______
61. ______
62. ______
63. ______
64. ______
65. ______
66. ______
67. ______
68. ______

37. $-7 - 14$

38. $-9 - 16$

39. $0 - (-5)$

40. $0 - (-1)$

41. $-8 - 0$

42. $-9 - 0$

43. $7 - (-5)$

44. $7 - (-4)$

45. $2 - 25$

46. $18 - 63$

47. $-42 - 26$

48. $-18 - 63$

49. $-71 - 2$

50. $-49 - 3$

51. $24 - (-92)$

52. $48 - (-73)$

53. $-50 - (-50)$

54. $-70 - (-70)$

55. $-\frac{3}{8} - \frac{5}{8}$

56. $\frac{3}{9} - \frac{9}{9}$

57. $\frac{3}{4} - \frac{2}{3}$

58. $\frac{5}{8} - \frac{3}{4}$

59. $-\frac{3}{4} - \frac{2}{3}$

60. $-\frac{5}{8} - \frac{3}{4}$

61. $-\frac{5}{8} - \left(-\frac{3}{4}\right)$

62. $-\frac{3}{4} - \left(-\frac{2}{3}\right)$

63. $6.1 - (-13.8)$

64. $1.5 - (-3.5)$

65. $-2.7 - 5.9$

66. $-3.2 - 5.8$

67. $0.99 - 1$

68. $0.87 - 1$

69. $-79 - 114$

70. $-197 - 216$

71. $0 - (-500)$

72. $500 - (-1000)$

73. $-2.8 - 0$

74. $6.04 - 1.1$

75. $7 - 10.53$

76. $8 - (-9.3)$

77. $\frac{1}{6} - \frac{2}{3}$

78. $-\frac{3}{8} - \left(-\frac{1}{2}\right)$

79. $-\frac{4}{7} - \left(-\frac{10}{7}\right)$

80. $\frac{12}{5} - \frac{12}{5}$

81. $-\frac{7}{10} - \frac{10}{15}$

82. $-\frac{4}{18} - \left(-\frac{2}{9}\right)$

83. $\frac{1}{5} - \frac{1}{3}$

84. $-\frac{1}{7} - \left(-\frac{1}{6}\right)$

Simplify.

85. $18 - (-15) - 3 - (-5) + 2$

86. $22 - (-18) + 7 + (-42) - 27$

87. $-31 + (-28) - (-14) - 17$

88. $-43 - (-19) - (-21) + 25$

89. $-34 - 28 + (-33) - 44$

90. $39 + (-88) - 29 - (-83)$

91. $-93 - (-84) - 41 - (-56)$

92. $84 + (-99) + 44 - (-18) - 43$

93. $-5 - (-30) + 30 + 40 - (-12)$

94. $14 - (-50) + 20 - (-32)$

95. $132 - (-21) + 45 - (-21)$

96. $81 - (-20) - 14 - (-50) + 53$

ANSWERS

69. ______
70. ______
71. ______
72. ______
73. ______
74. ______
75. ______
76. ______
77. ______
78. ______
79. ______
80. ______
81. ______
82. ______
83. ______
84. ______
85. ______
86. ______
87. ______
88. ______
89. ______
90. ______
91. ______
92. ______
93. ______
94. ______
95. ______
96. ______

ANSWERS

97. ______

98. ______

99. ______

100. ______

101. ______

102. ______

103. ______

104. ______

105. ______

106. ______

107. ______

108. ______

109. ______

110. ______

111. ______

112. ______

113. ______

114. ______

115. ______

116. ______

117. ______

b Solve.

97. Jose owes a relative \$530. The relative decides to cancel \$156 of the debt. How much does Jose owe now?

98. Lou has \$719.46 in her checking account. She wrote a check for \$1080 to pay for a sound system. By how much did she overdraw her checking account?

99. During a blizzard in Minneapolis, the temperature dropped from $-5°F$ to $-12°F$. How many degrees did it drop?

100. You are in debt \$476.89. How much money will you need to make your total assets \$213.80?

101. The deepest point in the Pacific Ocean is the Marianas Trench, with a depth of 34,370 ft. The deepest point in the Atlantic Ocean is the Puerto Rico Trench, with a depth of 28,538 ft. How much higher is the Puerto Rico Trench than the Marianas Trench?

102. The lowest point in Africa is Lake Assal, which is 515 ft below sea level. The lowest point in South America is the Valdes Peninsula, which is 132 ft below sea level. How much lower is Lake Assal than the Valdes Peninsula?

SKILL MAINTENANCE

103. Evaluate: 5^3.

104. Find the prime factorization of 864.

105. Simplify: $256 \div 64 \div 2^3 + 100$.

106. Simplify: $5 \cdot 6 + (7 \cdot 2)^2$.

107. Convert to decimal notation: 58.3%.

108. Simplify: $\frac{164}{256}$.

SYNTHESIS

Subtract.

109. 🖩 $123{,}907 - 433{,}789$

110. 🖩 $23{,}011 - (-60{,}432)$

Tell whether the statement is true or false for all integers a and b. If false, show why.

111. $a - 0 = 0 - a$

112. $0 - a = a$

113. If $a \neq b$, then $a - b \neq 0$.

114. If $a = -b$, then $a + b = 0$.

115. If $a + b = 0$, then a and b are opposites.

116. If $a - b = 0$, then $a = -b$.

117. Velma Quarles is a stockbroker. She kept track of the changes in the stock market over a period of 5 weeks. By how many points had the market risen or fallen over this time?

WEEK 1	WEEK 2	WEEK 3	WEEK 4	WEEK 5
Down 13 pts	Down 16 pts	Up 36 pts	Down 11 pts	Up 19 pts

1.5 *Multiplication of Real Numbers*

a MULTIPLICATION

Multiplication of real numbers is very much like multiplication of arithmetic numbers. The only difference is that we must determine whether the answer is positive or negative.

MULTIPLICATION OF A POSITIVE NUMBER AND A NEGATIVE NUMBER

To see how to multiply a positive number and a negative number, consider the pattern of the following.

This number decreases by 1 each time. This number decreases by 5 each time.

$$\begin{aligned} 4 \cdot 5 &= 20 \\ 3 \cdot 5 &= 15 \\ 2 \cdot 5 &= 10 \\ 1 \cdot 5 &= 5 \\ 0 \cdot 5 &= 0 \\ -1 \cdot 5 &= -5 \\ -2 \cdot 5 &= -10 \\ -3 \cdot 5 &= -15 \end{aligned}$$

Do Exercise 1.

According to this pattern, it looks as though the product of a negative number and a positive number is negative. That is the case, and we have the first part of the rule for multiplying numbers.

> To multiply a positive number and a negative number, multiply their absolute values. The answer is negative.

EXAMPLES Multiply.

1. $8(-5) = -40$

2. $-\frac{1}{3} \cdot \frac{5}{7} = -\frac{5}{21}$

3. $(-7.2)5 = -36$

Do Exercises 2–7.

MULTIPLICATION OF TWO NEGATIVE NUMBERS

How do we multiply two negative numbers? Again, we look for a pattern.

This number decreases by 1 each time. This number increases by 5 each time.

$$\begin{aligned} 4 \cdot (-5) &= -20 \\ 3 \cdot (-5) &= -15 \\ 2 \cdot (-5) &= -10 \\ 1 \cdot (-5) &= -5 \\ 0 \cdot (-5) &= 0 \\ -1 \cdot (-5) &= 5 \\ -2 \cdot (-5) &= 10 \\ -3 \cdot (-5) &= 15 \end{aligned}$$

Do Exercise 8.

OBJECTIVE

After finishing Section 1.5, you should be able to:

a Multiply real numbers.

FOR EXTRA HELP

TAPE 4 TAPE 3A MAC: 1B IBM: 1B

1. Complete, as in the example.

$$\begin{aligned} 4 \cdot 10 &= 40 \\ 3 \cdot 10 &= 30 \\ 2 \cdot 10 &= \\ 1 \cdot 10 &= \\ 0 \cdot 10 &= \\ -1 \cdot 10 &= \\ -2 \cdot 10 &= \\ -3 \cdot 10 &= \end{aligned}$$

Multiply.

2. $-3 \cdot 6$

3. $20 \cdot (-5)$

4. $4 \cdot (-20)$

5. $-\frac{2}{3} \cdot \frac{5}{6}$

6. $-4.23(7.1)$

7. $\frac{7}{8}\left(-\frac{4}{5}\right)$

8. Complete, as in the example.

$$\begin{aligned} 3 \cdot (-10) &= -30 \\ 2 \cdot (-10) &= -20 \\ 1 \cdot (-10) &= \\ 0 \cdot (-10) &= \\ -1 \cdot (-10) &= \\ -2 \cdot (-10) &= \\ -3 \cdot (-10) &= \end{aligned}$$

Answers on page A-2

Multiply.

9. $-3 \cdot (-4)$

10. $-16 \cdot (-2)$

11. $-7 \cdot (-5)$

12. $-\frac{4}{7}\left(-\frac{5}{9}\right)$

13. $-\frac{3}{2}\left(-\frac{4}{9}\right)$

14. $-3.25(-4.14)$

Multiply.

15. $5(-6)$

16. $(-5)(-6)$

17. $(-3.2) \cdot 0$

18. $\left(-\frac{4}{5}\right)\left(\frac{10}{3}\right)$

Answers on page A-2

According to the pattern, it appears that the product of two negative numbers is positive. That is actually so, and we have the second part of the rule for multiplying real numbers.

> To multiply two negative numbers, multiply their absolute values. The answer is positive.

Do Exercises 9–14.

The following is an alternative way to consider the rules we have for multiplication.

> To multiply two real numbers:
> 1. Multiply the absolute values.
> 2. If the signs are the same, the answer is positive.
> 3. If the signs are different, the answer is negative.

Multiplication by Zero

The only case that we have not considered is multiplying by zero. As with other numbers, the product of any real number and 0 is 0.

> **The Multiplication Property of Zero**
>
> For any real number a,
>
> $$a \cdot 0 = 0.$$
>
> (The product of 0 and any real number is 0.)

Examples Multiply.

4. $(-3)(-4) = 12$
5. $-1.6(2) = -3.2$
6. $-19 \cdot 0 = 0$
7. $\left(-\frac{5}{6}\right)\left(-\frac{1}{9}\right) = \frac{5}{54}$

Do Exercises 15–18.

MULTIPLYING MORE THAN TWO NUMBERS

When multiplying more than two real numbers, we can choose order and grouping as we please.

EXAMPLES Multiply.

8. $-8 \cdot 2(-3) = -16(-3)$ Multiplying the first two numbers
$= 48$ Multiplying the results

9. $-8 \cdot 2(-3) = 24 \cdot 2$ Multiplying the negatives. Every pair of negative numbers gives a positive product.
$= 48$

10. $-3(-2)(-5)(4) = 6(-5)(4)$ Multiplying the first two numbers
$= (-30)4$
$= -120$

11. $\left(-\frac{1}{2}\right)(8)\left(-\frac{2}{3}\right)(-6) = (-4)4$ Multiplying the first two numbers and the last two numbers
$= -16$

12. $-5 \cdot (-2) \cdot (-3) \cdot (-6) = 10 \cdot 18$
$= 180$

13. $(-3)(-5)(-2)(-3)(-6) = (-30)(18)$
$= -540$

We can see the following pattern in the results of Examples 12 and 13.

> The product of an even number of negative numbers is positive.
> The product of an odd number of negative numbers is negative.

Do Exercises 19–24.

EXAMPLE 14 Evaluate $(-x)^2$ and $-x^2$ when $x = 5$.

$(-x)^2 = (-5)^2 = (-5)(-5) = 25;$ Substitute 5 for x. Then evaluate the power.

$-x^2 = -(5)^2 = -25$ Substitute 5 for x. Evaluate the power. Then find the opposite.

Caution! Note that the expressions $(-x)^2$ and $-x^2$ are *not* equivalent. That is, they do not have the same value for every replacement of the variable by a real number. To find $(-x)^2$, we take the opposite and then square. To find $-x^2$, we find the square and then take the opposite.

EXAMPLE 15 Evaluate $2x^2$ when $x = 3$ and $x = -3$.

$2x^2 = 2(3)^2 = 2(9) = 18;$

$2x^2 = 2(-3)^2 = 2(9) = 18$

Do Exercises 25–27.

Multiply.

19. $5 \cdot (-3) \cdot 2$

20. $-3 \times (-4.1) \times (-2.5)$

21. $-\frac{1}{2} \cdot \left(-\frac{4}{3}\right) \cdot \left(-\frac{5}{2}\right)$

22. $-2 \cdot (-5) \cdot (-4) \cdot (-3)$

23. $(-4)(-5)(-2)(-3)(-1)$

24. $(-1)(-1)(-2)(-3)(-1)(-1)$

25. Evaluate $(-x)^2$ and $-x^2$ when $x = 2$.

26. Evaluate $(-x)^2$ and $-x^2$ when $x = 3$.

27. Evaluate $3x^2$ when $x = 4$ and $x = -4$.

Answers on page A-2

SIDELIGHTS

STUDY TIPS: STUDYING FOR TESTS AND MAKING THE MOST OF TUTORING SESSIONS

As has been stated, we will often present some tips and guidelines to enhance your learning abilities. Sometimes these tips will be focused on mathematics, but sometimes they will be more general, as is the case here where we consider test preparation and tutoring.

TEST-TAKING TIPS

- *Make up your own test questions as you study.* You have probably noted by now the section and objective codes that appear throughout the book. After you have done your homework over a particular objective, write one or two questions on your own that you think might be on a test. You will be amazed at the insight this will provide. You are actually carrying out a task similar to what a teacher does in preparing an exam.
- *Ask former students for old exams.* Working such exams can be very helpful and allows you to see what various professors think is important.
- *When taking a test, read each question carefully and try to do all the questions the first time through, but pace yourself.* Answer all the questions, and mark those to recheck if you have time at the end. Very often, your first hunch will be correct.
- *Try to write your test in a neat and orderly manner.* Very often, your instructor tries to give you partial credit when grading an exam. If your test paper is sloppy and disorderly, it is difficult to verify the partial credit. Doing your work neatly can ease such a task on an exam. Try using an erasable pen to make your writing darker and therefore more readable.

MAKING THE MOST OF TUTORING AND HELP SESSIONS

Often you will determine that a tutoring session would be helpful. The following comments may help you to make the most of such sessions.

- *Work on the topics before you go to the help or tutoring session. Do not go to such sessions viewing yourself as an empty cup and the tutor as a magician who will pour in the learning.* The primary source of your ability to learn is within you. We have seen so many students over the years go to help or tutoring sessions with no advanced preparation. You are often wasting your time and perhaps your money if you are paying for such sessions. Go to class, study the textbook, and mark trouble spots. Then use the help and tutoring sessions to deal with these difficulties most efficiently.
- *Do not be afraid to ask questions in these sessions*! The more you talk to your tutor, the more the tutor can help you with your difficulties.
- *Try being a "tutor" yourself.* Explaining a topic to someone else—a classmate, your instructor—is often the best way to learn it.
- *What about the student who says, "I could do the work at home, but on the test I made silly mistakes"*? Yes, all of us, including instructors, make silly computational mistakes in class, on homework, and on tests. But your instructor, if he or she has taught for some time, is probably aware that 90% of students who make such comments in truth do not have the depth of knowledge of the subject matter, and such silly mistakes often are a sign that the student has not mastered the material. There is no way we can make that analysis for you. It will have to be unraveled by some careful soul searching on your part or by a conference with your instructor.

NAME SECTION DATE

Exercise Set 1.5

a Multiply.

1. $-4 \cdot 2$

2. $-3 \cdot 5$

3. $-8 \cdot 6$

4. $-5 \cdot 2$

5. $8 \cdot (-3)$

6. $9 \cdot (-5)$

7. $-9 \cdot 8$

8. $-10 \cdot 3$

9. $-8 \cdot (-2)$

10. $-2 \cdot (-5)$

11. $-7 \cdot (-6)$

12. $-9 \cdot (-2)$

13. $15 \cdot (-8)$

14. $-12 \cdot (-10)$

15. $-14 \cdot 17$

16. $-13 \cdot (-15)$

17. $-25 \cdot (-48)$

18. $39 \cdot (-43)$

19. $-3.5 \cdot (-28)$

20. $97 \cdot (-2.1)$

21. $9 \cdot (-8)$

22. $7 \cdot (-9)$

23. $4 \cdot (-3.1)$

24. $3 \cdot (-2.2)$

25. $-5 \cdot (-6)$

26. $-6 \cdot (-4)$

27. $-7 \cdot (-3.1)$

28. $-4 \cdot (-3.2)$

29. $\frac{2}{3} \cdot \left(-\frac{3}{5}\right)$

30. $\frac{5}{7} \cdot \left(-\frac{2}{3}\right)$

31. $-\frac{3}{8} \cdot \left(-\frac{2}{9}\right)$

32. $-\frac{5}{8} \cdot \left(-\frac{2}{5}\right)$

33. -6.3×2.7

34. -4.1×9.5

35. $-\frac{5}{9} \cdot \frac{3}{4}$

36. $-\frac{8}{3} \cdot \frac{9}{4}$

37. $7 \cdot (-4) \cdot (-3) \cdot 5$

38. $9 \cdot (-2) \cdot (-6) \cdot 7$

39. $-\frac{2}{3} \cdot \frac{1}{2} \cdot \left(-\frac{6}{7}\right)$

40. $-\frac{1}{8} \cdot \left(-\frac{1}{4}\right) \cdot \left(-\frac{3}{5}\right)$

41. $-3 \cdot (-4) \cdot (-5)$

Answers

1. ______
2. ______
3. ______
4. ______
5. ______
6. ______
7. ______
8. ______
9. ______
10. ______
11. ______
12. ______
13. ______
14. ______
15. ______
16. ______
17. ______
18. ______
19. ______
20. ______
21. ______
22. ______
23. ______
24. ______
25. ______
26. ______
27. ______
28. ______
29. ______
30. ______
31. ______
32. ______
33. ______
34. ______
35. ______
36. ______
37. ______
38. ______
39. ______
40. ______
41. ______

ANSWERS

42. ______
43. ______
44. ______
45. ______
46. ______
47. ______
48. ______
49. ______
50. ______
51. ______
52. ______
53. ______
54. ______
55. ______
56. ______
57. ______
58. ______
59. ______
60. ______
61. ______
62. ______
63. ______
64. ______
65. ______
66. ______
67. ______
68. ______
69. ______
70. ______
71. ______
72. ______
73. ______
74. ______
75. a) ______
b) ______
c) ______
76. ______

42. $-2 \cdot (-5) \cdot (-7)$

43. $-2 \cdot (-5) \cdot (-3) \cdot (-5)$

44. $-3 \cdot (-5) \cdot (-2) \cdot (-1)$

45. $\frac{1}{5}\left(-\frac{2}{9}\right)$

46. $-\frac{3}{5}\left(-\frac{2}{7}\right)$

47. $-7 \cdot (-21) \cdot 13$

48. $-14 \cdot (34) \cdot 12$

49. $-4 \cdot (-1.8) \cdot 7$

50. $-8 \cdot (-1.3) \cdot (-5)$

51. $-\frac{1}{9}\left(-\frac{2}{3}\right)\left(\frac{5}{7}\right)$

52. $-\frac{7}{2}\left(-\frac{5}{7}\right)\left(-\frac{2}{5}\right)$

53. $4 \cdot (-4) \cdot (-5) \cdot (-12)$

54. $-2 \cdot (-3) \cdot (-4) \cdot (-5)$

55. $0.07 \cdot (-7) \cdot 6 \cdot (-6)$

56. $80 \cdot (-0.8) \cdot (-90) \cdot (-0.09)$

57. $\left(-\frac{5}{6}\right)\left(\frac{1}{8}\right)\left(-\frac{3}{7}\right)\left(-\frac{1}{7}\right)$

58. $\left(\frac{4}{5}\right)\left(-\frac{2}{3}\right)\left(-\frac{15}{7}\right)\left(\frac{1}{2}\right)$

59. $(-14) \cdot (-27) \cdot 0$

60. $7 \cdot (-6) \cdot 5 \cdot (-4) \cdot 3 \cdot (-2) \cdot 1 \cdot 0$

61. $(-8)(-9)(-10)$

62. $(-7)(-8)(-9)(-10)$

63. $(-6)(-7)(-8)(-9)(-10)$

64. $(-5)(-6)(-7)(-8)(-9)(-10)$

65. Evaluate $(-3x)^2$ and $-3x^2$ when $x = 7$.

66. Evaluate $(-2x)^2$ and $-2x^2$ when $x = 3$.

67. Evaluate $5x^2$ when $x = 2$ and $x = -2$.

68. Evaluate $2x^2$ when $x = 5$ and $x = -5$.

SYNTHESIS

Simplify. Keep in mind the rules for order of operations in Section R.5.

69. $-6[(-5) + (-7)]$

70. $-3[(-8) + (-6)]\left(-\frac{1}{7}\right)$

71. $-(3^5) \cdot [-(2^3)]$

72. $4(2^4) \cdot [-(3^3)] \cdot 6$

73. $|(-2)^3 + 4^2| - (2 - 7)^2$

74. $|-11(-3)^2 - 5^3 - 6^2 - (-4)^2|$

75. What must be true of a and b if $-ab$ is to be (a) positive? (b) zero? (c) negative?

76. Evaluate $-6(3x - 5y) + z$ when $x = -2$, $y = -4$, and $z = 5$.

1.6 *Division of Real Numbers*

We now consider division of real numbers. The definition of division results in rules for division that are the same as those for multiplication.

a DIVISION OF INTEGERS

The quotient $\frac{a}{b}$ (or $a \div b$) is the number, if there is one, that when multiplied by b gives a.

Let us use the definition to divide integers.

EXAMPLES Divide, if possible. Check your answer.

1. $14 \div (-7) = -2$ *Think:* What number multiplied by -7 gives 14? That number is -2. *Check:* $(-2)(-7) = 14$.

2. $\frac{-32}{-4} = 8$ *Think:* What number multiplied by -4 gives -32? That number is 8. *Check:* $8(-4) = -32$.

3. $\frac{-10}{7} = -\frac{10}{7}$ *Think:* What number multiplied by 7 gives -10? That number is $-\frac{10}{7}$. *Check:* $-\frac{10}{7} \cdot 7 = -10$.

4. $\frac{-17}{0}$ is **undefined.** *Think:* What number multiplied by 0 gives -17? There is no such number because the product of 0 and *any* number is 0.

The rules for division are the same as those for multiplication.

To multiply or divide two real numbers:
a) Multiply or divide the absolute values.
b) If the signs are the same, the answer is positive.
c) If the signs are different, the answer is negative.

Do Exercises 1–8.

DIVISION BY ZERO

Example 4 shows why we cannot divide -17 by 0. We can use the same argument to show why we cannot divide any nonzero number b by 0. Consider $b \div 0$. We look for a number that when multiplied by 0 gives b. There is no such number because the product of 0 and any number is 0. Thus we cannot divide a nonzero number b by 0.

On the other hand, if we divide 0 by 0, we look for a number r such that $0 \cdot r = 0$. But $0 \cdot r = 0$ for any number r. Thus it appears that $0 \div 0$ could be any number we choose. Getting any answer we want when we divide 0 by 0 would be very confusing. Thus we agree that division by zero is undefined.

Division by 0 is undefined.

$$a \div 0 \text{ is undefined for all real numbers } a.$$

0 divided by a nonzero number a is 0.

$$0 \div a = 0, \quad a \neq 0.$$

OBJECTIVES

After finishing Section 1.6, you should be able to:

- **a** Divide integers.
- **b** Find the reciprocal of a real number.
- **c** Divide real numbers.

FOR EXTRA HELP

TAPE 4 TAPE 3B MAC: 1B IBM: 1B

Divide.

1. $6 \div (-3)$

Think: What number multiplied by -3 gives 6?

2. $\frac{-15}{-3}$

Think: What number multiplied by -3 gives -15?

3. $-24 \div 8$

Think: What number multiplied by 8 gives -24?

4. $\frac{-48}{-6}$

5. $\frac{30}{-5}$

6. $\frac{30}{-7}$

7. $\frac{-5}{0}$

8. $\frac{0}{-3}$

Answers on page A-2

Find the reciprocal.

9. $\frac{2}{3}$

10. $-\frac{5}{4}$

11. -3

12. $-\frac{1}{5}$

13. 1.6

14. $\frac{1}{2/3}$

Answers on page A-2

b RECIPROCALS

When two numbers like $\frac{1}{2}$ and 2 are multiplied, the result is 1. Such numbers are called **reciprocals** of each other. Every nonzero real number has a reciprocal, also called a **multiplicative inverse.**

> Two numbers whose product is 1 are called *reciprocals* of each other.

EXAMPLES Find the reciprocal.

5. $\frac{7}{8}$ The reciprocal of $\frac{7}{8}$ is $\frac{8}{7}$ because $\frac{7}{8} \cdot \frac{8}{7} = 1$.

6. -5 The reciprocal of -5 is $-\frac{1}{5}$ because $-5\left(-\frac{1}{5}\right) = 1$.

7. 3.9 The reciprocal of 3.9 is $\frac{1}{3.9}$ because $3.9\left(\frac{1}{3.9}\right) = 1$.

8. $-\frac{1}{2}$ The reciprocal of $-\frac{1}{2}$ is -2 because $\left(-\frac{1}{2}\right)(-2) = 1$.

9. $-\frac{2}{3}$ The reciprocal of $-\frac{2}{3}$ is $-\frac{3}{2}$ because $\left(-\frac{2}{3}\right)\left(-\frac{3}{2}\right) = 1$.

10. $\frac{1}{3/4}$ The reciprocal of $\frac{1}{3/4}$ is $\frac{3}{4}$ because $\left(\frac{1}{3/4}\right)\left(\frac{3}{4}\right) = 1$.

> For $a \neq 0$, the reciprocal of a can be named $\frac{1}{a}$ and the reciprocal of $\frac{1}{a}$ is a.
>
> The reciprocal of a nonzero number $\frac{a}{b}$ can be named $\frac{b}{a}$.
>
> The number 0 has no reciprocal.

Do Exercises 9–14.

The reciprocal of a positive number is also a positive number, because their product must be the positive number 1. The reciprocal of a negative number is also a negative number, because their product must be the positive number 1.

> The reciprocal of a number has the same sign as the number itself.

It is important *not* to confuse *opposite* with *reciprocal.* Keep in mind that the opposite, or additive inverse, of a number is what we add to the number to get 0. A reciprocal, or multiplicative inverse, is what we multiply the number by to get 1.

Compare the following.

NUMBER	OPPOSITE (CHANGE THE SIGN.)	RECIPROCAL (INVERT BUT DO NOT CHANGE THE SIGN.)
$-\frac{3}{8}$	$\frac{3}{8}$	$-\frac{8}{3}$
19	-19	$\frac{1}{19}$
$\frac{18}{7}$	$-\frac{18}{7}$	$\frac{7}{18}$
-7.9	7.9	$-\frac{1}{7.9}$, or $-\frac{10}{79}$
0	0	Undefined

$\left(-\frac{3}{8}\right)\left(-\frac{8}{3}\right) = 1$

$-\frac{3}{8} + \frac{3}{8} = 0$

Do Exercise 15.

c DIVISION OF REAL NUMBERS

We know that we can subtract by adding an opposite. Similarly, we can divide by multiplying by a reciprocal.

For any real numbers a and b, $b \neq 0$,

$$a \div b = \frac{a}{b} = a \cdot \frac{1}{b}.$$

(To divide, we can multiply by the reciprocal of the divisor.)

EXAMPLES Rewrite the division as a multiplication.

11. $-4 \div 3$ $\qquad -4 \div 3$ is the same as $-4 \cdot \frac{1}{3}$

12. $\frac{6}{-7}$ $\qquad \frac{6}{-7} = 6\left(-\frac{1}{7}\right)$

13. $\frac{x+2}{5}$ $\qquad \frac{x+2}{5} = (x+2)\frac{1}{5}$ Parentheses are necessary here.

14. $\frac{-17}{1/b}$ $\qquad \frac{-17}{1/b} = -17 \cdot b$

15. $\frac{3}{5} \div \left(-\frac{9}{7}\right)$ $\qquad \frac{3}{5} \div \left(-\frac{9}{7}\right) = \frac{3}{5}\left(-\frac{7}{9}\right)$

Do Exercises 16–20.

When actually doing division calculations, we sometimes multiply by a reciprocal and we sometimes divide directly. With fractional notation, it is usually better to multiply by a reciprocal. With decimal notation, it is usually better to divide directly.

15. Complete the following table.

NUMBER	OPPOSITE	RECIPROCAL
$\frac{2}{3}$		
$-\frac{5}{4}$		
0		
1		
-8		
-4.5		

Rewrite the division as a multiplication.

16. $\frac{4}{7} \div \left(-\frac{3}{5}\right)$

17. $\frac{5}{-8}$

18. $\frac{a-b}{7}$

19. $\frac{-23}{1/a}$

20. $-5 \div 7$

Answers on page A-2

Divide by multiplying by the reciprocal of the divisor.

21. $\frac{4}{7} \div \left(-\frac{3}{5}\right)$

22. $-\frac{8}{5} \div \frac{2}{3}$

23. $-\frac{12}{7} \div \left(-\frac{3}{4}\right)$

24. Divide: $21.7 \div (-3.1)$.

Find two equal expressions for the number with negative signs in different places.

25. $\frac{-5}{6}$

26. $-\frac{8}{7}$

27. $\frac{10}{-3}$

Answers on page A-2

EXAMPLES Divide by multiplying by the reciprocal of the divisor.

16. $\frac{2}{3} \div \left(-\frac{5}{4}\right) = \frac{2}{3} \cdot \left(-\frac{4}{5}\right) = -\frac{8}{15}$

17. $-\frac{5}{6} \div \left(-\frac{3}{4}\right) = -\frac{5}{6} \cdot \left(-\frac{4}{3}\right) = \frac{20}{18} = \frac{10 \cdot 2}{9 \cdot 2} = \frac{10}{9} \cdot \frac{2}{2} = \frac{10}{9}$

Caution! Be careful not to change the sign when taking a reciprocal!

18. $-\frac{3}{4} \div \frac{3}{10} = -\frac{3}{4} \cdot \left(\frac{10}{3}\right) = -\frac{30}{12} = -\frac{5}{2} \cdot \frac{6}{6} = -\frac{5}{2}$

With decimal notation, it is easier to carry out long division than to multiply by the reciprocal.

EXAMPLES Divide.

19. $-27.9 \div (-3) = \frac{-27.9}{-3} = 9.3$ **Do the long division $3\overline{)27.9}$. The answer is positive.**

20. $-6.3 \div 2.1 = -3$ **Do the long division $2.1\overline{)6.3}$. The answer is negative.**

Do Exercises 21–24.

Consider the following:

1. $\frac{2}{3} = \frac{2}{3} \cdot 1 = \frac{2}{3} \cdot \frac{-1}{-1} = \frac{2(-1)}{3(-1)} = \frac{-2}{-3}$. Thus, $\frac{2}{3} = \frac{-2}{-3}$.

2. $-\frac{2}{3} = -1 \cdot \frac{2}{3} = \frac{-1}{1} \cdot \frac{2}{3} = \frac{-1 \cdot 2}{1 \cdot 3} = \frac{-2}{3}$. Thus, $-\frac{2}{3} = \frac{-2}{3}$.

$\frac{-2}{3} = \frac{-2}{3} \cdot 1 = \frac{-2}{3} \cdot \frac{-1}{-1} = \frac{-2(-1)}{3(-1)} = \frac{2}{-3}$. Thus, $\frac{-2}{3} = \frac{2}{-3}$.

We can use the following properties to make sign changes in fractional notation.

For any numbers a and b, $b \neq 0$:

1. $\frac{-a}{-b} = \frac{a}{b}$

(The opposite of a number a divided by the opposite of another number b is the same as the quotient of the two numbers a and b.)

2. $\frac{-a}{b} = \frac{a}{-b} = -\frac{a}{b}$

(The opposite of a number a divided by another number b is the same as the number a divided by the opposite of another number b, and both are the same as the opposite of a *divided by* b.)

Do Exercises 25–27.

NAME SECTION DATE

Exercise Set 1.6

a Divide, if possible. Check each answer.

1. $48 \div (-6)$

2. $\frac{42}{-7}$

3. $\frac{28}{-2}$

4. $24 \div (-12)$

5. $\frac{-24}{8}$

6. $-18 \div (-2)$

7. $\frac{-36}{-12}$

8. $-72 \div (-9)$

9. $\frac{-72}{9}$

10. $\frac{-50}{25}$

11. $-100 \div (-50)$

12. $\frac{-200}{8}$

13. $-108 \div 9$

14. $\frac{-63}{-7}$

15. $\frac{200}{-25}$

16. $-300 \div (-16)$

17. $\frac{75}{0}$

18. $\frac{0}{-5}$

19. $\frac{20}{-7}$

20. $\frac{-23}{-2}$

b Find the reciprocal.

21. $\frac{15}{7}$

22. $\frac{3}{8}$

23. $-\frac{47}{13}$

24. $-\frac{31}{12}$

25. 13

26. -10

27. 4.3

28. -8.5

29. $\frac{1}{-7.1}$

30. $\frac{1}{-4.9}$

31. $\frac{p}{q}$

32. $\frac{s}{t}$

33. $\frac{1}{4y}$

34. $\frac{-1}{8a}$

35. $\frac{2a}{3b}$

36. $\frac{-4y}{3x}$

Answers

1.
2.
3.
4.
5.
6.
7.
8.
9.
10.
11.
12.
13.
14.
15.
16.
17.
18.
19.
20.
21.
22.
23.
24.
25.
26.
27.
28.
29.
30.
31.
32.
33.
34.
35.
36.

ANSWERS

37. ______
38. ______
39. ______
40. ______
41. ______
42. ______
43. ______
44. ______
45. ______
46. ______
47. ______
48. ______
49. ______
50. ______
51. ______
52. ______
53. ______
54. ______
55. ______
56. ______
57. ______
58. ______
59. ______
60. ______
61. ______
62. ______
63. ______
64. ______
65. ______
66. ______
67. ______
68. ______
69. ______
70. ______
71. ______
72. ______
73. ______
74. ______
75. ______
76. ______
77. ______
78. ______
79. ______

c Rewrite the division as a multiplication.

37. $4 \div 17$

38. $5 \div (-8)$

39. $\dfrac{8}{-13}$

40. $-\dfrac{13}{47}$

41. $\dfrac{13.9}{-1.5}$

42. $-\dfrac{47.3}{21.4}$

43. $\dfrac{x}{\frac{1}{y}}$

44. $\dfrac{13}{x}$

45. $\dfrac{3x+4}{5}$

46. $\dfrac{4y-8}{-7}$

47. $\dfrac{5a-b}{5a+b}$

48. $\dfrac{2x+x^2}{x-5}$

Divide.

49. $\dfrac{3}{4} \div \left(-\dfrac{2}{3}\right)$

50. $\dfrac{7}{8} \div \left(-\dfrac{1}{2}\right)$

51. $-\dfrac{5}{4} \div \left(-\dfrac{3}{4}\right)$

52. $-\dfrac{5}{9} \div \left(-\dfrac{5}{6}\right)$

53. $-\dfrac{2}{7} \div \left(-\dfrac{4}{9}\right)$

54. $-\dfrac{3}{5} \div \left(-\dfrac{5}{8}\right)$

55. $-\dfrac{3}{8} \div \left(-\dfrac{8}{3}\right)$

56. $-\dfrac{5}{8} \div \left(-\dfrac{6}{5}\right)$

57. $-6.6 \div 3.3$

58. $-44.1 \div (-6.3)$

59. $\dfrac{-11}{-13}$

60. $\dfrac{-1.9}{20}$

61. $\dfrac{48.6}{-3}$

62. $\dfrac{-17.8}{3.2}$

63. $\dfrac{-9}{17-17}$

64. $\dfrac{-8}{-5+5}$

SKILL MAINTENANCE

65. Simplify: $\dfrac{264}{468}$.

66. Convert to decimal notation: 47.7%.

67. Simplify: $2^3 - 5 \cdot 3 + 8 \cdot 10 \div 2$.

68. Add and simplify: $\dfrac{2}{3} + \dfrac{5}{6}$.

69. Convert to percent notation: $\dfrac{7}{8}$.

70. Simplify: $\dfrac{40}{60}$.

SYNTHESIS

71. [calculator] Find the reciprocal of -10.5. Use the reciprocal key if one is available.

72. Determine those real numbers that are their own reciprocals.

73. Determine those real numbers a for which the opposite of a is the same as the reciprocal of a.

74. [calculator] What should happen if you enter a number on a calculator and press the reciprocal key twice? Why?

Tell whether the expression represents a positive number or a negative number when m and n are negative.

75. $\dfrac{-n}{m}$

76. $\dfrac{-n}{-m}$

77. $-\left(\dfrac{-n}{m}\right)$

78. $-\left(\dfrac{n}{-m}\right)$

79. $-\left(\dfrac{-n}{-m}\right)$

1.7 *Properties of Real Numbers*

a EQUIVALENT EXPRESSIONS

In solving equations and doing other kinds of work in algebra, we manipulate expressions in various ways. For example, instead of

$$x + x,$$

we might write

$$2x,$$

knowing that the two expressions represent the same number for any meaningful replacement of x. In that sense, the expressions $x + x$ and $2x$ are **equivalent**.

> Two expressions that have the same value for all meaningful replacements are called *equivalent*.

The expressions $x + 3x$ and $5x$ are *not* equivalent.

Do Exercises 1 and 2.

In this section, we will consider several laws of real numbers that will allow us to find equivalent expressions. The first two laws are the *identity properties of 0 and 1.*

> **THE IDENTITY PROPERTY OF 0**
>
> For any real number a,
>
> $$a + 0 = 0 + a = a.$$
>
> (The number 0 is the *additive identity.*)

> **THE IDENTITY PROPERTY OF 1**
>
> For any real number a,
>
> $$a \cdot 1 = 1 \cdot a = a.$$
>
> (The number 1 is the *multiplicative identity.*)

We often refer to the use of the identity property of 1 as "multiplying by 1." We do so to find equivalent fractional expressions.

EXAMPLE 1 Write a fractional expression equivalent to $\frac{2}{3}$ with a denominator of 15.

Note that $15 = 3 \cdot 5$. We want fractional notation for $\frac{2}{3}$ that has a denominator of 15, but the denominator 3 is missing a factor of 5. We multiply by 1, using $\frac{5}{5}$ as an equivalent expression for 1. Recall from arithmetic that to multiply with fractional notation, we multiply numerators and denominators:

OBJECTIVES

After finishing Section 1.7, you should be able to:

- **a** Find equivalent fractional expressions and simplify fractional expressions.
- **b** Use the commutative and associative laws to find equivalent expressions.
- **c** Use the distributive laws to multiply expressions like 8 and $x - y$.
- **d** Use the distributive laws to factor expressions like $4x - 12$.
- **e** Collect like terms.

FOR EXTRA HELP

TAPE 4 | TAPE 3B | MAC: 1B IBM: 1B

Complete the table by evaluating each expression for the given values.

1.

	$x + x$	$2x$
$x = 3$		
$x = -6$		
$x = 4.8$		

2.

	$x + 3x$	$5x$
$x = 2$		
$x = -6$		
$x = 4.8$		

Answers on page A-2

3. Write a fractional expression equivalent to $\frac{3}{4}$ with a denominator of 8.

4. Write a fractional expression equivalent to $\frac{3}{4}$ with a denominator of $4t$.

Simplify.

5. $\dfrac{3y}{4y}$

6. $-\dfrac{16m}{12m}$

Answers on page A-2

$$\frac{2}{3} = \frac{2}{3} \cdot 1 \qquad \text{Using the identity property of 1}$$

$$= \frac{2}{3} \cdot \frac{5}{5} \qquad \text{Using } \frac{5}{5} \text{ for 1}$$

$$= \frac{10}{15}. \qquad \text{Multiplying numerators and denominators}$$

Do Exercise 3.

EXAMPLE 2 Write a fractional expression equivalent to $\frac{2}{3}$ with a denominator of $3x$.

Note that $3x = 3 \cdot x$. We want fractional notation for $\frac{2}{3}$ that has a denominator of $3x$, but the denominator 3 is missing a factor of x. We multiply by 1, using x/x as an equivalent expression for 1:

$$\frac{2}{3} = \frac{2}{3} \cdot 1 = \frac{2}{3} \cdot \frac{x}{x} = \frac{2x}{3x}.$$

The expressions $2/3$ and $2x/3x$ are equivalent. They have the same value for any meaningful replacement. Note that $2x/3x$ is undefined for a replacement of 0, but for all nonzero real numbers, the expressions $2/3$ and $2x/3x$ have the same value.

Do Exercise 4.

In algebra, we consider an expression like $2/3$ to be "simplified" from $2x/3x$. To find such simplified expressions, we use the identity property of 1 to remove a factor equal to 1. (See also Section R.2.)

EXAMPLE 3 Simplify: $-\dfrac{20x}{12x}$.

$$-\frac{20x}{12x} = -\frac{5 \cdot 4x}{3 \cdot 4x} \qquad \text{We look for the largest factor common to both the numerator and the denominator and factor each.}$$

$$= -\frac{5}{3} \cdot \frac{4x}{4x} \qquad \text{Factoring the fractional expression}$$

$$= -\frac{5}{3} \cdot 1 \qquad \frac{4x}{4x} = 1$$

$$= -\frac{5}{3} \qquad \text{Removing a factor equal to 1 using the identity property of 1}$$

Do Exercises 5 and 6.

b THE COMMUTATIVE AND ASSOCIATIVE LAWS

THE COMMUTATIVE LAWS

Let us examine the expressions $x + y$ and $y + x$, as well as xy and yx.

EXAMPLE 4 Evaluate $x + y$ and $y + x$ when $x = 4$ and $y = 3$.

We substitute 4 for x and 3 for y in both expressions:

$$x + y = 4 + 3 = 7; \qquad y + x = 3 + 4 = 7.$$

EXAMPLE 5 Evaluate xy and yx when $x = 23$ and $y = 12$.

We substitute 23 for x and 12 for y in both expressions:

$$xy = 23 \cdot 12 = 276; \qquad yx = 12 \cdot 23 = 276.$$

Do Exercises 7 and 8.

Note that the expressions

$$x + y \quad \text{and} \quad y + x$$

have the same values no matter what the variables stand for. Thus they are equivalent. Therefore, when we add two numbers, the order in which we add does not matter. Similarly, the expressions xy and yx are equivalent. They also have the same values, no matter what the variables stand for. Therefore, when we multiply two numbers, the order in which we multiply does not matter.

The following are examples of general patterns or laws.

THE COMMUTATIVE LAWS

Addition. For any numbers a and b,

$$a + b = b + a.$$

(We can change the order when adding without affecting the answer.)

Multiplication. For any numbers a and b,

$$ab = ba.$$

(We can change the order when multiplying without affecting the answer.)

Using a commutative law, we know that $x + 2$ and $2 + x$ are equivalent. Similarly, $3x$ and $x(3)$ are equivalent. Thus, in an algebraic expression, we can replace one by the other and the result will be equivalent to the original expression.

EXAMPLE 6 Use the commutative laws to write an expression equivalent to $y + 5$, ab, and $7 + xy$.

An expression equivalent to $y + 5$ is $5 + y$ by the commutative law of addition.

An expression equivalent to ab is ba by the commutative law of multiplication.

An expression equivalent to $7 + xy$ is $xy + 7$ by the commutative law of addition. Another expression equivalent to $7 + xy$ is $7 + yx$ by the commutative law of multiplication.

Do Exercises 9–11.

THE ASSOCIATIVE LAWS

Now let us examine the expressions $a + (b + c)$ and $(a + b) + c$. Note that these expressions involve parentheses as *grouping* symbols, and they also involve three numbers. Calculations within parentheses are to be done first.

7. Evaluate $x + y$ and $y + x$ when $x = -2$ and $y = 3$.

8. Evaluate xy and yx when $x = -2$ and $y = 5$.

Use a commutative law to write an equivalent expression.

9. $x + 9$

10. pq

11. $xy + t$

Answers on page A-2

12. Calculate and compare:

$8 + (9 + 2)$ and $(8 + 9) + 2$.

13. Calculate and compare:

$10 \cdot (5 \cdot 3)$ and $(10 \cdot 5) \cdot 3$.

Use an associative law to write an equivalent expression.

14. $r + (s + 7)$

15. $9(ab)$

Answers on page A-2

EXAMPLE 7 Calculate and compare: $3 + (8 + 5)$ and $(3 + 8) + 5$.

$3 + (8 + 5) = 3 + 13$ **Calculating within parentheses first; adding the 8 and 5**

$= 16;$

$(3 + 8) + 5 = 11 + 5$ **Calculating within parentheses first; adding the 3 and 8**

$= 16$

The two expressions in Example 7 name the same number. Moving the parentheses to group the additions differently did not affect the value of the expression.

EXAMPLE 8 Calculate and compare: $3 \cdot (4 \cdot 2)$ and $(3 \cdot 4) \cdot 2$.

$3 \cdot (4 \cdot 2) = 3 \cdot 8 = 24;$ $(3 \cdot 4) \cdot 2 = 12 \cdot 2 = 24$

Do Exercises 12 and 13.

You may have noted that when only addition is involved, parentheses can be placed any way we please without affecting the answer. When only multiplication is involved, parentheses also can be placed any way we please without affecting the answer.

THE ASSOCIATIVE LAWS

Addition. For any numbers a, b, and c,

$$a + (b + c) = (a + b) + c.$$

(Numbers can be grouped in any manner for addition.)

Multiplication. For any numbers a, b, and c,

$$a \cdot (b \cdot c) = (a \cdot b) \cdot c.$$

(Numbers can be grouped in any manner for multiplication.)

EXAMPLE 9 Use an associative law to write an expression equivalent to $(y + z) + 3$.

An equivalent expression is

$$y + (z + 3)$$

by the associative law of addition.

EXAMPLE 10 Use an associative law to write an expression equivalent to $8(xy)$.

An equivalent expression is

$$(8x)y$$

by the associative law of multiplication.

Do Exercises 14 and 15.

The associative laws say parentheses may be placed any way we please when only additions or only multiplications are involved. Thus we often omit them. For example,

$x + (y + 2)$ means $x + y + 2$, and $(lw)h$ means lwh.

Using the Commutative and Associative Laws Together

Example 11 Use the commutative and associative laws to write at least three expressions equivalent to $(x + 5) + y$.

a) $(x + 5) + y = x + (5 + y)$ Using the associative law first and then using the commutative law
$= x + (y + 5)$

b) $(x + 5) + y = y + (x + 5)$ Using the commutative law first and then the commutative law again
$= y + (5 + x)$

c) $(x + 5) + y = (5 + x) + y$ Using the commutative law first and then the associative law
$= 5 + (x + y)$

Example 12 Use the commutative and associative laws to write at least three expressions equivalent to $(3x)y$.

a) $(3x)y = 3(xy)$ Using the associative law first and then using the commutative law
$= 3(yx)$

b) $(3x)y = y(3x)$ Using the commutative law twice
$= y(x3)$

c) $(3x)y = (x3)y$ Using the commutative law, and then the associative law, and then the commutative law again
$= x(3y)$
$= x(y3)$

Do Exercises 16 and 17.

Use the commutative and associative laws to write at least three equivalent expressions.

16. $4(tu)$

17. $r + (2 + s)$

c The Distributive Laws

The *distributive laws* are the basis of many procedures in both arithmetic and algebra. These are probably the most important laws that we use to manipulate algebraic expressions. The distributive law of multiplication over addition involves two operations: addition and multiplication.

Let us begin by considering a multiplication problem from arithmetic:

```
   4 5
 ×   7
 -----
   3 5 ← This is 7 · 5.
 2 8 0 ← This is 7 · 40.
 -----
 3 1 5 ← This is the sum 7 · 40 + 7 · 5.
```

To carry out the multiplication, we actually added two products. That is,

$$7 \cdot 45 = 7(40 + 5) = 7 \cdot 40 + 7 \cdot 5.$$

Let us examine this further. If we wish to multiply a sum of several numbers by a factor, we can either add and then multiply, or multiply and then add.

Example 13 Compute in two ways: $5 \cdot (4 + 8)$.

a) $5 \cdot (4 + 8)$ Adding within parentheses first, and then multiplying

$= 5 \cdot 12$
$= 60$

Answers on page A-2

Compute.

18. a) $7 \cdot (3 + 6)$

b) $(7 \cdot 3) + (7 \cdot 6)$

19. a) $2 \cdot (10 + 30)$

b) $(2 \cdot 10) + (2 \cdot 30)$

20. a) $(2 + 5) \cdot 4$

b) $(2 \cdot 4) + (5 \cdot 4)$

Calculate.

21. a) $4(5 - 3)$

b) $4 \cdot 5 - 4 \cdot 3$

22. a) $-2 \cdot (5 - 3)$

b) $-2 \cdot 5 - (-2) \cdot 3$

23. a) $5 \cdot (2 - 7)$

b) $5 \cdot 2 - 5 \cdot 7$

What are the terms of the expression?

24. $5x - 8y + 3$

25. $-4y - 2x + 3z$

Answers on page A-2

b) $(5 \cdot 4) + (5 \cdot 8)$ Distributing the multiplication to terms within parentheses first and then adding

$= 20 + 40$

$= 60$

Do Exercises 18–20.

THE DISTRIBUTIVE LAW OF MULTIPLICATION OVER ADDITION

For any numbers a, b, and c,

$$a(b + c) = ab + ac.$$

In the statement of the distributive law, we know that in an expression such as $ab + ac$, the multiplications are to be done first according to the rules for order of operations. (See Section R.5.) So, instead of writing $(4 \cdot 5) + (4 \cdot 7)$, we can write $4 \cdot 5 + 4 \cdot 7$. However, in $a(b + c)$, we cannot omit the parentheses. If we did, we would have $ab + c$, which means $(ab) + c$. For example, $3(4 + 2) = 18$, but $3 \cdot 4 + 2 = 14$.

There is another distributive law that relates multiplication and subtraction. This law says that to multiply by a difference, we can either subtract and then multiply, or multiply and then subtract.

THE DISTRIBUTIVE LAW OF MULTIPLICATION OVER SUBTRACTION

For any numbers a, b, and c,

$$a(b - c) = ab - ac.$$

We often refer to "*the* distributive law" when we mean *either* of these laws.

Do Exercises 21–23.

What do we mean by the *terms* of an expression? **Terms** are separated by addition signs. If there are subtraction signs, we can find an equivalent expression that uses addition signs.

EXAMPLE 14 What are the terms of $3x - 4y + 2z$?

We have

$$3x - 4y + 2z = 3x + (-4y) + 2z. \quad \text{Separating parts with + signs}$$

The terms are $3x$, $-4y$, and $2z$.

Do Exercises 24 and 25.

The distributive laws are a basis for a procedure in algebra called **multiplying**. In an expression like $8(a + 2b - 7)$, we multiply each term inside the parentheses by 8:

$$8(a + 2b - 7) = 8 \cdot a + 8 \cdot 2b - 8 \cdot 7 = 8a + 16b - 56.$$

EXAMPLES Multiply.

15. $9(x - 5) = 9x - 9(5)$ Using the distributive law of multiplication over subtraction

$= 9x - 45$

16. $2(w + 1) = 2 \cdot w + 2 \cdot 1$ Using the distributive law of multiplication over addition

$= 2w + 2$

17. $\frac{4}{3}(s - t + w) = \frac{4}{3}s - \frac{4}{3}t + \frac{4}{3}w$ Using both distributive laws

Do Exercises 26–28.

EXAMPLE 18 Multiply: $-4(x - 2y + 3z)$.

$$-4(x - 2y + 3z) = -4 \cdot x - (-4)(2y) + (-4)(3z)$$
$$= -4x - (-8y) + (-12z)$$
$$= -4x + 8y - 12z$$

We can also do this problem by first finding an equivalent expression with all plus signs and then multiplying:

$$-4(x - 2y + 3z) = -4[x + (-2y) + 3z]$$
$$= -4 \cdot x + (-4)(-2y) + (-4)(3z)$$
$$= -4x + 8y - 12z.$$

Do Exercises 29–31.

d FACTORING

Factoring is the reverse of multiplying. To factor, we can use the distributive laws in reverse:

$$ab + ac = a(b + c) \quad \text{and} \quad ab - ac = a(b - c).$$

> To *factor* an expression is to find an equivalent expression that is a product.

Look at Example 15. To *factor* $9x - 45$, we find an equivalent expression that is a product, $9(x - 5)$. When all the terms of an expression have a factor in common, we can "factor it out" using the distributive laws. Note the following.

$9x$ has the factors $9, -9, 3, -3, 1, -1, x, -x, 3x, -3x, 9x, -9x$;

-45 has the factors $1, -1, 3, -3, 5, -5, 9, -9, 15, -15, 45, -45$

We usually remove the largest common factor. In this case, that factor is 9. Thus,

$$9x - 45 = 9 \cdot x - 9 \cdot 5$$
$$= 9(x - 5).$$

Remember that an expression is factored when we find an equivalent expression that is a product.

Multiply.

26. $3(x - 5)$

27. $5(x + 1)$

28. $\frac{2}{3}(p + q - t)$

Multiply.

29. $-2(x - 3)$

30. $5(x - 2y + 4z)$

31. $-5(x - 2y + 4z)$

Answers on page A-2

Factor.

32. $6x - 12$

33. $3x - 6y + 9$

34. $bx + by - bz$

35. $16a - 36b + 42$

36. $\frac{3}{8}x - \frac{5}{8}y + \frac{7}{8}$

37. $-12x + 32y - 16z$

Collect like terms.

38. $6x - 3x$

39. $7x - x$

40. $x - 9x$

41. $x - 0.41x$

42. $5x + 4y - 2x - y$

43. $3x - 7x - 11 + 8y + 4 - 13y$

Answers on page A-2

EXAMPLES Factor.

19. $5x - 10 = 5 \cdot x - 5 \cdot 2$ **Try to do this step mentally.**

$= 5(x - 2)$ **You can check by multiplying.**

20. $ax - ay + az = a(x - y + z)$

21. $9x + 27y - 9 = 9 \cdot x + 9 \cdot 3y - 9 \cdot 1 = 9(x + 3y - 1)$

Caution! Note that although $3(3x + 9y - 3)$ is also equivalent to $9x + 27y - 9$, it is *not* the desired form. However, we can complete the process by factoring out another factor of 3:

$$9x + 27y - 9 = 3(3x + 9y - 3) = 3 \cdot 3(x + 3y - 1) = 9(x + 3y - 1).$$

Remember to factor out the *largest common factor.*

EXAMPLES Factor. Try to write just the answer, if you can.

22. $5x - 5y = 5(x - y)$

23. $-3x + 6y - 9z = -3(x - 2y + 3z)$

We usually factor out a negative when the first term is negative. The way we factor can depend on the situation in which we are working. We might also factor the expression in Example 23 as follows:

$$-3x + 6y - 9z = 3(-x + 2y - 3z).$$

24. $18z - 12x - 24 = 6(3z - 2x - 4)$

25. $\frac{1}{2}x + \frac{3}{2}y - \frac{1}{2} = \frac{1}{2}(x + 3y - 1)$

> *Remember:* An expression is factored when it is written as a product.

Do Exercises 32–37.

e COLLECTING LIKE TERMS

Terms such as $5x$ and $-4x$, whose variable factors are exactly the same, are called **like terms.** Similarly, numbers, such as -7 and 13, are like terms. Also, $3y^2$ and $9y^2$ are like terms because the variables are raised to the same power. Terms such as $4y$ and $5y^2$ are not like terms, and $7x$ and $2y$ are not like terms.

The process of **collecting like terms** is also based on the distributive laws. We can apply the distributive law when a factor is on the right because of the commutative law of multiplication.

EXAMPLES Collect like terms. Try to write just the answer, if you can.

26. $4x + 2x = (4 + 2)x = 6x$ **Factoring out the *x* using a distributive law**

27. $2x + 3y - 5x - 2y = 2x - 5x + 3y - 2y$

$= (2 - 5)x + (3 - 2)y = -3x + y$

28. $3x - x = (3 - 1)x = 2x$

29. $x - 0.24x = 1 \cdot x - 0.24x = (1 - 0.24)x = 0.76x$

30. $x - 6x = 1 \cdot x - 6 \cdot x = (1 - 6)x = -5x$

31. $4x - 7y + 9x - 5 + 3y - 8 = 13x - 4y - 13$

Do Exercises 38–43.

NAME SECTION DATE

Exercise Set 1.7

a Find an equivalent expression with the given denominator.

1. $\frac{3}{5}$; $5y$ **2.** $\frac{5}{8}$; $8t$ **3.** $\frac{2}{3}$; $15x$ **4.** $\frac{6}{7}$; $14y$

Simplify.

5. $-\frac{24a}{16a}$ **6.** $-\frac{42t}{18t}$ **7.** $-\frac{42ab}{36ab}$ **8.** $-\frac{64pq}{48pq}$

b Write an equivalent expression. Use a commutative law.

9. $y + 8$ **10.** $x + 3$ **11.** mn **12.** ab

13. $9 + xy$ **14.** $11 + ab$ **15.** $ab + c$ **16.** $rs + t$

Write an equivalent expression. Use an associative law.

17. $a + (b + 2)$ **18.** $3(vw)$ **19.** $(8x)y$ **20.** $(y + z) + 7$

21. $(a + b) + 3$ **22.** $(5 + x) + y$ **23.** $3(ab)$ **24.** $(6x)y$

Use the commutative and associative laws to write three equivalent expressions.

25. $(a + b) + 2$ **26.** $(3 + x) + y$ **27.** $5 + (v + w)$ **28.** $6 + (x + y)$

29. $(xy)3$ **30.** $(ab)5$ **31.** $7(ab)$ **32.** $5(xy)$

ANSWERS

1.
2.
3.
4.
5.
6.
7.
8.
9.
10.
11.
12.
13.
14.
15.
16.
17.
18.
19.
20.
21.
22.
23.
24.
25.
26.
27.
28.
29.
30.
31.
32.

ANSWERS

33. ____
34. ____
35. ____
36. ____
37. ____
38. ____
39. ____
40. ____
41. ____
42. ____
43. ____
44. ____
45. ____
46. ____
47. ____
48. ____
49. ____
50. ____
51. ____
52. ____
53. ____
54. ____
55. ____
56. ____
57. ____
58. ____
59. ____
60. ____
61. ____
62. ____
63. ____
64. ____

c Multiply.

33. $2(b + 5)$

34. $4(x + 3)$

35. $7(1 + t)$

36. $4(1 + y)$

37. $6(5x + 2)$

38. $9(6m + 7)$

39. $7(x + 4 + 6y)$

40. $4(5x + 8 + 3p)$

41. $7(4 - 3)$

42. $15(8 - 6)$

43. $-3(3 - 7)$

44. $1.2(5 - 2.1)$

45. $4.1(6.3 - 9.4)$

46. $-\frac{8}{9}\left(\frac{2}{3} - \frac{5}{3}\right)$

47. $7(x - 2)$

48. $5(x - 8)$

49. $-7(y - 2)$

50. $-9(y - 7)$

51. $-9(-5x - 6y + 8)$

52. $-7(-2x - 5y + 9)$

53. $-4(x - 3y - 2z)$

54. $8(2x - 5y - 8z)$

55. $3.1(-1.2x + 3.2y - 1.1)$

56. $-2.1(-4.2x - 4.3y - 2.2)$

List the terms of the expression.

57. $4x + 3z$

58. $8x - 1.4y$

59. $7x + 8y - 9z$

60. $8a + 10b - 18c$

d Factor. Check by multiplying.

61. $2x + 4$

62. $5y + 20$

63. $30 + 5y$

64. $7x + 28$

65. $14x + 21y$

66. $18a + 24b$

67. $5x + 10 + 15y$

68. $9a + 27b + 81$

69. $8x - 24$

70. $10x - 50$

71. $32 - 4y$

72. $24 - 6m$

73. $8x + 10y - 22$

74. $9a + 6b - 15$

75. $ax - a$

76. $by - 9b$

77. $ax - ay - az$

78. $cx + cy - cz$

79. $18x - 12y + 6$

80. $-14x + 21y + 7$

81. $3ax - 2ay + a$

82. $-5pq + 10pr - p$

e Collect like terms.

83. $9a + 10a$

84. $12x + 2x$

85. $10a - a$

86. $-16x + x$

87. $2x + 9z + 6x$

88. $3a - 5b + 7a$

89. $7x + 6y^2 + 9y^2$

90. $12m^2 + 6q + 9m^2$

91. $41a + 90 - 60a - 2$

92. $42x - 6 - 4x + 2$

93. $23 + 5t + 7y - t - y - 27$

94. $45 - 90d - 87 - 9d + 3 + 7d$

ANSWERS

65. ______
66. ______
67. ______
68. ______
69. ______
70. ______
71. ______
72. ______
73. ______
74. ______
75. ______
76. ______
77. ______
78. ______
79. ______
80. ______
81. ______
82. ______
83. ______
84. ______
85. ______
86. ______
87. ______
88. ______
89. ______
90. ______
91. ______
92. ______
93. ______
94. ______

ANSWERS

95. ____
96. ____
97. ____
98. ____
99. ____
100. ____
101. ____
102. ____
103. ____
104. ____
105. ____
106. ____
107. ____
108. ____
109. ____
110. ____
111. ____
112. ____
113. ____
114. ____
115. ____
116. ____
117. ____
118. ____
119. ____
120. ____
121. ____
122. ____
123. ____
124. ____
125. ____
126. ____

95. $\frac{1}{2}b + \frac{1}{2}b$

96. $\frac{2}{3}x + \frac{1}{3}x$

97. $2y + \frac{1}{4}y + y$

98. $\frac{1}{2}a + a + 5a$

99. $11x - 3x$

100. $9t - 17t$

101. $6n - n$

102. $10t - t$

103. $y - 17y$

104. $3m - 9m + 4$

105. $-8 + 11a - 5b + 6a - 7b + 7$

106. $8x - 5x + 6 + 3y - 2y - 4$

107. $9x + 2y - 5x$

108. $8y - 3z + 4y$

109. $11x + 2y - 4x - y$

110. $13a + 9b - 2a - 4b$

111. $2.7x + 2.3y - 1.9x - 1.8y$

112. $6.7a + 4.3b - 4.1a - 2.9b$

113. $\frac{1}{5}x + \frac{4}{5}y + \frac{2}{5}x - \frac{1}{5}y$

114. $\frac{7}{8}x + \frac{5}{8}y + \frac{1}{8}x - \frac{3}{8}y$

SKILL MAINTENANCE

115. Add and simplify: $\frac{11}{12} + \frac{15}{16}$.

116. Subtract and simplify: $\frac{7}{8} - \frac{2}{3}$.

117. Find the LCM of 16, 18, and 24.

118. Convert to percent notation: $\frac{3}{10}$.

119. Subtract and simplify: $\frac{1}{8} - \frac{1}{3}$.

120. Find the LCM of 12, 15, and 20.

SYNTHESIS

Tell whether the expressions are equivalent. Explain.

121. $3t + 5$ and $3 \cdot 5 + t$

122. $4x$ and $x + 4$

123. $5m + 6$ and $6 + 5m$

124. $(x + y) + z$ and $z + (x + y)$

Collect like terms, if possible, and factor the result.

125. $q + qr + qrs + qrst$

126. $21x + 44xy + 15y - 16x - 8y - 38xy + 2y + xy$

1.8 *Simplifying Expressions; Order of Operations*

We now expand our ability to manipulate expressions by first considering opposites of sums and differences. Then we simplify expressions involving parentheses.

Objectives

After finishing Section 1.8, you should be able to:

- **a** Find an equivalent expression for an opposite without parentheses, where an expression has several terms.
- **b** Simplify expressions by removing parentheses and collecting like terms.
- **c** Simplify expressions with parentheses inside parentheses.
- **d** Simplify expressions using rules for order of operations.

FOR EXTRA HELP

TAPE 5 TAPE 4A MAC: 1B IBM: 1B

a Opposites of Sums

What happens when we multiply a real number by -1? Consider the following products:

$$-1(7) = -7, \qquad -1(-5) = 5, \qquad -1(0) = 0.$$

From these examples, it appears that when we multiply a number by -1, we get the opposite, or additive inverse, of that number.

> **The Property of −1**
>
> For any real number a,
>
> $$-1 \cdot a = -a.$$
>
> (Negative one times a is the opposite, or additive inverse, of a.)

The property of -1 enables us to find certain expressions equivalent to opposites of sums.

Examples Find an equivalent expression without parentheses.

1. $-(3 + x) = -1(3 + x)$ — Using the property of -1

$= -1 \cdot 3 + (-1)x$ — Using a distributive law, multiplying each term by -1

$= -3 + (-x)$ — Using the property of -1

$= -3 - x$

2. $-(3x + 2y + 4) = -1(3x - 2y + 4)$ — Using the property of -1

$= -1(3x) + (-1)(2y) + (-1)4$ — Using a distributive law

$= -3x - 2y - 4$ — Using the property of -1

Do Exercises 1 and 2.

Find an equivalent expression without parentheses.

1. $-(x + 2)$

2. $-(5x + 2y + 8)$

Suppose we want to remove parentheses in an expression like

$$-(x - 2y + 5).$$

We can first find an equivalent expression in which the inside expression is separated by plus signs. Then we take the opposite of each term:

$$-(x - 2y + 5) = -[x + (-2y) + 5]$$
$$= -x + 2y - 5.$$

The most efficient method for this is to replace each term in the parentheses with its opposite ("change the sign of every term"). Doing so for $-(x - 2y + 5)$, we obtain $-x + 2y - 5$ as an equivalent expression.

Answers on page A-2

Find an equivalent expression without parentheses. Try to do this in one step.

3. $-(6 - t)$

4. $-(x - y)$

5. $-(-4a + 3t - 10)$

6. $-(18 - m - 2n + 4z)$

Remove parentheses and simplify.

7. $5x - (3x + 9)$

8. $5y - 2 - (2y - 4)$

Remove parentheses and simplify.

9. $6x - (4x + 7)$

10. $8y - 3 - (5y - 6)$

11. $(2a + 3b - c) - (4a - 5b + 2c)$

Answers on page A-2

EXAMPLES Find an equivalent expression without parentheses.

3. $-(5 - y) = -5 + y$ **Changing the sign of each term**

4. $-(2a - 7b - 6) = -2a + 7b + 6$

5. $-(-3x + 4y + z - 7w - 23) = 3x - 4y - z + 7w + 23$

Do Exercises 3–6.

b REMOVING PARENTHESES AND SIMPLIFYING

When a sum is added, as in $5x + (2x + 3)$, we can simply remove, or drop, the parentheses and collect like terms because of the associative law of addition:

$$5x + (2x + 3) = 5x + 2x + 3 = 7x + 3.$$

On the other hand, when a sum is subtracted, as in $3x - (4x + 2)$, no "associative" law applies. However, we can subtract by adding an opposite. We then remove parentheses by changing the sign of each term inside the parentheses and collecting like terms.

EXAMPLE 6 Remove parentheses and simplify.

$3x - (4x + 2) = 3x + [-(4x + 2)]$ **Adding the opposite of $(4x + 2)$**

$= 3x + (-4x - 2)$ **Changing the sign of each term inside the parentheses**

$= 3x - 4x - 2$

$= -x - 2$ **Collecting like terms**

Do Exercises 7 and 8.

In practice, the first three steps of Example 6 are usually combined by changing the sign of each term in parentheses and then collecting like terms.

EXAMPLES Remove parentheses and simplify.

7. $5y - (3y + 4) = 5y - 3y - 4$ **Removing parentheses by changing the sign of every term inside the parentheses**

$= 2y - 4$ **Collecting like terms**

8. $3y - 2 - (2y - 4) = 3y - 2 - 2y + 4 = y + 2$

9. $(3a + 4b - 5) - (2a - 7b + 4c - 8) = 3a + 4b - 5 - 2a + 7b - 4c + 8$

$= a + 11b - 4c + 3$

Do Exercises 9–11.

Next, consider subtracting an expression consisting of several terms preceded by a number other than 1 or -1.

EXAMPLE 10 Remove parentheses and simplify.

$x - 3(x + y) = x + [-3(x + y)]$ **Adding the opposite of $3(x + y)$**

$= x + [-3x - 3y]$ **Multiplying $x + y$ by -3**

$= x - 3x - 3y$

$= -2x - 3y$ **Collecting like terms**

EXAMPLES Remove parentheses and simplify.

11. $3y - 2(4y - 5) = 3y - 8y + 10$ Multiplying each term in parentheses by -2
$= -5y + 10$

12. $(2a + 3b - 7) - 4(-5a - 6b + 12) = 2a + 3b - 7 + 20a + 24b - 48$
$= 22a + 27b - 55$

13. $2y - \frac{1}{3}(9y - 12) = 2y - 3y + 4$
$= -y + 4$

Do Exercises 12–15.

c PARENTHESES WITHIN PARENTHESES

Some expressions contain more than one kind of grouping symbol such as brackets [ ] and braces { }.

> When more than one kind of grouping symbol occurs, do the computations in the innermost ones first. Then work from the inside out.

EXAMPLES Simplify.

14. $[3 - (7 + 3)] = [3 - 10]$ Computing $7 + 3$
$= -7$

15. $\{8 - [9 - (12 + 5)]\} = \{8 - [9 - 17]\}$ Computing $12 + 5$
$= \{8 - [-8]\}$ Computing $9 - 17$
$= 8 + 8$
$= 16$

16. $\left[(-4) \div \left(-\frac{1}{4}\right)\right] \div \frac{1}{4} = [(-4) \cdot (-4)] \div \frac{1}{4}$ Working within the brackets
$= 16 \div \frac{1}{4}$ Computing $(-4) \div \left(-\frac{1}{4}\right)$
$= 16 \cdot 4$
$= 64$

17. $4(2 + 3) - \{7 - [4 - (8 + 5)]\}$
$= 4 \cdot 5 - \{7 - [4 - 13]\}$ Working with the innermost parentheses first
$= 20 - \{7 - [-9]\}$ Computing $4 \cdot 5$ and $4 - 13$
$= 20 - 16$ Computing $7 - [-9]$
$= 4$

Do Exercises 16–19.

EXAMPLE 18 Simplify.

$[5(x + 2) - 3x] - [3(y + 2) - 7(y - 3)]$
$= [5x + 10 - 3x] - [3y + 6 - 7y + 21]$ Working with the innermost parentheses first
$= [2x + 10] - [-4y + 27]$ Collecting like terms within brackets
$= 2x + 10 + 4y - 27$ Removing brackets
$= 2x + 4y - 17$ Collecting like terms

Do Exercise 20.

Remove parentheses and simplify.

12. $y - 9(x + y)$

13. $5a - 3(7a - 6)$

14. $4a - b - 6(5a - 7b + 8c)$

15. $5x - \frac{1}{4}(8x + 28)$

Simplify.

16. $12 - (8 + 2)$

17. $\{9 - [10 - (13 + 6)]\}$

18. $[24 \div (-2)] \div (-2)$

19. $5(3 + 4) - \{8 - [5 - (9 + 6)]\}$

20. Simplify:
$[3(x + 2) + 2x] -$
$[4(y + 2) - 3(y - 2)]$.

Answers on page A-2

Simplify.

21. $23 - 42 \cdot 30$

22. $32 \div 8 \cdot 2$

23. $52 \cdot 5 + 5^3 - (4^2 - 48 \div 4)$

24. $\dfrac{5 - 10 - 5 \cdot 23}{2^3 + 3^2 - 7}$

Answers on page A-2

d ORDER OF OPERATIONS

When several operations are to be done in a calculation or a problem, we apply the same rules that we did in Section R.5. We repeat them here for review. (If you did not study that section earlier, you should do so now.)

RULES FOR ORDER OF OPERATIONS

1. Do all calculations within parentheses before operations outside.
2. Evaluate all exponential expressions.
3. Do all multiplications and divisions in order from left to right.
4. Do all additions and subtractions in order from left to right.

These rules are consistent with the way in which most computers perform calculations.

EXAMPLE 19 Simplify: $-34 \cdot 56 - 17$.

There are no parentheses or powers, so we start with the third step.

$-34 \cdot 56 - 17 = -1904 - 17$ — **Carrying out all multiplications and divisions in order from left to right**

$= -1921$ — **Carrying out all additions and subtractions in order from left to right**

EXAMPLE 20 Simplify: $2^4 + 51 \cdot 4 - (37 + 23 \cdot 2)$.

$2^4 + 51 \cdot 4 - (37 + 23 \cdot 2)$

$= 2^4 + 51 \cdot 4 - (37 + 46)$ — **Carrying out all operations inside parentheses first, multiplying 23 by 2, and following the rules for order of operations within the parentheses**

$= 2^4 + 51 \cdot 4 - 83$ — **Completing the addition inside parentheses**

$= 16 + 51 \cdot 4 - 83$ — **Evaluating exponential expressions**

$= 16 + 204 - 83$ — **Doing all multiplications**

$= 220 - 83$ — **Doing all additions and subtractions in order from left to right**

$= 137$

A fraction bar can play the role of a grouping symbol, although such a symbol is not as evident as the others.

EXAMPLE 21 Simplify: $\dfrac{-64 \div (-16) \div (-2)}{2^3 - 3^2}$.

An equivalent expression with brackets as grouping symbols is

$$[-64 \div (-16) \div (-2)] \div [2^3 - 3^2].$$

This shows, in effect, that we can do the calculations in the numerator and then in the denominator, and divide the results:

$$\frac{-64 \div (-16) \div (-2)}{2^3 - 3^2} = \frac{4 \div (-2)}{8 - 9} = \frac{-2}{-1} = 2.$$

Do Exercises 21–24.

NAME SECTION DATE

Exercise Set 1.8

a Find an equivalent expression without parentheses.

1. $-(2x + 7)$

2. $-(8x + 4)$

3. $-(5x - 8)$

4. $-(4x - 3)$

5. $-(4a - 3b + 7c)$

6. $-(x - 4y - 3z)$

7. $-(6x - 8y + 5)$

8. $-(4x + 9y + 7)$

9. $-(3x - 5y - 6)$

10. $-(6a - 4b - 7)$

11. $-(-8x - 6y - 43)$

12. $-(-2a + 9b - 5c)$

b Remove parentheses and simplify.

13. $9x - (4x + 3)$

14. $4y - (2y + 5)$

15. $2a - (5a - 9)$

16. $12m - (4m - 6)$

17. $2x + 7x - (4x + 6)$

18. $3a + 2a - (4a + 7)$

19. $2x - 4y - 3(7x - 2y)$

20. $3a - 9b - 1(4a - 8b)$

21. $15x - y - 5(3x - 2y + 5z)$

22. $4a - b - 4(5a - 7b + 8c)$

ANSWERS

1. ________
2. ________
3. ________
4. ________
5. ________
6. ________
7. ________
8. ________
9. ________
10. ________
11. ________
12. ________
13. ________
14. ________
15. ________
16. ________
17. ________
18. ________
19. ________
20. ________
21. ________
22. ________

ANSWERS

23. ______

24. ______

25. ______

26. ______

27. ______

28. ______

29. ______

30. ______

31. ______

32. ______

33. ______

34. ______

35. ______

36. ______

37. ______

38. ______

39. ______

40. ______

41. ______

42. ______

43. ______

44. ______

23. $(3x + 2y) - 2(5x - 4y)$

24. $(-6a - b) - 5(2b + a)$

25. $(12a - 3b + 5c) - 5(-5a + 4b - 6c)$

26. $(-8x + 5y - 12) - 6(2x - 4y - 10)$

c Simplify.

27. $[9 - 2(5 - 4)]$

28. $[6 - 5(8 - 4)]$

29. $8[7 - 6(4 - 2)]$

30. $10[7 - 4(7 - 5)]$

31. $[4(9 - 6) + 11] - [14 - (6 + 4)]$

32. $[7(8 - 4) + 16] - [15 - (7 + 8)]$

33. $[10(x + 3) - 4] + [2(x - 1) + 6]$

34. $[9(x + 5) - 7] + [4(x - 12) + 9]$

35. $[7(x + 5) - 19] - [4(x - 6) + 10]$

36. $[6(x + 4) - 12] - [5(x - 8) + 14]$

37. $3\{[7(x - 2) + 4] - [2(2x - 5) + 6]\}$

38. $4\{[8(x - 3) + 9] - [4(3x - 2) + 6]\}$

39. $4\{[5(x - 3) + 2] - 3[2(x + 5) - 9]\}$

40. $3\{[6(x - 4) + 5] - 2[5(x + 8) - 3]\}$

d Simplify.

41. $8 - 2 \cdot 3 - 9$

42. $8 - (2 \cdot 3 - 9)$

43. $(8 - 2 \cdot 3) - 9$

44. $(8 - 2)(3 - 9)$

45. $[(-24) \div (-3)] \div \left(-\frac{1}{2}\right)$

46. $[32 \div (-2)] \div (-2)$

47. $16 \cdot (-24) + 50$

48. $10 \cdot 20 - 15 \cdot 24$

49. $2^4 + 2^3 - 10$

50. $40 - 3^2 - 2^3$

51. $5^3 + 26 \cdot 71 - (16 + 25 \cdot 3)$

52. $4^3 + 10 \cdot 20 + 8^2 - 23$

53. $4 \cdot 5 - 2 \cdot 6 + 4$

54. $4 \cdot (6 + 8)/(4 + 3)$

55. $4^3/8$

56. $5^3 - 7^2$

57. $8(-7) + 6(-5)$

58. $10(-5) + 1(-1)$

59. $19 - 5(-3) + 3$

60. $14 - 2(-6) + 7$

61. $9 \div (-3) + 16 \div 8$

62. $-32 - 8 \div 4 - (-2)$

63. $6 - 4^2$

64. $(2 - 5)^2$

65. $(3 - 8)^2$

66. $3 - 3^2$

Answers

45. ______

46. ______

47. ______

48. ______

49. ______

50. ______

51. ______

52. ______

53. ______

54. ______

55. ______

56. ______

57. ______

58. ______

59. ______

60. ______

61. ______

62. ______

63. ______

64. ______

65. ______

66. ______

ANSWERS

67. ______
68. ______
69. ______
70. ______
71. ______
72. ______
73. ______
74. ______
75. ______
76. ______
77. ______
78. ______
79. ______
80. ______
81. ______
82. ______
83. ______
84. ______
85. ______
86. ______
87. ______
88. ______
89. ______
90. ______
91. ______
92. ______
93. ______
94. ______

67. $12 - 20^3$

68. $20 + 4^3 \div (-8)$

69. $2 \cdot 10^3 - 5000$

70. $-7(3^4) + 18$

71. $6[9 - (3 - 4)]$

72. $8[(6 - 13) - 11]$

73. $-1000 \div (-100) \div 10$

74. $256 \div (-32) \div (-4)$

75. $8 - (7 - 9)$

76. $(8 - 7) - 9$

77. $\dfrac{10 - 6^2}{9^2 + 3^2}$

78. $\dfrac{5^2 - 4^3 - 3}{9^2 - 2^2 - 1^5}$

79. $\dfrac{3(6 - 7) - 5 \cdot 4}{6 \cdot 7 - 8(4 - 1)}$

80. $\dfrac{20(8 - 3) - 4(10 - 3)}{10(2 - 6) - 2(5 + 2)}$

81. $\dfrac{2^3 - 3^2 + 12 \cdot 5}{-32 \div (-16) \div (-4)}$

82. $\dfrac{|3 - 5|^2 - |7 - 13|}{|12 - 9| + |11 - 14|}$

SKILL MAINTENANCE

83. Find the prime factorization of 236.

84. Find the LCM of 28 and 36.

85. Divide and simplify: $\dfrac{2}{3} \div \dfrac{5}{12}$.

Evaluate.

86. 3^4

87. 10^2

88. 15^2

SYNTHESIS

Find an equivalent expression by enclosing the last three terms in parentheses preceded by a minus sign.

89. $6y + 2x - 3a + c$

90. $x - y - a - b$

91. $6m + 3n - 5m + 4b$

Simplify.

92. $z - \{2z - [3z - (4z - 5z) - 6z] - 7z\} - 8z$

93. $\{x - [f - (f - x)] + [x - f]\} - 3x$

94. $x - \{x - 1 - [x - 2 - (x - 3 - \{x - 4 - [x - 5 - (x - 6)]\})]\}$

CRITICAL THINKING

These Critical Thinking exercises extend the concept of the Synthesis exercises, which have been introduced at the ends of exercise sets. They require you to synthesize objectives from several sections, thereby building your critical thinking skills. The exercises are of three types: Calculator Connections, which call for you to use your calculator; Extended Synthesis exercises, which are similar to the Synthesis exercises in the exercise sets, but may require more time; and Thinking and Writing exercises, which encourage you to both think and write about the ideas you have studied in this chapter.

CALCULATOR CONNECTION

In the following, use a calculator in any way you can to ease your work. Discussion of the general use of a calculator is presented in the section titled "Using a Scientific or Graphing Calculator" at the back of the book. If you are unfamiliar with the use of your particular calculator, consult its manual or your instructor.

1. Expressions like $x^2 + 3$ can be evaluated on a calculator with an exponent key or a square key. If your calculator does not have an exponent key, you can use the definition of an exponent and find the power by computing a product. Use your calculator to do the following.

a) Evaluate $x^2 + 3$ when $x = 7$, $x = -7$, and $x = -5.013$.

b) Evaluate $1 - x^2$ when $x = 5$, $x = -5$, and $x = -10.455$.

2. a) Look for a pattern in the following list, and find the missing numbers using your calculator:

1, 13, 25, 37, ___, ___, ___, ___, ___, ___, ___.

b) If the list in part (a) were to continue, at what position in the list would the number 1657 be located?

3. a) Look for a pattern in the following list, and find the missing numbers using your calculator:

2, 8, 32, 128, ___, ___, ___, ___, ___, ___, ___.

b) If the list in part (a) were to continue, at what position in the list would the number 131,072 be located?

4. The list of six numbers below is written by starting with two given numbers and multiplying them to get the next number in the list. Find the missing numbers using your calculator.

___; ___; ___; 1859; 265,837; 494,190,983

5. Consider the numbers 2, 4, 6, and 8. Assume that each can be placed in a blank in the following.

$\blacksquare + \blacksquare \cdot \blacksquare - \blacksquare = ?$

What placement of the numbers in the blanks yields the largest number? Explain why there are two answers.

6. Consider the numbers 3, 5, 7, and 9. Assume that each can be placed in a blank in the following.

$\blacksquare + \blacksquare^2 \cdot \blacksquare - \blacksquare = ?$

What placement of the numbers in the blanks yields the largest number? Explain why, compared to Exercise 5, there is just one answer.

In Exercises 7 and 8, place one of $+$, $-$, $\times$, and $\div$ in each blank to make a true sentence.

7. $-32 \blacksquare (88 \blacksquare 29) = -1888$

8. $3^5 \blacksquare 10^2 \blacksquare 5^2 = -2257$

9. Use your calculator as an aid. Try to characterize those denominators b in fractional notation a/b, where a and b are integers and $b \neq 0$, that result in terminating decimal notation.

10. Calculators do not always follow the rules for order of operations. To find out whether your calculator does, consider the computation

8 [+] 5 [·] 2 [=] → ?

If you get 18, the correct answer, then your calculator is programmed to follow the order of operations. If you get 26, then you know that you must always enter the operations in the correct order:

5 [·] 2 [+] 8 [=] → 18

to get the correct answer.

a) Does your calculator follow the rules for order of operations?

b) Use your calculator to find $6 \cdot 5 - 64 \div 16$.

(*continued*)

EXTENDED SYNTHESIS EXERCISES

1. Find two expressions that simplify to $\frac{5ab}{c}$.
2. Determine whether a commutative law holds for division of real numbers. That is, is it true that for all real numbers a and b, $a \div b = b \div a$?
3. Suppose you know that $p + q = 25$ and $p + (q + r) = 95$. Can you find $(p + q) + r$? r? $q + (r + p)$?
4. Determine whether $(a + b)^2$ and $a^2 + b^2$ are equivalent for any real numbers. Explain.

Write a formula for the shaded area.

5.

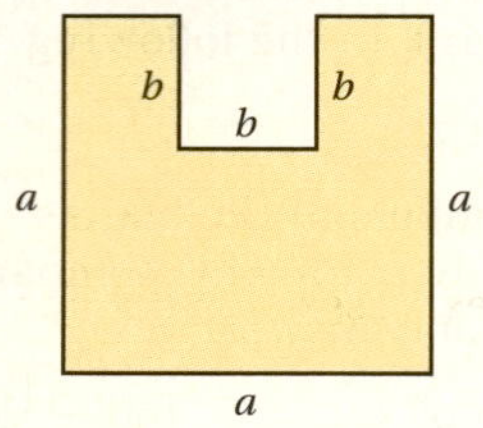

6.

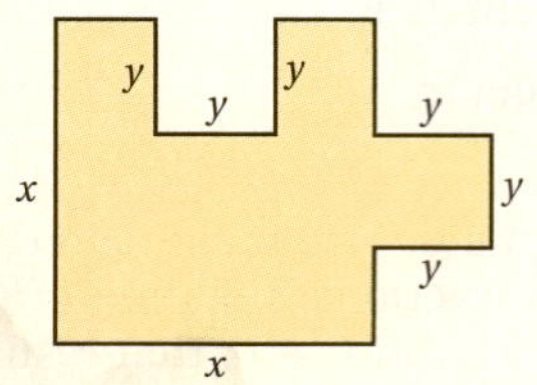

7. An **even number** is a number like 0, 2, 4, 6, 8, and so on, which can be expressed as $2 \cdot k$, where k is any whole number. For example, we know that 6 is an even number because we can express 6 as $6 = 2 \cdot 3$. An **odd number** is a number like 1, 3, 5, 7, 9, 11, and so on, which can be expressed as $2 \cdot k + 1$, where k is any whole number. For example, we know that 11 is odd because we can express it as $11 = 2 \cdot 5 + 1$.

   a) Show that 488 is even.

   b) Show that 0 is even.

   c) Show that 57 is odd.

   d) What can be said about the sum of two even numbers?

   e) What can be said about the sum of two odd numbers?

   f) What can be said about the sum of an even number and an odd number?

   g) What can be said about the product of an even number and an odd number?

   h) What can be said about the expression

   $$a^2 + a \cdot b,$$

   where a is odd and b is even?

Many of the problems that you will encounter in these critical thinking exercises involve making tables and looking for patterns. Such is the case in many of the following exercises.

8. The sum of two numbers is 800. The difference is 6. Find the numbers.
9. The sum of two numbers is 5. The product is -84. Find the numbers.
10. The sum of the digits of a two-digit number is 14. If the digits are reversed, the new number is 18 more than the original number. Find the original two-digit number.
11. Find the next three numbers in each sequence.

    a) 6, 5, 3, 0, ___, ___, ___

    b) 14, 10, 6, 2, ___, ___, ___

    c) -4, -6, -9, -13, ___, ___, ___

    d) 8, -4, 2, -1, 0.5, ___, ___, ___

12. The average attendance at the first six Poly-State Community College baseball games last year was 2035 people. The table below gives the change in attendance at each of this year's first six games compared to last year. By how much is the average attendance at this year's games above or below last year's?

GAME 1	GAME 2	GAME 3	GAME 4	GAME 5	GAME 6
+457	−244	−350	+447	−520	−288

EXERCISES FOR THINKING AND WRITING

1. Give three examples of rational numbers that are not integers.
2. Explain at least three uses of the distributive laws considered in this chapter.
3. One distributive law asserts that for any real numbers a, b, and c, $a(b + c) = ab + ac$. Use this fact and any other properties of real numbers to explain why the "right-hand" distributive law $(b + c)a = ba + ca$ holds for any real numbers.
4. Is it possible for a number to be its own reciprocal? Explain.
5. Write as many arguments as you can to convince a classmate that $-(-a) = a$ for all real numbers a.

Summary and Review: Chapter 1

Important Properties and Formulas

PROPERTIES OF THE REAL-NUMBER SYSTEM

The Commutative Laws: $a + b = b + a, \quad ab = ba$

The Associative Laws: $a + (b + c) = (a + b) + c, \quad a(bc) = (ab)c$

The Identity Properties: For every real number a, $a + 0 = a$ and $a \cdot 1 = a$.

The Inverse Properties: For each real number a, there is an opposite $-a$, such that $a + (-a) = 0$.

For each nonzero real number a, there is a reciprocal $\frac{1}{a}$, such that $a\left(\frac{1}{a}\right) = 1$.

The Distributive Laws: $a(b + c) = ab + ac, \quad a(b - c) = ab - ac$

Review Exercises

The review objectives to be tested in addition to the material in this chapter are [R.1a, b], [R.2b, c], [R.4a, d], and [R.5b, c].

1. Evaluate $\frac{x - y}{3}$ when $x = 17$ and $y = 5$.

2. Translate to an algebraic expression: Nineteen percent of some number.

3. Tell which integers correspond to this situation: Mike has a debt of \$45 and Joe has \$72 in his savings account.

4. Find: $|-38|$.

Graph the number on a number line.

5. -2.5

6. $\frac{8}{9}$

Use either $<$ or $>$ for $\blacksquare$ to write a true sentence.

7. $-3 \;\blacksquare\; 10$

8. $-1 \;\blacksquare\; -6$

9. $0.126 \;\blacksquare\; -12.6$

10. $-\frac{2}{3} \;\blacksquare\; -\frac{1}{10}$

Find the opposite.

11. 3.8

12. $-\frac{3}{4}$

Find the reciprocal.

13. $\frac{3}{8}$

14. -7

15. Find $-x$ when $x = -34$.

16. Find $-(-x)$ when $x = 5$.

Compute and simplify.

17. $4 + (-7)$

18. $6 + (-9) + (-8) + 7$

19. $-3.8 + 5.1 + (-12) + (-4.3) + 10$

20. $-3 - (-7)$

21. $-\frac{9}{10} - \frac{1}{2}$

22. $-3.8 - 4.1$

23. $-9 \cdot (-6)$

24. $-2.7(3.4)$

25. $\frac{2}{3} \cdot \left(-\frac{3}{7}\right)$

26. $3 \cdot (-7) \cdot (-2) \cdot (-5)$

27. $35 \div (-5)$

28. $-5.1 \div 1.7$

29. $-\frac{3}{11} \div \left(-\frac{4}{11}\right)$

30. $(-3.4 - 12.2) - 8(-7)$

31. $\dfrac{-12(-3) - 2^3 - (-9)(-10)}{3 \cdot 10 + 1}$

Solve.

32. On the first, second, and third downs, a football team had these gains and losses: 5-yd gain, 12-yd loss, and 15-yd gain, respectively. Find the total gain (or loss).

33. Your total assets are $170. You borrow $300. What are your total assets now?

Multiply.

34. $5(3x - 7)$

35. $-2(4x - 5)$

36. $10(0.4x + 1.5)$

37. $-8(3 - 6x)$

Factor.

38. $2x - 14$

39. $6x - 6$

40. $5x + 10$

41. $12 - 3x$

Collect like terms.

42. $11a + 2b - 4a - 5b$

43. $7x - 3y - 9x + 8y$

44. $6x + 3y - x - 4y$

45. $-3a + 9b + 2a - b$

Remove parentheses and simplify.

46. $2a - (5a - 9)$

47. $3(b + 7) - 5b$

48. $3[11 - 3(4 - 1)]$

49. $2[6(y - 4) + 7]$

50. $[8(x + 4) - 10] - [3(x - 2) + 4]$

51. $5\{[6(x - 1) + 7] - [3(3x - 4) + 8]\}$

Write true or false.

52. $-9 \leq 11$

53. $-11 \geq -3$

54. Write another inequality with the same meaning as $-3 < x$.

SKILL MAINTENANCE

55. Divide and simplify: $\frac{11}{12} \div \frac{7}{10}$.

56. Compute and simplify: $\dfrac{5^3 - 2^4}{5 \cdot 2 + 2^3}$.

57. Find the prime factorization of 648.

58. Convert to percent notation: $\frac{5}{8}$.

59. Convert to decimal notation: 5.67%.

60. Find the LCM of 15, 27, and 30.

SYNTHESIS

61. Simplify: $-\left|\frac{7}{8} - \left(-\frac{1}{2}\right) - \frac{3}{4}\right|$.

62. Simplify: $(|2.7 - 3| + 3^2 - |-3|) \div (-3)$.

NAME SECTION DATE

Test: Chapter 1

1. Evaluate $\frac{3x}{y}$ when $x = 10$ and $y = 5$.

2. Write an algebraic expression: Nine less than some number.

3. Find the area of a triangle when the height h is 30 ft and the base b is 16 ft.

Use either $<$ or $>$ for ■ to write a true sentence.

4. -4 ■ 0

5. -3 ■ -8

6. -0.78 ■ -0.87

7. $-\frac{1}{8}$ ■ $\frac{1}{2}$

Find the absolute value.

8. $|-7|$

9. $\left|\frac{9}{4}\right|$

10. $|-2.7|$

Find the opposite.

11. $\frac{2}{3}$

12. -1.4

13. Find $-x$ when $x = -8$.

Find the reciprocal.

14. -2

15. $\frac{4}{7}$

Compute and simplify.

16. $3.1 - (-4.7)$

17. $-8 + 4 + (-7) + 3$

18. $-\frac{1}{5} + \frac{3}{8}$

19. $2 - (-8)$

20. $3.2 - 5.7$

21. $\frac{1}{8} - \left(-\frac{3}{4}\right)$

ANSWERS

1. ______
2. ______
3. ______
4. ______
5. ______
6. ______
7. ______
8. ______
9. ______
10. ______
11. ______
12. ______
13. ______
14. ______
15. ______
16. ______
17. ______
18. ______
19. ______
20. ______
21. ______

ANSWERS

22. ______
23. ______
24. ______
25. ______
26. ______
27. ______
28. ______
29. ______
30. ______
31. ______
32. ______
33. ______
34. ______
35. ______
36. ______
37. ______
38. ______
39. ______
40. ______
41. ______
42. ______
43. ______
44. ______
45. ______

22. $4 \cdot (-12)$

23. $-\frac{1}{2} \cdot \left(-\frac{3}{8}\right)$

24. $-45 \div 5$

25. $-\frac{3}{5} \div \left(-\frac{4}{5}\right)$

26. $4.864 \div (-0.5)$

27. $-2(16) - |2(-8) - 5^3|$

28. Judy has $143 in her savings account. She withdraws $25. Then she makes a deposit of $30. How much is now in her savings account?

Multiply.

29. $3(6 - x)$

30. $-5(y - 1)$

Factor.

31. $12 - 22x$

32. $7x + 21 + 14y$

Simplify.

33. $6 + 7 - 4 - (-3)$

34. $5x - (3x - 7)$

35. $4(2a - 3b) + a - 7$

36. $4\{3[5(y - 3) + 9] + 2(y + 8)\}$

37. $256 \div (-16) \div 4$

38. $2^3 - 10[4 - (-2 + 18)3]$

39. Write an inequality with the same meaning as $x \leq -2$.

SKILL MAINTENANCE

40. Evaluate: $(1.2)^3$.

41. Convert to percent notation: $\frac{1}{8}$.

42. Find the prime factorization of 280.

43. Find the LCM of 16, 20, and 30.

SYNTHESIS

44. Simplify: $|-27 - 3(4)| - |-36| + |-12|$.

45. Simplify: $a - \{3a - [4a - (2a - 4a)]\}$.

2

Solving Equations and Inequalities

INTRODUCTION

In this chapter, we use the manipulations discussed in Chapter 1 to solve equations and inequalities. We then use equations and inequalities to solve problems.

AN APPLICATION

You see a flash of lightning. Your distance from the storm is M miles. You determine that distance by counting the number of seconds n that it takes the sound of the thunder to reach you and then multiplying by $\frac{1}{5}$. A formula relating M and n is $M = \frac{1}{5}n$.

Suppose we are 2 mi from the storm. How many seconds does it take for the sound of the thunder to reach us?

THE MATHEMATICS

We substitute 2 for M and solve for n:

$$\underbrace{2 = \frac{1}{5}n.}$$

To find n, we solve this equation.

Objectives for Review

The review objectives to be tested in addition to the material in this chapter are as follows.

[R.3a, b] Convert between decimal notation and fractional notation, and add, subtract, multiply, and divide using decimal notation.

[1.1a, b] Evaluate algebraic expressions by substitution, and translate phrases to algebraic expressions.

[1.3a] Add real numbers.

[1.8b] Simplify expressions by removing parentheses and collecting like terms.

Pretest: Chapter 2

Solve.

1. $-7x = 49$
2. $4y + 9 = 2y + 7$
3. $6a - 2 = 10$
4. $4 + x = 12$
5. $7 - 3(2x - 1) = 40$
6. $\frac{4}{9}x - 1 = \frac{7}{8}$
7. $1 + 2(a + 3) = 3(2a - 1) + 6$
8. $-3x \leq 18$
9. $y + 5 > 1$
10. $5 - 2a < 7$
11. $3x + 4 \geq 2x + 7$
12. $8y < -18$
13. Solve for G: $P = 3KG$.
14. Solve for a: $A = \dfrac{3a - b}{b}$.

Solve.

15. The perimeter of the ornate frame of an oil painting is 146 in. The width is 5 in. less than the length. Find the dimensions.
16. Money is invested in a savings account at 4.25% simple interest. After one year, there is $479.55 in the account. How much was originally invested?
17. The sum of three consecutive integers is 246. Find the integers.
18. When 18 is added to six times a number, the result is less than 120. For what numbers is this possible?

Graph on a number line.

19. $x > -3$
20. $x \leq 4$

2.1 *Solving Equations: The Addition Principle*

a EQUATIONS AND SOLUTIONS

In order to solve problems, we must learn to solve equations.

> An *equation* is a number sentence that says that the expressions on either side of the equals sign, =, represent the same number.

Here are some examples:

$$3 + 2 = 5, \quad 14 - 10 = 1 + 3, \quad x + 6 = 13, \quad 3x - 2 = 7 - x.$$

Equations have expressions on each side of the equals sign. The sentence "$14 - 10 = 1 + 3$" asserts that the expressions $14 - 10$ and $1 + 3$ name the same number.

Some equations are true. Some are false. Some are neither true nor false.

EXAMPLES Determine whether the equation is true, false, or neither.

1. $3 + 2 = 5$ The equation is *true.*

2. $7 - 2 = 4$ The equation is *false.*

3. $x + 6 = 13$ The equation is *neither* true nor false, because we do not know what number x represents.

Do Exercises 1–3.

> Any replacement for the variable that makes an equation true is called a *solution* of the equation. To solve an equation means to find *all* of its solutions.

One way to determine whether a number is a solution of an equation is to evaluate the algebraic expression on each side of the equation by substitution. If the values are the same, then the number is a solution.

EXAMPLE 4 Determine whether 7 is a solution of $x + 6 = 13$.

We have

$$\begin{array}{c|c} x + 6 = 13 & \\ \hline 7 + 6 & 13 \\ 13 & \end{array}$$

$x + 6 = 13$ **Writing the equation**
$7 + 6 \mid 13$ **Substituting 7 for x**
13 TRUE

Since the left-hand and the right-hand sides are the same, we have a solution. No other number makes the equation true, so the only solution is the number 7.

OBJECTIVES

After finishing Section 2.1, you should be able to:

a Determine whether a given number is a solution of a given equation.

b Solve equations using the addition principle.

FOR EXTRA HELP

TAPE 5 | TAPE 4A | MAC: 2A IBM: 2A

Determine whether the equation is true, false, or neither.

1. $5 - 8 = -4$

2. $12 + 6 = 18$

3. $x + 6 = 7 - x$

Answers on page A-2

Determine whether the given number is a solution of the given equation.

4. 8; $x + 4 = 12$

5. 0; $x + 4 = 12$

6. −3; $7 + x = -4$

7. Solve using the addition principle:

$x + 2 = 11.$

Answers on page A-2

EXAMPLE 5 Determine whether 19 is a solution of $7x = 141$.

We have

$7x$	$= 141$	Writing the equation
$7(19)$	141	Substituting 19 for x
133		FALSE

Since the left-hand and the right-hand sides are not the same, we do not have a solution.

Do Exercises 4–6.

b USING THE ADDITION PRINCIPLE

Consider the equation

$x = 7.$

We can easily see that the solution of this equation is 7. If we replace x by 7, we get

$7 = 7,$ which is true.

Now consider the equation of Example 4:

$x + 6 = 13.$

In Example 4, we discovered that the solution of this equation is also 7, but the fact that 7 is the solution is not as obvious. We now begin to consider principles that allow us to start with an equation and end up with an **equivalent equation,** like $x = 7$, in which the variable is alone on one side and for which the solution is easy to find.

> Equations with the same solutions are called *equivalent equations.*

One of the principles that we use in solving equations involves adding. An equation $a = b$ says that a and b stand for the same number. Suppose this is true, and we add a number c to the number a. We get the same answer if we add c to b, because a and b are the same number.

> **THE ADDITION PRINCIPLE**
>
> If an equation $a = b$ is true, then
>
> $a + c = b + c$
>
> is true for any number c.

Let us again solve the equation $x + 6 = 13$ using the addition principle. We want to get x alone on one side. To do so, we use the addition principle, adding −6 on both sides:

$x + 6 = 13$	
$x + 6 + (-6) = 13 + (-6)$	Using the addition principle: adding −6 on both sides
$x + 0 = 7$	Simplifying
$x = 7.$	Identity property of 0

Do Exercise 7.

Now we solve an equation with a subtraction using the addition principle.

EXAMPLE 6 Solve: $a - 4 = 10$.

We have

$$a - 4 = 10$$

$$a - 4 + 4 = 10 + 4 \quad \text{Using the addition principle: adding 4 on both sides}$$

$$a + 0 = 14 \quad \text{Simplifying}$$

$$a = 14. \quad \text{Identity property of 0}$$

CHECK:

$a - 4 = 10$	
$14 - 4$	10
10	TRUE

The solution is 14.

Do Exercise 8.

When we use the addition principle, we sometimes say that we "add the same number on both sides of the equation." This is also true for subtraction, since we can express every subtraction as an addition. That is, since

$$a - c = b - c \quad \text{means} \quad a + (-c) = b + (-c),$$

the addition principle tells us that we can "subtract the same number on both sides of the equation."

EXAMPLE 7 Solve: $x + 5 = -7$.

We have

$$x + 5 = -7$$

$$x + 5 - 5 = -7 - 5 \quad \text{Using the addition principle: adding } -5 \text{ on both sides or subtracting 5 on both sides}$$

$$x + 0 = -12 \quad \text{Simplifying}$$

$$x = -12. \quad \text{Identity property of 0}$$

We can see that the solution of $x = -12$ is the number -12. To check the answer, we substitute -12 in the original equation.

CHECK:

$x + 5 = -7$	
$-12 + 5$	-7
-7	TRUE

The solution of the original equation is -12.

In Example 7, to get x alone, we used the addition principle and subtracted 5 on both sides. This eliminated the 5 on the left. We started with $x + 5 = -7$, and, using the addition principle, we found a simpler equation $x = -12$ for which it was easy to "*see*" the solution. The equations $x + 5 = -7$ and $x = -12$ are equivalent.

Do Exercise 9.

8. Solve: $t - 3 = 19$.

9. Solve using the addition principle:

$$x + 7 = 2.$$

Answers on page A-2

Solve.

10. $8.7 = n - 4.5$

11. $y + 17.4 = 10.9$

Solve.

12. $x + \frac{1}{2} = -\frac{3}{2}$

13. $t - \frac{13}{4} = \frac{5}{8}$

Answers on page A-2

EXAMPLE 8 Solve: $-6.5 = y - 8.4$.

We have

$$-6.5 = y - 8.4$$

$$-6.5 + 8.4 = y - 8.4 + 8.4 \quad \text{Using the addition principle: adding 8.4 to eliminate } -8.4 \text{ on the right}$$

$$1.9 = y$$

CHECK:

$-6.5 = y - 8.4$	
-6.5	$1.9 - 8.4$
	-6.5 TRUE

The solution is 1.9.

Note that equations are reversible. That is, if $a = b$ is true, then $b = a$ is true. Thus when we solve $-6.5 = y - 8.4$, we can reverse it and solve $y - 8.4 = -6.5$ if we wish.

Do Exercises 10 and 11.

EXAMPLE 9 Solve: $-\frac{2}{3} + x = \frac{5}{2}$.

We have

$$-\frac{2}{3} + x = \frac{5}{2}$$

$$\frac{2}{3} - \frac{2}{3} + x = \frac{2}{3} + \frac{5}{2} \quad \text{Adding } \frac{2}{3}$$

$$x = \frac{2}{3} \cdot \frac{2}{2} + \frac{5}{2} \cdot \frac{3}{3} \quad \text{Multiplying by 1 to obtain equivalent fractional expressions with the least common denominator 6}$$

$$= \frac{4}{6} + \frac{15}{6}$$

$$= \frac{19}{6}.$$

CHECK:

$-\frac{2}{3} + x = \frac{5}{2}$	
$-\frac{2}{3} + \frac{19}{6}$	$\frac{5}{2}$
$-\frac{4}{6} + \frac{19}{6}$	
$\frac{15}{6}$	
$\frac{5}{2}$	TRUE

The solution is $\frac{19}{6}$.

Do Exercises 12 and 13.

NAME SECTION DATE

Exercise Set 2.1

a Determine whether the given number is a solution of the given equation.

1. 15; $x + 17 = 32$

2. 35; $t + 17 = 53$

3. 21; $x - 7 = 12$

4. 36; $a - 19 = 17$

5. -7; $6x = 54$

6. -9; $8y = -72$

7. 30; $\frac{x}{6} = 5$

8. 49; $\frac{y}{8} = 6$

9. 19; $5x + 7 = 107$

10. 9; $9x + 5 = 86$

11. -11; $7(y - 1) = 63$

12. -18; $x + 3 = 3 + x$

b Solve using the addition principle. Don't forget to check!

13. $x + 2 = 6$

14. $y + 4 = 11$

15. $x + 15 = -5$

16. $t + 10 = 44$

17. $x + 6 = -8$

18. $z + 9 = -14$

19. $x + 16 = -2$

20. $m + 18 = -13$

21. $x - 9 = 6$

22. $x - 11 = 12$

23. $x - 7 = -21$

24. $x - 3 = -14$

25. $5 + t = 7$

26. $8 + y = 12$

27. $-7 + y = 13$

ANSWERS

1. ______
2. ______
3. ______
4. ______
5. ______
6. ______
7. ______
8. ______
9. ______
10. ______
11. ______
12. ______
13. ______
14. ______
15. ______
16. ______
17. ______
18. ______
19. ______
20. ______
21. ______
22. ______
23. ______
24. ______
25. ______
26. ______
27. ______

ANSWERS

28. ______
29. ______
30. ______
31. ______
32. ______
33. ______
34. ______
35. ______
36. ______
37. ______
38. ______
39. ______
40. ______
41. ______
42. ______
43. ______
44. ______
45. ______
46. ______
47. ______
48. ______
49. ______
50. ______
51. ______
52. ______
53. ______
54. ______
55. ______
56. ______
57. ______
58. ______
59. ______
60. ______
61. ______
62. ______
63. ______
64. ______
65. ______

28. $-8 + y = 17$ **29.** $-3 + t = -9$ **30.** $-8 + t = -24$

31. $r + \frac{1}{3} = \frac{8}{3}$ **32.** $t + \frac{3}{8} = \frac{5}{8}$ **33.** $m + \frac{5}{6} = -\frac{11}{12}$

34. $x + \frac{2}{3} = -\frac{5}{6}$ **35.** $x - \frac{5}{6} = \frac{7}{8}$ **36.** $y - \frac{3}{4} = \frac{5}{6}$

37. $-\frac{1}{5} + z = -\frac{1}{4}$ **38.** $-\frac{1}{8} + y = -\frac{3}{4}$ **39.** $7.4 = x + 2.3$

40. $8.4 = 5.7 + y$ **41.** $7.6 = x - 4.8$ **42.** $8.6 = x - 7.4$

43. $-9.7 = -4.7 + y$ **44.** $-7.8 = 2.8 + x$ **45.** $5\frac{1}{6} + x = 7$

46. $5\frac{1}{4} = 4\frac{2}{3} + x$ **47.** $q + \frac{1}{3} = -\frac{1}{7}$ **48.** $52\frac{3}{8} = -84 + x$

SKILL MAINTENANCE

49. Add: $-3 + (-8)$.

50. Subtract: $-3 - (-8)$.

51. Multiply: $-\frac{2}{3} \cdot \frac{5}{8}$.

52. Divide: $-\frac{3}{7} \div \left(-\frac{9}{7}\right)$.

53. Divide: $\frac{2}{3} \div \left(-\frac{4}{9}\right)$.

54. Add: $-8.6 + 3.4$.

SYNTHESIS

Solve.

55. ▦ $-356.788 = -699.034 + t$

56. $-\frac{4}{5} + \frac{7}{10} = x - \frac{3}{4}$

57. $x + \frac{4}{5} = -\frac{2}{3} - \frac{4}{15}$ **58.** $8 - 25 = 8 + x - 21$ **59.** $16 + x - 22 = -16$

60. $x + x = x$ **61.** $x + 3 = 3 + x$ **62.** $x + 4 = 5 + x$

63. $-\frac{3}{2} + x = -\frac{5}{17} - \frac{3}{2}$ **64.** $|x| = 5$ **65.** $|x| + 6 = 19$

2.2 *Solving Equations: The Multiplication Principle*

OBJECTIVE

After finishing Section 2.2, you should be able to:

a Solve equations using the multiplication principle.

FOR EXTRA HELP

TAPE 5 | TAPE 4B | MAC: 2A IBM: 2A

a USING THE MULTIPLICATION PRINCIPLE

Suppose that $a = b$ is true, and we multiply a by some number c. We get the same answer if we multiply b by c, because a and b are the same number.

> **THE MULTIPLICATION PRINCIPLE**
>
> If an equation $a = b$ is true, then
>
> $$a \cdot c = b \cdot c$$
>
> is true for any number c.

When using the multiplication principle, we sometimes say that we "multiply on both sides of the equation by the same number."

EXAMPLE 1 Solve: $5x = 70$.

To get x alone, we multiply by the *multiplicative inverse,* or *reciprocal,* of 5. Then we get the *multiplicative identity* 1 times x, or $1 \cdot x$, which simplifies to x. This allows us to eliminate 5 on the left.

$5x = 70$ The reciprocal of 5 is $\frac{1}{5}$.

$\frac{1}{5} \cdot 5x = \frac{1}{5} \cdot 70$ Multiplying by $\frac{1}{5}$ to get $1 \cdot x$ and eliminate 5 on the left

$1 \cdot x = 14$ Simplifying

$x = 14$ Identity property of 1: $1 \cdot x = x$

CHECK:

$5x = 70$	
$5 \cdot 14$	70
70	

TRUE

The solution is 14.

Do Exercises 1 and 2.

Solve.

1. $6x = 90$

2. $4x = -7$

Answers on page A-2

3. Solve: $\frac{2}{3} = -\frac{5}{6}y$.

EXAMPLE 2 Solve: $\frac{3}{8} = -\frac{5}{4}x$.

We have

$$\frac{3}{8} = -\frac{5}{4}x$$

The reciprocal of $-\frac{5}{4}$ is $-\frac{4}{5}$. There is no sign change.

$$-\frac{4}{5} \cdot \frac{3}{8} = -\frac{4}{5} \cdot \left(-\frac{5}{4}x\right)$$ Multiplying by $-\frac{4}{5}$ to get $1 \cdot x$ and eliminate $-\frac{5}{4}$ on the right

$$-\frac{12}{40} = 1 \cdot x$$

$$-\frac{3}{10} = 1 \cdot x$$ Simplifying

$$-\frac{3}{10} = x$$ Identity property of 1: $1 \cdot x = x$

CHECK:

$\frac{3}{8} = -\frac{5}{4}x$		
$\frac{3}{8}$	$-\frac{5}{4}\left(-\frac{3}{10}\right)$	
	$\frac{3}{8}$	TRUE

The solution is $-\frac{3}{10}$.

Note that equations are reversible. That is, if $a = b$ is true, then $b = a$ is true. Thus, when we solve $\frac{3}{8} = -\frac{5}{4}x$, we can reverse it and solve $-\frac{5}{4}x = \frac{3}{8}$ if we wish.

Do Exercise 3.

The multiplication principle also tells us that we can "divide on both sides of the equation by a nonzero number." This is because division is the same as multiplying by a reciprocal. That is,

$$\frac{a}{c} = \frac{b}{c} \quad \text{means} \quad a \cdot \frac{1}{c} = b \cdot \frac{1}{c}, \quad \text{when } c \neq 0.$$

In an expression like $3x$, the number 3 is called the **coefficient**. In practice, it is usually more convenient to "divide" both sides of the equation if the coefficient of the variable is in decimal notation or is an integer. If the coefficient is in fractional notation, it is more convenient to "multiply" by a reciprocal.

Answer on page A-2

EXAMPLE 3 Solve: $3x = 9$.

We have

$$3x = 9$$

$$\frac{3x}{3} = \frac{9}{3} \qquad \text{Using the multiplication principle: multiplying by } \tfrac{1}{3} \text{ on both sides or dividing on both sides by 3}$$

$$1 \cdot x = 3 \qquad \text{Simplifying}$$

$$x = 3. \qquad \text{Identity property of 1}$$

CHECK:

$3x = 9$	
$3 \cdot 3$	9
9	TRUE

The solution is 3.

Do Exercise 4.

EXAMPLE 4 Solve: $-4x = 92$.

$$-4x = 92$$

$$\frac{-4x}{-4} = \frac{92}{-4} \qquad \text{Using the multiplication principle. Dividing on both sides by } -4 \text{ is the same as multiplying by } -\tfrac{1}{4}.$$

$$1 \cdot x = -23 \qquad \text{Simplifying}$$

$$x = -23 \qquad \text{Identity property of 1}$$

CHECK:

$-4x = 92$	
$-4(-23)$	92
92	TRUE

The solution is -23.

Do Exercise 5.

EXAMPLE 5 Solve: $-x = 9$.

$$-x = 9$$

$$-1 \cdot x = 9 \qquad \text{Using the property of } -1\text{: } -x = -1 \cdot x$$

$$-1 \cdot (-1 \cdot x) = -1 \cdot 9 \qquad \text{Multiplying on both sides by } -1 \text{, the reciprocal of itself, or dividing by } -1$$

$$1 \cdot x = -9$$

$$x = -9$$

CHECK:

$-x = 9$	
$-(-9)$	9
9	TRUE

The solution is -9.

Do Exercise 6.

4. Solve: $5x = 40$.

5. Solve: $-6x = 108$.

6. Solve: $-x = -10$.

Answers on page A-2

7. Solve: $-14 = \frac{-y}{2}$.

Solve.

8. $1.12x = 8736$

9. $6.3 = -2.1y$

Answers on page A-2

Now we solve an equation with a division using the multiplication principle. Consider an equation like $-y/9 = 14$. In Chapter 1, we learned that a division can be expressed as multiplication by the reciprocal of the divisor. Thus,

$$\frac{-y}{9} \text{ is equivalent to } \frac{1}{9}(-y).$$

The reciprocal of $\frac{1}{9}$ is 9. Then, using the multiplication principle, we multiply on both sides by 9. This is shown in the following example.

EXAMPLE 6 Solve: $\frac{-y}{9} = 14$.

$$\frac{-y}{9} = 14$$

$$\frac{1}{9}(-y) = 14$$

$$9 \cdot \frac{1}{9}(-y) = 9 \cdot 14 \quad \text{Multiplying by 9 on both sides}$$

$$-y = 126$$

$$-1 \cdot (-y) = -1 \cdot 126 \quad \text{Multiplying by } -1 \text{ on both sides}$$

$$y = -126$$

CHECK:

$\frac{-y}{9} = 14$	
$\frac{-(-126)}{9}$	14
$\frac{126}{9}$	
14	TRUE

The solution is -126.

Do Exercise 7.

EXAMPLE 7 Solve: $1.16y = 9744$.

$$1.16y = 9744$$

$$\frac{1.16y}{1.16} = \frac{9744}{1.16} \quad \text{Dividing by 1.16}$$

$$y = \frac{9744}{1.16}$$

$$= 8400$$

CHECK:

$1.16y = 9744$	
1.16(8400)	9744
9744	TRUE

The solution is 8400.

Do Exercises 8 and 9.

NAME SECTION DATE

Exercise Set 2.2

a Solve using the multiplication principle. Don't forget to check!

1. $6x = 36$

2. $3x = 51$

3. $5x = 45$

4. $8x = 72$

5. $84 = 7x$

6. $63 = 9x$

7. $-x = 40$

8. $53 = -x$

9. $-x = -1$

10. $-47 = -t$

11. $7x = -49$

12. $8x = -56$

13. $-12x = 72$

14. $-15x = 105$

15. $-21x = -126$

16. $-13x = -104$

17. $\frac{t}{7} = -9$

18. $\frac{y}{-8} = 11$

19. $\frac{3}{4}x = 27$

20. $\frac{4}{5}x = 16$

21. $\frac{-t}{3} = 7$

22. $\frac{-x}{6} = 9$

23. $-\frac{m}{3} = \frac{1}{5}$

24. $\frac{1}{8} = -\frac{y}{5}$

Answers

1. ______
2. ______
3. ______
4. ______
5. ______
6. ______
7. ______
8. ______
9. ______
10. ______
11. ______
12. ______
13. ______
14. ______
15. ______
16. ______
17. ______
18. ______
19. ______
20. ______
21. ______
22. ______
23. ______
24. ______

ANSWERS

25. ____
26. ____
27. ____
28. ____
29. ____
30. ____
31. ____
32. ____
33. ____
34. ____
35. ____
36. ____
37. ____
38. ____
39. ____
40. ____
41. ____
42. ____
43. ____
44. ____
45. ____
46. ____
47. ____
48. ____
49. ____
50. ____
51. ____
52. ____
53. ____
54. ____

25. $-\frac{3}{5}r = \frac{9}{10}$

26. $\frac{2}{5}y = -\frac{4}{15}$

27. $-\frac{3}{2}r = -\frac{27}{4}$

28. $-\frac{3}{8}x = -\frac{15}{16}$

29. $6.3x = 44.1$

30. $2.7y = 54$

31. $-3.1y = 21.7$

32. $-3.3y = 6.6$

33. $38.7m = 309.6$

34. $29.4m = 235.2$

35. $-\frac{2}{3}y = -10.6$

36. $-\frac{9}{7}y = 12.06$

SKILL MAINTENANCE

Collect like terms.

37. $3x + 4x$

38. $6x + 5 - 7x$

39. $-4x + 11 - 6x + 18x$

40. $8y - 16y - 24y$

Remove parentheses and simplify.

41. $3x - (4 + 2x)$

42. $2 - 5(x + 5)$

43. $8y - 6(3y + 7)$

44. $-2a - 4(5a - 1)$

SYNTHESIS

Solve.

45. 🖩 $-0.2344m = 2028.732$

46. $0 \cdot x = 0$

47. $0 \cdot x = 9$

48. $4|x| = 48$

49. $2|x| = -12$

Solve for x.

50. $ax = 5a$

51. $3x = \frac{b}{a}$

52. $cx = a^2 + 1$

53. $\frac{a}{b}x = 4$

54. A student makes a calculation and gets an answer of 22.5. On the last step, the student multiplies by 0.3 when a division by 0.3 should have been done. What is the correct answer?

2.3 *Using the Principles Together*

a APPLYING BOTH PRINCIPLES

Consider the equation $3x + 4 = 13$. It is more complicated than those we discussed in the preceding two sections. In order to solve such an equation, we first isolate the x-term, $3x$, using the addition principle. Then we apply the multiplication principle to get x by itself.

EXAMPLE 1 Solve: $3x + 4 = 13$.

$$3x + 4 = 13$$

$$3x + 4 - 4 = 13 - 4 \quad \text{Using the addition principle: subtracting 4 on both sides}$$

$$3x = 9 \quad \text{Simplifying}$$

$$\frac{3x}{3} = \frac{9}{3} \quad \text{Using the multiplication principle: dividing by 3 on both sides}$$

$$x = 3 \quad \text{Simplifying}$$

CHECK:

$3x + 4 = 13$	
$3 \cdot 3 + 4$	13
$9 + 4$	
13	TRUE

We use the rules for order of operations to carry out the check. We find the product $3 \cdot 3$. Then we add 4.

The solution is 3.

Do Exercise 1.

EXAMPLE 2 Solve: $-5x - 6 = 16$.

$$-5x - 6 = 16$$

$$-5x - 6 + 6 = 16 + 6 \quad \text{Adding 6 on both sides}$$

$$-5x = 22$$

$$\frac{-5x}{-5} = \frac{22}{-5} \quad \text{Dividing by } -5 \text{ on both sides}$$

$$x = -\frac{22}{5}, \text{ or } -4\frac{2}{5} \quad \text{Simplifying}$$

CHECK:

$-5x - 6 = 16$	
$-5\left(-\frac{22}{5}\right) - 6$	16
$22 - 6$	
16	TRUE

The solution is $-\frac{22}{5}$.

Do Exercises 2 and 3.

OBJECTIVES

After finishing Section 2.3, you should be able to:

a Solve equations using both the addition and the multiplication principles.

b Solve equations in which like terms may need to be collected.

c Solve equations by first removing parentheses and collecting like terms.

FOR EXTRA HELP

TAPE 6 — TAPE 4B — MAC: 2A, IBM: 2A

1. Solve: $9x + 6 = 51$.

Solve.

2. $8x - 4 = 28$

3. $-\frac{1}{2}x + 3 = 1$

Answers on page A-2

4. Solve: $-18 - x = -57$.

Solve.

5. $-4 - 8x = 8$

6. $41.68 = 4.7 - 8.6y$

Solve.

7. $4x + 3x = -21$

8. $x - 0.09x = 728$

Answers on page A-2

EXAMPLE 3 Solve: $45 - x = 13$.

$$45 - x = 13$$

$$-45 + 45 - x = -45 + 13 \quad \text{Adding } -45 \text{ on both sides}$$

$$-x = -32$$

$$-1 \cdot x = -32 \quad \text{Using the property of } -1\text{: } -x = -1 \cdot x$$

$$\frac{-1 \cdot x}{-1} = \frac{-32}{-1}$$

Dividing by -1 on both sides (You could have multiplied by -1 on both sides instead. That would also change the sign on both sides.)

$$x = 32$$

The number 32 checks and is the solution.

Do Exercise 4.

EXAMPLE 4 Solve: $16.3 - 7.2y = -8.18$.

$$16.3 - 7.2y = -8.18$$

$$-16.3 + 16.3 - 7.2y = -16.3 + (-8.18) \quad \text{Adding } -16.3 \text{ on both sides}$$

$$-7.2y = -24.48$$

$$\frac{-7.2y}{-7.2} = \frac{-24.48}{-7.2} \quad \text{Dividing by } -7.2 \text{ on both sides}$$

$$y = 3.4$$

CHECK:

$16.3 - 7.2y$	$= -8.18$
$16.3 - 7.2(3.4)$	-8.18
$16.3 - 24.48$	
-8.18	TRUE

The solution is 3.4.

Do Exercises 5 and 6.

b Collecting Like Terms

If there are like terms on one side of the equation, we collect them before using the addition or the multiplication principle.

EXAMPLE 5 Solve: $3x + 4x = -14$.

$$3x + 4x = -14$$

$$7x = -14 \quad \text{Collecting like terms}$$

$$\frac{7x}{7} = \frac{-14}{7} \quad \text{Dividing by 7 on both sides}$$

$$x = -2$$

The number -2 checks, so the solution is -2.

Do Exercises 7 and 8.

If there are like terms on opposite sides of the equation, we get them on the same side by using the addition principle. Then we collect them. In other words, we get all terms with a variable on one side and all numbers on the other.

EXAMPLE 6 Solve: $2x - 2 = -3x + 3$.

$$2x - 2 = -3x + 3$$
$$2x - 2 + 2 = -3x + 3 + 2 \quad \text{Adding 2}$$
$$2x = -3x + 5 \quad \text{Collecting like terms}$$
$$2x + 3x = -3x + 3x + 5 \quad \text{Adding } 3x$$
$$5x = 5 \quad \text{Simplifying}$$
$$\frac{5x}{5} = \frac{5}{5} \quad \text{Dividing by 5}$$
$$x = 1 \quad \text{Simplifying}$$

CHECK:

$2x - 2$	$= -3x + 3$	
$2 \cdot 1 - 2$	$-3 \cdot 1 + 3$	
$2 - 2$	$-3 + 3$	
0	0	TRUE

The solution is 1.

Do Exercise 9.

In Example 6, we used the addition principle to get all terms with a variable on one side and all numbers on the other side. Then we collected like terms and proceeded as before. If there are like terms on one side at the outset, they should be collected before proceeding.

EXAMPLE 7 Solve: $6x + 5 - 7x = 10 - 4x + 3$.

$$6x + 5 - 7x = 10 - 4x + 3$$
$$-x + 5 = 13 - 4x \quad \text{Collecting like terms}$$
$$4x - x + 5 = 13 - 4x + 4x \quad \text{Adding } 4x \text{ to get all terms with a variable on one side}$$
$$3x + 5 = 13 \quad \text{Simplifying; that is, collecting like terms}$$
$$3x + 5 - 5 = 13 - 5 \quad \text{Subtracting 5}$$
$$3x = 8 \quad \text{Simplifying}$$
$$\frac{3x}{3} = \frac{8}{3} \quad \text{Dividing by 3}$$
$$x = \frac{8}{3} \quad \text{Simplifying}$$

The number $\frac{8}{3}$ checks, so it is the solution.

Do Exercises 10–12.

9. Solve: $7y + 5 = 2y + 10$.

Solve.

10. $5 - 2y = 3y - 5$

11. $7x - 17 + 2x = 2 - 8x + 15$

12. $3x - 15 = 5x + 2 - 4x$

Answers on page A-2

13. Solve: $\frac{7}{8}x - \frac{1}{4} + \frac{1}{2}x = \frac{3}{4} + x.$

Answer on page A-2

Clearing Fractions and Decimals

For the types of equation considered thus far, we generally use the addition principle first. There are, however, some situations in which it is to our advantage to use the multiplication principle first. Consider, for example,

$$\frac{1}{2}x = \frac{3}{4}.$$

If we multiply by 4 on both sides, we get the equation $2x = 3$, which has no fractions. We have thus "cleared fractions." Consider

$$2.3x = 5.$$

If we multiply by 10 on both sides, we get the equation $23x = 50$, which has no decimal points. We have thus "cleared decimals." The equations are then easier to solve. *It is your choice* whether to clear the fractions or decimals, but doing so often eases computations.

In what follows, we use the multiplication principle first to clear, or eliminate, fractions or decimals. For fractions, the number by which we multiply is the **least common multiple of all the denominators.**

EXAMPLE 8 Solve: $\frac{2}{3}x - \frac{1}{6} + \frac{1}{2}x = \frac{7}{6} + 2x.$

The number 6 is the least common multiple of all the denominators. We multiply by 6 on both sides.

$$6\left(\frac{2}{3}x - \frac{1}{6} + \frac{1}{2}x\right) = 6\left(\frac{7}{6} + 2x\right) \qquad \text{Multiplying by 6 on both sides}$$

$$6 \cdot \frac{2}{3}x - 6 \cdot \frac{1}{6} + 6 \cdot \frac{1}{2}x = 6 \cdot \frac{7}{6} + 6 \cdot 2x \qquad \text{Using the distributive laws (*Caution!* Be sure to multiply *all* the terms by 6.)}$$

$$4x - 1 + 3x = 7 + 12x \qquad \text{Simplifying. Note that the fractions are cleared.}$$

$$7x - 1 = 7 + 12x \qquad \text{Collecting like terms}$$

$$7x - 1 - 12x = 7 + 12x - 12x \qquad \text{Subtracting } 12x$$

$$-5x - 1 = 7 \qquad \text{Collecting like terms}$$

$$-5x - 1 + 1 = 7 + 1 \qquad \text{Adding 1}$$

$$-5x = 8 \qquad \text{Collecting like terms}$$

$$\frac{-5x}{-5} = \frac{8}{-5} \qquad \text{Dividing by } -5$$

$$x = -\frac{8}{5}$$

The number $-\frac{8}{5}$ checks, so it is the solution.

Do Exercise 13.

To illustrate clearing decimals, we repeat Example 4, but this time we clear the equation of decimals first.

EXAMPLE 9 Solve: $16.3 - 7.2y = -8.18$.

The greatest number of decimal places in any one number is *two*. Multiplying by 100, which has *two* 0's, will clear decimals.

$$100(16.3 - 7.2y) = 100(-8.18) \quad \text{Multiplying by 100 on both sides}$$
$$100(16.3) - 100(7.2y) = 100(-8.18) \quad \text{Using a distributive law}$$
$$1630 - 720y = -818 \quad \text{Simplifying}$$
$$1630 - 720y - 1630 = -818 - 1630 \quad \text{Subtracting 1630 on both sides}$$
$$-720y = -2448 \quad \text{Collecting like terms}$$
$$\frac{-720y}{-720} = \frac{-2448}{-720} \quad \text{Dividing by } -720 \text{ on both sides}$$
$$y = 3.4$$

The number 3.4 checks, so it is the solution.

Do Exercise 14.

14. Solve: $41.68 = 4.7 - 8.6y$.

c EQUATIONS CONTAINING PARENTHESES

To solve certain kinds of equations that contain parentheses, we use the distributive laws to first remove the parentheses. Then we proceed as before.

EXAMPLE 10 Solve: $4x = 2(12 - 2x)$.

$$4x = 2(12 - 2x)$$
$$4x = 24 - 4x \quad \text{Using a distributive law to multiply and remove parentheses}$$
$$4x + 4x = 24 - 4x + 4x \quad \text{Adding } 4x \text{ to get all the } x\text{-terms on one side}$$
$$8x = 24 \quad \text{Collecting like terms}$$
$$\frac{8x}{8} = \frac{24}{8} \quad \text{Dividing by 8}$$
$$x = 3$$

CHECK:

$4x$	$= 2(12 - 2x)$
$4 \cdot 3$	$2(12 - 2 \cdot 3)$
12	$2(12 - 6)$
	$2 \cdot 6$
	12 TRUE

We use the rules for order of operations to carry out the calculations on each side of the equation.

The solution is 3.

Do Exercises 15 and 16.

Solve.

15. $2(2y + 3) = 14$

16. $5(3x - 2) = 35$

Answers on page A-2

Solve.

17. $3(7 + 2x) = 30 + 7(x - 1)$

18. $4(3 + 5x) - 4 = 3 + 2(x - 2)$

Answers on page A-2

Here is a procedure for solving the types of equation discussed in this section.

AN EQUATION-SOLVING PROCEDURE

1. Multiply on both sides to clear the equation of fractions or decimals. (This is optional, but it can ease computations.)
2. If parentheses occur, multiply to remove them using the *distributive laws.*
3. Collect like terms on each side, if necessary.
4. Get all terms with variables on one side and all constant terms on the other side, using the *addition principle.*
5. Collect like terms again, if necessary.
6. Multiply or divide to solve for the variable, using the *multiplication principle.*
7. Check all possible solutions in the original equation.

EXAMPLE 11 Solve: $2 - 5(x + 5) = 3(x - 2) - 1$.

$$2 - 5(x + 5) = 3(x - 2) - 1$$

$$2 - 5x - 25 = 3x - 6 - 1 \quad \text{Using the distributive laws to multiply and remove parentheses}$$

$$-5x - 23 = 3x - 7 \quad \text{Collecting like terms}$$

$$-5x - 23 + 5x = 3x - 7 + 5x \quad \text{Adding } 5x$$

$$-23 = 8x - 7 \quad \text{Collecting like terms}$$

$$-23 + 7 = 8x - 7 + 7 \quad \text{Adding 7}$$

$$-16 = 8x \quad \text{Collecting like terms}$$

$$\frac{-16}{8} = \frac{8x}{8} \quad \text{Dividing by 8}$$

$$-2 = x$$

CHECK:

$2 - 5(x + 5)$	$= 3(x - 2) - 1$	
$2 - 5(-2 + 5)$	$3(-2 - 2) - 1$	
$2 - 5(3)$	$3(-4) - 1$	
$2 - 15$	$-12 - 1$	
-13	-13	TRUE

The solution is -2.

Note that the solution of $-2 = x$ is -2, which is also the solution of $x = -2$.

Do Exercises 17 and 18.

NAME SECTION DATE

Exercise Set 2.3

a Solve. Don't forget to check!

1. $5x + 6 = 31$

2. $7x + 6 = 13$

3. $8x + 4 = 68$

4. $4y + 10 = 46$

5. $4x - 6 = 34$

6. $5y - 2 = 53$

7. $3x - 9 = 33$

8. $4x - 19 = 5$

9. $7x + 2 = -54$

10. $5x + 4 = -41$

11. $-45 = 3 + 6y$

12. $-91 = 9t + 8$

13. $-4x + 7 = 35$

14. $-5x - 7 = 108$

15. $-7x - 24 = -129$

16. $-6z - 18 = -132$

b Solve.

17. $5x + 7x = 72$

18. $8x + 3x = 55$

19. $8x + 7x = 60$

20. $8x + 5x = 104$

21. $4x + 3x = 42$

22. $7x + 18x = 125$

23. $-6y - 3y = 27$

24. $-5y - 7y = 144$

25. $-7y - 8y = -15$

Answers

1. ______
2. ______
3. ______
4. ______
5. ______
6. ______
7. ______
8. ______
9. ______
10. ______
11. ______
12. ______
13. ______
14. ______
15. ______
16. ______
17. ______
18. ______
19. ______
20. ______
21. ______
22. ______
23. ______
24. ______
25. ______

ANSWERS

26. ______
27. ______
28. ______
29. ______
30. ______
31. ______
32. ______
33. ______
34. ______
35. ______
36. ______
37. ______
38. ______
39. ______
40. ______
41. ______
42. ______
43. ______
44. ______
45. ______
46. ______
47. ______
48. ______
49. ______
50. ______

26. $-10y - 3y = -39$

27. $10.2y - 7.3y = -58$

28. $6.8y - 2.4y = -88$

29. $x + \frac{1}{3}x = 8$

30. $x + \frac{1}{4}x = 10$

31. $8y - 35 = 3y$

32. $4x - 6 = 6x$

33. $8x - 1 = 23 - 4x$

34. $5y - 2 = 28 - y$

35. $2x - 1 = 4 + x$

36. $4 - 3x = 6 - 7x$

37. $6x + 3 = 2x + 11$

38. $14 - 6a = -2a + 3$

39. $5 - 2x = 3x - 7x + 25$

40. $-7z + 2z - 3z - 7 = 17$

41. $4 + 3x - 6 = 3x + 2 - x$

42. $5 + 4x - 7 = 4x - 2 - x$

43. $4y - 4 + y + 24 = 6y + 20 - 4y$

44. $5y - 7 + y = 7y + 21 - 5y$

Solve. Clear fractions or decimals first.

45. $\frac{7}{2}x + \frac{1}{2}x = 3x + \frac{3}{2} + \frac{5}{2}x$

46. $\frac{7}{8}x - \frac{1}{4} + \frac{3}{4}x = \frac{1}{16} + x$

47. $\frac{2}{3} + \frac{1}{4}t = \frac{1}{3}$

48. $-\frac{3}{2} + x = -\frac{5}{6} - \frac{4}{3}$

49. $\frac{2}{3} + 3y = 5y - \frac{2}{15}$

50. $\frac{1}{2} + 4m = 3m - \frac{5}{2}$

51. $\frac{5}{3} + \frac{2}{3}x = \frac{25}{12} + \frac{5}{4}x + \frac{3}{4}$

52. $1 - \frac{2}{3}y = \frac{9}{5} - \frac{y}{5} + \frac{3}{5}$

53. $2.1x + 45.2 = 3.2 - 8.4x$

54. $0.96y - 0.79 = 0.21y + 0.46$

55. $1.03 - 0.62x = 0.71 - 0.22x$

56. $1.7t + 8 - 1.62t = 0.4t - 0.32 + 8$

57. $\frac{2}{7}x - \frac{1}{2}x = \frac{3}{4}x + 1$

58. $\frac{5}{16}y + \frac{3}{8}y = 2 + \frac{1}{4}y$

c Solve.

59. $3(2y - 3) = 27$

60. $8(3x + 2) = 30$

61. $40 = 5(3x + 2)$

62. $9 = 3(5x - 2)$

63. $2(3 + 4m) - 9 = 45$

64. $5x + 5(4x - 1) = 20$

65. $5r - (2r + 8) = 16$

66. $6b - (3b + 8) = 16$

67. $6 - 2(3x - 1) = 2$

68. $10 - 3(2x - 1) = 1$

69. $5(d + 4) = 7(d - 2)$

70. $3(t - 2) = 9(t + 2)$

71. $8(2t + 1) = 4(7t + 7)$

72. $7(5x - 2) = 6(6x - 1)$

ANSWERS

51. ______

52. ______

53. ______

54. ______

55. ______

56. ______

57. ______

58. ______

59. ______

60. ______

61. ______

62. ______

63. ______

64. ______

65. ______

66. ______

67. ______

68. ______

69. ______

70. ______

71. ______

72. ______

Answers

73. ____
74. ____
75. ____
76. ____
77. ____
78. ____
79. ____
80. ____
81. ____
82. ____
83. ____
84. ____
85. ____
86. ____
87. ____
88. ____
89. ____
90. ____
91. ____
92. ____
93. ____
94. ____
95. ____
96. ____
97. ____
98. ____
99. ____
100. ____
101. ____

73. $3(r - 6) + 2 = 4(r + 2) - 21$

74. $5(t + 3) + 9 = 3(t - 2) + 6$

75. $19 - (2x + 3) = 2(x + 3) + x$

76. $13 - (2c + 2) = 2(c + 2) + 3c$

77. $2[4 - 2(3 - x)] - 1 = 4[2(4x - 3) + 7] - 25$

78. $5[3(7 - t) - 4(8 + 2t)] - 20 = -6[2(6 + 3t) - 4]$

79. $0.7(3x + 6) = 1.1 - (x + 2)$

80. $0.9(2x + 8) = 20 - (x + 5)$

81. $a + (a - 3) = (a + 2) - (a + 1)$

82. $0.8 - 4(b - 1) = 0.2 + 3(4 - b)$

SKILL MAINTENANCE

83. Divide: $-22.1 \div 3.4$.

84. Factor: $7x - 21 - 14y$.

85. Use $<$ or $>$ for ▒ to write a true sentence:

-15 ▒ -13.

86. Find $-(-x)$ when $x = -14$.

87. Add: $-22.1 + 3.4$.

88. Subtract: $-22.1 - 3.4$.

SYNTHESIS

Solve.

89. 🖩 $0.008 + 9.62x - 42.8 = 0.944x + 0.0083 - x$

90. $\frac{y - 2}{3} = \frac{2 - y}{5}$

91. $0 = y - (-14) - (-3y)$

92. $3x = 4x$

93. $\frac{5 + 2y}{3} = \frac{25}{12} + \frac{5y + 3}{4}$

94. 🖩 $0.05y - 1.82 = 0.708y - 0.504$

95. $-2y + 5y = 6y$

96. $\frac{1}{4}(8y + 4) - 17 = -\frac{1}{2}(4y - 8)$

97. $\frac{1}{3}(6x + 24) - 20 = -\frac{1}{4}(12x - 72)$

98. $\frac{2}{3}\left(\frac{7}{8} - 4x\right) - \frac{5}{8} = \frac{3}{8}$

99. $\frac{3}{4}\left(3x - \frac{1}{2}\right) - \frac{2}{3} = \frac{1}{3}$

100. $\frac{4 - 3x}{7} = \frac{2 + 5x}{49} - \frac{x}{14}$

101. Solve the equation $4x - 8 = 32$ by first using the addition principle. Then solve it by first using the multiplication principle.

2.4 *Solving Problems*

a FIVE STEPS FOR SOLVING PROBLEMS

We have studied many new equation-solving tools in this chapter. We now apply them to problem solving. The following five-step strategy can be very helpful in solving problems.

FIVE STEPS FOR PROBLEM SOLVING IN ALGEBRA

1. *Familiarize* yourself with the problem situation.
2. *Translate* the problem to an equation.
3. *Solve* the equation.
4. *Check* the answer in the original problem.
5. *State* the answer to the problem clearly.

Of the five steps, the most important is probably the first one: becoming familiar with the problem situation. Here are some hints for familiarization.

To familiarize yourself with the problem situation:

1. If a problem is given in words, read it carefully.
2. Reread the problem, perhaps aloud. Try to verbalize the problem to yourself.
3. List the information given and the questions to be answered. Choose a variable (or variables) to represent the unknown and clearly state what the variable represents. Be descriptive! For example, let L = length, d = distance, and so on.
4. Make a drawing and label it with known information. Also, indicate unknown information, using specific units if given.
5. Find further information. Look up a formula on the inside back cover of this book or in a reference book. Talk to a reference librarian or an expert in the field.
6. Make a table of the given information and the information you have collected. Look for patterns that may help in the translation to an equation.
7. Guess or estimate the answer.

EXAMPLE 1 A 12-in. submarine sandwich is cut into two pieces. One piece is twice as long as the other. How long are the pieces?

1. ***Familiarize.*** We first draw a picture. We let

x = the length of the shorter piece.

Then $2x$ = the length of the longer piece.

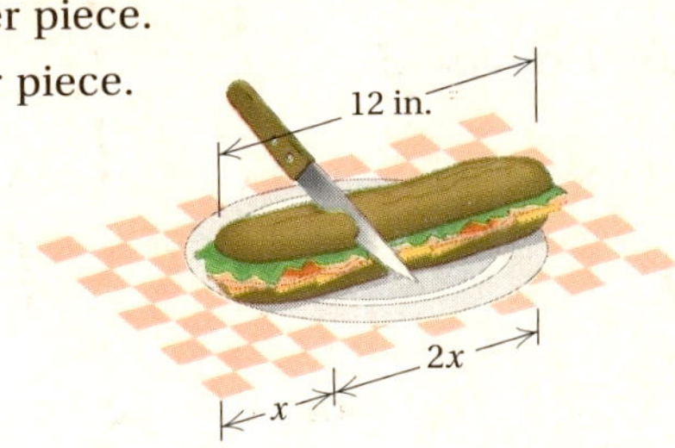

(We can also let y = the length of the longer piece. Then $\frac{1}{2}y$ = the length of the shorter piece. This introduces fractions, however, and will make the solution somewhat more difficult.)

OBJECTIVE

After finishing Section 2.4, you should be able to:

a Solve problems by translating to equations.

FOR EXTRA HELP

TAPE 6 | TAPE 5A | MAC: 2A IBM: 2A

4. Rule rents a more expensive color copier at a weekly rate of \$69.95 plus 8 cents per copy. How many copies can the business make on the same budget of \$300?

Answer on page A-2

EXAMPLE 4 Rule Office Supply rents a copy machine at a weekly rate of \$59.95 plus 4 cents a copy. A business decides that it needs an additional machine for one week to produce a special document, but its budget for a rental machine is \$300. How many copies can the business make on the \$300 budget?

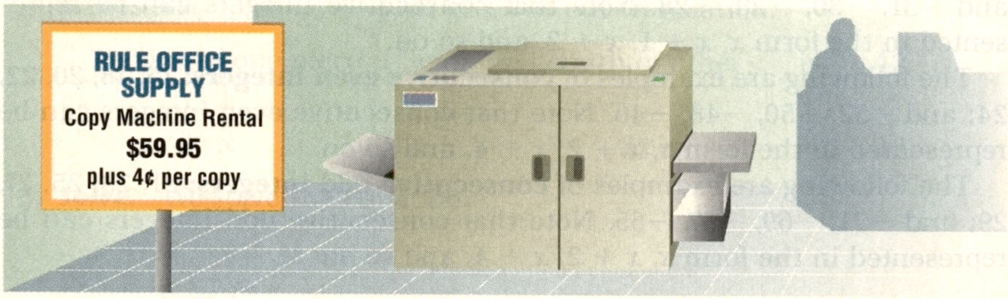

1. ***Familiarize.*** Suppose that the business makes 2000 copies. Then the cost is

Weekly charge plus Copy charges

or

$$\underbrace{\$59.95}_{\$59.95} \;\underbrace{\text{plus}}_{+}\; \underbrace{\text{Cost per copy}}_{\$0.04} \;\underbrace{\text{times}}_{\cdot}\; \underbrace{\text{Number of copies}}_{2000,}$$

which is \$59.95 + \$80.00, or \$139.95. This familiarizes us with the way in which a calculation is made. Note that we convert 4 cents to \$0.04 so that all information is in the same unit, dollars. Otherwise, we will not get the correct answer.

We let c = the number of copies that can be made for \$300.

2. ***Translate.*** We reword the problem and translate as follows.

$$\underbrace{\text{Weekly rate}}_{\$59.95} \;\underbrace{\text{plus}}_{+}\; \underbrace{\text{Cost per copy}}_{\$0.04} \;\underbrace{\text{times}}_{\cdot}\; \underbrace{\text{Number of copies}}_{c} \;\underbrace{\text{is}}_{=}\; \underbrace{\text{Cost}}_{\$300}$$

3. ***Solve.*** We solve the equation:

$$59.95 + 0.04c = 300$$

$$100(59.95 + 0.04c) = 100(300)$$ Multiplying by 100 on both sides to clear the decimals

$$100(59.95) + 100(0.04c) = 30{,}000$$ Using a distributive law

$$5995 + 4c = 30{,}000$$ Simplifying

$$5995 + 4c - 5995 = 30{,}000 - 5995$$ Subtracting 5995

$$4c = 24{,}005$$

$$\frac{4c}{4} = \frac{24{,}005}{4}$$ Dividing by 4

$$c \approx 6001.$$ Rounding to the nearest one. "≈" means "is approximately equal to."

4. ***Check.*** We check in the original problem. We first multiply 6001 by \$0.04, obtaining \$240.04. Then we add \$59.95 to \$240.04 and get \$299.99, which is just about the \$300 allotted.

5. ***State.*** The business can make about 6001 copies on the copy rental allotment of \$300.

Do Exercise 4.

EXAMPLE 5 The state of Wyoming is roughly in the shape of a rectangle whose perimeter is 1280 mi. The length is 90 mi more than the width. Find the dimensions.

1. ***Familiarize.*** We first draw a picture. We let

$$w = \text{the width of the rectangle.}$$

Then $w + 90 =$ the length.

(We can also let

$$l = \text{the length}$$

and $\quad l - 90 =$ the width.)

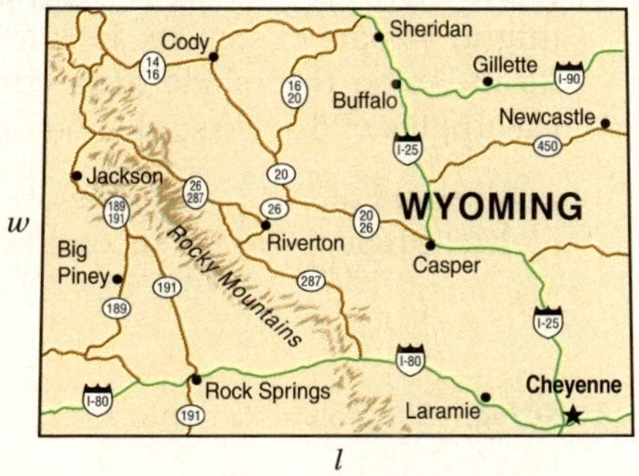

The perimeter P of a rectangle is the distance around it and is given by the formula $2l + 2w = P$, where $l =$ the length and $w =$ the width.

2. ***Translate.*** To translate the problem, we substitute $w + 90$ for l and 1280 for P:

$$2l + 2w = P$$
$$2(w + 90) + 2w = 1280.$$

3. ***Solve.*** We solve the equation:

$$2(w + 90) + 2w = 1280$$
$$2w + 180 + 2w = 1280 \qquad \text{Using a distributive law}$$
$$4w + 180 = 1280 \qquad \text{Collecting like terms}$$
$$4w + 180 - 180 = 1280 - 180 \qquad \text{Subtracting 180}$$
$$4w = 1100$$
$$\frac{4w}{4} = \frac{1100}{4} \qquad \text{Dividing by 4}$$
$$w = 275.$$

Possible dimensions are $w = 275$ mi and $w + 90 = 365$ mi.

4. ***Check.*** If the width is 275 mi and the length is 90 mi + 275 mi, or 365 mi, then the perimeter is 2(365 mi) + 2(275 mi), or 1280 mi. This checks.

5. ***State.*** The width is 275 mi, and the length is 365 mi.

Do Exercise 5.

EXAMPLE 6 The second angle of a triangle is twice as large as the first. The measure of the third angle is 20° greater than that of the first angle. How large are the angles?

1. ***Familiarize.*** We draw a picture. We let

$$\text{measure of first angle} = x.$$

Then $\quad$ measure of second angle $= 2x$

and $\quad$ measure of third angle $= x + 20$.

5. A standard-sized rug has a perimeter of 42 ft. The length is 3 ft more than the width. Find the dimensions of the rug.

Answer on page A-2

6. The second angle of a triangle is three times as large as the first. The third angle measures 30° more than the first angle. Find the measures of the angles.

Answer on page A-2

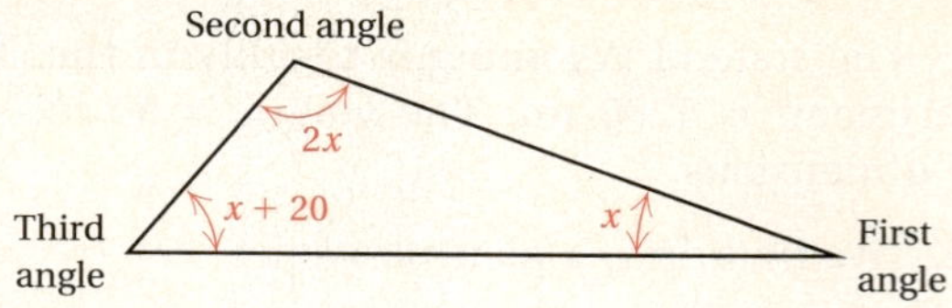

2. ***Translate.*** To translate, we need to recall a geometric fact. (You might, as part of step (1), look it up in a geometry book or in the list of formulas on the inside back cover.) Remember that the measures of the angles of any triangle add up to 180°.

Measure of first angle	+	Measure of second angle	+	Measure of third angle	=	180°
↓	↓	↓	↓	↓	↓	↓
x	$+$	$2x$	$+$	$(x + 20)$	$=$	180

3. ***Solve.*** We solve the equation:

$$
\begin{aligned}
x + 2x + (x + 20) &= 180 \\
4x + 20 &= 180 \\
4x + 20 - 20 &= 180 - 20 \\
4x &= 160 \\
\frac{4x}{4} &= \frac{160}{4} \\
x &= 40.
\end{aligned}
$$

Possible measures for the angles are as follows:

First angle: $x = 40°$;
Second angle: $2x = 2(40) = 80°$;
Third angle: $x + 20 = 40 + 20 = 60°$.

4. ***Check.*** Consider 40°, 80°, and 60°. The second is twice the first, and the third is 20° greater than the first. The sum is 180°. These numbers check.

5. ***State.*** The measures of the angles are 40°, 80°, and 60°.

Caution! Units are important in answers. Try to remember to include them, where appropriate.

Do Exercise 6.

We close this section with some other tips to aid you in problem solving.

PROBLEM-SOLVING TIPS

1. To be good at problem solving, work lots of problems.
2. Look for patterns when solving problems. Each time you study an example in a text, you may observe a pattern for problems that you will see later in the exercise sets or some other practical situation.
3. When translating to an equation, or some other mathematical language, consider the dimensions of the variables and constants in the equation. The variables that represent length should all be in the same unit, those that represent money should all be in dollars or all in cents, and so on.

NAME SECTION DATE

Exercise Set 2.4

a Solve.

1. The area of Lake Superior is about four times the area of Lake Ontario. The area of Lake Superior is 30,172 mi^2 (square miles). What is the area of Lake Ontario?

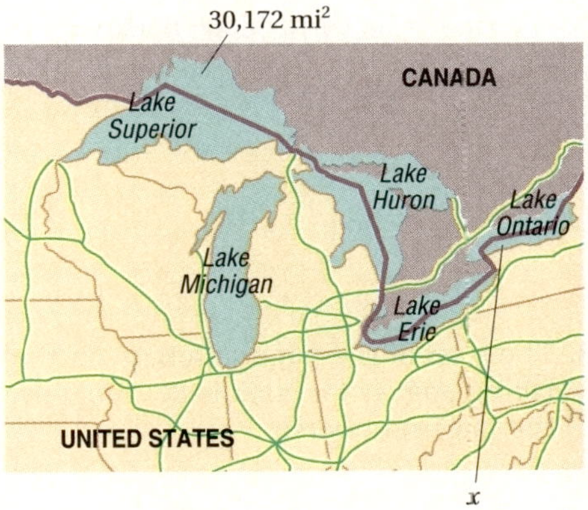

2. The height of the Eiffel Tower is 974 ft, which is about 669 ft higher than the Statue of Liberty. What is the height of the Statue of Liberty?

3. When 268 is added to a number, the result is 749. Find the number.

4. When 28 is multiplied by a number, the result is 1904. Find the number.

5. A free-lance editor charges $35 an hour. How many hours did she work to make $23,362.50?

6. A game board has 64 squares. If you win 39 squares and your opponent wins the rest, how many does your opponent get?

7. In a recent year, the cost of four 12-oz boxes of Frosted Mini-Wheats® was $13.20. How much did one box cost?

8. The total amount spent on women's blouses in a recent year was $6.5 billion. This was $0.2 billion more than was spent on women's dresses. How much was spent on women's dresses?

9. When 18 is subtracted from six times a certain number, the result is 96. What is the number?

10. When 28 is subtracted from five times a certain number, the result is 232. What is the number?

11. If you double a number and then add 85, you get three-fourths of the original number. What is the original number?

12. If you double a number and then add 16, you get two-fifths of the original number. What is the original number?

ANSWERS

1. ______
2. ______
3. ______
4. ______
5. ______
6. ______
7. ______
8. ______
9. ______
10. ______
11. ______
12. ______

ANSWERS

13. ________

14. ________

15. ________

16. ________

17. ________

18. ________

19. ________

20. ________

21. ________

22. ________

23. ________

24. ________

13. At breakfast, Francis drank 1 cup of low-fat milk and used $\frac{2}{5}$ cup on his cereal. Altogether, the milk provided him with 168 calories. How many calories are in a cup of low-fat milk?

14. On Monday, Karen filled the gas tank when it was $\frac{3}{4}$ empty. On Friday, she filled it when it was $\frac{1}{2}$ empty. Her total purchase of gasoline on those two days was $18\frac{3}{4}$ gal. How many gallons of gasoline does the tank hold?

15. A 480-m wire is cut into three pieces. The second piece is three times as long as the first. The third piece is four times as long as the second. How long is each piece?

16. A 180-m rope is cut into three pieces. The second piece is twice as long as the first. The third piece is three times as long as the second. How long is each piece of rope?

17. The sum of the page numbers on the facing pages of a book is 73. What are the page numbers?

18. The sum of the page numbers on the facing pages of a book is 81. What are the page numbers?

19. The sum of two consecutive even integers is 114. What are the integers?

20. The sum of two consecutive even integers is 106. What are the integers?

21. The ages of Edna, Ellie, and Elsa are consecutive integers. The sum of their ages is 126. What are their ages?

22. The ages of Wesley, Whitney, and Wiley are consecutive integers. The sum of their ages is 108. What are their ages?

23. The sum of three consecutive odd integers is 189. What are the integers?

24. Three consecutive integers are such that the first plus one-half the second plus three less than twice the third is 964. Find the integers.

25. The top of the John Hancock Building in Chicago is a rectangle whose length is 60 ft more than the width. The perimeter is 520 ft. Find the width and the length of the rectangle. Find the area of the rectangle.

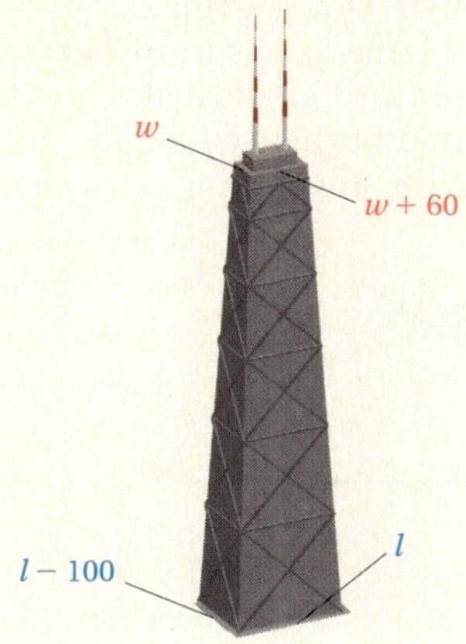

26. The ground floor of the John Hancock Building is a rectangle whose width is 100 ft less than the length. The perimeter is 860 ft. Find the width and the length of the rectangle. Find the area of the rectangle.

27. The perimeter of a cross section of a two-by-four piece of lumber is $10\frac{1}{2}$ in. The length is twice the width. Find the actual dimensions of the cross section of a two-by-four.

28. The state of Colorado is roughly in the shape of a rectangle with a perimeter of 1300 mi. The width is 110 mi less than the length. Find the width and the length.

29. The second angle of a triangle is four times as large as the first. The third angle is 45° less than the sum of the other two angles. Find the measure of the first angle.

30. The second angle of a triangle is three times as large as the first. The third angle is 25° less than the sum of the other two angles. Find the measure of the first angle.

31. Elite Rent-A-Car rents a compact-sized car at a daily rate of $34.95 plus 10 cents per mile. A businessperson is allotted $80 per day for car rental. How many miles can he travel on the $80 budget?

32. Elite Rent-A-Car rents intermediate cars at $43.95 plus 10 cents per mile. A businessperson has a car rental allotment of $90 per day. How many miles can she travel on the $90 budget?

ANSWERS

25. __________

26. __________

27. __________

28. __________

29. __________

30. __________

31. __________

32. __________

ANSWERS

33. ______

34. ______

35. a) ______

b) ______

36. ______

37. ______

38. ______

39. ______

40. ______

41. ______

42. ______

43. ______

44. ______

33. The second angle of a triangle is three times as large as the first. The measure of the third angle is 40° greater than that of the first angle. How large are the angles?

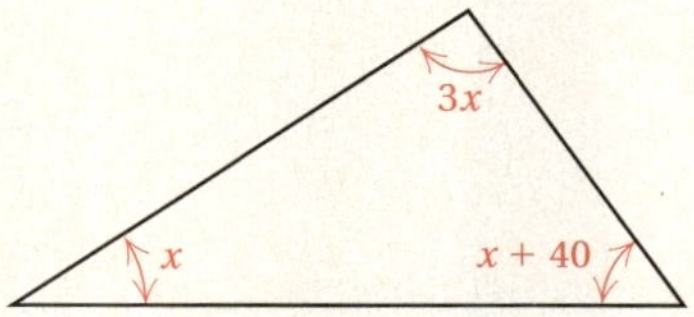

34. The second angle of a triangular parking lot is twice as large as the first. The third angle is 42° less than the sum of the other two angles. Find the measure of each angle.

35. The equation

$$P = 0.3444Y - 677$$

can be used to determine the average price P of a ticket to a major league baseball game in the year Y.

a) Find the average price of a ticket in 1996.

b) In what year will the price of a ticket be $11.80?

36. The equation

$$R = -0.028t + 20.8$$

can be used to predict the world record in the 200-m dash, where R stands for the record in seconds and t for the number of years since 1920. In what year will the record be 18.0 sec?

SYNTHESIS

37. Apples are collected in a basket for six people. One-third, one-fourth, one-eighth, and one-fifth are given to four people, respectively. The fifth person gets ten apples with one apple remaining for the sixth person. Find the original number of apples in the basket.

38. If the weekly rental price for a copy machine is $49.95 plus a certain price per copy and a business must make 5000 copies within a week and stay within a $300 budget, what is the highest price per copy that the business can afford?

39. A student scored 78 on a test that had 4 seven-point fill-ins and 24 three-point multiple-choice questions. The student had one fill-in wrong. How many multiple-choice questions did the student answer correctly?

40. The width of a rectangle is three-fourths of the length. The perimeter of the rectangle becomes 50 cm when the length and the width are each increased by 2 cm. Find the length and the width.

41. Abraham Lincoln's 1863 Gettysburg Address refers to the year 1776 as "Four *score* and seven years ago." Write an equation and find what number a score represents.

42. A student has an average score of 82 on three tests. The student's average score on the first two tests is 85. What was the score on the third test?

43. A storekeeper goes to the bank to get $10 worth of change. The storekeeper requests twice as many quarters as half dollars, twice as many dimes as quarters, three times as many nickels as dimes, and no pennies or dollars. How many of each coin did the storekeeper get?

44. The area of this triangle is 2.9047 in^2. Find x.

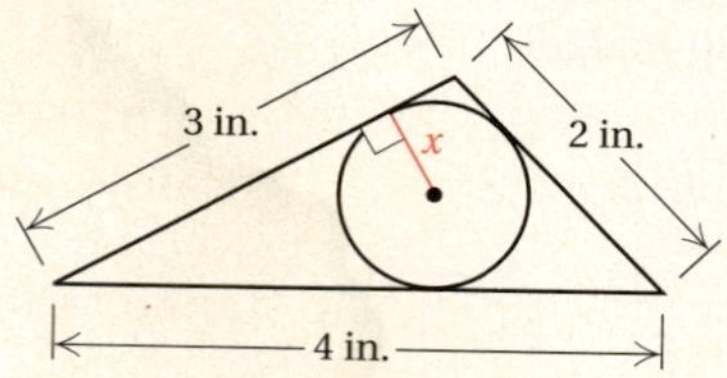

2.5 *Solving Percent Problems*

OBJECTIVE

After finishing Section 2.5, you should be able to:

a Solve problems involving percent.

FOR EXTRA HELP

TAPE 6 | TAPE 5A | MAC: 2A IBM: 2A

a Many problems involve percents. We can use our knowledge of equations and the problem-solving process to solve such problems.

EXAMPLE 1 What percent of 45 is 15?

1. ***Familiarize.*** This type of problem is stated so explicitly that we can proceed directly to the translation. We first let $x =$ the percent.

2. ***Translate.*** We translate as follows:

$$\underbrace{\text{What percent}}_{x\%} \quad \underset{\cdot}{\text{of}} \quad \underset{45}{45} \quad \underset{=}{\text{is}} \quad \underset{15.}{15?}$$

3. ***Solve.*** We solve the equation:

$$x\% \cdot 45 = 15$$
$$x \times 0.01 \times 45 = 15$$
$$x(0.45) = 15$$
$$\frac{x(0.45)}{0.45} = \frac{15}{0.45} \qquad \text{Dividing by 0.45}$$
$$x = \frac{15}{0.45} \times \frac{100}{100} = \frac{1500}{45}$$
$$= 33\frac{1}{3}.$$

4. ***Check.*** We check by finding $33\frac{1}{3}\%$ of 45:

$$33\frac{1}{3}\% \cdot 45 = \frac{1}{3} \cdot 45 = 15. \qquad \text{See Table 1 at the back of the book.}$$

5. ***State.*** The answer is $33\frac{1}{3}\%$.

Do Exercises 1 and 2.

Solve.

1. What percent of 50 is 16?

2. 15 is what percent of 60?

EXAMPLE 2 3 is 16% of what number?

1. ***Familiarize.*** This problem is stated so explicitly that we can proceed directly to the translation. We let $y =$ the number that we are taking 16% of.

2. ***Translate.*** The translation is as follows:

$$\underset{3}{3} \quad \underset{=}{\text{is}} \quad \underset{16\%}{16\%} \quad \underset{\cdot}{\text{of}} \quad \underset{y}{\text{what?}}$$

3. ***Solve.*** We solve the equation:

$$3 = 16\% \cdot y$$
$$3 = 0.16y \qquad \text{Converting to decimal notation}$$
$$0.16y = 3 \qquad \text{Reversing the equation}$$
$$\frac{0.16y}{0.16} = \frac{3}{0.16} \qquad \text{Dividing by 0.16}$$
$$y = 18.75.$$

Answers on page A-2

Solve.

3. 45 is 20 percent of what number?

4. 120 percent of what number is 60?

Solve.

5. What is 23% of 48?

6. The area of Arizona is 19% of the area of Alaska. The area of Alaska is 586,400 mi^2. What is the area of Arizona?

Answers on page A-2

4. *Check.* We check by finding 16% of 18.75:

$$16\% \times 18.75 = 0.16 \times 18.75 = 3.$$

5. *State.* The answer is 18.75.

Do Exercises 3 and 4.

Perhaps you have noticed that to handle percents in problems such as those in Examples 1 and 2, you can convert to decimal notation before continuing.

EXAMPLE 3 Blood is 90% water. The average adult has 5 quarts (qt) of blood. How much water is in the average adult's blood?

1. *Familiarize.* We first write down the given information.

Blood: 90% water

Adult: Body contains 5 qt of blood.

Water Content of Blood

We want to find the amount of water that is in the blood of an adult. We let $x =$ the amount of water in the blood of an adult. It seems reasonable that we take 90% of 5. This leads us to the rewording and translating of the problem.

2. *Translate.*

90%	of	5	is	what?	Rewording
↓	↓	↓	↓	↓	
90%	$\cdot$	5	$=$	x	Translating

3. *Solve.* We solve the equation:

$$90\% \cdot 5 = x$$

$$0.90 \times 5 = x \quad \text{Converting 90\% to decimal notation}$$

$$4.5 = x.$$

4. *Check.* The check is actually the computation we use to solve the equation:

$$90\% \cdot 5 = 0.90 \times 5 = 4.5.$$

5. *State.* There is 4.5 qt of water in an adult who has 5 qt of blood.

Do Exercises 5 and 6.

EXAMPLE 4 An investment is made at 6% simple interest for 1 year. It grows to \$768.50. How much was originally invested (the principal)?

1. ***Familiarize.*** Suppose that \$100 was invested. Recalling the formula for simple interest, $I = Prt$, we know that the interest for 1 year on \$100 at 6% simple interest is given by $I = \$100 \cdot 6\% \cdot 1 = \$6$. Then, at the end of the year, the amount in the account is found by adding the principal and the interest:

$$\begin{array}{ccccc} \text{Principal} & + & \text{Interest} & = & \text{Amount} \\ \downarrow & & \downarrow & & \downarrow \\ \$100 & + & \$6 & = & \$106. \end{array}$$

   In this problem, we are working backward. We are trying to find the principal, which is the original investment. We let $x =$ the principal.

2. ***Translate.*** We reword the problem and then translate.

$$\begin{array}{ccccc} \text{Principal} & + & \text{Interest} & = & \text{Amount} \\ \downarrow & & \downarrow & & \downarrow \\ x & + & 6\%x & = & 768.50 \end{array}$$

   Interest is 6% of the principal.

3. ***Solve.*** We solve the equation:

$$x + 6\%x = 768.50$$

$$x + 0.06x = 768.50 \quad \text{Converting to decimal notation}$$

$$1x + 0.06x = 768.50 \quad \text{Identity property of 1}$$

$$1.06x = 768.50 \quad \text{Collecting like terms}$$

$$\frac{1.06x}{1.06} = \frac{768.50}{1.06} \quad \text{Dividing by 1.06}$$

$$x = 725.$$

4. ***Check.*** We check by taking 6% of \$725 and adding it to \$725:

$$6\% \times \$725 = 0.06 \times 725 = \$43.50.$$

   Then \$725 + \$43.50 = \$768.50, so \$725 checks.

5. ***State.*** The original investment was \$725.

Do Exercise 7.

EXAMPLE 5 The price of a speedboat was decreased to a sale price of \$20,839. This was a 9% reduction. What was the former price?

7. An investment is made at 7% simple interest for 1 year. It grows to \$8988. How much was originally invested (the principal)?

Answer on page A-2

8. The price of a suit was decreased to a sale price of \$526.40. This was a 20% reduction. What was the former price?

Answer on page A-2

1. ***Familiarize.*** Suppose that the former price was \$24,000. A 9% reduction can be found by taking 9% of \$24,000—that is,

$$9\% \text{ of } \$24{,}000 = 0.09(\$24{,}000) = \$2160.$$

Then the sale price is found by subtracting the amount of the reduction:

Former price	$-$	Reduction	$=$	Sale price
\$24,000	$-$	\$2160	$=$	\$21,840.

Our guess of \$24,000 was too high, but we are becoming familiar with the problem. We let $x =$ the former price of the speedboat. It is reduced by 9%. So the sale price $= x - 9\%x$.

2. ***Translate.*** We reword and then translate:

Former price	$-$	Reduction	$=$	Sale price	Rewording
x	$-$	$9\%x$	$=$	\$20,839.	Translating

3. ***Solve.*** We solve the equation:

$$x - 9\%x = 20{,}839$$

$$x - 0.09x = 20{,}839 \quad \text{Converting to decimal notation}$$

$$1x - 0.09x = 20{,}839$$

$$0.91x = 20{,}839 \quad \text{Collecting like terms}$$

$$\frac{0.91x}{0.91} = \frac{20{,}839}{0.91} \quad \text{Dividing by 0.91}$$

$$x = 22{,}900.$$

4. ***Check.*** To check, we find 9% of \$22,900 and subtract:

$$9\% \times \$22{,}900 = 0.09 \times \$22{,}900 = \$2061;$$

$$\$22{,}900 - \$2061 = \$20{,}839.$$

Since we get the sale price, \$20,839, the answer \$22,900 checks.

5. ***State.*** The former price was \$22,900.

This problem is easy with algebra. Without algebra it is not. A common error in a problem like this is to take 9% of the sale price and subtract or add. Note that 9% of the original price ($9\% \cdot \$22{,}900 = \$2061$) is not equal to 9% of the sale price ($9\% \cdot \$20{,}839 = \$1875.51$)!

Do Exercise 8.

NAME SECTION DATE

Exercise Set 2.5

a Solve.

1. What percent of 180 is 36?

2. What percent of 76 is 19?

3. What percent of 125 is 30?

4. What percent of 300 is 57?

5. 45 is 30% of what number?

6. 20.4 is 24% of what number?

7. 0.3 is 12% of what number?

8. 7 is 175% of what number?

9. What number is 65% of 840?

10. What number is 1% of one million?

11. What percent of 80 is 100?

12. What percent of 10 is 205?

13. What is 2% of 40?

14. What is 40% of 2?

15. 2 is what percent of 40?

16. 40 is 2% of what number?

17. The premium on a $100,000 life insurance policy for a female nonsmoker, age 22, is about $166 per year. The premium for a smoker is 170% of the premium for a nonsmoker. What is the premium for a smoker?

18. The U.S. Postal Service reports that we open and read 78% of the junk mail that we receive. A business sends out 9500 advertising brochures. How many of them can it expect to be opened and read?

19. In a medical study, it was determined that if 800 people kiss someone else who has a cold, only 56 will actually catch the cold. What percent is this?

20. It has been determined by sociologists that 17% of the population is left-handed. Each week 160 men enter a tournament conducted by the Professional Bowlers Association. How many of them would you expect to be left-handed? Round to the nearest one.

21. On a test of 88 items, a student got 76 correct. What percent were correct?

22. One season, a basketball player made 36 out of 75 three-point shots. What percent did he make?

Answers

1. ______
2. ______
3. ______
4. ______
5. ______
6. ______
7. ______
8. ______
9. ______
10. ______
11. ______
12. ______
13. ______
14. ______
15. ______
16. ______
17. ______
18. ______
19. ______
20. ______
21. ______
22. ______

ANSWERS

23. ____________

24. ____________

25. ____________

26. ____________

27. ____________

28. ____________

29. ____________

30. ____________

31. ____________

32. ____________

33. ____________

34. ____________

35. ____________

36. ____________

37. ____________

38. ____________

39. ____________

23. The cost of a Navy F-14 Tomcat jet is \$24 million. Eventually each jet must be renovated at 2% of its original cost. How much will renovation cost?

24. The cost of a Navy A-6 Intruder attack bomber is \$30 million. Eventually each bomber must be renovated at 2% of its original cost. How much will renovation cost?

25. The sales tax rate in New York City is 8%. How much is charged on a purchase of \$428.86? How much is the total cost of the purchase?

26. A family spent \$224 one month for food. This was 28% of its income. What was its monthly income?

27. Water volume increases 9% when water freezes. If 400 cm^3 (cubic centimeters) of water is frozen, how much will its volume increase? What will be the volume of the ice?

28. An investment is made at 4% simple interest for 1 year. It grows to \$8112. How much was originally invested?

29. Due to inflation, the price of an item increased 12¢. This was an 8% increase. What was the old price? the new price?

30. Money is borrowed at 6.2% simple interest. After 1 year, \$6945.48 pays off the loan. How much was originally borrowed?

31. After a 40% price reduction, a shirt is on sale at \$19.20. What was the original price (that is, the price before reduction)?

32. After a 34% price reduction, a blouse is on sale at \$29.04. What was the original price?

SYNTHESIS

33. The weather report is "a 60% chance of showers during the day, 30% tonight, and 5% tomorrow morning." What are the chances that it will not rain during the day? tonight? tomorrow morning?

34. One number is 25% of another. The larger number is 12 more than the smaller. What are the numbers?

35. In one city, a sales tax of 9% was added to the price of gasoline as registered on the pump. Suppose a driver asked for \$10 worth of gas. The attendant filled the tank until the pump read \$9.10 and charged the driver \$10. Something was wrong. Use algebra to correct the error.

36. In a baseball league, the Mustangs won 15 of their first 20 games. How many more games will they have to play where they win only half the time in order to win 60% of the total number of games?

37. Twenty-seven people make a certain amount of money at a sale. What percentage does each receive if they share the profit equally?

38. If x is 160% of y, y is what percent of x?

39. Which of the following is higher, if either?

A. x is increased by 25%; then that amount is decreased by 25%.
B. x is decreased by 25%; then that amount is increased by 25%.
Explain.

2.6 Formulas

a SOLVING FORMULAS

A **formula** is a "recipe" for doing a certain type of calculation. Formulas are often given as equations. Here is an example of a formula that has to do with weather: $M = \frac{1}{5}n$. You see a flash of lightning. After a few seconds you hear the thunder associated with that flash. How far away was the lightning?

Your distance from the storm is M miles. You can find that distance by counting the number of seconds n that it takes the sound of the thunder to reach you and then multiplying by $\frac{1}{5}$.

EXAMPLE 1 Consider the formula $M = \frac{1}{5}n$. It takes 10 sec for the sound of thunder to reach you after you have seen a flash of lightning. How far away is the storm?

We substitute 10 for n and calculate M: $M = \frac{1}{5}n = \frac{1}{5}(10) = 2$. The storm is 2 mi away.

Do Exercise 1.

Suppose that we think we know how far we are from the storm and want to check by calculating the number of seconds it should take the sound of the thunder to reach us. We could substitute a number for M—say, 2—and solve for n:

$$2 = \frac{1}{5}n$$

$$10 = n. \quad \text{Multiplying by 5}$$

However, if we wanted to do this repeatedly, it might be easier to solve for n by getting it alone on one side. We "solve" the formula for n.

EXAMPLE 2 Solve for n: $M = \frac{1}{5}n$.

We have

$$M = \frac{1}{5}n \quad \text{We want this letter alone.}$$

$$5 \cdot M = 5 \cdot \frac{1}{5}n \quad \text{Multiplying on both sides by 5}$$

$$5M = n.$$

In the above situation for $M = 2$, $n = 5(2)$, or 10.

Do Exercise 2.

To see how the addition and multiplication principles apply to formulas, compare the following.

A. Solve.

$$5x + 2 = 12$$
$$5x = 12 - 2$$
$$5x = 10$$
$$x = \frac{10}{5} = 2$$

B. Solve.

$$5x + 2 = 12$$
$$5x = 12 - 2$$
$$x = \frac{12 - 2}{5}$$

C. Solve for x.

$$ax + b = c$$
$$ax = c - b$$
$$x = \frac{c - b}{a}$$

In (A), we solved as we did before. In (B), we did not carry out the calculations. In (C), we could not carry out the calculations because we had unknown numbers.

OBJECTIVE

After finishing Section 2.6, you should be able to:

a Solve a formula for a specified letter.

FOR EXTRA HELP

TAPE 6 | TAPE 5B | MAC: 2A IBM: 2A

1. Suppose that it takes the sound of thunder 14 sec to reach you. How far away is the storm?

2. Solve for I: $E = IR$.

(This is a formula from electricity relating voltage E, current I, and resistance R.)

Answers on page A-2

3. Solve for D: $C = \pi D$.
(This is a formula for the circumference C of a circle of diameter D.)

EXAMPLE 3 Solve for r: $C = 2\pi r$.

This is a formula for the circumference C of a circle of radius r.

$$C = 2\pi r \qquad \text{We want this letter alone.}$$

$$\frac{C}{2\pi} = \frac{2\pi r}{2\pi} \qquad \text{Dividing by } 2\pi$$

$$\frac{C}{2\pi} = r$$

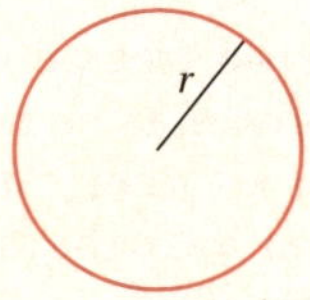

Do Exercise 3.

To solve a formula for a given letter, identify the letter and:

1. Multiply on both sides to clear fractions or decimals, if that is needed.
2. Collect like terms on each side, if necessary.
3. Get all terms with the letter to be solved for on one side of the equation and all other terms on the other side.
4. Collect like terms again, if necessary.
5. Solve for the letter in question.

4. Solve for c:

$$A = \frac{a + b + c + d}{4}.$$

5. Solve for I:

$$E = \frac{9R}{I}.$$

(This is a formula for computing the earned run average E of a pitcher who has given up R earned runs in I innings of pitching.)

EXAMPLE 4 Solve for a: $A = \dfrac{a + b + c}{3}$.

This is a formula for the average A of three numbers a, b, and c.

$$A = \frac{a + b + c}{3} \qquad \text{We want the letter } a \text{ alone.}$$

$$3A = 3 \cdot \frac{a + b + c}{3} \qquad \text{Multiplying by 3 to clear the fraction}$$

$$3A = a + b + c$$

$$3A - b = a + b + c - b \qquad \text{Subtracting } b$$

$$3A - b = a + c$$

$$3A - b - c = a + c - c \qquad \text{Subtracting } c$$

$$3A - b - c = a$$

EXAMPLE 5 Solve for C: $Q = \dfrac{100M}{C}$.

This is a formula used in psychology for intelligence quotient Q, where M is mental age and C is chronological, or actual, age.

$$Q = \frac{100M}{C} \qquad \text{We want the letter } C \text{ alone.}$$

$$CQ = C \cdot \frac{100M}{C} \qquad \text{Multiplying by } C \text{ to clear the fraction}$$

$$CQ = 100M$$

$$\frac{CQ}{Q} = \frac{100M}{Q} \qquad \text{Dividing by } Q$$

$$C = \frac{100M}{Q}$$

Do Exercises 4 and 5.

Answers on page A-2

NAME SECTION DATE

Exercise Set 2.6

a Solve for the given letter.

1. $A = bh$, for b
(Area of a parallelogram with base b and height h)

2. $A = bh$, for h

3. $d = rt$, for r
(A distance formula, where d is distance, r is speed, and t is time)

4. $d = rt$, for t

5. $I = Prt$, for P
(Simple-interest formula, where I is interest, P is principal, r is interest rate, and t is time)

6. $I = Prt$, for t

7. $F = ma$, for a
(A physics formula, where F is force, m is mass, and a is acceleration)

8. $F = ma$, for m

9. $P = 2l + 2w$, for w
(Perimeter of a rectangle of length l and width w)

10. $P = 2l + 2w$, for l

11. $A = \pi r^2$, for r^2
(Area of a circle with radius r)

12. $A = \pi r^2$, for π

13. $A = \frac{1}{2}bh$, for b
(Area of a triangle with base b and height h)

14. $A = \frac{1}{2}bh$, for h

15. $E = mc^2$, for m
(A relativity formula)

16. $E = mc^2$, for c^2

17. $Q = \frac{c + d}{2}$, for d

18. $Q = \frac{p - q}{2}$, for p

19. $A = \frac{a + b + c}{3}$, for b

20. $A = \frac{a + b + c}{3}$, for c

21. $Ax + By = C$, for y

22. $Ax + By = C$, for x

ANSWERS

1. ______
2. ______
3. ______
4. ______
5. ______
6. ______
7. ______
8. ______
9. ______
10. ______
11. ______
12. ______
13. ______
14. ______
15. ______
16. ______
17. ______
18. ______
19. ______
20. ______
21. ______
22. ______

ANSWERS

23. ________
24. ________
25. ________
26. ________
27. ________
28. ________
29. ________
30. ________
31. ________
32. ________
33. ________
34. ________
35. ________
36. ________
37. ________
38. ________
39. ________
40. ________
41. ________
42. ________
43. ________
44. ________

23. $v = \dfrac{3k}{t}$, for t

24. $P = \dfrac{ab}{c}$, for c

25. The formula

$$H = \frac{D^2N}{2.5}$$

is used to find the horsepower H of an N-cylinder engine. Solve for D^2.

26. Solve for N:

$$H = \frac{D^2N}{2.5}.$$

27. The area of a sector of a circle is given by

$$A = \frac{\pi r^2 S}{360},$$

where r is the radius and S is the angle measure of the sector. Solve for S.

28. Solve for r^2:

$$A = \frac{\pi r^2 S}{360}.$$

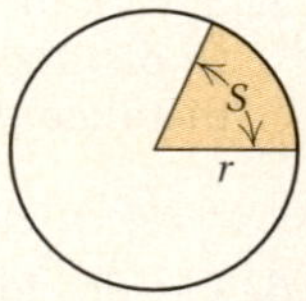

29. The formula

$$R = -0.0075t + 3.85$$

can be used to estimate the world record in the 1500-m run t years after 1930. Solve for t.

30. The formula

$$F = \frac{9}{5}C + 32$$

can be used to convert from Celsius, or Centigrade, temperature C to Fahrenheit temperature F. Solve for C.

SKILL MAINTENANCE

31. Convert to decimal notation: $\dfrac{23}{25}$.

32. Add: $-23 + (-67)$.

33. Subtract: $-45.8 - (-32.6)$.

34. Remove parentheses and simplify:

$$4a - 8b - 5(5a - 4b).$$

35. Add: $-\dfrac{2}{3} + \dfrac{5}{6}$.

36. Subtract: $-\dfrac{2}{3} - \dfrac{5}{6}$.

SYNTHESIS

Solve.

37. $A = \dfrac{1}{2}ah + \dfrac{1}{2}bh$, for b; for h

38. $A = \dfrac{1}{2}ah - \dfrac{1}{2}bh$, for a; for h

39. $Q = 3a + 5ca$, for a

40. $P = 4m + 7mn$, for m

41. In $A = lw$, l and w both double. What is the effect on A?

42. In $P = 2a + 2b$, P doubles. Do a and b necessarily both double?

43. In $A = \frac{1}{2}bh$, b increases by 4 units and h does not change. What happens to A?

44. Solve for F:

$$D = \frac{1}{E + F}.$$

2.7 *Solving Inequalities*

We now extend our equation-solving principles to the solving of inequalities.

a SOLUTIONS OF INEQUALITIES

In Section 1.2, we defined the symbols $>$ (greater than), $<$ (less than), $\geq$ (greater than or equal to), and $\leq$ (less than or equal to). For example, $3 \leq 4$ and $3 \leq 3$ are both true, but $-3 \leq -4$ and $0 \geq 2$ are both false.

An **inequality** is a number sentence with $>$, $<$, $\geq$, or $\leq$ as its verb—for example,

$$-4 > t, \qquad x < 3, \qquad 2x + 5 \geq 0, \quad \text{and} \quad -3y + 7 \leq -8.$$

Some replacements for a variable in an inequality make it true and some make it false.

> A replacement that makes an inequality true is called a *solution.* The set of all solutions is called the *solution set.* When we have found the set of all solutions of an inequality, we say that we have *solved* the inequality.

EXAMPLES Determine whether the number is a solution of $x < 2$.

1. -2 Since $-2 < 2$ is true, -2 is a solution.

2. 2 Since $2 < 2$ is false, 2 is not a solution.

EXAMPLES Determine whether the number is a solution of $y \geq 6$.

3. 6 Since $6 \geq 6$ is true, 6 is a solution.

4. -4 Since $-4 \geq 6$ is false -4 is not a solution.

Do Exercises 1 and 2.

b GRAPHS OF INEQUALITIES

Some solutions of $x < 2$ are 0.45, -8.9, $-\pi$, and so on. In fact, there are infinitely many real numbers that are solutions. Because we cannot list them all individually, it is helpful to make a drawing that represents all the solutions.

A **graph** of an inequality is a drawing that represents its solutions. An inequality in one variable can be graphed on a number line. An inequality in two variables can be graphed on a coordinate plane; we will study such graphs in Chapter 6.

We first graph inequalities in one variable on a number line.

EXAMPLE 5 Graph: $x < 2$.

The solutions of $x < 2$ are all those numbers less than 2. They are shown on the graph by shading all points to the left of 2. The open circle at 2 indicates that 2 is not part of the graph.

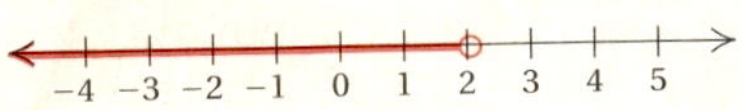

OBJECTIVES

After finishing Section 2.7, you should be able to:

- **a** Determine whether a given number is a solution of an inequality.
- **b** Graph an inequality on a number line.
- **c** Solve inequalities using the addition principle.
- **d** Solve inequalities using the multiplication principle.
- **e** Solve inequalities using the addition and multiplication principles together.

FOR EXTRA HELP

TAPE 7 TAPE 5B MAC: 2B IBM: 2B

Determine whether each number is a solution of the inequality.

1. $x > 3$

a) 2 b) 0

c) -5 d) 15

e) 3

2. $x \leq 6$

a) 6 b) 0

c) -4 d) 25

e) -6

Answers on page A-2

Graph.

3. $x < 4$

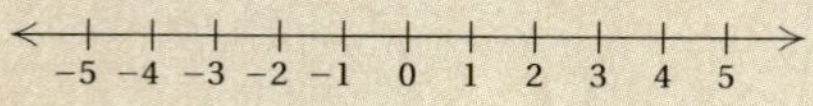

4. $y \geq -2$

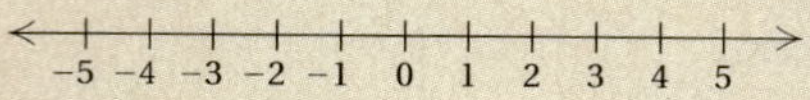

5. $-2 \leq x < 4$

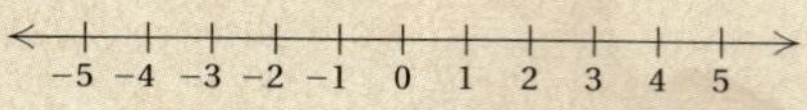

Answers on page A-2

EXAMPLE 6 Graph: $y \geq -3$.

The solutions of $y \geq -3$ are shown on the number line by shading the point for -3 and all points to the right of -3. The closed circle at -3 indicates that -3 *is* part of the graph.

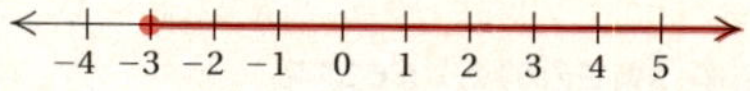

EXAMPLE 7 Graph: $-2 < x \leq 3$.

The inequality $-2 < x \leq 3$ is read "-2 is less than x *and* x is less than or equal to 3," or "x is greater than -2 *and* less than or equal to 3." To be a solution of this inequality, a number must be a solution of both $-2 < x$ and $x \leq 3$. The number 1 is a solution, as are -0.5, 2, 2.5, and 3. The solution is graphed as follows:

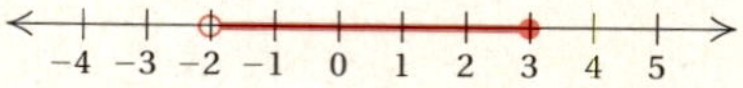

The open circle at -2 means that -2 is not part of the graph. The closed circle at 3 means that 3 is part of the graph. The other solutions are shaded.

Do Exercises 3–5.

c SOLVING INEQUALITIES USING THE ADDITION PRINCIPLE

Consider the true inequality

$$3 < 7.$$

If we add 2 on both sides, we get another true inequality:

$$3 + 2 < 7 + 2, \quad \text{or} \quad 5 < 9.$$

Similarly, if we add -3 on both sides, we get another true inequality:

$$3 + (-3) < 7 + (-3), \quad \text{or} \quad 0 < 4.$$

> **THE ADDITION PRINCIPLE FOR INEQUALITIES**
> If the same number is added (or subtracted) on both sides of a true inequality, we get another true inequality.

Let us see how we use the addition principle to solve inequalities.

EXAMPLE 8 Solve: $x + 2 > 8$. Then graph.

We use the addition principle, subtracting 2 on both sides:

$$x + 2 - 2 > 8 - 2$$
$$x > 6.$$

Using the addition principle, we get an inequality for which we can determine the solutions easily.

Any number greater than 6 makes the last sentence true and is a solution of that sentence. Any such number is also a solution of the original sentence. Thus the inequality is solved. The graph is as follows:

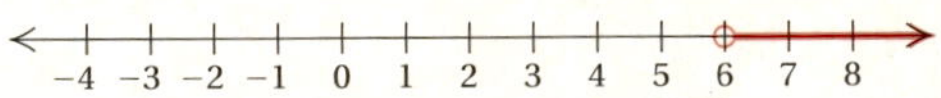

We cannot check all the solutions of an inequality by substitution, as we can check solutions of equations, because there are too many of them. A partial check can be done by substituting a number greater than 6—say, 7—into the original inequality:

$x + 2 > 8$	
$7 + 2$	8
9	TRUE

Since $9 > 8$ is true, 7 is a solution. Any number greater than 6 is a solution.

When two inequalities have the same solutions, we say that they are **equivalent**. Whenever we use the addition principle with inequalities, the first and last sentences will be equivalent.

EXAMPLE 9 Solve: $3x + 1 \leq 2x - 3$. Then graph.

We have

$$3x + 1 \leq 2x - 3$$

$$3x + 1 - 1 \leq 2x - 3 - 1 \quad \text{Subtracting 1}$$

$$3x \leq 2x - 4 \quad \text{Simplifying}$$

$$3x - 2x \leq 2x - 4 - 2x \quad \text{Subtracting } 2x$$

$$x \leq -4. \quad \text{Simplifying}$$

The graph is as follows:

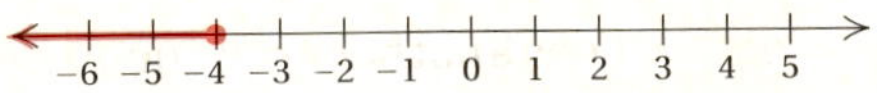

Any number less than or equal to -4 is a solution. The following are some solutions:

$$-4, \quad -5, \quad -6, \quad -4.1, \quad -2045, \quad \text{and} \quad -18\pi.$$

Besides drawing a graph, we can also describe all the solutions of an inequality using **set notation.** We could just begin to list them in a set using roster notation (see p. 49), as follows:

$$\{-4, -5, -6, -4.1, -2045, -18\pi, \ldots\}.$$

We can never list them all this way, however. Seeing this set without knowing the inequality makes it difficult for us to know what real numbers we are considering. There is, however, another kind of notation that we can use. It is

$$\{x \mid x \leq -4\},$$

which is read

"The set of all x such that x is less than or equal to -4."

This shorter notation for sets is called **set-builder notation** (see Section 1.2). From now on, you should use this notation when solving inequalities.

Do Exercises 6–8.

Solve. Then graph.

6. $x + 3 > 5$

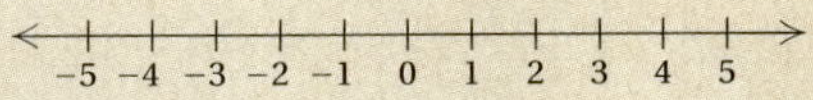

7. $x - 1 \leq 2$

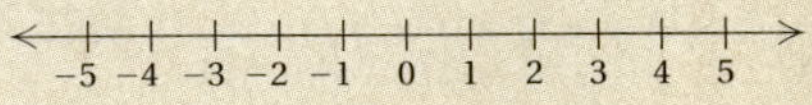

8. $5x + 1 < 4x - 2$

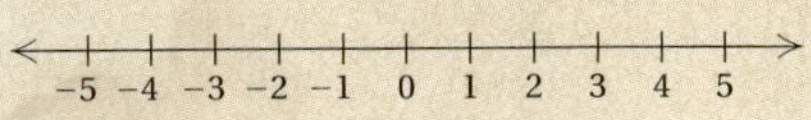

Answers on page A-2

Solve.

9. $x + \frac{2}{3} \geq \frac{4}{5}$

10. $5y + 2 \leq -1 + 4y$

Answers on page A-2

EXAMPLE 10 Solve: $x + \frac{1}{3} > \frac{5}{4}$.

We have

$$x + \tfrac{1}{3} > \tfrac{5}{4}$$

$$x + \tfrac{1}{3} - \tfrac{1}{3} > \tfrac{5}{4} - \tfrac{1}{3} \qquad \text{Subtracting } \tfrac{1}{3}$$

$$x > \tfrac{5}{4} \cdot \tfrac{3}{3} - \tfrac{1}{3} \cdot \tfrac{4}{4} \qquad \text{Multiplying by 1 to obtain a common denominator}$$

$$x > \tfrac{15}{12} - \tfrac{4}{12}$$

$$x > \tfrac{11}{12}.$$

Any number greater than $\frac{11}{12}$ is a solution. The solution set is

$$\{x \mid x > \tfrac{11}{12}\},$$

which is read

"The set of all x such that x is greater than $\frac{11}{12}$."

When solving inequalities, you may obtain an answer like $7 < x$. Recall from Chapter 1 that this has the same meaning as $x > 7$. Thus the solution set can be described as $\{x \mid 7 < x\}$ or as $\{x \mid x > 7\}$. The latter is used most often.

Do Exercises 9 and 10.

d Solving Inequalities Using the Multiplication Principle

There is a multiplication principle for inequalities similar to that for equations, but it must be modified when multiplying on both sides by a negative number. Consider the true inequality

$$3 < 7.$$

If we multiply on both sides by a positive number 2, we get another true inequality:

$$3 \cdot 2 < 7 \cdot 2,$$

or $\quad 6 < 14. \quad$ True

If we multiply on both sides by a negative number -3, we get the false inequality

$$3 \cdot (-3) < 7 \cdot (-3),$$

or $\quad -9 < -21. \quad$ False

However, if we reverse the inequality symbol, we get a true inequality:

$-9 > -21. \quad$ True

Summarizing these results, we obtain the multiplication principle for inequalities.

The Multiplication Principle for Inequalities

If we multiply (or divide) on both sides of a true inequality by a positive number, we get another true inequality. If we multiply (or divide) by a *negative* number and *reverse* the inequality symbol, we get another true inequality.

EXAMPLE 11 Solve: $4x < 28$. Then graph.

We have

$$4x < 28$$

$$\frac{4x}{4} < \frac{28}{4} \quad \text{Dividing by 4}$$

The symbol stays the same.

$$x < 7. \quad \text{Simplifying}$$

The solution set is $\{x \mid x < 7\}$. The graph is as follows:

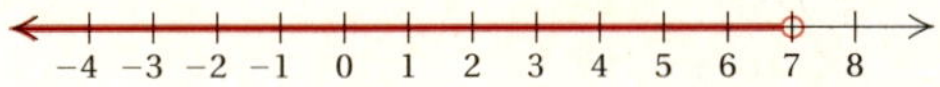

Do Exercises 11 and 12.

EXAMPLE 12 Solve: $-2y < 18$. Then graph.

We have

$$-2y < 18$$

$$\frac{-2y}{-2} > \frac{18}{-2} \quad \text{Dividing by } -2$$

The symbol must be reversed!

$$y > -9. \quad \text{Simplifying}$$

The solution set is $\{y \mid y > -9\}$. The graph is as follows:

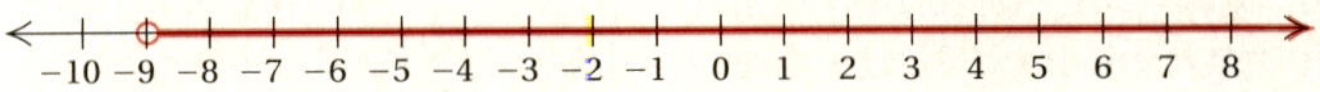

Do Exercises 13 and 14.

e USING THE PRINCIPLES TOGETHER

We use the addition and multiplication principles together in solving inequalities in much the same way as in solving equations. We generally use the addition principle first.

EXAMPLE 13 Solve: $6 - 5y > 7$.

We have

$$6 - 5y > 7$$

$$-6 + 6 - 5y > -6 + 7 \quad \text{Adding } -6\text{. The symbol stays the same.}$$

$$-5y > 1 \quad \text{Simplifying}$$

$$\frac{-5y}{-5} < \frac{1}{-5} \quad \text{Dividing by } -5$$

The symbol must be reversed.

$$y < -\frac{1}{5}. \quad \text{Simplifying}$$

The solution set is $\{y \mid y < -\frac{1}{5}\}$.

Do Exercise 15.

Solve. Then graph.

11. $8x < 64$

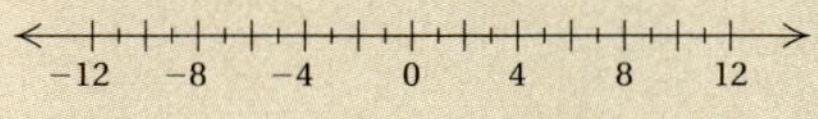

12. $5y \geq 160$

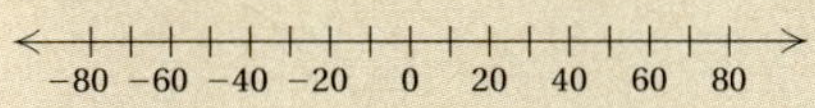

Solve.

13. $-4x \leq 24$

14. $-5y > 13$

15. Solve: $7 - 4x < 8$.

Answers on page A-2

16. Solve: $24 - 7y \leq 11y - 14$.

17. Solve. Use a method like the one used in Example 15.

$24 - 7y \leq 11y - 14$

18. Solve:

$3(7 + 2x) \leq 30 + 7(x - 1)$.

Answers on page A-2

EXAMPLE 14 Solve: $8y - 5 > 17 - 5y$.

$-17 + 8y - 5 > -17 + 17 - 5y$ — **Adding −17. The symbol stays the same.**

$8y - 22 > -5y$ — **Simplifying**

$-8y + 8y - 22 > -8y - 5y$ — **Adding −8y**

$-22 > -13y$ — **Simplifying**

$\frac{-22}{-13} < \frac{-13y}{-13}$ — **Dividing by −13**

The symbol must be reversed.

$\frac{22}{13} < y$.

The solution set is $\{y \mid \frac{22}{13} < y\}$. Since $\frac{22}{13} < y$ has the same meaning as $y > \frac{22}{13}$, we can also describe the solution set as $\{y \mid y > \frac{22}{13}\}$. That is, $\frac{22}{13} < y$ and $y > \frac{22}{13}$ are equivalent. Answers are generally written, however, with the variable on the left.

We can often solve inequalities in such a way as to avoid having to reverse the inequality symbol. We add so that after like terms have been collected, the coefficient of the variable term is positive. We show this by solving the inequality in Example 14 a different way.

EXAMPLE 15 Solve: $8y - 5 > 17 - 5y$.

We note that if we add $5y$ on both sides, the coefficient of the y-term will be positive after like terms have been collected.

$8y - 5 + 5y > 17 - 5y + 5y$ — **Adding 5y on both sides**

$13y - 5 > 17$ — **Simplifying**

$13y - 5 + 5 > 17 + 5$ — **Adding 5 on both sides**

$13y > 22$ — **Simplifying**

$\frac{13y}{13} > \frac{22}{13}$ — **Dividing by 13 on both sides**

$y > \frac{22}{13}$

The solution set is $\{y \mid y > \frac{22}{13}\}$.

Do Exercises 16 and 17.

EXAMPLE 16 Solve: $3(x - 2) - 1 < 2 - 5(x + 6)$.

$3(x - 2) - 1 < 2 - 5(x + 6)$

$3x - 6 - 1 < 2 - 5x - 30$ — **Using the distributive laws to multiply and remove parentheses**

$3x - 7 < -5x - 28$ — **Simplifying**

$3x + 5x < -28 + 7$ — **Adding 5x and 7 to get all x-terms on one side and all other terms on the other side**

$8x < -21$ — **Simplifying**

$x < \frac{-21}{8}$, or $-\frac{21}{8}$ — **Dividing by 8**

The solution set is $\{x \mid x < -\frac{21}{8}\}$.

Do Exercise 18.

NAME SECTION DATE

Exercise Set 2.7

a Determine whether each number is a solution of the given inequality.

1. $x > -4$
a) 4
b) 0
c) -4
d) 6
e) 5.6

2. $x \leq 5$
a) 0
b) 5
c) -1
d) -5
e) $7\frac{1}{4}$

3. $x \geq 6$
a) -6
b) 0
c) 6
d) 8
e) $-3\frac{1}{2}$

4. $x < 8$
a) 8
b) -10
c) 0
d) 11
e) -4.7

b Graph on a number line.

5. $x > 4$

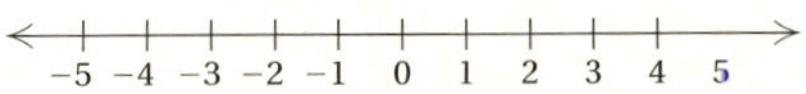

6. $x < 0$

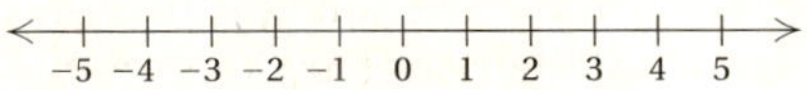

7. $t < -3$

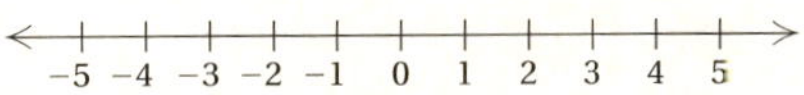

8. $y > 5$

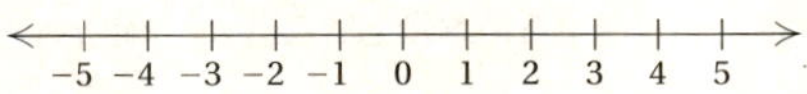

9. $m \geq -1$

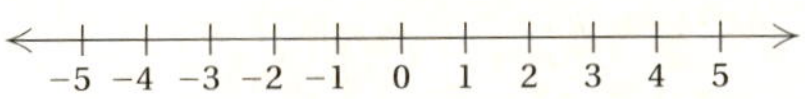

10. $x \leq -2$

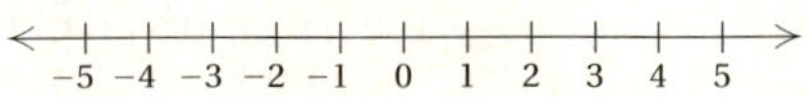

11. $-3 < x \leq 4$

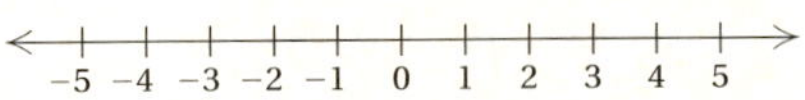

12. $-5 \leq x < 2$

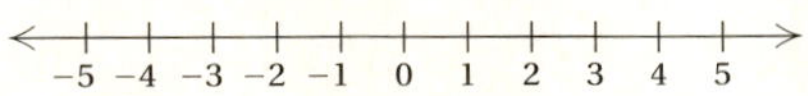

13. $0 < x < 3$

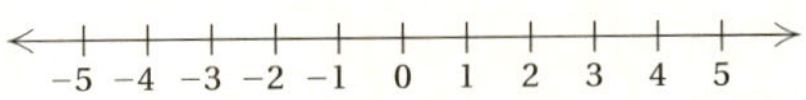

14. $-5 \leq x \leq 0$

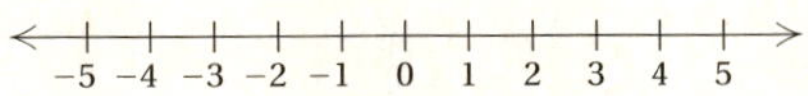

c Solve using the addition principle. Then graph.

15. $x + 7 > 2$

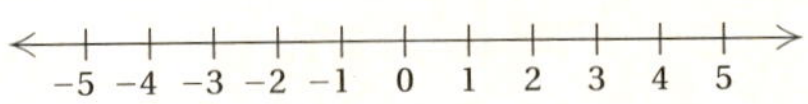

16. $x + 5 > 2$

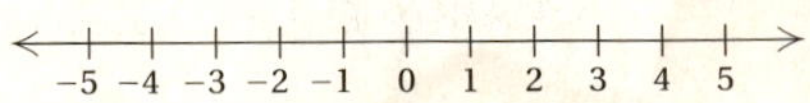

17. $x + 8 \leq -10$

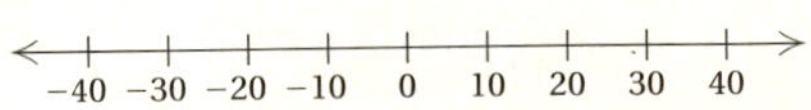

18. $x + 8 \leq -11$

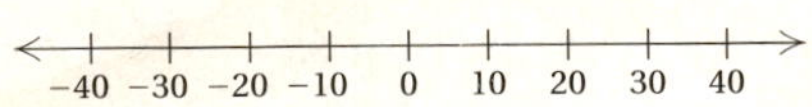

Solve using the addition principle.

19. $y - 7 > -12$

20. $y - 9 > -15$

21. $2x + 3 > x + 5$

22. $2x + 4 > x + 7$

23. $3x + 9 \leq 2x + 6$

24. $3x + 18 \leq 2x + 16$

25. $5x - 6 < 4x - 2$

26. $9x - 8 < 8x - 9$

27. $-9 + t > 5$

ANSWERS

1.
2.
3.
4.
5. See graph.
6. See graph.
7. See graph.
8. See graph.
9. See graph.
10. See graph.
11. See graph.
12. See graph.
13. See graph.
14. See graph.
15.
16.
17.
18.
19.
20.
21.
22.
23.
24.
25.
26.
27.

ANSWERS

28. ________
29. ________
30. ________
31. ________
32. ________
33. ________
34. ________
35. ________
36. ________
37. ________
38. ________
39. ________
40. ________
41. ________
42. ________
43. ________
44. ________
45. ________
46. ________
47. ________
48. ________
49. ________
50. ________
51. ________
52. ________
53. ________
54. ________

28. $-8 + p > 10$ **29.** $y + \frac{1}{4} \leq \frac{1}{2}$ **30.** $x - \frac{1}{3} \leq \frac{5}{6}$

31. $x - \frac{1}{3} > \frac{1}{4}$ **32.** $x + \frac{1}{8} > \frac{1}{2}$

d Solve using the multiplication principle. Then graph.

33. $5x < 35$

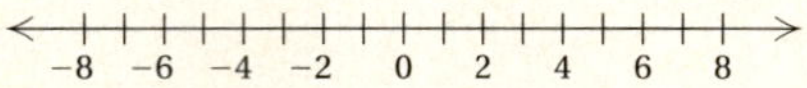

34. $8x \geq 32$

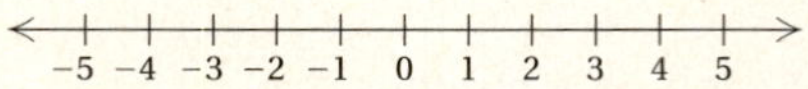

35. $-12x > -36$

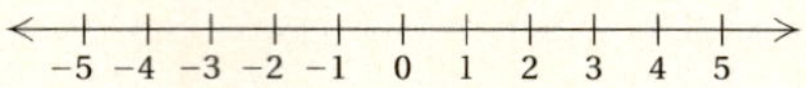

36. $-16x > -64$

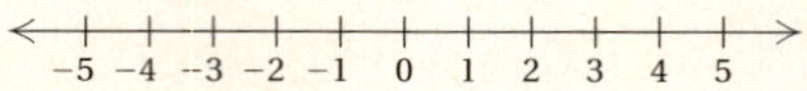

Solve using the multiplication principle.

37. $5y \geq -2$ **38.** $3x < -4$ **39.** $-2x \leq 12$ **40.** $-3x \leq 15$

41. $-4y \geq -16$ **42.** $-7x < -21$ **43.** $-3x < -17$ **44.** $-5y > -23$

45. $-2y > \frac{1}{7}$ **46.** $-4x \leq \frac{1}{9}$ **47.** $-\frac{6}{5} \leq -4x$ **48.** $-\frac{7}{9} > 63x$

e Solve using the addition and multiplication principles.

49. $4 + 3x < 28$ **50.** $3 + 4y < 35$ **51.** $3x - 5 \leq 13$

52. $5y - 9 \leq 21$ **53.** $13x - 7 < -46$ **54.** $8y - 6 < -54$

55. $30 > 3 - 9x$

56. $48 > 13 - 7y$

57. $4x + 2 - 3x \leq 9$

58. $15x + 5 - 14x \leq 9$

59. $-3 < 8x + 7 - 7x$

60. $-8 < 9x + 8 - 8x - 3$

61. $6 - 4y > 4 - 3y$

62. $9 - 8y > 5 - 7y + 2$

63. $5 - 9y \leq 2 - 8y$

64. $6 - 18x \leq 4 - 12x - 5x$

65. $19 - 7y - 3y < 39$

66. $18 - 6y - 4y < 63 + 5y$

67. $2.1x + 45.2 > 3.2 - 8.4x$

68. $0.96y - 0.79 \leq 0.21y + 0.46$

69. $\frac{x}{3} - 2 \leq 1$

70. $\frac{2}{3} + \frac{x}{5} < \frac{4}{15}$

71. $\frac{y}{5} + 1 \leq \frac{2}{5}$

72. $\frac{3x}{4} - \frac{7}{8} \geq -15$

73. $3(2y - 3) < 27$

74. $4(2y - 3) > 28$

75. $2(3 + 4m) - 9 \geq 45$

76. $3(5 + 3m) - 8 \leq 88$

Answers

55. ______

56. ______

57. ______

58. ______

59. ______

60. ______

61. ______

62. ______

63. ______

64. ______

65. ______

66. ______

67. ______

68. ______

69. ______

70. ______

71. ______

72. ______

73. ______

74. ______

75. ______

76. ______

ANSWERS

77. ______
78. ______
79. ______
80. ______
81. ______
82. ______
83. ______
84. ______
85. ______
86. ______
87. ______
88. ______
89. ______
90. ______
91. ______
92. ______
93. ______
94. ______
95. See graph.
96. ______
97. ______
98. ______
99. ______

77. $8(2t + 1) > 4(7t + 7)$

78. $7(5y - 2) > 6(6y - 1)$

79. $3(r - 6) + 2 < 4(r + 2) - 21$

80. $5(x + 3) + 9 \leq 3(x - 2) + 6$

81. $0.8(3x + 6) \geq 1.1 - (x + 2)$

82. $0.4(2x + 8) \geq 20 - (x + 5)$

83. $a + (a - 1) < (a + 2) - (a + 1)$

84. $0.8 - 4(b - 1) > 0.2 + 3(4 - b)$

SKILL MAINTENANCE

Add or subtract.

85. $-56 + (-18)$

86. $-2.3 + 7.1$

87. $-\frac{3}{4} + \frac{1}{8}$

88. $8.12 - 9.23$

89. $-56 - (-18)$

90. $-\frac{3}{4} - \frac{1}{8}$

91. $-2.3 - 7.1$

92. $-8.12 + 9.23$

SYNTHESIS

93. Suppose that $2x - 5 \geq 9$ is true for some value of x. Determine whether $2x - 5 \geq 8$ is true for that same value of x.

94. Determine whether each number is a solution of the inequality $|x| < 3$.

a) 0
b) -2
c) -3
d) 4
e) 3
f) 1.7
g) -2.8

95. Graph $|x| < 3$ on a number line.

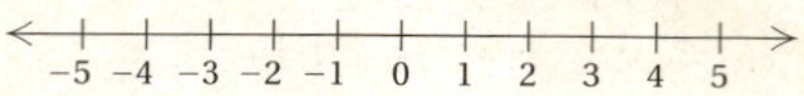

96. Determine whether each number is a solution of the inequality $|x| \geq 4$.

a) 0
b) -5
c) 6
d) -3
e) 3
f) -8
g) 9.7

Solve.

97. $x + 3 \leq 3 + x$

98. $x + 4 < 3 + x$

99. Suppose that we are considering *only* integer solutions to $x > 5$. Find an equivalent inequality involving $\geq$.

2.8 Solving Problems Using Inequalities

We can use inequalities to solve certain types of problems.

OBJECTIVES

After finishing Section 2.8, you should be able to:

a Translate number sentences to inequalities.

b Solve problems using inequalities.

FOR EXTRA HELP

TAPE 7 | TAPE 6A | MAC: 2B IBM: 2B

a TRANSLATING TO INEQUALITIES

First let us practice translating sentences to inequalities.

EXAMPLES Translate to an inequality.

1. A number is less than 5.

$x < 5$

2. A number is greater than or equal to $3\frac{1}{2}$.

$y \geq 3\frac{1}{2}$

3. He can earn, at most, \$34,000.

$E \leq \$34,000$

4. The number of compact disc players sold in this city in a year is at least 2700.

$C \geq 2700$

5. 12 more than twice a number is less than 37.

$2x + 12 < 37$

Do Exercises 1–5.

Translate.

1. A number is less than or equal to 8.

2. A number is greater than −2.

3. The speed of that car is at most 180 mph.

4. The price of that car is at least \$5800.

5. Twice a number minus 32 is greater than 5.

Answers on page A-2

b SOLVING PROBLEMS

EXAMPLE 6 A pre-med student is taking a chemistry course in which four tests are to be given. To get an A, she must average at least 90 on the four tests. The student got scores of 91, 86, and 89 on the first three tests. Determine (in terms of an inequality) what scores on the last test will allow her to get an A.

1. ***Familiarize.*** Let us try some guessing. Suppose the student gets a 92 on the last test. The average of the four scores is their sum divided by the number of tests, 4, and is given by

$$\frac{91 + 86 + 89 + 92}{4} = 89.5.$$

For this average to be *at least* 90, it must be greater than or equal to 90. In this case, we have $89.5 \geq 90$, which is not true. But there are scores that will give the A. To find them, we translate to an inequality and solve. Let x = the student's score on the last test.

2. ***Translate.*** The average of the four scores must be *at least* 90. This means that it must be greater than or equal to 90. Thus we can translate the problem to the inequality

$$\frac{91 + 86 + 89 + x}{4} \geq 90.$$

6. A student is taking a literature course in which four tests are to be given. To get a B, he must average at least 80 on the four tests. The student got scores of 82, 76, and 78 on the first three tests. Determine (in terms of an inequality) what scores on the last test will allow him to get at least a B.

7. Gold stays solid at Fahrenheit temperatures below 1945.4°. Determine (in terms of an inequality) those Celsius temperatures for which gold stays solid. Use the formula given in Example 7.

Answers on page A-2

3. ***Solve.*** We solve the inequality. We first multiply by 4 to clear the fraction.

$$4\left(\frac{91 + 86 + 89 + x}{4}\right) \geq 4 \cdot 90 \qquad \text{Multiplying by 4}$$
$$91 + 86 + 89 + x \geq 360$$
$$266 + x \geq 360 \qquad \text{Collecting like terms}$$
$$x \geq 94 \qquad \text{Subtracting 266}$$

The solution set is $\{x \mid x \geq 94\}$.

4. ***Check.*** We can obtain a partial check by substituting a number greater than or equal to 94. We leave it to the student to try 95 in a manner similar to what was done in the familiarization step.

5. ***State.*** Any score that is at least 94 will give the student an A.

Do Exercise 6.

EXAMPLE 7 Butter stays solid at Fahrenheit temperatures below 88°. The formula

$$F = \tfrac{9}{5}C + 32$$

can be used to convert Celsius temperatures C to Fahrenheit temperatures F. Determine (in terms of an inequality) those Celsius temperatures for which butter stays solid.

1. ***Familiarize.*** Suppose we guess to see how we might consider a solution. We try a Celsius temperature of 40°. We substitute and find F:

$$F = \tfrac{9}{5}C + 32 = \tfrac{9}{5}(40) + 32 = 72 + 32 = 104°.$$

This is higher than 88°, so 40° is *not* a solution. To find the solutions, we need to solve an inequality.

2. ***Translate.*** The Fahrenheit temperature F is to be less than 88. We have the inequality

$$F < 88.$$

To find the Celsius temperatures C that satisfy this condition, we substitute $\tfrac{9}{5}C + 32$ for F, which gives us the following inequality:

$$\tfrac{9}{5}C + 32 < 88.$$

3. ***Solve.*** We solve the inequality:

$$\tfrac{9}{5}C + 32 < 88$$
$$5\left(\tfrac{9}{5}C + 32\right) < 5(88) \qquad \text{Multiplying by 5 to clear the fraction}$$
$$5\left(\tfrac{9}{5}C\right) + 5(32) < 440 \qquad \text{Using a distributive law}$$
$$9C + 160 < 440 \qquad \text{Simplifying}$$
$$9C < 280 \qquad \text{Subtracting 160}$$
$$C < \frac{280}{9} \qquad \text{Dividing by 9}$$
$$C < 31.1. \qquad \text{Dividing and rounding to the nearest tenth}$$

The solution set of the inequality is $\{C \mid C < 31.1°\}$.

4. ***Check.*** The check is left to the student.

5. ***State.*** Butter stays solid at Celsius temperatures below 31.1°.

Do Exercise 7.

NAME SECTION DATE

Exercise Set 2.8

a Translate to an inequality.

1. A number is greater than 8.

2. A number is less than 5.

3. A number is less than or equal to -4.

4. A number is greater than or equal to 18.

5. The number of people is at least 1300.

6. The cost is at most \$4857.95.

7. The amount of acid is not to exceed 500 liters.

8. The cost of gasoline is no less than 94 cents per gallon.

9. Two more than three times a number is less than 13.

10. Five less than one-half a number is greater than 17.

b Solve.

11. Your quiz grades are 73, 75, 89, and 91. Determine (in terms of an inequality) those scores that you can obtain on the last quiz in order to receive an average quiz grade of at least 85.

12. A human body is considered to be fevered when its temperature is higher than 98.6°F. Using the formula given in Example 7, determine (in terms of an inequality) those Celsius temperatures for which the body is fevered.

13. The formula

$$R = -0.075t + 3.85$$

can be used to predict the world record in the 1500-m run t years after 1930. Determine (in terms of an inequality) those years for which the world record will be less than 3.5 min.

14. The formula

$$R = -0.028t + 20.8$$

can be used to predict the world record in the 200-m dash t years after 1920. Determine (in terms of an inequality) those years for which the world record will be less than 19.0 sec.

15. Black Angus calves weigh about 75 lb at birth and gain about 2 lb per day for the first few weeks. Determine (in terms of an inequality) those days for which the calf's weight is more than 125 lb.

16. Rule Office Supply rents a certain kind of copy machine for \$59.95 per week plus 4 cents per copy. A business decides to rent such a machine for a week, but must stay within a budget of \$400. Determine (in terms of an inequality) the number of copies that the business can make and stay within the budget.

ANSWERS

1. ______
2. ______
3. ______
4. ______
5. ______
6. ______
7. ______
8. ______
9. ______
10. ______
11. ______
12. ______
13. ______
14. ______
15. ______
16. ______

ANSWERS

17. ______

18. ______

19. ______

20. ______

21. ______

22. ______

23. ______

24. ______

25. ______

26. ______

27. ______

28. ______

29. ______

30. ______

17. Find all numbers such that the sum of the number and 15 is less than four times the number.

18. Find all numbers such that three times the number minus ten times the number is greater than or equal to eight times the number.

19. The width of a rectangle is fixed at 4 cm. Determine (in terms of an inequality) those lengths for which the area will be less than 86 cm^2.

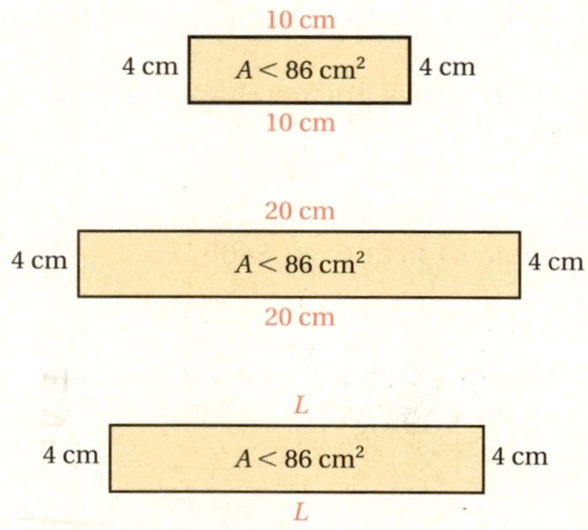

20. An overnight delivery service accepts packages up to 165 in. in length and girth combined. (Girth is the distance around the package.) A package has a fixed girth of 11 in. Determine (in terms of an inequality) those lengths for which a package is acceptable.

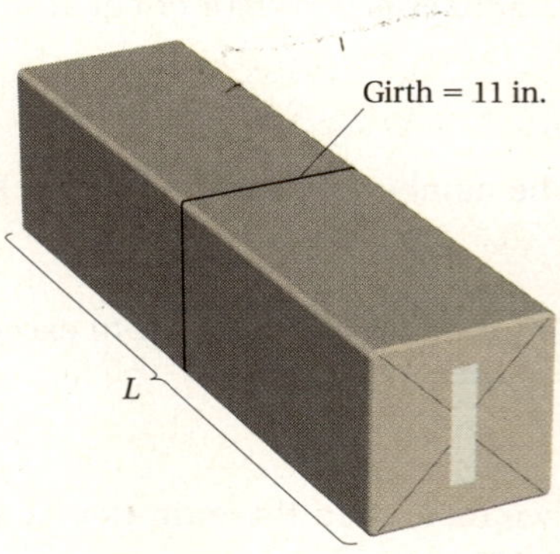

21. One side of a triangle is 2 cm shorter than the base. The other side is 3 cm longer than the base. What lengths of the base will allow the perimeter to be greater than 19 cm?

22. The perimeter of a rectangular swimming pool is not to exceed 70 ft. The length is to be twice the width. What widths will meet these conditions?

23. Dirk's Electric made 17 customer calls last week and 22 calls this week. How many calls must be made next week in order to maintain an average of at least 20 for the three-week period?

24. Ginny and Jill do volunteer work at a hospital. Jill worked 3 hr more than Ginny, and together they worked more than 27 hr. What possible number of hours did each work?

25. A family's air conditioner needs freon. The charge for a service call is a flat fee of \$70 plus \$60 an hour. The freon costs \$35. The family has at most \$150 to pay for the service call. Determine (in terms of an inequality) those lengths of time of the call that will allow the family to stay within its \$150 budget.

26. A student is shopping for a new pair of jeans and two sweaters of the same kind. He is determined to spend no more than \$120.00 for the outfit. He buys jeans for \$21.95. What is the most that the student can spend for each sweater?

27. A landscaping company is laying out a triangular flower bed. The height of the triangle is 16 ft. What lengths of the base will make the area at least 200 ft^2?

28. Rhetoric Advertising is a direct-mail company. It determines that for a particular campaign, it can use any letter with a fixed width of $3\frac{1}{2}$ in. and an area of at least $17\frac{1}{2}$ in^2. Determine (in terms of an inequality) those lengths that will satisfy the company constraints.

SYNTHESIS

29. The sum of two consecutive odd integers is less than 100. What is the largest possible pair of such integers?

30. The area of a square can be no more than 64 cm^2. What lengths of a side will allow this?

CALCULATOR CONNECTION

One way to use percents with calculators is to mentally convert from the percent notation to decimal notation as you enter a number in the calculator. For example, 27% of 350 = 0.27 × 350 = 94.5. On some calculators, there is a percent key. Consult your manual for its usage. Find each of the following.

1. 46% of 550
2. 67.3% of 12,456
3. 8.24% of $5800

Solve using a calculator. Express the answer to three decimal places.

4. $35.6x - 567.889 = 78{,}945.667$
5. $23.4(x - 7.009) + 18.2(10.3 + 5x) = 88.902$

EXTENDED SYNTHESIS EXERCISES

1. Write an equation that has the solution $-\frac{8}{15}$ and for which that solution is found using only the multiplication principle.
2. Write an equation that has the solution $-\frac{8}{15}$ and for which that solution is found using only the addition principle.
3. Write an equation that has the solution $-\frac{8}{15}$ and for which that solution is found using only the addition and the multiplication principles one time each.
4. Find an equation for which 5 and −5 are both solutions.
5. Find an equation for which there are no real-number solutions.
6. Find an equation for which all real numbers are solutions.
7. A rocket is divided into three sections: the payload and the navigation section in the top, the fuel section in the middle, and the rocket section in the bottom. The top section is one-sixth as long as the rocket section. The rocket section is one-half the total length of the rocket. The total length is 240 ft. Find the length of each section.

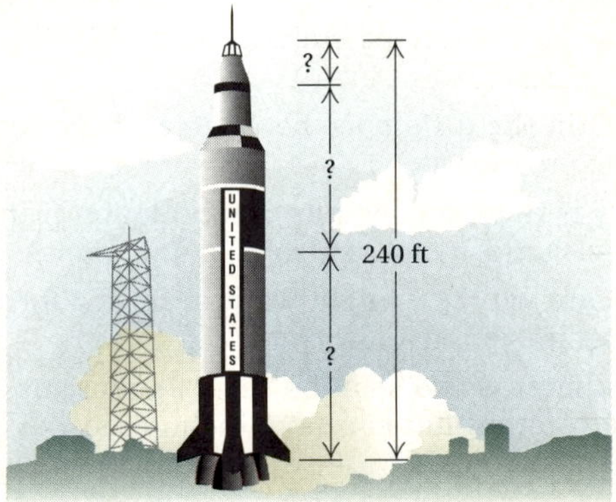

8. Sam and Becky went bowling at Stardust Lanes. The total cost was $9. Sam bowled three games and Becky bowled two games. Shoe rental was $0.75. What was the cost of each game?
9. Kyoko shared a package of engineering graph paper with three of his friends. He gave $\frac{1}{4}$ of the pack to Raul. Sarah got $\frac{1}{3}$ of what was left. After that, Teresa took $\frac{1}{6}$ of what was left in the package. Kyoko kept the remaining 30 sheets. How many sheets of paper did Kyoko have originally?
10. For what values of x is $|x| > x$?
11. Which is better, a discount of 40% or successive discounts of 35% and 5%?
12. Which is better, successive discounts of 10%, 10%, and 20% or 20%, 10%, and 10%?
13. Express the number 5 as the sum of 10 consecutive integers. Find the product of these integers.
14. There are 9 teams in a softball league. During the season, each team will play every other team in the league twice. How many games will be played by the entire league?
15. The number 64 is expressed as a sum of four numbers. Of these four numbers, when 3 is added to the first, 3 is subtracted from the second, the third is multiplied by 3, and the fourth is divided by 3, the resulting four numbers are the same. Find the numbers.

(*continued*)

CRITICAL THINKING

EXERCISES FOR THINKING AND WRITING

Explain all the errors in each of the following.

1. $4 - 3x = 5$
$3x = 9$
$x = 3$

2. $2(x - 5) = 7$
$2x - 5 = 7$
$2x = 12$
$x = 6$

3. Explain the difference in using the multiplication principle for solving equations and for solving inequalities.

4. Explain the difference between an equation and an expression.

5. Why is there no need for a subtraction principle or a division principle for solving equations?

6. Explain why you might want to solve a formula for a specific letter.

7. In what ways do solution sets of equations seem to differ from solution sets of inequalities?

8. Use the inequality $-5 < 8$ to explain why an inequality symbol must be reversed when multiplying by a negative number.

Summary and Review: Chapter 2

Important Properties and Formulas

The Addition Principles:	If $a = b$ is true, then $a + c = b + c$ is true for any real number c. If the same number is added on both sides of an inequality, we get another true inequality.
The Multiplication Principles:	If $a = b$ is true, then $ac = bc$ is true for any real number c. If we multiply on both sides of a true inequality by a positive number, we get another true inequality. If we multiply by a negative number and reverse the inequality symbol, we get another true inequality.

Review Exercises

The review objectives to be tested in addition to the material in this chapter are [R.3a, b], [1.1a, b], [1.3a], and [1.8b].

Solve.

1. $x + 5 = -17$

2. $-8x = -56$

3. $-\frac{x}{4} = 48$

4. $n - 7 = -6$

5. $15x = -35$

6. $x - 11 = 14$

7. $-\frac{2}{3} + x = -\frac{1}{6}$

8. $\frac{4}{5}y = -\frac{3}{16}$

9. $y - 0.9 = 9.09$

10. $5 - x = 13$

11. $5t + 9 = 3t - 1$

12. $7x - 6 = 25x$

13. $\frac{1}{4}x - \frac{5}{8} = \frac{3}{8}$

14. $14y = 23y - 17 - 10$

15. $0.22y - 0.6 = 0.12y + 3 - 0.8y$

16. $\frac{1}{4}x - \frac{1}{8}x = 3 - \frac{1}{16}x$

17. $4(x + 3) = 36$

18. $3(5x - 7) = -66$

19. $8(x - 2) = 5(x + 4)$

20. $-5x + 3(x + 8) = 16$

Determine whether the given number is a solution of the inequality $x \leq 4$.

21. -3

22. 7

23. 4

Solve. Write set notation for the answers.

24. $y + \frac{2}{3} \geq \frac{1}{6}$

25. $9x \geq 63$

26. $2 + 6y > 14$

27. $7 - 3y \geq 27 + 2y$

28. $3x + 5 < 2x - 6$

29. $-4y < 28$

30. $3 - 4x < 27$

31. $4 - 8x < 13 + 3x$

32. $-3y \geq -21$

33. $-4x \leq \frac{1}{3}$

Graph on a number line.

34. $4x - 6 < x + 3$

35. $-2 < x \leq 5$

36. $y > 0$

Solve.

37. $C = \pi d$, for d

38. $V = \frac{1}{3}Bh$, for B

39. $A = \frac{a + b}{2}$, for a

40. An entertainment center sold for $2449 in June. This was $332 more than the cost in February. Find the cost in February.

41. Selma is paid a $4 commission for each appliance that she sells. One week, she received $108 in commissions. How many appliances did she sell?

42. An 8-m board is cut into two pieces. One piece is 2 m longer than the other. How long are the pieces?

43. If 14 is added to three times a certain number, the result is 41. Find the number.

44. The sum of two consecutive odd integers is 116. Find the integers.

45. The perimeter of a rectangle is 56 cm. The width is 6 cm less than the length. Find the width and the length.

46. After a 30% reduction, a bread maker is on sale for $154. What was the marked price (the price before reducing)?

47. A hotel manager's salary is $30,000. That is a 15% increase over the previous year's salary. What was the previous salary (to the nearest dollar)?

48. The measure of the second angle of a triangle is 50° more than that of the first. The measure of the third angle is 10° less than twice the first. Find the measures of the angles.

49. Your quiz grades are 71, 75, 82, and 86. What is the lowest grade that you can get on the next quiz and still have an average of at least 80?

50. The length of a rectangle is 43 cm. What widths will make the perimeter greater than 120 cm?

SKILL MAINTENANCE

51. Convert to decimal notation: $\frac{17}{12}$.

52. Divide: $12.42 \div 5.4$.

53. Add: $-12 + 10 + (-19) + (-24)$.

54. Remove parentheses and simplify: $5x - 8(6x - y)$.

SYNTHESIS

55. Consumer experts advise us never to pay the sticker price for a car. A rule of thumb is to pay the sticker price minus 20% of the sticker price, plus $200. A car is purchased for $11,520 using the rule. What was the sticker price?

56. The total length of the Nile and Amazon Rivers is 13,108 km. If the Amazon were 234 km longer, it would be as long as the Nile. Find the length of each river.

Solve.

57. $2|n| + 4 = 50$

58. $|3n| = 60$

59. $y = 2a - ab + 3$, for a

NAME SECTION DATE

Test: Chapter 2

Solve.

1. $x + 7 = 15$

2. $t - 9 = 17$

3. $3x = -18$

4. $-\frac{4}{7}x = -28$

5. $3t + 7 = 2t - 5$

6. $\frac{1}{2}x - \frac{3}{5} = \frac{2}{5}$

7. $8 - y = 16$

8. $-\frac{2}{5} + x = -\frac{3}{4}$

9. $3(x + 2) = 27$

10. $-3x - 6(x - 4) = 9$

11. $0.4p + 0.2 = 4.2p - 7.8 - 0.6p$

Solve. Write set notation for the answers.

12. $x + 6 \leq 2$

13. $14x + 9 > 13x - 4$

14. $12x \leq 60$

15. $-2y \geq 26$

16. $-4y \leq -32$

17. $-5x \geq \frac{1}{4}$

18. $4 - 6x > 40$

19. $5 - 9x \geq 19 + 5x$

ANSWERS

1. ____________
2. ____________
3. ____________
4. ____________
5. ____________
6. ____________
7. ____________
8. ____________
9. ____________
10. ____________
11. ____________
12. ____________
13. ____________
14. ____________
15. ____________
16. ____________
17. ____________
18. ____________
19. ____________

ANSWERS

20. ____
21. ____
22. ____
23. ____
24. ____
25. ____
26. ____
27. ____
28. ____
29. ____
30. ____
31. ____
32. ____
33. ____
34. ____
35. ____
36. ____
37. ____

Graph on a number line.

20. $y \leq 9$

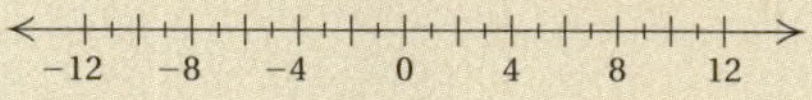

21. $6x - 3 < x + 2$

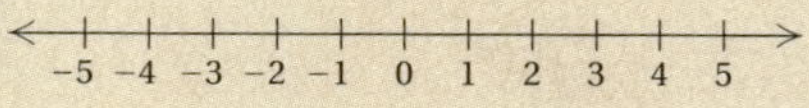

22. $-2 \leq x \leq 2$

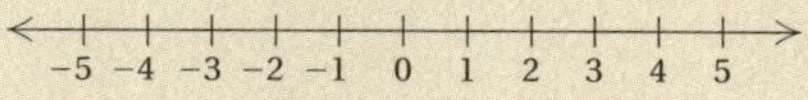

Solve.

23. The perimeter of a rectangular photograph is 36 cm. The length is 4 cm greater than the width. Find the width and the length.

24. If you triple a number and then subtract 14, you get two-thirds of the original number. What is the original number?

25. The sum of three consecutive odd integers is 249. Find the integers.

26. Money is invested in a savings account at 5% simple interest. After 1 year, there is \$924 in the account. How much was originally invested?

Solve the formula for the given letter.

27. $A = 2\pi rh$, for r

28. $w = \dfrac{P - 2l}{2}$, for l

Solve.

29. Find all numbers such that six times the number is greater than the number plus 30.

30. The width of a rectangle is 96 yd. Find all possible lengths so that the perimeter of the rectangle will be at least 540 yd.

SKILL MAINTENANCE

31. Add: $\dfrac{2}{3} + \left(-\dfrac{8}{9}\right)$.

32. Evaluate $\dfrac{4x}{y}$ when $x = 2$ and $y = 3$.

33. Translate to an algebraic expression: Seventy-three percent of p.

34. Simplify: $2x - 3y - 5(4x - 8y)$.

SYNTHESIS

35. Solve $c = \dfrac{1}{a - d}$, for d.

36. Solve: $3|w| - 8 = 37$.

37. A movie theater had a certain number of tickets to give away. Five people got the tickets. The first got one-third of the tickets, the second got one-fourth of the tickets, and the third got one-fifth of the tickets. The fourth person got eight tickets, and there were five tickets left for the fifth person. Find the total number of tickets given away.

Cumulative Review: Chapters 1–2

Evaluate.

1. $\frac{y - x}{4}$ when $y = 12$ and $x = 6$

2. $\frac{3x}{y}$ when $x = 5$ and $y = 4$

3. $x - 3$ when $x = 3$

4. Translate to an algebraic expression: Four less than twice w.

Use $<$ or $>$ for ▆ to write a true sentence.

5. -4 ▆ -6

6. 0 ▆ -5

7. -8 ▆ 7

8. Find the opposite and the reciprocal of $\frac{2}{5}$.

Find the absolute value.

9. $|3|$

10. $\left|-\frac{3}{4}\right|$

11. $|0|$

Compute and simplify.

12. $-6.7 + 2.3$

13. $-\frac{1}{6} - \frac{7}{3}$

14. $-\frac{5}{8}\left(-\frac{4}{3}\right)$

15. $(-7)(5)(-6)(-0.5)$

16. $81 \div (-9)$

17. $-10.8 \div 3.6$

18. $-\frac{4}{5} \div -\frac{25}{8}$

Multiply.

19. $5(3x + 5y + 2z)$

20. $4(-3x - 2)$

21. $-6(2y - 4x)$

Factor.

22. $64 + 18x + 24y$

23. $16y - 56$

24. $5a - 15b + 25$

Collect like terms.

25. $9b + 18y + 6b + 4y$

26. $3y + 4 + 6z + 6y$

27. $-4d - 6a + 3a - 5d + 1$

28. $3.2x + 2.9y - 5.8x - 8.1y$

Simplify.

29. $7 - 2x - (-5x) - 8$

30. $-3x - (-x + y)$

31. $-3(x - 2) - 4x$

32. $10 - 2(5 - 4x)$

33. $[3(x + 6) - 10] - [5 - 2(x - 8)]$

Solve.

34. $x + 1.75 = 6.25$

35. $\frac{5}{2}y = \frac{2}{5}$

36. $-2.6 + x = 8.3$

37. $4\frac{1}{2} + y = 8\frac{1}{3}$

38. $-\frac{3}{4}x = 36$

39. $-2.2y = -26.4$

40. $5.8x = -35.96$

41. $-4x + 3 = 15$

42. $-3x + 5 = -8x - 7$

43. $4y - 4 + y = 6y + 20 - 4y$

44. $-3(x - 2) = -15$

45. $\frac{1}{3}x - \frac{5}{6} = \frac{1}{2} + 2x$

46. $-3.7x + 6.2 = -7.3x - 5.8$

47. $3x - 1 < 2x + 1$

48. $5 - y \leq 2y - 7$

49. $3y + 7 > 5y + 13$

50. $A = \frac{1}{2}h(b + c)$, for h

51. $Q = \frac{p - q}{2}$, for q

Solve.

52. If 25 is subtracted from a certain number, the result is 129. Find the number.

53. Lance and Rocky purchased rollerblades for a total of $107. Lance paid $17 more for his rollerblades than Rocky did. What did Rocky pay?

54. Money is invested in a savings account at 8% simple interest. After 1 year, there is $1134 in the account. How much was originally invested?

55. A 143-m wire is cut into three pieces. The second is 3 m longer than the first. The third is four-fifths as long as the first. How long is each piece?

56. Your test grades are 75, 82, 86, and 79. Determine (in terms of an inequality) what scores on the last test will allow you an average test score of at least 80.

57. After a 25% reduction, a tie is on sale for $18.45. What was the price before reduction?

SYNTHESIS

58. An engineer's salary at the end of a year is $38,563.20. This reflects a 4% salary increase and a later 3% cost-of-living adjustment during the year. What was the salary at the beginning of the year?

Solve.

59. $4|x| - 13 = 3$

60. $4(x + 2) = 4(x - 2) + 16$

61. $0(x + 3) + 4 = 0$

62. $\frac{2 + 5x}{4} = \frac{11}{28} + \frac{8x + 3}{7}$

63. $5(7 + x) = (x + 7)5$

64. $p = \frac{2}{m + Q}$, for Q

3

Polynomials: Operations

Introduction

Algebraic expressions like

$$-0.05t^2 + 2t + 2$$

and

$$x^3 - 5x^2 + 4x - 11$$

are called *polynomials.* One of the most important parts of introductory algebra is the study of polynomials. In this chapter, we learn to add, subtract, multiply, and divide polynomials.

Of particular importance here is the study of the fast ways to multiply certain polynomials. These *special products* will be helpful not only in this text but also in more advanced mathematics.

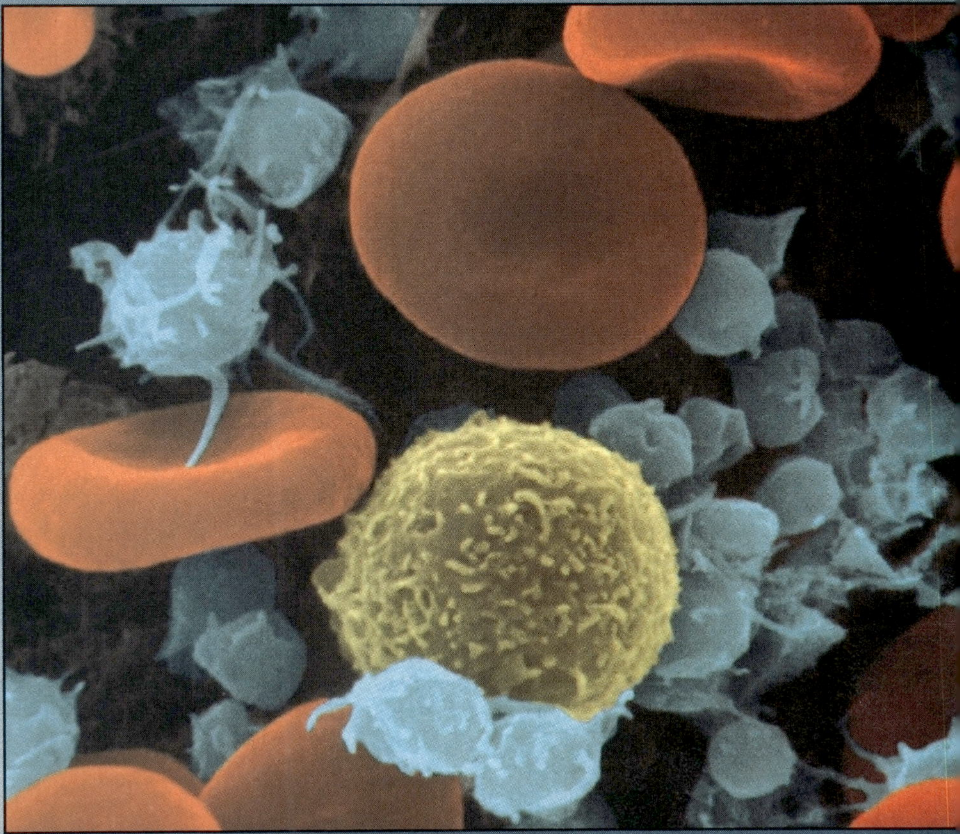

An Application

The concentration, in parts per million, of a certain medication in the bloodstream after time t, in hours, is given by the polynomial

$$-0.05t^2 + 2t + 2.$$

Find the concentration after 2 hr.

The Mathematics

To solve the problem, we substitute 2 for t and evaluate:

$$\underbrace{-0.05t^2 + 2t + 2}_{\uparrow} = -0.05(2)^2 + 2 \cdot 2 + 2$$

$$= 5.8 \text{ parts per million.}$$

This is a polynomial.

Objectives for Review

The review objectives to be tested in addition to the material in this chapter are as follows.

[1.4a] Subtract real numbers and simplify combinations of additions and subtractions.

[1.7d] Use the distributive laws to factor expressions like $4x - 12$.

[2.3b, c] Solve equations in which terms may need to be collected, and solve equations by first removing parentheses and collecting like terms.

[2.4a] Solve problems by translating to equations.

Pretest: Chapter 3

1. Multiply: $x^{-3} \cdot x^5$.

2. Divide: $\frac{x^{-2}}{x^5}$.

3. Simplify: $(-4x^2y^{-3})^2$.

4. Express using a positive exponent: p^{-3}.

5. Convert to scientific notation: 0.000347.

6. Convert to decimal notation: 3.4×10^6.

7. Identify the degree of each term and the degree of the polynomial:

$2x^3 - 4x^2 + 3x - 5$.

8. Collect like terms:

$2a^3b - a^2b^2 + ab^3 + 9 - 5a^3b - a^2b^2 + 12b^3$.

9. Add:

$(5x^2 - 7x + 8) + (6x^2 + 11x - 19)$.

10. Subtract:

$(5x^2 - 7x + 8) - (6x^2 + 11x - 19)$.

Multiply.

11. $5x^2(3x^2 - 4x + 1)$

12. $(x + 5)^2$

13. $(x - 5)(x + 5)$

14. $(x^3 + 6)(4x^3 - 5)$

15. $(2x - 3y)(2x - 3y)$

16. Divide: $(x^3 - x^2 + x + 2) \div (x - 2)$.

3.1 *Integers as Exponents*

We introduced integer exponents of 2 or higher in Section R.5. Here we consider 0 and 1, as well as negative integers, as exponents.

a EXPONENTIAL NOTATION

An exponent of 2 or greater tells how many times the base is used as a factor. For example,

$$a \cdot a \cdot a \cdot a = a^4.$$

In this case, the **exponent** is 4 and the **base** is a. An expression for a power is called **exponential notation.**

a^n ← This is the exponent.
↑
This is the base.

EXAMPLE 1 What is the meaning of 3^5? of n^4? of $(2n)^3$? of $50x^2$?

3^5 means $3 \cdot 3 \cdot 3 \cdot 3 \cdot 3$; $(2n)^3$ means $2n \cdot 2n \cdot 2n$;

n^4 means $n \cdot n \cdot n \cdot n$; $50x^2$ means $50 \cdot x \cdot x$

Do Exercises 1–4.

We read exponential notation as follows:

a^n is read the ***n*th power of *a*,** or simply ***a* to the *n*th,** or ***a* to the *n*.**

We often read x^2 as "***x*-squared.**" The reason for this is that the area of a square of side x is $x \cdot x$, or x^2. We often read x^3 as "***x*-cubed.**" The reason for this is that the volume of a cube with length, width, and height x is $x \cdot x \cdot x$, or x^3.

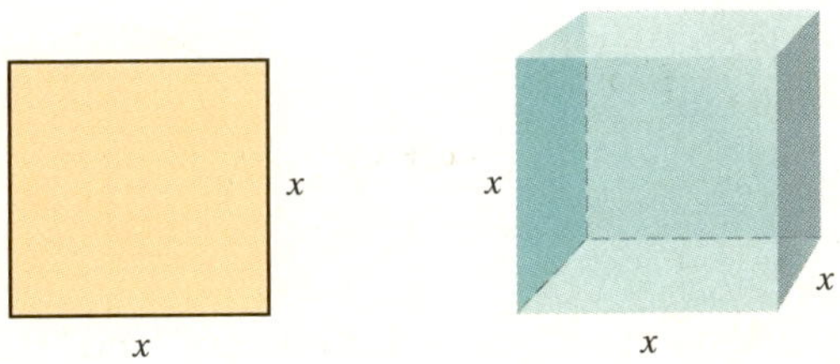

b ONE AND ZERO AS EXPONENTS

Look for a pattern in the following:

On each side, we divide by 8 at each step.

$$8 \cdot 8 \cdot 8 \cdot 8 = 8^4$$
$$8 \cdot 8 \cdot 8 = 8^3$$
$$8 \cdot 8 = 8^2$$
$$8 = 8^?$$
$$1 = 8^?.$$

On this side, the exponents decrease by 1.

To continue the pattern, we would say that

$$8 = 8^1$$

and $$1 = 8^0.$$

OBJECTIVES

After finishing Section 3.1, you should be able to:

- **a** Tell the meaning of exponential notation.
- **b** Evaluate exponential expressions with exponents of 0 and 1.
- **c** Evaluate algebraic expressions containing exponents.
- **d** Use the product rule to multiply exponential expressions with like bases.
- **e** Use the quotient rule to divide exponential expressions with like bases.
- **f** Express an exponential expression involving negative exponents with positive exponents.

FOR EXTRA HELP

TAPE 7 | TAPE 6A | MAC: 3A IBM: 3A

What is the meaning of each of the following?

1. 5^4

2. x^5

3. $(3t)^2$

4. $3t^2$

Answers on page A-3

Evaluate.

5. 6^1

6. 7^0

7. $(8.4)^1$

8. 8654^0

Answers on page A-3

We make the following definition.

> $a^1 = a$, for any number a;
> $a^0 = 1$, for any nonzero number a.

We consider 0^0 to be undefined. We will explain why later in this section.

EXAMPLE 2 Evaluate 5^1, 8^1, 3^0, $(-7.3)^0$, and $(186{,}892{,}046)^0$.

$5^1 = 5; \quad 8^1 = 8; \quad 3^0 = 1;$

$(-7.3)^0 = 1; \quad (186{,}892{,}046)^0 = 1$

Do Exercises 5–8.

c EVALUATING ALGEBRAIC EXPRESSIONS

Algebraic expressions can involve exponential notation. For example, the following are algebraic expressions:

$$x^4, \quad (3x)^3 - 2, \quad a^2 + 2ab + b^2.$$

We evaluate algebraic expressions by replacing variables with numbers and following the rules for order of operations.

EXAMPLE 3 Evaluate x^4 when $x = 2$.

$x^4 = 2^4$ Substituting

$= 2 \cdot 2 \cdot 2 \cdot 2 = 16$

EXAMPLE 4 The area of a circle is given by $A = \pi r^2$, where r is the radius. Find the area of a circle with a radius of 10 cm. Use 3.14 as an approximation for π.

$A = \pi r^2 \approx 3.14 \times (10\text{ cm})^2$

$\approx 3.14 \times 100\text{ cm}^2$

$\approx 314\text{ cm}^2$

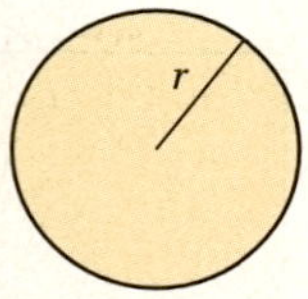

In Example 4, "cm^2" means "square centimeters" and "$\approx$" means "is approximately equal to."

EXAMPLE 5 Evaluate $m^1 + 5$ and $m^0 + 5$ when $m = 4$.

$m^1 + 5 = 4^1 + 5 = 4 + 5 = 9;$

$m^0 + 5 = 4^0 + 5 = 1 + 5 = 6$

EXAMPLE 6 Evaluate $(5x)^3$ when $x = -2$.

When we evaluate with a negative number, we often use extra parentheses to show the substitution.

$(5x)^3 = [5 \cdot (-2)]^3$ Substituting

$= [-10]^3$ Multiplying within brackets first

$= -1000$ Evaluating the power

EXAMPLE 7 Evaluate $5x^3$ when $x = -2$.

$$5x^3 = 5 \cdot (-2)^3 \quad \text{Substituting}$$
$$= 5(-8) \quad \text{Evaluating the power first}$$
$$= -40$$

Recall that two expressions are equivalent if they have the same value for all meaningful replacements. Note that Examples 6 and 7 show that $(5x)^3$ and $5x^3$ are *not* equivalent—that is, $(5x)^3 \neq 5x^3$.

Do Exercises 9–13.

d MULTIPLYING POWERS WITH LIKE BASES

There are several rules for manipulating exponential notation to obtain equivalent expressions. We first consider multiplying powers with like bases:

$$a^3 \cdot a^2 = \underbrace{(a \cdot a \cdot a)}_{3 \text{ factors}}\underbrace{(a \cdot a)}_{2 \text{ factors}} = \underbrace{a \cdot a \cdot a \cdot a \cdot a}_{5 \text{ factors}} = a^5.$$

Since an integer exponent greater than 1 tells how many times we use a base as a factor, then $(a \cdot a \cdot a)(a \cdot a) = a \cdot a \cdot a \cdot a \cdot a = a^5$ by the associative law. Note that the exponent in a^5 is the sum of those in $a^3 \cdot a^2$. That is, $3 + 2 = 5$. Likewise,

$$b^4 \cdot b^3 = (b \cdot b \cdot b \cdot b)(b \cdot b \cdot b) = b^7, \quad \text{where} \quad 4 + 3 = 7.$$

Adding the exponents gives the correct result.

THE PRODUCT RULE

For any number a and any positive integers m and n,

$$a^m \cdot a^n = a^{m+n}.$$

(When multiplying with exponential notation, if the bases are the same, keep the base and add the exponents.)

EXAMPLES Multiply and simplify. By simplify, we mean write the expression as one number to a nonnegative power.

8. $8^4 \cdot 8^3 = 8^{4+3}$ **Adding exponents:** $a^m \cdot a^n = a^{m+n}$
$= 8^7$

9. $x^2 \cdot x^9 = x^{2+9}$
$= x^{11}$

10. $m^5m^{10}m^3 = m^{5+10+3}$
$= m^{18}$

11. $x \cdot x^8 = x^1 \cdot x^8 = x^{1+8}$
$= x^9$

12. $(a^3b^2)(a^3b^5) = (a^3a^3)(b^2b^5)$
$= a^6b^7$

Do Exercises 14–18.

9. Evaluate t^3 when $t = 5$.

10. Find the area of a circle when $r = 32$ cm. Use 3.14 for π.

11. Evaluate $200 - a^4$ when $a = 3$.

12. Evaluate $t^1 - 4$ and $t^0 - 4$ when $t = 7$.

13. a) Evaluate $(4t)^2$ when $t = -3$.

b) Evaluate $4t^2$ when $t = -3$.

c) Determine whether $(4t)^2$ and $4t^2$ are equivalent.

Multiply and simplify.

14. $3^5 \cdot 3^5$

15. $x^4 \cdot x^6$

16. $p^4p^{12}p^8$

17. $x \cdot x^4$

18. $(a^2b^3)(a^7b^5)$

Answers on page A-3

Divide and simplify.

19. $\dfrac{4^5}{4^2}$

20. $\dfrac{y^6}{y^2}$

21. $\dfrac{p^{10}}{p}$

22. $\dfrac{a^7b^6}{a^3b^4}$

Answers on page A-3

e DIVIDING POWERS WITH LIKE BASES

The following suggests a rule for dividing powers with like bases, such as a^5/a^2:

$$\frac{a^5}{a^2} = \frac{a \cdot a \cdot a \cdot a \cdot a}{a \cdot a} = \frac{a \cdot a \cdot a \cdot a \cdot a}{1 \cdot a \cdot a} = \frac{a \cdot a \cdot a}{1} \cdot \frac{a \cdot a}{a \cdot a} = \frac{a \cdot a \cdot a}{1} \cdot 1$$
$$= a \cdot a \cdot a = a^3.$$

Note that the exponent in a^3 is the difference of those in $a^5 \div a^2$. If we subtract exponents, we get $5 - 2$, which is 3.

THE QUOTIENT RULE

For any nonzero number a and any positive integers m and n,

$$\frac{a^m}{a^n} = a^{m-n}.$$

(When dividing with exponential notation, if the bases are the same, keep the base and subtract the exponent of the denominator from the exponent of the numerator.)

EXAMPLES Divide and simplify. By simplify, we mean write the expression as one number to a nonnegative power.

13. $\dfrac{6^5}{6^3} = 6^{5-3}$ Subtracting exponents
$= 6^2$

14. $\dfrac{x^8}{x^2} = x^{8-2}$
$= x^6$

15. $\dfrac{t^{12}}{t} = \dfrac{t^{12}}{t^1} = t^{12-1}$
$= t^{11}$

16. $\dfrac{p^5q^7}{p^2q^5} = \dfrac{p^5}{p^2} \cdot \dfrac{q^7}{q^5} = p^{5-2}q^{7-5}$
$= p^3q^2$

The quotient rule can also be used to explain the definition of 0 as an exponent. Consider the expression a^4/a^4, where a is nonzero:

$$\frac{a^4}{a^4} = \frac{a \cdot a \cdot a \cdot a}{a \cdot a \cdot a \cdot a} = 1.$$

This is true because the numerator and the denominator are the same. Now suppose we apply the rule for dividing powers with the same base:

$$\frac{a^4}{a^4} = a^{4-4} = a^0 = 1.$$

Since both expressions a^4/a^4 and a^{4-4} are equivalent to 1, it follows that $a^0 = 1$, when $a \neq 0$.

We can explain why we do not define 0^0 using the quotient rule. We know that 0^0 is 0^{1-1}. But 0^{1-1} is also equal to $0/0$. We have already seen that division by 0 is undefined, so 0^0 is also undefined.

Do Exercises 19–22.

f NEGATIVE INTEGERS AS EXPONENTS

We can use the rule for dividing powers with like bases to lead us to a definition of exponential notation when the exponent is a negative integer. Consider $5^3/5^7$ and first simplify it using procedures we have learned for working with fractions:

$$\frac{5^3}{5^7} = \frac{5 \cdot 5 \cdot 5}{5 \cdot 5 \cdot 5 \cdot 5 \cdot 5 \cdot 5 \cdot 5} = \frac{5 \cdot 5 \cdot 5 \cdot 1}{5 \cdot 5 \cdot 5 \cdot 5 \cdot 5 \cdot 5 \cdot 5}$$
$$= \frac{5 \cdot 5 \cdot 5}{5 \cdot 5 \cdot 5} \cdot \frac{1}{5 \cdot 5 \cdot 5 \cdot 5} = \frac{1}{5^4}.$$

Now we apply the rule for dividing powers with the same bases. Then

$$\frac{5^3}{5^7} = 5^{3-7} = 5^{-4}.$$

From these two expressions for $5^3/5^7$, it follows that

$$5^{-4} = \frac{1}{5^4}.$$

This leads to our definition of negative exponents:

> For any real number a that is nonzero and any integer n,
>
> $$\frac{1}{a^n} = a^{-n}.$$
>
> (To take the reciprocal of a^n, change the sign of the exponent.)

The numbers a^n and a^{-n} are reciprocals because

$$a^n \cdot a^{-n} = a^n \cdot \frac{1}{a^n} = \frac{a^n}{a^n} = 1.$$

EXAMPLES Express using positive exponents. Then simplify.

17. $4^{-2} = \frac{1}{4^2} = \frac{1}{16}$

18. $(-3)^{-2} = \frac{1}{(-3)^2} = \frac{1}{(-3)(-3)} = \frac{1}{9}$

19. $m^{-3} = \frac{1}{m^3}$

20. $ab^{-1} = a\left(\frac{1}{b^1}\right) = a\left(\frac{1}{b}\right) = \frac{a}{b}$

21. $\frac{1}{x^{-3}} = x^{-(-3)} = x^3$

22. $3c^{-5} = 3\left(\frac{1}{c^5}\right) = \frac{3}{c^5}$

Caution! Note in Example 17 that

$$4^{-2} \neq -16 \quad \text{and} \quad 4^{-2} \neq -\frac{1}{16}.$$

Do Exercises 23–28.

The rules for multiplying and dividing powers with like bases still hold when exponents are 0 or negative. We will state them in a summary at the end of this section.

Express with positive exponents. Then simplify.

23. 4^{-3}

24. 5^{-2}

25. 2^{-4}

26. $(-2)^{-3}$

27. $4p^{-3}$

28. $\frac{1}{x^{-2}}$

Answers on page A-3

Simplify.

29. $5^{-2} \cdot 5^4$

30. $x^{-3} \cdot x^{-4}$

31. $\dfrac{7^{-2}}{7^3}$

32. $\dfrac{b^{-2}}{b^{-3}}$

33. $\dfrac{t}{t^{-5}}$

Answers on page A-3

EXAMPLES Simplify. By simplify, we mean write the expression as one number to a nonnegative power.

23. $7^{-3} \cdot 7^6 = 7^{-3+6} = 7^3$ **Adding exponents**

24. $x^4 \cdot x^{-3} = x^{4+(-3)} = x^1 = x$

25. $\dfrac{5^4}{5^{-2}} = 5^{4-(-2)} = 5^{4+2} = 5^6$ **Subtracting exponents**

26. $\dfrac{x}{x^7} = x^{1-7} = x^{-6} = \dfrac{1}{x^6}$

27. $\dfrac{b^{-4}}{b^{-5}} = b^{-4-(-5)} = b^1 = b$

28. $y^{-4} \cdot y^{-8} = y^{-4+(-8)} = y^{-12} = \dfrac{1}{y^{12}}$

In Examples 25–27 (division with exponents), it may help to think as follows: After writing the base, write the top exponent. Then write a subtraction sign. Next write the bottom exponent. Then do the subtraction by adding the opposite. For example,

$$\frac{x^{-3}}{x^{-5}} = x^{-3-(-5)} = x^{-3+5} = x^2$$

(1) Write the base.
(2) Write the top exponent.
(3) Write a subtraction sign.
(4) Write the bottom exponent.

Do Exercises 29–33.

The following is another way to arrive at the definition of negative exponents.

On each side, we divide by 5 at each step.

$$5 \cdot 5 \cdot 5 \cdot 5 = 5^4$$
$$5 \cdot 5 \cdot 5 = 5^3$$
$$5 \cdot 5 = 5^2$$
$$5 = 5^1$$
$$1 = 5^0$$
$$\frac{1}{5} = 5^?$$
$$\frac{1}{25} = 5^?$$

On this side, the exponents decrease by 1.

To continue the pattern, it should follow that

$$\frac{1}{5} = \frac{1}{5^1} = 5^{-1} \quad \text{and} \quad \frac{1}{25} = \frac{1}{5^2} = 5^{-2}.$$

The following is a summary of the definitions and rules for exponents that we have considered in this section.

DEFINITIONS AND RULES FOR EXPONENTS

For any integers m and n,

1 as an exponent:	$a^1 = a$;
0 as an exponent:	$a^0 = 1,\ a \neq 0$;
Negative integers as exponents:	$\dfrac{1}{a^n} = a^{-n},\ \dfrac{1}{a^{-n}} = a^n$;
Product Rule:	$a^m \cdot a^n = a^{m+n}$;
Quotient Rule:	$\dfrac{a^m}{a^n} = a^{m-n}$

53. $(3y)^4(3y)^8$

54. $(2t)^8(2t)^{17}$

55. $(7y)^1(7y)^{16}$

56. $(8x)^0(8x)^1$

57. $3^{-5} \cdot 3^8$

58. $5^{-8} \cdot 5^9$

59. $x^{-2} \cdot x$

60. $x \cdot x^{-1}$

61. $x^{14} \cdot x^3$

62. $x^9 \cdot x^4$

63. $x^{-7} \cdot x^{-6}$

64. $y^{-5} \cdot y^{-8}$

65. $t^8 \cdot t^{-8}$

66. $m^{10} \cdot m^{-10}$

e, **f** Divide and simplify.

67. $\dfrac{7^5}{7^2}$

68. $\dfrac{5^8}{5^6}$

69. $\dfrac{8^{12}}{8^6}$

70. $\dfrac{8^{13}}{8^2}$

71. $\dfrac{y^9}{y^5}$

72. $\dfrac{x^{11}}{x^9}$

73. $\dfrac{16^2}{16^8}$

74. $\dfrac{7^2}{7^9}$

75. $\dfrac{m^6}{m^{12}}$

76. $\dfrac{a^3}{a^4}$

77. $\dfrac{(8x)^6}{(8x)^{10}}$

78. $\dfrac{(8t)^4}{(8t)^{11}}$

ANSWERS

53. ______
54. ______
55. ______
56. ______
57. ______
58. ______
59. ______
60. ______
61. ______
62. ______
63. ______
64. ______
65. ______
66. ______
67. ______
68. ______
69. ______
70. ______
71. ______
72. ______
73. ______
74. ______
75. ______
76. ______
77. ______
78. ______

ANSWERS

79. ____
80. ____
81. ____
82. ____
83. ____
84. ____
85. ____
86. ____
87. ____
88. ____
89. ____
90. ____
91. ____
92. ____
93. ____
94. ____
95. ____
96. ____
97. ____
98. ____
99. ____
100. ____
101. ____
102. ____
103. ____
104. ____
105. ____
106. ____
107. ____
108. ____
109. ____
110. ____
111. ____

79. $\frac{(2y)^9}{(2y)^9}$ **80.** $\frac{(6y)^7}{(6y)^7}$ **81.** $\frac{x}{x^{-1}}$ **82.** $\frac{y^8}{y}$

83. $\frac{x^7}{x^{-2}}$ **84.** $\frac{t^8}{t^{-3}}$ **85.** $\frac{z^{-6}}{z^{-2}}$ **86.** $\frac{x^{-9}}{x^{-3}}$

87. $\frac{x^{-5}}{x^{-8}}$ **88.** $\frac{y^{-2}}{y^{-9}}$ **89.** $\frac{m^{-9}}{m^{-9}}$ **90.** $\frac{x^{-7}}{x^{-7}}$

Simplify.

91. 5^2, 5^{-2}, $\left(\frac{1}{5}\right)^2$, $\left(\frac{1}{5}\right)^{-2}$, -5^2, and $(-5)^2$

92. 8^2, 8^{-2}, $\left(\frac{1}{8}\right)^2$, $\left(\frac{1}{8}\right)^{-2}$, -8^2, and $(-8)^2$

SKILL MAINTENANCE

93. Translate to an algebraic expression: Sixty-four percent of t.

94. Evaluate $3x/y$ when $x = 4$ and $y = 12$.

95. Divide: $1555.2 \div 24.3$.

96. Add: $1555.2 + 24.3$.

97. Solve: $3x - 4 + 5x - 10x = x - 8$.

98. Factor: $8x - 56$.

SYNTHESIS

99. Determine whether $(5y)^0$ and $5y^0$ are equivalent expressions.

Simplify.

100. $(y^{2x})(y^{3x})$ **101.** $a^{5k} \div a^{3k}$ **102.** $\frac{a^{6t}(a^{7t})}{a^{9t}}$

103. $\frac{\left(\frac{1}{2}\right)^4}{\left(\frac{1}{2}\right)^5}$ **104.** $\frac{(0.8)^5}{(0.8)^3(0.8)^2}$

105. Determine whether $(a + b)^2$ and $a^2 + b^2$ are equivalent. (*Hint:* Choose values for a and b and evaluate.)

Use $>$, $<$, or $=$ for ▒ to write a true sentence.

106. 3^5 ▒ 3^4 **107.** 4^2 ▒ 4^3 **108.** 4^3 ▒ 5^3 **109.** 4^3 ▒ 3^4

Find a value of the variable that shows that the two expressions are *not* equivalent.

110. $3x^2$; $(3x)^2$

111. $\frac{x + 2}{2}$; x

3.2 *Exponents and Scientific Notation*

We now enhance our ability to manipulate exponential expressions by considering three more rules. The rules are also applied to a new way to name numbers called *scientific notation.*

OBJECTIVES

After finishing Section 3.2, you should be able to:

- **a** Use the power rule to raise powers to powers.
- **b** Raise a product to a power and a quotient to a power.
- **c** Convert between scientific notation and decimal notation.
- **d** Multiply and divide using scientific notation.
- **e** Solve problems using scientific notation.

FOR EXTRA HELP

TAPE 8 | TAPE 6B | MAC: 3A IBM: 3A

a RAISING POWERS TO POWERS

Consider an expression like $(3^2)^4$. We are raising 3^2 to the fourth power:

$$\begin{aligned}(3^2)^4 &= (3^2)(3^2)(3^2)(3^2)\\ &= (3\cdot 3)(3\cdot 3)(3\cdot 3)(3\cdot 3)\\ &= 3\cdot 3\cdot 3\cdot 3\cdot 3\cdot 3\cdot 3\cdot 3\\ &= 3^8.\end{aligned}$$

Note that in this case we could have multiplied the exponents:

$$(3^2)^4 = 3^{2\cdot 4} = 3^8.$$

Likewise, $(y^8)^3 = (y^8)(y^8)(y^8) = y^{24}$. Once again, we get the same result if we multiply the exponents:

$$(y^8)^3 = y^{8\cdot 3} = y^{24}.$$

THE POWER RULE

For any real number a and any integers m and n,

$$(a^m)^n = a^{mn}.$$

(To raise a power to a power, multiply the exponents.)

EXAMPLES Simplify. Express the answers using positive exponents.

1. $(3^5)^4 = 3^{5\cdot 4} = 3^{20}$ — Multiplying exponents

2. $(2^2)^5 = 2^{2\cdot 5} = 2^{10}$

3. $(y^{-5})^7 = y^{-5\cdot 7} = y^{-35} = \dfrac{1}{y^{35}}$

4. $(x^4)^{-2} = x^{4(-2)} = x^{-8} = \dfrac{1}{x^8}$

5. $(a^{-4})^{-6} = a^{(-4)(-6)} = a^{24}$

Do Exercises 1–4.

Simplify. Express the answers using positive exponents.

1. $(3^4)^5$

2. $(x^{-3})^4$

3. $(y^{-5})^{-3}$

4. $(x^{-4})^8$

Answers on page A-3

b RAISING A PRODUCT OR A QUOTIENT TO A POWER

When an expression inside parentheses is raised to a power, the inside expression is the base. Let us compare $2a^3$ and $(2a)^3$:

$2a^3 = 2\cdot a\cdot a\cdot a;$ — The base is a.

$$\begin{aligned}(2a)^3 &= (2a)(2a)(2a) && \text{The base is } 2a.\\ &= (2\cdot 2\cdot 2)(a\cdot a\cdot a) && \text{Using the associative law of multiplication to regroup the factors}\\ &= 2^3a^3\\ &= 8a^3.\end{aligned}$$

We see that $2a^3$ and $(2a)^3$ are *not* equivalent. We also see that we can evaluate the power $(2a)^3$ by raising each factor to the power 3. This leads us to the following rule for raising a product to a power.

Simplify.

5. $(2x^5y^{-3})^4$

6. $(5x^5y^{-6}z^{-3})^2$

7. $[(-x)^{37}]^2$

8. $(3y^{-2}x^{-5}z^8)^3$

Simplify.

9. $\left(\frac{x^6}{5}\right)^2$

10. $\left(\frac{2t^5}{w^4}\right)^3$

11. $\left(\frac{x^4}{3}\right)^{-2}$

Answers on page A-3

RAISING A PRODUCT TO A POWER

For any real numbers a and b and any integer n,

$$(ab)^n = a^n b^n.$$

(To raise a product to the nth power, raise each factor to the nth power.)

EXAMPLES

6. $(4x^2)^3 = 4^3 \cdot (x^2)^3$ Raising each factor to the third power
$= 64x^6$

7. $(5x^3y^5z^2)^4 = 5^4(x^3)^4(y^5)^4(z^2)^4$ Raising each factor to the fourth power
$= 625x^{12}y^{20}z^8$

8. $(-5x^4y^3)^3 = (-5)^3(x^4)^3(y^3)^3$
$= -125x^{12}y^9$

9. $[(-x)^{25}]^2 = (-x)^{50}$
$= (-1 \cdot x)^{50}$ Using the property of -1 (Section 1.8)
$= (-1)^{50}x^{50}$
$= 1 \cdot x^{50}$ The product of an even number of negative factors is positive.
$= x^{50}$

10. $(5x^2y^{-2})^3 = 5^3(x^2)^3(y^{-2})^3 = 125x^6y^{-6}$ Be careful to raise *each* factor to the third power.
$= \frac{125x^6}{y^6}$

11. $(3x^3y^{-5}z^2)^4 = 3^4(x^3)^4(y^{-5})^4(z^2)^4$
$= 81x^{12}y^{-20}z^8 = \frac{81x^{12}z^8}{y^{20}}$

Do Exercises 5–8.

There is a similar rule for raising a quotient to a power.

RAISING A QUOTIENT TO A POWER

For any real numbers a and b, $b \neq 0$, and any integer n,

$$\left(\frac{a}{b}\right)^n = \frac{a^n}{b^n}.$$

(To raise a quotient to a power, raise both the numerator and the denominator to the power.)

EXAMPLES Simplify.

12. $\left(\frac{x^2}{4}\right)^3 = \frac{(x^2)^3}{4^3} = \frac{x^6}{64}$

13. $\left(\frac{3a^4}{b^3}\right)^2 = \frac{(3a^4)^2}{(b^3)^2} = \frac{3^2(a^4)^2}{b^{3\cdot 2}} = \frac{9a^8}{b^6}$

14. $\left(\frac{y^3}{5}\right)^{-2} = \frac{(y^3)^{-2}}{5^{-2}} = \frac{y^{-6}}{5^{-2}} = \frac{\frac{1}{y^6}}{\frac{1}{5^2}} = \frac{1}{y^6} \div \frac{1}{5^2} = \frac{1}{y^6} \cdot \frac{5^2}{1} = \frac{25}{y^6}$

Do Exercises 9–11.

c SCIENTIFIC NOTATION

There are many kinds of symbols, or notation, for numbers. You are already familiar with fractional notation, decimal notation, and percent notation. Now we study another, **scientific notation,** which is especially useful when calculations involve very large or very small numbers. The following are examples of scientific notation:

The distance from the earth to the planet Neptune:

$$2.79 \times 10^9 \text{ mi} = 2{,}790{,}000{,}000 \text{ mi};$$

The mass of a hydrogen atom:

$$1.7 \times 10^{-24} \text{ g} = 0.0000000000000000000000017 \text{ g}.$$

> *Scientific notation* for a number is an expression of the type
>
> $$N \times 10^n,$$
>
> where N is greater than or equal to 1 and less than 10 ($1 \le N < 10$) and N is expressed in decimal notation. 10^n is also considered to be scientific notation when $N = 1$.

Convert to scientific notation.

12. 0.000517

13. 523,000,000

You should try to make conversions to scientific notation mentally as much as possible. Here is a handy mental device.

> A positive exponent in scientific notation indicates a large number (greater than 1) and a negative exponent indicates a small number (less than 1).

EXAMPLES

15. $78{,}000 = 7.8 \times 10^4$ 7.8,000. (4 places)

Large number, so the exponent is positive.

16. $0.0000057 = 5.7 \times 10^{-6}$ 0.000005.7 (6 places)

Small number, so the exponent is negative.

Each of the following is *not* scientific notation.

12.46×10^7 — This number is greater than 10.

0.347×10^{-5} — This number is less than 1.

Do Exercises 12 and 13.

EXAMPLES Convert mentally to decimal notation.

17. $7.893 \times 10^5 = 789{,}300$ 7.89300. (5 places)

Positive exponent, so the answer is a large number.

Answers on page A-3

Convert to decimal notation.

14. 6.893×10^{11}

15. 5.67×10^{-5}

18. $4.7 \times 10^{-8} = 0.000000047$ $\qquad 0.00000004.7$ (8 places)

Negative exponent, so the answer is a small number.

Do Exercises 14 and 15.

When using a calculator, we might express a number like 260,000,000,000 using scientific notation in a form like

2.6 E 11, or 2.6 11,

or perhaps in other forms, depending on your calculator.

d MULTIPLYING AND DIVIDING USING SCIENTIFIC NOTATION

MULTIPLYING

Consider the product

$$400 \cdot 2000 = 800{,}000.$$

In scientific notation, this is

$$(4 \times 10^2) \cdot (2 \times 10^3) = (4 \cdot 2)(10^2 \cdot 10^3) = 8 \times 10^5.$$

By applying the commutative and associative laws, we can find this product by multiplying $4 \cdot 2$, to get 8, and $10^2 \cdot 10^3$, to get 10^5 (we do this by adding the exponents).

Multiply and write scientific notation for the result.

16. $(1.12 \times 10^{-8})(5 \times 10^{-7})$

17. $(9.1 \times 10^{-17})(8.2 \times 10^3)$

EXAMPLE 19 Multiply: $(1.8 \times 10^6) \cdot (2.3 \times 10^{-4})$.

We apply the commutative and associative laws to get

$$\begin{aligned}(1.8 \times 10^6) \cdot (2.3 \times 10^{-4}) &= (1.8 \cdot 2.3) \times (10^6 \cdot 10^{-4}) \\ &= 4.14 \times 10^{6+(-4)} \quad \text{Adding exponents} \\ &= 4.14 \times 10^2.\end{aligned}$$

EXAMPLE 20 Multiply: $(3.1 \times 10^5) \cdot (4.5 \times 10^{-3})$.

We have

$$\begin{aligned}(3.1 \times 10^5) \cdot (4.5 \times 10^{-3}) &= (3.1 \times 4.5)(10^5 \cdot 10^{-3}) \\ &= 13.95 \times 10^2.\end{aligned}$$

The answer at this stage is 13.95×10^2, but this is *not* scientific notation, because 13.95 is not a number between 1 and 10. To find scientific notation, we convert 13.95 to scientific notation and simplify:

$$\begin{aligned}13.95 \times 10^2 &= (1.395 \times 10^1) \times 10^2 \quad \text{Substituting } 1.395 \times 10^1 \text{ for } 13.95 \\ &= 1.395 \times (10^1 \times 10^2) \quad \text{Associative law} \\ &= 1.395 \times 10^3. \quad \text{Adding exponents}\end{aligned}$$

The answer is

$$1.395 \times 10^3.$$

Do Exercises 16 and 17.

Answers on page A-3

DIVIDING

Consider the quotient

$$800{,}000 \div 400 = 2000.$$

In scientific notation, this is

$$(8 \times 10^5) \div (4 \times 10^2) = \frac{8 \times 10^5}{4 \times 10^2} = \frac{8}{4} \times \frac{10^5}{10^2} = 2 \times 10^3.$$

We can find this product by dividing 8 by 4, to get 2, and 10^5 by 10^2, to get 10^3 (we do this by subtracting the exponents).

EXAMPLE 21 Divide: $(3.41 \times 10^5) \div (1.1 \times 10^{-3})$.

$$\begin{aligned}(3.41 \times 10^5) \div (1.1 \times 10^{-3}) &= \frac{3.41 \times 10^5}{1.1 \times 10^{-3}}\\ &= \frac{3.41}{1.1} \times \frac{10^5}{10^{-3}}\\ &= 3.1 \times 10^{5-(-3)}\\ &= 3.1 \times 10^8\end{aligned}$$

EXAMPLE 22 Divide: $(6.4 \times 10^{-7}) \div (8.0 \times 10^6)$.

We have

$$\begin{aligned}(6.4 \times 10^{-7}) \div (8.0 \times 10^6) &= \frac{6.4 \times 10^{-7}}{8.0 \times 10^6}\\ &= \frac{6.4}{8.0} \times \frac{10^{-7}}{10^6}\\ &= 0.8 \times 10^{-7-6}\\ &= 0.8 \times 10^{-13}.\end{aligned}$$

The answer at this stage is

$$0.8 \times 10^{-13},$$

but this is *not* scientific notation, because 0.8 is not a number between 1 and 10. To find scientific notation, we convert 0.8 to scientific notation and simplify:

$$\begin{aligned}0.8 \times 10^{-13} &= (8.0 \times 10^{-1}) \times 10^{-13} && \text{Substituting } 8.0 \times 10^{-1} \text{ for } 0.8\\ &= 8.0 \times (10^{-1} \times 10^{-13}) && \text{Associative law}\\ &= 8.0 \times 10^{-14}. && \text{Adding exponents}\end{aligned}$$

The answer is

$$8.0 \times 10^{-14}.$$

Do Exercises 18 and 19.

e SOLVING PROBLEMS WITH SCIENTIFIC NOTATION

EXAMPLE 23 There are 4400 members in the Professional Bowlers Association. There are 261 million people in the United States. What part of the population are members of the Professional Bowlers Association? Write scientific notation for the answer.

Divide and write scientific notation for the result.

18. $\dfrac{4.2 \times 10^5}{2.1 \times 10^2}$

19. $\dfrac{1.1 \times 10^{-4}}{2.0 \times 10^{-7}}$

Answers on page A-3

20. There are 300,000 words in the English language. The above-average person knows about 20,000 of them. What part of the total number of words does the above-average person know? Write scientific notation for the answer.

21. Americans eat 6.5 million gallons of popcorn each day. How much popcorn do they eat in one year? Write scientific notation for the answer.

Answers on page A-3

The part of the population that belongs to the Professional Bowlers Association is

$$\frac{4400}{261 \text{ million}}.$$

We know that 1 million = 1,000,000 = 10^6, so 261 million = 261×10^6, or 2.61×10^8. We also have 4400 = 4.4×10^3. We can now divide and write scientific notation for the answer:

$$\frac{4400}{261 \text{ million}} = \frac{4.4 \times 10^3}{2.61 \times 10^8}$$
$$\approx 1.6858 \times 10^{-5}.$$

Do Exercise 20.

EXAMPLE 24 Americans drink 3 million gallons of orange juice in one day. How much orange juice is consumed in this country in one year? Write scientific notation for the answer.

There are 365 days in a year, so the amount of orange juice consumed is

$$(365 \text{ days}) \cdot (3 \text{ million}) = (3.65 \times 10^2)(3 \times 10^6)$$
$$= 10.95 \times 10^8$$
$$= (1.095 \times 10^1) \times 10^8$$
$$= 1.095 \times 10^9.$$

There are 1.095×10^9 gallons of orange juice consumed in this country in one year.

Do Exercise 21.

The following is a summary of the definitions and rules for exponents that we have considered in this section and the preceding one.

DEFINITIONS AND RULES FOR EXPONENTS

Exponent of 1:	$a^1 = a$
Exponent of 0:	$a^0 = 1, a \neq 0$
Negative exponents:	$\frac{1}{a^n} = a^{-n}, a \neq 0$
Product Rule:	$a^m \cdot a^n = a^{m+n}$
Quotient Rule:	$\frac{a^m}{a^n} = a^{m-n}$
Power Rule:	$(a^m)^n = a^{mn}$
Raising a product to a power:	$(ab)^n = a^n b^n$
Raising a quotient to a power:	$\left(\frac{a}{b}\right)^n = \frac{a^n}{b^n}$
Scientific notation:	$N \times 10^n$, or 10^n, where $1 \leq N < 10$

NAME SECTION DATE

Exercise Set 3.2

a, **b** Simplify.

1. $(2^3)^2$

2. $(5^2)^4$

3. $(5^2)^{-3}$

4. $(7^{-3})^5$

5. $(x^{-3})^{-4}$

6. $(a^{-5})^{-6}$

7. $(4x^3)^2$

8. $4(x^3)^2$

9. $(x^4y^5)^{-3}$

10. $(t^5x^3)^{-4}$

11. $(x^{-6}y^{-2})^{-4}$

12. $(x^{-2}y^{-7})^{-5}$

13. $(3x^3y^{-8}z^{-3})^2$

14. $(2a^2y^{-4}z^{-5})^3$

15. $\left(\frac{a^2}{b^3}\right)^4$

16. $\left(\frac{x^3}{y^4}\right)^5$

17. $\left(\frac{y^3}{2}\right)^2$

18. $\left(\frac{a^5}{3}\right)^3$

19. $\left(\frac{y^2}{2}\right)^{-3}$

20. $\left(\frac{a^4}{3}\right)^{-2}$

21. $\left(\frac{x^2y}{z}\right)^3$

22. $\left(\frac{m}{n^4p}\right)^3$

23. $\left(\frac{a^2b}{cd^3}\right)^{-2}$

24. $\left(\frac{2a^2}{3b^4}\right)^{-3}$

ANSWERS

1. ______
2. ______
3. ______
4. ______
5. ______
6. ______
7. ______
8. ______
9. ______
10. ______
11. ______
12. ______
13. ______
14. ______
15. ______
16. ______
17. ______
18. ______
19. ______
20. ______
21. ______
22. ______
23. ______
24. ______

ANSWERS

25. ______

26. ______

27. ______

28. ______

29. ______

30. ______

31. ______

32. ______

33. ______

34. ______

35. ______

36. ______

37. ______

38. ______

39. ______

40. ______

41. ______

42. ______

c Convert to scientific notation.

25. 28,000,000,000

26. 4,900,000,000,000

27. 907,000,000,000,000,000

28. 168,000,000,000,000

29. 0.00000304

30. 0.000000000865

31. 0.000000018

32. 0.00000000002

33. 100,000,000,000

34. 0.0000001

Convert to decimal notation.

35. 8.74×10^7

36. 1.85×10^8

37. 5.704×10^{-8}

38. 8.043×10^{-4}

39. 10^7

40. 10^6

41. 10^{-5}

42. 10^{-8}

d Multiply or divide and write scientific notation for the result.

43. $(3 \times 10^4)(2 \times 10^5)$

44. $(3.9 \times 10^8)(8.4 \times 10^{-3})$

45. $(5.2 \times 10^5)(6.5 \times 10^{-2})$

46. $(7.1 \times 10^{-7})(8.6 \times 10^{-5})$

47. $(9.9 \times 10^{-6})(8.23 \times 10^{-8})$

48. $(1.123 \times 10^4) \times 10^{-9}$

49. $\dfrac{8.5 \times 10^8}{3.4 \times 10^{-5}}$

50. $\dfrac{5.6 \times 10^{-2}}{2.5 \times 10^5}$

51. $(3.0 \times 10^6) \div (6.0 \times 10^9)$

52. $(1.5 \times 10^{-3}) \div (1.6 \times 10^{-6})$

53. $\dfrac{7.5 \times 10^{-9}}{2.5 \times 10^{12}}$

54. $\dfrac{4.0 \times 10^{-3}}{8.0 \times 10^{20}}$

e Solve. Write scientific notation for the answer.

55. The average discharge at the mouth of the Amazon River is 4,200,000 ft^3 (cubic feet) per second. How much water is discharged from the Amazon River in one hour? in one year?

56. About 250,000 people die each day in the world. How many die in one year?

57. The mass of the earth is about 5.98×10^{24} kilograms (kg). The mass of the planet Saturn is about 95 times the mass of the earth. Write scientific notation for the mass of Saturn.

58. There are 300,000 words in the English language. The average person knows about 10,000 of them. What part of the total number of words does the average person know?

Answers

43. ______

44. ______

45. ______

46. ______

47. ______

48. ______

49. ______

50. ______

51. ______

52. ______

53. ______

54. ______

55. ______

56. ______

57. ______

58. ______

ANSWERS

59. ______

60. ______

61. ______

62. ______

63. ______

64. ______

65. ______

66. ______

67. ______

68. ______

69. ______

70. ______

71. ______

72. ______

73. ______

74. ______

75. ______

76. ______

77. ______

59. Americans spend $6,800,000 in vending machines each day. How much do they spend in one year?

60. The wavelength of light is given by the velocity divided by the frequency. The velocity of light is 300,000,000 m/sec. Red light has a frequency of 400,000,000,000,000 cycles per second. What is the wavelength of red light?

SKILL MAINTENANCE

Factor.

61. $9x - 36$

62. $4x - 2y + 16$

Solve.

63. $2x - 4 - 5x + 8 = x - 3$

64. $4(x - 3) + 5 = 6(x + 2) - 8$

65. $8(2x + 3) - 2(x - 5) = 10$

66. $8x + 7 - 9x = 12 - 6x + 5$

SYNTHESIS

67. Carry out the indicated operations. Write scientific notation for the result.

$$\frac{(5.2 \times 10^{6})(6.1 \times 10^{-11})}{1.28 \times 10^{-3}}$$

68. Find the reciprocal and express in scientific notation.

$$(6.25 \times 10^{-3})$$

Simplify.

69. $\dfrac{(5^{12})^2}{5^{25}}$

70. $\dfrac{a^{22}}{(a^2)^{11}}$

71. $\dfrac{(3^5)^4}{3^5 \cdot 3^4}$

72. $\dfrac{49^{18}}{7^{35}}$

(*Hint:* Study Exercise 70.)

73. $\left(\dfrac{1}{a}\right)^{-n}$

74. $\dfrac{(0.4)^5}{[(0.4)^3]^2}$

Determine whether each of the following is true for any pairs of integers m and n and any positive numbers x and y.

75. $x^m \cdot y^n = (xy)^{mn}$

76. $x^m \cdot y^m = (xy)^{2m}$

77. $(x - y)^m = x^m - y^m$

3.3 *Introduction to Polynomials*

We have already learned to evaluate and to manipulate certain kinds of algebraic expressions. We will now consider algebraic expressions called *polynomials*.

The following are examples of *monomials in one variable:*

$$3x^2, \quad 2x, \quad -5, \quad 37p^4, \quad 0.$$

Each expression is a constant or a constant times some variable to a nonnegative integer power. More formally, a **monomial** is an expression of the type ax^n, where a is a real-number constant and n is a nonnegative integer.

Algebraic expressions like the following are **polynomials:**

$$\tfrac{3}{4}y^5, \quad -2, \quad 5y + 3, \quad 3x^2 + 2x - 5, \quad -7a^3 + \tfrac{1}{2}a, \quad 6x, \quad 37p^4, \quad x, \quad 0.$$

> A *polynomial* is a monomial or a combination of sums and/or differences of monomials.

The following algebraic expressions are *not* polynomials:

$$\textbf{(1)}\ \frac{x+3}{x-4}, \qquad \textbf{(2)}\ 5x^3 - 2x^2 + \frac{1}{x}, \qquad \textbf{(3)}\ \frac{1}{x^3 - 2}.$$

Expressions (1) and (3) are not polynomials because they represent quotients, not sums. Expression (2) is not a polynomial because

$$\frac{1}{x} = x^{-1},$$

and this is not a monomial because the exponent is negative.

Do Exercise 1.

a EVALUATING POLYNOMIALS AND APPLICATIONS

When we replace the variable in a polynomial by a number, the polynomial then represents a number called a **value** of the polynomial. Finding that number, or value, is called **evaluating the polynomial.** We evaluate a polynomial using the rules for order of operations (Section 1.8).

EXAMPLE 1 Evaluate the polynomial when $x = 2$.

a) $3x + 5 = 3 \cdot 2 + 5$
$= 6 + 5$
$= 11$

b) $2x^2 - 7x + 3 = 2 \cdot 2^2 - 7 \cdot 2 + 3$
$= 2 \cdot 4 - 14 + 3$
$= 8 - 14 + 3$
$= -3$

OBJECTIVES

After finishing Section 3.3, you should be able to:

- **a** Evaluate a polynomial for a given value of the variable.
- **b** Identify the terms of a polynomial.
- **c** Identify the like terms of a polynomial.
- **d** Identify the coefficients of a polynomial.
- **e** Collect the like terms of a polynomial.
- **f** Arrange a polynomial in descending order, or collect the like terms and then arrange in descending order.
- **g** Identify the degree of each term of a polynomial and the degree of the polynomial.
- **h** Identify the missing terms of a polynomial.
- **i** Classify a polynomial as a monomial, binomial, trinomial, or none of these.

FOR EXTRA HELP

TAPE 8 | TAPE 6B | MAC: 3A IBM: 3A

1. Write three polynomials.

Answer on page A-3

EXAMPLE 2 Evaluate the polynomial when $x = -5$.

a) $2 - x^3 = 2 - (-5)^3 = 2 - (-125) = 2 + 125 = 127$

b) $-x^2 - 3x + 1 = -(-5)^2 - 3(-5) + 1$
$= -25 + 15 + 1$
$= -9$

Do Exercises 2–5.

Polynomials occur in many real-world situations. The following examples are two such applications.

EXAMPLE 3 *Medical dosage.* The concentration, in parts per million, of a certain medication in the bloodstream after time t, in hours, is given by the polynomial

$$-0.05t^2 + 2t + 2.$$

Find the concentration after 2 hr.

To find the concentration after 2 hr, we evaluate the polynomial for $t = 2$:

$$\begin{aligned} -0.05t^2 + 2t + 2 &= -0.05(2)^2 + 2(2) + 2 \\ &= -0.05(4) + 2(2) + 2 \\ &= -0.2 + 4 + 2 \\ &= -0.2 + 6 \\ &= 5.8. \end{aligned}$$

Carrying out the calculation using the rules for order of operations

The concentration after 2 hr is 5.8 parts per million.

EXAMPLE 4 *Games in a sports league.* In a sports league of n teams in which each team plays every other team twice, the total number of games to be played is given by the polynomial

$$n^2 - n.$$

A women's slow-pitch softball league has 10 teams. What is the total number of games to be played?

We evaluate the polynomial for $n = 10$:

$$n^2 - n = 10^2 - 10 = 100 - 10 = 90.$$

The league plays 90 games.

Do Exercises 6–8.

b IDENTIFYING TERMS

As we saw in Section 1.4, subtractions can be rewritten as additions. For any polynomial that has some subtractions, we can find an equivalent polynomial using only additions.

EXAMPLES Find an equivalent polynomial using only additions.

5. $-5x^2 - x = -5x^2 + (-x)$

6. $4x^5 - 2x^6 - 4x + 7 = 4x^5 + (-2x^6) + (-4x) + 7$

Do Exercises 9 and 10.

Evaluate the polynomial when $x = 3$.

2. $-4x - 7$

3. $-5x^3 + 7x + 10$

Evaluate the polynomial when $x = -4$.

4. $5x + 7$

5. $2x^2 + 5x - 4$

6. In the situation of Example 4, what is the total number of games to be played in a league of 12 teams?

7. The perimeter of a square of side x is given by the polynomial $4x$.

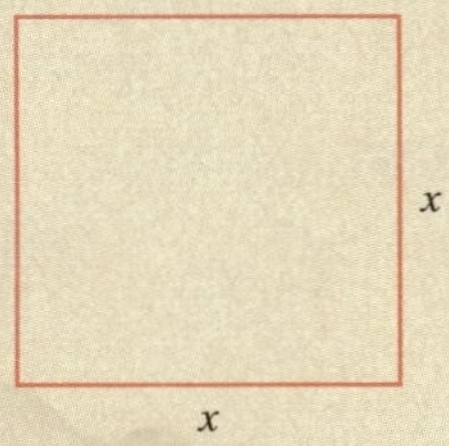

A baseball diamond is a square 90 ft on a side. Find the perimeter of a baseball diamond.

8. In the situation of Example 3, find the concentration after 3 hr.

Find an equivalent polynomial using only additions.

9. $-9x^3 - 4x^5$

10. $-2y^3 + 3y^7 - 7y$

Answers on page A-3

When a polynomial has only additions, the monomials being added are called **terms**. In Example 5, the terms are $-5x^2$ and $-x$. In Example 6, the terms are $4x^5$, $-2x^6$, $-4x$, and 7.

EXAMPLE 7 Identify the terms of the polynomial

$$4x^7 + 3x + 12 + 8x^3 + 5x.$$

Terms: $4x^7$, $3x$, 12, $8x^3$, and $5x$.

If there are subtractions, you can *think* of them as additions without rewriting.

EXAMPLE 8 Identify the terms of the polynomial

$$3t^4 - 5t^6 - 4t + 2.$$

Terms: $3t^4$, $-5t^6$, $-4t$, and 2.

Do Exercises 11 and 12.

c LIKE TERMS

When terms have the same variable and the variable is raised to the same power, we say that they are **like terms,** or **similar terms.**

EXAMPLES Identify the like terms in the polynomials.

9. $4x^3 + 5x - 4x^2 + 2x^3 + x^2$

Like terms: $4x^3$ and $2x^3$ — **Same variable and exponent**

Like terms: $-4x^2$ and x^2 — **Same variable and exponent**

10. $6 - 3a^2 + 8 - a - 5a$

Like terms: 6 and 8 — **Constant terms are like terms because $6 = 6x^0$ and $8 = 8x^0$.**

Like terms: $-a$ and $-5a$

Do Exercises 13–15.

d COEFFICIENTS

The coefficient of the term $5x^3$ is 5. In the following polynomial, the color numbers are the **coefficients**:

$$3x^5 - 2x^3 + 5x + 4.$$

EXAMPLE 11 Identify the coefficient of each term in the polynomial

$$3x^4 - 4x^3 + 7x^2 + x - 8.$$

The coefficient of the first term is 3.

The coefficient of the second term is -4.

The coefficient of the third term is 7.

The coefficient of the fourth term is 1.

The coefficient of the fifth term is -8.

Do Exercise 16.

Identify the terms of the polynomial.

11. $3x^2 + 6x + \dfrac{1}{2}$

12. $-4y^5 + 7y^2 - 3y - 2$

Identify the like terms in the polynomial.

13. $4x^3 - x^3 + 2$

14. $4t^4 - 9t^3 - 7t^4 + 10t^3$

15. $5x^2 + 3x - 10 + 7x^2 - 8x + 11$

16. Identify the coefficient of each term in the polynomial

$$2x^4 - 7x^3 - 8.5x^2 + 10x - 4.$$

Answers on page A-3

Collect like terms.

17. $3x^2 + 5x^2$

18. $4x^3 - 2x^3 + 2 + 5$

19. $\frac{1}{2}x^5 - \frac{3}{4}x^5 + 4x^2 - 2x^2$

20. $24 - 4x^3 - 24$

21. $5x^3 - 8x^5 + 8x^5$

22. $-2x^4 + 16 + 2x^4 + 9 - 3x^5$

Collect like terms.

23. $7x - x$

24. $5x^3 - x^3 + 4$

25. $\frac{3}{4}x^3 + 4x^2 - x^3 + 7$

26. $8x^2 - x^2 + x^3 - 1 - 4x^2 + 10$

Answers on page A-3

e Collecting Like Terms

We can often simplify polynomials by **collecting like terms,** or **combining similar terms.** To do this, we use the distributive laws. We factor out the exponential expression and add or subtract the coefficients. We try to do this mentally as much as possible.

EXAMPLES Collect like terms.

12. $2x^3 - 6x^3 = (2 - 6)x^3 = -4x^3$ Using a distributive law

13. $5x^2 + 7 + 4x^4 + 2x^2 - 11 - 2x^4 = (5 + 2)x^2 + (4 - 2)x^4 + (7 - 11)$
$= 7x^2 + 2x^4 - 4$

Note that using the distributive laws in this manner allows us to collect like terms by adding or subtracting the coefficients. Often the middle step is omitted and we add or subtract mentally, writing just the answer. In collecting like terms, we may get 0.

EXAMPLES Collect like terms.

14. $5x^3 - 5x^3 = (5 - 5)x^3 = 0x^3 = 0$

15. $3x^4 + 2x^2 - 3x^4 + 8 = (3 - 3)x^4 + 2x^2 + 8$
$= 0x^4 + 2x^2 + 8$
$= 2x^2 + 8$

Do Exercises 17–22.

Multiplying a term of a polynomial by 1 does not change the term, but it may make the polynomial easier to factor or add and subtract.

EXAMPLES Collect like terms.

16. $5x^2 + x^2 = 5x^2 + 1x^2$ Replacing x^2 by $1x^2$
$= (5 + 1)x^2$ Using a distributive law
$= 6x^2$

17. $5x^4 - 6x^3 - x^4 = 5x^4 - 6x^3 - 1x^4$ $x^4 = 1x^4$
$= (5 - 1)x^4 - 6x^3$
$= 4x^4 - 6x^3$

18. $\frac{2}{3}x^4 - x^3 - \frac{1}{6}x^4 + \frac{2}{5}x^3 - \frac{3}{10}x^3 = \left(\frac{2}{3} - \frac{1}{6}\right)x^4 + \left(-1 + \frac{2}{5} - \frac{3}{10}\right)x^3$
$= \left(\frac{4}{6} - \frac{1}{6}\right)x^4 + \left(-\frac{10}{10} + \frac{4}{10} - \frac{3}{10}\right)x^3$
$= \frac{3}{6}x^4 - \frac{9}{10}x^3$
$= \frac{1}{2}x^4 - \frac{9}{10}x^3$

Do Exercises 23–26.

f Descending and Ascending Order

Note in the following polynomial that the exponents decrease. We say that the polynomial is arranged in **descending order:**

$$2x^4 - 8x^3 + 5x^2 - x + 3.$$

The term with the largest exponent is first. The term with the next largest exponent is second, and so on. The associative and commutative laws allow us to arrange the terms of a polynomial in descending order.

EXAMPLES Arrange the polynomial in descending order.

19. $6x^5 + 4x^7 + x^2 + 2x^3 = 4x^7 + 6x^5 + 2x^3 + x^2$

20. $\frac{2}{3} + 4x^5 - 8x^2 + 5x - 3x^3 = 4x^5 - 3x^3 - 8x^2 + 5x + \frac{2}{3}$

We usually arrange polynomials in descending order, but not always. The opposite order is called **ascending order.** Generally, if an exercise is written in a certain order, we give the answer in that same order.

Do Exercises 27–29.

EXAMPLE 21 Collect like terms and then arrange in descending order:

$2x^2 - 4x^3 + 3 - x^2 - 2x^3$.

We have

$$2x^2 - 4x^3 + 3 - x^2 - 2x^3 = x^2 - 6x^3 + 3 \quad \text{Collecting like terms}$$
$$= -6x^3 + x^2 + 3 \quad \text{Arranging in descending order}$$

Do Exercises 30 and 31.

g DEGREES

The **degree** of a term is the exponent of the variable. The degree of the term $5x^3$ is 3.

EXAMPLE 22 Identify the degree of each term of $8x^4 + 3x + 7$.

The degree of $8x^4$ is 4.

The degree of $3x$ is 1. Recall that $x = x^1$.

The degree of 7 is 0. Think of 7 as $7x^0$. Recall that $x^0 = 1$.

The **degree of a polynomial** is the largest of the degrees of the terms, unless it is the polynomial 0. The polynomial 0 is a special case. We agree that it has *no* degree either as a term or as a polynomial. This is because we can express 0 as $0 = 0x^5 = 0x^7$, and so on, using any exponent we wish.

EXAMPLE 23 Identify the degree of the polynomial $5x^3 - 6x^4 + 7$.

We have

$5x^3 - 6x^4 + 7$. The largest exponent is 4.

The degree of the polynomial is 4.

Do Exercise 32.

Arrange the polynomial in descending order.

27. $x + 3x^5 + 4x^3 + 5x^2 + 6x^7 - 2x^4$

28. $4x^2 - 3 + 7x^5 + 2x^3 - 5x^4$

29. $-14 + 7t^2 - 10t^5 + 14t^7$

Collect like terms and then arrange in descending order.

30. $3x^2 - 2x + 3 - 5x^2 - 1 - x$

31. $-x + \dfrac{1}{2} + 14x^4 - 7x - 1 - 4x^4$

32. Identify the degree of each term and the degree of the polynomial

$-6x^4 + 8x^2 - 2x + 9$.

Answers on page A-3

Identify the missing terms in the polynomial.

33. $2x^3 + 4x^2 - 2$

34. $-3x^4$

35. $x^3 + 1$

36. $x^4 - x^2 + 3x + 0.25$

Classify the polynomial as a monomial, binomial, trinomial, or none of these.

37. $5x^4$

38. $4x^3 - 3x^2 + 4x + 2$

39. $3x^2 + x$

40. $3x^2 + 2x - 4$

Answers on page A-3

Let us summarize the terminology that we have learned, using the polynomial

$$3x^4 - 8x^3 + 5x^2 + 7x - 6.$$

TERM	COEFFICIENT	DEGREE OF THE TERM	DEGREE OF THE POLYNOMIAL
$3x^4$	3	4	
$-8x^3$	−8	3	
$5x^2$	5	2	4
$7x$	7	1	
−6	−6	0	

h Missing Terms

If a coefficient is 0, we usually do not write the term. We say that we have a **missing term.**

Example 24 Identify the missing terms in the polynomial

$$8x^5 - 2x^3 + 5x^2 + 7x + 8.$$

There is no term with x^4. We say that the x^4-term (or the *fourth-degree term*) is missing.

For certain skills or manipulations, we can write missing terms with zero coefficients or leave space. For example, we can write the polynomial $3x^2 + 9$ as

$$3x^2 + 0x + 9 \quad \text{or} \quad 3x^2 + \qquad 9.$$

Do Exercises 33–36.

i Classifying Polynomials

Polynomials with just one term are called **monomials**. Polynomials with just two terms are called **binomials**. Those with just three terms are called **trinomials**. Those with more than three terms are usually not specified with a name.

Example 25

MONOMIALS	BINOMIALS	TRINOMIALS	NONE OF THESE
$4x^2$	$2x + 4$	$3x^3 + 4x + 7$	$4x^3 - 5x^2 + x - 8$
9	$3x^5 + 6x$	$6x^7 - 7x^2 + 4$	
$-23x^{19}$	$-9x^7 - 6$	$4x^2 - 6x - \frac{1}{2}$	

Do Exercises 37–40.

NAME SECTION DATE

Exercise Set 3.3

a Evaluate the polynomial when $x = 4$.

1. $-5x + 2$

2. $-8x + 1$

3. $2x^2 - 5x + 7$

4. $3x^2 + x - 7$

5. $x^3 - 5x^2 + x$

6. $7 - x + 3x^2$

Evaluate the polynomial when $x = -1$.

7. $3x + 5$

8. $8 - 4x$

9. $x^2 - 2x + 1$

10. $5x + 6 - x^2$

11. $-3x^3 + 7x^2 - 3x - 2$

12. $-2x^3 + 5x^2 - 4x + 3$

Falling distance. The distance s, in feet, traveled by a body falling freely from rest in t seconds is approximated by the polynomial

$$s = 16t^2.$$

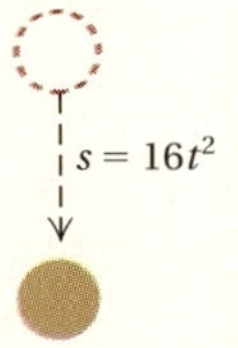

13. A brick is dropped from a building and takes 3 sec to hit the ground. How high is the building?

14. A stone is dropped from a cliff and takes 8 sec to hit the ground. How high is the cliff?

Daily accidents. The daily number of accidents N (average number of accidents per day) involving drivers of age a is approximated by the polynomial

$$N = 0.4a^2 - 40a + 1039.$$

15. Evaluate the polynomial when $a = 20$ to find the daily number of accidents involving 20-year-old drivers.

16. Evaluate the polynomial when $a = 18$ to find the daily number of accidents involving 18-year-old drivers.

Total cost. An electronics firm is marketing a new kind of stereo. It is determined that the total cost of producing x stereos is given by

$$5000 + 0.6x^2 \text{ dollars.}$$

17. What is the total cost of producing 650 stereos?

18. What is the total cost of producing 500 stereos?

Total revenue. Total revenue is the total amount of money taken in. The electronics firm determines that when it sells x stereos, it will take in

$$280x - 0.4x^2 \text{ dollars.}$$

19. What is the total revenue from the sale of 100 stereos?

20. What is the total revenue from the sale of 75 stereos?

ANSWERS

1. ______
2. ______
3. ______
4. ______
5. ______
6. ______
7. ______
8. ______
9. ______
10. ______
11. ______
12. ______
13. ______
14. ______
15. ______
16. ______
17. ______
18. ______
19. ______
20. ______

ANSWERS

21. ______
22. ______
23. ______
24. ______
25. ______
26. ______
27. ______
28. ______
29. ______
30. ______
31. ______
32. ______
33. ______
34. ______
35. ______
36. ______
37. ______
38. ______
39. ______
40. ______
41. ______
42. ______
43. ______
44. ______
45. ______
46. ______
47. ______
48. ______
49. ______
50. ______

b Identify the terms of the polynomial.

21. $2 - 3x + x^2$

22. $2x^2 + 3x - 4$

c Identify the like terms in the polynomial.

23. $5x^3 + 6x^2 - 3x^2$

24. $3x^2 + 4x^3 - 2x^2$

25. $2x^4 + 5x - 7x - 3x^4$

26. $-3t + t^3 - 2t - 5t^3$

27. $3x^5 - 7x + 8 + 14x^5 - 2x - 9$

28. $8x^3 + 7x^2 - 11 - 4x^3 - 8x^2 - 29$

d Identify the coefficient of each term of the polynomial.

29. $-3x + 6$

30. $2x - 4$

31. $5x^2 + 3x + 3$

32. $3x^2 - 5x + 2$

33. $-5x^4 + 6x^3 - 3x^2 + 8x - 2$

34. $7x^3 - 4x^2 - 4x + 5$

e Collect like terms.

35. $2x - 5x$

36. $2x^2 + 8x^2$

37. $x - 9x$

38. $x - 5x$

39. $5x^3 + 6x^3 + 4$

40. $6x^4 - 2x^4 + 5$

41. $5x^3 + 6x - 4x^3 - 7x$

42. $3a^4 - 2a + 2a + a^4$

43. $6b^5 + 3b^2 - 2b^5 - 3b^2$

44. $2x^2 - 6x + 3x + 4x^2$

45. $\frac{1}{4}x^5 - 5 + \frac{1}{2}x^5 - 2x - 37$

46. $\frac{1}{3}x^3 + 2x - \frac{1}{6}x^3 + 4 - 16$

47. $6x^2 + 2x^4 - 2x^2 - x^4 - 4x^2$

48. $8x^2 + 2x^3 - 3x^3 - 4x^2 - 4x^2$

49. $\frac{1}{4}x^3 - x^2 - \frac{1}{6}x^2 + \frac{3}{8}x^3 + \frac{5}{16}x^3$

50. $\frac{1}{5}x^4 + \frac{1}{5} - 2x^2 + \frac{1}{10} - \frac{3}{15}x^4 + 2x^2 - \frac{3}{10}$

f Arrange the polynomial in descending order.

51. $x^5 + x + 6x^3 + 1 + 2x^2$

52. $3 + 2x^2 - 5x^6 - 2x^3 + 3x$

53. $5y^3 + 15y^9 + y - y^2 + 7y^8$

54. $9p - 5 + 6p^3 - 5p^4 + p^5$

Collect like terms and then arrange in descending order.

55. $3x^4 - 5x^6 - 2x^4 + 6x^6$

56. $-1 + 5x^3 - 3 - 7x^3 + x^4 + 5$

57. $-2x + 4x^3 - 7x + 9x^3 + 8$

58. $-6x^2 + x - 5x + 7x^2 + 1$

59. $3x + 3x + 3x - x^2 - 4x^2$

60. $-2x - 2x - 2x + x^3 - 5x^3$

61. $-x + \frac{3}{4} + 15x^4 - x - \frac{1}{2} - 3x^4$

62. $2x - \frac{5}{6} + 4x^3 + x + \frac{1}{3} - 2x$

g Identify the degree of each term of the polynomial and the degree of the polynomial.

63. $2x - 4$

64. $6 - 3x$

65. $3x^2 - 5x + 2$

66. $5x^3 - 2x^2 + 3$

67. $-7x^3 + 6x^2 + 3x + 7$

68. $5x^4 + x^2 - x + 2$

69. $x^2 - 3x + x^6 - 9x^4$

70. $8x - 3x^2 + 9 - 8x^3$

71. For the polynomial $-7x^4 + 6x^3 - 3x^2 + 8x - 2$, complete this table.

TERM	COEFFICIENT	DEGREE OF THE TERM	DEGREE OF THE POLYNOMIAL
$6x^3$	6		
		2	
$8x$		1	
	−2		

ANSWERS

51. ____

52. ____

53. ____

54. ____

55. ____

56. ____

57. ____

58. ____

59. ____

60. ____

61. ____

62. ____

63. ____

64. ____

65. ____

66. ____

67. ____

68. ____

69. ____

70. ____

71. See table.

ANSWERS

72. See table.
73.
74.
75.
76.
77.
78.
79.
80.
81.
82.
83.
84.
85.
86.
87.
88.
89.
90.
91.
92.
93.
94.
95.
96.

72. For the polynomial $3x^2 + 8x^5 - 46x^3 + 6x - 2.4 - \frac{1}{2}x^4$, complete this table.

TERM	COEFFICIENT	DEGREE OF THE TERM	DEGREE OF THE POLYNOMIAL
		5	
$-\frac{1}{2}x^4$		4	
	-46		
$3x^2$		2	
	6		
-2.4			

h Identify the missing terms in the polynomial.

73. $x^3 - 27$

74. $x^5 + x$

75. $x^4 - x$

76. $5x^4 - 7x + 2$

77. $2x^3 - 5x^2 + x - 3$

78. $-6x^3$

i Classify the polynomial as a monomial, binomial, trinomial, or none of these.

79. $x^2 - 10x + 25$

80. $-6x^4$

81. $x^3 - 7x^2 + 2x - 4$

82. $x^2 - 9$

83. $4x^2 - 25$

84. $2x^4 - 7x^3 + x^2 + x - 6$

85. $40x$

86. $4x^2 + 12x + 9$

SKILL MAINTENANCE

87. Three tired campers stopped for the night. All they had to eat was a bag of apples. During the night, one awoke and ate one-third of the apples. Later, a second camper awoke and ate one-third of the apples that remained. Much later, the third camper awoke and ate one-third of those apples yet remaining after the other two had eaten. When they got up the next morning, 8 apples were left. How many apples did they begin with?

Subtract.

88. $1 - 20$

89. $\frac{1}{8} - \frac{5}{6}$

90. $\frac{3}{8} - \left(-\frac{1}{4}\right)$

91. $5.6 - 8.2$

SYNTHESIS

Collect like terms.

92. $\frac{9}{2}x^8 + \frac{1}{9}x^2 + \frac{1}{2}x^9 + \frac{9}{2}x^1 + \frac{9}{2}x^9 + \frac{8}{9}x^2 + \frac{1}{2}x - \frac{1}{2}x^8$

93. $(3x^2)^3 + 4x^2 \cdot 4x^4 - x^4(2x)^2 + ((2x)^2)^3 - 100x^2(x^2)^2$

94. Construct a polynomial in x (meaning that x is the variable) of degree 5 with four terms and coefficients that are integers.

95. What is the degree of $(5m^5)^2$?

96. A polynomial in x has degree 3. The coefficient of x^2 is three less than the coefficient of x^3. The coefficient of x is three times the coefficient of x^2. The remaining coefficient is two more than the coefficient of x^3. The sum of the coefficients is -4. Find the polynomial.

3.4 *Addition and Subtraction of Polynomials*

a ADDITION OF POLYNOMIALS

To add two polynomials, we can write a plus sign between them and then collect like terms. Depending on the situation, you may see polynomials written in descending order, ascending order, or neither. Generally, if an exercise is written in a particular order, we write the answer in that same order.

EXAMPLE 1 Add: $(-3x^3 + 2x - 4) + (4x^3 + 3x^2 + 2)$.

$$(-3x^3 + 2x - 4) + (4x^3 + 3x^2 + 2)$$
$$= (-3 + 4)x^3 + 3x^2 + 2x + (-4 + 2) \quad \text{Collecting like terms (No signs are changed.)}$$
$$= x^3 + 3x^2 + 2x - 2$$

EXAMPLE 2 Add:

$$\left(\tfrac{2}{3}x^4 + 3x^2 - 2x + \tfrac{1}{2}\right) + \left(-\tfrac{1}{3}x^4 + 5x^3 - 3x^2 + 3x - \tfrac{1}{2}\right).$$

We have

$$\left(\tfrac{2}{3}x^4 + 3x^2 - 2x + \tfrac{1}{2}\right) + \left(-\tfrac{1}{3}x^4 + 5x^3 - 3x^2 + 3x - \tfrac{1}{2}\right)$$
$$= \left(\tfrac{2}{3} - \tfrac{1}{3}\right)x^4 + 5x^3 + (3 - 3)x^2 + (-2 + 3)x + \left(\tfrac{1}{2} - \tfrac{1}{2}\right) \quad \text{Collecting like terms}$$
$$= \tfrac{1}{3}x^4 + 5x^3 + x.$$

We can add polynomials as we do because they represent numbers. After some practice, you will be able to add mentally.

Do Exercises 1–4.

EXAMPLE 3 Add: $(3x^2 - 2x + 2) + (5x^3 - 2x^2 + 3x - 4)$.

$$(3x^2 - 2x + 2) + (5x^3 - 2x^2 + 3x - 4)$$
$$= 5x^3 + (3 - 2)x^2 + (-2 + 3)x + (2 - 4) \quad \text{You might do this step mentally.}$$
$$= 5x^3 + x^2 + x - 2 \quad \text{Then you would write only this.}$$

Do Exercises 5 and 6.

We can also add polynomials by writing like terms in columns.

EXAMPLE 4 Add: $9x^5 - 2x^3 + 6x^2 + 3$ and $5x^4 - 7x^2 + 6$ and $3x^6 - 5x^5 + x^2 + 5$.

We arrange the polynomials with the like terms in columns.

$$\begin{array}{rrrrrr} & 9x^5 & & -\,2x^3 & +\,6x^2 & +\ \ 3 \\ & & 5x^4 & & -\,7x^2 & +\ \ 6 \\ 3x^6 & -\,5x^5 & & & +\ \ x^2 & +\ \ 5 \\ \hline 3x^6 & +\,4x^5 & +\,5x^4 & -\,2x^3 & & +\,14 \end{array}$$

We leave spaces for missing terms.

Adding

We write the answer as $3x^6 + 4x^5 + 5x^4 - 2x^3 + 14$ without the space.

OBJECTIVES

After finishing Section 3.4, you should be able to:

- **a** Add polynomials.
- **b** Find the opposite of a polynomial.
- **c** Subtract polynomials.
- **d** Solve problems using addition and subtraction of polynomials.

FOR EXTRA HELP

TAPE 8 | TAPE 7A | MAC: 3A IBM: 3A

Add.

1. $(3x^2 + 2x - 2) + (-2x^2 + 5x + 5)$

2. $(-4x^5 + x^3 + 4) + (7x^4 + 2x^2)$

3. $(31x^4 + x^2 + 2x - 1) + (-7x^4 + 5x^3 - 2x + 2)$

4. $(17x^3 - x^2 + 3x + 4) + \left(-15x^3 + x^2 - 3x - \dfrac{2}{3}\right)$

Add mentally. Try to write just the answer.

5. $(4x^2 - 5x + 3) + (-2x^2 + 2x - 4)$

6. $(3x^3 - 4x^2 - 5x + 3) + \left(5x^3 + 2x^2 - 3x - \dfrac{1}{2}\right)$

Answers on page A-3

Add.

7. $\begin{array}{r} -2x^3+5x^2-2x+4 \\ x^4 \qquad +6x^2+7x-10 \\ -9x^4+6x^3+x^2 \qquad -2 \\ \hline \end{array}$

8. $-3x^3 + 5x + 2$ and
$x^3 + x^2 + 5$ and
$x^3 - 2x - 4$

Find two equivalent expressions for the opposite of the polynomial.

9. $12x^4 - 3x^2 + 4x$

10. $-4x^4 + 3x^2 - 4x$

11. $-13x^6 + 2x^4 - 3x^2 + x - \frac{5}{13}$

12. $-7y^3 + 2y^2 - y + 3$

Simplify.

13. $-(4x^3 - 6x + 3)$

14. $-(5x^4 + 3x^2 + 7x - 5)$

15. $-\left(14x^{10} - \frac{1}{2}x^5 + 5x^3 - x^2 + 3x\right)$

Answers on page A-3

Do Exercises 7 and 8.

b Opposites of Polynomials

We now look at subtraction of polynomials. To do so, we first consider the opposite, or additive inverse, of a polynomial.

We know that two numbers are opposites of each other if their sum is zero. For example, 5 and -5 are opposites, since $5 + (-5) = 0$. The same definition holds for polynomials.

> Two polynomials are *opposites,* or *additive inverses,* of each other if their sum is zero.

To find a way to determine an opposite, look for a pattern in the following examples:

a) $2x + (-2x) = 0$;

b) $-6x^2 + 6x^2 = 0$;

c) $(5t^3 - 2) + (-5t^3 + 2) = 0$;

d) $(7x^3 - 6x^2 - x + 4) + (-7x^3 + 6x^2 + x - 4) = 0$.

Since $(5t^3 - 2) + (-5t^3 + 2) = 0$, we know that the opposite of $(5t^3 - 2)$ is $(-5t^3 + 2)$. To say the same thing with purely algebraic symbolism, consider

$$\underbrace{\text{The opposite of}}_{\downarrow \; -} \quad \underbrace{(5t^3 - 2)}_{\downarrow \; (5t^3-2)} \quad \underset{=}{\text{is}} \quad \underbrace{-5t^3 + 2.}_{\downarrow \; -5t^3+2.}$$

> We can find an equivalent polynomial for the opposite, or additive inverse, of a polynomial by replacing each term by its opposite—that is, *changing the sign of every term.*

EXAMPLE 5 Find two equivalent expressions for the opposite of

$$4x^5 - 7x^3 - 8x + \tfrac{5}{6}.$$

a) $-\left(4x^5 - 7x^3 - 8x + \frac{5}{6}\right)$

b) $-4x^5 + 7x^3 + 8x - \frac{5}{6}$ Changing the sign of every term

Thus, $-\left(4x^5 - 7x^3 - 8x + \frac{5}{6}\right)$ is equivalent to $-4x^5 + 7x^3 + 8x - \frac{5}{6}$, and each is the opposite of the original polynomial $4x^5 - 7x^3 - 8x + \frac{5}{6}$.

Do Exercises 9–12.

EXAMPLE 6 Simplify: $-\left(-7x^4 - \frac{5}{9}x^3 + 8x^2 - x + 67\right)$.

$$-\left(-7x^4 - \tfrac{5}{9}x^3 + 8x^2 - x + 67\right) = 7x^4 + \tfrac{5}{9}x^3 - 8x^2 + x - 67$$

Do Exercises 13–15.

C SUBTRACTION OF POLYNOMIALS

Recall that we can subtract a real number by adding its opposite, or additive inverse: $a - b = a + (-b)$. This allows us to find an equivalent expression for the difference of two polynomials.

EXAMPLE 7 Subtract:

$$(9x^5 + x^3 - 2x^2 + 4) - (2x^5 + x^4 - 4x^3 - 3x^2).$$

We have

$$(9x^5 + x^3 - 2x^2 + 4) - (2x^5 + x^4 - 4x^3 - 3x^2)$$

$$= 9x^5 + x^3 - 2x^2 + 4 + [-(2x^5 + x^4 - 4x^3 - 3x^2)]$$ Adding the opposite

$$= 9x^5 + x^3 - 2x^2 + 4 - 2x^5 - x^4 + 4x^3 + 3x^2$$ Finding the opposite by changing the sign of *each* term

$$= 7x^5 - x^4 + 5x^3 + x^2 + 4$$ Collecting like terms

Do Exercises 16 and 17.

As with similar work in Section 1.8, we combine steps by changing the sign of each term of the polynomial being subtracted and collecting like terms. Try to do this mentally as much as possible.

EXAMPLE 8 Subtract: $(9x^5 + x^3 - 2x) - (-2x^5 + 5x^3 + 6)$.

$$(9x^5 + x^3 - 2x) - (-2x^5 + 5x^3 + 6)$$

$$= 9x^5 + x^3 - 2x + 2x^5 - 5x^3 - 6$$ Finding the opposite by changing the sign of each term

$$= 11x^5 - 4x^3 - 2x - 6$$ Collecting like terms

Do Exercises 18 and 19.

We can use columns to subtract. We replace coefficients by their opposites, as shown in Example 7.

EXAMPLE 9 Write in columns and subtract:

$$(5x^2 - 3x + 6) - (9x^2 - 5x - 3).$$

a) $$\begin{array}{l} 5x^2 - 3x + 6 \\ \underline{-(9x^2 - 5x - 3)} \end{array}$$ Writing similar terms in columns

b) $$\begin{array}{l} 5x^2 - 3x + 6 \\ \underline{-9x^2 + 5x + 3} \end{array}$$ Changing signs

c) $$\begin{array}{l} 5x^2 - 3x + 6 \\ \underline{-9x^2 + 5x + 3} \\ -4x^2 + 2x + 9 \end{array}$$ Adding

If you can do so without error, you can arrange the polynomials in columns and write just the answer.

Subtract.

16. $(7x^3 + 2x + 4) - (5x^3 - 4)$

17. $(-3x^2 + 5x - 4) - (-4x^2 + 11x - 2)$

Subtract.

18. $(-6x^4 + 3x^2 + 6) - (2x^4 + 5x^3 - 5x^2 + 7)$

19. $\left(\frac{3}{2}x^3 - \frac{1}{2}x^2 + 0.3\right) - \left(\frac{1}{2}x^3 + \frac{1}{2}x^2 + \frac{4}{3}x + 1.2\right)$

Answers on page A-3

Write in columns and subtract.

20. $(4x^3 + 2x^2 - 2x - 3) - (2x^3 - 3x^2 + 2)$

21. $(2x^3 + x^2 - 6x + 2) - (x^5 + 4x^3 - 2x^2 - 4x)$

22. Find a polynomial for the sum of the areas of the rectangles.

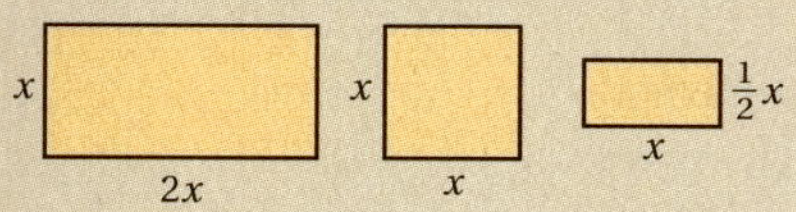

23. Find a polynomial for the shaded area.

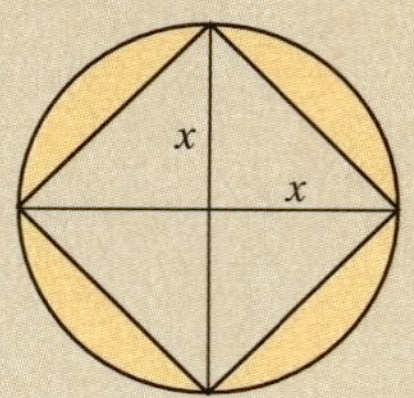

Answers on page A-3

EXAMPLE 10 Write in columns and subtract:

$$(x^3 + x^2 + 2x - 12) - (-2x^3 + x^2 - 3x).$$

We have

$$\begin{array}{r} x^3 + x^2 + 2x - 12 \\ -2x^3 + x^2 - 3x \phantom{{}-12} \\ \hline 3x^3 \phantom{{}+x^2} + 5x - 12. \end{array}$$

Do Exercises 20 and 21.

d SOLVING PROBLEMS

EXAMPLE 11 Find a polynomial for the sum of the areas of these rectangles.

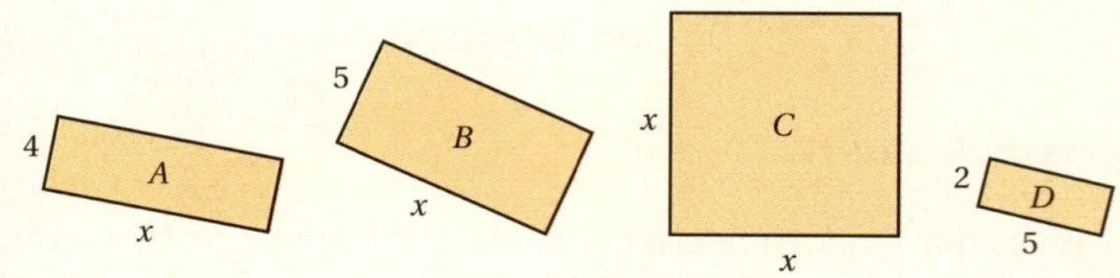

Recall that the area of a rectangle is the product of the length and the width. The sum of the areas is a sum of products. We find these products and then collect like terms.

Area of A	plus	Area of B	plus	Area of C	plus	Area of D
$4x$	$+$	$5x$	$+$	$x \cdot x$	$+$	$2 \cdot 5$

We collect like terms:

$$4x + 5x + x^2 + 10 = x^2 + 9x + 10.$$

Do Exercise 22.

EXAMPLE 12 A 4-ft by 4-ft sandbox is placed on a square lawn x ft on a side. To determine the amount of grass seed needed for the lawn, find a polynomial for the remaining area.

We draw a picture of the situation as shown here.

We reword the problem and write the polynomial as follows.

Area of lawn	$-$	Area of sandbox	$=$	Area left over
$x \cdot x$	$-$	$4 \cdot 4$	$=$	Area left over

Then $x^2 - 16$ = Area left over.

Do Exercise 23.

NAME SECTION DATE

Exercise Set 3.4

a Add.

1. $(3x + 2) + (-4x + 3)$

2. $(6x + 1) + (-7x + 2)$

3. $(-6x + 2) + (x^2 + x - 3)$

4. $(x^2 - 5x + 4) + (8x - 9)$

5. $(x^2 - 9) + (x^2 + 9)$

6. $(x^3 + x^2) + (2x^3 - 5x^2)$

7. $(3x^2 - 5x + 10) + (2x^2 + 8x - 40)$

8. $(6x^4 + 3x^3 - 1) + (4x^2 - 3x + 3)$

9. $(1.2x^3 + 4.5x^2 - 3.8x) + (-3.4x^3 - 4.7x^2 + 23)$

10. $(0.5x^4 - 0.6x^2 + 0.7) + (2.3x^4 + 1.8x - 3.9)$

11. $(1 + 4x + 6x^2 + 7x^3) + (5 - 4x + 6x^2 - 7x^3)$

12. $(3x^4 - 6x - 5x^2 + 5) + (6x^2 - 4x^3 - 1 + 7x)$

13. $\left(\frac{1}{4}x^4 + \frac{2}{3}x^3 + \frac{5}{8}x^2 + 7\right) + \left(-\frac{3}{4}x^4 + \frac{3}{8}x^2 - 7\right)$

14. $\left(\frac{1}{3}x^9 + \frac{1}{5}x^5 - \frac{1}{2}x^2 + 7\right) + \left(-\frac{1}{5}x^9 + \frac{1}{4}x^4 - \frac{3}{5}x^5 + \frac{3}{4}x^2 + \frac{1}{2}\right)$

15. $(0.02x^5 - 0.2x^3 + x + 0.08) + (-0.01x^5 + x^4 - 0.8x - 0.02)$

16. $(0.03x^6 + 0.05x^3 + 0.22x + 0.05) + \left(\frac{7}{100}x^6 - \frac{3}{100}x^3 + 0.5\right)$

17. $(9x^8 - 7x^4 + 2x^2 + 5) + (8x^7 + 4x^4 - 2x) + (-3x^4 + 6x^2 + 2x - 1)$

18. $(4x^5 - 6x^3 - 9x + 1) + (6x^3 + 9x^2 + 9x) + (-4x^3 + 8x^2 + 3x - 2)$

ANSWERS

1. ______
2. ______
3. ______
4. ______
5. ______
6. ______
7. ______
8. ______
9. ______
10. ______
11. ______
12. ______
13. ______
14. ______
15. ______
16. ______
17. ______
18. ______

ANSWERS

19. __________

20. __________

21. __________

22. __________

23. __________

24. __________

25. __________

26. __________

27. __________

28. __________

29. __________

30. __________

31. __________

32. __________

33. __________

34. __________

35. __________

36. __________

37. __________

38. __________

39. __________

40. __________

41. __________

42. __________

19.
$$\begin{array}{rrrrr} 0.15x^4 & +\ 0.10x^3 & -\ 0.9x^2 & & \\ & -\ 0.01x^3 & +\ 0.01x^2 & +\ x & \\ 1.25x^4 & & +\ 0.11x^2 & & +\ 0.01 \\ & 0.27x^3 & & & +\ 0.99 \\ -0.35x^4 & & +\ 15x^2 & & -\ 0.03 \\ \hline \end{array}$$

20.
$$\begin{array}{rrrrr} 0.05x^4 & +\ 0.12x^3 & -\ 0.5x^2 & & \\ & -\ 0.02x^3 & +\ 0.02x^2 & +\ 2x & \\ 1.5x^4 & & +\ 0.01x^2 & & +\ 0.15 \\ & 0.25x^3 & & & +\ 0.85 \\ -0.25x^4 & & +\ 10x^2 & & -\ 0.04 \\ \hline \end{array}$$

b Find two equivalent expressions for the opposite of the polynomial.

21. $-5x$

22. $x^2 - 3x$

23. $-x^2 + 10x - 2$

24. $-4x^3 - x^2 - x$

25. $12x^4 - 3x^3 + 3$

26. $4x^3 - 6x^2 - 8x + 1$

Simplify.

27. $-(3x - 7)$

28. $-(-2x + 4)$

29. $-(4x^2 - 3x + 2)$

30. $-(-6a^3 + 2a^2 - 9a + 1)$

31. $-(-4x^4 + 6x^2 + \frac{3}{4}x - 8)$

32. $-(-5x^4 + 4x^3 - x^2 + 0.9)$

c Subtract.

33. $(3x + 2) - (-4x + 3)$

34. $(6x + 1) - (-7x + 2)$

35. $(-6x + 2) - (x^2 + x - 3)$

36. $(x^2 - 5x + 4) - (8x - 9)$

37. $(x^2 - 9) - (x^2 + 9)$

38. $(x^3 + x^2) - (2x^3 - 5x^2)$

39. $(6x^4 + 3x^3 - 1) - (4x^2 - 3x + 3)$

40. $(-4x^2 + 2x) - (3x^3 - 5x^2 + 3)$

41. $(1.2x^3 + 4.5x^2 - 3.8x) -$
$(-3.4x^3 - 4.7x^2 + 23)$

42. $(0.5x^4 - 0.6x^2 + 0.7) -$
$(2.3x^4 + 1.8x - 3.9)$

43. $\left(\frac{5}{8}x^3 - \frac{1}{4}x - \frac{1}{3}\right) - \left(-\frac{1}{8}x^3 + \frac{1}{4}x - \frac{1}{3}\right)$

44. $\left(\frac{1}{5}x^3 + 2x^2 - 0.1\right) - \left(-\frac{2}{5}x^3 + 2x^2 + 0.01\right)$

45. $(0.08x^3 - 0.02x^2 + 0.01x) - (0.02x^3 + 0.03x^2 - 1)$

46. $(0.8x^4 + 0.2x - 1) - \left(\frac{7}{10}x^4 + \frac{1}{5}x - 0.1\right)$

Subtract.

47. $\begin{array}{r} x^2 + 5x + 6 \\ \underline{x^2 + 2x \phantom{{}+6}} \end{array}$

48. $\begin{array}{l} x^3 \phantom{{}+x^2} + 1 \\ \underline{x^3 + x^2 \phantom{{}+1}} \end{array}$

49. $\begin{array}{r} 5x^4 + 6x^3 - 9x^2 \phantom{{}+8x+9} \\ \underline{-6x^4 - 6x^3 \phantom{{}-9x^2} + 8x + 9} \end{array}$

50. $\begin{array}{r} 5x^4 \phantom{{}+6x^3} + 6x^2 - 3x + 6 \\ \underline{6x^3 + 7x^2 - 8x - 9} \end{array}$

51. $\begin{array}{r} x^5 \phantom{{}- x^4 + x^3 - x^2 + x} - 1 \\ \underline{x^5 - x^4 + x^3 - x^2 + x - 1} \end{array}$

52. $\begin{array}{r} x^5 + x^4 - x^3 + x^2 - x + 2 \\ \underline{x^5 - x^4 + x^3 - x^2 - x + 2} \end{array}$

d Solve.

53. Find a polynomial for the sum of the areas of these rectangles.

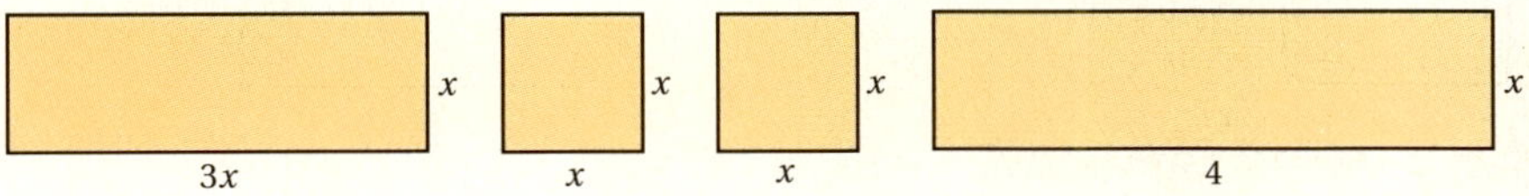

Find a polynomial for the perimeter of the figure.

54.

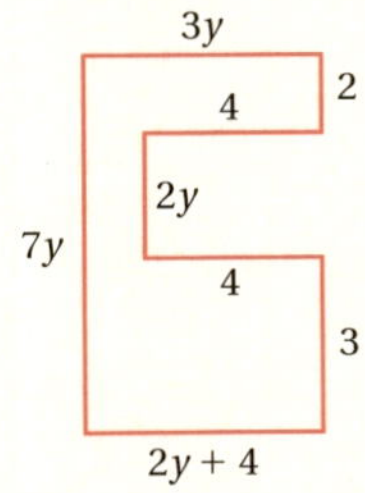

55.

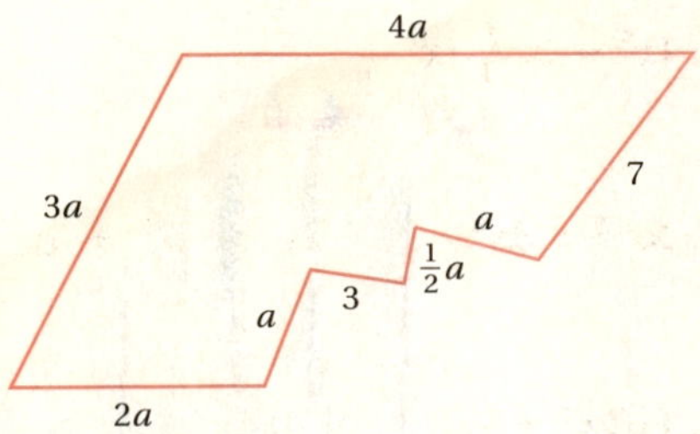

ANSWERS

43. ______

44. ______

45. ______

46. ______

47. ______

48. ______

49. ______

50. ______

51. ______

52. ______

53. ______

54. ______

55. ______

ANSWERS

56. ______

57. ______

58. ______

59. ______

60. ______

61. ______

62. ______

63. ______

64. ______

65. ______

66. ______

67. ______

68. ______

69. ______

SKILL MAINTENANCE

Solve.

56. $5x - 7x = 38$

57. $8x + 3x = 66$

58. $5x - 4 = 26 - x$

59. $\frac{3}{8}x + \frac{1}{4} - \frac{3}{4}x = \frac{11}{16} + x$

60. $8(5x + 2) = 7(6x - 3)$

61. $6(y - 3) - 8 = 4(y + 2) + 5$

SYNTHESIS

Find two algebraic expressions for the area of the figure.

62.

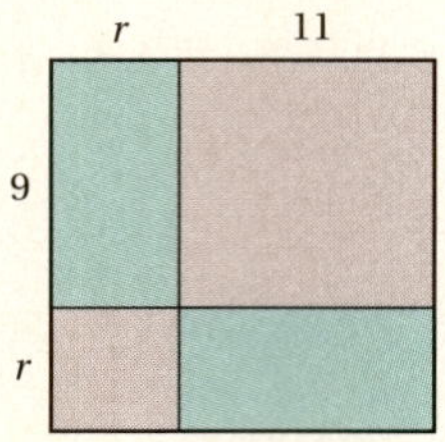

63.

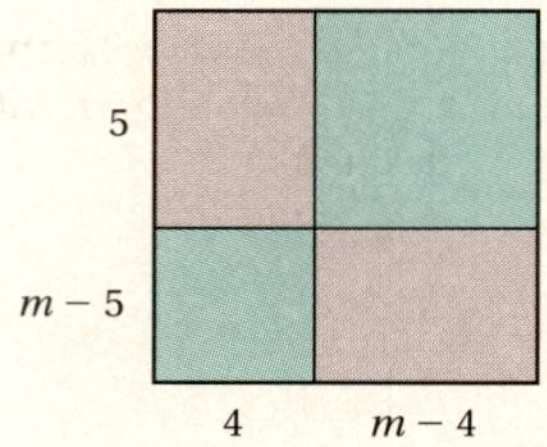

Find a polynomial for the shaded area.

64.

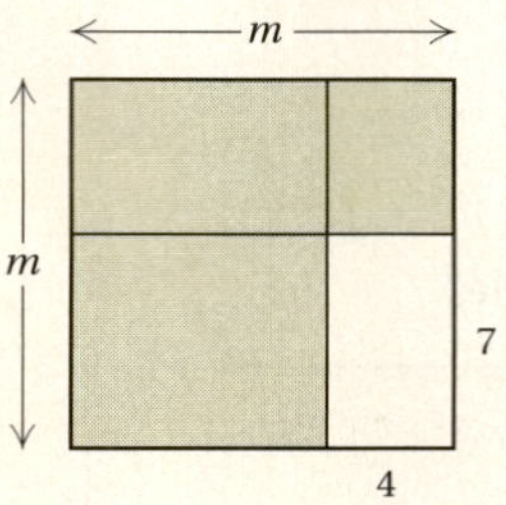

65.

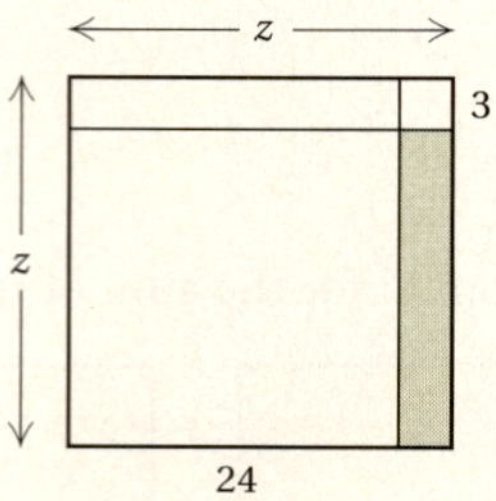

Simplify.

66. $(7y^2 - 5y + 6) - (3y^2 + 8y - 12) + (8y^2 - 10y + 3)$

67. $(3x^2 - 4x + 6) - (-2x^2 + 4) + (-5x - 3)$

68. $(-y^4 - 7y^3 + y^2) + (-2y^4 + 5y - 2) - (-6y^3 + y^2)$

69. $(-4 + x^2 + 2x^3) - (-6 - x + 3x^3) - (-x^2 - 5x^3)$

3.5 *Multiplication of Polynomials*

We now multiply polynomials using techniques based, for the most part, on the distributive laws, but also on the associative and commutative laws. As we proceed in this chapter, we will develop special ways to find certain products.

OBJECTIVES

After finishing Section 3.5, you should be able to:

- **a** Multiply monomials.
- **b** Multiply a monomial and any polynomial.
- **c** Multiply two binomials.
- **d** Multiply any two polynomials.

FOR EXTRA HELP

TAPE 9 TAPE 7A MAC: 3B IBM: 3B

a MULTIPLYING MONOMIALS

Consider $(3x)(4x)$. We multiply as follows:

$$\begin{aligned} (3x)(4x) &= 3 \cdot x \cdot 4 \cdot x && \text{By the associative law of multiplication} \\ &= 3 \cdot 4 \cdot x \cdot x && \text{By the commutative law of multiplication} \\ &= (3 \cdot 4)(x \cdot x) && \text{By the associative law} \\ &= 12x^2. && \text{Using the product rule for exponents} \end{aligned}$$

> To find an equivalent expression for the product of two monomials, multiply the coefficients and then multiply the variables using the product rule for exponents.

EXAMPLES Multiply.

1. $$\begin{aligned} 5x \cdot 6x &= (5 \cdot 6)(x \cdot x) && \text{Multiplying the coefficients} \\ &= 30x^2 && \text{Simplifying} \end{aligned}$$

2. $$\begin{aligned} (3x)(-x) &= (3x)(-1x) \\ &= (3)(-1)(x \cdot x) \\ &= -3x^2 \end{aligned}$$

3. $$\begin{aligned} (-7x^5)(4x^3) &= (-7 \cdot 4)(x^5 \cdot x^3) \\ &= -28x^{5+3} && \text{Adding the exponents} \\ &= -28x^8 && \text{Simplifying} \end{aligned}$$

After some practice, you can do this mentally. Multiply the coefficients and then the variables by keeping the base and adding the exponents. Write only the answer.

Do Exercises 1–8.

Multiply.

1. $(3x)(-5)$
2. $(-x) \cdot x$
3. $(-x)(-x)$
4. $(-x^2)(x^3)$
5. $3x^5 \cdot 4x^2$
6. $(4y^5)(-2y^6)$
7. $(-7y^4)(-y)$
8. $7x^5 \cdot 0$

Answers on page A-3

b MULTIPLYING A MONOMIAL AND ANY POLYNOMIAL

To find an equivalent expression for the product of a monomial, such as $2x$, and a binomial, such as $5x + 3$, we use a distributive law and multiply each term of $5x + 3$ by $2x$.

EXAMPLE 4 Multiply: $2x(5x + 3)$.

$$\begin{aligned} 2x(5x + 3) &= (2x)(5x) - (2x)(3) && \text{Using a distributive law} \\ &= 10x^2 + 6x && \text{Multiplying the monomials} \end{aligned}$$

Multiply.

9. $4x(2x + 4)$

10. $3t^2(-5t + 2)$

11. $5x^3(x^3 + 5x^2 - 6x + 8)$

Multiply.

12. $(x + 8)(x + 5)$

13. $(x + 5)(x - 4)$

Multiply.

14. $(5x + 3)(x - 4)$

15. $(2x - 3)(3x - 5)$

Answers on page A-3

EXAMPLE 5 Multiply: $5x(2x^2 - 3x + 4)$.

$$5x(2x^2 - 3x + 4) = (5x)(2x^2) - (5x)(3x) + (5x)(4)$$
$$= 10x^3 - 15x^2 + 20x$$

To multiply a monomial and a polynomial, multiply each term of the polynomial by the monomial.

EXAMPLE 6 Multiply: $2x^2(x^3 - 7x^2 + 10x - 4)$.

$$2x^2(x^3 - 7x^2 + 10x - 4) = 2x^5 - 14x^4 + 20x^3 - 8x^2$$

Do Exercises 9–11.

c MULTIPLYING TWO BINOMIALS

To find an equivalent expression for the product of two binomials, we use the distributive laws more than once. In Example 7, we use a distributive law three times.

EXAMPLE 7 Multiply: $(x + 5)(x + 4)$.

$(x + 5)(x + 4) = x(x + 4) + 5(x + 4)$ **Using a distributive law**

$= x \cdot x + x \cdot 4 + 5 \cdot x + 5 \cdot 4$ **Using a distributive law on each part**

$= x^2 + 4x + 5x + 20$ **Multiplying the monomials**

$= x^2 + 9x + 20$ **Collecting like terms**

Do Exercises 12 and 13.

EXAMPLE 8 Multiply: $(4x + 3)(x - 2)$.

$(4x + 3)(x - 2) = 4x(x - 2) + 3(x - 2)$ **Using a distributive law**

$= 4x \cdot x - 4x \cdot 2 + 3 \cdot x - 3 \cdot 2$ **Using a distributive law on each part**

$= 4x^2 - 8x + 3x - 6$ **Multiplying the monomials**

$= 4x^2 - 5x - 6$ **Collecting like terms**

Do Exercises 14 and 15.

d Multiplying Any Two Polynomials

Let us consider the product of a binomial and a trinomial. We again use a distributive law three times. You may see ways to skip some steps and do the work mentally.

EXAMPLE 9 Multiply: $(x^2 + 2x - 3)(x^2 + 4)$.

$$
\begin{aligned}
(x^2 + 2x - 3)(x^2 + 4) &= (x^2 + 2x - 3)x^2 + (x^2 + 2x - 3)4 \\
&= x^2 \cdot x^2 + 2x \cdot x^2 - 3 \cdot (x^2) + x^2(4) + 2x \cdot 4 \\
&\quad - 3 \cdot 4 \\
&= x^4 + 2x^3 - 3x^2 + 4x^2 + 8x - 12 \\
&= x^4 + 2x^3 + x^2 + 8x - 12
\end{aligned}
$$

Do Exercises 16 and 17.

Perhaps you have discovered the following in the preceding examples.

> To multiply two polynomials P and Q, select one of the polynomials—say, P. Then multiply each term of P by every term of Q and collect like terms.

We can use columns for long multiplications. We multiply each term at the top by every term at the bottom. We write like terms in columns, and then we add the results. Such multiplication is like multiplying with whole numbers:

4 5 7	4 5 7	$= 400 + 50 + 7$
× 6 3	× 6 3	$= 60 + 3$
1 3 7 1	1200 + 150 + 21	$= 3(457) = 3(400 + 50 + 7)$
2 7 4 2 0	24000 + 3000 + 420	$= 60(457) = 60(400 + 50 + 7)$
2 8 7 9 1	24000 + 4200 + 570 + 21	$= 28{,}791$

EXAMPLE 10 Multiply: $(4x^2 - 2x + 3)(x + 2)$.

$$
\begin{array}{rrrrl}
 & 4x^2 & -\ 2x & +\ 3 & \\
 & & x & +\ 2 & \\
\hline
 & 8x^2 & -\ 4x & +\ 6 & \text{Multiplying the top row by 2} \\
4x^3 & -\ 2x^2 & +\ 3x & & \text{Multiplying the top row by } x \\
\hline
4x^3 & +\ 6x^2 & -\ x & +\ 6 & \text{Collecting like terms} \\
\uparrow & \uparrow & \uparrow & \uparrow & \text{Line up like terms in columns.}
\end{array}
$$

Multiply.

16. $(x^2 + 3x - 4)(x^2 + 5)$

17. $(3y^2 - 7)(2y^3 - 2y + 5)$

Answers on page A-3

Multiply.

18.
$$\begin{array}{r} 3x^2 - 2x + 4 \\ x + 5 \\ \hline \end{array}$$

19.
$$\begin{array}{r} -5x^2 + 4x + 2 \\ -4x^2 - 8 \\ \hline \end{array}$$

20. Multiply.
$$\begin{array}{r} 3x^2 - 2x - 5 \\ 2x^2 + x - 2 \\ \hline \end{array}$$

Answers on page A-3

EXAMPLE 11 Multiply: $(5x^3 - 3x + 4)(-2x^2 - 3)$.

When missing terms occur, it helps to leave spaces for them and align like terms as we multiply.

$$\begin{array}{rrrrrl} & 5x^3 & & -3x & +4 & \\ & & -2x^2 & & -3 & \\ \hline & -15x^3 & & +9x & -12 & \text{Multiplying by } -3 \\ -10x^5 & +6x^3 & -8x^2 & & & \text{Multiplying by } -2x^2 \\ \hline -10x^5 & -9x^3 & -8x^2 & +9x & -12 & \text{Collecting like terms} \end{array}$$

Do Exercises 18 and 19.

EXAMPLE 12 Multiply: $(2x^2 + 3x - 4)(2x^2 - x + 3)$.

$$\begin{array}{rrrrrl} & & 2x^2 & +3x & -4 & \\ & & 2x^2 & -x & +3 & \\ \hline & & 6x^2 & +9x & -12 & \text{Multiplying by } 3 \\ & -2x^3 & -3x^2 & +4x & & \text{Multiplying by } -x \\ 4x^4 & +6x^3 & -8x^2 & & & \text{Multiplying by } 2x^2 \\ \hline 4x^4 & +4x^3 & -5x^2 & +13x & -12 & \text{Collecting like terms} \end{array}$$

Do Exercise 20.

SIDELIGHTS

FACTORS AND SUMS

To *factor* a number is to express it as a product. Since $12 = 4 \cdot 3$, we say that 12 is *factored* and that 4 and 3 are *factors* of 12. In the table below, the top number has been factored in such a way that the sum of the factors is the bottom number. For example, in the first column, 40 has been factored as $5 \cdot 8$, and $5 + 8 = 13$, the bottom number. Such thinking is important in algebra when we factor trinomials of the type $x^2 + bx + c$.

PRODUCT	40	63	36	72	−140	−96	48	168	110			
FACTOR	5									−9	−24	−3
FACTOR	8									−10	18	
SUM	13	16	−20	−38	−4	4	−14	−29	−21			18

EXERCISES

Find the missing numbers in the table.

NAME SECTION DATE

Exercise Set 3.5

a Multiply.

1. $(8x^2)(5)$

2. $(4x^2)(-2)$

3. $(-x^2)(-x)$

4. $(-x^3)(x^2)$

5. $(8x^5)(4x^3)$

6. $(10a^2)(2a^2)$

7. $(0.1x^6)(0.3x^5)$

8. $(0.3x^4)(-0.8x^6)$

9. $\left(-\frac{1}{5}x^3\right)\left(-\frac{1}{3}x\right)$

10. $\left(-\frac{1}{4}x^4\right)\left(\frac{1}{5}x^8\right)$

11. $(-4x^2)(0)$

12. $(-4m^5)(-1)$

13. $(3x^2)(-4x^3)(2x^6)$

14. $(-2y^5)(10y^4)(-3y^3)$

b Multiply.

15. $2x(-x + 5)$

16. $3x(4x - 6)$

17. $-5x(x - 1)$

18. $-3x(-x - 1)$

19. $x^2(x^3 + 1)$

20. $-2x^3(x^2 - 1)$

21. $3x(2x^2 - 6x + 1)$

22. $-4x(2x^3 - 6x^2 - 5x + 1)$

23. $(-6x^2)(x^2 + x)$

24. $(-4x^2)(x^2 - x)$

25. $(3y^2)(6y^4 + 8y^3)$

26. $(4y^4)(y^3 - 6y^2)$

c Multiply.

27. $(x + 6)(x + 3)$

28. $(x + 5)(x + 2)$

29. $(x + 5)(x - 2)$

30. $(x + 6)(x - 2)$

31. $(x - 4)(x - 3)$

32. $(x - 7)(x - 3)$

33. $(x + 3)(x - 3)$

34. $(x + 6)(x - 6)$

35. $(5 - x)(5 - 2x)$

36. $(3 + x)(6 + 2x)$

37. $(2x + 5)(2x + 5)$

38. $(3x - 4)(3x - 4)$

39. $\left(x - \frac{5}{2}\right)\left(x + \frac{2}{5}\right)$

40. $\left(x + \frac{4}{3}\right)\left(x + \frac{3}{2}\right)$

ANSWERS

1.
2.
3.
4.
5.
6.
7.
8.
9.
10.
11.
12.
13.
14.
15.
16.
17.
18.
19.
20.
21.
22.
23.
24.
25.
26.
27.
28.
29.
30.
31.
32.
33.
34.
35.
36.
37.
38.
39.
40.

ANSWERS

41. ______
42. ______
43. ______
44. ______
45. ______
46. ______
47. ______
48. ______
49. ______
50. ______
51. ______
52. ______
53. ______
54. ______
55. ______
56. ______
57. ______
58. ______
59. ______
60. ______
61. ______
62. ______
63. ______
64. ______
65. ______
66. ______
67. ______

d Multiply.

41. $(x^2 + x + 1)(x - 1)$

42. $(x^2 + x - 2)(x + 2)$

43. $(2x + 1)(2x^2 + 6x + 1)$

44. $(3x - 1)(4x^2 - 2x - 1)$

45. $(y^2 - 3)(3y^2 - 6y + 2)$

46. $(3y^2 - 3)(y^2 + 6y + 1)$

47. $(x^3 + x^2)(x^3 + x^2 - x)$

48. $(x^3 - x^2)(x^3 - x^2 + x)$

49. $(-5x^3 - 7x^2 + 1)(2x^2 - x)$

50. $(-4x^3 + 5x^2 - 2)(5x^2 + 1)$

51. $(1 + x + x^2)(-1 - x + x^2)$

52. $(1 - x + x^2)(1 - x + x^2)$

53. $(2t^2 - t - 4)(3t^2 + 2t - 1)$

54. $(3a^2 - 5a + 2)(2a^2 - 3a + 4)$

55. $(x - x^3 + x^5)(x^2 - 1 + x^4)$

56. $(x - x^3 + x^5)(3x^2 + 3x^6 + 3x^4)$

57. $(x^3 + x^2 + x + 1)(x - 1)$

58. $(x + 2)(x^3 - x^2 + x - 2)$

SKILL MAINTENANCE

59. Subtract: $-\frac{1}{4} - \frac{1}{2}$.

60. Factor: $16x - 24y + 36$.

61. Factor: $9x - 45y + 15$.

62. Subtract: $-3.8 - (-10.2)$.

SYNTHESIS

63. Find a polynomial for the shaded area.

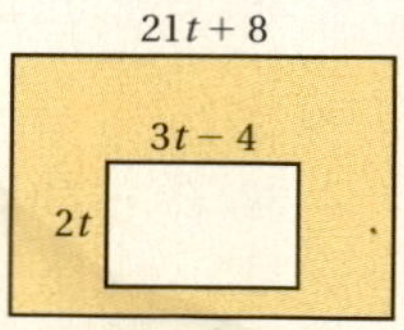

64. A box with a square bottom is to be made from a 12-in.-square piece of cardboard. Squares with side x are cut out of the corners and the sides are folded up. Find polynomials for the volume and the outside surface area of the box.

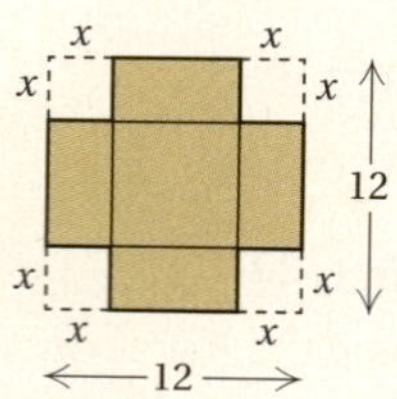

65. The height of a triangle is 4 ft longer than its base. Find a polynomial for the area.

Compute and simplify.

66. $(x + 3)(x + 6) + (x + 3)(x + 6)$

67. $(x - 2)(x - 7) - (x - 2)(x - 7)$

We can look at the rule for finding $(A + B)^2$ geometrically as follows. The area of the large square is

$$(A + B)(A + B) = (A + B)^2.$$

This is equal to the sum of the areas of the smaller rectangles:

$$A^2 + AB + AB + B^2 = A^2 + 2AB + B^2.$$

Thus,

$$(A + B)^2 = A^2 + 2AB + B^2.$$

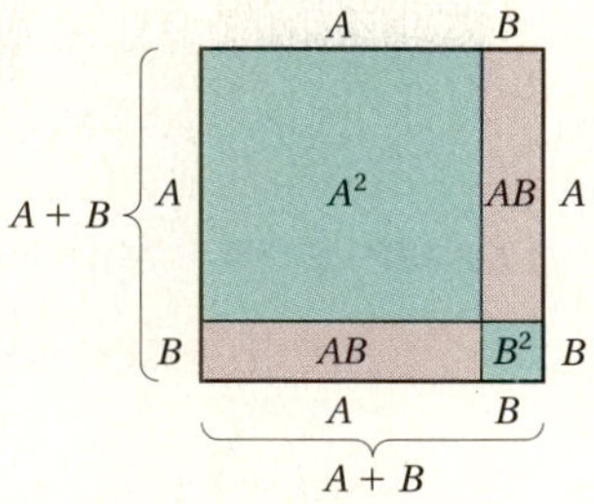

d MULTIPLICATIONS OF VARIOUS TYPES

We have considered how to quickly multiply certain kinds of polynomials. Let us now try several types of multiplications mixed together so that we can learn to sort them out. When you multiply, first see what kind of multiplication you have. Then use the best method. The formulas you should know and the questions you should ask yourself are as follows.

MULTIPLYING TWO POLYNOMIALS

1. Is the product the square of a binomial? If so, use the following:

   $$(A + B)(A + B) = (A + B)^2 = A^2 + 2AB + B^2,$$

   or $$(A - B)(A - B) = (A - B)^2 = A^2 - 2AB + B^2.$$

   The square of a binomial is the square of the first term, plus or minus *twice* the product of the two terms, plus the square of the last term.

   [The answer has 3 terms.]

2. Is it the product of the sum and the difference of the *same* two terms? If so, use the following:

   $$(A + B)(A - B) = A^2 - B^2.$$

   The product of the sum and the difference of the same two terms is the difference of the squares.

   [The answer has 2 terms.]

3. Is it the product of two binomials other than those above? If so, use FOIL.

   [The answer will have 3 or 4 terms.]

4. Is it the product of a monomial and a polynomial? If so, multiply each term of the polynomial by the monomial.

5. Is it the product of two polynomials other than those above? If so, multiply each term of one by every term of the other. Use columns if you wish.

   [The answer will have 2 or more terms, usually more than 2 terms.]

Remember that FOIL will *always* work for two binomials. You may use it instead of either of the first two rules, but those rules will make your work go faster.

Multiply.

27. $(x + 5)(x + 6)$

28. $(t - 4)(t + 4)$

29. $4x^2(-2x^3 + 5x^2 + 10)$

30. $(9x^2 + 1)^2$

31. $(2a - 5)(2a + 8)$

32. $\left(5x + \frac{1}{2}\right)^2$

33. $\left(2x - \frac{1}{2}\right)^2$

34. $(x^2 - x + 4)(x - 2)$

Answers on page A-3

EXAMPLE 19 Multiply: $(x + 3)(x - 3)$.

$(x + 3)(x - 3) = x^2 - 9$ Using method 2 (the product of the sum and the difference of two terms)

EXAMPLE 20 Multiply: $(t + 7)(t - 5)$.

$(t + 7)(t - 5) = t^2 + 2t - 35$ Using method 3 (the product of two binomials, but neither the square of a binomial nor the product of the sum and the difference of two terms)

EXAMPLE 21 Multiply: $(x + 7)(x + 7)$.

$(x + 7)(x + 7) = x^2 + 2(7)x + 49$ Using method 1 (the square of a binomial sum)

$= x^2 + 14x + 49$

EXAMPLE 22 Multiply: $2x^3(9x^2 + x - 7)$.

$2x^3(9x^2 + x - 7) = 18x^5 + 2x^4 - 14x^3$ Using method 4 (the product of a monomial and a trinomial; multiplying each term of the trinomial by the monomial)

EXAMPLE 23 Multiply: $(5x^3 - 7x)^2$.

$(5x^3 - 7x)^2 = 25x^6 - 2(5x^3)(7x) + 49x^2$ Using method 1 (the square of a binomial difference)

$= 25x^6 - 70x^4 + 49x^2$

EXAMPLE 24 Multiply: $\left(3x + \frac{1}{4}\right)^2$.

$\left(3x + \frac{1}{4}\right)^2 = 9x^2 + 2(3x)\left(\frac{1}{4}\right) + \frac{1}{16}$ Using method 1 (the square of a binomial sum. To get the middle term, we multiply $3x$ by $\frac{1}{4}$ and double.)

$= 9x^2 + \frac{3}{2}x + \frac{1}{16}$

EXAMPLE 25 Multiply: $\left(4x - \frac{3}{4}\right)^2$.

$\left(4x - \frac{3}{4}\right)^2 = 16x^2 - 2(4x)\left(\frac{3}{4}\right) + \frac{9}{16}$ Using method 1 (the square of a binomial difference)

$= 16x^2 - 6x + \frac{9}{16}$

EXAMPLE 26 Multiply: $(p + 3)(p^2 + 2p - 1)$.

$$
\begin{array}{r}
p^2 + 2p - 1 \\
p + 3 \\
\hline
3p^2 + 6p - 3 \\
p^3 + 2p^2 - p \phantom{{}-3} \\
\hline
p^3 + 5p^2 + 5p - 3
\end{array}
$$

Using method 5 (the product of two polynomials)

Multiplying by 3

Multiplying by p

Do Exercises 27–34.

NAME SECTION DATE

Exercise Set 3.6

a Multiply. Try to write only the answer. If you need more steps, be sure to use them.

1. $(x + 1)(x^2 + 3)$

2. $(x^2 - 3)(x - 1)$

3. $(x^3 + 2)(x + 1)$

4. $(x^4 + 2)(x + 10)$

5. $(y + 2)(y - 3)$

6. $(a + 2)(a + 3)$

7. $(3x + 2)(3x + 2)$

8. $(4x + 1)(4x + 1)$

9. $(5x - 6)(x + 2)$

10. $(x - 8)(x + 8)$

11. $(3t - 1)(3t + 1)$

12. $(2m + 3)(2m + 3)$

13. $(4x - 2)(x - 1)$

14. $(2x - 1)(3x + 1)$

15. $\left(p - \frac{1}{4}\right)\left(p + \frac{1}{4}\right)$

16. $\left(q + \frac{3}{4}\right)\left(q + \frac{3}{4}\right)$

17. $(x - 0.1)(x + 0.1)$

18. $(x + 0.3)(x - 0.4)$

19. $(2x^2 + 6)(x + 1)$

20. $(2x^2 + 3)(2x - 1)$

21. $(-2x + 1)(x + 6)$

22. $(3x + 4)(2x - 4)$

23. $(a + 7)(a + 7)$

24. $(2y + 5)(2y + 5)$

25. $(1 + 2x)(1 - 3x)$

26. $(-3x - 2)(x + 1)$

27. $(x^2 + 3)(x^3 - 1)$

28. $(x^4 - 3)(2x + 1)$

29. $(3x^2 - 2)(x^4 - 2)$

30. $(x^{10} + 3)(x^{10} - 3)$

ANSWERS

1. ______
2. ______
3. ______
4. ______
5. ______
6. ______
7. ______
8. ______
9. ______
10. ______
11. ______
12. ______
13. ______
14. ______
15. ______
16. ______
17. ______
18. ______
19. ______
20. ______
21. ______
22. ______
23. ______
24. ______
25. ______
26. ______
27. ______
28. ______
29. ______
30. ______

ANSWERS

31. ______
32. ______
33. ______
34. ______
35. ______
36. ______
37. ______
38. ______
39. ______
40. ______
41. ______
42. ______
43. ______
44. ______
45. ______
46. ______
47. ______
48. ______
49. ______
50. ______
51. ______
52. ______
53. ______
54. ______
55. ______
56. ______
57. ______
58. ______
59. ______
60. ______
61. ______
62. ______
63. ______
64. ______

31. $(3x^5 + 2)(2x^2 + 6)$ **32.** $(1 - 2x)(1 + 3x^2)$ **33.** $(8x^3 + 1)(x^3 + 8)$

34. $(4 - 2x)(5 - 2x^2)$ **35.** $(4x^2 + 3)(x - 3)$ **36.** $(7x - 2)(2x - 7)$

37. $(4y^4 + y^2)(y^2 + y)$ **38.** $(5y^6 + 3y^3)(2y^6 + 2y^3)$

b Multiply mentally, if possible. If you need extra steps, be sure to use them.

39. $(x + 4)(x - 4)$ **40.** $(x + 1)(x - 1)$ **41.** $(2x + 1)(2x - 1)$

42. $(x^2 + 1)(x^2 - 1)$ **43.** $(5m - 2)(5m + 2)$ **44.** $(3x^4 + 2)(3x^4 - 2)$

45. $(2x^2 + 3)(2x^2 - 3)$ **46.** $(6x^5 - 5)(6x^5 + 5)$ **47.** $(3x^4 - 4)(3x^4 + 4)$

48. $(t^2 - 0.2)(t^2 + 0.2)$ **49.** $(x^6 - x^2)(x^6 + x^2)$ **50.** $(2x^3 - 0.3)(2x^3 + 0.3)$

51. $(x^4 + 3x)(x^4 - 3x)$ **52.** $\left(\frac{3}{4} + 2x^3\right)\left(\frac{3}{4} - 2x^3\right)$ **53.** $(x^{12} - 3)(x^{12} + 3)$

54. $(12 - 3x^2)(12 + 3x^2)$ **55.** $(2y^8 + 3)(2y^8 - 3)$ **56.** $\left(m - \frac{2}{3}\right)\left(m + \frac{2}{3}\right)$

c Multiply mentally, if possible. If you need extra steps, be sure to use them.

57. $(x + 2)^2$ **58.** $(2x - 1)^2$ **59.** $(3x^2 + 1)^2$ **60.** $\left(3x + \frac{3}{4}\right)^2$

61. $\left(a - \frac{1}{2}\right)^2$ **62.** $\left(2a - \frac{1}{5}\right)^2$ **63.** $(3 + x)^2$ **64.** $(x^3 - 1)^2$

65. $(x^2 + 1)^2$

66. $(8x - x^2)^2$

67. $(2 - 3x^4)^2$

68. $(6x^3 - 2)^2$

69. $(5 + 6t^2)^2$

70. $(3p^2 - p)^2$

d Multiply mentally, if possible.

71. $(3 - 2x^3)^2$

72. $(x - 4x^3)^2$

73. $4x(x^2 + 6x - 3)$

74. $8x(-x^5 + 6x^2 + 9)$

75. $\left(2x^2 - \frac{1}{2}\right)\left(2x^2 - \frac{1}{2}\right)$

76. $(-x^2 + 1)^2$

77. $(-1 + 3p)(1 + 3p)$

78. $(-3q + 2)(3q + 2)$

79. $3t^2(5t^3 - t^2 + t)$

80. $-6x^2(x^3 + 8x - 9)$

81. $(6x^4 + 4)^2$

82. $(8a + 5)^2$

83. $(3x + 2)(4x^2 + 5)$

84. $(2x^2 - 7)(3x^2 + 9)$

85. $(8 - 6x^4)^2$

86. $\left(\frac{1}{5}x^2 + 9\right)\left(\frac{3}{5}x^2 - 7\right)$

87. $(t - 1)(t^2 + t + 1)$

88. $(y + 5)(y^2 - 5y + 25)$

Compute each of the following and compare.

89. $3^2 + 4^2$; $(3 + 4)^2$

90. $6^2 - 7^2$; $(6 + 7)^2$

91. $9^2 - 5^2$; $(9 - 5)^2$

92. $11^2 - 4^2$; $(11 - 4)^2$

ANSWERS

65. __________
66. __________
67. __________
68. __________
69. __________
70. __________
71. __________
72. __________
73. __________
74. __________
75. __________
76. __________
77. __________
78. __________
79. __________
80. __________
81. __________
82. __________
83. __________
84. __________
85. __________
86. __________
87. __________
88. __________
89. __________
90. __________
91. __________
92. __________

ANSWERS

93. ______
94. ______
95. ______
96. ______
97. ______
98. ______
99. ______
100. ______
101. ______
102. ______
103. ______
104. ______
105. ______
106. ______
107. ______
108. ______
109. ______
110. ______
111. a) ______
b) ______
c) ______
d) ______
112. ______

SKILL MAINTENANCE

93. In apartment 3B, lamps, an air conditioner, and a television set are all operating at the same time. The lamps use 10 times as many watts of electricity as the television set, and the air conditioner uses 40 times as many watts as the television set. The total wattage used in the apartment is 2550. How many watts are used by each appliance?

Solve.

94. $3x - 8x = 4(7 - 8x)$

95. $3(x - 2) = 5(2x + 7)$

96. $5(2x - 3) - 2(3x - 4) = 20$

SYNTHESIS

Multiply.

97. $4y(y + 5)(2y + 8)$

98. $8x(2x - 3)(5x + 9)$

99. $[(x + 1) - x^2][(x - 2) + 2x^2]$

100. $[(2x - 1)(2x + 1)](4x^2 + 1)$

Solve.

101. $(x + 2)(x - 5) = (x + 1)(x - 3)$

102. $(2x + 5)(x - 4) = (x + 5)(2x - 4)$

The height of a box is 1 more than its length l, and the length is 1 more than its width w. Find a polynomial for the volume V in terms of the following.

103. The width w

104. The length l

105. The height h

Calculate as the difference of squares.

106. 18×22 [*Hint:* $(20 - 2)(20 + 2)$.]

107. 93×107

Multiply. (Do not collect like terms before multiplying.)

108. $[(3x - 2)(3x + 2)](9x^2 + 4)$

109. $[3a - (2a - 3)][3a + (2a - 3)]$

110. $(5t^2 - 3)^2(5t^2 + 3)^2$

111. A factored polynomial for the shaded area in this rectangle is $(A + B)(A - B)$.

a) Find a polynomial for the area of the entire rectangle.

b) Find a polynomial for the sum of the areas of the two small unshaded rectangles.

c) Find a polynomial for the area in part (a) minus the area in part (b).

d) Find a polynomial for the area of the shaded region and compare this with the polynomial found in part (c).

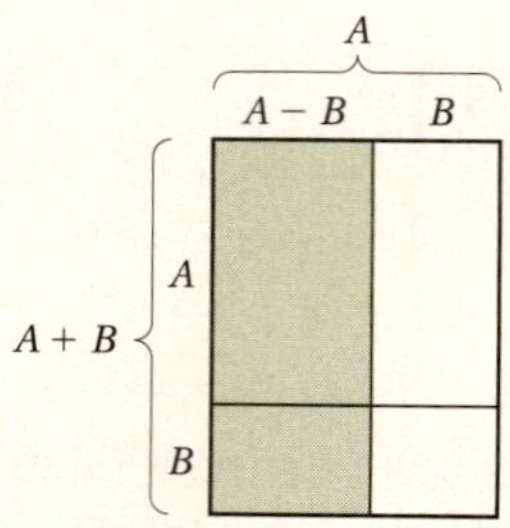

112. Find $(10x + 5)^2$. Use your result to show how to mentally square any two-digit number ending in 5.

3.7 Operations with Polynomials in Several Variables

The polynomials that we have been studying have only one variable. A **polynomial in several variables** is an expression like those you have already seen, but with more than one variable. Here are some examples:

$$3x + xy^2 + 5y + 4, \qquad 8xy^2z - 2x^3z - 13x^4y^2 + 15.$$

a Evaluating Polynomials

Example 1 Evaluate the polynomial $4 + 3x + xy^2 + 8x^3y^3$ when $x = -2$ and $y = 5$.

We replace x by -2 and y by 5:

$$\begin{aligned} 4 + 3x + xy^2 + 8x^3y^3 &= 4 + 3(-2) + (-2) \cdot 5^2 + 8(-2)^3 \cdot 5^3 \\ &= 4 - 6 - 50 - 8000 \\ &= -8052. \end{aligned}$$

Example 2 *Surface area of a right circular cylinder.* The surface area of a right circular cylinder is given by the polynomial

$$2\pi rh + 2\pi r^2,$$

where h is the height and r is the radius of the base. (This formula can be derived by cutting the cylinder apart, as shown on the right below, and adding the areas of the parts.)

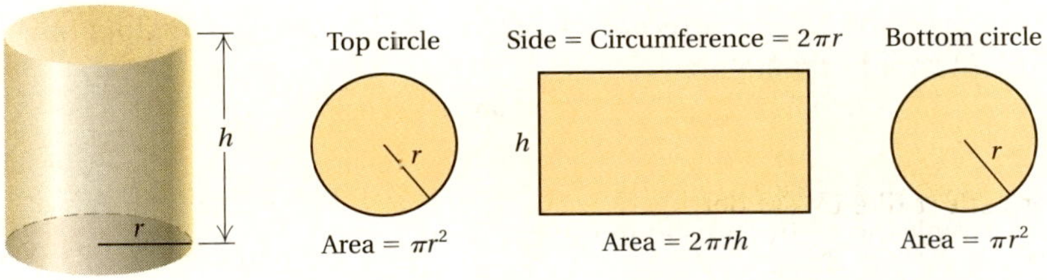

A 12-oz beverage can has a height of 4.7 in. and a radius of 1.2 in. To find the surface area, we can evaluate the polynomial when $h = 4.7$ and $r = 1.2$. Use 3.14 as an approximation for π.

We evaluate the polynomial when $h = 4.7$, $r = 1.2$, and $\pi \approx 3.14$:

$$\begin{aligned} 2\pi rh + 2\pi r^2 &\approx 2(3.14)(1.2)(4.7) + 2(3.14)(1.2)^2 \\ &\approx 2(3.14)(1.2)(4.7) + 2(3.14)(1.44) \\ &\approx 35.4192 + 9.0432 \\ &\approx 44.4624. \end{aligned}$$

The surface area is about 44.4624 in^2.

Do Exercises 1–3.

Objectives

After finishing Section 3.7, you should be able to:

- a Evaluate a polynomial in several variables for given values of the variables.
- b Identify the coefficients and the degrees of the terms of a polynomial and the degree of a polynomial.
- c Collect like terms of a polynomial.
- d Add polynomials.
- e Subtract polynomials.
- f Multiply polynomials.

For Extra Help

TAPE 9 | TAPE 7B | MAC: 3B IBM: 3B

1. Evaluate the polynomial

$$4 + 3x + xy^2 + 8x^3y^3$$

when $x = 2$ and $y = -5$.

2. Evaluate the polynomial

$$8xy^2 - 2x^3z - 13x^4y^2 + 5$$

when $x = -1$, $y = 3$, and $z = 4$.

3. For the situation of Example 2, find the surface area of a tank with $h = 20$ ft and $r = 3$ ft. Use 3.14 for π.

Answers on page A-3

4. Identify the coefficient of each term:

$-3xy^2 + 3x^2y - 2y^3 + xy + 2.$

5. Identify the degree of each term and the degree of the polynomial

$4xy^2 + 7x^2y^3z^2 - 5x + 2y + 4.$

Collect like terms.

6. $4x^2y + 3xy - 2x^2y$

7. $-3pq - 5pqr^3 - 12 + 8pq + 5pqr^3 + 4$

Answers on page A-3

b COEFFICIENTS AND DEGREES

The **degree** of a term is the sum of the exponents of the variables. The **degree of a polynomial** is the degree of the term of highest degree.

EXAMPLE 3 Identify the coefficient and the degree of each term and the degree of the polynomial

$$9x^2y^3 - 14xy^2z^3 + xy + 4y + 5x^2 + 7.$$

TERM	COEFFICIENT	DEGREE	DEGREE OF THE POLYNOMIAL
$9x^2y^3$	9	5	
$-14xy^2z^3$	−14	6	6
xy	1	2	
$4y$	4	1	Think: $4y = 4y^1$.
$5x^2$	5	2	
7	7	0	Think: $7 = 7x^0$, or $7x^0y^0z^0$.

Do Exercises 4 and 5.

c COLLECTING LIKE TERMS

Like terms (or **similar terms**) have exactly the same variables with exactly the same exponents. For example,

$3x^2y^3$ and $-7x^2y^3$ are like terms;

$9x^4z^7$ and $12x^4z^7$ are like terms.

But

$13xy^5$ and $-2x^2y^5$ are *not* like terms, because the x-factors have different exponents;

and

$3xyz^2$ and $4xy$ are *not* like terms, because there is no factor involving z^2 in the second expression.

Collecting like terms is based on the distributive laws.

EXAMPLES Collect like terms.

4. $5x^2y + 3xy^2 - 5x^2y - xy^2 = (5 - 5)x^2y + (3 - 1)xy^2 = 2xy^2$

5. $7xy - 5xy^2 + 3xy^2 - 7 + 6x^3 + 9xy - 11x^3 + y - 1$
$= -2xy^2 + 16xy - 5x^3 + y - 8$

Do Exercises 6 and 7.

d ADDITION

We can find the sum of two polynomials in several variables by writing a plus sign between them and then collecting like terms.

EXAMPLE 6 Add: $(-5x^3 + 3y - 5y^2) + (8x^3 + 4x^2 + 7y^2)$.

$$(-5x^3 + 3y - 5y^2) - (8x^3 + 4x^2 + 7y^2)$$
$$= (-5 + 8)x^3 + 4x^2 + 3y + (-5 + 7)y^2$$
$$= 3x^3 + 4x^2 + 3y + 2y^2$$

EXAMPLE 7 Add:

$$(5xy^2 - 4x^2y + 5x^3 + 2) + (3xy^2 - 2x^2y + 3x^3y - 5).$$

We first look for like terms. They are $5xy^2$ and $3xy^2$, $-4x^2y$ and $-2x^2y$, and 2 and -5. We collect these. Since there are no more like terms, the answer is

$$8xy^2 - 6x^2y + 5x^3 + 3x^3y - 3.$$

Do Exercises 8–10.

e SUBTRACTION

We subtract a polynomial by adding its opposite, or additive inverse. The opposite of the polynomial

$$4x^2y - 6x^3y^2 + x^2y^2 - 5y$$

can be represented by

$$-(4x^2y - 6x^3y^2 + x^2y^2 - 5y).$$

We find an equivalent expression for the opposite of a polynomial by replacing each coefficient by its opposite, or by changing the sign of each term. Thus,

$$-(4x^2y - 6x^3y^2 + x^2y^2 - 5y) = -4x^2y + 6x^3y^2 - x^2y^2 + 5y.$$

EXAMPLE 8 Subtract:

$$(4x^2y + x^3y^2 + 3x^2y^3 + 6y + 10) - (4x^2y - 6x^3y^2 + x^2y^2 - 5y - 8).$$

We have

$$(4x^2y + x^3y^2 + 3x^2y^3 + 6y + 10) - (4x^2y - 6x^3y^2 + x^2y^2 - 5y - 8)$$
$$= 4x^2y + x^3y^2 + 3x^2y^3 + 6y + 10 - 4x^2y + 6x^3y^2 - x^2y^2 + 5y + 8$$

Adding the opposite

$$= 7x^3y^2 + 3x^2y^3 - x^2y^2 + 11y + 18$$

Collecting like terms. (Try to write just the answer!)

Do Exercises 11 and 12.

Add.

8. $(4x^3 + 4x^2 - 8y - 3) + (-8x^3 - 2x^2 + 4y + 5)$

9. $(13x^3y + 3x^2y - 5y) + (x^3y + 4x^2y - 3xy + 3y)$

10. $(-5p^2q^4 + 2p^2q^2 + 3q) + (6pq^2 + 3p^2q + 5)$

Subtract.

11. $(-4s^4t + s^3t^2 + 2s^2t^3) - (4s^4t - 5s^3t^2 + s^2t^2)$

12. $(-5p^4q + 5p^3q^2 - 3p^2q^3 - 7q^4 - 2) - (4p^4q - 4p^3q^2 + p^2q^3 + 2q^4 - 7)$

Answers on page A-3

Multiply.

13. $(x^2y^3 + 2x)(x^3y^2 + 3x)$

14. $(p^4q - 2p^3q^2 + 3q^3)(p + 2q)$

Multiply.

15. $(3xy + 2x)(x^2 + 2xy^2)$

16. $(x - 3y)(2x - 5y)$

17. $(4x + 5y)^2$

18. $(3x^2 - 2xy^2)^2$

19. $(2xy^2 + 3x)(2xy^2 - 3x)$

20. $(3xy^2 + 4y)(-3xy^2 + 4y)$

21. $(3y + 4 - 3x)(3y + 4 + 3x)$

22. $(2a + 5b + c)(2a - 5b - c)$

Answers on page A-3

f MULTIPLICATION

To multiply polynomials in several variables, we can multiply each term of one by every term of the other. Where appropriate, we use the special products that we have learned.

EXAMPLE 9 Multiply: $(3x^2y - 2xy + 3y)(xy + 2y)$.

$$\begin{array}{rrrrl} & 3x^2y & - 2xy & + 3y & \\ & & xy & + 2y & \\ \hline & 6x^2y^2 & - 4xy^2 & + 6y^2 & \text{Multiplying by } 2y \\ 3x^3y^2 & - 2x^2y^2 & + 3xy^2 & & \text{Multiplying by } xy \\ \hline 3x^3y^2 & + 4x^2y^2 & - xy^2 & + 6y^2 & \text{Adding} \end{array}$$

Do Exercises 13 and 14.

EXAMPLES Multiply.

10. $(x^2y + 2x)(xy^2 + y^2) = x^3y^3 + x^2y^3 + 2x^2y^2 + 2xy^2$ (F O I L)

11. $(p + 5q)(2p - 3q) = 2p^2 - 3pq + 10pq - 15q^2$
$= 2p^2 + 7pq - 15q^2$

$(A + B)^2 = A^2 + 2 \cdot A \cdot B + B^2$

12. $(3x + 2y)^2 = (3x)^2 + 2(3x)(2y) + (2y)^2$
$= 9x^2 + 12xy + 4y^2$

$(A - B)^2 = A^2 - 2 \cdot A \cdot B + B^2$

13. $(2y^2 - 5x^2y)^2 = (2y^2)^2 - 2(2y^2)(5x^2y) + (5x^2y)^2$
$= 4y^4 - 20x^2y^3 + 25x^4y^2$

$(A + B)(A - B) = A^2 - B^2$

14. $(3x^2y + 2y)(3x^2y - 2y) = (3x^2y)^2 - (2y)^2$
$= 9x^4y^2 - 4y^2$

15. $(-2x^3y^2 + 5t)(2x^3y^2 + 5t) = (5t - 2x^3y^2)(5t + 2x^3y^2)$
$= (5t)^2 - (2x^3y^2)^2$
$= 25t^2 - 4x^6y^4$

$(A - B)(A + B) = A^2 - B^2$

16. $(2x + 3 - 2y)(2x + 3 + 2y) = (2x + 3)^2 - (2y)^2$
$= 4x^2 + 12x + 9 - 4y^2$

Do Exercises 15–22.

NAME SECTION DATE

Exercise Set 3.7

ANSWERS

1. ______
2. ______
3. ______
4. ______
5. ______
6. ______
7. ______
8. ______
9. ______
10. ______
11. ______
12. ______
13. ______
14. ______

a Evaluate the polynomial when $x = 3$ and $y = -2$.

1. $x^2 - y^2 + xy$

2. $x^2 + y^2 - xy$

Evaluate the polynomial when $x = 2$, $y = -3$, and $z = -1$.

3. $xyz^2 + z$

4. $xy - xz + yz$

Interest compounded annually for two years. An amount of money P is invested at interest rate i. In 2 years, it will grow to an amount given by the polynomial

$$P(1 + i)^2.$$

5. Evaluate the polynomial when $P = 10{,}000$ and $i = 0.06$ to find the amount to which \$10,000 will grow at 6% interest for 2 years.

6. Evaluate the polynomial when $P = 10{,}000$ and $i = 0.04$ to find the amount to which \$10,000 will grow at 4% interest for 2 years.

Interest compounded annually for three years. An amount of money P is invested at interest rate i. In 3 years, it will grow to an amount given by the polynomial

$$P(1 + i)^3.$$

7. Evaluate the polynomial when $P = 10{,}000$ and $i = 0.06$ to find the amount to which \$10,000 will grow at 6% interest for 3 years.

8. Evaluate the polynomial when $P = 10{,}000$ and $i = 0.04$ to find the amount to which \$10,000 will grow at 4% interest for 3 years.

Surface area of a right circular cylinder. The area of a right circular cylinder is given by the polynomial

$$2\pi rh + 2\pi r^2,$$

where h is the height and r is the radius of the base.

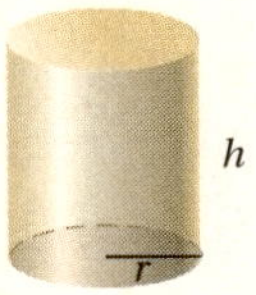

9. A 16-oz beverage can has a height of 6.3 in. and a radius of 1.2 in. Evaluate the polynomial when $h = 6.3$ and $r = 1.2$ to find the area of the can. Use 3.14 for π.

10. A 26-oz coffee can has a height of 6.5 in. and a radius of 2.5 in. Evaluate the polynomial when $h = 6.5$ and $r = 2.5$ to find the area of the can. Use 3.14 for π.

b Identify the coefficient and the degree of each term of the polynomial. Then find the degree of the polynomial

11. $x^3y - 2xy + 3x^2 - 5$

12. $5y^3 - y^2 + 15y + 1$

13. $17x^2y^3 - 3x^3yz - 7$

14. $6 - xy + 8x^2y^2 - y^5$

ANSWERS

15. ________

16. ________

17. ________

18. ________

19. ________

20. ________

21. ________

22. ________

23. ________

24. ________

25. ________

26. ________

27. ________

28. ________

29. ________

30. ________

31. ________

32. ________

33. ________

34. ________

35. ________

36. ________

37. ________

38. ________

c Collect like terms.

15. $a + b - 2a - 3b$

16. $y^2 - 1 + y - 6 - y^2$

17. $3x^2y - 2xy^2 + x^2$

18. $m^3 + 2m^2n - 3m^2 + 3mn^2$

19. $2u^2v - 3uv^2 + 6u^2v - 2uv^2$

20. $3x^2 + 6xy + 3y^2 - 5x^2 - 10xy - 5y^2$

21. $6au + 3av + 14au + 7av$

22. $3x^2y - 2z^2y + 3xy^2 + 5z^2y$

d Add.

23. $(2x^2 - xy + y^2) + (-x^2 - 3xy + 2y^2)$

24. $(2z - z^2 + 5) + (z^2 - 3z + 1)$

25. $(r - 2s + 3) + (2r + s) + (s + 4)$

26. $(b^3a^2 - 2b^2a^3 + 3ba + 4) + (b^2a^3 - 4b^3a^2 + 2ba - 1)$

27. $(2x^2 - 3xy + y^2) + (-4x^2 - 6xy - y^2) + (x^2 + xy - y^2)$

e Subtract.

28. $(x^3 - y^3) - (-2x^3 + x^2y - xy^2 + 2y^3)$

29. $(xy - ab - 8) - (xy - 3ab - 6)$

30. $(3y^4x^2 + 2y^3x - 3y - 7) - (2y^4x^2 + 2y^3x - 4y - 2x + 5)$

31. $(-2a + 7b - c) - (-3b + 4c - 8d)$

32. Find the sum of $2a + b$ and $3a - b$. Then subtract $5a + 2b$.

f Multiply.

33. $(3z - u)(2z + 3u)$

34. $(a - b)(a^2 + b^2 + 2ab)$

35. $(a^2b - 2)(a^2b - 5)$

36. $(xy + 7)(xy - 4)$

37. $(a + a^2 - 1)(a^2 + 1 - y)$

38. $(r + tx)(vx + s)$

39. $(a^3 + bc)(a^3 - bc)$

40. $(m^2 + n^2 - mn)(m^2 + mn + n^2)$

41. $(y^4x + y^2 + 1)(y^2 + 1)$

42. $(a - b)(a^2 + ab + b^2)$

43. $(3xy - 1)(4xy + 2)$

44. $(m^3n + 8)(m^3n - 6)$

45. $(3 - c^2d^2)(4 + c^2d^2)$

46. $(6x - 2y)(5x - 3y)$

47. $(m^2 - n^2)(m + n)$

48. $(pq + 0.2)(0.4pq - 0.1)$

49. $(xy + x^5y^5)(x^4y^4 - xy)$

50. $(x - y^3)(2y^3 + x)$

51. $(x + h)^2$

52. $(3a + 2b)^2$

53. $(r^3t^2 - 4)^2$

54. $(3a^2b - b^2)^2$

55. $(p^4 + m^2n^2)^2$

56. $(2ab - cd)^2$

57. $\left(2a^3 - \frac{1}{2}b^3\right)^2$

58. $-3x(x + 8y)^2$

59. $3a(a - 2b)^2$

60. $(a^2 + b + 2)^2$

61. $(2a - b)(2a + b)$

62. $(x - y)(x + y)$

63. $(c^2 - d)(c^2 + d)$

64. $(p^3 - 5q)(p^3 + 5q)$

ANSWERS

39. ________

40. ________

41. ________

42. ________

43. ________

44. ________

45. ________

46. ________

47. ________

48. ________

49. ________

50. ________

51. ________

52. ________

53. ________

54. ________

55. ________

56. ________

57. ________

58. ________

59. ________

60. ________

61. ________

62. ________

63. ________

64. ________

ANSWERS

65. ______

66. ______

67. ______

68. ______

69. ______

70. ______

71. ______

72. ______

73. ______

74. ______

75. ______

76. ______

77. ______

78. ______

79. ______

65. $(ab + cd^2)(ab - cd^2)$

66. $(xy + pq)(xy - pq)$

67. $(x + y - 3)(x + y + 3)$

68. $(p + q + 4)(p + q - 4)$

69. $[x + y + z][x - (y + z)]$

70. $[a + b + c][a - (b + c)]$

71. $(a + b + c)(a - b - c)$

72. $(3x + 2 - 5y)(3x + 2 + 5y)$

SYNTHESIS

Find a polynomial for the shaded area. (Leave results in terms of π where appropriate.)

73.

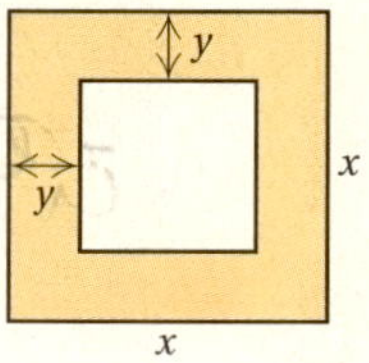

74.

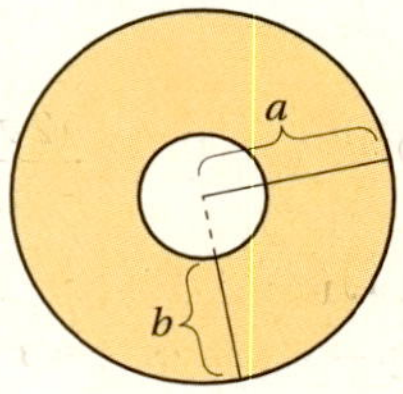

75.

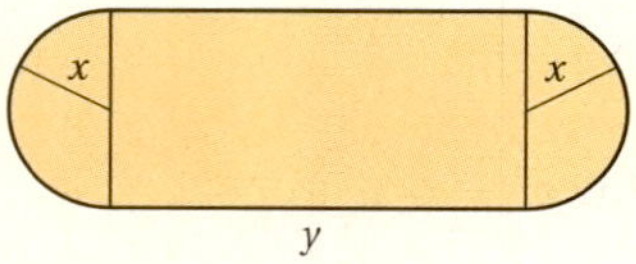

76.

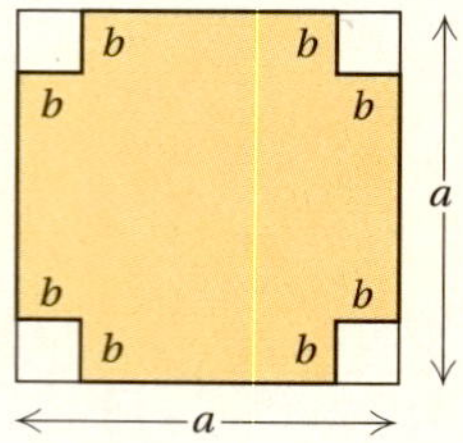

77. *The magic number.* The Boston Red Sox are leading the Detroit Tigers for the Eastern Division championship of the American League. The magic number is 8. This means that any combination of Red Sox wins and Tiger losses that totals 8 will ensure the championship for the Red Sox. The magic number is given by the polynomial

$$G - W_1 - L_2 + 1,$$

where G is the number of games in the season, W_1 is the number of wins by the leading team, and L_2 is the number of losses by the second-place team.

Given the situation shown in the table and assuming a 162-game season, what is the magic number for the Philadelphia Phillies?

EASTERN DIVISION

	W	L	PCT.	GB
Philadelphia	77	40	.658	—
Pittsburgh.........	65	53	.551	$12\frac{1}{2}$
Florida............	61	60	.504	18
Chicago	55	67	.451	$24\frac{1}{2}$
St. Louis	51	65	.440	$25\frac{1}{2}$
Montreal..........	41	73	.360	$34\frac{1}{2}$
New York	35	79	.307	$40\frac{1}{2}$

78. *Lung capacity.* The polynomial

$$0.041h - 0.018A - 2.69$$

can be used to estimate the lung capacity, in liters, of a female with height h, in centimeters, and age A, in years. Find the lung capacity of a 29-year-old woman who is 138.7 cm tall.

79. Find a formula for $(A + B)^3$.

3.8 *Division of Polynomials*

In this section, we consider division of polynomials. You will see that such division is similar to what is done in arithmetic.

OBJECTIVES

After finishing Section 3.8, you should be able to:

a Divide a polynomial by a monomial.

b Divide a polynomial by a divisor that is not a monomial and, if there is a remainder, express the result in two ways.

FOR EXTRA HELP

TAPE 10 TAPE 8A MAC: 3B IBM: 3B

a DIVISOR A MONOMIAL

We first consider division by a monomial. When we are dividing a monomial by a monomial, we can use the rules of exponents and subtract exponents when bases are the same. We studied this in Section 3.1. For example,

$$\frac{15x^{10}}{3x^4} = \frac{15}{3}x^{10-4} = 5x^6; \qquad \frac{42a^2b^5}{-3ab^2} = \frac{42}{-3}a^{2-1}b^{5-2} = -14ab^3.$$

When we divide a polynomial by a monomial, we break up the division into an addition of quotients of monomials. To do this, we use the rule for addition using fractional notation. That is, since

$$\frac{A}{C} + \frac{B}{C} = \frac{A+B}{C},$$

we know that

$$\frac{A+B}{C} = \frac{A}{C} + \frac{B}{C}.$$

EXAMPLE 1 Divide $x^3 + 10x^2 + 8x$ by $2x$.

We write the division as follows:

$$\frac{x^3 + 10x^2 + 8x}{2x}.$$

This is equivalent to

$$\frac{x^3}{2x} + \frac{10x^2}{2x} + \frac{8x}{2x}.$$ To see this, add and get the original expression.

Next, we do the separate divisions:

$$\frac{x^3}{2x} + \frac{10x^2}{2x} + \frac{8x}{2x} = \frac{1}{2}x^{3-1} + \frac{10}{2}x^{2-1} + \frac{8}{2}x^{1-1} = \frac{1}{2}x^2 + 5x + 4.$$

CHECK: We multiply the quotient by $2x$:

$$\begin{array}{r} \frac{1}{2}x^2 + 5x + 4 \\ 2x \\ \hline x^3 + 10x^2 + 8x \end{array}$$

We multiply. The answer checks.

Do Exercises 1 and 2.

1. Divide $2x^3 + 6x^2 + 4x$ by $2x$. Check the result.

2. Divide. Check the result.

$$(6x^2 + 3x - 2) \div 3$$

Answers on page A-3

Divide and check.

3. $(8x^2 - 3x + 1) \div 2$

4. $\dfrac{2x^4y^6 - 3x^3y^4 + 5x^2y^3}{x^2y^2}$

Answers on page A-3

EXAMPLE 2 Divide and check: $(10a^5b^4 - 2a^3b^2 + 6a^2b) \div (2a^2b)$.

$$\frac{10a^5b^4 - 2a^3b^2 + 6a^2b}{2a^2b} = \frac{10a^5b^4}{2a^2b} - \frac{2a^3b^2}{2a^2b} + \frac{6a^2b}{2a^2b}$$
$$= \frac{10}{2}a^{5-2}b^{4-1} - \frac{2}{2}a^{3-2}b^{2-1} + \frac{6}{2}$$
$$= 5a^3b^3 - ab + 3$$

CHECK:

$$\begin{array}{r} 5a^3b^3 - ab + 3 \\ 2a^2b \\ \hline 10a^5b^4 - 2a^3b^2 + 6a^2b \end{array}$$

We multiply.

The answer checks.

> To divide a polynomial by a monomial, divide each term by the monomial.

Do Exercises 3 and 4.

b DIVISOR NOT A MONOMIAL

Let us consider long division in arithmetic. We divide 3711 by 8.

$$\begin{array}{r} 463 \\ 8\overline{)3711} \\ \underline{32} \\ 51 \\ \underline{48} \\ 31 \\ \underline{24} \\ 7 \end{array}$$

The quotient is 463. The remainder is 7, expressed as R = 7. We write the answer as

463 R 7

or $463 + \dfrac{7}{8} = 463\dfrac{7}{8}$.

We check by multiplying the quotient, 463, by the divisor, 8, and adding the remainder, 7:

$8 \cdot 463 + 7 = 3704 + 7 = 3711.$

Now let us look at long division with polynomials. We use this procedure when the divisor is not a monomial. We write polynomials in descending order and then write in missing terms.

EXAMPLE 3 Divide $x^2 + 5x + 6$ by $x + 2$.

We have

$$
\begin{array}{rll}
 & x & \leftarrow \text{Divide the first term by the first term: } x^2/x = x. \text{ Ignore the term 2.} \\
x + 2 \enclose{longdiv}{} & x^2 + 5x + 6 & \\
 & \underline{x^2 + 2x} & \leftarrow \text{Multiply } x \text{ above by the divisor, } x + 2. \\
 & 3x & \leftarrow \text{Subtract: } (x^2 + 5x) - (x^2 + 2x) = x^2 + 5x - x^2 - 2x = 3x.
\end{array}
$$

We now "bring down" the next term of the dividend—in this case, 6.

$$
\begin{array}{rll}
 & x + 3 & \leftarrow \text{Divide the first term by the first term: } 3x/x = 3. \\
x + 2 \enclose{longdiv}{} & x^2 + 5x + 6 & \\
 & \underline{x^2 + 2x} & \\
 & 3x + 6 & \leftarrow \text{The 6 has been "brought down."} \\
 & \underline{3x + 6} & \leftarrow \text{Multiply 3 by the divisor, } x + 2. \\
 & 0 & \leftarrow \text{Subtract: } (3x + 6) - (3x + 6) = 3x + 6 - 3x - 6 = 0.
\end{array}
$$

The quotient is $x + 3$. The remainder is 0, expressed as $R = 0$. A remainder of 0 is generally not listed in an answer.

To check, we multiply the quotient by the divisor and add the remainder, if any, to see if we get the dividend:

$$\underbrace{(x + 2)}_{\text{Divisor}} \cdot \underbrace{(x + 3)}_{\text{Quotient}} + \underbrace{0}_{\text{Remainder}} = \underbrace{x^2 + 5x + 6}_{\text{Dividend}}. \quad \text{The division checks.}$$

Do Exercise 5.

EXAMPLE 4 Divide and check: $(x^2 + 2x - 12) \div (x - 3)$.

We have

$$
\begin{array}{rll}
 & x & \leftarrow \text{Divide the first term by the first term: } x^2/x = x. \\
x - 3 \enclose{longdiv}{} & x^2 + 2x - 12 & \\
 & \underline{x^2 - 3x} & \leftarrow \text{Multiply } x \text{ above by the divisor, } x - 3. \\
 & 5x & \leftarrow \text{Subtract: } (x^2 + 2x) - (x^2 - 3x) = x^2 + 2x - x^2 + 3x = 5x.
\end{array}
$$

We now "bring down" the next term of the dividend—in this case, -12.

$$
\begin{array}{rll}
 & x + 5 & \leftarrow \text{Divide the first term by the first term: } 5x/x = 5. \\
x - 3 \enclose{longdiv}{} & x^2 + 2x - 12 & \\
 & \underline{x^2 - 3x} & \\
 & 5x - 12 & \leftarrow \text{Bring down the } -12. \\
 & \underline{5x - 15} & \leftarrow \text{Multiply 5 above by the divisor, } x - 3. \\
 & 3 & \leftarrow \text{Subtract: } (5x - 12) - (5x - 15) = 5x - 12 - 5x + 15 = 3.
\end{array}
$$

The answer is $x + 5$ with $R = 3$, or

$$\underbrace{x + 5}_{\text{Quotient}} + \frac{3 \;\rightarrow \text{Remainder}}{\underbrace{x - 3}_{\text{Divisor}}}$$

(This is the way answers will be given at the back of the book.)

5. Divide and check:

$(x^2 + x - 6) \div (x + 3)$.

Answer on page A-3

6. Divide and check:

$$x - 2\overline{)x^2 + 2x - 8}.$$

Divide and check.

7. $x + 3\overline{)x^2 + 7x + 10}$

8. $(x^3 - 1) \div (x - 1)$

Answers on page A-3

CHECK: When the answer is given in the preceding form, we can check by multiplying the divisor by the quotient and adding the remainder, as follows:

$$\begin{aligned}(x - 3)(x + 5) + 3 &= x^2 + 2x - 15 + 3 \\ &= x^2 + 2x - 12.\end{aligned}$$

When dividing, an answer may "come out even" (that is, have a remainder of 0, as in Example 3), or it may not (as in Example 4). If a remainder is not 0, we continue dividing until the degree of the remainder is less than the degree of the divisor. Check this in each of Examples 3 and 4.

Do Exercise 6.

EXAMPLE 5 Divide and check: $(x^3 + 1) \div (x + 1)$.

$$\begin{array}{r l}
 & x^2 - x + 1 \\
x + 1 & \overline{)x^3 + 0x^2 + 0x + 1} \\
 & \underline{x^3 + x^2} \\
 & - x^2 + 0x \\
 & \underline{- x^2 - x} \\
 & x + 1 \\
 & \underline{x + 1} \\
 & 0
\end{array}$$

← Fill in the missing terms (see Section 3.3h).

This subtraction is $x^3 - (x^3 + x^2)$.

This subtraction is $-x^2 - (-x^2 - x)$.

The answer is $x^2 - x + 1$.

CHECK:
$$\begin{aligned}(x + 1)(x^2 - x + 1) + 0 &= x^2 - x + 1 + x^3 - x^2 + x + 0 \\ &= x^3 + 1.\end{aligned}$$

EXAMPLE 6 Divide and check: $(x^4 - 3x^2 + 1) \div (x - 4)$.

$$\begin{array}{r l}
 & x^3 + 4x^2 + 13x + 52 \\
x - 4 & \overline{)x^4 + 0x^3 - 3x^2 + 0x + 1} \\
 & \underline{x^4 - 4x^3} \\
 & 4x^3 - 3x^2 \\
 & \underline{4x^3 - 16x^2} \\
 & 13x^2 + 0x \\
 & \underline{13x^2 - 52x} \\
 & 52x + 1 \\
 & \underline{52x - 208} \\
 & 209
\end{array}$$

← Fill in the missing terms.

$x^4 - (x^4 - 4x^3)$

$(4x^3 - 3x^2) - (4x^3 - 16x^2)$

The answer is $x^3 + 4x^2 + 13x + 52$, with R = 209, or

$$x^3 + 4x^2 + 13x + 52 + \frac{209}{x - 4}.$$

CHECK:
$$\begin{aligned}&(x - 4)(x^3 + 4x^2 + 13x + 52) + 209 \\ &= -4x^3 - 16x^2 - 52x - 208 + x^4 + 4x^3 + 13x^2 + 52x + 209 \\ &= x^4 - 3x^2 + 1\end{aligned}$$

Do Exercises 7 and 8.

NAME SECTION DATE

Exercise Set 3.8

a Divide and check.

1. $\dfrac{24x^4 - 4x^3 + x^2 - 16}{8}$

2. $\dfrac{12a^4 - 3a^2 + a - 6}{6}$

3. $\dfrac{u - 2u^2 - u^5}{u}$

4. $\dfrac{50x^5 - 7x^4 + x^2}{x}$

5. $(15t^3 + 24t^2 - 6t) \div (3t)$

6. $(25t^3 + 15t^2 - 30t) \div (5t)$

7. $(20x^6 - 20x^4 - 5x^2) \div (-5x^2)$

8. $(24x^6 + 32x^5 - 8x^2) \div (-8x^2)$

9. $(24x^5 - 40x^4 + 6x^3) \div (4x^3)$

10. $(18x^6 - 27x^5 - 3x^3) \div (9x^3)$

11. $\dfrac{18x^2 - 5x + 2}{2}$

12. $\dfrac{15x^2 - 30x + 6}{3}$

13. $\dfrac{12x^3 + 26x^2 + 8x}{2x}$

14. $\dfrac{2x^4 - 3x^3 + 5x^2}{x^2}$

15. $\dfrac{9r^2s^2 + 3r^2s - 6rs^2}{3rs}$

16. $\dfrac{4x^4y - 8x^6y^2 + 12x^8y^6}{4x^4y}$

b Divide.

17. $(x^2 + 4x + 4) \div (x + 2)$

18. $(x^2 - 6x + 9) \div (x - 3)$

19. $(x^2 - 10x - 25) \div (x - 5)$

20. $(x^2 + 8x - 16) \div (x + 4)$

21. $(x^2 + 4x - 14) \div (x + 6)$

22. $(x^2 + 5x - 9) \div (x - 2)$

ANSWERS

1. ______
2. ______
3. ______
4. ______
5. ______
6. ______
7. ______
8. ______
9. ______
10. ______
11. ______
12. ______
13. ______
14. ______
15. ______
16. ______
17. ______
18. ______
19. ______
20. ______
21. ______
22. ______

ANSWERS

23. ______

24. ______

25. ______

26. ______

27. ______

28. ______

29. ______

30. ______

31. ______

32. ______

33. ______

34. ______

35. ______

36. ______

37. ______

38. ______

39. ______

40. ______

41. ______

42. ______

43. ______

44. ______

45. ______

46. ______

47. ______

48. ______

49. ______

50. ______

51. ______

23. $\dfrac{x^2 - 9}{x + 3}$

24. $\dfrac{x^2 - 25}{x - 5}$

25. $\dfrac{x^5 + 1}{x + 1}$

26. $\dfrac{x^5 - 1}{x - 1}$

27. $\dfrac{8x^3 - 22x^2 - 5x + 12}{4x + 3}$

28. $\dfrac{2x^3 - 9x^2 + 11x - 3}{2x - 3}$

29. $(x^6 - 13x^3 + 42) \div (x^3 - 7)$

30. $(x^6 + 5x^3 - 24) \div (x^3 - 3)$

31. $(x^4 - 16) \div (x - 2)$

32. $(x^4 - 81) \div (x - 3)$

33. $(t^3 - t^2 + t - 1) \div (t - 1)$

34. $(t^3 - t^2 + t - 1) \div (t + 1)$

SKILL MAINTENANCE

35. Subtract: $-2.3 - (-9.1)$.

36. Factor: $4x - 12 + 24y$.

37. The perimeter of a rectangle is 640 ft. The length is 15 ft more than the width. Find the area of the rectangle.

38. Solve: $-6(2 - x) + 10(5x - 7) = 10$.

39. Solve: $-10(x - 4) = 5(2x + 5) - 7$.

40. Subtract: $-\dfrac{5}{8} - \dfrac{3}{4}$.

SYNTHESIS

Divide.

41. $(x^4 + 9x^2 + 20) \div (x^2 + 4)$

42. $(y^4 + a^2) \div (y + a)$

43. $(5a^3 + 8a^2 - 23a - 1) \div (5a^2 - 7a - 2)$

44. $(15y^3 - 30y + 7 - 19y^2) \div (3y^2 - 2 - 5y)$

45. $(6x^5 - 13x^3 + 5x + 3 - 4x^2 + 3x^4) \div (3x^3 - 2x - 1)$

46. $(5x^7 - 3x^4 + 2x^2 - 10x + 2) \div (x^2 - x + 1)$

47. $(a^6 - b^6) \div (a - b)$

48. $(x^5 + y^5) \div (x + y)$

If the remainder is 0 when one polynomial is divided by another, the divisor is a *factor* of the dividend. Find the value(s) of c for which $x - 1$ is a factor of the polynomial.

49. $x^2 + 4x + c$

50. $2x^2 + 3cx - 8$

51. $c^2x^2 - 2cx + 1$

CRITICAL THINKING

CALCULATOR CONNECTION

Use a calculator in any way that you can to do the following operations with polynomials.

1. Add:

$(-20.344x^6 - 70.789x^5 + 890x) + (68.888x^6 + 69.994x^5)$.

2. Subtract:

$$(345.099x^3 - 6.178x) - (-224.508x^3 + 8.99x).$$

Multiply.

3. $(1.206y + 3.42)(2.401y - 8.992)$

4. $(67.58x + 3.225)^2$

5. $(0.127x - 56.88)^2$

6. $(789.02x - 12.34)(789.02x + 12.34)$

Some calculators allow the use of scientific notation. For example, a key like [E] or [EXP] is used to tell the calculator that a power of ten is being entered. For example, 240,000,000 is entered as 2.4 [E] 8, and 0.0068 is entered as 6.8 [E] −3. Enter each of the following products in scientific notation and find scientific notation for the answer.

7. (24,000,000)(58,000,000)

8. 240,000,000 ÷ 74,000,000,000

9. (0.000000233)(0.0000000707)

10. 0.08989 ÷ 786,564.67

We can evaluate polynomials with a calculator by using the memory key [STO] and the memory recall key [RCL] to save steps. (On some calculators, these keys might be represented by [M] and [MR].) To evaluate

$$2x^3 - 4x^2 + 5 \quad \text{when } x = 14,$$

we proceed as follows:

14 [STO]   This stores 14 in the memory.

2 × [RCL] [y^x] 3 [−] 4 × [RCL] [y^x] 2 [+] 5 [=] → 4709.

11. Evaluate $5x^3 + 16x^2 - 4x - 23$ when $x = 12$.

12. Evaluate $6x^4 - 28.2x^3 + 16x^2 - 281.2$ when $x = -6.814$.

EXTENDED SYNTHESIS EXERCISES

1. Express $3 \cdot 9 \cdot 27 \cdot 81$ as a power of 3.

2. Express $8 \cdot 4^3 \cdot 16$ as a power of 2.

3. Express $64 \cdot 2^8 \cdot 16$ as a power of 4.

4. Suppose it takes about 3 days for a rocket to travel from the earth to the moon. About how long would it take the same rocket traveling at the same speed to reach Mars? Give scientific notation for your answer.

APPROXIMATE DISTANCE FROM EARTH TO:	
Moon	240,000 mi
Mars	35,000,000 mi
Pluto	2,670,000,000 mi

5. Use the information in Exercise 4. Suppose it takes about 5 days for a rocket to travel from the earth to the moon. About how long would it take the same rocket traveling at the same speed to reach Pluto? Give scientific notation for your answer.

6. The sum of two polynomials is $3x^2 - x + 18$. One of the polynomials is $x^2 - 7$. Find the other.

7. The difference of two polynomials is $3x^2 - x + 18$. One of the polynomials is $x^2 - 7$. Find the other.

8. The product of two polynomials is $3x^2 + 17x - 28$. One of the polynomials is $x + 7$. Find the other.

9. The product of two polynomials is $x^5 - 1$. One of the polynomials is $x - 1$. Find the other.

10. Show that the difference of the squares of two consecutive integers is the same as the sum of the integers. (*Hint:* Let x = the smallest integer and $x + 1$ = the larger.)

(*continued*)

CRITICAL THINKING

11. Write $(x + 3)^2$ using the four areas of the square shown here.

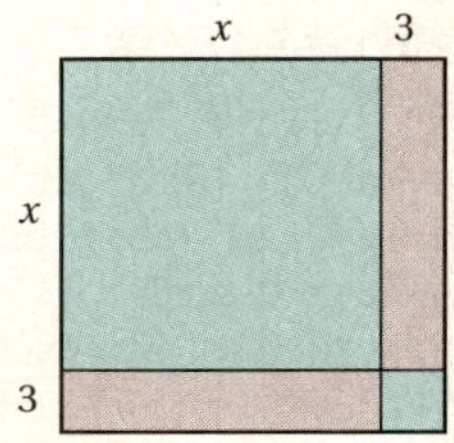

12. Write $(y - 2)^2$ using the four areas of the square shown here.

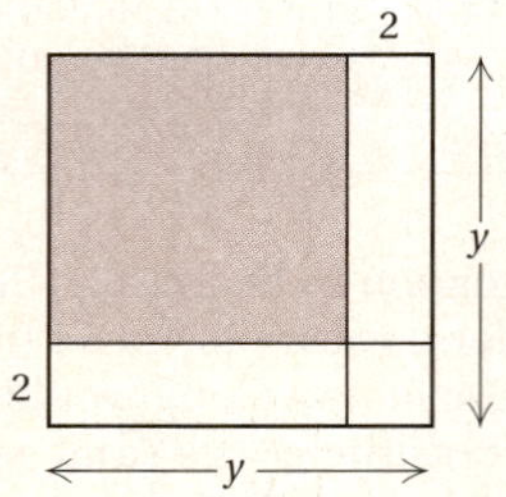

13. Find a polynomial for the shaded area.

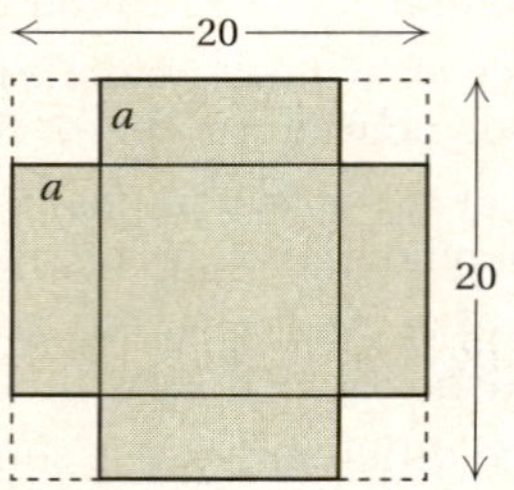

Find two expressions for the shaded area.

14.

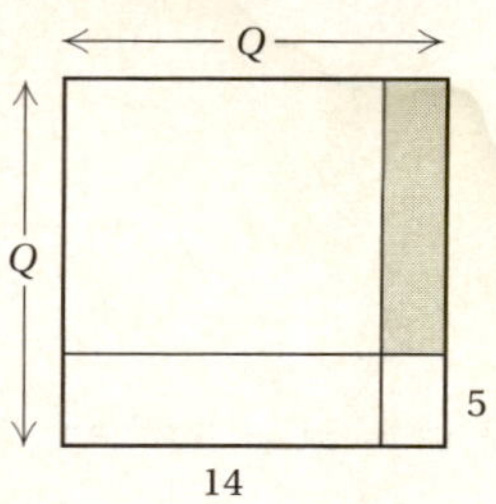

15.

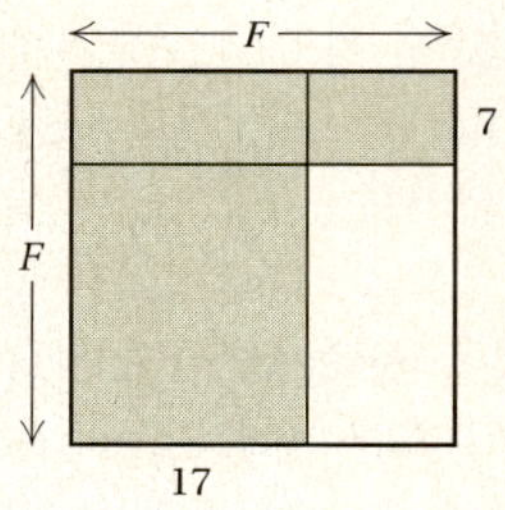

16.

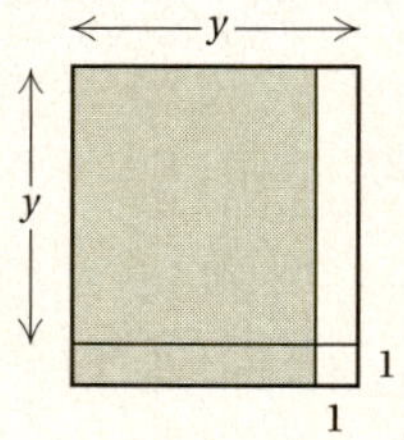

EXERCISES FOR THINKING AND WRITING

Explain the errors in each of the following.

1. $(a + 2)^2 = a^2 + 4$
2. $(p + 7)(p - 7) = p^2 + 14p - 49$
3. $(t - 3)^2 = t^2 - 9$
4. $2^{-3} = -6$
5. $2^{-3} = \dfrac{1}{-8}$
6. $\dfrac{a^2}{a^5} = a^3$
7. $m^{-2}m^5 = m^{-10}$
8. $b^8b^5 = b^3$
9. Explain why 578.6×10^{-7} is not scientific notation.
10. Write a short explanation of the difference between a monomial, a binomial, a trinomial, and a general polynomial.
11. Discuss the ways in which the distributive law is used in this chapter.
12. Why is scientific notation useful?
13. For what values of m and n is m^{-n} negative?

Summary and Review: Chapter 3

IMPORTANT PROPERTIES AND FORMULAS

FOIL:	$(A + B)(C + D) = AC + AD + BC + BD$
Square of a Sum:	$(A + B)(A + B) = (A + B)^2 = A^2 + 2AB + B^2$
Square of a Difference:	$(A - B)(A - B) = (A - B)^2 = A^2 - 2AB + B^2$
Product of a Sum and a Difference:	$(A + B)(A - B) = A^2 - B^2$

DEFINITIONS AND RULES FOR EXPONENTS

See p. 184.

Review Exercises

The review objectives to be tested in addition to the material in this chapter are [1.4a], [1.7d], [2.3b, c], and [2.4a].

Multiply and simplify.

1. $7^2 \cdot 7^{-4}$

2. $y^7 \cdot y^3 \cdot y$

3. $(3x)^5 \cdot (3x)^9$

4. $t^8 \cdot t^0$

Divide and simplify.

5. $\frac{4^5}{4^2}$

6. $\frac{a^5}{a^8}$

7. $\frac{(7x)^4}{(7x)^4}$

Simplify.

8. $(3t^4)^2$

9. $(2x^3)^2(-3x)^2$

10. $\left(\frac{2x}{y}\right)^{-3}$

11. Express using a negative exponent: $\frac{1}{t^5}$.

12. Express using a positive exponent: y^{-4}.

13. Convert to scientific notation: 0.0000328.

14. Convert to decimal notation: 8.3×10^6.

Multiply or divide and write scientific notation for the result.

15. $(3.8 \times 10^4)(5.5 \times 10^{-1})$

16. $\frac{1.28 \times 10^{-8}}{2.5 \times 10^{-4}}$

17. Each day Americans eat 170 million eggs. How many eggs are eaten in one year? Write scientific notation for the answer.

18. Evaluate the polynomial $x^2 - 3x + 6$ when $x = -1$.

19. Identify the terms of the polynomial $-4y^5 + 7y^2 - 3y - 2$.

20. Identify the missing terms in $x^3 + x$.

21. Identify the degree of each term and the degree of the polynomial $4x^3 + 6x^2 - 5x + \frac{5}{3}$.

Classify the polynomial as a monomial, binomial, trinomial, or none of these.

22. $4x^3 - 1$

23. $4 - 9t^3 - 7t^4 + 10t^2$

24. $7y^2$

Collect like terms and then arrange in descending order.

25. $3x^2 - 2x + 3 - 5x^2 - 1 - x$

26. $-x + \frac{1}{2} + 14x^4 - 7x^2 - 1 - 4x^4$

Add.

27. $(3x^4 - x^3 + x - 4) + (x^5 + 7x^3 - 3x^2 - 5) + (-5x^4 + 6x^2 - x)$

28. $(3x^5 - 4x^4 + x^3 - 3) + (3x^4 - 5x^3 + 3x^2) + (4x^5 + 4x^3) + (-5x^5 - 5x^2) + (-5x^4 + 2x^3 + 5)$

Subtract.

29. $(5x^2 - 4x + 1) - (3x^2 + 7)$

30. $(3x^5 - 4x^4 + 3x^2 + 3) - (2x^5 - 4x^4 + 3x^3 + 4x^2 - 5)$

31. The length of a rectangle is 4 m greater than its width. Find a polynomial for the perimeter and a polynomial for the area.

Multiply.

32. $\left(x + \frac{2}{3}\right)\left(x + \frac{1}{2}\right)$

33. $(7x + 1)^2$

34. $(4x^2 - 5x + 1)(3x - 2)$

35. $(3x^2 + 4)(3x^2 - 4)$

36. $5x^4(3x^3 - 8x^2 + 10x + 2)$

37. $(x + 4)(x - 7)$

38. $(3y^2 - 2y)^2$

39. $(2t^2 + 3)(t^2 - 7)$

40. Evaluate the polynomial $2 - 5xy + y^2 - 4xy^3 + x^6$ when $x = -1$ and $y = 2$.

41. Identify the coefficient and degree of each term of the polynomial $x^5y - 7xy + 9x^2 - 8$. Then find the degree of the polynomial.

Collect like terms.

42. $y + w - 2y + 8w - 5$

43. $m^6 - 2m^2n + m^2n^2 + n^2m - 6m^3 + m^2n^2 + 7n^2m$

44. Add:

$(5x^2 - 7xy + y^2) + (-6x^2 - 3xy - y^2) + (x^2 + xy - 2y^2)$.

45. Subtract:

$(6x^3y^2 - 4x^2y - 6x) - (-5x^3y^2 + 4x^2y + 6x^2 - 6)$.

Multiply.

46. $(p - q)(p^2 + pq + q^2)$

47. $\left(3a^4 - \frac{1}{3}b^3\right)^2$

Divide.

48. $(10x^3 - x^2 + 6x) \div (2x)$

49. $(6x^3 - 5x^2 - 13x + 13) \div (2x + 3)$

SKILL MAINTENANCE

50. Factor: $25t - 50 + 100m$.

51. Solve: $7x + 6 - 8x = 11 - 5x + 4$.

52. Subtract: $-3.4 - 7.8$.

53. The perimeter of a rectangle is 540 m. The width is 19 m less than the length. Find the width and the length.

SYNTHESIS

54. Collect like terms:

$$-3x^5 \cdot 3x^3 - x^6(2x)^2 + (3x^4)^2 + (2x^2)^4 - 40x^2(x^3)^2.$$

55. Solve:

$$(x - 7)(x + 10) = (x - 4)(x - 6).$$

NAME SECTION DATE

Test: Chapter 3

Multiply and simplify.

1. $6^{-2} \cdot 6^{-3}$

2. $x^6 \cdot x^2 \cdot x$

3. $(4a)^3 \cdot (4a)^8$

Divide and simplify.

4. $\frac{3^5}{3^2}$

5. $\frac{x^3}{x^8}$

6. $\frac{(2x)^5}{(2x)^5}$

Simplify.

7. $(x^3)^2$

8. $(-3y^2)^3$

9. $(2a^3b)^4$

10. $\left(\frac{ab}{c}\right)^3$

11. $(3x^2)^3(-2x^5)^3$

12. $3(x^2)^3(-2x^5)^3$

13. $2x^2(-3x^2)^4$

14. $(2x)^2(-3x^2)^4$

15. Express using a positive exponent:

5^{-3}.

16. Express using a negative exponent:

$\frac{1}{y^8}$.

17. Convert to scientific notation:

3,900,000,000.

18. Convert to decimal notation:

5×10^{-8}.

Multiply or divide and write scientific notation for the answer.

19. $\frac{5.6 \times 10^6}{3.2 \times 10^{-11}}$

20. $(2.4 \times 10^5)(5.4 \times 10^{16})$

21. Each day Americans eat 170 million eggs. There are 261 million people in this country. How many eggs does each person eat in one year? Write scientific notation for the answer.

22. Evaluate the polynomial $x^5 + 5x - 1$ when $x = -2$.

23. Identify the coefficient of each term of the polynomial $\frac{1}{3}x^5 - x + 7$.

24. Identify the degree of each term and the degree of the polynomial $2x^3 - 4 + 5x + 3x^6$.

25. Classify the polynomial $7 - x$ as a monomial, binomial, trinomial, or none of these.

Answers

1. ______
2. ______
3. ______
4. ______
5. ______
6. ______
7. ______
8. ______
9. ______
10. ______
11. ______
12. ______
13. ______
14. ______
15. ______
16. ______
17. ______
18. ______
19. ______
20. ______
21. ______
22. ______
23. ______
24. ______
25. ______

ANSWERS

26. ______
27. ______
28. ______
29. ______
30. ______
31. ______
32. ______
33. ______
34. ______
35. ______
36. ______
37. ______
38. ______
39. ______
40. ______
41. ______
42. ______
43. ______
44. ______
45. ______
46. ______
47. ______
48. ______
49. ______
50. ______
51. ______

Collect like terms.

26. $4a^2 - 6 + a^2$

27. $y^2 - 3y - y + \frac{3}{4}y^2$

28. Collect like terms and then arrange in descending order:

$$3 - x^2 + 2x^3 + 5x^2 - 6x - 2x + x^5.$$

Add.

29. $(3x^5 + 5x^3 - 5x^2 - 3) +$
$(x^5 + x^4 - 3x^3 - 3x^2 + 2x - 4)$

30. $\left(x^4 + \frac{2}{3}x + 5\right) + \left(4x^4 + 5x^2 + \frac{1}{3}x\right)$

Subtract.

31. $(2x^4 + x^3 - 8x^2 - 6x - 3) -$
$(6x^4 - 8x^2 + 2x)$

32. $(x^3 - 0.4x^2 - 12) -$
$(x^5 + 0.3x^3 + 0.4x^2 + 9)$

Multiply.

33. $-3x^2(4x^2 - 3x - 5)$

34. $\left(x - \frac{1}{3}\right)^2$

35. $(3x + 10)(3x - 10)$

36. $(3b + 5)(b - 3)$

37. $(x^6 - 4)(x^8 + 4)$

38. $(8 - y)(6 + 5y)$

39. $(2x + 1)(3x^2 - 5x - 3)$

40. $(5t + 2)^2$

41. Collect like terms: $x^3y - y^3 + xy^3 + 8 - 6x^3y - x^2y^2 + 11.$

42. Subtract: $(8a^2b^2 - ab + b^3) - (-6ab^2 - 7ab - ab^3 + 5b^3).$

43. Multiply: $(3x^5 - 4y^5)(3x^5 + 4y^5).$

Divide.

44. $(12x^4 + 9x^3 - 15x^2) \div (3x^2)$

45. $(6x^3 - 8x^2 - 14x + 13) \div (3x + 2)$

SKILL MAINTENANCE

46. Solve: $7x - 4x - 2 = 37.$

47. Factor: $64t - 32m + 16.$

48. Subtract: $\frac{2}{5} - \left(-\frac{3}{4}\right).$

49. The first angle of a triangle is four times as large as the second. The measure of the third angle is 30° greater than that of the second. How large are the angles?

SYNTHESIS

50. The height of a box is 1 less than its length, and the length is 2 more than its width. Find the volume in terms of the length.

51. Solve: $(x - 5)(x + 5) = (x + 6)^2.$

Cumulative Review: Chapters 1–3

1. Evaluate $\frac{x}{2y}$ when $x = 10$ and $y = 2$.

2. Evaluate $2x^3 + x^2 - 3$ when $x = -1$.

3. Evaluate $x^3y^2 + xy + 2xy^2$ when $x = -1$ and $y = 2$.

4. Find the absolute value: $|-4|$.

5. Find the reciprocal of 5.

Compute and simplify.

6. $-\frac{3}{5} + \frac{5}{12}$

7. $3.4 - (-0.8)$

8. $(-2)(-1.4)(2.6)$

9. $\frac{3}{8} \div \left(-\frac{9}{10}\right)$

10. $(1.1 \times 10^{10})(2 \times 10^{12})$

11. $(3.2 \times 10^{-10}) \div (8 \times 10^{-6})$

Simplify.

12. $\frac{-9x}{3x}$

13. $y - (3y + 7)$

14. $3(x - 1) - 2[x - (2x + 7)]$

15. $2 - [32 \div (4 + 2^2)]$

Add.

16. $(x^4 + 3x^3 - x + 7) + (2x^5 - 3x^4 + x - 5)$

17. $(x^2 + 2xy) + (y^2 - xy) + (2x^2 - 3y^2)$

Subtract.

18. $(x^3 + 3x^2 - 4) - (-2x^2 + x + 3)$

19. $\left(\frac{1}{3}x^2 - \frac{1}{4}x - \frac{1}{5}\right) - \left(\frac{2}{3}x^2 + \frac{1}{2}x - \frac{1}{5}\right)$

Multiply.

20. $3(4x - 5y + 7)$

21. $(-2x^3)(-3x^5)$

22. $2x^2(x^3 - 2x^2 + 4x - 5)$

23. $(y^2 - 2)(3y^2 + 5y + 6)$

24. $(2p^3 + p^2q + pq^2)(p - pq + q)$

25. $(2x + 3)(3x + 2)$

26. $(3x^2 + 1)^2$

27. $\left(t + \frac{1}{2}\right)\left(t - \frac{1}{2}\right)$

28. $(2y^2 + 5)(2y^2 - 5)$

29. $(2x^4 - 3)(2x^2 + 3)$

30. $(t - 2t^2)^2$

31. $(3p + q)(5p - 2q)$

Divide.

32. $(18x^3 + 6x^2 - 9x) \div 3x$

33. $(3x^3 + 7x^2 - 13x - 21) \div (x + 3)$

Solve.

34. $1.5 = 2.7 + x$

35. $\frac{2}{7}x = -6$

36. $5x - 9 = 36$

37. $\frac{2}{3} = \frac{-m}{10}$

38. $5.4 - 1.9x = 0.8x$

39. $x - \frac{7}{8} = \frac{3}{4}$

40. $2(2 - 3x) = 3(5x + 7)$

41. $\frac{1}{4}x - \frac{2}{3} = \frac{3}{4} + \frac{1}{3}x$

42. $y + 5 - 3y = 5y - 9$

43. $\frac{1}{4}x - 7 < 5 - \frac{1}{2}x$

44. $2(x + 2) \geq 5(2x + 3)$

45. $A = 2\pi rh + \pi r^2$, for h

46. A 6-ft by 3-ft raft is floating in a swimming pool of radius r. Find a polynomial for the area of the surface of the pool not covered by the raft.

Solve.

47. The sum of the page numbers on the facing pages of a book is 37. What are the page numbers?

48. The perimeter of a room is 88 ft. The width is 4 ft less than the length. Find the width and the length.

49. The second angle of a triangle is five times as large as the first. The third angle is twice the sum of the other two angles. Find the measure of the first angle.

50. If you triple a number and then add 99, you get $\frac{4}{5}$ of the original number. What is the original number?

51. A bookstore sells books at a price that is 80% higher than the price the store pays for the books. A book is priced for sale at \$6.30. How much did the store pay for the book?

Simplify.

52. $y^2 \cdot y^{-6} \cdot y^8$

53. $\frac{x^6}{x^7}$

54. $(-3x^3y^{-2})^3$

55. $\frac{x^3x^{-4}}{x^{-5}x}$

56. Identify the coefficient of each term of the polynomial $\frac{2}{3}x^2 + 4x - 6$.

57. Identify the degree of each term and the degree of the polynomial $2x^4 + 3x^2 + 2x + 1$.

Classify the polynomial as a monomial, binomial, trinomial, or none of these.

58. $2x^2 + 1$

59. $2x^2 + x + 1$

SYNTHESIS

60. A picture frame is x inches square. The picture that it frames is 2 in. shorter than the frame in both length and width. Find a polynomial for the area of the frame.

Add.

61. $[(2x)^2 - (3x)^3 + 2x^2x^3 + (x^2)^2] + [5x^2(2x^3) - ((2x)^2)^2]$

62. $(x - 3)^2 + (2x + 1)^2$

63. $[(3x^3 + 11x^2 + 11x + 15) \div (x + 3)] + [(2x^3 - 7x^2 + 2) \div (2x + 1)]$

Solve.

64. $(x + 3)(2x - 5) + (x - 1)^2 = (3x + 1)(x - 3)$

65. $(2x^2 + x - 6) \div (2x - 3) = (2x^2 - 9x - 5) \div (x - 5)$

66. $20 - 3|x| = 5$

67. $(x + 2)^2 = (x + 1)(x + 3)$

68. $(x - 3)(x + 4) = (x^3 - 4x^2 - 17x + 60) \div (x - 5)$

4

Polynomials: Factoring

INTRODUCTION

Factoring is the reverse of multiplying. To *factor* a polynomial, or other algebraic expression, is to find an equivalent expression that is a product. In this chapter, we study factoring polynomials. To learn to factor quickly, we use the fast methods for multiplication that we learned in Chapter 3.

At the end of the chapter, we find the payoff for learning to factor. We can now solve certain new equations containing second-degree polynomials. This in turn allows us to solve problems that we could not have solved before.

AN APPLICATION

Sharks' teeth are shaped like triangles. The height of a tooth of a great white shark is 1 cm longer than the base. The area is 15 cm^2. Find the height and the base.

THE MATHEMATICS

We let $b =$ the base. Then the height is $b + 1$ and the problem translates to the equation

$$\underbrace{\tfrac{1}{2}b(b + 1) = 15.}$$

This is a second-degree, or quadratic equation.

Objectives for Review

The review objectives to be tested in addition to the material in this chapter are as follows.

[1.6a, c] Divide real numbers.

[2.6a] Solve a formula for a specified letter.

[2.7e] Solve inequalities using the addition and multiplication principles together.

[3.6d] Find special products of polynomials.

Pretest: Chapter 4

1. Find three factorizations of $-20x^6$.

Factor.

2. $2x^2 + 4x + 2$

3. $x^2 + 6x + 8$

4. $8a^5 + 4a^3 - 20a$

5. $-6 + 5x^2 - 13x$

6. $81 - z^4$

7. $y^6 - 4y^3 + 4$

8. $3x^3 + 2x^2 + 12x + 8$

9. $p^2 - p - 30$

10. $x^4y^2 - 64$

11. $2p^2 + 7pq - 4q^2$

Solve.

12. $x^2 - 5x = 0$

13. $(x - 4)(5x - 3) = 0$

14. $3x^2 + 10x - 8 = 0$

Solve.

15. Six less than the square of a number is five times the number. Find all such numbers.

16. The height of a triangle is 3 cm longer than the base. The area of the triangle is 44 cm^2. Find the base and the height.

4.1 *Introduction to Factoring*

To solve certain types of algebraic equations involving polynomials of second degree, we must learn to factor polynomials.

Consider the product $15 = 3 \cdot 5$. We say that 3 and 5 are **factors** of 15 and that $3 \cdot 5$ is a **factorization** of 15. Since $15 = 15 \cdot 1$, we also know that 15 and 1 are factors of 15 and that $15 \cdot 1$ is a factorization of 15.

> To *factor* a polynomial is to express it as a product.
>
> A *factor* of a polynomial P is a polynomial that can be used to express P as a product.
>
> A *factorization* of a polynomial is an expression that names that polynomial as a product.

OBJECTIVES

After finishing Section 4.1, you should be able to:

- **a** Factor monomials.
- **b** Factor polynomials when the terms have a common factor, factoring out the largest common factor.
- **c** Factor certain expressions with four terms using factoring by grouping.

FOR EXTRA HELP

TAPE 10 | TAPE 8A | MAC: 4A IBM: 4A

a FACTORING MONOMIALS

To factor a monomial, we find two monomials whose product is equivalent to the original monomial. Compare.

	Multiplying	*Factoring*
a)	$(4x)(5x) = 20x^2$	$20x^2 = (4x)(5x)$
b)	$(2x)(10x) = 20x^2$	$20x^2 = (2x)(10x)$
c)	$(-4x)(-5x) = 20x^2$	$20x^2 = (-4x)(-5x)$
d)	$(x)(20x) = 20x^2$	$20x^2 = (x)(20x)$

You can see that the monomial $20x^2$ has many factorizations. There are still other ways to factor $20x^2$.

Do Exercises 1 and 2.

EXAMPLE 1 Find three factorizations of $15x^3$.

a) $15x^3 = (3 \cdot 5)(x \cdot x^2)$
$\quad = (3x)(5x^2)$

b) $15x^3 = (3 \cdot 5)(x^2 \cdot x)$
$\quad = (3x^2)(5x)$

c) $15x^3 = (-15)(-1)x^3$
$\quad = (-15)(-x^3)$

Do Exercises 3–5.

1. a) Multiply: $(3x)(4x)$.

b) Factor: $12x^2$.

2. a) Multiply: $(2x)(8x^2)$.

b) Factor: $16x^3$.

Find three factorizations of the monomial.

3. $8x^4$

4. $21x^2$

5. $6x^5$

Answers on page A-3

6. a) Multiply: $3(x+2)$.

b) Factor: $3x+6$.

7. a) Multiply: $2x(x^2+5x+4)$.

b) Factor: $2x^3+10x^2+8x$.

Answers on page A-4

b FACTORING WHEN TERMS HAVE A COMMON FACTOR

To factor polynomials quickly, we consider the special-product rules learned in Chapter 3, but we first factor out the largest common factor.

To multiply a monomial and a polynomial with more than one term, we multiply each term of the polynomial by the monomial using the distributive laws, $a(b+c)=ab+ac$ and $a(b-c)=ab-ac$. To factor, we do the reverse. We express a polynomial as a product using the distributive laws in reverse: $ab+ac=a(b+c)$ and $ab-ac=a(b-c)$. Compare.

Multiply	*Factor*
$3x(x^2+2x-4)$	$3x^3+6x^2-12x$
$=3x\cdot x^2+3x\cdot 2x-3x\cdot 4$	$=3x\cdot x^2+3x\cdot 2x-3x\cdot 4$
$=3x^3+6x^2-12x$	$=3x(x^2+2x-4)$

Do Exercises 6 and 7.

Caution! Consider the following:

$$3x^3+6x^2-12x=3\cdot x\cdot x\cdot x+2\cdot 3\cdot x\cdot x-2\cdot 2\cdot 3\cdot x.$$

The terms of the polynomial, $3x^3$, $6x^2$, and $-12x$, have been factored but the polynomial itself has not been factored. This is not what we mean by a factorization of the polynomial. The *factorization* is

$$3x(x^2+2x-4).$$

The expressions $3x$ and x^2+2x-4 are *factors.*

To factor, we first try to find a factor common to all terms. There may not always be one other than 1. When there is, we generally use the factor with the largest possible coefficient and the largest possible exponent.

EXAMPLE 2 Factor: $3x^2+6$.

We have

$$3x^2+6=3\cdot x^2+3\cdot 2 \quad \text{Factoring each term}$$
$$=3(x^2+2). \quad \text{Factoring out the common factor 3}$$

We can check by multiplying: $3(x^2+2)=3\cdot x^2+3\cdot 2=3x^2+6$.

EXAMPLE 3 Factor: $16x^3+20x^2$.

$$16x^3+20x^2=(4x^2)(4x)+(4x^2)(5) \quad \text{Factoring each term}$$
$$=4x^2(4x+5) \quad \text{Factoring out the common factor } 4x^2$$

Suppose in Example 3 that you had not recognized the largest common factor and removed only part of it, as follows:

$$16x^3+20x^2=(2x^2)(8x)+(2x^2)(10)$$
$$=2x^2(8x+10).$$

Note that $8x+10$ still has a common factor of 2. You need not begin again. Just continue factoring out common factors, as follows, until finished:

$$=2x^2[2(4x+5)]$$
$$=4x^2(4x+5).$$

EXAMPLE 4 Factor: $15x^5 - 12x^4 + 27x^3 - 3x^2$.

$$15x^5 - 12x^4 + 27x^3 - 3x^2 = (3x^2)(5x^3) - (3x^2)(4x^2) + (3x^2)(9x) - (3x^2)(1)$$
$$= 3x^2(5x^3 - 4x^2 + 9x - 1) \quad \text{Factoring out } 3x^2$$

Caution! Don't forget the term -1.

CHECK: We multiply to check.

$$3x^2(5x^3 - 4x^2 + 9x - 1)$$
$$= (3x^2)(5x^3) - (3x^2)(4x^2) + (3x^2)(9x) - (3x^2)(1)$$
$$= 15x^5 - 12x^4 + 27x^3 - 3x^2$$

As you become more familiar with factoring, you will be able to spot the largest common factor without factoring each term. Then you can write just the answer.

EXAMPLES Factor.

5. $8m^3 - 16m = 8m(m^2 - 2)$

6. $14p^2y^3 - 8py^2 + 2py = 2py(7py^2 - 4y + 1)$

7. $\frac{4}{5}x^2 + \frac{1}{5}x + \frac{2}{5} = \frac{1}{5}(4x^2 + x + 2)$

Do Exercises 8–12.

There are two important points to keep in mind as we study this chapter.

1. Before doing any other kind of factoring, first try to factor out the largest common factor.
2. Always check the result of factoring by multiplying.

c FACTORING BY GROUPING

Certain polynomials with four terms can be factored using a method called *factoring by grouping*.

EXAMPLE 8 Factor: $x^2(x + 1) + 2(x + 1)$.

The binomial $x + 1$ is common to both terms:

$$x^2(x + 1) + 2(x + 1) = (x^2 + 2)(x + 1).$$

The factorization is $(x^2 + 2)(x + 1)$.

Do Exercises 13 and 14.

Factor. Check by multiplying.

8. $x^2 + 3x$

9. $3y^6 - 5y^3 + 2y^2$

10. $9x^4 - 15x^3 + 3x^2$

11. $\frac{3}{4}t^3 + \frac{5}{4}t^2 + \frac{7}{4}t + \frac{1}{4}$

12. $35x^7 - 49x^6 + 14x^5 - 63x^3$

Factor.

13. $x^2(x + 7) + 3(x + 7)$

14. $x^2(a + b) + 2(a + b)$

Answers on page A-4

Consider the four-term polynomial

$$x^3 + x^2 + 2x + 2.$$

There is no factor other than 1 that is common to all the terms. We can, however, factor $x^3 + x^2$ and $2x + 2$ separately:

$$x^3 + x^2 = x^2(x + 1); \qquad \text{Factoring } x^3 + x^2$$

$$2x + 2 = 2(x + 1). \qquad \text{Factoring } 2x + 2$$

We have grouped certain terms and factored each polynomial separately:

$$\begin{aligned} x^3 + x^2 + 2x + 2 &= (x^3 + x^2) + (2x + 2) \\ &= x^2(x + 1) + 2(x + 1) \\ &= (x^2 + 2)(x + 1), \end{aligned}$$

as in Example 8. This method is called **factoring by grouping.** We began with a polynomial with four terms. After grouping and removing common factors, we obtained a polynomial with two parts, each having a common factor $x + 1$. Not all polynomials with four terms can be factored by this method, but it does give us a method to try.

EXAMPLES Factor by grouping.

9. $6x^3 - 9x^2 + 4x - 6$

$= (6x^3 - 9x^2) + (4x - 6)$

$= 3x^2(2x - 3) + 2(2x - 3)$ — Factoring each binomial

$= (3x^2 + 2)(2x - 3)$ — Factoring out the common factor $2x - 3$

10. $x^3 + x^2 + x + 1 = (x^3 + x^2) + (x + 1)$

$= x^2(x + 1) + 1(x + 1)$ — Factoring each binomial

$= (x^2 + 1)(x + 1)$ — Factoring out the common factor $x + 1$

11. $2x^3 - 6x^2 - x + 3$

$= (2x^3 - 6x^2) + (-x + 3)$

$= 2x^2(x - 3) - 1(x - 3)$ — *Check:* $-1(x - 3) = -x + 3$.

$= (2x^2 - 1)(x - 3)$ — Factoring out the common factor $x - 3$

12. $12x^5 + 20x^2 - 21x^3 - 35 = 4x^2(3x^3 + 5) - 7(3x^3 + 5)$

$= (4x^2 - 7)(3x^3 + 5)$

13. $x^3 + x^2 + 2x - 2 = x^2(x + 1) + 2(x - 1)$

This polynomial is not factorable using factoring by grouping. It may be factorable, but not by methods that we will consider in this text.

Do Exercises 15–20.

Factor by grouping.

15. $x^3 + 7x^2 + 3x + 21$

16. $8t^3 + 2t^2 + 12t + 3$

17. $3m^5 - 15m^3 + 2m^2 - 10$

18. $3x^3 - 6x^2 - x + 2$

19. $4x^3 - 6x^2 - 6x + 9$

20. $y^4 - 2y^3 - 2y - 10$

Answers on page A-4

NAME SECTION DATE

Exercise Set 4.1

a Find three factorizations for the monomial.

1. $8x^3$ **2.** $6x^4$ **3.** $-10a^6$ **4.** $-8y^5$ **5.** $24x^4$ **6.** $15x^5$

b Factor. Check by multiplying.

7. $x^2 - 6x$

8. $x^2 + 5x$

9. $2x^2 + 6x$

10. $8y^2 - 8y$

11. $x^3 + 6x^2$

12. $3x^4 - x^2$

13. $8x^4 - 24x^2$

14. $5x^5 + 10x^3$

15. $2x^2 + 2x - 8$

16. $8x^2 - 4x - 20$

17. $17x^5y^3 + 34x^3y^2 + 51xy$

18. $16p^6q^4 + 32p^5q^3 - 48pq^2$

19. $6x^4 - 10x^3 + 3x^2$

20. $5x^5 + 10x^2 - 8x$

21. $x^5y^5 + x^4y^3 + x^3y^3 - x^2y^2$

22. $x^9y^6 - x^7y^5 + x^4y^4 + x^3y^3$

23. $2x^7 - 2x^6 - 64x^5 + 4x^3$

24. $8y^2 - 20y^2 + 12y - 16$

25. $1.6x^4 - 2.4x^3 + 3.2x^2 + 6.4x$

ANSWERS

1. ______
2. ______
3. ______
4. ______
5. ______
6. ______
7. ______
8. ______
9. ______
10. ______
11. ______
12. ______
13. ______
14. ______
15. ______
16. ______
17. ______
18. ______
19. ______
20. ______
21. ______
22. ______
23. ______
24. ______
25. ______

ANSWERS

26. ______
27. ______
28. ______
29. ______
30. ______
31. ______
32. ______
33. ______
34. ______
35. ______
36. ______
37. ______
38. ______
39. ______
40. ______
41. ______
42. ______
43. ______
44. ______
45. ______
46. ______
47. ______
48. ______
49. ______
50. ______
51. ______
52. ______
53. ______
54. ______
55. ______
56. ______
57. ______

26. $2.5x^6 - 0.5x^4 + 5x^3 + 10x^2$

27. $\frac{5}{3}x^6 + \frac{4}{3}x^5 + \frac{1}{3}x^4 + \frac{1}{3}x^3$

28. $\frac{5}{9}x^7 + \frac{2}{9}x^5 - \frac{4}{9}x^3 - \frac{1}{9}x$

c Factor.

29. $x^2(x + 3) + 2(x + 3)$

30. $3z^2(2z + 1) + (2z + 1)$

Factor by grouping.

31. $x^3 + 3x^2 + 2x + 6$

32. $6z^3 + 3z^2 + 2z + 1$

33. $2x^3 + 6x^2 + x + 3$

34. $3x^3 + 2x^2 + 3x + 2$

35. $8x^3 - 12x^2 + 6x - 9$

36. $10x^3 - 25x^2 + 4x - 10$

37. $12x^3 - 16x^2 + 3x - 4$

38. $18x^3 - 21x^2 + 30x - 35$

39. $5x^3 - 5x^2 - x + 1$

40. $7x^3 - 14x^2 - x + 2$

41. $x^3 + 8x^2 - 3x - 24$

42. $2x^3 + 12x^2 - 5x - 30$

43. $2x^3 - 8x^2 - 9x + 36$

44. $20g^3 - 4g^2 - 25g + 5$

SKILL MAINTENANCE

Solve.

45. $-2x < 48$

46. $4x - 8x + 16 \geq 6(x - 2)$

47. Divide: $\frac{-108}{-4}$.

48. Solve $A = \frac{p + q}{2}$ for p.

Multiply.

49. $(y + 5)(y + 7)$

50. $(y + 7)^2$

51. $(y + 7)(y - 7)$

52. $(y - 7)^2$

SYNTHESIS

Factor.

53. $4x^5 + 6x^3 + 6x^2 + 9$

54. $x^6 + x^4 + x^2 + 1$

55. $x^{12} + x^7 + x^5 + 1$

56. $x^3 - x^2 - 2x + 5$

57. $p^3 + p^2 - 3p + 10$

4.2 *Factoring Trinomials of the Type $x^2 + bx + c$*

a We now begin a study of the factoring of trinomials. We first try to factor trinomials like

$$x^2 + 5x + 6 \quad \text{and} \quad x^2 + 3x - 10$$

by a refined *trial-and-error* process. In this section, we restrict our attention to trinomials of the type $ax^2 + bx + c$, where $a = 1$. The coefficient a is often called the **leading coefficient.**

Constant Term Positive

Recall the FOIL method of multiplying two binomials:

$$(x + 2)(x + 5) = \overset{F}{x^2} + \overset{O}{5x} + \overset{I}{2x} + \overset{L}{10}$$
$$= x^2 + 7x + 10.$$

The product above is a trinomial. The term of highest degree, x^2, called the leading term, has a coefficient of 1. The constant term, 10, is positive. To factor $x^2 + 7x + 10$, we think of FOIL in reverse. We multiplied x times x to get the first term of the trinomial, so we know that the first term of each binomial factor is x. Next we look for numbers p and q such that

$$x^2 + 7x + 10 = (x + p)(x + q).$$

To get the middle term and the last term of the trinomial, we look for two numbers p and q whose product is 10 and whose sum is 7. Those numbers are 2 and 5. Thus the factorization is

$$(x + 2)(x + 5).$$

EXAMPLE 1 Factor: $x^2 + 5x + 6$.

Think of FOIL in reverse. The first term of each factor is x:

$$(x + p)(x + q).$$

We then look for two numbers p and q whose product is 6 and whose sum is 5. Since both 5 and 6 are positive, we need consider only positive factors.

PAIRS OF FACTORS	SUMS OF FACTORS
1, 6	7
2, 3	5 ← The numbers we need are 2 and 3.

The factorization is $(x + 2)(x + 3)$. We can check by multiplying to see whether we get the original trinomial.

CHECK: $(x + 2)(x + 3) = x^2 + 3x + 2x + 6 = x^2 + 5x + 6.$

Do Exercises 1 and 2.

OBJECTIVE

After finishing Section 4.2, you should be able to:

a Factor trinomials of the type $x^2 + bx + c$ by examining the constant term c.

FOR EXTRA HELP

TAPE 10 | TAPE 8B | MAC: 4A IBM: 4A

Factor.

1. $x^2 + 7x + 12$

2. $x^2 + 13x + 36$

Answers on page A-4

Factor.

3. $x^2 - 8x + 15$

4. $t^2 - 9t + 20$

Answers on page A-4

Consider this multiplication:

$$(x-2)(x-5) = \overset{F}{x^2} \underbrace{\overset{O}{-\,5x} \overset{I}{-\,2x}} \overset{L}{+\,10}$$
$$= x^2 \quad - 7x \quad + 10.$$

> When the constant term of a trinomial is positive, we look for two numbers with the same sign (both negative or both positive). The sign is that of the middle term:
>
> $x^2 - 7x + 10 = (x - 2)(x - 5)$, or $x^2 + 7x + 10 = (x + 2)(x + 5)$.

EXAMPLE 2 Factor: $y^2 - 8y + 12$.

Since the constant term, 12, is positive and the coefficient of the middle term, −8, is negative, we look for a factorization of 12 in which both factors are negative. Their sum must be −8.

PAIRS OF FACTORS	SUMS OF FACTORS
−1, −12	−13
−2, −6	−8 ← The numbers we need are −2 and −6.
−3, −4	−7

The factorization is $(y - 2)(y - 6)$.

Do Exercises 3 and 4.

Constant Term Negative

Sometimes when we use FOIL, the product has a negative constant term. Consider these multiplications:

a) $(x-5)(x+2) = \overset{F}{x^2} \underbrace{\overset{O}{+\,2x} \overset{I}{-\,5x}} \overset{L}{-\,10}$
$$= x^2 \quad - 3x \quad - 10;$$

b) $(x+5)(x-2) = \overset{F}{x^2} \underbrace{\overset{O}{-\,2x} \overset{I}{+\,5x}} \overset{L}{-\,10}$
$$= x^2 \quad + 3x \quad - 10.$$

Reversing the signs of the factors changes the sign of the middle term.

> When the constant term of a trinomial is negative, we look for two factors whose product is negative. One of them must be positive and the other negative. Their sum must be the coefficient of the middle term:
>
> $x^2 - 3x - 10 = (x - 5)(x + 2)$, or $x^2 + 3x - 10 = (x + 5)(x - 2)$.

EXAMPLE 3 Factor: $x^3 - 8x^2 - 20x$.

Always look first for a common factor. This time there is one, x. We first factor it out: $x^3 - 8x^2 - 20x = x(x^2 - 8x - 20)$. Now consider the expression $x^2 - 8x - 20$. Since the constant term, -20, is negative, we look for a factorization of -20 in which one factor is positive and one factor is negative. The sum must be -8, so the negative factor must have the larger absolute value. Thus we consider only pairs of factors in which the negative factor has the larger absolute value.

PAIRS OF FACTORS	SUMS OF FACTORS
1, −20	−19
2, −10	−8
4, −5	−1

← The numbers we need are 2 and −10.

The factorization of $x^2 - 8x - 20$ is $(x + 2)(x - 10)$. But we must also remember to include the common factor. The factorization of the original polynomial is

$$x(x + 2)(x - 10).$$

Do Exercise 5.

EXAMPLE 4 Factor: $t^2 - 24 + 5t$.

It helps to first write the trinomial in descending order: $t^2 + 5t - 24$. Since the constant term, -24, is negative, we look for a factorization of -24 in which one factor is positive and one factor is negative. Their sum must be 5, so the positive factor must have the larger absolute value. Thus we consider only pairs of factors in which the positive term has the larger absolute value.

PAIRS OF FACTORS	SUMS OF FACTORS
−1, 24	23
−2, 12	10
−3, 8	5
−4, 6	2

← The numbers we need are −3 and 8.

The factorization is $(t - 3)(t + 8)$.

Do Exercise 6.

EXAMPLE 5 Factor: $x^4 - x^2 - 110$.

Consider this trinomial as $(x^2)^2 - x^2 - 110$. We look for numbers p and q such that

$$x^4 - x^2 - 110 = (x^2 + p)(x^2 + q).$$

Since the constant term, -110, is negative, we look for a factorization of -110 in which one factor is positive and one factor is negative. Their sum must be -1. The middle-term coefficient, -1, is small compared to -110. This tells us that the desired factors are close to each other in absolute value. The numbers we want are 10 and -11. The factorization is

$$(x^2 + 10)(x^2 - 11).$$

5. Explain why you would not consider these pairs of factors in factoring $x^2 - 8x - 20$.

PAIRS OF FACTORS	SUMS OF FACTORS
−1, 20	
−2, 10	
−4, 5	

6. Explain why you would not consider these pairs of factors in factoring $t^2 + 5t - 24$.

PAIRS OF FACTORS	SUMS OF FACTORS
1, −24	
2, −12	
3, −8	
4, −6	

Answers on page A-4

Factor.

7. $x^3 + 4x^2 - 12x$

8. $y^2 - 12 - 4y$

9. $t^4 + 5t^2 - 14$

10. $p^2 - pq - 3pq^2$

11. $x^2 + 2x + 7$

12. Factor: $x^2 + 8x + 16$.

Answers on page A-4

EXAMPLE 6 Factor: $a^2 + 4ab - 21b^2$.

We consider the trinomial in the equivalent form

$$a^2 + 4ba - 21b^2.$$

We think of $4b$ as a "coefficient" of a. Then we look for factors of $-21b^2$ whose sum is $4b$. Those factors are $-3b$ and $7b$. The factorization is

$$(a - 3b)(a + 7b).$$

There are polynomials that are not factorable.

EXAMPLE 7 Factor: $x^2 - x + 5$.

Since 5 has very few factors, we can easily check all possibilities.

PAIRS OF FACTORS	SUMS OF FACTORS
5, 1	6
−5, −1	−6

There are no factors whose sum is −1. Thus the polynomial is *not* factorable into binomials.

Do Exercises 7–11.

Can we factor a trinomial that is a perfect square using this method? The answer is yes.

EXAMPLE 8 Factor: $x^2 - 10x + 25$.

Since the constant term, 25, is positive and the coefficient of the middle term, −10, is negative, we look for a factorization of 25 in which both factors are negative. Their sum must be −10.

PAIRS OF FACTORS	SUMS OF FACTORS
−25, −1	−26
−5, −5	−10 ← The numbers we need are −5 and −5.

The factorization is $(x - 5)(x - 5)$, or $(x - 5)^2$.

Do Exercise 12.

The following is a summary of our procedure for factoring $x^2 + bx + c$.

> To factor $x^2 + bx + c$:
>
> 1. First arrange in descending order.
> 2. Use a trial-and-error process that looks for factors of c whose sum is b.
> 3. If c is positive, the signs of the factors are the same as the sign of b.
> 4. If c is negative, one factor is positive and the other is negative. If the sum of two factors is the opposite of b, changing the sign of each factor will give the desired factors whose sum is b.
> 5. Check by multiplying.

NAME SECTION DATE

Exercise Set 4.2

a Factor. Remember that you can check by multiplying.

1. $x^2 + 8x + 15$
2. $x^2 + 5x + 6$
3. $x^2 + 7x + 12$
4. $x^2 + 9x + 8$
5. $x^2 - 6x + 9$
6. $y^2 - 11y + 28$
7. $x^2 + 9x + 14$
8. $a^2 + 11a + 30$
9. $b^2 + 5b + 4$
10. $z^2 - 8z + 7$
11. $x^2 + \frac{2}{3}x + \frac{1}{9}$
12. $x^2 - \frac{2}{5}x + \frac{1}{25}$
13. $d^2 - 7d + 10$
14. $t^2 - 12t + 35$
15. $y^2 - 11y + 10$
16. $x^2 - 4x - 21$
17. $x^2 + x - 42$
18. $x^2 + 2x - 15$
19. $x^2 - 7x - 18$
20. $y^2 - 3y - 28$
21. $x^3 - 6x^2 - 16x$
22. $x^3 - x^2 - 42x$
23. $y^2 - 4y - 45$
24. $x^2 - 7x - 60$
25. $-2x - 99 + x^2$
26. $x^2 - 72 + 6x$
27. $c^4 + c^2 - 56$
28. $b^4 + 5b^2 - 24$
29. $a^4 + 2a^2 - 35$
30. $x^4 - x^2 - 6$
31. $x^2 + x + 1$
32. $x^2 + 5x + 3$
33. $7 - 2p + p^2$
34. $11 - 3w + w^2$
35. $x^2 + 20x + 100$
36. $a^2 + 19a + 88$

ANSWERS

1.
2.
3.
4.
5.
6.
7.
8.
9.
10.
11.
12.
13.
14.
15.
16.
17.
18.
19.
20.
21.
22.
23.
24.
25.
26.
27.
28.
29.
30.
31.
32.
33.
34.
35.
36.

ANSWERS

37. ________
38. ________
39. ________
40. ________
41. ________
42. ________
43. ________
44. ________
45. ________
46. ________
47. ________
48. ________
49. ________
50. ________
51. ________
52. ________
53. ________
54. ________
55. ________
56. ________
57. ________
58. ________
59. ________
60. ________
61. ________
62. ________
63. ________
64. ________
65. ________
66. ________
67. ________
68. ________
69. ________
70. ________
71. ________
72. ________

37. $x^2 - 21x - 100$ **38.** $x^2 - 20x + 96$ **39.** $x^2 - 21x - 72$

40. $4x^2 + 40x + 100$ **41.** $x^2 - 25x + 144$ **42.** $y^2 - 21y + 108$

43. $a^2 + a - 132$ **44.** $a^2 + 9a - 90$ **45.** $120 - 23x + x^2$

46. $96 + 22d + d^2$ **47.** $108 - 3x - x^2$ **48.** $112 + 9y - y^2$

49. $y^2 - 0.2y - 0.08$ **50.** $t^2 - 0.3t - 0.10$ **51.** $p^2 + 3pq - 10q^2$

52. $a^2 + 2ab - 3b^2$ **53.** $m^2 + 5mn + 4n^2$ **54.** $x^2 + 11xy + 24y^2$

55. $s^2 - 2st - 15t^2$ **56.** $p^2 + 5pq - 24q^2$

SKILL MAINTENANCE

Multiply.

57. $8x(2x^2 - 6x + 1)$ **58.** $(7w + 6)(4w - 11)$ **59.** $(7w + 6)^2$

60. $(4w - 11)^2$ **61.** $(4w - 11)(4w + 11)$

62. Simplify: $(3x^4)^3$.

SYNTHESIS

63. Find all integers m for which $y^2 + my + 50$ can be factored.

64. Find all integers b for which $a^2 + ba - 50$ can be factored.

Factor completely.

65. $x^2 - \frac{1}{2}x - \frac{3}{16}$ **66.** $x^2 - \frac{1}{4}x - \frac{1}{8}$ **67.** $x^2 + \frac{30}{7}x - \frac{25}{7}$

68. $\frac{1}{3}x^3 + \frac{1}{3}x^2 - 2x$ **69.** $b^{2n} + 7b^n + 10$ **70.** $a^{2m} - 11a^m + 28$

Find a polynomial in factored form for the shaded area. (Leave answers in terms of π.)

71.

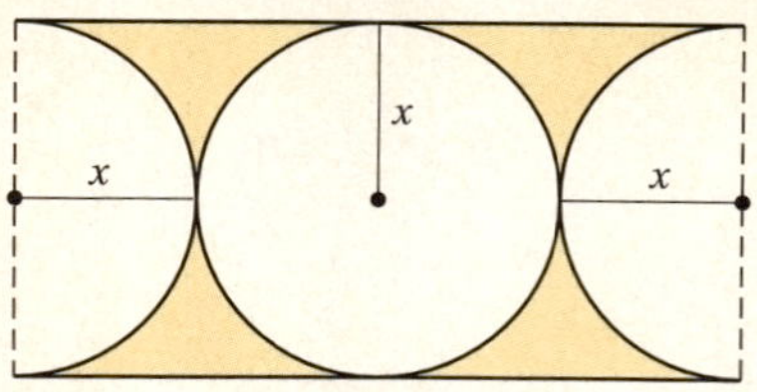

72.

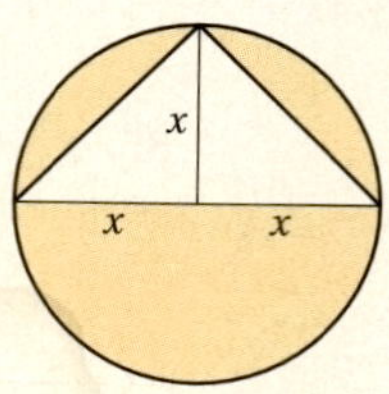

4.3 Factoring Trinomials of the Type $ax^2 + bx + c,\ a \neq 1$

OBJECTIVE

After finishing Section 4.3, you should be able to:

a Factor trinomials of the type $ax^2 + bx + c,\ a \neq 1$.

FOR EXTRA HELP

TAPE 10 | TAPE 8B | MAC: 4A IBM: 4A

In Section 4.2, we learned a trial-and-error method to factor trinomials of the type $x^2 + bx + c$. In this section, we factor trinomials in which the coefficient of the leading term x^2 is not 1. The procedure we learn is a refined trial-and-error method. (In Section 4.4, we will consider an alternative method for the same kind of factoring. It involves *factoring by grouping.*)

a We want to factor trinomials of the type $ax^2 + bx + c$. Consider the following multiplication:

$$\begin{aligned}(2x + 5)(3x + 4) &= \overset{F}{6x^2} + \overset{O}{8x} + \overset{I}{15x} + \overset{L}{20}\\ &= 6x^2 + 23x + 20\end{aligned}$$

F	O + I		L
$2 \cdot 3$	$2 \cdot 4$	$5 \cdot 3$	$5 \cdot 4$

To factor $6x^2 + 23x + 20$, we reverse the above multiplication, using what we might call an "unFOIL" process. We look for two binomials $rx + p$ and $sx + q$ whose product is this trinomial. The product of the First terms must be $6x^2$. The product of the Outside terms plus the product of the Inside terms must be $23x$. The product of the Last terms must be 20. We know from the preceding discussion that the answer is

$$(2x + 5)(3x + 4).$$

Generally, however, finding such an answer is a refined trial-and-error process. It turns out that $(-2x - 5)(-3x - 4)$ is also a correct answer, but we usually choose an answer in which the first coefficients are positive.

We will use the following trial-and-error method.

> To factor $ax^2 + bx + c,\ a \neq 1$, using the FOIL method:
>
> 1. Factor out a common factor, if any.
> 2. Factor the term ax^2. This gives these possibilities for r and s:
>
> $$(rx + p)(sx + q).$$
> $$rx \cdot sx = ax^2$$
>
> 3. Factor the last term c. This gives these possibilities for p and q:
>
> $$(rx + p)(sx + q).$$
> $$p \cdot q = c$$
>
> 4. Look for combinations of factors from steps (2) and (3) for which the sum of their products is the middle term, bx:
>
> $$rx \cdot q$$
> $$(rx + p)(sx + q). \qquad rx \cdot q + p \cdot sx \stackrel{?}{=} bx$$
> $$p \cdot sx$$

Factor.

1. $2x^2 - x - 15$

EXAMPLE 1 Factor: $3x^2 - 10x - 8$.

1) First, factor out a common factor, if any. There is none (other than 1 or -1).

2) Factor the first term, $3x^2$. The only possibility is $3x \cdot x$. The desired factorization is then of the form

$$(3x + \square)(x + \square),$$

where we must determine the numbers in the blanks.

3) Factor the last term, -8, which is negative. The possibilities are

$$(-8)(1), \quad 8(-1), \quad (-2)(4), \quad \text{and} \quad 2(-4).$$

4) We look for combinations of factors from steps (2) and (3) such that the sum of their products is the middle term, $-10x$:

$3x$
$(3x - 8)(x + 1) = 3x^2 - 5x - 8$;
$-8x$ Wrong middle term

$-3x$
$(3x + 8)(x - 1) = 3x^2 + 5x - 8$;
$8x$ Wrong middle term

$12x$
$(3x - 2)(x + 4) = 3x^2 + 10x - 8$;
$-2x$ Wrong middle term

$-12x$
$(3x + 2)(x - 4) = 3x^2 - 10x - 8$;
$2x$ Correct middle term!

There are four other possibilities that we could try, but we need not since we have found a factorization.

2. $12x^2 - 17x - 5$

The factorization is $(3x + 2)(x - 4)$.

Do Exercises 1 and 2.

EXAMPLE 2 Factor: $24x^2 - 76x + 40$.

1) Factor out a common factor, if any. This time there is one, 4. We factor it out:

$$4(6x^2 - 19x + 10).$$

Now we factor the trinomial $6x^2 - 19x + 10$.

2) Factor the first term, $6x^2$. The possibilities are $3x, 2x$ and $6x, x$. Then we have these as possibilities for factorizations:

$$(3x + \square)(2x + \square) \quad \text{or} \quad (6x + \square)(x + \square).$$

3) Factor the last term, 10, which is positive. The possibilities are

$$10, 1 \quad \text{and} \quad -10, -1 \quad \text{and} \quad 5, 2 \quad \text{and} \quad -5, -2.$$

4) We look for combinations of factors from steps (2) and (3) such that the sum of their products is the middle term, $-19x$. The sign of the middle term, $-19x$, is negative, but the sign of the last term, 10, is positive. Thus the signs of both factors of the last term, 10, must be negative. From our list of factors in step (3), we can use only $-10, -1$ and $-5, -2$ as possibilities. This reduces the possibilities for factorizations to 8. We start by using these factors with $(3x + \square)(2x + \square)$.

Answers on page A-4

Should we not find the correct factorization, we will consider $(6x + \square)(x + \square)$.

$-3x$

$(3x - 10)(2x - 1) = 6x^2 - 23x + 10;$

$-20x$ ↑ Wrong middle term

$-30x$

$(3x - 1)(2x - 10) = 6x^2 - 32x + 10;$

$-2x$ ↑ Wrong middle term

$-6x$

$(3x - 5)(2x - 2) = 6x^2 - 16x + 10;$

$-10x$ ↑ Wrong middle term

$-15x$

$(3x - 2)(2x - 5) = 6x^2 - 19x + 10;$

$-4x$ ↑ Correct middle term!

We have a correct answer. We need not consider $(6x + \square)(x + \square)$.

Look again at the possibility $(3x - 5)(2x - 2)$. Without multiplying, we can reject such a possibility. Look at the following:

$$(3x - 5)(2x - 2) = 2(3x - 5)(x - 1).$$

The expression $2x - 2$ has a common factor, 2. But we removed the largest common factor before we began. If this expression were a factorization, then 2 would have to be a common factor in addition to the original 4. Thus, as we saw when we multiplied, $(3x - 5)(2x - 2)$ cannot be part of the factorization of the original trinomial.

> Given that we factored out the largest common factor at the outset, we can eliminate factorizations that have a common factor.

The factorization of $6x^2 - 19x + 10$ is $(3x - 2)(2x - 5)$. But do not forget the common factor! We must include it in order to get a factorization of the original trinomial:

$$24x^2 - 76x + 40 = 4(3x - 2)(2x - 5).$$

Do Exercises 3 and 4.

EXAMPLE 3 Factor: $10x^2 + 37x + 7$.

1) First, factor out a common factor, if any. There is none (other than 1 or -1).

2) Factor the first term, $10x^2$. The possibilities are $10x$, x and $5x$, $2x$. We have these as possibilities for factorizations:

$$(10x + \square)(x + \square) \quad \text{and} \quad (5x + \square)(2x + \square).$$

Factor.

3. $3x^2 - 19x + 20$

4. $20x^2 - 46x + 24$

Answers on page A-4

5. Factor: $6x^2 + 7x + 2$.

Factor.

6. $6a^2 - 5ab + b^2$

7. $6x^2 + 15xy + 9y^2$

Answers on page A-4

3) Factor the last term, 7. The possibilities are 1, 7 and −1, −7.

4) We look for factors from steps (2) and (3) such that the sum of their products is the middle term. In this case, all signs are positive, so we need consider only plus signs. The possibilities are

$$(10x + 1)(x + 7) = 10x^2 + 71x + 7,$$
$$(10x + 7)(x + 1) = 10x^2 + 17x + 7,$$
$$(5x + 7)(2x + 1) = 10x^2 + 19x + 7,$$
$$(5x + 1)(2x + 7) = 10x^2 + 37x + 7.$$

The factorization is $(5x + 1)(2x + 7)$.

Tips for factoring $ax^2 + bx + c$, $a \neq 1$:

1. If the largest common factor has been factored out of the original trinomial, then no binomial factor can have a common factor (other than 1 or −1).
2. If all the signs of all the terms are positive, then the signs of all the terms of the binomial factors are positive.
3. Be systematic about your trials. Keep track of those you have tried and those you have not.

Do Exercise 5.

Keep in mind that this method of factoring trinomials of the type $ax^2 + bx + c$ involves trial and error. As you practice, you will find that you can make better and better guesses.

Caution! When factoring any polynomial, always look first for a common factor. Failure to do so is such a common error that this caution bears repeating.

EXAMPLE 4 Factor: $6p^2 - 13pq - 28q^2$.

1) Factor out a common factor, if any. There is none (other than 1 or −1).

2) Factor the first term, $6p^2$. Possibilities are $2p$, $3p$ and $6p$, p. We have these as possibilities for factorizations:

$$(2p + \square)(3p + \square) \quad \text{and} \quad (6p + \square)(p + \square).$$

3) Factor the last term, $-28q^2$, which has a negative coefficient. The possibilities are $-14q$, $2q$ and $14q$, $-2q$; $-28q$, q and $28q$, $-q$; and $-7q$, $4q$ and $7q$, $-4q$.

4) The coefficient of the middle term is negative, so we look for combinations of factors from steps (2) and (3) such that the sum of their products has a negative coefficient. We try some possibilities:

$$(2p - 14q)(3p + 2q) = 6p^2 - 38pq - 28q^2,$$
$$(2p - 28q)(3p + q) = 6p^2 - 82pq - 28q^2,$$
$$(2p - 7q)(3p + 4q) = 6p^2 - 13pq - 28q^2.$$

The factorization of $6p^2 - 13pq - 28q^2$ is $(2p - 7q)(3p + 4q)$.

Do Exercises 6 and 7.

NAME SECTION DATE

Exercise Set 4.3

a Factor.

1. $2x^2 - 7x - 4$

2. $3x^2 - x - 4$

3. $5x^2 - x - 18$

4. $4x^2 - 17x + 15$

5. $6x^2 + 23x + 7$

6. $6x^2 - 23x + 7$

7. $3x^2 + 4x + 1$

8. $7x^2 + 15x + 2$

9. $4x^2 + 4x - 15$

10. $9x^2 + 6x - 8$

11. $2x^2 - x - 1$

12. $15x^2 - 19x - 10$

13. $9x^2 + 18x - 16$

14. $2x^2 + 5x + 2$

15. $3x^2 - 5x - 2$

16. $18x^2 - 3x - 10$

17. $12x^2 + 31x + 20$

18. $15x^2 + 19x - 10$

19. $14x^2 + 19x - 3$

20. $35x^2 + 34x + 8$

21. $9x^2 + 18x + 8$

22. $6 - 13x + 6x^2$

23. $49 - 42x + 9x^2$

24. $16 + 36x^2 + 48x$

25. $24x^2 + 47x - 2$

26. $16p^2 - 78p + 27$

27. $35x^2 - 57x - 44$

28. $9a^2 + 12a - 5$

29. $20 + 6x - 2x^2$

30. $15 + x - 2x^2$

31. $12x^2 + 28x - 24$

32. $6x^2 + 33x + 15$

33. $30x^2 - 24x - 54$

34. $18t^2 - 24t + 6$

35. $4x + 6x^2 - 10$

36. $-9 + 18x^2 - 21x$

ANSWERS

1. ______
2. ______
3. ______
4. ______
5. ______
6. ______
7. ______
8. ______
9. ______
10. ______
11. ______
12. ______
13. ______
14. ______
15. ______
16. ______
17. ______
18. ______
19. ______
20. ______
21. ______
22. ______
23. ______
24. ______
25. ______
26. ______
27. ______
28. ______
29. ______
30. ______
31. ______
32. ______
33. ______
34. ______
35. ______
36. ______

ANSWERS

37. ______
38. ______
39. ______
40. ______
41. ______
42. ______
43. ______
44. ______
45. ______
46. ______
47. ______
48. ______
49. ______
50. ______
51. ______
52. ______
53. ______
54. ______
55. ______
56. ______
57. ______
58. ______
59. ______
60. ______
61. ______
62. ______
63. ______
64. ______
65. ______
66. ______
67. ______
68. ______
69. ______
70. ______
71. ______
72. ______
73. ______
74. ______
75. ______
76. ______
77. ______
78. ______
79. ______
80. ______
81. ______
82. ______

37. $3x^2 - 4x + 1$

38. $6x^2 + 13x + 6$

39. $12x^2 - 28x - 24$

40. $6x^2 - 33x + 15$

41. $-1 + 2x^2 - x$

42. $-19x + 15x^2 + 6$

43. $9x^2 - 18x - 16$

44. $14x^2 + 35x + 14$

45. $15x^2 - 25x - 10$

46. $18x^2 + 3x - 10$

47. $12x^3 + 31x^2 + 20x$

48. $15x^3 + 19x^2 - 10x$

49. $14x^4 + 19x^3 - 3x^2$

50. $70x^4 + 68x^3 + 16x^2$

51. $168x^3 - 45x^2 + 3x$

52. $144x^5 + 168x^4 + 48x^3$

53. $15x^4 - 19x^2 + 6$

54. $9x^4 + 18x^2 + 8$

55. $25t^2 + 80t + 64$

56. $9x^2 - 42x + 49$

57. $6x^3 + 4x^2 - 10x$

58. $18x^3 - 21x^2 - 9x$

59. $25x^2 + 79x + 64$

60. $9y^2 + 42y + 47$

61. $x^2 + 15x - 11$

62. $x^2 + 4x - 8$

63. $12m^2 - mn - 20n^2$

64. $12a^2 - 17ab + 6b^2$

65. $6a^2 - ab - 15b^2$

66. $3p^2 - 16pq - 12q^2$

67. $9a^2 + 18ab + 8b^2$

68. $10s^2 + 4st - 6t^2$

69. $35p^2 + 34pq + 8q^2$

70. $30a^2 + 87ab + 30b^2$

71. $18x^2 - 6xy - 24y^2$

72. $15a^2 - 5ab - 20b^2$

SKILL MAINTENANCE

73. Solve $A = pq - 7$ for q.

74. Solve $y = mx + b$ for x.

75. Solve $3x + 2y = 6$ for y.

76. Solve $p - q + r = 2$ for q.

Solve.

77. $5 - 4x < -11$

78. $2x - 4(x + 3x) \geq 6x - 8 - 9x$

SYNTHESIS

Factor.

79. $20x^{2n} + 16x^n + 3$

80. $-15x^{2m} + 26x^m - 8$

81. $3x^{6a} - 2x^{3a} - 1$

82. $x^{2n+1} - 2x^{n+1} + x$

4.4 Factoring $ax^2 + bx + c$, $a \neq 1$, Using Grouping

OBJECTIVE

After finishing Section 4.4, you should be able to:

a Factor trinomials of the type $ax^2 + bx + c$, $a \neq 1$, by splitting the middle term and using grouping.

FOR EXTRA HELP

TAPE 11 | TAPE 9A | MAC: 4A IBM: 4A

a Another method of factoring trinomials of the type $ax^2 + bx + c$, $a \neq 1$, is known as the **grouping method.** It involves factoring by grouping. We know how to factor the trinomial $x^2 + 5x + 6$. We look for factors of the constant term, 6, whose sum is the coefficient of the middle term, 5:

$$x^2 + 5x + 6.$$

(1) Factor: $6 = 2 \cdot 3$

(2) Sum: $2 + 3 = 5$

What happens when the leading coefficient is not 1? Consider the trinomial $3x^2 - 10x - 8$. The method we use is similar to what we used for the preceding trinomial, but we need two more steps. The method is outlined as follows.

> To factor $ax^2 + bx + c$, $a \neq 1$, using the grouping method:
>
> 1. Factor out a common factor, if any.
> 2. Multiply the leading coefficient a and the constant c.
> 3. Try to factor the product ac so that the sum of the factors is b. That is, find integers p and q such that $pq = ac$ and $p + q = b$.
> 4. Split the middle term. That is, write it as a sum using the factors found in step (3).
> 5. Then factor by grouping.

EXAMPLE 1 Factor: $3x^2 - 10x - 8$.

1) First, factor out a common factor, if any. There is none (other than 1 or -1).

2) Multiply the leading coefficient, 3, and the constant, -8:

$$3(-8) = -24.$$

3) Then look for a factorization of -24 in which the sum of the factors is the coefficient of the middle term, -10.

PAIRS OF FACTORS	SUMS OF FACTORS
-1, 24	23
-2, 12	10
-3, 8	5
-4, 6	2
-6, 4	-2
-8, 3	-5
-12, 2	-10 ← $-12 + 2 = -10$
-24, 1	-23

4) Next, split the middle term as a sum or a difference using the factors found in step (3):

$$-10x = -12x + 2x.$$

Factor.

1. $6x^2 + 7x + 2$

2. $12x^2 - 17x - 5$

Factor.

3. $6x^2 + 15x + 9$

4. $20x^2 - 46x + 24$

Answers on page A-4

5) Factor by grouping, as follows:

$$3x^2 - 10x - 8 = 3x^2 - 12x + 2x - 8 \quad \text{Substituting } -12x + 2x \text{ for } -10x$$
$$= 3x(x - 4) + 2(x - 4) \quad \text{Factoring by grouping; see Section 4.1}$$
$$= (3x + 2)(x - 4).$$

We can also split the middle term as $2x - 12x$. We still get the same factorization, although the factors may be in a different order. Note the following:

$$3x^2 - 10x - 8 = 3x^2 + 2x - 12x - 8$$
$$= x(3x + 2) - 4(3x + 2)$$
$$= (x - 4)(3x + 2).$$

Check by multiplying: $(x - 4)(3x + 2) = 3x^2 - 10x - 8.$

Do Exercises 1 and 2.

EXAMPLE 2 Factor: $8x^2 + 8x - 6$.

1) First, factor out a common factor, if any. The number 2 is common to all three terms, so we factor it out:

$$2(4x^2 + 4x - 3).$$

2) Next, factor the trinomial $4x^2 + 4x - 3$. Multiply the leading coefficient and the constant, 4 and -3:

$$4(-3) = -12.$$

3) Try to factor -12 so that the sum of the factors is 4.

PAIRS OF FACTORS	SUMS OF FACTORS
−3, 4	1
3, −4	−1
−12, 1	−11
12, −1	11
−6, 2	−4
6, −2	4 ← 6 + (−2) = 4

4) Split the middle term, $4x$, as follows:

$$4x = 6x - 2x.$$

5) Factor by grouping:

$$4x^2 + 4x - 3 = 4x^2 + 6x - 2x - 3 \quad \text{Substituting } 6x - 2x \text{ for } 4x$$
$$= 2x(2x + 3) - 1(2x + 3) \quad \text{Factoring by grouping}$$
$$= (2x - 1)(2x + 3).$$

The factorization of $4x^2 + 4x - 3$ is $(2x - 1)(2x + 3)$. But don't forget the common factor! We must include it to get a factorization of the original trinomial:

$$8x^2 + 8x - 6 = 2(2x - 1)(2x + 3).$$

Do Exercises 3 and 4.

NAME SECTION DATE

Exercise Set 4.4

a Factor. Note that the middle term has already been split.

1. $x^2 + 2x + 7x + 14$

2. $x^2 + 3x + x + 3$

3. $x^2 - 4x - x + 4$

4. $a^2 + 5a - 2a - 10$

5. $6x^2 + 4x + 9x + 6$

6. $3x^2 - 2x + 3x - 2$

7. $3x^2 - 4x - 12x + 16$

8. $24 - 18y - 20y + 15y^2$

9. $35x^2 - 40x + 21x - 24$

10. $8x^2 - 6x - 28x + 21$

11. $4x^2 + 6x - 6x - 9$

12. $2x^4 - 6x^2 - 5x^2 + 15$

13. $2x^4 + 6x^2 + 5x^2 + 15$

14. $9x^4 - 6x^2 - 6x^2 + 4$

Factor by grouping.

15. $2x^2 - 7x - 4$

16. $5x^2 - x - 18$

17. $3x^2 + 4x - 15$

(3x-5)(x+3)

18. $3x^2 + x - 4$

19. $6x^2 + 23x + 7$

20. $6x^2 + 13x + 6$

21. $3x^2 + 4x + 1$

22. $7x^2 + 15x + 2$

23. $4x^2 + 4x - 15$

Answers

1. ______
2. ______
3. ______
4. ______
5. ______
6. ______
7. ______
8. ______
9. ______
10. ______
11. ______
12. ______
13. ______
14. ______
15. ______
16. ______
17. ______
18. ______
19. ______
20. ______
21. ______
22. ______
23. ______

ANSWERS

24. ______
25. ______
26. ______
27. ______
28. ______
29. ______
30. ______
31. ______
32. ______
33. ______
34. ______
35. ______
36. ______
37. ______
38. ______
39. ______
40. ______
41. ______
42. ______
43. ______
44. ______
45. ______
46. ______
47. ______
48. ______
49. ______
50. ______
51. ______
52. ______
53. ______
54. ______

24. $9x^2 + 6x - 8$

25. $2x^2 + x - 1$

26. $15x^2 + 19x - 10$

27. $9x^2 - 18x - 16$

28. $2x^2 - 5x + 2$

29. $3x^2 + 5x - 2$

30. $18x^2 + 3x - 10$

31. $12x^2 - 31x + 20$

32. $15x^2 - 19x - 10$

33. $14x^2 + 19x - 3$

34. $35x^2 + 34x + 8$

35. $9x^2 + 18x + 8$

36. $6 - 13x + 6x^2$

37. $49 - 42x + 9x^2$

38. $25x^2 + 40x + 16$

39. $24x^2 + 47x - 2$

40. $16a^2 + 78a + 27$

41. $35x^5 - 57x^4 - 44x^3$

42. $18a^3 + 24a^2 - 10a$

43. $60x + 18x^2 - 6x^3$

44. $60x + 4x^2 - 8x^3$

SKILL MAINTENANCE

Solve.

45. $6 - 3x \geq -18$

46. $3 - 2x - 4x > -9$

47. $\frac{1}{2}x - 6x + 10 \leq x - 5x$

48. $-2(x + 7) > -4(x - 5)$

49. $3x - 6x + 2(x - 4) > 2(9 - 4x)$

50. $-6(x - 4) + 8(4 - x) \leq 3(x - 7)$

SYNTHESIS

Factor.

51. $9x^{10} - 12x^5 + 4$

52. $24x^{2n} + 22x^n + 3$

53. $16x^{10} + 8x^5 + 1$

54. $(a + 4)^2 - 2(a + 4) + 1$

4.5 *Factoring Trinomial Squares and Differences of Squares*

OBJECTIVES

After finishing Section 4.5, you should be able to:

a Recognize trinomial squares.

b Factor trinomial squares.

c Recognize differences of squares.

d Factor differences of squares, being careful to factor completely.

FOR EXTRA HELP

TAPE 11 | TAPE 9A | MAC: 4A IBM: 4A

In this section, we first learn to factor trinomials that are squares of binomials. Then we factor binomials that are differences of squares.

a RECOGNIZING TRINOMIAL SQUARES

Some trinomials are squares of binomials. For example, the trinomial $x^2 + 10x + 25$ is the square of the binomial $x + 5$. To see this, we can calculate $(x + 5)^2$. It is $x^2 + 2 \cdot x \cdot 5 + 5^2$, or $x^2 + 10x + 25$. A trinomial that is the square of a binomial is called a **trinomial square.**

In Chapter 3, we considered squaring binomials as special-product rules:

$$(A + B)^2 = A^2 + 2AB + B^2;$$
$$(A - B)^2 = A^2 - 2AB + B^2.$$

We can use these equations in reverse to factor trinomial squares.

$$A^2 + 2AB + B^2 = (A + B)^2;$$
$$A^2 - 2AB + B^2 = (A - B)^2$$

How can we recognize when an expression to be factored is a trinomial square? Look at $A^2 + 2AB + B^2$ and $A^2 - 2AB + B^2$. In order for an expression to be a trinomial square:

a) Two terms, A^2 and B^2, must be squares, such as

$$4, \quad x^2, \quad 25x^4, \quad 16t^2.$$

b) There must be no minus sign before A^2 or B^2.

c) If we multiply A and B (the square roots of A^2 and B^2) and double the result, we get either the remaining term $2 \cdot A \cdot B$, or its opposite, $-2 \cdot A \cdot B$.

EXAMPLE 1 Determine whether $x^2 + 6x + 9$ is a trinomial square.

a) We know that x^2 and 9 are squares.

b) There is no minus sign before x^2 or 9.

c) If we multiply the square roots, x and 3, and double the product, we get the remaining term: $2 \cdot x \cdot 3 = 6x$.

Thus, $x^2 + 6x + 9$ is the square of a binomial. In fact, $x^2 + 6x + 9 = (x + 3)^2$.

EXAMPLE 2 Determine whether $x^2 + 6x + 11$ is a trinomial square.

The answer is no, because only one term is a square.

Determine whether each is a trinomial square. Write "yes" or "no."

1. $x^2 + 8x + 16$

2. $25 - x^2 + 10x$

3. $t^2 - 12t + 4$

4. $25 + 20y + 4y^2$

5. $5x^2 + 16 - 14x$

6. $16x^2 + 40x + 25$

7. $p^2 + 6p - 9$

8. $25a^2 + 9 - 30a$

Factor.

9. $x^2 + 2x + 1$

10. $1 - 2x + x^2$

11. $4 + t^2 + 4t$

12. $25x^2 - 70x + 49$

13. $49 - 56y + 16y^2$

Answers on page A-4

EXAMPLE 3 Determine whether $16x^2 + 49 - 56x$ is a trinomial square.

It helps to first write the trinomial in descending order:

$$16x^2 - 56x + 49.$$

a) We know that $16x^2$ and 49 are squares.

b) There is no minus sign before $16x^2$ or 49.

c) If we multiply the square roots, $4x$ and 7, and double the product, we get the opposite of the remaining term: $2 \cdot 4x \cdot 7 = 56x$; $56x$ is the opposite of $-56x$.

Thus, $16x^2 + 49 - 56x$ is a trinomial square. In fact, $16x^2 - 56x + 49 = (4x - 7)^2$.

Do Exercises 1–8.

b FACTORING TRINOMIAL SQUARES

We can use the trial-and-error or grouping methods from Sections 4.2–4.4 to factor such trinomial squares, but there is a faster method using the following equations:

$$A^2 + 2AB + B^2 = (A + B)^2;$$
$$A^2 - 2AB + B^2 = (A - B)^2.$$

We consider 3 to be a square root of 9 because $3^2 = 9$. Similarly, A is a square root of A^2. We use square roots of the squared terms and the sign of the remaining term to factor a trinomial square.

EXAMPLE 4 Factor: $x^2 + 6x + 9$.

$$x^2 + 6x + 9 = x^2 + 2 \cdot x \cdot 3 + 3^2 = (x + 3)^2$$
$$A^2 + 2\ A\ B + B^2 = (A + B)^2$$

The sign of the middle term is positive.

EXAMPLE 5 Factor: $x^2 + 49 - 14x$.

$$x^2 + 49 - 14x = x^2 - 14x + 49 \quad \text{Changing order}$$
$$= x^2 - 2 \cdot x \cdot 7 + 7^2 \quad \text{The sign of the middle term is negative.}$$
$$= (x - 7)^2$$

EXAMPLE 6 Factor: $16x^2 - 40x + 25$.

$$16x^2 - 40x + 25 = (4x)^2 - 2 \cdot 4x \cdot 5 + 5^2 = (4x - 5)^2$$
$$A^2 - 2\ A\ B + B^2 = (A - B)^2$$

Do Exercises 9–13.

EXAMPLE 7 Factor: $t^4 + 20t^2 + 100$.

$$t^4 + 20t^2 + 100 = (t^2)^2 + 2(t^2)(10) + 10^2$$
$$= (t^2 + 10)^2$$

EXAMPLE 8 Factor: $75m^3 + 210m^2 + 147m$.

Always look first for a common factor. This time there is one, $3m$:

$$75m^3 + 210m^2 + 147m = 3m[25m^2 + 70m + 49]$$
$$= 3m[(5m)^2 + 2(5m)(7) + 7^2]$$
$$= 3m(5m + 7)^2.$$

EXAMPLE 9 Factor: $4p^2 - 12pq + 9q^2$.

$$4p^2 - 12pq + 9q^2 = (2p)^2 - 2(2p)(3q) + (3q)^2$$
$$= (2p - 3q)^2$$

Do Exercises 14–17.

c RECOGNIZING DIFFERENCES OF SQUARES

The following polynomials are *differences of squares:*

$x^2 - 9$, $4t^2 - 49$, $a^2 - 25b^2$.

To factor a difference of squares such as $x^2 - 9$, think about the formula we used in Chapter 3:

$$(A + B)(A - B) = A^2 - B^2.$$

Equations are reversible, so we also know that

$$A^2 - B^2 = (A + B)(A - B).$$

Thus,

$$x^2 - 9 = (x + 3)(x - 3).$$

To use this formula, we must be able to recognize when it applies. A **difference of squares** is an expression like the following:

$A^2 - B^2$.

How can we recognize such expressions? Look at $A^2 - B^2$. In order for a binomial to be a difference of squares:

a) There must be two expressions, both squares, such as

$4x^2$, 9, $25t^4$, 1, x^6, $49y^8$.

When the coefficient is a perfect square and the power(s) of the variable(s) is (are) even, then the expression is a perfect square.

b) The terms must have different signs.

Factor.

14. $48m^2 + 75 + 120m$

15. $p^4 + 18p^2 + 81$

16. $4z^5 - 20z^4 + 25z^3$

17. $9a^2 + 30ab + 25b^2$

Answers on page A-4

Determine whether each is a difference of squares. Write "yes" or "no."

18. $x^2 - 25$

19. $t^2 - 24$

20. $y^2 + 36$

21. $4x^2 - 15$

22. $16x^4 - 49$

23. $9w^6 - 1$

24. $-49 + 25t^2$

Answers on page A-4

EXAMPLE 10 Is $9x^2 - 64$ a difference of squares?

a) The first expression is a square: $9x^2 = (3x)^2$.
   The second expression is a square: $64 = 8^2$.

b) The terms have different signs.

Thus we have a difference of squares, $(3x)^2 - 8^2$.

EXAMPLE 11 Is $25 - t^3$ a difference of squares?

a) The expression t^3 is not a square.

The expression is not a difference of squares.

EXAMPLE 12 Is $-4x^2 + 16$ a difference of squares?

a) The expressions $4x^2$ and 16 are squares: $4x^2 = (2x)^2$ and $16 = 4^2$.

b) The terms have different signs.

Thus we have a difference of squares. We can also see this by rewriting in the equivalent form: $16 - 4x^2$.

Do Exercises 18–24.

d FACTORING DIFFERENCES OF SQUARES

To factor a difference of squares, we use the following equation:

$$A^2 - B^2 = (A + B)(A - B).$$

To factor a difference of squares $A^2 - B^2$, we find A and B, which are square roots of the expressions A^2 and B^2. We then use A and B to form two factors. One is the sum $A + B$, and the other is the difference $A - B$.

EXAMPLE 13 Factor: $x^2 - 4$.

$$x^2 - 4 = x^2 - 2^2 = (x + 2)(x - 2)$$

$$A^2 - B^2 = (A + B)(A - B)$$

EXAMPLE 14 Factor: $9 - 16t^4$.

$$9 - 16t^4 = 3^2 - (4t^2)^2 = (3 + 4t^2)(3 - 4t^2)$$

$$A^2 - B^2 = (A + B)(A - B)$$

EXAMPLE 15 Factor: $m^2 - 4p^2$.

$$m^2 - 4p^2 = m^2 - (2p)^2 = (m + 2p)(m - 2p)$$

EXAMPLE 16 Factor: $x^2 - \frac{1}{9}$.

$$x^2 - \frac{1}{9} = x^2 - \left(\frac{1}{3}\right)^2 = \left(x + \frac{1}{3}\right)\left(x - \frac{1}{3}\right)$$

EXAMPLE 17 Factor: $18x^2 - 50x^6$.

Always look first for a factor common to all terms. This time there is one, $2x^2$.

$$\begin{aligned} 18x^2 - 50x^6 &= 2x^2(9 - 25x^4) \\ &= 2x^2[3^2 - (5x^2)^2] \\ &= 2x^2(3 + 5x^2)(3 - 5x^2) \end{aligned}$$

EXAMPLE 18 Factor: $49x^4 - 9x^6$.

$$49x^4 - 9x^6 = x^4(49 - 9x^2) = x^4(7 + 3x)(7 - 3x)$$

Do Exercises 25–29.

Caution! Note carefully in these examples that a difference of squares is *not* the square of the difference; that is,

$$A^2 - B^2 \neq (A - B)^2 = A^2 - 2AB + B^2.$$

For example,

$$(45 - 5)^2 = 40^2 = 1600,$$

but

$$45^2 - 5^2 = 2025 - 25 = 2000.$$

FACTORING COMPLETELY

If a factor with more than one term can still be factored, you should do so. When no factor can be factored further, you have **factored completely.** Always factor completely whenever told to factor.

EXAMPLE 19 Factor: $p^4 - 16$.

$$\begin{aligned} p^4 - 16 &= (p^2)^2 - 4^2 \\ &= (p^2 + 4)(p^2 - 4) && \text{Factoring a difference of squares} \\ &= (p^2 + 4)(p + 2)(p - 2) && \text{Factoring further. The factor } p^2 - 4 \text{ is a difference of squares.} \end{aligned}$$

The polynomial $p^2 + 4$ cannot be factored further into polynomials with real coefficients.

EXAMPLE 20 Factor: $y^4 - 16x^{12}$.

$$\begin{aligned} y^4 - 16x^{12} &= (y^2 + 4x^6)(y^2 - 4x^6) && \text{Factoring a difference of squares} \\ &= (y^2 + 4x^6)(y + 2x^3)(y - 2x^3) && \text{Factoring further. The factor } y^2 - 4x^6 \text{ is a difference of squares.} \end{aligned}$$

Factor.

25. $x^2 - 9$

26. $64 - 4t^2$

27. $a^2 - 25b^2$

28. $64x^4 - 25x^6$

29. $5 - 20t^6$
[*Hint*: $1 = 1^2$, $t^6 = (t^3)^2$.]

Answers on page A-4

Factor completely.

30. $81x^4 - 1$

31. $49p^4 - 25q^6$

Answers on page A-4

FACTORING HINTS

1. Always look first for a common factor. If there is one, factor it out!
2. Always factor completely.
3. Check by multiplying.

Caution! If the greatest common factor has been removed, then you cannot factor a sum of squares further. In particular,

$$(A + B)^2 \neq A^2 + B^2.$$

Consider $25x^2 + 100$. This is a case in which we have a sum of squares, but there is a common factor, 25. Factoring, we get $25(x^2 + 4)$. Now $x^2 + 4$ cannot be factored further.

Do Exercises 30 and 31.

NAME ______ SECTION ______ DATE ______

Exercise Set 4.5

a Determine whether each of the following is a trinomial square.

1. $x^2 - 14x + 49$

2. $x^2 - 16x + 64$

3. $x^2 + 16x - 64$

4. $x^2 - 14x - 49$

5. $x^2 - 2x + 4$

6. $x^2 + 3x + 9$

7. $9x^2 - 36x + 24$

8. $36x^2 - 24x + 16$

b Factor completely. Remember to look first for a common factor and to check by multiplying.

9. $x^2 - 14x + 49$

10. $x^2 - 20x + 100$

11. $x^2 + 16x + 64$

12. $x^2 + 20x + 100$

13. $x^2 - 2x + 1$

14. $x^2 + 2x + 1$

15. $4 + 4x + x^2$

16. $4 + x^2 - 4x$

17. $q^4 - 6q^2 + 9$

18. $64 + 16a^2 + a^4$

19. $49 + 56y + 16y^2$

20. $75 + 48a^2 - 120a$

21. $2x^2 - 4x + 2$

22. $2x^2 - 40x + 200$

23. $x^3 - 18x^2 + 81x$

24. $x^3 + 24x^2 + 144x$

25. $12q^2 - 36q + 27$

26. $20p^2 + 100p + 125$

ANSWERS

1. ______
2. ______
3. ______
4. ______
5. ______
6. ______
7. ______
8. ______
9. ______
10. ______
11. ______
12. ______
13. ______
14. ______
15. ______
16. ______
17. ______
18. ______
19. ______
20. ______
21. ______
22. ______
23. ______
24. ______
25. ______
26. ______

ANSWERS

27. ______

28. ______

29. ______

30. ______

31. ______

32. ______

33. ______

34. ______

35. ______

36. ______

37. ______

38. ______

39. ______

40. ______

41. ______

42. ______

43. ______

44. ______

45. ______

46. ______

47. ______

48. ______

27. $49 - 42x + 9x^2$

28. $64 - 112x + 49x^2$

29. $5y^4 + 10y^2 + 5$

30. $a^4 + 14a^2 + 49$

31. $1 + 4x^4 + 4x^2$

32. $1 - 2a^5 + a^{10}$

33. $4p^2 + 12pq + 9q^2$

34. $25m^2 + 20mn + 4n^2$

35. $a^2 - 6ab + 9b^2$

36. $x^2 - 14xy + 49y^2$

37. $81a^2 - 18ab + b^2$

38. $64p^2 + 16pq + q^2$

39. $36a^2 + 96ab + 64b^2$

40. $16m^2 - 40mn + 25n^2$

c Determine whether each of the following is a difference of squares.

41. $x^2 - 4$

42. $x^2 - 36$

43. $x^2 + 25$

44. $x^2 + 9$

45. $x^2 - 45$

46. $x^2 - 80y^2$

47. $16x^2 - 25y^2$

48. $-1 + 36x^2$

d Factor completely. Remember to look first for a common factor.

49. $y^2 - 4$

50. $q^2 - 1$

51. $p^2 - 9$

52. $x^2 - 36$

53. $-49 + t^2$

54. $-64 + m^2$

55. $a^2 - b^2$

56. $p^2 - q^2$

57. $25t^2 - m^2$

58. $w^2 - 49z^2$

59. $100 - k^2$

60. $81 - w^2$

61. $16a^2 - 9$

62. $25x^2 - 4$

63. $4x^2 - 25y^2$

64. $9a^2 - 16b^2$

65. $8x^2 - 98$

66. $24x^2 - 54$

67. $36x - 49x^3$

68. $16x - 81x^3$

69. $49a^4 - 81$

70. $25a^4 - 9$

71. $a^4 - 16$

72. $y^4 - 1$

73. $5x^4 - 405$

74. $4x^4 - 64$

75. $1 - y^8$

76. $x^8 - 1$

ANSWERS

49. ________

50. ________

51. ________

52. ________

53. ________

54. ________

55. ________

56. ________

57. ________

58. ________

59. ________

60. ________

61. ________

62. ________

63. ________

64. ________

65. ________

66. ________

67. ________

68. ________

69. ________

70. ________

71. ________

72. ________

73. ________

74. ________

75. ________

76. ________

ANSWERS

77.
78.
79.
80.
81.
82.
83.
84.
85.
86.
87.
88.
89.
90.
91.
92.
93.
94.
95.
96.
97.
98.
99.
100.
101.
102.
103.
104.
105.
106.
107.
108.
109.
110.
111.
112.
113.
114.
115.
116.
117.
118.

77. $x^{12} - 16$ **78.** $x^8 - 81$ **79.** $y^2 - \frac{1}{16}$ **80.** $x^2 - \frac{1}{25}$

81. $25 - \frac{1}{49}x^2$ **82.** $\frac{1}{4} - 9q^2$ **83.** $16m^4 - t^4$ **84.** $p^4q^4 - 1$

SKILL MAINTENANCE

Divide.

85. $(-110) \div 10$ **86.** $-1000 \div (-2.5)$ **87.** $\left(-\frac{2}{3}\right) \div \frac{4}{5}$

88. $8.1 \div (-9)$ **89.** $-64 \div (-32)$ **90.** $-256 \div 1.6$

SYNTHESIS

Factor completely, if possible.

91. $49x^2 - 216$ **92.** $27x^3 - 13x$ **93.** $x^2 + 22x + 121$

94. $x^2 - 5x + 25$ **95.** $18x^3 + 12x^2 + 2x$ **96.** $162x^2 - 82$

97. $x^8 - 2^8$ **98.** $4x^4 - 4x^2$ **99.** $3x^5 - 12x^3$

100. $3x^2 - \frac{1}{3}$ **101.** $18x^3 - \frac{8}{25}x$ **102.** $x^2 - 2.25$

103. $0.49p - p^3$ **104.** $3.24x^2 - 0.81$ **105.** $0.64x^2 - 1.21$

106. $1.28x^2 - 2$ **107.** $(x + 3)^2 - 9$ **108.** $(y - 5)^2 - 36q^2$

109. $x^2 - \left(\frac{1}{x}\right)^2$ **110.** $a^{2n} - 49b^{2n}$ **111.** $81 - b^{4k}$

112. $9x^{18} + 48x^9 + 64$ **113.** $9b^{2n} + 12b^n + 4$ **114.** $(x + 7)^2 - 4x - 24$

115. $(y + 3)^2 + 2(y + 3) + 1$ **116.** $49(x + 1)^2 - 42(x + 1) + 9$

Find c so that the polynomial will be the square of a binomial.

117. $cy^2 + 6y + 1$ **118.** $cy^2 - 24y + 9$

4.6 *Factoring: A General Strategy*

Objective

After finishing Section 4.6, you should be able to:

a Factor polynomials completely using any of the methods considered in this chapter.

FOR EXTRA HELP

TAPE 11 | TAPE 9B | MAC: 4B IBM: 4B

a We now combine all of our factoring techniques and consider a general strategy for factoring polynomials. Here we will encounter polynomials of all the types we have considered, in random order, so you will have the opportunity to determine which method to use.

> To factor a polynomial:
>
> **a)** Always look first for a common factor. If there is one, factor out the largest common factor.
>
> **b)** Then look at the number of terms.
>
> *Two terms:* Determine whether you have a difference of squares. Do not try to factor a sum of squares: $A^2 + B^2$.
>
> *Three terms:* Determine whether the trinomial is a square. If it is, you know how to factor. If not, try trial and error, using the standard method or grouping.
>
> *Four terms:* Try factoring by grouping.
>
> **c)** Always *factor completely.* If a factor with more than one term can still be factored, you should factor it. When no factor can be factored further, you have finished.

EXAMPLE 1 Factor: $5t^4 - 80$.

a) We look for a common factor:

$$5t^4 - 80 = 5(t^4 - 16).$$

b) The factor $t^4 - 16$ has only two terms. It is a difference of squares: $(t^2)^2 - 4^2$. We factor it, being careful to include the common factor:

$$5(t^2 + 4)(t^2 - 4).$$

c) We see that one of the factors is again a difference of squares. We factor it:

$$5(t^2 + 4)(t + 2)(t - 2).$$

↑ This is a sum of squares. It cannot be factored!

We have factored completely because no factor with more than one term can be factored further.

EXAMPLE 2 Factor: $2x^3 + 10x^2 + x + 5$.

a) We look for a common factor. There isn't one.

b) There are four terms. We try factoring by grouping:

$$2x^3 + 10x^2 + x + 5$$

$= (2x^3 + 10x^2) + (x + 5)$ Separating into two binomials

$= 2x^2(x + 5) + 1(x + 5)$ Factoring each binomial

$= (2x^2 + 1)(x + 5).$ Factoring out the common factor, $x + 5$

c) None of these factors can be factored further, so we have factored completely.

Factor.

1. $3m^4 - 3$

2. $x^6 + 8x^3 + 16$

3. $2x^4 + 8x^3 + 6x^2$

4. $3x^3 + 12x^2 - 2x - 8$

5. $8x^3 - 200x$

Answers on page A-4

EXAMPLE 3 Factor: $x^5 - 2x^4 - 35x^3$.

a) We look first for a common factor. This time there is one, x^3:

$$x^5 - 2x^4 - 35x^3 = x^3(x^2 - 2x - 35).$$

b) The factor $x^2 - 2x - 35$ has three terms, but it is not a trinomial square. We factor it using trial and error:

$$x^5 - 2x^4 - 35x^3 = x^3(x^2 - 2x - 35) = x^3(x - 7)(x + 5).$$

Don't forget to include the common factor in the final answer!

c) No factor with more than one term can be factored further, so we have factored completely.

EXAMPLE 4 Factor: $x^4 - 10x^2 + 25$.

a) We look first for a common factor. There isn't one.

b) There are three terms. We see that this polynomial is a trinomial square. We factor it:

$$x^4 - 10x^2 + 25 = (x^2)^2 - 2 \cdot x^2 \cdot 5 + 5^2 = (x^2 - 5)^2.$$

c) Since $x^2 - 5$ cannot be factored further, we have factored completely.

Do Exercises 1–5.

EXAMPLE 5 Factor: $6x^2y^4 - 21x^3y^5 + 3x^2y^6$.

a) We look first for a common factor:

$$6x^2y^4 - 21x^3y^5 + 3x^2y^6 = 3x^2y^4(2 - 7xy + y^2).$$

b) There are three terms in $2 - 7xy + y^2$. We determine whether the trinomial is a square. Since only y^2 is a square, we do not have a trinomial square. Can the trinomial be factored by trial and error? A key to the answer is that x is only in the term $-7xy$. The polynomial might be in a form like $(1 - y)(2 + y)$, but there would be no x in the middle term. Thus, $2 - 7xy + y^2$ cannot be factored.

c) Have we factored completely? Yes, because no factor with more than one term can be factored further.

EXAMPLE 6 Factor: $(p + q)(x + 2) + (p + q)(x + y)$.

a) We look for a common factor:

$$\begin{aligned}(p + q)(x + 2) + (p + q)(x + y) &= (p + q)[(x + 2) + (x + y)] \\ &= (p + q)(2x + y + 2).\end{aligned}$$

b) There are three terms in $2x + y + 2$, but this trinomial cannot be factored further.

c) Neither factor can be factored further, so we have factored completely.

EXAMPLE 7 Factor: $px + py + qx + qy$.

a) We look first for a common factor. There isn't one.

b) There are four terms. We try factoring by grouping:

$$\begin{aligned}px + py + qx + qy &= p(x + y) + q(x + y) \\ &= (p + q)(x + y).\end{aligned}$$

c) Have we factored completely? Since neither factor can be factored further, we have factored completely.

EXAMPLE 8 Factor: $25x^2 + 20xy + 4y^2$.

a) We look first for a common factor. There isn't one.

b) There are three terms. We determine whether the trinomial is a square. The first term and the last term are squares:

$$25x^2 = (5x)^2 \quad \text{and} \quad 4y^2 = (2y)^2.$$

Since twice the product of $5x$ and $2y$ is the other term,

$$2 \cdot 5x \cdot 2y = 20xy,$$

the trinomial is a perfect square.

We factor by writing the square roots of the square terms and the sign of the middle term:

$$25x^2 + 20xy + 4y^2 = (5x + 2y)^2.$$

We can check by squaring $5x + 2y$.

c) Since $5x + 2y$ cannot be factored further, we have factored completely.

Factor.

6. $x^4y^2 + 2x^3y + 3x^2y$

7. $10p^6q^2 + 4p^5q^3 + 2p^4q^4$

8. $(a - b)(x + 5) + (a - b)(x + y^2)$

9. $ax^2 + ay + bx^2 + by$

10. $x^4 + 2x^2y^2 + y^4$

11. $x^2y^2 + 5xy + 4$

12. $p^4 - 81q^4$

Answers on page A-4

EXAMPLE 9 Factor: $p^2q^2 + 7pq + 12$.

a) We look first for a common factor. There isn't one.

b) There are three terms. We determine whether the trinomial is a square. The first term is a square, but neither of the other terms is a square, so we do not have a trinomial square. We use the trial-and-error or grouping method, thinking of the product pq as a single variable. We consider this possibility for factorization:

$$(pq + \blacksquare)(pq + \blacksquare).$$

We factor the last term, 12. All the signs are positive, so we consider only positive factors. Possibilities are 1, 12 and 2, 6 and 3, 4. The pair 3, 4 gives a sum of 7 for the coefficient of the middle term. Thus,

$$p^2q^2 + 7pq + 12 = (pq + 3)(pq + 4).$$

c) No factor with more than one term can be factored further, so we have factored completely.

EXAMPLE 10 Factor: $8x^4 - 20x^2y - 12y^2$.

a) We look first for a common factor:

$$8x^4 - 20x^2y - 12y^2 = 4(2x^4 - 5x^2y - 3y^2).$$

b) There are three terms in $2x^4 - 5x^2y - 3y^2$. We determine whether the trinomial is a square. Since none of the terms is a square, we do not have a trinomial square. We factor $2x^4$. Possibilities are $2x^2$, x^2 and $2x$, x^3 and others. We also factor the last term, $-3y^2$. Possibilities are $3y$, $-y$ and $-3y$, y and others. We look for factors such that the sum of their products is the middle term. We try some possibilities:

$$(2x - y)(x^3 + 3y) = 2x^4 + 6xy - x^3y - 3y^2,$$
$$(2x^2 - y)(x^2 + 3y) = 2x^4 + 5x^2y - 3y^2,$$
$$(2x^2 + y)(x^2 - 3y) = 2x^4 - 5x^2y - 3y^2.$$

c) No factor with more than one term can be factored further, so we have factored completely. The factorization, including the common factor, is

$$4(2x^2 + y)(x^2 - 3y).$$

EXAMPLE 11 Factor: $a^4 - 16b^4$.

a) We look first for a common factor. There isn't one.

b) There are two terms. Since $a^4 = (a^2)^2$ and $16b^4 = (4b^2)^2$, we see that we do have a difference of squares. Thus,

$$a^4 - 16b^4 = (a^2 + 4b^2)(a^2 - 4b^2).$$

c) The last factor can be factored further. It is also a difference of squares. Thus,

$$a^4 - 16b^4 = (a^2 + 4b^2)(a + 2b)(a - 2b).$$

Do Exercises 6–12.

NAME SECTION DATE

Exercise Set 4.6

a Factor completely.

1. $3x^2 - 192$

2. $2t^2 - 18$

3. $a^2 + 25 - 10a$

4. $y^2 + 49 + 14y$

5. $2x^2 - 11x + 12$

6. $8y^2 - 18y - 5$

7. $x^3 + 24x^2 + 144x$

8. $x^3 - 18x^2 + 81x$

9. $x^3 + 3x^2 - 4x - 12$

10. $x^3 - 5x^2 - 25x + 125$

11. $48x^2 - 3$

12. $50x^2 - 32$

13. $9x^3 + 12x^2 - 45x$

14. $20x^3 - 4x^2 - 72x$

15. $x^2 + 4$

16. $t^2 + 25$

17. $x^4 + 7x^2 - 3x^3 - 21x$

18. $m^4 + 8m^3 + 8m^2 + 64m$

19. $x^5 - 14x^4 + 49x^3$

20. $2x^6 + 8x^5 + 8x^4$

21. $20 - 6x - 2x^2$

22. $45 - 3x - 6x^2$

ANSWERS

1. ______
2. ______
3. ______
4. ______
5. ______
6. ______
7. ______
8. ______
9. ______
10. ______
11. ______
12. ______
13. ______
14. ______
15. ______
16. ______
17. ______
18. ______
19. ______
20. ______
21. ______
22. ______

Answers

23. ______

24. ______

25. ______

26. ______

27. ______

28. ______

29. ______

30. ______

31. ______

32. ______

33. ______

34. ______

35. ______

36. ______

37. ______

38. ______

39. ______

40. ______

41. ______

42. ______

43. ______

44. ______

23. $x^2 - 6x + 1$

24. $x^2 + 8x + 5$

25. $4x^4 - 64$

26. $5x^5 - 80x$

27. $1 - y^8$

28. $t^8 - 1$

29. $x^5 - 4x^4 + 3x^3$

30. $x^6 - 2x^5 + 7x^4$

31. $\frac{1}{81}x^6 - \frac{8}{27}x^3 + \frac{16}{9}$

32. $36a^2 - 15a + \frac{25}{16}$

33. $mx^2 + my^2$

34. $12p^2 + 24q^3$

35. $9x^2y^2 - 36xy$

36. $x^2y - xy^2$

37. $2\pi rh + 2\pi r^2$

38. $10p^4q^4 + 35p^3q^3 + 10p^2q^2$

39. $(a + b)(x - 3) + (a + b)(x + 4)$

40. $5c(a^3 + b) - (a^3 + b)$

41. $(x - 1)(x + 1) - y(x + 1)$

42. $3(p - q) - q^2(p - q)$

43. $n^2 + 2n + np + 2p$

44. $a^2 - 3a + ay - 3y$

45. $6q^2 - 3q + 2pq - p$

46. $2x^2 - 4x + xy - 2y$

47. $4b^2 + a^2 - 4ab$

48. $x^2 + y^2 - 2xy$

49. $16x^2 + 24xy + 9y^2$

50. $9c^2 + 6cd + d^2$

51. $49m^4 - 112m^2n + 64n^2$

52. $4x^2y^2 + 12xyz + 9z^2$

53. $y^4 + 10y^2z^2 + 25z^4$

54. $0.01x^4 - 0.1x^2y^2 + 0.25y^4$

55. $\frac{1}{4}a^2 + \frac{1}{3}ab + \frac{1}{9}b^2$

56. $4p^2q + pq^2 + 4p^3$

57. $a^2 - ab - 2b^2$

58. $3b^2 - 17ab - 6a^2$

59. $2mn - 360n^2 + m^2$

60. $15 + x^2y^2 + 8xy$

61. $m^2n^2 - 4mn - 32$

62. $p^2q^2 + 7pq + 6$

63. $a^2b^6 + 4ab^5 - 32b^4$

64. $p^5q^2 + 3p^4q - 10p^3$

65. $a^5 + 4a^4b - 5a^3b^2$

66. $2s^6t^2 + 10s^3t^3 + 12t^4$

67. $a^2 - \frac{1}{25}b^2$

68. $p^2 - \frac{1}{49}b^2$

Answers

45. ____________

46. ____________

47. ____________

48. ____________

49. ____________

50. ____________

51. ____________

52. ____________

53. ____________

54. ____________

55. ____________

56. ____________

57. ____________

58. ____________

59. ____________

60. ____________

61. ____________

62. ____________

63. ____________

64. ____________

65. ____________

66. ____________

67. ____________

68. ____________

ANSWERS

69. ________
70. ________
71. ________
72. ________
73. ________
74. ________
75. ________
76. ________
77. ________
78. ________
79. ________
80. ________
81. ________
82. ________
83. ________
84. ________
85. ________
86. ________
87. ________
88. ________
89. ________
90. ________
91. ________
92. ________
93. ________
94. ________
95. ________
96. ________
97. ________
98. ________
99. ________
100. ________

69. $a^4 + a^2bc - 2b^2c^2$

70. $x^6 + x^3y - 2y^2$

71. $x^2 - y^2$

72. $p^2q^2 - r^2$

73. $16 - p^4q^4$

74. $15a^4 - 15b^4$

75. $1 - 16x^{12}y^{12}$

76. $81a^4 - b^4$

77. $q^3 + 8q^2 - q - 8$

78. $m^3 - 7m^2 - 4m + 28$

SKILL MAINTENANCE

79. Divide: $\frac{7}{5} \div \left(-\frac{11}{10}\right)$.

80. Multiply: $(5x - t)^2$.

81. Solve $A = aX + bX - 7$ for X.

82. Solve: $4(x - 9) - 2(x + 7) < 14$.

SYNTHESIS

Factor completely.

83. $a^4 - 2a^2 + 1$

84. $x^4 + 9$

85. $12.25x^2 - 7x + 1$

86. $\frac{1}{5}x^2 - x + \frac{4}{5}$

87. $5x^2 + 13x + 7.2$

88. $x^3 - (x - 3x^2) - 3$

89. $18 + y^3 - 9y - 2y^2$

90. $-(x^4 - 7x^2 - 18)$

91. $a^3 + 4a^2 + a + 4$

92. $x^3 + x^2 - (4x + 4)$

93. $x^4 - 7x^2 - 18$

94. $3x^4 - 15x^2 + 12$

95. $x^3 - x^2 - 4x + 4$

96. $y^2(y + 1) - 4y(y + 1) - 21(y + 1)$

97. $y^2(y - 1) - 2y(y - 1) + (y - 1)$

98. $6(x - 1)^2 + 7y(x - 1) - 3y^2$

99. $(y + 4)^2 + 2x(y + 4) + x^2$

100. $a^4 - 81$

4.7 *Solving Quadratic Equations by Factoring*

OBJECTIVES

After finishing Section 4.7, you should be able to:

a Solve equations (already factored) using the principle of zero products.

b Solve quadratic equations by factoring and then using the principle of zero products.

FOR EXTRA HELP

TAPE 11 TAPE 9B MAC: 4B IBM: 4B

In this section, we introduce a new equation-solving method and use it along with factoring to solve certain equations like $x^2 + x - 156 = 0$.

> A *quadratic equation* is an equation equivalent to an equation of the type
>
> $$ax^2 + bx + c = 0, \quad \text{where } a > 0.$$
>
> The trinomial on the left is of second degree.

a The Principle of Zero Products

The product of two numbers is 0 if one or both of the numbers is 0. Furthermore, *if any product is* 0, *then a factor must be* 0. For example, if $7x = 0$, then we know that $x = 0$. If $x(2x - 9) = 0$, we know that $x = 0$ or $2x - 9 = 0$. If $(x + 3)(x - 2) = 0$, we know that $x + 3 = 0$ or $x - 2 = 0$. In a product such as $ab = 24$, we cannot conclude with certainty that a is 24 or that b is 24, but if $ab = 0$, we can conclude that $a = 0$ or $b = 0$.

EXAMPLE 1 Solve: $(x + 3)(x - 2) = 0$.

We have a product of 0. This equation will be true when either factor is 0. Thus it is true when

$$x + 3 = 0 \quad \text{or} \quad x - 2 = 0.$$

Here we have two simple equations that we know how to solve:

$$x = -3 \quad \text{or} \quad x = 2.$$

Each of the numbers -3 and 2 is a solution of the original equation, as we can see in the following checks.

CHECK: For -3:

$$\begin{array}{r|l} (x+3)(x-2) & = 0 \\ \hline (-3+3)(-3-2) & 0 \\ 0(-5) & \\ 0 & \quad \text{TRUE} \end{array}$$

For 2:

$$\begin{array}{r|l} (x+3)(x-2) & = 0 \\ \hline (2+3)(2-2) & 0 \\ 5(0) & \\ 0 & \quad \text{TRUE} \end{array}$$

We now have a principle to help in solving quadratic equations.

> **THE PRINCIPLE OF ZERO PRODUCTS**
>
> An equation $ab = 0$ is true if and only if $a = 0$ is true or $b = 0$ is true, or both are true. (A product is 0 if and only if one or both of the factors is 0.)

Solve using the principle of zero products.

1. $(x-3)(x+4)=0$

2. $(x-7)(x-3)=0$

3. $(4t+1)(3t-2)=0$

4. Solve: $y(3y-17)=0$.

Answers on page A-4

EXAMPLE 2 Solve: $(5x+1)(x-7)=0$.

$$
\begin{aligned}
(5x+1)(x-7)&=0 \\
5x+1=0 \quad &\text{or} \quad x-7=0 && \text{Using the principle of zero products} \\
5x=-1 \quad &\text{or} \quad x=7 && \text{Solving the two equations separately} \\
x=-\tfrac{1}{5} \quad &\text{or} \quad x=7
\end{aligned}
$$

CHECK:

For $-\frac{1}{5}$:

$(5x+1)(x-7)=0$	
$\left(5\left(-\frac{1}{5}\right)+1\right)\left(-\frac{1}{5}-7\right)$	0
$(-1+1)\left(-7\frac{1}{5}\right)$	
$0\left(-7\frac{1}{5}\right)$	
0	TRUE

For 7:

$(5x+1)(x-7)=0$	
$(5(7)+1)(7-7)$	0
$(35+1)\cdot 0$	
$36\cdot 0$	
0	TRUE

The solutions are $-\frac{1}{5}$ and 7.

When you solve an equation using the principle of zero products, you may wish to check by substitution as in Examples 1 and 2. Such a check will detect errors in solving.

Do Exercises 1–3.

When some factors have only one term, you can still use the principle of zero products.

EXAMPLE 3 Solve: $x(2x-9)=0$.

$$
\begin{aligned}
x(2x-9)&=0 \\
x=0 \quad &\text{or} \quad 2x-9=0 && \text{Using the principle of zero products} \\
x=0 \quad &\text{or} \quad 2x=9 \\
x=0 \quad &\text{or} \quad x=\frac{9}{2}
\end{aligned}
$$

The solutions are 0 and $\frac{9}{2}$. The check is left to the student.

Do Exercise 4.

b Using Factoring to Solve Equations

Using factoring and the principle of zero products, we can solve some new kinds of equations. Thus we have extended our equation-solving abilities.

EXAMPLE 4 Solve: $x^2+5x+6=0$.

Compare this equation to those that we know how to solve from Chapter 2. There are no like terms to collect, and we have a squared term. We first factor the polynomial. Then we use the principle of zero products.

$$
\begin{aligned}
x^2+5x+6&=0 \\
(x+2)(x+3)&=0 && \text{Factoring} \\
x+2=0 \quad &\text{or} \quad x+3=0 && \text{Using the principle of zero products} \\
x=-2 \quad &\text{or} \quad x=-3
\end{aligned}
$$

CHECK: For -2:

$$\begin{array}{r|l} x^2 + 5x + 6 = 0 & \\ \hline (-2)^2 + 5(-2) + 6 & 0 \\ 4 - 10 + 6 & \\ -6 + 6 & \\ 0 & \end{array} \quad \text{TRUE}$$

For -3:

$$\begin{array}{r|l} x^2 + 5x + 6 = 0 & \\ \hline (-3)^2 + 5(-3) + 6 & 0 \\ 9 - 15 + 6 & \\ -6 + 6 & \\ 0 & \end{array} \quad \text{TRUE}$$

The solutions are -2 and -3.

Caution! Keep in mind that you *must* have 0 on one side of the equation before you can use the principle of zero products. Get all nonzero terms on one side and 0 on the other.

Do Exercise 5.

EXAMPLE 5 Solve: $x^2 - 8x = -16$.

We first add 16 to get a 0 on one side:

$$x^2 - 8x = -16$$

$$x^2 - 8x + 16 = 0 \qquad \text{Adding 16}$$

$$(x - 4)(x - 4) = 0 \qquad \text{Factoring}$$

$$x - 4 = 0 \quad \text{or} \quad x - 4 = 0 \qquad \text{Using the principle of zero products}$$

$$x = 4 \quad \text{or} \quad x = 4.$$

There is only one solution, 4. The check is left to the student.

Do Exercises 6 and 7.

EXAMPLE 6 Solve: $x^2 + 5x = 0$.

$$x^2 + 5x = 0$$

$$x(x + 5) = 0 \qquad \text{Factoring out a common factor}$$

$$x = 0 \quad \text{or} \quad x + 5 = 0 \qquad \text{Using the principle of zero products}$$

$$x = 0 \quad \text{or} \quad x = -5$$

The solutions are 0 and -5. The check is left to the student.

EXAMPLE 7 Solve: $4x^2 = 25$.

$$4x^2 = 25$$

$$4x^2 - 25 = 0 \qquad \text{Subtracting 25 on both sides to get 0 on one side}$$

$$(2x - 5)(2x + 5) = 0 \qquad \text{Factoring a difference of squares}$$

$$2x - 5 = 0 \quad \text{or} \quad 2x + 5 = 0$$

$$2x = 5 \quad \text{or} \quad 2x = -5$$

$$x = \frac{5}{2} \quad \text{or} \quad x = -\frac{5}{2}$$

The solutions are $\frac{5}{2}$ and $-\frac{5}{2}$.

Do Exercises 8 and 9.

5. Solve: $x^2 - x - 6 = 0$.

Solve.

6. $x^2 - 3x = 28$

7. $x^2 = 6x - 9$

Solve.

8. $x^2 - 4x = 0$

9. $9x^2 = 16$

Answers on page A-4

10. Solve: $(x + 1)(x - 1) = 8$.

Answer on page A-4

EXAMPLE 8 Solve: $(x + 2)(x - 2) = 5$.

Be careful with an equation like this one! It might be tempting to set each factor equal to 5. Remember: We must have a 0 on one side. We first carry out the product on the left. Then we subtract 5 on both sides to get 0 on one side. Then we proceed with the principle of zero products.

$$(x + 2)(x - 2) = 5$$

$$x^2 - 4 = 5 \quad \text{Multiplying on the left}$$

$$x^2 - 4 - 5 = 5 - 5 \quad \text{Subtracting 5}$$

$$x^2 - 9 = 0 \quad \text{Simplifying}$$

$$(x + 3)(x - 3) = 0 \quad \text{Factoring}$$

$$x + 3 = 0 \quad \text{or} \quad x - 3 = 0 \quad \text{Using the principle of zero products}$$

$$x = -3 \quad \text{or} \quad x = 3$$

CHECK: For -3:

$$\begin{array}{r|l} (x + 2)(x - 2) & 5 \\ \hline (-3 + 2)(-3 - 2) & 5 \\ (-1)(-5) & \\ 5 & \end{array} \quad \text{TRUE}$$

For 3:

$$\begin{array}{r|l} (x + 2)(x - 2) & 5 \\ \hline (3 + 2)(3 - 2) & 5 \\ (5)(1) & \\ 5 & \end{array} \quad \text{TRUE}$$

The solutions are -3 and 3.

Do Exercise 10.

NAME SECTION DATE

Exercise Set 4.7

a Solve using the principle of zero products.

1. $(x + 4)(x + 9) = 0$

2. $(x + 2)(x - 7) = 0$

3. $(x + 3)(x - 8) = 0$

4. $(x + 6)(x - 8) = 0$

5. $(x + 12)(x - 11) = 0$

6. $(x - 13)(x + 53) = 0$

7. $x(x + 3) = 0$

8. $y(y + 5) = 0$

9. $0 = y(y + 18)$

10. $0 = x(x - 19)$

11. $(2x + 5)(x + 4) = 0$

12. $(2x + 9)(x + 8) = 0$

13. $(5x + 1)(4x - 12) = 0$

14. $(4x + 9)(14x - 7) = 0$

15. $(7x - 28)(28x - 7) = 0$

16. $(13x + 14)(6x - 5) = 0$

17. $2x(3x - 2) = 0$

18. $55x(8x - 9) = 0$

19. $\frac{1}{2}x\left(\frac{2}{3}x - 12\right) = 0$

20. $\frac{5}{8}x\left(\frac{3}{5}x - 6\right) = 0$

21. $\left(\frac{1}{5} + 2x\right)\left(\frac{1}{9} - 3x\right) = 0$

22. $\left(\frac{7}{4}x - \frac{1}{16}\right)\left(\frac{2}{3}x - \frac{16}{15}\right) = 0$

23. $(0.3x - 0.1)(0.05x + 1) = 0$

24. $(0.1x + 0.3)(0.4x - 20) = 0$

25. $9x(3x - 2)(2x - 1) = 0$

26. $(x + 5)(x - 75)(5x - 1) = 0$

b Solve by factoring and using the principle of zero products. Remember to check.

27. $x^2 + 6x + 5 = 0$

28. $x^2 + 7x + 6 = 0$

29. $x^2 + 7x - 18 = 0$

30. $x^2 + 4x - 21 = 0$

31. $x^2 - 8x + 15 = 0$

32. $x^2 - 9x + 14 = 0$

33. $x^2 - 8x = 0$

34. $x^2 - 3x = 0$

35. $x^2 + 18x = 0$

36. $x^2 + 16x = 0$

37. $x^2 = 16$

38. $100 = x^2$

Answers

1.
2.
3.
4.
5.
6.
7.
8.
9.
10.
11.
12.
13.
14.
15.
16.
17.
18.
19.
20.
21.
22.
23.
24.
25.
26.
27.
28.
29.
30.
31.
32.
33.
34.
35.
36.
37.
38.

ANSWERS

39. ______
40. ______
41. ______
42. ______
43. ______
44. ______
45. ______
46. ______
47. ______
48. ______
49. ______
50. ______
51. ______
52. ______
53. ______
54. ______
55. ______
56. ______
57. ______
58. ______
59. ______
60. ______
61. ______
62. ______
63. ______
64. ______
65. ______
66. ______
67. ______
68. ______
69. ______
70. ______
71. ______
72. ______
73. a) ______
b) ______
c) ______
d) ______
e) ______
74. a) ______
b) ______

39. $9x^2 - 4 = 0$ **40.** $4x^2 - 9 = 0$ **41.** $0 = 6x + x^2 + 9$

42. $0 = 25 + x^2 + 10x$ **43.** $x^2 + 16 = 8x$ **44.** $1 + x^2 = 2x$

45. $5x^2 = 6x$ **46.** $7x^2 = 8x$ **47.** $6x^2 - 4x = 10$

48. $3x^2 - 7x = 20$ **49.** $12y^2 - 5y = 2$ **50.** $2y^2 + 12y = -10$

51. $t(3t + 1) = 2$ **52.** $x(x - 5) = 14$ **53.** $100y^2 = 49$

54. $64a^2 = 81$ **55.** $x^2 - 5x = 18 + 2x$ **56.** $3x^2 + 8x = 9 + 2x$

57. $10x^2 - 23x + 12 = 0$ **58.** $12x^2 + 17x - 5 = 0$

SKILL MAINTENANCE

Translate to an algebraic expression.

59. The square of the sum of a and b

60. The sum of the squares of a and b

Divide.

61. $144 \div (-9)$ **62.** $-24.3 \div 5.4$ **63.** $-\frac{5}{8} \div \frac{3}{16}$ **64.** $-\frac{3}{16} \div \left(-\frac{5}{8}\right)$

SYNTHESIS

Solve.

65. $b(b + 9) = 4(5 + 2b)$ **66.** $y(y + 8) = 16(y - 1)$ **67.** $(t - 3)^2 = 36$

68. $(t - 5)^2 = 2(5 - t)$ **69.** $x^2 - \frac{1}{64} = 0$ **70.** $x^2 - \frac{25}{36} = 0$

71. $\frac{5}{16}x^2 = 5$ **72.** $\frac{27}{25}x^2 = \frac{1}{3}$

73. Find an equation that has the given numbers as solutions. For example, 3 and -2 are solutions to $x^2 - x - 6 = 0$.

a) $-3, 4$ b) $-3, -4$ c) $\frac{1}{2}, \frac{1}{2}$ d) $5, -5$ e) $0, 0.1, \frac{1}{4}$

74. Check the numbers found below. What's wrong with the methods used? Try to find the solutions.

a) $(x - 3)(x + 4) = 0$
$x = -3$ or $x = 4$

b) $(x - 3)(x + 4) = 8$
$x - 3 = 8$ or $x + 4 = 8$
$x = 11$ or $x = 4$

4.8 *Solving Problems*

a We can now use our new method for solving quadratic equations and the five steps for solving problems.

EXAMPLE 1 One more than a number times one less than the number is 8. Find all such numbers.

1. ***Familiarize.*** Let us make a guess. Try 5. One more than 5 is 6. One less than the number is 4. The product of one more than the number and one less than the number is 6(4), or 24, which is too large. We could continue to guess, but let us use our algebraic skills to find the numbers. Let $x =$ the number (there could be more than one).
2. ***Translate.*** From the familiarization, we can translate as follows:

One more than a number	times	One less than that number	is	8.
↓	↓	↓	↓	↓
$(x + 1)$	$\cdot$	$(x - 1)$	$=$	8

3. ***Solve.*** We solve the equation as follows:

$$(x + 1)(x - 1) = 8$$

$$x^2 - 1 = 8 \qquad \text{Multiplying}$$

$$x^2 - 1 - 8 = 8 - 8 \qquad \text{Subtracting 8 on both sides to get 0 on one side}$$

$$x^2 - 9 = 0 \qquad \text{Simplifying}$$

$$(x - 3)(x + 3) = 0 \qquad \text{Factoring}$$

$$x - 3 = 0 \quad \text{or} \quad x + 3 = 0 \qquad \text{Using the principle of zero products}$$

$$x = 3 \quad \text{or} \quad x = -3.$$

4. ***Check.*** One more than 3 (this is 4) times one less than 3 (this is 2) is 8. Thus, 3 checks. One more than -3 (this is -2) times one less than -3 (this is -4) is 8. Thus, -3 also checks.
5. ***State.*** There are two such numbers, 3 and -3.

Do Exercises 1 and 2.

EXAMPLE 2 The square of a number minus twice the number is 48. Find all such numbers.

1. ***Familiarize.*** Let us make a guess to help understand the problem and ease the translation. Try 6. The square of 6 is 36, and twice the number is 12. Then $36 - 12 = 24$, so 6 is not a number we want. We find the numbers using our algebraic skills. Let $x =$ the number, or numbers.
2. ***Translate.*** We translate as follows:

The square of a number	minus	Twice the number	is	48.
↓	↓	↓	↓	↓
x^2	$-$	$2x$	$=$	48

OBJECTIVE

After finishing Section 4.8, you should be able to:

a Solve problems involving quadratic equations that can be solved by factoring.

FOR EXTRA HELP

TAPE 12 TAPE 10A MAC: 4B IBM: 4B

1. One more than a number times one less than the number is 24. Find all such numbers.

2. Seven less than a number times eight less than the number is 0. Find all such numbers.

Answers on page A-4

3. The square of a number minus the number is 20. Find all such numbers.

4. The width of a rectangle is 2 cm less than the length. The area is 15 cm^2. Find the length and the width.

Answers on page A-4

3. ***Solve.*** We solve the equation as follows:

$$x^2 - 2x = 48$$
$$x^2 - 2x - 48 = 48 - 48 \quad \text{Subtracting 48 to get 0 on one side}$$
$$x^2 - 2x - 48 = 0 \quad \text{Simplifying}$$
$$(x - 8)(x + 6) = 0 \quad \text{Factoring}$$
$$x - 8 = 0 \quad \text{or} \quad x + 6 = 0 \quad \text{Using the principle of zero products}$$
$$x = 8 \quad \text{or} \quad x = -6.$$

4. ***Check.*** The square of 8 is 64, and twice the number 8 is 16. Then 64 − 16 is 48, so 8 checks. The square of −6 is $(-6)^2$, or 36, and twice −6 is −12. Then 36 − (−12) is 48, so −6 checks.

5. ***State.*** There are two such numbers, 8 and −6.

Do Exercise 3.

EXAMPLE 3 The height of a triangular foresail on a racing yacht is 7 ft more than the base. The area of the triangle is 30 ft^2. Find the height and the base.

1. ***Familiarize.*** We first make a drawing. If you don't remember the formula for the area of a triangle, look it up on the inside back cover of this book or in a geometry book. The area is $\frac{1}{2}$(base)(height).

   We let b = the base of the triangle. Then $b + 7$ = the height.

2. ***Translate.*** It helps to reword this problem before translating:

$\frac{1}{2}$	times	Base	times	Base plus 7	is	30.	Rewording
↓	↓	↓	↓	↓	↓	↓	
$\frac{1}{2}$	$\cdot$	b	$\cdot$	$(b + 7)$	$=$	30	Translating

3. ***Solve.*** We solve the equation as follows:

$$\frac{1}{2} \cdot b \cdot (b + 7) = 30$$
$$\frac{1}{2}(b^2 + 7b) = 30 \quad \text{Multiplying}$$
$$2 \cdot \frac{1}{2}(b^2 + 7b) = 2 \cdot 30 \quad \text{Multiplying by 2}$$
$$b^2 + 7b = 60 \quad \text{Simplifying}$$
$$b^2 + 7b - 60 = 60 - 60 \quad \text{Subtracting 60 to get 0 on one side}$$
$$b^2 + 7b - 60 = 0$$
$$(b + 12)(b - 5) = 0 \quad \text{Factoring}$$
$$b + 12 = 0 \quad \text{or} \quad b - 5 = 0 \quad \text{Using the principle of zero products}$$
$$b = -12 \quad \text{or} \quad b = 5.$$

4. ***Check.*** The base of a triangle cannot have a negative length, so −12 cannot be a solution. Suppose the base is 5 ft. Then the height is 7 ft more than the base, so the height is 12 ft and the area is $\frac{1}{2}(5)(12)$, or 30 ft^2. These numbers check in the original problem.

5. ***State.*** The height is 12 ft and the base is 5 ft.

Do Exercise 4.

EXAMPLE 4 In a sports league of n teams in which each team plays every other team twice, the total number N of games to be played is given by

$$n^2 - n = N.$$

If a basketball league plays a total of 240 games, how many teams are in the league?

1., 2. ***Familiarize*** and ***Translate.*** We are given that $n =$ the number of teams in a league and $N =$ the number of games. To familiarize yourself with this problem, reread Example 4 in Section 3.3 where we first considered it. To find the number of teams n in a league in which 240 games are played, we substitute 240 for N in the equation:

$$n^2 - n = 240. \quad \text{Substituting 240 for } N$$

3. ***Solve.*** We solve the equation as follows:

$$\begin{aligned} n^2 - n &= 240 \\ n^2 - n - 240 &= 240 - 240 && \text{Subtracting 240 to get 0 on one side} \\ n^2 - n - 240 &= 0 \\ (n - 16)(n + 15) &= 0 && \text{Factoring} \\ n - 16 = 0 \quad &\text{or} \quad n + 15 = 0 && \text{Using the principle of zero products} \\ n = 16 \quad &\text{or} \quad n = -15. \end{aligned}$$

4. ***Check.*** The solutions of the equation are 16 and -15. Since the number of teams cannot be negative, -15 cannot be a solution. But 16 checks, since $16^2 - 16 = 256 - 16 = 240$.
5. ***State.*** There are 16 teams in the league.

Do Exercise 5.

EXAMPLE 5 The product of the page numbers on two consecutive pages in a book is 156. Find the page numbers.

1. ***Familiarize.*** The page numbers are consecutive integers. Recall that consecutive integers are next to each other, such as 49 and 50, or -6 and -5. Let $x =$ the smaller integer; then $x + 1 =$ the larger integer.
2. ***Translate.*** It helps to reword the problem before translating:

First integer	times	Second integer	is	156.	Rewording
↓	↓	↓	↓	↓	
x	$\cdot$	$(x + 1)$	$=$	156	Translating

3. ***Solve.*** We solve the equation as follows:

$$\begin{aligned} x(x + 1) &= 156 \\ x^2 + x &= 156 && \text{Multiplying} \\ x^2 + x - 156 &= 156 - 156 && \text{Subtracting 156 to get 0 on one side} \\ x^2 + x - 156 &= 0 && \text{Simplifying} \\ (x - 12)(x + 13) &= 0 && \text{Factoring} \\ x - 12 = 0 \quad &\text{or} \quad x + 13 = 0 && \text{Using the principle of zero products} \\ x = 12 \quad &\text{or} \quad x = -13. \end{aligned}$$

4. ***Check.*** The solutions of the equation are 12 and -13. When x is 12, then $x + 1$ is 13, and $12 \cdot 13 = 156$. The numbers 12 and 13 are consecutive integers that are solutions to the problem. When x is -13, then $x + 1$ is -12, and $(-13)(-12) = 156$. The numbers -13 and -12

5. Use $N = n^2 - n$ for the following.

a) A women's volleyball league has 19 teams. What is the total number of games to be played?

b) A slow-pitch softball league plays a total of 72 games. How many teams are in the league?

Answers on page A-4

6. The product of the page numbers on two facing pages of a book is 506. Find the page numbers.

7. The length of one leg of a right triangle is 1 m longer than the other. The length of the hypotenuse is 5 m. Find the lengths of the legs.

Answers on page A-4

are also consecutive integers, but they are not solutions of the problem because negative numbers are not used as page numbers.

5. ***State.*** The page numbers are 12 and 13.

Do Exercise 6.

The following problem involves the Pythagorean theorem, which relates the lengths of the sides of a right triangle. A **right triangle** has a 90° angle. The side opposite the 90° angle is called the **hypotenuse**. The other sides are called **legs**.

THE PYTHAGOREAN THEOREM

The sum of the squares of the legs of a right triangle is equal to the square of the hypotenuse:

$$a^2 + b^2 = c^2.$$

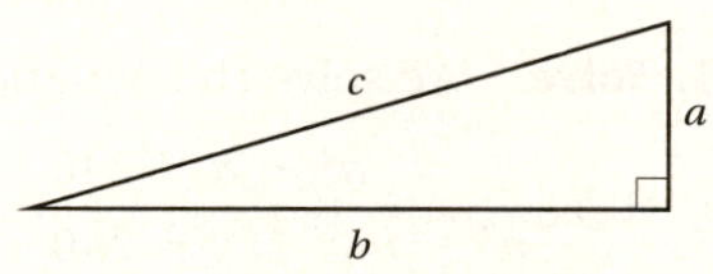

EXAMPLE 6 The length of one leg of a right triangle is 7 ft longer than the other. The length of the hypotenuse is 13 ft. Find the lengths of the legs.

1. ***Familiarize.*** We first make a drawing. We let

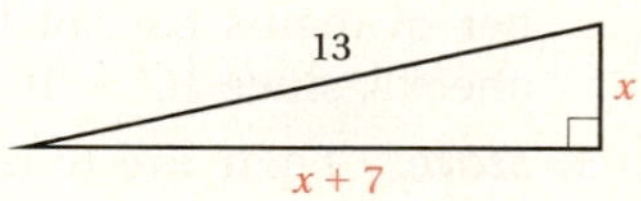

x = the length of one leg.

Since the other leg is 7 ft longer, we know that

$x + 7$ = the length of the other leg.

The hypotenuse has length 13 ft.

2. ***Translate.*** Applying the Pythagorean theorem, we obtain the following translation:

$$a^2 + b^2 = c^2$$
$$x^2 + (x + 7)^2 = 13^2.$$

3. ***Solve.*** We solve the equation as follows:

$x^2 + (x^2 + 14x + 49) = 169$ — Squaring the binomial and 13

$2x^2 + 14x + 49 = 169$ — Collecting like terms

$2x^2 + 14x + 49 - 169 = 169 - 169$ — Subtracting 169 to get 0 on one side

$2x^2 + 14x - 120 = 0$ — Simplifying

$2(x^2 + 7x - 60) = 0$ — Factoring out a common factor

$x^2 + 7x - 60 = 0$ — Dividing by 2 on both sides, or multiplying by $\frac{1}{2}$

$(x + 12)(x - 5) = 0$ — Factoring

$x + 12 = 0$ or $x - 5 = 0$

$x = -12$ or $x = 5$.

4. ***Check.*** The integer -12 cannot be a length of a side because it is negative. When $x = 5$, $x + 7 = 12$, and $5^2 + 12^2 = 13^2$. So 5 and 12 check.

5. ***State.*** The lengths of the legs are 5 ft and 12 ft.

Do Exercise 7.

NAME SECTION DATE

Exercise Set 4.8

a Solve.

1. If 7 is added to the square of a number, the result is 32. Find all such numbers.

2. If you subtract a number from four times its square, the result is 3. Find all such numbers.

3. Fifteen more than the square of a number is eight times the number. Find all such numbers.

4. Eight more than the square of a number is six times the number. Find all such numbers.

5. The product of the page numbers on two facing pages of a book is 210. Find the page numbers.

6. The product of the page numbers on two facing pages of a book is 420. Find the page numbers.

7. The product of two consecutive even integers is 168. Find the integers.

8. The product of two consecutive even integers is 224. Find the integers.

9. The product of two consecutive odd integers is 255. Find the integers.

10. The product of two consecutive odd integers is 143. Find the integers.

ANSWERS

1. ____
2. ____
3. ____
4. ____
5. ____
6. ____
7. ____
8. ____
9. ____
10. ____

ANSWERS

11. ______

12. ______

13. ______

14. ______

15. ______

16. ______

17. ______

18. ______

11. The length of a rectangular calculator is 5 cm greater than the width. The area of the calculator is 84 cm^2. Find the length and the width.

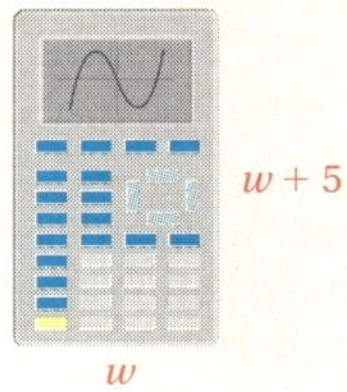

12. The length of a rectangular garden is 4 m greater than the width. The area of the garden is 96 m^2. Find the length and the width.

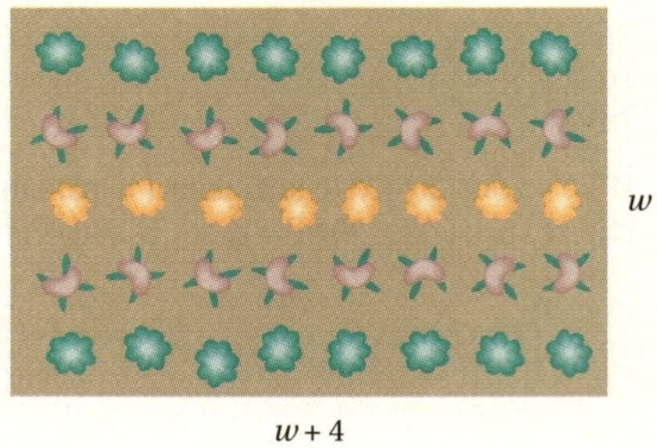

13. The area of a square bookcase is 5 more than the perimeter. Find the length of a side.

14. The perimeter of a square porch is 3 more than the area. Find the length of a side.

15. Sharks' teeth are shaped like triangles. The height of a tooth of a great white shark is 1 cm longer than the base. The area is 15 cm^2. Find the height and the base.

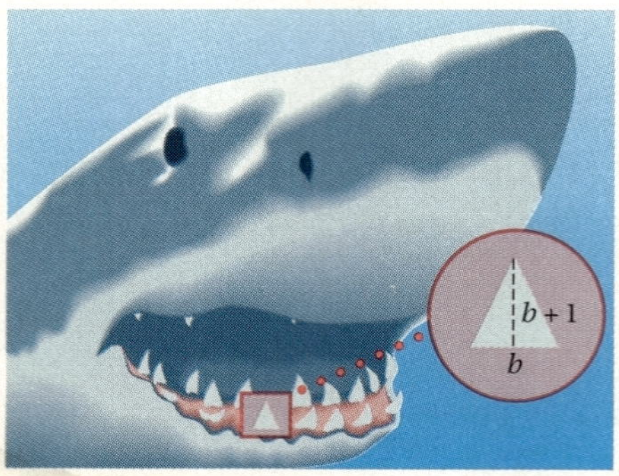

16. The base of a triangle is 10 cm greater than the height. The area is 28 cm^2. Find the height and the base.

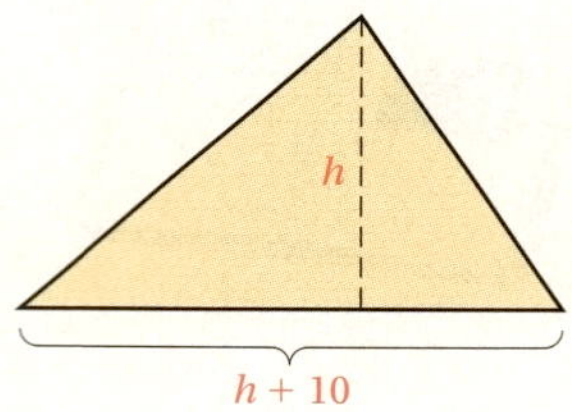

17. If the sides of a square are lengthened by 3 km, the area becomes 81 km^2. Find the length of a side of the original square.

18. The base and the height of a triangle are the same length. If the length of the base is increased by 4 in., the area becomes 96 in^2. Find the length of the base of the original triangle.

19. The sum of the squares of two consecutive odd positive integers is 74. Find the integers.

20. The sum of the squares of two consecutive odd positive integers is 130. Find the integers.

Use $n^2 - n = N$ for Exercises 21–24.

21. A chess league has 14 teams. What is the total number of games to be played?

22. A women's volleyball league has 23 teams. What is the total number of games to be played?

23. A slow-pitch softball league plays a total of 132 games. How many teams are in the league?

24. A basketball league plays a total of 90 games. How many teams are in the league?

A researcher wants to investigate the potential spread of germs by contact. She knows that the number of possible handshakes within a group of n people is given by $N = \frac{1}{2}(n^2 - n)$.

25. There are 100 people at a party. How many handshakes are possible?

26. There are 40 people at a meeting. How many handshakes are possible?

27. Everyone at a meeting shook hands. There were 300 handshakes in all. How many people were at the meeting?

28. Everyone at a party shook hands. There were 190 handshakes in all. How many people were at the party?

ANSWERS

19. ______

20. ______

21. ______

22. ______

23. ______

24. ______

25. ______

26. ______

27. ______

28. ______

ANSWERS

29. ______

30. ______

31. ______

32. ______

33. ______

34. ______

35. ______

36. ______

29. The length of one leg of a right triangle is 8 ft. The length of the hypotenuse is 2 ft longer than the other leg. Find the length of the hypotenuse and the other leg.

30. The length of one leg of a right triangle is 24 ft. The length of the other leg is 16 ft shorter than the hypotenuse. Find the length of the hypotenuse and the other leg.

SYNTHESIS

31. A cement walk of constant width is built around a 20-ft by 40-ft rectangular pool. The total area of the pool and the walk is 1500 ft^2. Find the width of the walk.

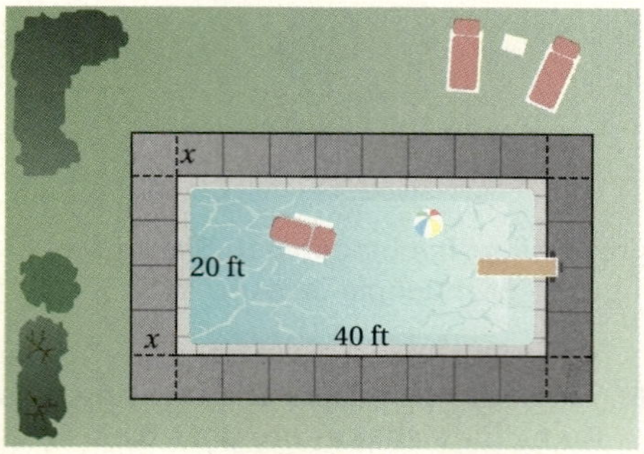

32. An open rectangular gutter is made by turning up the sides of a piece of metal 20 in. wide. The area of the cross-section of the gutter is 50 in^2. Find the depth of the gutter.

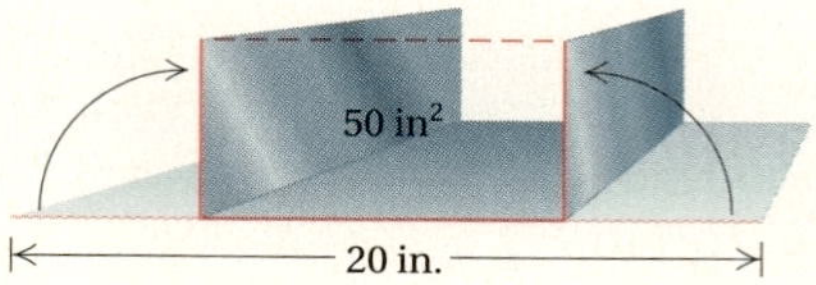

33. The ones digit of a number less than 100 is 4 greater than the tens digit. The sum of the number and the product of the digits is 58. Find the number.

34. The total surface area of a closed box is 350 m^2. The box is 9 m high and has a square base and lid. Find the length of the side of the base.

35. A rectangular piece of cardboard is twice as long as it is wide. A 4-cm square is cut out of each corner, and the sides are turned up to make a box with an open top. The volume of the box is 616 cm^3. Find the original dimensions of the cardboard.

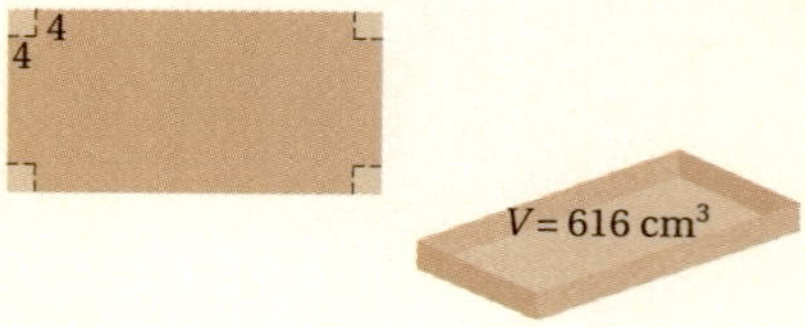

36. The length of each side of a square is increased by 5 cm to form a new square. The area of the new square is $2\frac{1}{4}$ times the area of the original square. Find the area of each square.

CRITICAL THINKING

CALCULATOR CONNECTION

Solve using a calculator as much as possible.

1. $(0.00005x + 0.1)(0.0097x + 0.5) = 0$
2. $x^2 - 0.3x - 0.1 = 0$
3. $x^2 - 0.9x - 3.22 = 0$
4. Evaluate 3^2, 33^2, 333^2, and 3333^2 using a calculator. Look for a pattern and find $33{,}333^2$ without using a calculator.

EXTENDED SYNTHESIS EXERCISES

1. Find a polynomial with two factors, one of which is $x^2 - 2$.
2. Find a fourth-degree polynomial with three factors, one of which is $x^2 - 3$.
3. Suppose that $ax^2 + 12x + c$ is a trinomial square. What values can a and c have if they are both integers and $a > 0$?
4. Find all the second-degree polynomials with integer coefficients that have x^2 as a first term and -16 as a last term, that are factorable, and whose factors have integer coefficients.
5. Determine which integers between 1 and 10, inclusive, can be expressed as a difference of squares of whole numbers. Then factor the differences.
6. Factor: $3(x + 1)^{n+1}(x + 3)^2 - 5(x + 1)^n(x + 3)^3$.

Factor part or parts of each polynomial as the square of a binomial. Then factor the polynomial as a difference of two squares.

7. $x^2 + 2x + 1 - 9$
8. $y^2 - 6y + 9 - x^2 + 8x - 16$
9. $a^2 + 2ab + b^2 - 1$
10. $4 - a^2 + 6ab - 9b^2$
11. Find a common binomial factor of the polynomials $x^2 + 7x + 10$ and $x^4 - 3x^2 - 4$.
12. Find two consecutive positive integers such that the product of the sum and the difference of the integers plus 8 is the sum of their squares.
13. The sum of the two factors of $x^2 - 28x + 171$ is subtracted from the sum of the two factors of $x^2 - 27x + 170$. What is the result?
14. If $ab = 2$ and $(a + b)^2 = 10$, then find the value of $a^2 + b^2$.
15. If $ab = 7$ and $(a - b)^2 = 38$, then find the value of $a^2 + b^2$.
16. If $x + y = a$, $x - y = 1/a$, and $a \neq 0$, find the value of $x^2 - y^2$.
17. If $a - 2 = 4$ and $ab - 2b + 7a - 14 = 12$, then find the value of $b + 7$.
18. *Pythagorean triples* are sets of three positive integers that can be the lengths of the sides of a right triangle. One example of such a set is 3, 4, and 5, because $3^2 + 4^2 = 5^2$. Find two more sets of Pythagorean triples.
19. A model rocket is launched with an initial velocity of 180 ft/sec. Its height h, in feet, after t seconds is given by the formula $h = 180t - 16t^2$.

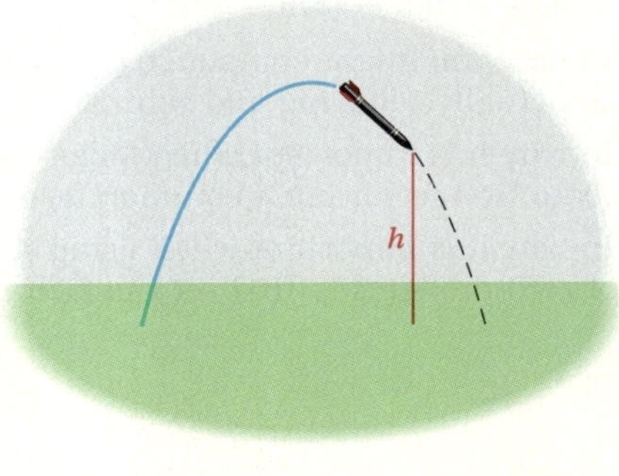

a) After how many seconds will the rocket first reach a height of 464 ft?

b) After how many seconds will the rocket be at that height again?

(*continued*)

CRITICAL THINKING

EXERCISES FOR THINKING AND WRITING

1. Compare the type of equations we are able to solve after studying this chapter with those we studied earlier.
2. Describe a procedure that can always be used to check the result of factoring a polynomial, and explain why it works.
3. Suppose we know that $(x - 5)(x + 2)$ would be a correct factorization of a trinomial $x^2 + bx + c$ except the sign of the middle term is incorrect. What is a correct factorization?

Explain the error, if any, in each of the following when the direction is "Factor completely."

4. $x^2 + 9 = (x + 3)^2$
5. $x^2 - 6x + 8 = (x + 4)(x - 2)$
6. $p^2 - 9 = (p - 3)^2$
7. $a^2 + 6a - 9 = (a - 3)^2$
8. $16a^4 - 81 = (4a^2 - 9)(4a^2 + 9)$
9. $16m^2 - 80m + 100 = (4m - 10)^2$
10. Explain how you might compute $0.87^2 - 0.86^2$ without actually squaring and subtracting.
11. Explain why it is important when factoring polynomials to first factor out a common factor.
12. Explain why it is important when using the principle of zero products to first set the factors equal to 0.

Summary and Review: Chapter 4

Important Properties and Formulas

Factoring Formulas: $A^2 - B^2 = (A + B)(A - B)$,
$A^2 + 2AB + B^2 = (A + B)^2$,
$A^2 - 2AB + B^2 = (A - B)^2$

The Principle of Zero Products: **An equation $ab = 0$ is true if and only if $a = 0$ is true or $b = 0$ is true, or both are true.**

Review Exercises

The review objectives to be tested in addition to the material in this chapter are [1.6a, c], [2.6a], [2.7e], and [3.6d].

Find three factorizations of the monomial.

1. $-10x^2$

2. $36x^5$

Factor completely.

3. $5 - 20x^6$

4. $x^2 - 3x$

5. $9x^2 - 4$

6. $x^2 + 4x - 12$

7. $x^2 + 14x + 49$

8. $6x^3 + 12x^2 + 3x$

9. $x^3 + x^2 + 3x + 3$

10. $6x^2 - 5x + 1$

11. $x^4 - 81$

12. $9x^3 + 12x^2 - 45x$

13. $2x^2 - 50$

14. $x^4 + 4x^3 - 2x - 8$

15. $16x^4 - 1$

16. $8x^6 - 32x^5 + 4x^4$

17. $75 + 12x^2 + 60x$

18. $x^2 + 9$

19. $x^3 - x^2 - 30x$

20. $4x^2 - 25$

21. $9x^2 + 25 - 30x$

22. $6x^2 - 28x - 48$

23. $x^2 - 6x + 9$

24. $2x^2 - 7x - 4$

25. $18x^2 - 12x + 2$

26. $3x^2 - 27$

27. $15 - 8x + x^2$

28. $25x^2 - 20x + 4$

29. $49b^{10} + 4a^8 - 28a^4b^5$

30. $x^2y^2 + xy - 12$

31. $12a^2 + 84ab + 147b^2$

32. $m^2 + 5m + mt + 5t$

33. $32x^4 - 128y^4z^4$

Solve.

34. $(x - 1)(x + 3) = 0$

35. $x^2 + 2x - 35 = 0$

36. $x^2 + x - 12 = 0$

37. $3x^2 + 2 = 5x$

38. $2x^2 + 5x = 12$

39. $16 = x(x - 6)$

Solve.

40. The square of a number is 6 more than the number. Find all such numbers.

41. The product of two consecutive even integers is 288. Find the integers.

42. The product of two consecutive odd integers is 323. Find the integers.

43. Twice the square of a number is 10 more than the number. Find all such numbers.

44. The width of a rectangle is 7 ft less than the length. The area of the rectangle is 228 ft^2. Find the length and the width.

45. The height of a triangle is 6 more than the base. The area is 216 m^2. Find the base and the height.

SKILL MAINTENANCE

46. Divide: $-\frac{12}{25} \div \left(-\frac{21}{10}\right)$.

47. Solve: $20 - (3x + 2) \geq 2(x + 5) + x$.

48. Solve $A = a + 2b$ for b.

49. Multiply: $(2a + 3)^2$.

50. Multiply: $(2a - 3)(2a + 3)$.

51. Multiply: $(2a - 3)(5a + 7)$.

SYNTHESIS

Solve.

52. The pages of a book measure 15 cm by 20 cm. Margins of equal width surround the printing on each page and constitute one-half of the area of the page. Find the width of the margins.

53. The cube of a number is the same as twice the square of the number. Find all such numbers.

54. The length of a rectangle is two times its width. When the length is increased by 20 and the width decreased by 1, the area is 160. Find the original length and width.

Solve.

55. $x^2 + 25 = 0$

56. $(x - 2)(x + 3)(2x - 5) = 0$

57. For each equation on the left, find an equivalent equation on the right.

a) $3x^2 - 4x + 8 = 0$
b) $(x - 6)(x + 3) = 0$
c) $x^2 + 2x + 9 = 0$
d) $(2x - 5)(x + 4) = 0$
e) $5x^2 - 5 = 0$
f) $x^2 + 10x - 2 = 0$

g) $4x^2 + 8x + 36 = 0$
h) $(2x + 8)(2x - 5) = 0$
i) $9x^2 - 12x + 24 = 0$
j) $(x + 1)(5x - 5) = 0$
k) $x^2 - 3x - 18 = 0$
l) $2x^2 + 20x - 4 = 0$

NAME SECTION DATE

Test: Chapter 4

1. Find three factorizations of $4x^3$.

Factor completely.

2. $x^2 - 7x + 10$

3. $x^2 + 25 - 10x$

4. $6y^2 - 8y^3 + 4y^4$

5. $x^3 + x^2 + 2x + 2$

6. $x^2 - 5x$

7. $x^3 + 2x^2 - 3x$

8. $28x - 48 + 10x^2$

9. $4x^2 - 9$

10. $x^2 - x - 12$

11. $6m^3 + 9m^2 + 3m$

12. $3w^2 - 75$

13. $60x + 45x^2 + 20$

$45x^2 + 60x + 20$

$(15x + 10)(3x + 2)$

14. $3x^4 - 48$

15. $49x^2 - 84x + 36$

$(7x - 6)^2$

16. $5x^2 - 26x + 5$

$(5x - 1)(x - 5)$

Answers

1. ______
2. ______
3. ______
4. ______
5. ______
6. ______
7. ______
8. ______
9. ______
10. ______
11. ______
12. ______
13. ______
14. ______
15. ______
16. ______

ANSWERS

17. ____
18. ____
19. ____
20. ____
21. ____
22. ____
23. ____
24. ____
25. ____
26. ____
27. ____
28. ____
29. ____
30. ____
31. ____
32. ____
33. ____
34. ____

17. $x^4 + 2x^3 - 3x - 6$

18. $80 - 5x^4$

19. $4x^2 - 4x - 15$

20. $6t^3 + 9t^2 - 15t$

21. $3m^2 - 9mn - 30n^2$

Solve.

22. $x^2 - x - 20 = 0$

23. $2x^2 + 7x = 15$

24. $x(x - 3) = 28$

Solve.

25. The square of a number is 24 more than five times the number. Find all such numbers.

26. The length of a rectangle is 6 m more than the width. The area of the rectangle is 40 m^2. Find the length and the width.

SKILL MAINTENANCE

27. Divide: $\frac{5}{8} \div \left(-\frac{11}{16}\right)$.

28. Solve: $10(x - 3) < 4(x + 2)$.

29. Solve $I = PRT$ for T.

30. Multiply: $(5x^2 - 7)^2$.

SYNTHESIS

31. The length of a rectangle is five times its width. When the length is decreased by 3 and the width is increased by 2, the area of the new rectangle is 60. Find the original length and width.

32. Factor: $(a + 3)^2 - 2(a + 3) - 35$.

33. If $x^2 - 4 = (14)(18)$, then one possibility for x is which of the following?

a) 12 b) 14
c) 16 d) 18

34. If $x + y = 4$ and $x - y = 6$, then $x^2 - y^2 = ?$

a) 2 b) 10
c) 34 d) 24

Cumulative Review: Chapters 1–4

Use either < or > for $\blacksquare$ to write a true sentence.

1. $\frac{2}{3} \blacksquare \frac{5}{7}$

2. $-\frac{4}{7} \blacksquare -\frac{8}{11}$

Compute and simplify.

3. $2.06 + (-4.79) - (-3.08)$

4. $5.652 \div (-3.6)$

5. $\left(\frac{2}{9}\right)\left(-\frac{3}{8}\right)\left(\frac{6}{7}\right)$

6. $\frac{21}{5} \div \left(-\frac{7}{2}\right)$

Simplify.

7. $[3x + 2(x - 1)] - [2x - (x + 3)]$

8. $1 - [14 + 28 \div 7 - (6 + 9 \div 3)]$

9. $(2x^2y^{-1})^3$

10. $\frac{3x^5}{4x^3} \cdot \frac{-2x^{-3}}{9x^2}$

11. Add:

$(2x^2 - 3x^3 + x - 4) + (x^4 - x - 5x^2)$.

12. Subtract:

$(2x^2y^2 + xy - 2xy^2) - (2xy - 2xy^2 + x^2y)$.

13. Divide: $(x^3 + 2x^2 - x + 1) \div (x - 1)$.

Multiply.

14. $(2t - 3)^2$

15. $(x^2 - 3)(x^2 + 3)$

16. $(2x + 4)(3x - 4)$

17. $2x(x^3 + 3x^2 + 4x)$

18. $(2y - 1)(2y^2 + 3y + 4)$

19. $\left(x + \frac{2}{3}\right)\left(x - \frac{2}{3}\right)$

Factor.

20. $x^2 + 2x - 8$

21. $4x^2 - 25$

22. $3x^3 - 4x^2 + 3x - 4$

23. $x^2 - 26x + 169$

24. $75x^2 - 108y^2$

25. $6x^2 - 13x - 63$

26. $x^4 - 2x^2 - 3$

27. $4y^3 - 6y^2 - 4y + 6$

28. $6p^2 + pq - q^2$

29. $10x^3 + 52x^2 + 10x$

30. $49x^3 - 42x^2 + 9x$

31. $3x^2 + 5x - 4$

32. $75x^3 + 27x$

33. $3x^8 - 48y^8$

34. $14x^2 + 28 + 42x$

35. $x^5 - x^3 + x^2 - 1$

Solve.

36. $3x - 5 = 2x + 10$

37. $3y + 4 > 5y - 8$

38. $(x - 15)\left(x + \frac{1}{4}\right) = 0$

39. $-98x(x + 37) = 0$

40. $x^3 + x^2 = 25x + 25$

41. $2x^2 = 72$

42. $9x^2 + 1 = 6x$

43. $x^2 + 17x + 70 = 0$

44. $14y^2 = 21y$

45. $1.6 - 3.5x = 0.9$

46. $(x + 3)(x - 4) = 8$

47. $1.5x - 3.6 \leq 1.3x + 0.4$

48. $2x - [3x - (2x + 3)] = 3x + [4 - (2x + 1)]$

49. $y = mx + b$, for m

Solve.

50. The sum of two consecutive even integers is 102. Find the integers.

51. The product of two consecutive even integers is 360. Find the integers.

52. The length of a rectangular window is 3 ft longer than the height. The area of the window is 18 ft^2. Find the length and the height.

53. The length of a rectangular lot is 200 m longer than the width. The perimeter of the lot is 1000 m. Find the dimensions of the lot.

54. Money is borrowed at 12% simple interest. After 1 year, $7280 pays off the loan. How much was originally borrowed?

55. The length of one leg of a right triangle is 15 m. The length of the other leg is 9 m shorter than the length of the hypotenuse. Find the length of the hypotenuse.

56. A 100-m wire is cut into three pieces. The second piece is twice as long as the first piece. The third piece is one-third as long as the first piece. How long is each piece?

57. After a 25% price reduction, a pair of shoes is on sale for $21.75. What was the price before reduction?

58. The height of a triangle is 2 cm more than the base. The area of the triangle is 144 cm^2. Find the height and the base.

SYNTHESIS

Solve.

59. $(x + 3)(x - 5) \leq (x + 2)(x - 1)$

60. $\frac{x - 3}{2} - \frac{2x + 5}{26} = \frac{4x + 11}{13}$

61. $(x + 1)^2 = 25$

Factor.

62. $x^2(x - 3) - x(x - 3) - 2(x - 3)$

63. $4a^2 - 4a + 1 - 9b^2 - 24b - 16$

Solve.

64. Find c so that the polynomial will be the square of a binomial: $cx^2 - 40x + 16$.

65. The length of the radius of a circle is increased by 2 cm to form a new circle. The area of the new circle is four times the area of the original circle. Find the length of the radius of the original circle.

5

Rational Expressions and Equations

INTRODUCTION

In this chapter, we learn to manipulate rational expressions. We learn how to simplify rational expressions as well as to add, subtract, multiply, and divide them. Then we use these skills to solve equations, formulas, and problems.

AN APPLICATION

A cheetah can run 35 km/h faster than a greyhound. If a cheetah can run 19 km in the same time that a greyhound can run 12 km, find each of their speeds.

THE MATHEMATICS

We let $r =$ the speed of the greyhound. The problem then translates to this equation:

These are *rational expressions.*

$$\frac{12}{r} = \frac{19}{r + 35}.$$

This is a *rational equation.*

Objectives for Review

The review objectives to be tested in addition to the material in this chapter are as follows.

[3.2a, b] Use the power rule to raise powers to powers, and raise a product or a quotient to a power.

[3.4a, c] Add and subtract polynomials.

[4.6a] Factor polynomials completely.

[4.8a] Solve problems involving quadratic equations that can be solved by factoring.

Pretest: Chapter 5

1. Find the LCM of $x^2 + 5x + 6$ and $x^2 + 6x + 9$.

Perform the indicated operations and simplify.

2. $\dfrac{b-1}{2-b} + \dfrac{b^2-3}{b^2-4}$

3. $\dfrac{4y-4}{y^2-y-2} - \dfrac{3y-5}{y^2-y-2}$

4. $\dfrac{4}{a+2} + \dfrac{3}{a}$

5. $\dfrac{x}{x+1} - \dfrac{x}{x-1} + \dfrac{2x^2}{x^2-1}$

6. $\dfrac{4x+8}{x+1} \cdot \dfrac{x^2-2x-3}{2x^2-8}$

7. $\dfrac{x+3}{x^2-9} \div \dfrac{x+3}{x^2-6x+9}$

8. Simplify: $\dfrac{\dfrac{1}{x}+\dfrac{1}{y}}{\dfrac{1}{x}-\dfrac{1}{y}}$.

Solve.

9. $\dfrac{1}{x+4} = \dfrac{5}{x}$

10. $\dfrac{3}{x-2} + \dfrac{x}{2} = \dfrac{6}{2x-4}$

11. Solve $R = \dfrac{1}{3}M(a-b)$ for M.

12. It takes 6 hr for a paper carrier to deliver 200 papers. At this rate, how long would it take to deliver 350 papers?

13. One data-entry clerk can key in a report in 6 hr. Another can key in the same report in 5 hr. How long would it take them to key in the same report working together?

14. One car travels 20 mph faster than another. While one car travels 300 mi, the other travels 400 mi. Find their speeds.

5.1 Multiplying and Simplifying Rational Expressions

a Rational Expressions and Replacements

Rational numbers are quotients of integers. Some examples are

$$\frac{2}{3}, \quad \frac{4}{-5}, \quad \frac{-8}{17}, \quad \frac{563}{1}.$$

The following are called **rational expressions** or **fractional expressions.** They are quotients of polynomials:

$$\frac{3}{4}, \quad \frac{5}{x+2}, \quad \frac{t^2+3t-10}{7t^2-4}.$$

A rational expression is also a division. For example,

$$\frac{3}{4} \text{ means } 3 \div 4 \quad \text{and} \quad \frac{x-8}{x+2} \text{ means } (x-8) \div (x+2).$$

Because rational expressions indicate division, we must be careful to avoid denominators of zero. When a variable is replaced with a number that produces a denominator equal to zero, the rational expression is undefined. For example, in the expression

$$\frac{x-8}{x+2},$$

when x is replaced with -2, the denominator is 0, and the expression is undefined:

$$\frac{x-8}{x+2} = \frac{-2-8}{-2+2} = \frac{-10}{0} \leftarrow \text{Undefined}$$

When x is replaced with a number other than -2, such as 3, the expression *is* defined because the denominator is nonzero:

$$\frac{x-8}{x+2} = \frac{3-8}{3+2} = \frac{-5}{5} = -1.$$

Objectives

After finishing Section 5.1, you should be able to:

- **a** Find all numbers for which a rational expression is undefined.
- **b** Multiply a rational expression by 1, using an expression such as A/A.
- **c** Simplify rational expressions by factoring the numerator and the denominator and removing factors of 1.
- **d** Multiply rational expressions and simplify.

FOR EXTRA HELP

TAPE 12 | TAPE 10A | MAC: 5A IBM: 5A

Find all numbers for which the rational expression is undefined.

1. $\dfrac{16}{x - 3}$

2. $\dfrac{2x - 7}{x^2 + 5x - 24}$

3. $\dfrac{x + 5}{8}$

Answers on page A-4

EXAMPLE 1 Find all numbers for which the rational expression

$$\frac{x + 4}{x^2 - 3x - 10}$$

is undefined.

To determine which numbers make the rational expression undefined, we set the denominator equal to 0 and solve:

$$x^2 - 3x - 10 = 0$$

$$(x - 5)(x + 2) = 0 \quad \text{Factoring}$$

$$x - 5 = 0 \quad \text{or} \quad x + 2 = 0 \quad \text{Using the principle of zero products}$$

$$x = 5 \quad \text{or} \quad x = -2.$$

The expression is undefined for the replacement numbers 5 and -2.

Do Exercises 1–3.

b MULTIPLYING BY 1

We multiply rational expressions in the same way that we multiply fractional notation in arithmetic. For a review, see Section R.2.

> To multiply rational expressions, multiply the numerators and multiply the denominators.

For example,

$$\frac{x - 2}{3} \cdot \frac{x + 2}{x + 7} = \frac{(x - 2)(x + 2)}{3(x + 7)}. \quad \text{Multiplying the numerators and the denominators}$$

Note that we leave the numerator, $(x - 2)(x + 2)$, and the denominator, $3(x + 7)$, in factored form because it is easier to simplify if we do not multiply. In order to learn to simplify, we first need to consider multiplying the rational expression by 1.

Any rational expression with the same numerator and denominator is a symbol for 1:

$$\frac{x + 8}{x + 8} = 1, \quad \frac{3x^2 - 4}{3x^2 - 4} = 1, \quad \frac{-1}{-1} = 1.$$

> Expressions that have the same value for all meaningful replacements are called *equivalent expressions*.

We can multiply by 1 to obtain an equivalent expression.

EXAMPLES Multiply.

2. $\dfrac{3x+2}{x+1} \cdot 1 = \dfrac{3x+2}{x+1} \cdot \dfrac{2x}{2x} = \dfrac{(3x+2)2x}{(x+1)2x}$

3. $\dfrac{x+2}{x-7} \cdot \dfrac{x+3}{x+3} = \dfrac{(x+2)(x+3)}{(x-7)(x+3)}$

4. $\dfrac{2+x}{2-x} \cdot \dfrac{-1}{-1} = \dfrac{(2+x)(-1)}{(2-x)(-1)}$

Do Exercises 4–6.

Multiply.

4. $\dfrac{2x+1}{3x-2} \cdot \dfrac{x}{x}$

5. $\dfrac{x+1}{x-2} \cdot \dfrac{x+2}{x+2}$

6. $\dfrac{x-8}{x-y} \cdot \dfrac{-1}{-1}$

c SIMPLIFYING RATIONAL EXPRESSIONS

Simplifying rational expressions is similar to simplifying fractional expressions in arithmetic. For a review, see Section R.2. We saw there, for example, that an expression like $\frac{15}{40}$ can be simplified as follows:

$$\frac{15}{40} = \frac{3 \cdot 5}{8 \cdot 5}$$ Factoring the numerator and the denominator. Note the common factor of 5.

$$= \frac{3}{8} \cdot \frac{5}{5}$$ Factoring the fractional expression

$$= \frac{3}{8} \cdot 1$$ $\frac{5}{5} = 1$

$$= \frac{3}{8}.$$ Using the identity property of 1. We call this "removing a factor equal to 1," or "removing a factor of 1."

In algebra, instead of simplifying

$$\frac{15}{40},$$

we may simplify an expression like

$$\frac{x^2-16}{x+4}.$$

Just as factoring is important in simplifying in arithmetic, so too is it important in simplifying rational expressions. The factoring we use most is the factoring of polynomials, which we studied in Chapter 4.

To simplify, we can do the reverse of multiplying. We factor the numerator and the denominator and "remove" a factor of 1.

EXAMPLE 5 Simplify by removing a factor of 1: $\dfrac{8x^2}{24x}$.

$$\frac{8x^2}{24x} = \frac{8 \cdot x \cdot x}{3 \cdot 8 \cdot x}$$ Factoring the numerator and the denominator

$$= \frac{8x}{8x} \cdot \frac{x}{3}$$ Factoring the rational expression

$$= 1 \cdot \frac{x}{3}$$ $\frac{8x}{8x} = 1$

$$= \frac{x}{3}$$ We removed a factor of 1.

Do Exercises 7 and 8.

Simplify by removing a factor of 1.

7. $\dfrac{5y}{y}$

8. $\dfrac{9x^2}{36x}$

Answers on page A-4

EXAMPLES Simplify by removing a factor of 1.

6. $\dfrac{5a + 15}{10} = \dfrac{5(a + 3)}{5 \cdot 2}$ Factoring the numerator and the denominator

$= \dfrac{5}{5} \cdot \dfrac{a + 3}{2}$ Factoring the rational expression

$= 1 \cdot \dfrac{a + 3}{2}$ $\dfrac{5}{5} = 1$

$= \dfrac{a + 3}{2}$ Removing a factor of 1

7. $\dfrac{6a + 12}{7a + 14} = \dfrac{6(a + 2)}{7(a + 2)}$ Factoring the numerator and the denominator

$= \dfrac{6}{7} \cdot \dfrac{a + 2}{a + 2}$ Factoring the rational expression

$= \dfrac{6}{7} \cdot 1$ $\dfrac{a + 2}{a + 2} = 1$

$= \dfrac{6}{7}$ Removing a factor of 1

8. $\dfrac{6x^2 + 4x}{2x^2 + 2x} = \dfrac{2x(3x + 2)}{2x(x + 1)}$ Factoring the numerator and the denominator

$= \dfrac{2x}{2x} \cdot \dfrac{3x + 2}{x + 1}$ Factoring the rational expression

$= 1 \cdot \dfrac{3x + 2}{x + 1}$ $\dfrac{2x}{2x} = 1$

$= \dfrac{3x + 2}{x + 1}$ Removing a factor of 1. Note in this step that you *cannot* remove the x's because x is not a factor of the entire numerator and the entire denominator.

9. $\dfrac{x^2 + 3x + 2}{x^2 - 1} = \dfrac{(x + 2)(x + 1)}{(x + 1)(x - 1)}$

$= \dfrac{x + 1}{x + 1} \cdot \dfrac{x + 2}{x - 1}$

$= 1 \cdot \dfrac{x + 2}{x - 1}$

$= \dfrac{x + 2}{x - 1}$

CANCELING

You may have encountered canceling when working with rational expressions. With great concern, we mention it as a possible way to speed up your work. Our concern is that canceling be done with care and understanding. Example 9 might have been done faster as follows:

$$\frac{x^2 + 3x + 2}{x^2 - 1} = \frac{(x + 2)(x + 1)}{(x + 1)(x - 1)} \quad \text{Factoring the numerator and the denominator}$$

$$= \frac{(x + 2)\cancel{(x + 1)}}{\cancel{(x + 1)}(x - 1)} \quad \text{When a factor of 1 is noted, it is canceled, as shown: } \frac{x+1}{x+1} = 1.$$

$$= \frac{x + 2}{x - 1}. \quad \text{Simplifying}$$

Caution! The difficulty with canceling is that it is often applied incorrectly, as in the following situations:

$$\frac{\cancel{x} + 3}{\cancel{x}} = 3; \qquad \frac{\cancel{4} + 1}{\cancel{4} + 2} = \frac{1}{2}; \qquad \frac{1\cancel{5}}{\cancel{5}4} = \frac{1}{4}.$$

Wrong! Wrong! Wrong!

In each of these situations, the expressions canceled were *not* factors of 1. Factors are parts of products. For example, in $2 \cdot 3$, 2 and 3 are factors, but in $2 + 3$, 2 and 3 are *not* factors. **If you can't factor, you can't cancel. If in doubt, don't cancel!**

Do Exercises 9–12.

FACTORS THAT ARE OPPOSITES

Consider

$$\frac{x - 4}{4 - x}.$$

At first glance, the numerator and the denominator do not appear to have any common factors other than 1. But $x - 4$ and $4 - x$ are opposites, or additive inverses, of each other. Thus we can rewrite one as the opposite of the other.

EXAMPLE 10 Simplify: $\frac{x - 4}{4 - x}$.

$$\frac{x - 4}{4 - x} = \frac{x - 4}{-(-4 + x)}$$

$$= \frac{x - 4}{-1(x - 4)}$$

$$= -1 \cdot \frac{x - 4}{x - 4}$$

$$= -1 \cdot 1$$

$$= -1$$

Do Exercises 13–15.

Simplify by removing a factor of 1.

9. $\frac{2x^2 + x}{3x^2 + 2x}$

10. $\frac{x^2 - 1}{2x^2 - x - 1}$

11. $\frac{7x + 14}{7}$

12. $\frac{12y + 24}{48}$

Simplify.

13. $\frac{x - 8}{8 - x}$

14. $\frac{c - d}{d - c}$

15. $\frac{-x - 7}{x + 7}$

Answers on page A-4

Multiply and simplify.

16. $\frac{a^2 - 4a + 4}{a^2 - 9} \cdot \frac{a + 3}{a - 2}$

17. $\frac{x^2 - 25}{6} \cdot \frac{3}{x + 5}$

Answers on page A-4

d MULTIPLYING AND SIMPLIFYING

We try to simplify after we multiply. That is why we do not multiply out the numerator and the denominator too soon. We would need to factor them again anyway in order to simplify.

EXAMPLE 11 Multiply and simplify: $\frac{5a^3}{4} \cdot \frac{2}{5a}$.

$$\frac{5a^3}{4} \cdot \frac{2}{5a} = \frac{5a^3(2)}{4(5a)}$$ Multiplying the numerators and the denominators

$$= \frac{2 \cdot 5 \cdot a \cdot a \cdot a}{2 \cdot 2 \cdot 5 \cdot a}$$ Factoring the numerator and the denominator

$$= \frac{\cancel{2} \cdot \cancel{5} \cdot \cancel{a} \cdot a \cdot a}{\cancel{2} \cdot 2 \cdot \cancel{5} \cdot \cancel{a}}$$ Removing a factor of 1: $\frac{2 \cdot 5 \cdot a}{2 \cdot 5 \cdot a} = 1$

$$= \frac{a^2}{2}$$ Simplifying

EXAMPLE 12 Multiply and simplify: $\frac{x^2 + 6x + 9}{x^2 - 4} \cdot \frac{x - 2}{x + 3}$.

$$\frac{x^2 + 6x + 9}{x^2 - 4} \cdot \frac{x - 2}{x + 3} = \frac{(x^2 + 6x + 9)(x - 2)}{(x^2 - 4)(x + 3)}$$ Multiplying the numerators and the denominators

$$= \frac{(x + 3)(x + 3)(x - 2)}{(x + 2)(x - 2)(x + 3)}$$ Factoring the numerator and the denominator

$$= \frac{\cancel{(x + 3)}(x + 3)\cancel{(x - 2)}}{(x + 2)\cancel{(x - 2)}\cancel{(x + 3)}}$$ Removing a factor of 1: $\frac{(x + 3)(x - 2)}{(x + 3)(x - 2)} = 1$

$$= \frac{x + 3}{x + 2}$$ Simplifying

EXAMPLE 13 Multiply and simplify: $\frac{x^2 + x - 2}{15} \cdot \frac{5}{2x^2 - 3x + 1}$.

$$\frac{x^2 + x - 2}{15} \cdot \frac{5}{2x^2 - 3x + 1} = \frac{(x^2 + x - 2)5}{15(2x^2 - 3x + 1)}$$ Multiplying the numerators and the denominators

$$= \frac{(x + 2)(x - 1)5}{5(3)(x - 1)(2x - 1)}$$ Factoring the numerator and the denominator

$$= \frac{(x + 2)\cancel{(x - 1)5}}{\cancel{5}(3)\cancel{(x - 1)}(2x - 1)}$$ Removing a factor of 1: $\frac{(x - 1)5}{(x - 1)5} = 1$

$$= \frac{x + 2}{3(2x - 1)}$$ Simplifying

You need not carry out this multiplication.

Do Exercises 16 and 17.

NAME SECTION DATE

Exercise Set 5.1

a Find all numbers for which the rational expression is undefined.

1. $\frac{-3}{2x}$

2. $\frac{24}{-8y}$

3. $\frac{5}{x-8}$

4. $\frac{y-4}{y+6}$

5. $\frac{3}{2y+5}$

6. $\frac{x^2-9}{4x-12}$

7. $\frac{x^2+11}{x^2-3x-28}$

8. $\frac{p^2-9}{p^2-7p+10}$

9. $\frac{m^3-2m}{m^2-25}$

10. $\frac{7-3x+x^2}{49-x^2}$

11. $\frac{x-4}{3}$

12. $\frac{x^2-25}{14}$

b Multiply. Do not simplify. Note that in each case you are multiplying by 1.

13. $\frac{4x}{4x} \cdot \frac{3x^2}{5y}$

14. $\frac{5x^2}{5x^2} \cdot \frac{6y^3}{3z^4}$

15. $\frac{2x}{2x} \cdot \frac{x-1}{x+4}$

16. $\frac{2a-3}{5a+2} \cdot \frac{a}{a}$

17. $\frac{3-x}{4-x} \cdot \frac{-1}{-1}$

18. $\frac{x-5}{5-x} \cdot \frac{-1}{-1}$

19. $\frac{y+6}{y+6} \cdot \frac{y-7}{y+2}$

20. $\frac{x^2+1}{x^3-2} \cdot \frac{x-4}{x-4}$

Answers

1. ____
2. ____
3. ____
4. ____
5. ____
6. ____
7. ____
8. ____
9. ____
10. ____
11. ____
12. ____
13. ____
14. ____
15. ____
16. ____
17. ____
18. ____
19. ____
20. ____

Answers

21. ______

22. ______

23. ______

24. ______

25. ______

26. ______

27. ______

28. ______

29. ______

30. ______

31. ______

32. ______

33. ______

34. ______

35. ______

36. ______

37. ______

38. ______

39. ______

40. ______

c Simplify.

21. $\frac{8x^3}{32x}$

22. $\frac{4x^2}{20x}$

23. $\frac{48p^7q^5}{18p^5q^4}$

24. $\frac{-76x^8y^3}{-24x^4y^3}$

25. $\frac{4x - 12}{4x}$

26. $\frac{5a - 40}{5}$

27. $\frac{3m^2 + 3m}{6m^2 + 9m}$

28. $\frac{4y^2 - 2y}{5y^2 - 5y}$

29. $\frac{a^2 - 9}{a^2 + 5a + 6}$

30. $\frac{t^2 - 25}{t^2 + t - 20}$

31. $\frac{a^2 - 10a + 21}{a^2 - 11a + 28}$

32. $\frac{x^2 - 2x - 8}{x^2 - x - 6}$

33. $\frac{x^2 - 25}{x^2 - 10x + 25}$

34. $\frac{x^2 + 8x + 16}{x^2 - 16}$

35. $\frac{a^2 - 1}{a - 1}$

36. $\frac{t^2 - 1}{t + 1}$

37. $\frac{x^2 + 1}{x + 1}$

38. $\frac{m^2 + 9}{m + 3}$

39. $\frac{6x^2 - 54}{4x^2 - 36}$

40. $\frac{8x^2 - 32}{4x^2 - 16}$

41. $\dfrac{6t + 12}{t^2 - t - 6}$

42. $\dfrac{4x + 32}{x^2 + 9x + 8}$

43. $\dfrac{2t^2 + 6t + 4}{4t^2 - 12t - 16}$

44. $\dfrac{3a^2 - 9a - 12}{6a^2 + 30a + 24}$

45. $\dfrac{t^2 - 4}{(t + 2)^2}$

46. $\dfrac{m^2 - 10m + 25}{m^2 - 25}$

47. $\dfrac{6 - x}{x - 6}$

48. $\dfrac{t - 3}{3 - t}$

49. $\dfrac{a - b}{b - a}$

50. $\dfrac{y - x}{-x + y}$

51. $\dfrac{6t - 12}{2 - t}$

52. $\dfrac{5a - 15}{3 - a}$

53. $\dfrac{x^2 - 1}{1 - x}$

54. $\dfrac{a^2 - b^2}{b^2 - a^2}$

d Multiply and simplify.

55. $\dfrac{4x^3}{3x} \cdot \dfrac{14}{x}$

56. $\dfrac{18}{x^3} \cdot \dfrac{5x^2}{6}$

57. $\dfrac{3c}{d^2} \cdot \dfrac{4d}{6c^3}$

58. $\dfrac{3x^2y}{2} \cdot \dfrac{4}{xy^3}$

59. $\dfrac{x^2 - 3x - 10}{x^2 - 4x + 4} \cdot \dfrac{x - 2}{x - 5}$

60. $\dfrac{t^2}{t^2 - 4} \cdot \dfrac{t^2 - 5t + 6}{t^2 - 3t}$

ANSWERS

41. ______

42. ______

43. ______

44. ______

45. ______

46. ______

47. ______

48. ______

49. ______

50. ______

51. ______

52. ______

53. ______

54. ______

55. ______

56. ______

57. ______

58. ______

59. ______

60. ______

ANSWERS

61. ______

62. ______

63. ______

64. ______

65. ______

66. ______

67. ______

68. ______

69. ______

70. ______

71. ______

72. ______

73. ______

74. ______

75. ______

76. ______

77. ______

78. ______

79. ______

80. ______

81. ______

82. ______

61. $\dfrac{a^2-9}{a^2} \cdot \dfrac{a^2-3a}{a^2+a-12}$

62. $\dfrac{x^2+10x-11}{x^2-1} \cdot \dfrac{x+1}{x+11}$

63. $\dfrac{4a^2}{3a^2-12a+12} \cdot \dfrac{3a-6}{2a}$

64. $\dfrac{5v+5}{v-2} \cdot \dfrac{v^2-4v+4}{v^2-1}$

65. $\dfrac{t^4-16}{t^4-1} \cdot \dfrac{t^2+1}{t^2+4}$

66. $\dfrac{x^4-1}{x^4-81} \cdot \dfrac{x^2+9}{x^2+1}$

67. $\dfrac{(x+4)^3}{(x+2)^3} \cdot \dfrac{x^2+4x+4}{x^2+8x+16}$

68. $\dfrac{(t-2)^3}{(t-1)^3} \cdot \dfrac{t^2-2t+1}{t^2-4t+4}$

69. $\dfrac{5a^2-180}{10a^2-10} \cdot \dfrac{20a+20}{2a-12}$

70. $\dfrac{2t^2-98}{4t^2-4} \cdot \dfrac{8t+8}{16t-112}$

SKILL MAINTENANCE

71. The product of two consecutive even integers is 360. Find the integers.

Factor.

72. $16 - t^4$

73. $2y^3 - 10y^2 + y - 5$

74. $x^5 - 2x^4 - 35x^3$

75. $a^2 - 16a + 64$

76. $x^2 - x - 56$

SYNTHESIS

Simplify.

77. $\dfrac{x^4-16y^4}{(x^2+4y^2)(x-2y)}$

78. $\dfrac{(a-b)^2}{b^2-a^2}$

79. $\dfrac{t^4-1}{t^4-81} \cdot \dfrac{t^2-9}{t^2+1} \cdot \dfrac{(t-9)^2}{(t+1)^2}$

80. $\dfrac{(t+2)^3}{(t+1)^3} \cdot \dfrac{t^2+2t+1}{t^2+4t+4} \cdot \dfrac{t+1}{t+2}$

81. $\dfrac{x^2-y^2}{(x-y)^2} \cdot \dfrac{x^2-2xy+y^2}{x^2-4xy-5y^2}$

82. $\dfrac{x-1}{x^2+1} \cdot \dfrac{x^4-1}{(x-1)^2} \cdot \dfrac{x^2-1}{x^4-2x^2+1}$

5.2 *Division and Reciprocals*

OBJECTIVES

After finishing Section 5.2, you should be able to:

a Find the reciprocal of a rational expression.

b Divide rational expressions and simplify.

FOR EXTRA HELP

TAPE 12 | TAPE 10B | MAC: 5A IBM: 5A

There is a similarity throughout this chapter between what we do with rational expressions and what we do with rational numbers. In fact, after variables have been replaced with rational numbers, a rational expression represents a rational number.

a FINDING RECIPROCALS

Two expressions are reciprocals of each other if their product is 1. The reciprocal of a rational expression is found by interchanging the numerator and the denominator.

EXAMPLES

1. The reciprocal of $\frac{2}{5}$ is $\frac{5}{2}$. $\left(\text{This is because } \frac{2}{5}\cdot\frac{5}{2}=\frac{10}{10}=1.\right)$

2. The reciprocal of $\frac{2x^2-3}{x+4}$ is $\frac{x+4}{2x^2-3}$.

3. The reciprocal of $x+2$ is $\frac{1}{x+2}$. $\left(\text{Think of } x+2 \text{ as } \frac{x+2}{1}.\right)$

Do Exercises 1–4.

Find the reciprocal.

1. $\frac{7}{2}$

2. $\frac{x^2+5}{2x^3-1}$

3. $x-5$

4. $\frac{1}{x^2-3}$

b DIVISION

We divide rational expressions in the same way that we divide fractional notation in arithmetic. For a review, see Section R.2.

> To divide rational expressions, multiply by the reciprocal of the divisor. Then factor and simplify the result.

EXAMPLES Divide.

4.
$$\frac{3}{4}\div\frac{2}{5}=\frac{3}{4}\cdot\frac{5}{2}\qquad\text{Multiplying by the reciprocal of the divisor}$$
$$=\frac{3\cdot 5}{4\cdot 2}$$
$$=\frac{15}{8}$$

5.
$$\frac{2}{x}\div\frac{x}{3}=\frac{2}{x}\cdot\frac{3}{x}\qquad\text{Multiplying by the reciprocal of the divisor}$$
$$=\frac{2\cdot 3}{x\cdot x}$$
$$=\frac{6}{x^2}$$

Do Exercises 5 and 6.

Divide.

5. $\frac{3}{5}\div\frac{7}{2}$

6. $\frac{x}{8}\div\frac{5}{x}$

Answers on page A-4

7. Divide:

$$\frac{x-3}{x+5} \div \frac{x+5}{x-2}.$$

Divide and simplify.

8. $\dfrac{x-3}{x+5} \div \dfrac{x+2}{x+5}$

9. $\dfrac{x^2-5x+6}{x+5} \div \dfrac{x+2}{x+5}$

10. $\dfrac{y^2-1}{y+1} \div \dfrac{y^2-2y+1}{y+1}$

Answers on page A-4

EXAMPLE 6 Divide: $\dfrac{x+1}{x+2} \div \dfrac{x-1}{x+3}$.

$$\frac{x+1}{x+2} \div \frac{x-1}{x+3} = \frac{x+1}{x+2} \cdot \frac{x+3}{x-1}$$ Multiplying by the reciprocal of the divisor

$$= \frac{(x+1)(x+3)}{(x+2)(x-1)}$$ ← You need not carry out the multiplications in the numerator and the denominator.

Do Exercise 7.

EXAMPLE 7 Divide and simplify: $\dfrac{x+1}{x^2-1} \div \dfrac{x+1}{x^2-2x+1}$.

$$\frac{x+1}{x^2-1} \div \frac{x+1}{x^2-2x+1}$$

$$= \frac{x+1}{x^2-1} \cdot \frac{x^2-2x+1}{x+1}$$ Multiplying by the reciprocal

$$= \frac{(x+1)(x^2-2x+1)}{(x^2-1)(x+1)}$$

$$= \frac{(x+1)(x-1)(x-1)}{(x-1)(x+1)(x+1)}$$ Factoring the numerator and the denominator

$$= \frac{\cancel{(x+1)}\cancel{(x-1)}(x-1)}{\cancel{(x-1)}\cancel{(x+1)}(x+1)}$$ Removing a factor of 1: $\dfrac{(x+1)(x-1)}{(x+1)(x-1)} = 1$

$$= \frac{x-1}{x+1}$$

EXAMPLE 8 Divide and simplify: $\dfrac{x^2-2x-3}{x^2-4} \div \dfrac{x+1}{x+5}$.

$$\frac{x^2-2x-3}{x^2-4} \div \frac{x+1}{x+5}$$

$$= \frac{x^2-2x-3}{x^2-4} \cdot \frac{x+5}{x+1}$$ Multiplying by the reciprocal

$$= \frac{(x^2-2x-3)(x+5)}{(x^2-4)(x+1)}$$

$$= \frac{(x-3)(x+1)(x+5)}{(x-2)(x+2)(x+1)}$$ Factoring the numerator and the denominator

$$= \frac{(x-3)\cancel{(x+1)}(x+5)}{(x-2)(x+2)\cancel{(x+1)}}$$ Removing a factor of 1: $\dfrac{x+1}{x+1} = 1$

$$= \frac{(x-3)(x+5)}{(x-2)(x+2)}$$ ← You need not carry out the multiplications in the numerator and the denominator.

Do Exercises 8–10.

NAME SECTION DATE

Exercise Set 5.2

a Find the reciprocal.

1. $\dfrac{4}{x}$

2. $\dfrac{a+3}{a-1}$

3. $x^2 - y^2$

4. $\dfrac{1}{a+b}$

5. $\dfrac{x^2+2x-5}{x^2-4x+7}$

6. $\dfrac{(a-b)(a+b)}{(a+4)(a-5)}$

b Divide and simplify.

7. $\dfrac{2}{5} \div \dfrac{4}{3}$

8. $\dfrac{3}{10} \div \dfrac{3}{2}$

9. $\dfrac{2}{x} \div \dfrac{8}{x}$

10. $\dfrac{t}{3} \div \dfrac{t}{15}$

11. $\dfrac{a}{b^2} \div \dfrac{a^2}{b^3}$

12. $\dfrac{x^2}{y} \div \dfrac{x^3}{y^3}$

13. $\dfrac{a+2}{a-3} \div \dfrac{a-1}{a+3}$

14. $\dfrac{x-8}{x+9} \div \dfrac{x+2}{x-1}$

15. $\dfrac{x^2-1}{x} \div \dfrac{x+1}{x-1}$

16. $\dfrac{4y-8}{y+2} \div \dfrac{y-2}{y^2-4}$

17. $\dfrac{x+1}{6} \div \dfrac{x+1}{3}$

18. $\dfrac{a}{a-b} \div \dfrac{b}{a-b}$

19. $\dfrac{5x-5}{16} \div \dfrac{x-1}{6}$

20. $\dfrac{4y-12}{12} \div \dfrac{y-3}{3}$

21. $\dfrac{-6+3x}{5} \div \dfrac{4x-8}{25}$

ANSWERS

1. ________
2. ________
3. ________
4. ________
5. ________
6. ________
7. ________
8. ________
9. ________
10. ________
11. ________
12. ________
13. ________
14. ________
15. ________
16. ________
17. ________
18. ________
19. ________
20. ________
21. ________

ANSWERS

22. ______
23. ______
24. ______
25. ______
26. ______
27. ______
28. ______
29. ______
30. ______
31. ______
32. ______
33. ______
34. ______
35. ______
36. ______
37. ______
38. ______
39. ______
40. ______
41. ______
42. ______
43. ______
44. ______

22. $\dfrac{-12 + 4x}{4} \div \dfrac{-6 + 2x}{6}$

23. $\dfrac{a + 2}{a - 1} \div \dfrac{3a + 6}{a - 5}$

24. $\dfrac{t - 3}{t + 2} \div \dfrac{4t - 12}{t + 1}$

25. $\dfrac{x^2 - 4}{x} \div \dfrac{x - 2}{x + 2}$

26. $\dfrac{x + y}{x - y} \div \dfrac{x^2 + y}{x^2 - y^2}$

27. $\dfrac{x^2 - 9}{4x + 12} \div \dfrac{x - 3}{6}$

28. $\dfrac{a - b}{2a} \div \dfrac{a^2 - b^2}{8a^3}$

29. $\dfrac{c^2 + 3c}{c^2 + 2c - 3} \div \dfrac{c}{c + 1}$

30. $\dfrac{y + 5}{2y} \div \dfrac{y^2 - 25}{4y^2}$

31. $\dfrac{2y^2 - 7y + 3}{2y^2 + 3y - 2} \div \dfrac{6y^2 - 5y + 1}{3y^2 + 5y - 2}$

32. $\dfrac{x^2 + x - 20}{x^2 - 7x + 12} \div \dfrac{x^2 + 10x + 25}{x^2 - 6x + 9}$

33. $\dfrac{x^2 - 1}{4x + 4} \div \dfrac{2x^2 - 4x + 2}{8x + 8}$

34. $\dfrac{5t^2 + 5t - 30}{10t + 30} \div \dfrac{2t^2 - 8}{6t^2 + 36t + 54}$

SKILL MAINTENANCE

35. Sixteen more than the square of a number is eight times the number. Find the number.

36. Subtract:
$(8x^3 - 3x^2 + 7) - (8x^2 + 3x - 5).$

Simplify.

37. $(2x^{-3}y^4)^2$

38. $(5x^6y^{-4})^3$

39. $\left(\dfrac{2x^3}{y^5}\right)^2$

40. $\left(\dfrac{a^{-3}}{b^4}\right)^5$

SYNTHESIS

Simplify.

41. $\dfrac{3a^2 - 5ab - 12b^2}{3ab + 4b^2} \div (3b^2 - ab)$

42. $\dfrac{3x^2 - 2xy - y^2}{x^2 - y^2} \div 3x^2 + 4xy + y^2$

43. $\dfrac{3x + 3y + 3}{9x} \div \left(\dfrac{x^2 + 2xy + y^2 - 1}{x^4 + x^2}\right)$

44. $\left(\dfrac{y^2 + 5y + 6}{y^2} \cdot \dfrac{3y^3 + 6y^2}{y^2 - y - 12}\right) \div \dfrac{y^2 - y}{y^2 - 2y - 8}$

5.3 Least Common Multiples and Denominators

OBJECTIVES

After finishing Section 5.3, you should be able to:

a Find the LCM of several numbers by factoring.

b Add fractions, first finding the LCD.

c Find the LCM of algebraic expressions by factoring.

FOR EXTRA HELP

TAPE 13 | TAPE 10B | MAC: 5A IBM: 5A

a LEAST COMMON MULTIPLES

To add when denominators are different, we first find a common denominator. For a review, see Sections R.1 and R.2. We saw there, for example, that to add $\frac{5}{12}$ and $\frac{7}{30}$, we first look for the **least common multiple, LCM,** of both 12 and 30. That number becomes the **least common denominator, LCD.** To find the LCM of 12 and 30, we factor:

$$12 = 2 \cdot 2 \cdot 3;$$
$$30 = 2 \cdot 3 \cdot 5.$$

The LCM is the number that has 2 as a factor twice, 3 as a factor once, and 5 as a factor once:

$$\text{LCM} = 2 \cdot 2 \cdot 3 \cdot 5, \text{ or } 60.$$

> To find the LCM, use each factor the greatest number of times that it appears in any one factorization.

EXAMPLE 1 Find the LCM of 24 and 36.

$$\left.\begin{aligned} 24 &= 2 \cdot 2 \cdot 2 \cdot 3 \\ 36 &= 2 \cdot 2 \cdot 3 \cdot 3 \end{aligned}\right\} \quad \text{LCM} = 2 \cdot 2 \cdot 2 \cdot 3 \cdot 3, \text{ or } 72$$

Do Exercises 1–4.

Find the LCM by factoring.

1. 16, 18

2. 6, 12

3. 2, 5

4. 24, 30, 20

Answers on page A-4

b ADDING USING THE LCD

Let us finish adding $\frac{5}{12}$ and $\frac{7}{30}$:

$$\frac{5}{12} + \frac{7}{30} = \frac{5}{2 \cdot 2 \cdot 3} + \frac{7}{2 \cdot 3 \cdot 5}.$$

The least common denominator, LCD, is $2 \cdot 2 \cdot 3 \cdot 5$. To get the LCD in the first denominator, we need a 5. To get the LCD in the second denominator, we need another 2. We get these numbers by multiplying by 1:

$$\frac{5}{12} + \frac{7}{30} = \frac{5}{2 \cdot 2 \cdot 3} \cdot \frac{5}{5} + \frac{7}{2 \cdot 3 \cdot 5} \cdot \frac{2}{2} \qquad \text{Multiplying by 1}$$

$$= \frac{25}{2 \cdot 2 \cdot 3 \cdot 5} + \frac{14}{2 \cdot 3 \cdot 5 \cdot 2} \qquad \text{The denominators are now the LCD.}$$

$$= \frac{39}{2 \cdot 2 \cdot 3 \cdot 5} \qquad \text{Adding the numerators and keeping the LCD}$$

$$= \frac{\cancel{3} \cdot 13}{2 \cdot 2 \cdot \cancel{3} \cdot 5} \qquad \text{Factoring the numerator and removing a factor of 1: } \frac{3}{3} = 1$$

$$= \frac{13}{20}. \qquad \text{Simplifying}$$

Add, first finding the LCD. Simplify, if possible.

5. $\frac{3}{16} + \frac{1}{18}$

6. $\frac{1}{6} + \frac{1}{12}$

7. $\frac{1}{2} + \frac{3}{5}$

8. $\frac{1}{24} + \frac{1}{30} + \frac{3}{20}$

Find the LCM.

9. $12xy^2, \quad 15x^3y$

10. $y^2 + 5y + 4, \quad y^2 + 2y + 1$

11. $t^2 + 16, \quad t - 2, \quad 7$

12. $x^2 + 2x + 1, \quad 3x^2 - 3x, \quad x^2 - 1$

Answers on page A-4

EXAMPLE 2 Add: $\frac{5}{12} + \frac{11}{18}$.

$$\left.\begin{aligned} 12 &= 2 \cdot 2 \cdot 3 \\ 18 &= 2 \cdot 3 \cdot 3 \end{aligned}\right\} \quad \text{LCD} = 2 \cdot 2 \cdot 3 \cdot 3, \text{ or } 36$$

$$\frac{5}{12} + \frac{11}{18} = \frac{5}{2 \cdot 2 \cdot 3} \cdot \frac{3}{3} + \frac{11}{2 \cdot 3 \cdot 3} \cdot \frac{2}{2} = \frac{37}{2 \cdot 2 \cdot 3 \cdot 3} = \frac{37}{36}$$

Do Exercises 5–8.

c LCMs of Algebraic Expressions

To find the LCM of two or more algebraic expressions, we factor them. Then we use each factor the greatest number of times that it occurs in any one expression.

EXAMPLE 3 Find the LCM of $12x$, $16y$, and $8xyz$.

$$\left.\begin{aligned} 12x &= 2 \cdot 2 \cdot 3 \cdot x \\ 16y &= 2 \cdot 2 \cdot 2 \cdot 2 \cdot y \\ 8xyz &= 2 \cdot 2 \cdot 2 \cdot x \cdot y \cdot z \end{aligned}\right\} \quad \begin{aligned} \text{LCM} &= 2 \cdot 2 \cdot 2 \cdot 2 \cdot 3 \cdot x \cdot y \cdot z \\ &= 48xyz \end{aligned}$$

EXAMPLE 4 Find the LCM of $x^2 + 5x - 6$ and $x^2 - 1$.

$$\left.\begin{aligned} x^2 + 5x - 6 &= (x + 6)(x - 1) \\ x^2 - 1 &= (x + 1)(x - 1) \end{aligned}\right\} \quad \text{LCM} = (x + 6)(x - 1)(x + 1)$$

EXAMPLE 5 Find the LCM of $x^2 + 4$, $x + 1$, and 5.

These expressions are not factorable, so the LCM is their product:

$5(x^2 + 4)(x + 1)$.

EXAMPLE 6 Find the LCM of $x^2 - 25$ and $2x - 10$.

$$\left.\begin{aligned} x^2 - 25 &= (x + 5)(x - 5) \\ 2x - 10 &= 2(x - 5) \end{aligned}\right\} \quad \text{LCM} = 2(x + 5)(x - 5)$$

EXAMPLE 7 Find the LCM of $x^2 - 4y^2$, $x^2 - 4xy + 4y^2$, and $x - 2y$.

$$\left.\begin{aligned} x^2 - 4y^2 &= (x - 2y)(x + 2y) \\ x^2 - 4xy + 4y^2 &= (x - 2y)(x - 2y) \\ x - 2y &= x - 2y \end{aligned}\right\} \quad \begin{aligned} \text{LCM} &= (x + 2y)(x - 2y)(x - 2y) \\ &= (x + 2y)(x - 2y)^2 \end{aligned}$$

Do Exercises 9–12.

NAME SECTION DATE

Exercise Set 5.3

a Find the LCM.

1. 12, 27

2. 10, 15

3. 8, 9

4. 12, 18

5. 6, 9, 21

6. 8, 36, 40

7. 24, 36, 40

8. 4, 5, 20

9. 10, 100, 500

10. 28, 42, 60

b Add, first finding the LCD. Simplify, if possible.

11. $\frac{7}{24} + \frac{11}{18}$

12. $\frac{7}{60} + \frac{2}{25}$

13. $\frac{1}{6} + \frac{3}{40}$

14. $\frac{5}{24} + \frac{3}{20}$

15. $\frac{1}{20} + \frac{1}{30} + \frac{2}{45}$

16. $\frac{2}{15} + \frac{5}{9} + \frac{3}{20}$

c Find the LCM.

17. $6x^2$, $12x^3$

18. $2a^2b$, $8ab^3$

19. $2x^2$, $6xy$, $18y^2$

20. p^3q, p^2q, pq^2

21. $2(y - 3)$, $6(y - 3)$

22. $5(m + 2)$, $15(m + 2)$

23. t, $t + 2$, $t - 2$

24. y, $y - 5$, $y + 5$

25. $x^2 - 4$, $x^2 + 5x + 6$

26. $x^2 - 4$, $x^2 - x - 2$

ANSWERS

1. ______
2. ______
3. ______
4. ______
5. ______
6. ______
7. ______
8. ______
9. ______
10. ______
11. ______
12. ______
13. ______
14. ______
15. ______
16. ______
17. ______
18. ______
19. ______
20. ______
21. ______
22. ______
23. ______
24. ______
25. ______
26. ______

ANSWERS

27.
28.
29.
30.
31.
32.
33.
34.
35.
36.
37.
38.
39.
40.
41.
42.
43.
44.
45.
46.
47.
48.
49.
50.

27. $t^3 + 4t^2 + 4t, \quad t^2 - 4t$

28. $m^4 - m^2, \quad m^3 - m^2$

29. $a + 1, \quad (a - 1)^2, \quad a^2 - 1$

30. $a^2 - 2ab + b^2, \quad a^2 - b^2, \quad 3a + 3b$

31. $m^2 - 5m + 6, \quad m^2 - 4m + 4$

32. $2x^2 + 5x + 2, \quad 2x^2 - x - 1$

33. $2 + 3x, \quad 4 - 9x^2, \quad 2 - 3x$

34. $9 - 4x^2, \quad 3 + 2x, \quad 3 - 2x$

35. $10v^2 + 30v, \quad 5v^2 + 35v + 60$

36. $12a^2 + 24a, \quad 4a^2 + 20a + 24$

37. $9x^3 - 9x^2 - 18x, \quad 6x^5 - 24x^4 + 24x^3$

38. $x^5 - 4x^3, \quad x^3 + 4x^2 + 4x$

39. $x^5 + 4x^4 + 4x^3, \quad 3x^2 - 12, \quad 2x + 4$

40. $x^5 + 2x^4 + x^3, \quad 2x^3 - 2x, \quad 5x - 5$

SKILL MAINTENANCE

Factor.

41. $x^2 - 6x + 9$

42. $6x^2 + 4x$

43. $x^2 - 9$

44. $x^2 + 4x - 21$

45. $x^2 + 6x + 9$

46. $x^2 - 4x - 21$

SYNTHESIS

Find the LCM.

47. 72, 90, 96

48. $8x^2 - 8, \quad 6x^2 - 12x + 6, \quad 10x - 10$

49. Two joggers leave the starting point of a fitness loop at the same time and jog in the same direction. One jogger completes a lap in 6 min and the second jogger in 8 min. Assuming they continue to run at the same pace, when will they next meet at the starting place?

50. If the LCM of two expressions is the same as one of the expressions, what relationship exists between the two expressions?

5.4 *Adding Rational Expressions*

OBJECTIVE

After finishing Section 5.4, you should be able to:

a Add rational expressions.

FOR EXTRA HELP

TAPE 13 | TAPE 11A | MAC: 5A IBM: 5A

a We add rational expressions as we do rational numbers.

To add when the denominators are the same, add the numerators and keep the same denominator.

EXAMPLES Add.

1. $\dfrac{x}{x+1} + \dfrac{2}{x+1} = \dfrac{x+2}{x+1}$

2. $\dfrac{2x^2+3x-7}{2x+1} + \dfrac{x^2+x-8}{2x+1} = \dfrac{(2x^2+3x-7)+(x^2+x-8)}{2x+1}$

$$= \frac{3x^2+4x-15}{2x+1}$$

3. $\dfrac{x-5}{x^2-9} + \dfrac{2}{x^2-9} = \dfrac{(x-5)+2}{x^2-9} = \dfrac{x-3}{x^2-9}$

$= \dfrac{x-3}{(x-3)(x+3)}$ Factoring

$= \dfrac{\cancel{x-3}}{\cancel{(x-3)}(x+3)}$ Removing a factor of 1: $\dfrac{x-3}{x-3} = 1$

$= \dfrac{1}{x+3}$ Simplifying

As in Example 3, simplifying should be done if possible after adding.

Do Exercises 1–3.

Add.

1. $\dfrac{5}{9} + \dfrac{2}{9}$

2. $\dfrac{3}{x-2} + \dfrac{x}{x-2}$

3. $\dfrac{4x+5}{x-1} + \dfrac{2x-1}{x-1}$

When denominators are not the same, we multiply by 1 to obtain equivalent expressions with the same denominator. When one denominator is the opposite of the other, we can first multiply either expression by 1 using $-1/-1$.

EXAMPLES

4. $\dfrac{x}{2} + \dfrac{3}{-2} = \dfrac{x}{2} + \dfrac{3}{-2}\cdot\dfrac{-1}{-1}$ Multiplying by 1 using $\dfrac{-1}{-1}$

$= \dfrac{x}{2} + \dfrac{-3}{2}$ The denominators are now the same.

$= \dfrac{x+(-3)}{2} = \dfrac{x-3}{2}$

5. $\dfrac{3x+4}{x-2} + \dfrac{x-7}{2-x} = \dfrac{3x+4}{x-2} + \dfrac{x-7}{2-x}\cdot\dfrac{-1}{-1}$

We could have chosen to multiply this expression by $-1/-1$. We multiply only one expression, *not* both.

$= \dfrac{3x+4}{x-2} + \dfrac{-x+7}{x-2}$ *Note:* $(2-x)(-1) = -2+x = x-2.$

$= \dfrac{(3x+4)+(-x+7)}{x-2} = \dfrac{2x+11}{x-2}$

Do Exercises 4 and 5.

Add.

4. $\dfrac{x}{4} + \dfrac{5}{-4}$

5. $\dfrac{2x+1}{x-3} + \dfrac{x+2}{3-x}$

Answers on page A-4

Add.

6. $\dfrac{3x}{16} + \dfrac{5x^2}{24}$

7. $\dfrac{3}{16x} + \dfrac{5}{24x^2}$

Answers on page A-4

When denominators are different, we find the least common denominator, LCD. The procedure we will use is as follows.

> To add rational expressions with different denominators:
>
> 1. Find the LCM of the denominators. This is the least common denominator (LCD).
> 2. For each rational expression, find an equivalent expression with the LCD. To do so, multiply by 1 using an expression for 1 made up of factors of the LCD that are missing from the original denominator.
> 3. Add the numerators. Write the sum over the LCD.
> 4. Simplify, if possible.

EXAMPLE 6 Add: $\dfrac{5x^2}{8} + \dfrac{7x}{12}$.

First, we find the LCD:

$$\left.\begin{aligned} 8 &= 2 \cdot 2 \cdot 2 \\ 12 &= 2 \cdot 2 \cdot 3 \end{aligned}\right\} \quad \text{LCD} = 2 \cdot 2 \cdot 2 \cdot 3, \text{ or } 24.$$

Compare the factorization $8 = 2 \cdot 2 \cdot 2$ with the factorization of the LCD, $24 = 2 \cdot 2 \cdot 2 \cdot 3$. The factor of the LCD missing from 8 is 3. Compare $12 = 2 \cdot 2 \cdot 3$ and $24 = 2 \cdot 2 \cdot 2 \cdot 3$. The factor of the LCD missing from 12 is 2. We multiply by 1 to get the LCD in each expression, and then add and simplify, if possible:

$$\begin{aligned} \frac{5x^2}{8} + \frac{7x}{12} &= \frac{5x^2}{2 \cdot 2 \cdot 2} + \frac{7x}{2 \cdot 2 \cdot 3} \\ &= \frac{5x^2}{2 \cdot 2 \cdot 2} \cdot \frac{3}{3} + \frac{7x}{2 \cdot 2 \cdot 3} \cdot \frac{2}{2} \qquad \text{Multiplying by 1 to get the same denominators} \\ &= \frac{15x^2}{24} + \frac{14x}{24} \\ &= \frac{15x^2 + 14x}{24}. \end{aligned}$$

EXAMPLE 7 Add: $\dfrac{3}{8x} + \dfrac{5}{12x^2}$.

First, we find the LCD:

$$\left.\begin{aligned} 8x &= 2 \cdot 2 \cdot 2 \cdot x \\ 12x^2 &= 2 \cdot 2 \cdot 3 \cdot x \cdot x \end{aligned}\right\} \quad \text{LCD} = 2 \cdot 2 \cdot 2 \cdot 3 \cdot x \cdot x, \text{ or } 24x^2.$$

The factors of the LCD missing from $8x$ are 3 and x. The factor of the LCD missing from $12x^2$ is 2. We multiply by 1 to get the LCD in each expression, and then add and simplify, if possible:

$$\begin{aligned} \frac{3}{8x} + \frac{5}{12x^2} &= \frac{3}{8x} \cdot \frac{3 \cdot x}{3 \cdot x} + \frac{5}{12x^2} \cdot \frac{2}{2} \\ &= \frac{9x}{24x^2} + \frac{10}{24x^2} \\ &= \frac{9x + 10}{24x^2}. \end{aligned}$$

Do Exercises 6 and 7.

EXAMPLE 8 Add: $\frac{2a}{a^2 - 1} + \frac{1}{a^2 + a}$.

First, we find the LCD:

$$\left.\begin{aligned} a^2 - 1 &= (a - 1)(a + 1) \\ a^2 + a &= a(a + 1) \end{aligned}\right\} \quad \text{LCD} = a(a - 1)(a + 1).$$

We multiply by 1 to get the LCD in each expression, and then add and simplify:

$$\begin{aligned}
&\frac{2a}{(a - 1)(a + 1)} \cdot \frac{a}{a} + \frac{1}{a(a + 1)} \cdot \frac{a - 1}{a - 1} \\
&= \frac{2a^2}{a(a - 1)(a + 1)} + \frac{a - 1}{a(a - 1)(a + 1)} \\
&= \frac{2a^2 + a - 1}{a(a - 1)(a + 1)} \\
&= \frac{(a + 1)(2a - 1)}{a(a - 1)(a + 1)} \qquad \text{Factoring the numerator in order to simplify} \\
&= \frac{(a + 1)(2a - 1)}{a(a - 1)(a + 1)} \qquad \text{Removing a factor of 1: } \frac{a + 1}{a + 1} = 1 \\
&= \frac{2a - 1}{a(a - 1)}.
\end{aligned}$$

Do Exercise 8.

EXAMPLE 9 Add: $\frac{x + 4}{x - 2} + \frac{x - 7}{x + 5}$.

First, we find the LCD. It is just the product of the denominators:

$$\text{LCD} = (x - 2)(x + 5).$$

We multiply by 1 to get the LCD in each expression, and then add and simplify:

$$\begin{aligned}
\frac{x + 4}{x - 2} \cdot \frac{x + 5}{x + 5} + \frac{x - 7}{x + 5} \cdot \frac{x - 2}{x - 2} &= \frac{(x + 4)(x + 5)}{(x - 2)(x + 5)} + \frac{(x - 7)(x - 2)}{(x - 2)(x + 5)} \\
&= \frac{x^2 + 9x + 20}{(x - 2)(x + 5)} + \frac{x^2 - 9x + 14}{(x - 2)(x + 5)} \\
&= \frac{x^2 + 9x + 20 + x^2 - 9x + 14}{(x - 2)(x + 5)} \\
&= \frac{2x^2 + 34}{(x - 2)(x + 5)}.
\end{aligned}$$

Do Exercise 9.

8. Add:

$$\frac{3}{x^3 - x} + \frac{4}{x^2 + 2x + 1}.$$

9. Add:

$$\frac{x - 2}{x + 3} + \frac{x + 7}{x + 8}.$$

Answers on page A-4

10. Add:

$$\frac{5}{x^2 + 17x + 16} + \frac{3}{x^2 + 9x + 8}.$$

11. Add:

$$\frac{x + 3}{x^2 - 16} + \frac{5}{12 - 3x}.$$

Answers on page A-4

EXAMPLE 10 Add: $\dfrac{x}{x^2 + 11x + 30} + \dfrac{-5}{x^2 + 9x + 20}$.

$$\frac{x}{x^2 + 11x + 30} + \frac{-5}{x^2 + 9x + 20}$$

$$= \frac{x}{(x + 5)(x + 6)} + \frac{-5}{(x + 5)(x + 4)}$$

Factoring the denominators in order to find the LCD. The LCD is $(x + 4)(x + 5)(x + 6)$.

$$= \frac{x}{(x + 5)(x + 6)} \cdot \frac{x + 4}{x + 4} + \frac{-5}{(x + 5)(x + 4)} \cdot \frac{x + 6}{x + 6}$$

Multiplying by 1

$$= \frac{x(x + 4) + (-5)(x + 6)}{(x + 4)(x + 5)(x + 6)} = \frac{x^2 + 4x - 5x - 30}{(x + 4)(x + 5)(x + 6)}$$

$$= \frac{x^2 - x - 30}{(x + 4)(x + 5)(x + 6)}$$

$$= \frac{(x - 6)\cancel{(x + 5)}}{(x + 4)\cancel{(x + 5)}(x + 6)}$$

$$= \frac{(x - 6)}{(x + 4)(x + 6)}$$

Always simplify at the end if possible: $\dfrac{x + 5}{x + 5} = 1$.

Do Exercise 10.

Suppose that after we factor to find the LCD, we find factors that are opposites. There are several ways to handle this, but the easiest is to first go back and multiply by $-1/-1$ appropriately to change factors so they are not opposites.

EXAMPLE 11 Add: $\dfrac{x}{x^2 - 25} + \dfrac{3}{10 - 2x}$.

First, we factor as though we are going to find the LCD:

$$x^2 - 25 = (x - 5)(x + 5);$$
$$10 - 2x = 2(5 - x).$$

We note that there is an $x - 5$ as one factor and a $5 - x$ as another factor. If the denominator of the second expression were $2x - 10$, this situation would not occur. To avoid $10 - 2x$, we first multiply by 1 using $-1/-1$, and then continue as before:

$$\frac{x}{x^2 - 25} + \frac{3}{10 - 2x} = \frac{x}{(x - 5)(x + 5)} + \frac{3}{10 - 2x} \cdot \frac{-1}{-1}$$

$$= \frac{x}{(x - 5)(x + 5)} + \frac{-3}{2x - 10}$$

$$= \frac{x}{(x - 5)(x + 5)} + \frac{-3}{2(x - 5)}$$

LCD = $2(x - 5)(x + 5)$

$$= \frac{x}{(x - 5)(x + 5)} \cdot \frac{2}{2} + \frac{-3}{2(x - 5)} \cdot \frac{x + 5}{x + 5}$$

$$= \frac{2x - 3(x + 5)}{2(x - 5)(x + 5)} = \frac{2x - 3x - 15}{2(x - 5)(x + 5)}$$

$$= \frac{-x - 15}{2(x - 5)(x + 5)}.$$

Collecting like terms

Do Exercise 11.

NAME SECTION DATE

Exercise Set 5.4

a Add. Simplify, if possible.

1. $\frac{5}{8} + \frac{3}{8}$

2. $\frac{3}{16} + \frac{5}{16}$

3. $\frac{1}{3 + x} + \frac{5}{3 + x}$

4. $\frac{4x + 6}{2x - 1} + \frac{5 - 8x}{-1 + 2x}$

5. $\frac{x^2 + 7x}{x^2 - 5x} + \frac{x^2 - 4x}{x^2 - 5x}$

6. $\frac{4}{x + y} + \frac{9}{y + x}$

7. $\frac{7}{8} + \frac{5}{-8}$

8. $\frac{5}{-3} + \frac{11}{3}$

9. $\frac{3}{t} + \frac{4}{-t}$

10. $\frac{5}{-a} + \frac{8}{a}$

11. $\frac{2x + 7}{x - 6} + \frac{3x}{6 - x}$

12. $\frac{2x - 7}{5x - 8} + \frac{6 + 10x}{8 - 5x}$

13. $\frac{y^2}{y - 3} + \frac{9}{3 - y}$

14. $\frac{t^2}{t - 2} + \frac{4}{2 - t}$

15. $\frac{b - 7}{b^2 - 16} + \frac{7 - b}{16 - b^2}$

16. $\frac{a - 3}{a^2 - 25} + \frac{a - 3}{25 - a^2}$

17. $\frac{a^2}{a - b} + \frac{b^2}{b - a}$

18. $\frac{x^2}{x - 7} + \frac{49}{7 - x}$

ANSWERS

1. ________
2. ________
3. ________
4. ________
5. ________
6. ________
7. ________
8. ________
9. ________
10. ________
11. ________
12. ________
13. ________
14. ________
15. ________
16. ________
17. ________
18. ________

ANSWERS

19. ______

20. ______

21. ______

22. ______

23. ______

24. ______

25. ______

26. ______

27. ______

28. ______

29. ______

30. ______

31. ______

32. ______

33. ______

34. ______

35. ______

36. ______

19. $\frac{x+3}{x-5} + \frac{2x-1}{5-x} + \frac{2(3x-1)}{x-5}$

20. $\frac{3(x-2)}{2x-3} + \frac{5(2x+1)}{2x-3} + \frac{3(x+1)}{3-2x}$

21. $\frac{2(4x+1)}{5x-7} + \frac{3(x-2)}{7-5x} + \frac{-10x-1}{5x-7}$

22. $\frac{5(x-2)}{3x-4} + \frac{2(x-3)}{4-3x} + \frac{3(5x+1)}{4-3x}$

23. $\frac{x+1}{(x+3)(x-3)} + \frac{4(x-3)}{(x-3)(x+3)} + \frac{(x-1)(x-3)}{(3-x)(x+3)}$

24. $\frac{2(x+5)}{(2x-3)(x-1)} + \frac{3x+4}{(2x-3)(1-x)} + \frac{x-5}{(3-2x)(x-1)}$

25. $\frac{2}{x} + \frac{5}{x^2}$

26. $\frac{3}{y^2} + \frac{6}{y}$

27. $\frac{5}{6r} + \frac{7}{8r}$

28. $\frac{13}{18x} + \frac{7}{24x}$

29. $\frac{4}{xy^2} + \frac{6}{x^2y}$

30. $\frac{8}{ab^3} + \frac{3}{a^2b}$

31. $\frac{2}{9t^3} + \frac{1}{6t^2}$

32. $\frac{5}{c^2d^3} + \frac{-4}{7cd^2}$

33. $\frac{x+y}{xy^2} + \frac{3x+y}{x^2y}$

34. $\frac{2c-d}{c^2d} + \frac{c+d}{cd^2}$

35. $\frac{3}{x-2} + \frac{3}{x+2}$

36. $\frac{2}{y+1} + \frac{2}{y-1}$

37. $\frac{3}{x+1} + \frac{2}{3x}$

38. $\frac{4}{5y} + \frac{7}{y-2}$

39. $\frac{2x}{x^2-16} + \frac{x}{x-4}$

40. $\frac{4x}{x^2-25} + \frac{x}{x+5}$

41. $\frac{5}{z+4} + \frac{3}{3z+12}$

42. $\frac{t}{t-3} + \frac{5}{4t-12}$

43. $\frac{3}{x-1} + \frac{2}{(x-1)^2}$

44. $\frac{8}{(y+3)^2} + \frac{5}{y+3}$

45. $\frac{4a}{5a-10} + \frac{3a}{10a-20}$

46. $\frac{9x}{6x-30} + \frac{3x}{4x-20}$

47. $\frac{x+4}{x} + \frac{x}{x+4}$

48. $\frac{a}{a-3} + \frac{a-3}{a}$

49. $\frac{4}{a^2-a-2} + \frac{3}{a^2+4a+3}$

50. $\frac{a}{a^2-2a+1} + \frac{1}{a^2-5a+4}$

51. $\frac{x+3}{x-5} + \frac{x-5}{x+3}$

52. $\frac{3x}{2y-3} + \frac{2x}{3y-2}$

53. $\frac{a}{a^2-1} + \frac{2a}{a^2-a}$

54. $\frac{3x+2}{3x+6} + \frac{x-2}{x^2-4}$

ANSWERS

37. ______

38. ______

39. ______

40. ______

41. ______

42. ______

43. ______

44. ______

45. ______

46. ______

47. ______

48. ______

49. ______

50. ______

51. ______

52. ______

53. ______

54. ______

ANSWERS

55. ______

56. ______

57. ______

58. ______

59. ______

60. ______

61. ______

62. ______

63. ______

64. ______

65. ______

66. ______

67. ______

68. ______

69. ______

70. ______

71. ______

72. ______

55. $\dfrac{6}{x-y} + \dfrac{4x}{y^2 - x^2}$

56. $\dfrac{a-2}{3-a} + \dfrac{4-a^2}{a^2-9}$

57. $\dfrac{4-a}{25-a^2} + \dfrac{a+1}{a-5}$

58. $\dfrac{x+2}{x-7} + \dfrac{3-x}{49-x^2}$

59. $\dfrac{2}{t^2+t-6} + \dfrac{3}{t^2-9}$

60. $\dfrac{10}{a^2-a-6} + \dfrac{3a}{a^2+4a+4}$

SKILL MAINTENANCE

Subtract.

61. $(x^2 + x) - (x + 1)$

62. $(4y^3 - 5y^2 + 7y - 24) - (-9y^3 + 9y^2 - 5y + 49)$

Simplify.

63. $(2x^4y^3)^{-3}$

64. $\left(\dfrac{x^3}{5y}\right)^2$

65. $\left(\dfrac{x^{-4}}{y^7}\right)^3$

66. $(5x^{-2}y^{-3})^2$

SYNTHESIS

Find the perimeter and the area of the figure.

67.

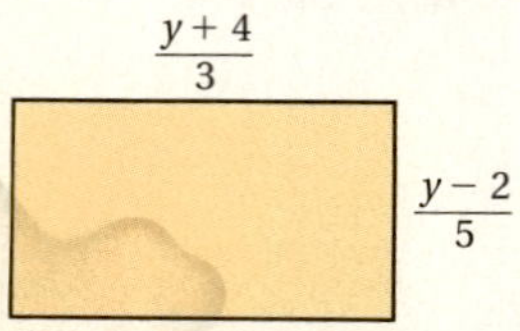

68.

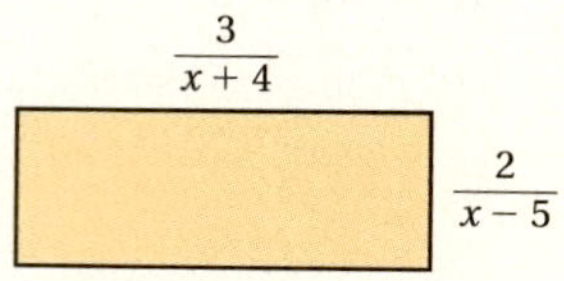

Add. Simplify, if possible.

69. $\dfrac{5}{z+2} + \dfrac{4z}{z^2-4} + 2$

70. $\dfrac{-2}{y^2-9} + \dfrac{4y}{(y-3)^2} + \dfrac{6}{3-y}$

71. $\dfrac{3z^2}{z^4-4} + \dfrac{5z^2-3}{2z^4+z^2-6}$

72. Find an expression equivalent to

$$\frac{a-3b}{a-b}$$

that is a sum of two fractional expressions. Answers can vary.

5.5 Subtracting Rational Expressions

a We subtract rational expressions as we do rational numbers.

> To subtract when the denominators are the same, subtract the numerators and keep the same denominator.

EXAMPLE 1 Subtract: $\dfrac{8}{x} - \dfrac{3}{x}$.

$$\frac{8}{x} - \frac{3}{x} = \frac{8-3}{x} = \frac{5}{x}$$

EXAMPLE 2 Subtract: $\dfrac{3x}{x+2} - \dfrac{x-2}{x+2}$.

$$\frac{3x}{x+2} - \frac{x-2}{x+2} = \frac{3x-(x-2)}{x+2}$$

The parentheses are important to make sure that you subtract the entire numerator.

$$= \frac{3x-x+2}{x+2} = \frac{2x+2}{x+2}$$

Do Exercises 1–3.

When one denominator is the additive inverse of the other, we can first multiply one expression by $-1/-1$ to obtain a common denominator.

EXAMPLE 3 Subtract: $\dfrac{x}{5} - \dfrac{3x-4}{-5}$.

$$\frac{x}{5} - \frac{3x-4}{-5} = \frac{x}{5} - \frac{3x-4}{-5}\cdot\frac{-1}{-1}$$

Multiplying by $\frac{-1}{-1}$

This is equal to 1 (not −1).

$$= \frac{x}{5} - \frac{(3x-4)(-1)}{(-5)(-1)}$$

$$= \frac{x}{5} - \frac{4-3x}{5}$$

Remember the parentheses!

$$= \frac{x-(4-3x)}{5}$$

$$= \frac{x-4+3x}{5} = \frac{4x-4}{5}$$

EXAMPLE 4 Subtract: $\dfrac{5y}{y-5} - \dfrac{2y-3}{5-y}$.

$$\frac{5y}{y-5} - \frac{2y-3}{5-y} = \frac{5y}{y-5} - \frac{2y-3}{5-y}\cdot\frac{-1}{-1}$$

$$= \frac{5y}{y-5} - \frac{(2y-3)(-1)}{(5-y)(-1)} = \frac{5y}{y-5} - \frac{3-2y}{y-5}$$

Remember the parentheses!

$$= \frac{5y-(3-2y)}{y-5}$$

OBJECTIVES

After finishing Section 5.5, you should be able to:

a Subtract rational expressions.

b Simplify combined additions and subtractions of rational expressions.

FOR EXTRA HELP

TAPE 13

TAPE 11A

MAC: 5A
IBM: 5A

Subtract.

1. $\dfrac{7}{11} - \dfrac{3}{11}$

2. $\dfrac{7}{y} - \dfrac{2}{y}$

3. $\dfrac{2x^2+3x-7}{2x+1} - \dfrac{x^2+x-8}{2x+1}$

Answers on page A-4

Subtract.

4. $\frac{x}{3} - \frac{2x-1}{-3}$

5. $\frac{3x}{x-2} - \frac{x-3}{2-x}$

6. Subtract:

$$\frac{x-2}{3x} - \frac{2x-1}{5x}.$$

Answers on page A-4

Then

$$= \frac{5y - 3 + 2y}{y - 5}$$

$$= \frac{7y-3}{y-5}.$$

Do Exercises 4 and 5.

To subtract rational expressions with different denominators, we use a procedure similar to what we used for addition, except that we subtract numerators and write the difference over the LCD.

> To subtract rational expressions with different denominators:
>
> 1. Find the LCM of the denominators. This is the least common denominator (LCD).
> 2. For each rational expression, find an equivalent expression with the LCD. To do so, multiply by 1 using a symbol for 1 made up of factors of the LCD that are missing from the original denominator.
> 3. Subtract the numerators. Write the difference over the LCD.
> 4. Simplify, if possible.

EXAMPLE 5 Subtract: $\frac{x+2}{x-4} - \frac{x+1}{x+4}$.

The LCD $= (x-4)(x+4)$.

$$\frac{x+2}{x-4} \cdot \frac{x+4}{x+4} - \frac{x+1}{x+4} \cdot \frac{x-4}{x-4}$$

$$= \frac{(x+2)(x+4)}{(x-4)(x+4)} - \frac{(x+1)(x-4)}{(x-4)(x+4)}$$

$$= \frac{x^2+6x+8}{(x-4)(x+4)} - \frac{x^2-3x-4}{(x-4)(x+4)}$$

$$= \frac{x^2+6x+8-(x^2-3x-4)}{(x-4)(x+4)}$$

Subtracting this numerator. Don't forget the parentheses.

$$= \frac{x^2+6x+8-x^2+3x+4}{(x-4)(x+4)}$$

$$= \frac{9x+12}{(x-4)(x+4)}$$

Do Exercise 6.

EXAMPLE 6 Subtract: $\frac{x}{x^2+5x+6} - \frac{2}{x^2+3x+2}$.

$$\frac{x}{x^2+5x+6} - \frac{2}{x^2+3x+2}$$

$$= \frac{x}{(x+2)(x+3)} - \frac{2}{(x+2)(x+1)}$$

LCD $= (x+1)(x+2)(x+3)$

$$= \frac{x}{(x+2)(x+3)} \cdot \frac{x+1}{x+1} - \frac{2}{(x+2)(x+1)} \cdot \frac{x+3}{x+3}$$

$$= \frac{x^2+x}{(x+1)(x+2)(x+3)} - \frac{2x+6}{(x+1)(x+2)(x+3)}$$

Then

Subtracting this numerator. Don't forget the parentheses.

$$= \frac{x^2 + x - (2x + 6)}{(x + 1)(x + 2)(x + 3)}$$

$$= \frac{x^2 + x - 2x - 6}{(x + 1)(x + 2)(x + 3)}$$

$$= \frac{x^2 - x - 6}{(x + 1)(x + 2)(x + 3)}$$

$$= \frac{(x + 2)(x - 3)}{(x + 1)(x + 2)(x + 3)}$$

$$= \frac{\cancel{(x + 2)}(x - 3)}{(x + 1)\cancel{(x + 2)}(x + 3)}$$ Simplifying by removing a factor of 1: $\frac{x + 2}{x + 2} = 1$

$$= \frac{x - 3}{(x + 1)(x + 3)}.$$

Do Exercise 7.

Suppose that after we factor to find the LCD, we find factors that are opposites. Then we multiply by $-1/-1$ appropriately to change factors so that they are not opposites.

EXAMPLE 7 Subtract: $\frac{p}{64 - p^2} - \frac{5}{p - 8}$.

Factoring $64 - p^2$, we get $(8 - p)(8 + p)$. Note that the factor $8 - p$ in the first denominator and $p - 8$ in the second denominator are opposites. We multiply the first expression by $-1/-1$ to avoid this situation. Then we proceed as before.

$$\frac{p}{64 - p^2} - \frac{5}{p - 8} = \frac{p}{64 - p^2} \cdot \frac{-1}{-1} - \frac{5}{p - 8}$$

$$= \frac{-p}{p^2 - 64} - \frac{5}{p - 8}$$

$$= \frac{-p}{(p - 8)(p + 8)} - \frac{5}{p - 8}$$ LCD = $(p - 8)(p + 8)$

$$= \frac{-p}{(p - 8)(p + 8)} - \frac{5}{p - 8} \cdot \frac{p + 8}{p + 8}$$

$$= \frac{-p}{(p - 8)(p + 8)} - \frac{5p + 40}{(p - 8)(p + 8)}$$

Subtracting this numerator. Don't forget the parentheses.

$$= \frac{-p - (5p + 40)}{(p - 8)(p + 8)}$$

$$= \frac{-p - 5p - 40}{(p - 8)(p + 8)}$$

$$= \frac{-6p - 40}{(p - 8)(p + 8)}$$

Do Exercise 8.

7. Subtract:

$$\frac{x}{x^2 + 15x + 56} - \frac{6}{x^2 + 13x + 42}.$$

8. Subtract:

$$\frac{y}{16 - y^2} - \frac{7}{y - 4}.$$

Answers on page A-4

9. Perform the indicated operations and simplify:

$$\frac{x+2}{x^2-9}-\frac{x-7}{9-x^2}+\frac{-8-x}{x^2-9}.$$

10. Perform the indicated operations and simplify:

$$\frac{1}{x}-\frac{5}{3x}+\frac{2x}{x+1}.$$

Answers on page A-4

b COMBINED ADDITIONS AND SUBTRACTIONS

Now let us look at some combined additions and subtractions.

EXAMPLE 8 Perform the indicated operations and simplify:

$$\frac{x+9}{x^2-4}+\frac{5-x}{4-x^2}-\frac{2+x}{x^2-4}.$$

We have

$$\begin{aligned}\frac{x+9}{x^2-4}+\frac{5-x}{4-x^2}-\frac{2+x}{x^2-4} &= \frac{x+9}{x^2-4}+\frac{5-x}{4-x^2}\cdot\frac{-1}{-1}-\frac{2+x}{x^2-4}\\ &= \frac{x+9}{x^2-4}+\frac{x-5}{x^2-4}-\frac{2+x}{x^2-4}\\ &= \frac{(x+9)+(x-5)-(2+x)}{x^2-4}\\ &= \frac{x+9+x-5-2-x}{x^2-4}\\ &= \frac{x+2}{x^2-4}\\ &= \frac{\cancel{(x+2)}\cdot 1}{\cancel{(x+2)}(x-2)} \qquad \frac{x+2}{x+2}=1\\ &= \frac{1}{x-2}.\end{aligned}$$

Do Exercise 9.

EXAMPLE 9 Perform the indicated operations and simplify:

$$\frac{1}{x}-\frac{1}{x^2}+\frac{2}{x+1}.$$

The LCD $= x\cdot x(x+1)$, or $x^2(x+1)$.

$$\begin{aligned}&\frac{1}{x}\cdot\frac{x(x+1)}{x(x+1)}-\frac{1}{x^2}\cdot\frac{(x+1)}{(x+1)}+\frac{2}{x+1}\cdot\frac{x^2}{x^2}\\ &= \frac{x(x+1)}{x^2(x+1)}-\frac{x+1}{x^2(x+1)}+\frac{2x^2}{x^2(x+1)}\\ &= \frac{x(x+1)-(x+1)+2x^2}{x^2(x+1)} \qquad \text{Subtracting this numerator. Don't forget the parentheses.}\\ &= \frac{x^2+x-x-1+2x^2}{x^2(x+1)}\\ &= \frac{3x^2-1}{x^2(x+1)}\end{aligned}$$

Do Exercise 10.

NAME SECTION DATE

Exercise Set 5.5

a Subtract. Simplify, if possible.

1. $\dfrac{7}{x} - \dfrac{3}{x}$

2. $\dfrac{5}{a} - \dfrac{8}{a}$

3. $\dfrac{y}{y-4} - \dfrac{4}{y-4}$

4. $\dfrac{t^2}{t+5} - \dfrac{25}{t+5}$

5. $\dfrac{2x-3}{x^2+3x-4} - \dfrac{x-7}{x^2+3x-4}$

6. $\dfrac{x+1}{x^2-2x+1} - \dfrac{5-3x}{x^2-2x+1}$

7. $\dfrac{11}{6} - \dfrac{5}{-6}$

8. $\dfrac{5}{9} - \dfrac{7}{-9}$

9. $\dfrac{5}{a} - \dfrac{8}{-a}$

10. $\dfrac{8}{x} - \dfrac{3}{-x}$

11. $\dfrac{4}{y-1} - \dfrac{4}{1-y}$

12. $\dfrac{5}{a-2} - \dfrac{3}{2-a}$

13. $\dfrac{3-x}{x-7} - \dfrac{2x-5}{7-x}$

14. $\dfrac{t^2}{t-2} - \dfrac{4}{2-t}$

ANSWERS

1. ______
2. ______
3. ______
4. ______
5. ______
6. ______
7. ______
8. ______
9. ______
10. ______
11. ______
12. ______
13. ______
14. ______

ANSWERS

15. ______

16. ______

17. ______

18. ______

19. ______

20. ______

21. ______

22. ______

23. ______

24. ______

25. ______

26. ______

27. ______

28. ______

15. $\frac{a-2}{a^2-25} - \frac{6-a}{25-a^2}$

16. $\frac{x-8}{x^2-16} - \frac{x-8}{16-x^2}$

17. $\frac{4-x}{x-9} - \frac{3x-8}{9-x}$

18. $\frac{4x-6}{x-5} - \frac{7-2x}{5-x}$

19. $\frac{2(x-1)}{2x-3} - \frac{3(x+2)}{2x-3} - \frac{x-1}{3-2x}$

20. $\frac{5(2y+1)}{2y-3} - \frac{3(y-1)}{3-2y} - \frac{3(y-2)}{2y-3}$

21. $\frac{a-2}{10} - \frac{a+1}{5}$

22. $\frac{y+3}{2} - \frac{y-4}{4}$

23. $\frac{4z-9}{3z} - \frac{3z-8}{4z}$

24. $\frac{a-1}{4a} - \frac{2a+3}{a}$

25. $\frac{4x+2t}{3xt^2} - \frac{5x-3t}{x^2t}$

26. $\frac{5x+3y}{2x^2y} - \frac{3x+4y}{xy^2}$

27. $\frac{5}{x+5} - \frac{3}{x-5}$

28. $\frac{3t}{t-1} - \frac{8t}{t+1}$

29. $\dfrac{3}{2t^2 - 2t} - \dfrac{5}{2t - 2}$

30. $\dfrac{11}{x^2 - 4} - \dfrac{8}{x + 2}$

31. $\dfrac{2s}{t^2 - s^2} - \dfrac{s}{t - s}$

32. $\dfrac{3}{12 + x - x^2} - \dfrac{2}{x^2 - 9}$

33. $\dfrac{y - 5}{y} - \dfrac{3y - 1}{4y}$

34. $\dfrac{3x - 2}{4x} - \dfrac{3x + 1}{6x}$

35. $\dfrac{a}{x + a} - \dfrac{a}{x - a}$

36. $\dfrac{a}{a - b} - \dfrac{a}{a + b}$

37. $\dfrac{5x}{x^2 - 9} - \dfrac{4}{3 - x}$

38. $\dfrac{8x}{16 - x^2} - \dfrac{5}{x - 4}$

39. $\dfrac{t^2}{2t^2 - 2t} - \dfrac{1}{2t - 2}$

40. $\dfrac{4}{5a^2 - 5a} - \dfrac{2}{5a - 5}$

41. $\dfrac{x}{x^2 + 5x + 6} - \dfrac{2}{x^2 + 3x + 2}$

42. $\dfrac{a}{a^2 + 11a + 30} - \dfrac{5}{a^2 + 9a + 20}$

ANSWERS

29. __________

30. __________

31. __________

32. __________

33. __________

34. __________

35. __________

36. __________

37. __________

38. __________

39. __________

40. __________

41. __________

42. __________

ANSWERS

43. ______

44. ______

45. ______

46. ______

47. ______

48. ______

49. ______

50. ______

51. ______

52. ______

53. ______

54. ______

55. ______

56. ______

57. ______

58. ______

59. ______

60. ______

61. ______

62. ______

b Perform the indicated operations and simplify.

43. $\dfrac{3(2x+5)}{x-1} - \dfrac{3(2x-3)}{1-x} + \dfrac{6x-1}{x-1}$

44. $\dfrac{a-2b}{b-a} - \dfrac{3a-3b}{a-b} + \dfrac{2a-b}{a-b}$

45. $\dfrac{x-y}{x^2-y^2} + \dfrac{x+y}{x^2-y^2} - \dfrac{2x}{x^2-y^2}$

46. $\dfrac{x-3y}{2(y-x)} + \dfrac{x+y}{2(x-y)} - \dfrac{2x-2y}{2(x-y)}$

47. $\dfrac{10}{2y-1} - \dfrac{6}{1-2y} + \dfrac{y}{2y-1} + \dfrac{y-4}{1-2y}$

48. $\dfrac{(x+1)(2x-1)}{(2x-3)(x-3)} - \dfrac{(x-3)(x+1)}{(3-x)(3-2x)} + \dfrac{(2x+1)(x+3)}{(3-2x)(x-3)}$

49. $\dfrac{a+6}{4-a^2} - \dfrac{a+3}{a+2} + \dfrac{a-3}{2-a}$

50. $\dfrac{4t}{t^2-1} - \dfrac{2}{t} - \dfrac{2}{t+1}$

51. $\dfrac{2z}{1-2z} + \dfrac{3z}{2z+1} - \dfrac{3}{4z^2-1}$

52. $\dfrac{1}{x-y} - \dfrac{2x}{x^2-y^2} + \dfrac{1}{x+y}$

53. $\dfrac{1}{x+y} - \dfrac{1}{x-y} + \dfrac{2x}{x^2-y^2}$

54. $\dfrac{2b}{a^2-b^2} - \dfrac{1}{a+b} + \dfrac{1}{a-b}$

SKILL MAINTENANCE

Simplify.

55. $\dfrac{x^8}{x^3}$

56. $3x^4 \cdot 10x^8$

57. $(a^2b^{-5})^{-4}$

58. $\dfrac{54x^{10}}{3x^7}$

59. $\dfrac{66x^2}{11x^5}$

60. $5x^{-7} \cdot 2x^4$

SYNTHESIS

Subtract. Simplify, if possible.

61. $\dfrac{5}{3-2x} + \dfrac{3}{2x-3} - \dfrac{x-3}{2x^2-x-3}$

62. $\dfrac{2r}{r^2-s^2} + \dfrac{1}{r+s} - \dfrac{1}{r-s}$

5.6 Solving Rational Equations

a In Sections 5.1–5.5 we studied operations with rational *expressions*. These are expressions without an equals sign. We cannot clear expressions of fractions other than occasionally when simplifying by removing a factor of 1. In this section, we study rational *equations*. Equations do have an equals sign, and we can clear them of fractions as we did in Section 2.3.

A **rational**, or **fractional, equation** is an equation containing one or more rational expressions. Here are some examples:

$$\frac{2}{3} + \frac{5}{6} = \frac{x}{9}, \qquad x - \frac{6}{x} = -5, \qquad \frac{x^2}{x-1} = \frac{1}{x-1}.$$

> To solve a rational equation, the first step is to clear the equation of fractions. To do this, multiply both sides of the equation by the LCM of all the denominators. Then carry out the equation-solving process as we learned it in Chapter 2.

When clearing an equation of fractions, we use the terminology LCM instead of LCD because we are *not* adding or subtracting rational expressions.

EXAMPLE 1 Solve: $\frac{2}{3} + \frac{5}{6} = \frac{x}{9}$.

The LCM of all denominators is $2 \cdot 3 \cdot 3$, or 18. We multiply on both sides by 18:

$$\frac{2}{3} + \frac{5}{6} = \frac{x}{9}$$

$$18\left(\frac{2}{3} + \frac{5}{6}\right) = 18 \cdot \frac{x}{9} \qquad \text{Multiplying on both sides by the LCM}$$

$$18 \cdot \frac{2}{3} + 18 \cdot \frac{5}{6} = 18 \cdot \frac{x}{9} \qquad \text{Multiplying to remove parentheses}$$

When clearing an equation of fractions, be sure to multiply *each* term by the LCM.

$$12 + 15 = 2x \qquad \text{Simplifying. Note that we have now cleared fractions.}$$

$$27 = 2x$$

$$\frac{27}{2} = x.$$

Do Exercise 1.

OBJECTIVE

After finishing Section 5.6, you should be able to:

a Solve rational equations.

FOR EXTRA HELP

TAPE 13 | TAPE 11B | MAC: 5A IBM: 5A

1. Solve: $\frac{3}{4} + \frac{5}{8} = \frac{x}{12}$.

Answer on page A-4

2. Solve: $\frac{1}{x} = \frac{1}{6 - x}$.

3. Solve: $\frac{x}{4} - \frac{x}{6} = \frac{1}{8}$.

Answers on page A-4

EXAMPLE 2 Solve: $\frac{1}{x} = \frac{1}{4 - x}$.

The LCM is $x(4 - x)$. We multiply on both sides by $x(4 - x)$:

$$\frac{1}{x} = \frac{1}{4 - x}$$

$$x(4 - x) \cdot \frac{1}{x} = x(4 - x) \cdot \frac{1}{4 - x} \quad \text{Multiplying on both sides by the LCM}$$

$$4 - x = x \quad \text{Simplifying}$$

$$4 = 2x$$

$$x = 2.$$

CHECK:

$\frac{1}{x} = \frac{1}{4 - x}$	
$\frac{1}{2}$	$\frac{1}{4 - 2}$
	$\frac{1}{2}$ TRUE

This checks, so 2 is the solution.

Do Exercise 2.

EXAMPLE 3 Solve: $\frac{x}{6} - \frac{x}{8} = \frac{1}{12}$.

The LCM is 24. We multiply on both sides by 24:

$$\frac{x}{6} - \frac{x}{8} = \frac{1}{12}$$

$$24\left(\frac{x}{6} - \frac{x}{8}\right) = 24 \cdot \frac{1}{12} \quad \text{Multiplying on both sides by the LCM}$$

$$24 \cdot \frac{x}{6} - 24 \cdot \frac{x}{8} = 24 \cdot \frac{1}{12} \quad \text{Multiplying to remove parentheses}$$

Be sure to multiply *each* term by the LCM.

$$4x - 3x = 2 \quad \text{Simplifying}$$

$$x = 2.$$

CHECK:

$\frac{x}{6} - \frac{x}{8} = \frac{1}{12}$	
$\frac{2}{6} - \frac{2}{8}$	$\frac{1}{12}$
$\frac{1}{3} - \frac{1}{4}$	
$\frac{4}{12} - \frac{3}{12}$	
$\frac{1}{12}$	TRUE

This checks, so the solution is 2.

Do Exercise 3.

EXAMPLE 4 Solve: $\frac{2}{3x} + \frac{1}{x} = 10$.

The LCM is $3x$. We multiply on both sides by $3x$:

$$\frac{2}{3x} + \frac{1}{x} = 10$$

$$3x\left(\frac{2}{3x} + \frac{1}{x}\right) = 3x \cdot 10 \qquad \text{Multiplying on both sides by the LCM}$$

$$3x \cdot \frac{2}{3x} + 3x \cdot \frac{1}{x} = 3x \cdot 10 \qquad \text{Multiplying to remove parentheses}$$

$$2 + 3 = 30x \qquad \text{Simplifying}$$

$$5 = 30x$$

$$\frac{5}{30} = x$$

$$\frac{1}{6} = x.$$

We leave the check to the student. The solution is $\frac{1}{6}$.

Do Exercise 4.

4. Solve: $\frac{1}{2x} + \frac{1}{x} = -12$.

EXAMPLE 5 Solve: $x + \frac{6}{x} = -5$.

The LCM is x. We multiply on both sides by x:

$$x + \frac{6}{x} = -5$$

$$x\left(x + \frac{6}{x}\right) = -5x \qquad \text{Multiplying on both sides by } x$$

$$x \cdot x + x \cdot \frac{6}{x} = -5x \qquad \text{Note that each rational expression on the left is now multiplied by } x.$$

$$x^2 + 6 = -5x \qquad \text{Simplifying}$$

$$x^2 + 5x + 6 = 0 \qquad \text{Adding } 5x \text{ to get a 0 on one side}$$

$$(x + 3)(x + 2) = 0 \qquad \text{Factoring}$$

$$x + 3 = 0 \quad \text{or} \quad x + 2 = 0 \qquad \text{Using the principle of zero products}$$

$$x = -3 \quad \text{or} \quad x = -2.$$

CHECK:

For -3:

$x + \frac{6}{x} = -5$	
$-3 + \frac{6}{-3}$	-5
$-3 - 2$	
-5	TRUE

For -2:

$x + \frac{6}{x} = -5$	
$-2 + \frac{6}{-2}$	-5
$-2 - 3$	
-5	TRUE

Both of these check, so there are two solutions, -3 and -2.

Do Exercise 5.

5. Solve: $x + \frac{1}{x} = 2$.

Answers on page A-4

6. Solve: $\dfrac{x^2}{x+2} = \dfrac{4}{x+2}$.

7. Solve: $\dfrac{4}{x-2} + \dfrac{1}{x+2} = \dfrac{26}{x^2-4}$.

Caution! We have introduced a new use of the LCM in this section. You previously used the LCM in adding or subtracting rational expressions. *Now* we have equations with equals signs. We clear fractions by multiplying on both sides of the equation by the LCM. This eliminates the denominators. Do *not* make the mistake of trying to clear fractions when you do not have an equation.

Answers on page A-4

When we multiply on both sides of an equation by the LCM, the resulting equation might have solutions that are *not* solutions of the original equation. Thus we must *always* check possible solutions in the original equation.

1. If you have carried out all algebraic procedures correctly, you need only check to see if a number makes a denominator 0 in the original equation.
2. To be sure that no computational errors have been made and that you indeed have a solution, a complete check is necessary, as we did in Chapter 2.

The next example illustrates the importance of checking all possible solutions.

EXAMPLE 6 Solve: $\dfrac{x^2}{x-1} = \dfrac{1}{x-1}$.

The LCM is $x - 1$. We multiply on both sides by $x - 1$:

$$\frac{x^2}{x-1} = \frac{1}{x-1}$$

$$(x-1)\cdot\frac{x^2}{x-1} = (x-1)\cdot\frac{1}{x-1} \quad \text{Multiplying on both sides by } x-1$$

$$x^2 = 1 \quad \text{Simplifying}$$

$$x^2 - 1 = 0 \quad \text{Subtracting 1 to get a 0 on one side}$$

$$(x-1)(x+1) = 0 \quad \text{Factoring}$$

$$x - 1 = 0 \quad \text{or} \quad x + 1 = 0 \quad \text{Using the principle of zero products}$$

$$x = 1 \quad \text{or} \quad x = -1.$$

The numbers 1 and -1 are possible solutions. We look at the original equation and see that 1 makes a denominator 0 and is therefore not a solution. The number -1 checks and is a solution.

Do Exercise 6.

EXAMPLE 7 Solve: $\dfrac{3}{x-5} + \dfrac{1}{x+5} = \dfrac{2}{x^2-25}$.

The LCM is $(x-5)(x+5)$. We multiply on both sides by $(x-5)(x+5)$:

$$(x-5)(x+5)\left(\frac{3}{x-5} + \frac{1}{x+5}\right) = (x-5)(x+5)\left(\frac{2}{x^2-25}\right)$$

Multiplying on both sides by the LCM

$$(x-5)(x+5)\cdot\frac{3}{x-5} + (x-5)(x+5)\cdot\frac{1}{x+5} = (x-5)(x+5)\cdot\frac{2}{x^2-25}$$

$$3(x+5) + (x-5) = 2 \quad \text{Simplifying}$$

$$3x + 15 + x - 5 = 2 \quad \text{Removing parentheses}$$

$$4x + 10 = 2$$

$$4x = -8$$

$$x = -2.$$

The check is left to the student. The number -2 checks and is the solution.

Do Exercise 7.

NAME SECTION DATE

Exercise Set 5.6

a Solve. Don't forget to check!

1. $\frac{3}{8} + \frac{4}{5} = \frac{x}{20}$

2. $\frac{3}{5} + \frac{2}{3} = \frac{x}{9}$

3. $\frac{1}{x} = \frac{2}{3} - \frac{5}{6}$

4. $\frac{1}{x} = \frac{1}{8} - \frac{3}{5}$

5. $\frac{1}{6} + \frac{1}{8} = \frac{1}{t}$

6. $\frac{1}{8} + \frac{1}{12} = \frac{1}{t}$

7. $x + \frac{4}{x} = -5$

8. $\frac{10}{x} - x = 3$

9. $\frac{x}{4} - \frac{4}{x} = 0$

10. $\frac{x}{5} - \frac{5}{x} = 0$

11. $\frac{5}{x} = \frac{6}{x} - \frac{1}{3}$

12. $\frac{4}{x} = \frac{5}{x} - \frac{1}{2}$

ANSWERS

1. ______
2. ______
3. ______
4. ______
5. ______
6. ______
7. ______
8. ______
9. ______
10. ______
11. ______
12. ______

ANSWERS

13. ________

14. ________

15. ________

16. ________

17. ________

18. ________

19. ________

20. ________

21. ________

22. ________

13. $\frac{5}{3x} + \frac{3}{x} = 1$

14. $\frac{5}{2y} + \frac{8}{y} = 1$

15. $\frac{t-2}{t+3} = \frac{3}{8}$

16. $\frac{x-7}{x+2} = \frac{1}{4}$

17. $\frac{2}{x+1} = \frac{1}{x-2}$

18. $\frac{8}{y-3} = \frac{6}{y+4}$

19. $\frac{x}{6} - \frac{x}{10} = \frac{1}{6}$

20. $\frac{x}{8} - \frac{x}{12} = \frac{1}{8}$

21. $\frac{t+2}{5} - \frac{t-2}{4} = 1$

22. $\frac{x+1}{3} - \frac{x-1}{2} = 1$

23. $\dfrac{a-3}{3a+2}=\dfrac{1}{5}$

24. $\dfrac{x+7}{8x-5}=\dfrac{2}{3}$

25. $\dfrac{x-1}{x-5}=\dfrac{4}{x-5}$

26. $\dfrac{x-7}{x-9}=\dfrac{2}{x-9}$

27. $\dfrac{2}{x+3}=\dfrac{5}{x}$

28. $\dfrac{6}{y}=\dfrac{5}{y-8}$

29. $\dfrac{x-2}{x-3}=\dfrac{x-1}{x+1}$

30. $\dfrac{t+5}{t-2}=\dfrac{t-2}{t+4}$

31. $\dfrac{1}{x+3}+\dfrac{1}{x-3}=\dfrac{1}{x^2-9}$

32. $\dfrac{4}{x-3}+\dfrac{2x}{x^2-9}=\dfrac{1}{x+3}$

ANSWERS

23. ______

24. ______

25. ______

26. ______

27. ______

28. ______

29. ______

30. ______

31. ______

32. ______

ANSWERS

33. ______

34. ______

35. ______

36. ______

37. ______

38. ______

39. ______

40. ______

41. ______

42. ______

43. ______

44. ______

45. ______

46. ______

47. ______

48. ______

49. ______

50. ______

33. $\dfrac{x}{x+4} - \dfrac{4}{x-4} = \dfrac{x^2+16}{x^2-16}$

34. $\dfrac{5}{y-3} - \dfrac{30}{y^2-9} = 1$

35. $\dfrac{4-a}{8-a} = \dfrac{4}{a-8}$

36. $\dfrac{-3}{x-7} = \dfrac{-10-x}{7-x}$

SKILL MAINTENANCE

Simplify.

37. $(a^2b^5)^{-3}$

38. $(x^{-2}y^{-3})^{-4}$

39. $\left(\dfrac{2x}{t^2}\right)^4$

40. $\left(\dfrac{y^3}{w^2}\right)^{-2}$

41. $4x^{-5} \cdot 8x^{11}$

42. $(8x^5y^{-4})^2$

SYNTHESIS

Solve.

43. $\dfrac{4}{y-2} - \dfrac{2y-3}{y^2-4} = \dfrac{5}{y+2}$

44. $\dfrac{x}{x^2+3x-4} + \dfrac{x+1}{x^2+6x+8} = \dfrac{2x}{x^2+x-2}$

45. $\dfrac{x+1}{x+2} = \dfrac{x+3}{x+4}$

46. $\dfrac{x^2}{x^2-4} = \dfrac{x}{x+2} - \dfrac{2x}{2-x}$

47. $4a - 3 = \dfrac{a+13}{a+1}$

48. $\dfrac{3x-9}{x-3} = \dfrac{5x-4}{2}$

49. $\dfrac{y^2-4}{y+3} = 2 - \dfrac{y-2}{y+3}$

50. $\dfrac{3a-5}{a^2+4a+3} + \dfrac{2a+2}{a+3} = \dfrac{a-3}{a+1}$

5.7 *Solving Problems and Proportions*

a SOLVING PROBLEMS

EXAMPLE 1 If 2 is subtracted from a number and then the reciprocal is found, the result is twice the reciprocal of the number itself. What is the number?

1. ***Familiarize.*** Let us try to guess such a number. Try 10: $10 - 2$ is 8, and the reciprocal of 8 is $\frac{1}{8}$. Two times the reciprocal of 10 is $2\left(\frac{1}{10}\right)$, or $\frac{1}{5}$. Since $\frac{1}{8} \neq \frac{1}{5}$, the number 10 does not check, but the process helps us understand the translation. Let x = the number.

2. ***Translate.*** From the familiarization step, we get the following translation. Subtracting 2 from the number gives us $x - 2$. Twice the reciprocal of the original number is $2(1/x)$.

The reciprocal of 2 less than the number **is** twice the reciprocal of the number.

$$\frac{1}{x-2} = 2 \cdot \frac{1}{x} \qquad \text{Translating}$$

3. ***Solve.*** We solve the equation. The LCM is $x(x - 2)$.

$$x(x-2) \cdot \frac{1}{x-2} = x(x-2) \cdot \frac{2}{x} \qquad \text{Multiplying by the LCM}$$
$$x = 2(x-2) \qquad \text{Simplifying}$$
$$x = 2x - 4$$
$$-x = -4$$
$$x = 4$$

4. ***Check.*** We go back to the original problem. The number to be checked is 4. Two from 4 is 2. The reciprocal of 2 is $\frac{1}{2}$. The reciprocal of the number itself is $\frac{1}{4}$. Since $\frac{1}{2}$ is twice $\frac{1}{4}$, the conditions are satisfied.

5. ***State.*** The number is 4.

Do Exercise 1.

EXAMPLE 2 A cheetah can run 35 km/h faster than a greyhound. A cheetah can run 19 km in the same time that a greyhound can run 12 km. Find the speed of each animal.

1. ***Familiarize.*** We first make a drawing. Let r = the speed of the greyhound. Then $r + 35$ = the speed of the cheetah.

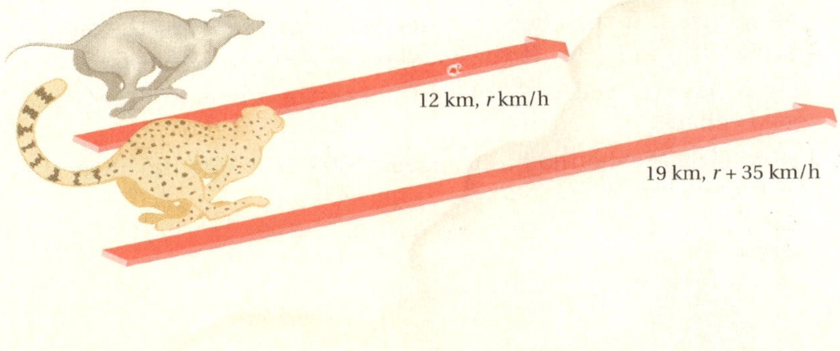

OBJECTIVES

After finishing Section 5.7, you should be able to:

a Solve applied problems using rational equations.

b Solve proportion problems.

FOR EXTRA HELP

TAPE 14 | TAPE 11B | MAC: 5B IBM: 5B

1. The reciprocal of 2 more than a number is three times the reciprocal of the number. Find the number.

Answer on page A-4

2. One car travels 20 km/h faster than another. While one car travels 240 km, the other travels 160 km. Find the speed of each car.

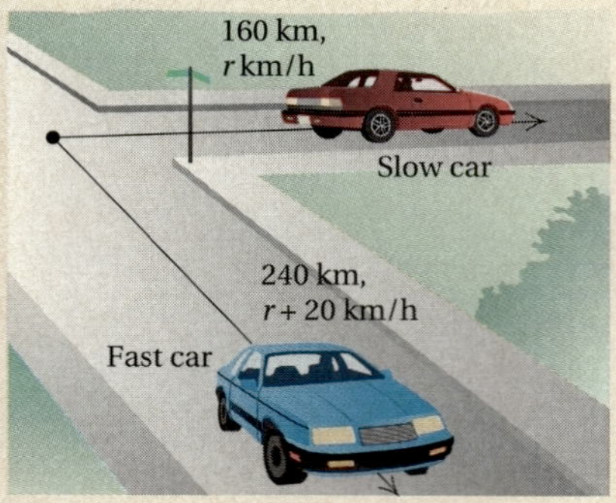

Recall that we may need to find a formula that relates the notions of distance, speed, and time; it is $d = rt$, or *Distance = Speed · Time.* (Indeed, you may need to look up such a formula.) Each animal travels the same length of time, so we can use just t for time. We organize the information in a chart.

$$d \quad = \quad r \quad \cdot \quad t$$

	DISTANCE	SPEED	TIME	
GREYHOUND	12	r	t	→ $12 = rt$
CHEETAH	19	$r + 35$	t	→ $19 = (r + 35)t$

2. ***Translate.*** We can apply the formula $d = rt$ along the rows of the table to obtain two equations:

$$12 = rt, \qquad (1)$$
$$19 = (r + 35)t. \qquad (2)$$

The animals travel for the same length of time. Thus, if we solve each equation for t and set the results equal to each other, we get an equation in terms of r.

Solving $12 = rt$ for t: $\quad t = \dfrac{12}{r}$

Solving $19 = (r + 35)t$ for t: $\quad t = \dfrac{19}{r + 35}$

Since the times are the same, we have the following equation:

$$\frac{12}{r} = \frac{19}{r + 35}.$$

3. ***Solve.*** To solve the equation, we first multiply on both sides by the LCM, which is $r(r + 35)$:

$$r(r + 35) \cdot \frac{12}{r} = r(r + 35) \cdot \frac{19}{r + 35} \qquad \text{Multiplying on both sides by the LCM, which is } r(r + 35)$$
$$12(r + 35) = 19r \qquad \text{Simplifying}$$
$$12r + 420 = 19r \qquad \text{Removing parentheses}$$
$$420 = 7r$$
$$\frac{420}{7} = r$$
$$60 = r.$$

We now have a possible solution. The speed of the greyhound is 60 km/h, and the speed of the cheetah is $r = 60 + 35$, or 95 km/h.

4. ***Check.*** We first reread the problem to see what we were to find. We check the speeds of 60 km/h for the greyhound and 95 km/h for the cheetah. The cheetah does travel 35 km/h faster than the greyhound. The cheetah will travel farther than the greyhound, which runs at a slower speed. If the cheetah runs 19 km at 95 km/h, the time it has traveled is $\frac{19}{95}$, or $\frac{1}{5}$ hr. If the greyhound runs 12 km at 60 km/h, the time it has traveled is $\frac{12}{60}$, or $\frac{1}{5}$ hr. Since the times are the same, the speeds check.

5. ***State.*** The greyhound has a speed of 60 km/h, and the cheetah has a speed of 95 km/h.

Answer on page A-4

Do Exercise 2.

EXAMPLE 3 Erin and Nick work as volunteers at a community recycling depot. Erin can sort a morning's accumulation of recyclables in 4 hr, while Nick requires 6 hr to do the same job. How long would it take them, working together, to sort the recyclables?

1. ***Familiarize.*** We familiarize ourselves with the problem by considering two *incorrect* ways of translating the problem to mathematical language.

   a) A common *incorrect* way to translate the problem is to add the two times: 4 hr + 6 hr = 10 hr. Let us think about this. Erin can do the job alone in 4 hr. If Erin and Nick work together, whatever time it takes them should be *less* than 4 hr. Thus we reject 10 hr as a solution, but we do have a partial check on any answer we get. The answer should be less than 4 hr.

   b) Another *incorrect* way to translate the problem is as follows. Suppose the two people split up the sorting job in such a way that Erin does half the sorting and Nick does the other half. Then

   Erin sorts $\frac{1}{2}$ the recyclables in $\frac{1}{2}$(4 hr), or 2 hr,

   and Nick sorts $\frac{1}{2}$ the recyclables in $\frac{1}{2}$(6 hr), or 3 hr.

   But time is wasted since Erin would finish 1 hr earlier than Nick. In effect, they have not worked together to get the job done as fast as possible. If Erin helps Nick after completing her half, the entire job could be done in a time somewhere between 2 hr and 3 hr.

   We proceed to a translation by considering how much of the job is finished in 1 hr, 2 hr, 3 hr, and so on. It takes Erin 4 hr to do the sorting job alone. Then, in 1 hr, she can do $\frac{1}{4}$ of the job. It takes Nick 6 hr to do the job alone. Then, in 1 hr, he can do $\frac{1}{6}$ of the job. Working together, they can do

   $\frac{1}{4} + \frac{1}{6}$, or $\frac{5}{12}$ of the job in 1 hr.

   In 2 hr, Erin can do $2\left(\frac{1}{4}\right)$ of the job and Nick can do $2\left(\frac{1}{6}\right)$ of the job. Working together, they can do

   $2\left(\frac{1}{4}\right) + 2\left(\frac{1}{6}\right)$, or $\frac{5}{6}$ of the job in 2 hr.

   Continuing this reasoning, we can create a table like the following one.

TIME	FRACTION OF THE JOB COMPLETED		
	Erin	Nick	Together
1 hr	$\frac{1}{4}$	$\frac{1}{6}$	$\frac{1}{4} + \frac{1}{6}$, or $\frac{5}{12}$
2 hr	$2\left(\frac{1}{4}\right)$	$2\left(\frac{1}{6}\right)$	$2\left(\frac{1}{4}\right) + 2\left(\frac{1}{6}\right)$, or $\frac{5}{6}$
3 hr	$3\left(\frac{1}{4}\right)$	$3\left(\frac{1}{6}\right)$	$3\left(\frac{1}{4}\right) + 3\left(\frac{1}{6}\right)$, or $1\frac{1}{4}$
t hr	$t\left(\frac{1}{4}\right)$	$t\left(\frac{1}{6}\right)$	$t\left(\frac{1}{4}\right) + t\left(\frac{1}{6}\right)$

3. By checking work records, a contractor finds that it takes Eduardo 6 hr to construct a wall of a certain size. It takes Yolanda 8 hr to construct the same wall. How long would it take if they worked together?

4. Find the ratio of 145 km to 2.5 liters (L).

5. Recently, a baseball player got 7 hits in 25 times at bat. What was the rate, or batting average, in hits per times at bat?

6. Impulses in nerve fibers travel 310 km in 2.5 hr. What is the rate, or speed, in kilometers per hour?

7. A lake of area 550 yd^2 contains 1320 fish. What is the population density of the lake in fish per square yard?

Answers on page A-4

From the table, we see that if they work 3 hr, the fraction of the job completed is $1\frac{1}{4}$, which is more of the job than needs to be done. We also see that the answer is somewhere between 2 hr and 3 hr. What we want is a number t such that the fraction of the job that gets completed is 1; that is, the job is just completed—more than $\frac{5}{6}$ and less than $1\frac{1}{4}$.

2. *Translate.* From the table, we see that the time we want is some number t for which

$$t\left(\frac{1}{4}\right) + t\left(\frac{1}{6}\right) = 1, \quad \text{or} \quad \frac{t}{4} + \frac{t}{6} = 1,$$

where 1 represents the idea that the entire job is completed in time t.

3. *Solve.* We solve the equation:

$$12\left(\frac{t}{4} + \frac{t}{6}\right) = 12 \cdot 1 \qquad \text{Multiplying by the LCM, which is } 2 \cdot 2 \cdot 3\text{, or 12.}$$

$$12 \cdot \frac{t}{4} + 12 \cdot \frac{t}{6} = 12$$

$$3t + 2t = 12$$

$$5t = 12$$

$$t = \frac{12}{5}, \text{ or } 2\frac{2}{5} \text{ hr.}$$

4. *Check.* The check can be done by recalculating:

$$\frac{12}{5}\left(\frac{1}{4}\right) + \frac{12}{5}\left(\frac{1}{6}\right) = \frac{3}{5} + \frac{2}{5} = \frac{5}{5} = 1.$$

We also have another check in what we learned from our familiarization. The answer, $2\frac{2}{5}$ hr, is between 2 hr and 3 hr (see the table), and it is less than 4 hr, the time it takes Erin working alone.

5. *State.* It takes $2\frac{2}{5}$ hr for them to do the sorting, working together.

THE WORK PRINCIPLE

Suppose a = the time it takes A to do a job, b = the time it takes B to do the same job, and t = the time it takes them to do the same job working together. Then

$$\frac{t}{a} + \frac{t}{b} = 1.$$

Do Exercise 3.

b SOLVING PROPORTION PROBLEMS

We now consider problems concerning proportions. A **proportion** involves ratios. A **ratio** of two quantities is their quotient. For example, 37% is the ratio of 37 to 100, $\frac{37}{100}$. The ratio of two different kinds of measure is called a **rate**. If you travel 400 mi in 7 hr, your rate, or **speed**, is

$$\frac{400 \text{ mi}}{7 \text{ hr}} \approx 57.1 \frac{\text{mi}}{\text{hr}}, \quad \text{or } 57.1 \text{ mph.}$$

Do Exercises 4–7.

An equality of ratios, $A/B = C/D$, is called a *proportion*. The numbers named in a proportion are said to be *proportional*.

Proportions can be used to solve applied problems by expressing a single ratio in two ways. For example, suppose it takes 9 gal of gas to drive 120 mi, and we wish to find the amount of gas required to travel 550 mi. If we assume that the car uses gas at the same rate throughout the trip, the ratios are the same, and we can write a proportion. Note that the units of *gasoline* are in both numerators and the units of *mileage* are in both denominators.

$$\begin{matrix}\text{Gas} \rightarrow \\ \text{Miles} \rightarrow\end{matrix} \frac{9}{120} = \frac{x}{550} \begin{matrix}\leftarrow \text{Gas} \\ \leftarrow \text{Miles}\end{matrix}$$

To solve this proportion, we multiply by 550 to get x alone on one side:

$$550 \cdot \frac{9}{120} = 550 \cdot \frac{x}{550}$$

$$\frac{550 \cdot 9}{120} = x$$

$$41.25 = x.$$

Thus, 41.25 gal is required to travel 550 mi. (Note that we could have multiplied by the LCM of 120 and 550, which is 6600, but in this case, that would have been more complicated.) We can also use **cross-products** to solve proportions:

$$\frac{9}{120} = \frac{x}{550}$$

$$9 \cdot 550 = 120 \cdot x \qquad \textbf{9 · 550 and 120 · } x \textbf{ are called } \textit{cross-products.}$$

$$\frac{9 \cdot 550}{120} = x$$

$$41.25 = x.$$

This method can be verified using the multiplication principle, multiplying on both sides by 550 and then by 120, but we will not do so here.

EXAMPLE 4 An industrial supervisor read 32 employee evaluations. It takes the supervisor 48 min to read 5 evaluations. At this rate, how long did it take to read all 32 evaluations?

1. ***Familiarize.*** The supervisor reads 5 evaluations in 48 min, and we wish to find how long it will take to read all 32 evaluations. We can set up ratios, letting t = the total time to read 32 evaluations.
2. ***Translate.*** If we assume that the supervisor continues to read at the same rate, the ratios are the same, and we have an equation:

$$\begin{matrix}\text{Number of evaluations} \rightarrow \\ \text{Amount of time} \rightarrow\end{matrix} \frac{5}{48} = \frac{32}{t} \begin{matrix}\leftarrow \text{Number of evaluations} \\ \leftarrow \text{Amount of time}\end{matrix}$$

3. ***Solve.*** We solve the equation:

$$48t \cdot \frac{5}{48} = 48t \cdot \frac{32}{t} \qquad \textbf{Multiplying by the LCM, } 48t$$

$$5t = 48 \cdot 32$$

$$t = \frac{48 \cdot 32}{5} = 307.2 \text{ min.}$$

8. It takes 60 oz of grass seed to seed 3000 ft^2 of lawn. At this rate, how much would be needed for 5000 ft^2 of lawn?

9. A sample of 184 light bulbs contained 6 defective bulbs. How many would you expect to find in a sample of 1288 bulbs?

10. To determine the number of deer in a forest, a conservationist catches 612 deer, tags them, and lets them loose. Later, 244 deer are caught, 72 of which are tagged. Estimate how many deer are in the forest.

Answers on page A-4

4. ***Check.*** We leave the check to the student.
5. ***State.*** It will take the supervisor 307.2 min, or 5.12 hr, to read 32 evaluations.

Do Exercises 8 and 9.

EXAMPLE 5 *Estimating wildlife populations.* To determine the number of fish in a lake, a park ranger catches 225 fish, tags them, and throws them back into the lake. Later, 108 fish are caught, and 15 of them are found to be tagged. Estimate how many fish are in the lake.

1. ***Familiarize.*** The ratio of fish tagged to the total number of fish in the lake, F, is $\frac{225}{F}$. Of the 108 fish caught later, 15 fish were tagged. The ratio of fish tagged to fish caught is $\frac{15}{108}$.
2. ***Translate.*** Assuming that the two ratios are the same, we can translate to a proportion.

$$\text{Fish tagged originally} \rightarrow \frac{225}{F} = \frac{15}{108} \leftarrow \text{Tagged fish caught later}$$
(denominators: Fish in lake → F; 108 ← Fish caught later)

3. ***Solve.*** We solve the proportion. We multiply by the LCM, which is $108F$:

$$108F \cdot \frac{225}{F} = 108F \cdot \frac{15}{108} \quad \text{Multiplying by } 108F$$
$$108 \cdot 225 = F \cdot 15$$
$$\frac{108 \cdot 225}{15} = F \quad \text{Dividing by 15}$$
$$1620 = F.$$

4. ***Check.*** We leave the check to the student.
5. ***State.*** We estimate that there are about 1620 fish in the lake.

Do Exercise 10.

NAME SECTION DATE

Exercise Set 5.7

a Solve.

1. The reciprocal of 6 plus the reciprocal of 8 is the reciprocal of what number?

2. The reciprocal of 5 plus the reciprocal of 4 is the reciprocal of what number?

3. One number is 5 more than another. The quotient of the larger divided by the smaller is $\frac{4}{3}$. Find the numbers.

4. One number is 4 more than another. The quotient of the larger divided by the smaller is $\frac{5}{2}$. Find the numbers.

5. Rick drives his four-wheel drive truck 40 km/h faster than Sarah drives her Saturn. While Sarah travels 150 km, Rick travels 350 km. Find their speeds.

   Complete this table and the equations as part of the familiarization.

$d = r \cdot t$

	DISTANCE	SPEED	TIME	
CAR	150	r		$\rightarrow 150 = r(\quad)$
TRUCK	350		t	$\rightarrow 350 = (\qquad)t$

6. A passenger car travels 30 km/h faster than a delivery truck. While the car goes 400 km, the truck goes 250 km. Find their speeds.

7. The speed of a freight train is 14 mph slower than the speed of a passenger train. The freight train travels 330 mi in the same time that it takes the passenger train to travel 400 mi. Find the speed of each train.

   Complete this table and the equations as part of the familiarization.

$d = r \cdot t$

	DISTANCE	SPEED	TIME	
FREIGHT	330		t	$\rightarrow 330 = (\qquad)t$
PASSENGER	400	r		$\rightarrow 400 = r(\quad)$

8. The speed of a freight train is 15 mph slower than the speed of a passenger train. The freight train travels 390 mi in the same time that it takes the passenger train to travel 480 mi. Find the speed of each train.

ANSWERS

1.

2.

3.

4.

5.

6.

7.

8.

ANSWERS

9. ____________

10. ____________

11. ____________

12. ____________

13. ____________

14. ____________

15. ____________

16. ____________

17. ____________

18. ____________

9. A long-distance trucker traveled 120 mi in one direction during a snowstorm. The return trip in rainy weather was accomplished at double the speed and took 3 hr less time. Find the speed going.

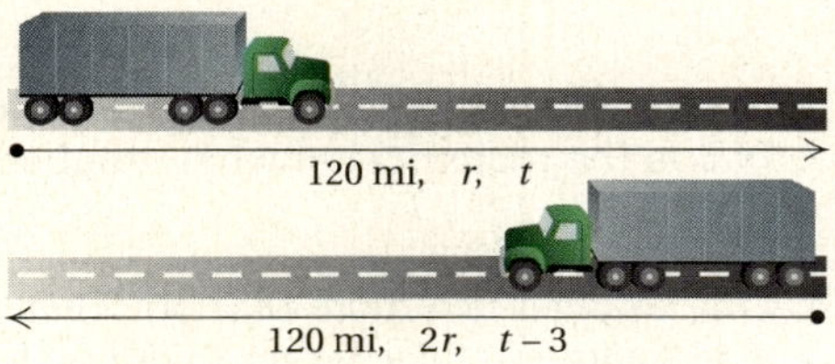

10. After making a trip of 126 mi, a person found that the trip would have taken 1 hr less time by increasing the speed by 8 mph. What was the actual speed?

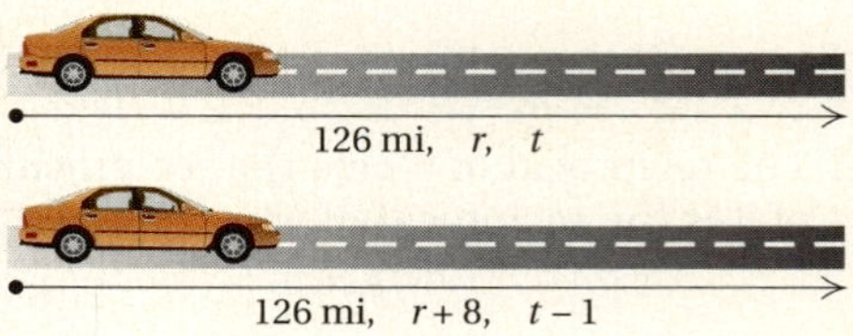

11. By checking work records, a carpenter finds that Juanita can build a garage in 12 hr. Antoine can do the same job in 16 hr. How long would it take if they worked together?

12. It takes David 4 hr to paint a family room. It takes Sierra 5 hr to do the same job. How long would it take them, working together, to do the painting job?

13. By checking work records, a plumber finds that Rory can fit a kitchen in 12 hr. Mira can do the same job in 9 hr. How long would it take if they worked together?

14. Morgan can proofread 25 pages in 40 min. Shelby can proofread the same 25 pages in 30 min. How long would it take them, working together, to proofread 25 pages?

b Find the ratio of the following. Simplify, if possible.

15. 54 days, 6 days

16. 800 mi, 50 gal

17. A black racer snake travels 4.6 km in 2 hr. What is the speed in kilometers per hour?

18. Light travels 558,000 mi in 3 sec. What is the speed in miles per second?

Solve.

19. A 120-lb person should eat a minimum of 44 g of protein each day. How much protein should a 180-lb person eat each day?

20. The coffee beans from 14 trees are required to produce 7.7 kg of coffee (this is the average amount that each person in the United States drinks each year). How many trees are required to produce 320 kg of coffee?

21. A student traveled 234 km in 14 days. At this same rate, how far would the student travel in 42 days?

22. In a potato bread recipe, the ratio of milk to flour is $\frac{3}{13}$. If 5 cups of milk are used, how many cups of flour are used?

23. It is known that 10 cm^3 of a normal specimen of human blood contains 1.2 g of hemoglobin. How many grams of hemoglobin would 16 cm^3 of the same blood contain?

24. A sample of 144 firecrackers contained 9 "duds." How many duds would you expect in a sample of 320 firecrackers?

25. To determine the number of blue whales in the world's oceans, marine biologists tag 500 blue whales in various parts of the world. Later, 400 blue whales are checked, and it is found that 20 of them are tagged. Estimate the blue whale population.

26. To determine the number of trout in a lake, a conservationist catches 112 trout, tags them, and throws them back into the lake. Later, 82 trout are caught; 32 of them are tagged. Estimate the number of trout in the lake.

Answers

19. ______

20. ______

21. ______

22. ______

23. ______

24. ______

25. ______

26. ______

ANSWERS

27. a) ______

b) ______

28. a) ______

b) ______

29. ______

30. ______

31. ______

32. ______

33. ______

34. ______

35. ______

36. ______

37. ______

38. ______

39. ______

40. ______

27. The ratio of the weight of an object on Mars to the weight of an object on earth is 0.4 to 1.

a) How much would a 12-ton rocket weigh on Mars?
b) How much would a 120-lb astronaut weigh on Mars?

28. The ratio of the weight of an object on the moon to the weight of an object on earth is 0.16 to 1.

a) How much would a 12-ton rocket weigh on the moon?
b) How much would a 180-lb astronaut weigh on the moon?

29. A basketball team has 12 more games to play. They have won 25 of the 36 games they have played. How many more games must they win in order to finish with a 0.750 record?

30. Simplest fractional notation for a rational number is $\frac{9}{17}$. Find an equal ratio where the sum of the numerator and the denominator is 104.

SYNTHESIS

31. Larry, Moe, and Curly are accountants who can complete a financial report in 3 days. Larry can do the job in 8 days and Moe can do it in 10 days. How many days will it take Curly to complete the job?

32. Together, Whitney, Michael, and Garth are bakers who prepare a batch of donuts in 1 hr and 20 min. To do the job alone, Whitney needs twice the time that Michael needs and 2 hr more than Garth needs. How long would it take each to complete the job working alone?

33. The denominator of a fraction is 1 more than the numerator. If 2 is subtracted from both the numerator and the denominator, the resulting fraction is $\frac{1}{2}$. Find the original fraction.

34. Express 100 as the sum of two numbers for which the ratio of one number, increased by 5, to the other number, decreased by 5, is 4.

35. In a proportion

$$\frac{A}{B} = \frac{C}{D},$$

the numbers A and D are often called extremes, whereas the numbers B and C are called the means. Write four true proportions.

36. Compare

$$\frac{A + B}{B} = \frac{C + D}{D}$$

with the proportion

$$\frac{A}{B} = \frac{C}{D}.$$

37. Ann and Betty work together and complete a sales report in 4 hr. It would take Betty 6 hr longer, working alone, to do the job than it would Ann. How long would it take each of them to do the job working alone?

38. The speed of a boat in still water is 10 mph. It travels 24 mi upstream and 24 mi downstream in a total time of 5 hr. What is the speed of the current?

39. How soon after 5 o'clock will the hands on a clock first be together?

40. Rachel allows herself 1 hr to reach a sales appointment 50 mi away. After she has driven 30 mi, she realizes that she must increase her speed by 15 mph in order to get there on time. What was her speed for the first 30 mi?

5.8 Formulas

OBJECTIVE

After finishing Section 5.8, you should be able to:

a Solve a formula for a letter.

FOR EXTRA HELP

TAPE 14 | TAPE 11B | MAC: 5B IBM: 5B

a The use of formulas is important in many applications of mathematics. We use the following procedure to solve a rational formula for a letter.

> To solve a rational formula for a given letter, identify the letter, and:
>
> 1. Multiply on both sides to clear fractions or decimals, if that is needed.
> 2. Multiply to remove parentheses, if necessary.
> 3. Get all terms with the letter to be solved for on one side of the equation and all other terms on the other side, using the addition principle.
> 4. Factor out the unknown.
> 5. Solve for the letter in question, using the multiplication principle.

EXAMPLE 1 *Gravitational force.* The gravitational force f between planets of mass M and m, at a distance d from each other, is given by

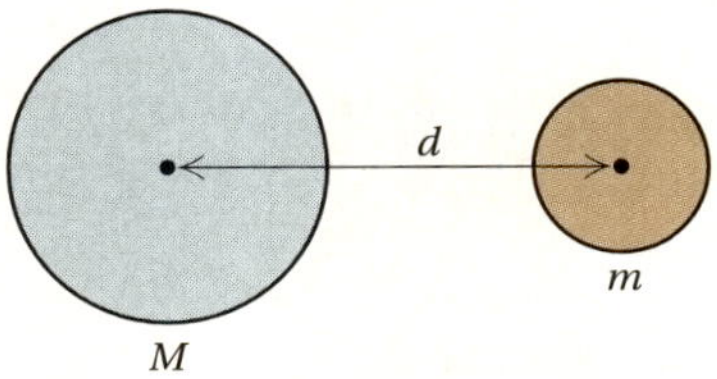

$$f = \frac{kMm}{d^2},$$

where k represents a fixed number constant. Solve for m.

We have

$$f \cdot d^2 = \frac{kMm}{d^2} \cdot d^2 \quad \text{Multiplying by the LCM, } d^2$$

$$fd^2 = kMm \quad \text{Simplifying}$$

$$\frac{fd^2}{kM} = m. \quad \text{Dividing by } kM$$

Do Exercise 1.

1. Solve for M: $f = \dfrac{kMm}{d^2}$.

EXAMPLE 2 *The area of a trapezoid.* The area A of a trapezoid is half the product of the height h and the sum of the lengths b_1 and b_2 of the parallel sides. Solve for b_2.

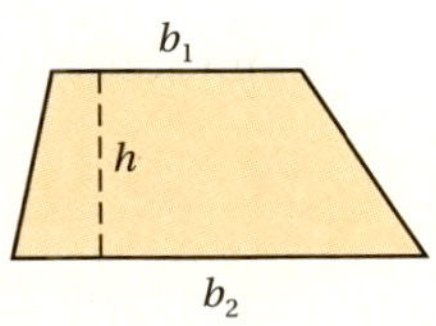

$$A = \frac{1}{2}h(b_1 + b_2)$$

We consider b_1 and b_2 to be different variables (or constants). The letter b_1 represents the length of the first parallel side and b_2 represents the length of the second parallel side. The small numbers 1 and 2 are called **subscripts**. Subscripts are used to identify different variables with related meanings.

$$2 \cdot A = 2 \cdot \frac{1}{2}h(b_1 + b_2) \quad \text{Multiplying by 2 to clear fractions}$$

$$2A = h(b_1 + b_2) \quad \text{Simplifying}$$

Answer on page A-5

2. Solve for b_1: $A = \frac{1}{2}h(b_1 + b_2)$.

3. Solve for f: $\frac{1}{p} + \frac{1}{q} = \frac{1}{f}$.
(This is an optics formula.)

4. Solve for b: $Q = \frac{a - b}{2b}$.

Answers on page A-5

Then $2A = hb_1 + hb_2$ — Using a distributive law to remove parentheses

$2A - hb_1 = hb_2$ — Subtracting hb_1

$\frac{2A - hb_1}{h} = b_2$. — Dividing by h

Do Exercise 2.

EXAMPLE 3 *A work formula.* The following work formula was considered in Section 5.7. Solve it for t.

$$\frac{t}{a} + \frac{t}{b} = 1$$

We multiply by the LCM, which is ab:

$ab \cdot \left(\frac{t}{a} + \frac{t}{b}\right) = ab \cdot 1$ — Multiplying by ab

$ab \cdot \frac{t}{a} + ab \cdot \frac{t}{b} = ab$ — Using a distributive law to remove parentheses

$bt + at = ab$ — Simplifying

$(b + a)t = ab$ — Factoring out t

$t = \frac{ab}{b + a}$. — Dividing by $b + a$

Do Exercise 3.

In Examples 1 and 2, the letter for which we solved was on the right side of the equation. In Example 3, the letter was on the left. The location of the letter is a matter of choice, since all equations are reversible.

> **TIP FOR FORMULA SOLVING**
>
> The variable to be solved for should be alone on one side of the equation, with *no* occurrence of that variable on the other side.

EXAMPLE 4 Solve for b: $S = \frac{a + b}{3b}$.

We multiply by the LCM, which is $3b$:

$3b \cdot S = 3b \cdot \frac{a + b}{3b}$ — Multiplying by $3b$

$3bS = a + b$ — Simplifying

If we divide by $3S$, we will have b alone on the left, but we will still have a term with b on the right.

$3bS - b = a$ — Subtracting b to get all terms involving b on one side

$b(3S - 1) = a$

$b = \frac{a}{3S - 1}$. — Dividing by $3S - 1$

Do Exercise 4.

NAME SECTION DATE

Exercise Set 5.8

a Solve.

1. $S = 2\pi rh$ for r

2. $A = P(1 + rt)$ for t
(An interest formula)

3. $A = \frac{1}{2}bh$ for b
(The area of a triangle)

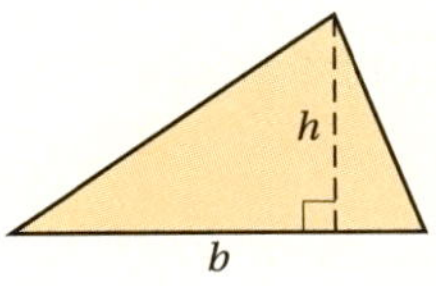

4. $s = \frac{1}{2}gt^2$ for g

5. $S = 180(n - 2)$ for n

6. $S = \frac{n}{2}(a + l)$ for a

7. $V = \frac{1}{3}k(B + b + 4M)$ for b

8. $A = P + Prt$ for P
(*Hint:* Factor the right-hand side.)

9. $S(r - 1) = rl - a$ for r

10. $T = mg - mf$ for m
(*Hint:* Factor the right-hand side.)

11. $A = \frac{1}{2}h(b_1 + b_2)$ for h

12. $S = 2\pi r(r + h)$ for h
(The area of a right circular cylinder)

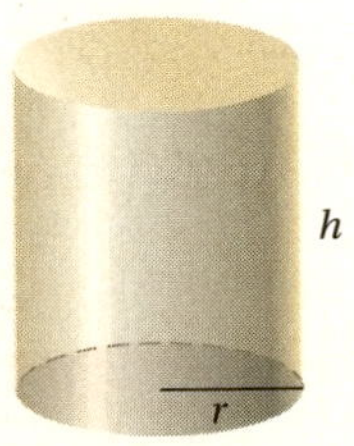

13. $\frac{A - B}{AB} = Q$ for B

14. $L = \frac{Mt + g}{t}$ for t

15. $\frac{1}{p} + \frac{1}{q} = \frac{1}{f}$ for p

16. $\frac{1}{a} + \frac{1}{b} = \frac{1}{t}$ for b

ANSWERS

1. ______
2. ______
3. ______
4. ______
5. ______
6. ______
7. ______
8. ______
9. ______
10. ______
11. ______
12. ______
13. ______
14. ______
15. ______
16. ______

ANSWERS

17. ______

18. ______

19. ______

20. ______

21. ______

22. ______

23. ______

24. ______

25. ______

26. ______

27. ______

28. ______

29. ______

30. ______

31. ______

32. ______

33. ______

34. ______

35. ______

36. ______

17. $\frac{A}{P} = 1 + r$ for A

18. $\frac{2A}{h} = a + b$ for h

19. $\frac{1}{R} = \frac{1}{r_1} + \frac{1}{r_2}$ for R
(An electricity formula)

20. $\frac{1}{R} = \frac{1}{r_1} + \frac{1}{r_2}$ for r_1

21. $\frac{A}{B} = \frac{C}{D}$ for D

22. $q = \frac{VQ}{I}$ for I
(An engineering formula)

23. $h_1 = q\left(1 + \frac{h_2}{p}\right)$ for h_2

24. $S = \frac{a - ar^n}{1 - r}$ for a

25. $C = \frac{Ka - b}{a}$ for a

26. $Q = \frac{Pt - h}{t}$ for t

SKILL MAINTENANCE

27. Subtract: $(5x^4 - 6x^3 + 23x^2 - 79x + 24) - (-18x^4 - 56x^3 + 84x - 17)$.

Factor.

28. $x^2 - 4$

29. $30y^4 + 9y^2 - 12$

30. $49m^2 - 112mn + 64n^2$

31. $y^2 + 2y - 35$

32. $y^4 - 1$

SYNTHESIS

Solve.

33. $u = -F\left(E - \frac{P}{T}\right)$ for T

34. $l = a + (n - 1)d$ for d

35. The formula

$$C = \frac{5}{9}(F - 32)$$

is used to convert Fahrenheit temperatures to Celsius temperatures. At what temperature are the Fahrenheit and Celsius readings the same?

36. In

$$N = \frac{a}{c},$$

what is the effect on N when c increases? when c decreases? Assume that a, c, and N are positive.

5.9 Complex Rational Expressions

a A **complex rational expression,** or **complex fractional expression,** is a rational expression that has one or more rational expressions within its numerator or denominator. Here are some examples:

$$\frac{1+\dfrac{2}{x}}{3},\quad \frac{\dfrac{x+y}{2}}{\dfrac{2x}{x+1}},\quad \frac{\dfrac{1}{3}+\dfrac{1}{5}}{\dfrac{2}{x}-\dfrac{x}{y}}.$$

These are rational expressions within the complex rational expression.

There are two methods to simplify complex rational expressions. We will consider them both. Use the one that works best for you or the one that your instructor directs you to use.

Multiplying by the LCM of All the Denominators: Method 1

Method 1

To simplify a complex rational expression:

1. First, find the LCM of all the denominators of all the rational expressions occurring *within* both the numerator and the denominator of the complex rational expression.
2. Then multiply by 1 using LCM/LCM.
3. If possible, simplify by removing a factor of 1.

Example 1 Simplify: $\dfrac{\dfrac{1}{2}+\dfrac{3}{4}}{\dfrac{5}{6}-\dfrac{3}{8}}$.

We have

$$\frac{\dfrac{1}{2}+\dfrac{3}{4}}{\dfrac{5}{6}-\dfrac{3}{8}}$$

The denominators *within* the complex rational expression are 2, 4, 6, and 8. The LCM of these denominators is 24. We multiply by 1 using $\frac{24}{24}$.

$$=\frac{\dfrac{1}{2}+\dfrac{3}{4}}{\dfrac{5}{6}-\dfrac{3}{8}}\cdot\frac{24}{24}\qquad\text{Multiplying by 1}$$

$$=\frac{\left(\dfrac{1}{2}+\dfrac{3}{4}\right)24}{\left(\dfrac{5}{6}-\dfrac{3}{8}\right)24}$$

← Multiplying the numerator by 24

← Multiplying the denominator by 24

Objective

After finishing Section 5.9, you should be able to:

a Simplify complex rational expressions.

For Extra Help

TAPE 14 — TAPE 12A — MAC: 5B IBM: 5B

1. Simplify. Use Method 1.

$$\frac{\frac{1}{3}+\frac{4}{5}}{\frac{7}{8}-\frac{5}{6}}$$

2. Simplify. Use Method 1.

$$\frac{\frac{x}{2}+\frac{2x}{3}}{\frac{1}{x}-\frac{x}{2}}$$

3. Simplify. Use Method 1.

$$\frac{1+\frac{1}{x}}{1-\frac{1}{x^2}}$$

Answers on page A-5

Using the distributive laws, we carry out the multiplications:

$$= \frac{\frac{1}{2}(24)+\frac{3}{4}(24)}{\frac{5}{6}(24)-\frac{3}{8}(24)}$$

$$= \frac{12+18}{20-9} \quad \text{Simplifying}$$

$$= \frac{30}{11}.$$

Multiplying in this manner has the effect of clearing fractions in both the top and the bottom of the complex rational expression.

Do Exercise 1.

EXAMPLE 2 Simplify: $\dfrac{\frac{3}{x}+\frac{1}{2x}}{\frac{1}{3x}-\frac{3}{4x}}$.

The denominators within the complex expression are x, $2x$, $3x$, and $4x$. The LCM of these denominators is $12x$. We multiply by 1 using $12x/12x$.

$$\frac{\frac{3}{x}+\frac{1}{2x}}{\frac{1}{3x}-\frac{3}{4x}} \cdot \frac{12x}{12x} = \frac{\left(\frac{3}{x}+\frac{1}{2x}\right)12x}{\left(\frac{1}{3x}-\frac{3}{4x}\right)12x} = \frac{\frac{3}{x}(12x)+\frac{1}{2x}(12x)}{\frac{1}{3x}(12x)-\frac{3}{4x}(12x)}$$

$$= \frac{36+6}{4-9} = -\frac{42}{5}$$

Do Exercise 2.

EXAMPLE 3 Simplify: $\dfrac{1-\frac{1}{x}}{1-\frac{1}{x^2}}$.

The denominators within the complex expression are x and x^2. The LCM of these denominators is x^2. We multiply by 1 using x^2/x^2. Then, after obtaining a single rational expression, we simplify:

$$\frac{1-\frac{1}{x}}{1-\frac{1}{x^2}} \cdot \frac{x^2}{x^2} = \frac{\left(1-\frac{1}{x}\right)x^2}{\left(1-\frac{1}{x^2}\right)x^2} = \frac{1(x^2)-\frac{1}{x}(x^2)}{1(x^2)-\frac{1}{x^2}(x^2)} = \frac{x^2-x}{x^2-1}$$

$$= \frac{x(x-1)}{(x+1)(x-1)} = \frac{x}{x+1}.$$

Do Exercise 3.

Adding in the Numerator and the Denominator: Method 2

Method 2

To simplify a complex rational expression:

1. Add or subtract, as necessary, to get a single rational expression in the numerator.
2. Add or subtract, as necessary, to get a single rational expression in the denominator.
3. Divide the numerator by the denominator.
4. If possible, simplify by removing a factor of 1.

We will redo Examples 1–3 using this method.

Example 4 Simplify: $\dfrac{\frac{1}{2}+\frac{3}{4}}{\frac{5}{6}-\frac{3}{8}}$.

We have

$$\frac{\frac{1}{2}+\frac{3}{4}}{\frac{5}{6}-\frac{3}{8}} = \frac{\frac{1}{2}\cdot\frac{2}{2}+\frac{3}{4}}{\frac{5}{6}\cdot\frac{4}{4}-\frac{3}{8}\cdot\frac{3}{3}}$$

Multiplying the $\frac{1}{2}$ by 1 to get a common denominator

Multiplying the $\frac{5}{6}$ and the $\frac{3}{8}$ by 1 to get a common denominator

$$= \frac{\frac{2}{4}+\frac{3}{4}}{\frac{20}{24}-\frac{9}{24}}$$

$$= \frac{\frac{5}{4}}{\frac{11}{24}}$$

Adding in the numerator; subtracting in the denominator

$$= \frac{5}{4}\cdot\frac{24}{11}$$

Multiplying by the reciprocal of the divisor

$$= \frac{5\cdot 3\cdot 2\cdot 2\cdot 2}{2\cdot 2\cdot 11}$$

Factoring

$$= \frac{5\cdot 3\cdot 2\cdot \cancel{2}\cdot \cancel{2}}{\cancel{2}\cdot \cancel{2}\cdot 11}$$

Removing a factor of 1: $\frac{2\cdot 2}{2\cdot 2} = 1$

$$= \frac{30}{11}.$$

Do Exercise 4.

4. Simplify. Use Method 2.

$$\frac{\frac{1}{3}+\frac{4}{5}}{\frac{7}{8}-\frac{5}{6}}$$

Answer on page A-5

5. Simplify. Use Method 2.

$$\frac{\frac{x}{2} + \frac{2x}{3}}{\frac{1}{x} - \frac{x}{2}}$$

6. Simplify. Use Method 2.

$$\frac{1 + \frac{1}{x}}{1 - \frac{1}{x^2}}$$

Answers on page A-5

EXAMPLE 5 Simplify: $\dfrac{\frac{3}{x} + \frac{1}{2x}}{\frac{1}{3x} - \frac{3}{4x}}$.

We have

$$\frac{\frac{3}{x} + \frac{1}{2x}}{\frac{1}{3x} - \frac{3}{4x}} = \frac{\frac{3}{x} \cdot \frac{2}{2} + \frac{1}{2x}}{\frac{1}{3x} \cdot \frac{4}{4} - \frac{3}{4x} \cdot \frac{3}{3}}$$

Finding the LCM, $2x$, and multiplying by 1 in the numerator

Finding the LCM, $12x$, and multiplying by 1 in the denominator

$$= \frac{\frac{6}{2x} + \frac{1}{2x}}{\frac{4}{12x} - \frac{9}{12x}} = \frac{\frac{7}{2x}}{\frac{-5}{12x}}$$

Adding in the numerator and subtracting in the denominator

$$= \frac{7}{2x} \cdot \frac{12x}{-5}$$

Multiplying by the reciprocal of the divisor

$$= \frac{7}{2x} \cdot \frac{6(2x)}{-5}$$

Factoring

$$= \frac{7}{\cancel{2x}} \cdot \frac{6(\cancel{2x})}{-5}$$

Removing a factor of 1: $\frac{2x}{2x} = 1$

$$= \frac{42}{-5} = -\frac{42}{5}.$$

Do Exercise 5.

EXAMPLE 6 Simplify: $\dfrac{1 - \frac{1}{x}}{1 - \frac{1}{x^2}}$.

We have

$$\frac{1 - \frac{1}{x}}{1 - \frac{1}{x^2}} = \frac{\frac{x}{x} - \frac{1}{x}}{\frac{x^2}{x^2} - \frac{1}{x^2}}$$

Finding the LCM, x, and multiplying by 1 in the numerator

Finding the LCM, x^2, and multiplying by 1 in the denominator

$$= \frac{\frac{x-1}{x}}{\frac{x^2-1}{x^2}}$$

Subtracting in the numerator and subtracting in the denominator

$$= \frac{x-1}{x} \cdot \frac{x^2}{x^2-1}$$

Multiplying by the reciprocal of the divisor

$$= \frac{(x-1)x \cdot x}{x(x-1)(x+1)}$$

Factoring

$$= \frac{\cancel{(x-1)}\cancel{x} \cdot x}{\cancel{x}\cancel{(x-1)}(x+1)}$$

Removing a factor of 1: $\frac{x(x-1)}{x(x-1)} = 1$

$$= \frac{x}{x+1}.$$

Do Exercise 6.

NAME SECTION DATE

Exercise Set 5.9

a Simplify.

1. $\dfrac{1+\dfrac{9}{16}}{1-\dfrac{3}{4}}$

2. $\dfrac{6-\dfrac{3}{8}}{4+\dfrac{5}{6}}$

3. $\dfrac{1-\dfrac{3}{5}}{1+\dfrac{1}{5}}$

4. $\dfrac{2+\dfrac{2}{3}}{2-\dfrac{2}{3}}$

5. $\dfrac{\dfrac{1}{2}+\dfrac{3}{4}}{\dfrac{5}{8}-\dfrac{5}{6}}$

6. $\dfrac{\dfrac{3}{4}+\dfrac{7}{8}}{\dfrac{2}{3}-\dfrac{5}{6}}$

7. $\dfrac{\dfrac{1}{x}+3}{\dfrac{1}{x}-5}$

8. $\dfrac{2-\dfrac{1}{a}}{4+\dfrac{1}{a}}$

9. $\dfrac{4-\dfrac{1}{x^2}}{2-\dfrac{1}{x}}$

10. $\dfrac{\dfrac{2}{y}+\dfrac{1}{2y}}{y+\dfrac{y}{2}}$

11. $\dfrac{8+\dfrac{8}{d}}{1+\dfrac{1}{d}}$

12. $\dfrac{3+\dfrac{2}{t}}{3-\dfrac{2}{t}}$

13. $\dfrac{\dfrac{x}{8}-\dfrac{8}{x}}{\dfrac{1}{8}+\dfrac{1}{x}}$

14. $\dfrac{\dfrac{2}{m}+\dfrac{m}{2}}{\dfrac{m}{3}-\dfrac{3}{m}}$

15. $\dfrac{1+\dfrac{1}{y}}{1-\dfrac{1}{y^2}}$

ANSWERS

1. ______
2. ______
3. ______
4. ______
5. ______
6. ______
7. ______
8. ______
9. ______
10. ______
11. ______
12. ______
13. ______
14. ______
15. ______

ANSWERS

16. ______
17. ______
18. ______
19. ______
20. ______
21. ______
22. ______
23. ______
24. ______
25. ______
26. ______
27. ______
28. ______
29. ______
30. ______
31. ______
32. ______
33. ______
34. ______

16. $\dfrac{\dfrac{1}{q^2} - 1}{\dfrac{1}{q} + 1}$

17. $\dfrac{\dfrac{1}{5} - \dfrac{1}{a}}{\dfrac{5 - a}{5}}$

18. $\dfrac{\dfrac{4}{t}}{4 + \dfrac{1}{t}}$

19. $\dfrac{\dfrac{1}{a} + \dfrac{1}{b}}{\dfrac{1}{a^2} - \dfrac{1}{b^2}}$

20. $\dfrac{\dfrac{1}{x^2} - \dfrac{1}{y^2}}{\dfrac{2}{x} - \dfrac{2}{y}}$

21. $\dfrac{\dfrac{p}{q} + \dfrac{q}{p}}{\dfrac{1}{p} + \dfrac{1}{q}}$

22. $\dfrac{x - 3 + \dfrac{2}{x}}{x - 4 + \dfrac{3}{x}}$

SKILL MAINTENANCE

23. Add: $(2x^3 - 4x^2 + x - 7) + (4x^4 + x^3 + 4x^2 + x)$.

24. The length of a rectangle is 3 yd greater than the width. The area of the rectangle is 10 yd^2. Find the perimeter.

Factor.

25. $p^2 - 10p + 25$

26. $p^2 + 10p + 25$

27. $50p^2 - 100$

28. $5p^2 - 40p - 100$

SYNTHESIS

29. Find the reciprocal of $\dfrac{2}{x - 1} - \dfrac{1}{3x - 2}$.

Simplify.

30. $\dfrac{\dfrac{a}{b} + \dfrac{c}{d}}{\dfrac{b}{a} + \dfrac{d}{c}}$

31. $\dfrac{\dfrac{a}{b} - \dfrac{c}{d}}{\dfrac{b}{a} - \dfrac{d}{c}}$

32. $\left[\dfrac{\dfrac{x + 1}{x - 1} + 1}{\dfrac{x + 1}{x - 1} - 1}\right]^5$

33. $1 + \dfrac{1}{1 + \dfrac{1}{1 + \dfrac{1}{1 + \dfrac{1}{x}}}}$

34. $\dfrac{\dfrac{z}{1 - \dfrac{z}{2 + 2z}} - 2z}{\dfrac{2z}{5z - 2} - 3}$

CRITICAL THINKING

CALCULATOR CONNECTION

Many formulas involve rational equations. Using the reciprocal key [1/x] can make it easier to work with these formulas.

Example It takes Bianca 6.75 hr, working alone, to keyboard a financial statement. It takes Arturo 7 hr, working alone, and Alani 8 hr, working alone, to produce the same statement. How long does it take if they all work together?

Let t = the time it takes them to do the job working together. Then

$$\frac{t}{6.75} + \frac{t}{7} + \frac{t}{8} = 1, \quad \text{or} \quad \frac{1}{t} = \frac{1}{6.75} + \frac{1}{7} + \frac{1}{8}.$$

We solve for $1/t$ using the reciprocal key:

$\frac{1}{t}$ = 6.75 [1/x] [+] 7 [1/x] [+] 8 [1/x]

$= 0.416005291.$

This is $1/t$, so we press the reciprocal key [1/x] again to find

$$t = 2.40381558.$$

The financial statement will take about 2.4 hr if they all work together.

1. Irwin can edge a lawn in 45 min if he works alone. It takes Hannah 38 min to edge the same lawn if she works alone. How long would it take if they work together?
2. Five sorority sisters work on sending out invitations to a dance. Working alone, it would take each of them, respectively, 2.8, 3.1, 2.2, 3.4, and 2.4 hr to do the job. How long will it take if they work together?

EXTENDED SYNTHESIS EXERCISES

Evaluate the expression before and after simplifying.

1. $\dfrac{x^2 + 3x}{x}$ when $x = 13$
2. $\dfrac{x^2 - 9}{(x + 3)^2}$ when $x = 4$
3. $\dfrac{3b - 6}{2 - b}$ when $b = 2$
4. $\dfrac{a^2 + b^2}{(a + b)^2}$ when $a = 3$ and $b = 4$
5. Find two different pairs of rational expressions whose product is

$$\frac{8x^2 + 16x - 24}{x^2 + 13x + 40}.$$

6. The volume of this rectangular solid is $x - 3$. What is its height?

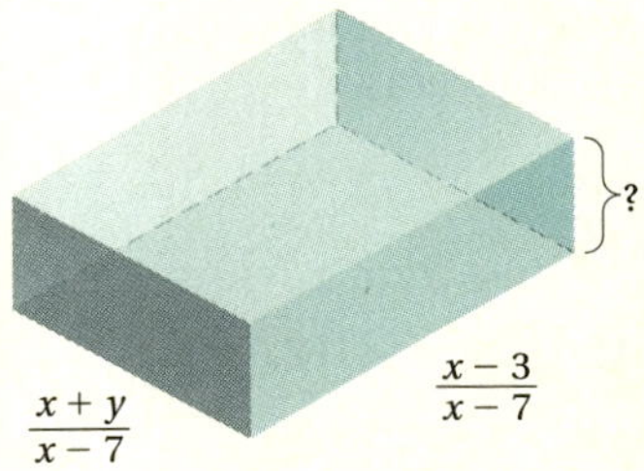

Find the perimeter and the area of the rectangle.

7.

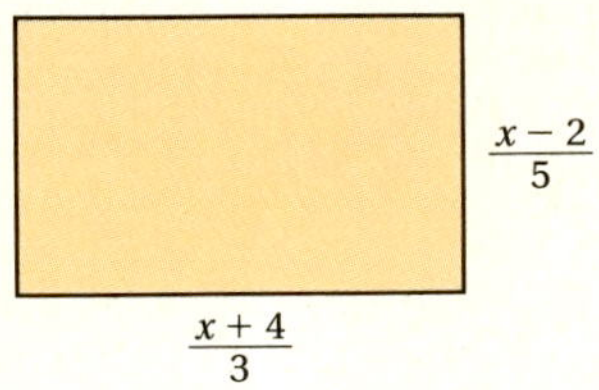

8.

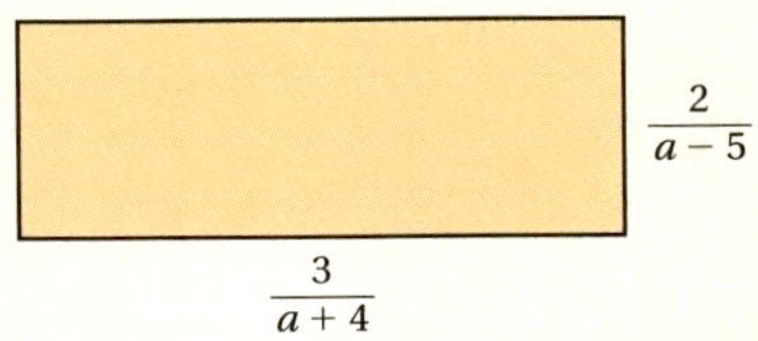

9. The perimeter of the right triangle is $2a + 5$. Find the length of the missing side and the area.

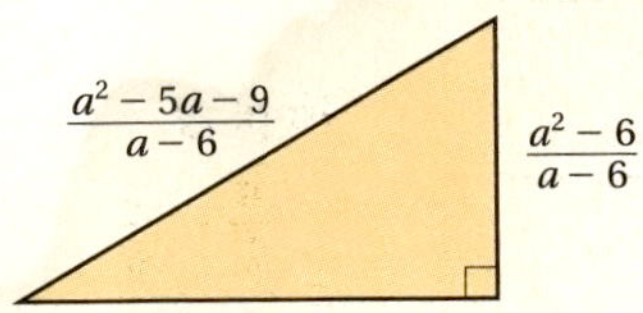

(*continued*)

Critical Thinking

10. *Batting averages of 0.400 or better.* A baseball player's batting average is defined to be the number of hits divided by the number of official at-bats. (An official at-bat does not include walks, sacrifices, or being hit by a pitch.) The last major-league baseball player to attain a 0.400 batting average was Ted Williams in 1941. He got 185 hits in 456 at-bats to bat 0.406.

    Almost every year, there is a great clamor in the press when a player appears to have a chance at a 0.400 average. Nevertheless, the last player to come close was George Brett in 1980 when he got 175 hits in 449 at-bats for a 0.390 average. How many more hits would Brett have had to get to achieve a 0.400 average?

Solve.

11. $\dfrac{\dfrac{2x+3}{x+1}}{\dfrac{x-2}{x+1}} = 12$

12. $\dfrac{\dfrac{x+1}{x-1}+1}{\dfrac{x+1}{x-1}-1} = 10$

Exercises for Thinking and Writing

Carry out the direction for each of the following. Explain the use of the LCM in each exercise.

1. Add: $\dfrac{4}{x-2} + \dfrac{1}{x+2}$.

2. Subtract: $\dfrac{4}{x-2} - \dfrac{1}{x+2}$.

3. Solve: $\dfrac{4}{x-2} + \dfrac{1}{x+2} = \dfrac{26}{x^2-4}$.

4. Simplify: $\dfrac{1-\dfrac{2}{x}}{1+\dfrac{x}{4}}$. Give two answers.

Tell whether the rational expression has been simplified correctly. If it has not, describe the error.

5. $\dfrac{10x^4y^2}{2x^3y^2} = 5x$

6. $\dfrac{5a-8}{5a} = -8$

7. $\dfrac{(p-2)^2}{p^2-4} = p+2$

8. $\dfrac{8a^2-32}{4a^2-16} = 2a^2-2$

9. $\dfrac{a^2+b^2}{(a+b)^2} = 1$

10. Carry out each of the following tasks. Then compare the procedures.

    a) Solve: $\dfrac{1}{t} = \dfrac{1}{3} + \dfrac{1}{4}$.

    b) Solve the following for t: $\dfrac{1}{t} = \dfrac{1}{a} + \dfrac{1}{b}$.

11. Explain why the use of parentheses is important when subtracting rational expressions.

Summary and Review Exercises: Chapter 5

The review objectives to be tested in addition to the material in this chapter are [3.2a, b], [3.4a, c], [4.6a], and [4.8a].

Find all numbers for which the rational expression is undefined.

1. $\frac{3}{x}$

2. $\frac{4}{x - 6}$

3. $\frac{x + 5}{x^2 - 36}$

4. $\frac{x^2 - 3x + 2}{x^2 + x - 30}$

5. $\frac{-4}{(x + 2)^2}$

6. $\frac{x - 5}{x^3 - 8x^2 + 15x}$

Simplify.

7. $\frac{4x^2 - 8x}{4x^2 + 4x}$

8. $\frac{14x^2 - x - 3}{2x^2 - 7x + 3}$

9. $\frac{(y - 5)^2}{y^2 - 25}$

Multiply or divide and simplify.

10. $\frac{a^2 - 36}{10a} \cdot \frac{2a}{a + 6}$

11. $\frac{6t - 6}{2t^2 + t - 1} \cdot \frac{t^2 - 1}{t^2 - 2t + 1}$

12. $\frac{10 - 5t}{3} \div \frac{t - 2}{12t}$

13. $\frac{4x^4}{x^2 - 1} \div \frac{2x^3}{x^2 - 2x + 1}$

Find the LCM.

14. $3x^2, \quad 10xy, \quad 15y^2$

15. $a - 2, \quad 4a - 8$

16. $y^2 - y - 2, \quad y^2 - 4$

Add or subtract and simplify.

17. $\frac{x + 8}{x + 7} + \frac{10 - 4x}{x + 7}$

18. $\frac{3}{3x - 9} + \frac{x - 2}{3 - x}$

19. $\frac{6x - 3}{x^2 - x - 12} - \frac{2x - 15}{x^2 - x - 12}$

20. $\frac{3x - 1}{2x} - \frac{x - 3}{x}$

21. $\frac{x + 3}{x - 2} - \frac{x}{2 - x}$

22. $\frac{2a}{a + 1} + \frac{4a}{a^2 - 1}$

23. $\frac{d^2}{d - c} + \frac{c^2}{c - d}$

24. $\frac{1}{x^2 - 25} - \frac{x - 5}{x^2 - 4x - 5}$

25. $\frac{3x}{x + 2} - \frac{x}{x - 2} + \frac{8}{x^2 - 4}$

Simplify.

26. $\dfrac{\dfrac{1}{z}+1}{\dfrac{1}{z^2}-1}$

27. $\dfrac{\dfrac{c}{d}-\dfrac{d}{c}}{\dfrac{1}{c}+\dfrac{1}{d}}$

Solve.

28. $\dfrac{3}{y}-\dfrac{1}{4}=\dfrac{1}{y}$

29. $\dfrac{15}{x}-\dfrac{15}{x+2}=2$

Solve.

30. In checking records, a contractor finds that crew A can pave a certain length of highway in 9 hr. Crew B can do the same job in 12 hr. How long would it take if they worked together?

31. A manufacturer is testing two high-speed trains. One train travels 40 km/h faster than the other. While one train travels 70 km, the other travels 60 km. Find their speeds.

32. The reciprocal of 1 more than a number is twice the reciprocal of the number itself. What is the number?

33. A sample of 250 calculators contained 8 defective calculators. How many defective calculators would you expect to find in a sample of 5000?

34. One plane travels 80 mph faster than another. While one of them travels 1750 mi, the other travels 950 mi. Find the speed of each plane.

Solve for the letter indicated.

35. $\dfrac{1}{r}+\dfrac{1}{s}=\dfrac{1}{t}$, for s

36. $F=\dfrac{9C+160}{5}$, for C

SKILL MAINTENANCE

37. Factor: $5x^3+20x^2-3x-12$.

38. Simplify: $(5x^3y^2)^{-3}$.

39. Subtract:
$(5x^3-4x^2+3x-4)-(7x^3-7x^2-9x+14)$.

40. The width of a rectangle is 2 cm less than the length. The area is 15 cm^2. Find the dimensions and the perimeter of the rectangle.

SYNTHESIS

Simplify.

41. $\dfrac{2a^2+5a-3}{a^2}\cdot\dfrac{5a^3+30a^2}{2a^2+7a-4}\div\dfrac{a^2+6a}{a^2+7a+12}$

42. $\dfrac{12a}{(a-b)(b-c)}-\dfrac{2a}{(b-a)(c-b)}$

43. a) Add: $\dfrac{6}{x-3}+\dfrac{4}{x+3}$.

b) Solve: $\dfrac{6}{x-3}+\dfrac{4}{x+3}=\dfrac{7}{x^2-9}$.

c) Explain the difference in the use of the LCM in parts (a) and (b).

NAME SECTION DATE

Test: Chapter 5

Find all numbers for which the rational expression is undefined.

1. $\frac{8}{2x}$

2. $\frac{5}{x + 8}$

3. $\frac{x - 7}{x^2 - 49}$

4. $\frac{x^2 + x - 30}{x^2 - 3x + 2}$

5. $\frac{11}{(x - 1)^2}$

6. $\frac{x + 2}{x^3 + 8x^2 + 15x}$

7. Simplify:

$$\frac{6x^2 + 17x + 7}{2x^2 + 7x + 3}.$$

8. Multiply and simplify:

$$\frac{a^2 - 25}{6a} \cdot \frac{3a}{a - 5}.$$

9. Divide and simplify:

$$\frac{25x^2 - 1}{9x^2 - 6x} \div \frac{5x^2 + 9x - 2}{3x^2 + x - 2}.$$

10. Find the LCM:

$y^2 - 9,\ y^2 + 10y + 21,\ y^2 + 4y - 21.$

Add or subtract. Simplify, if possible.

11. $\frac{16 + x}{x^3} + \frac{7 - 4x}{x^3}$

12. $\frac{5 - t}{t^2 + 1} - \frac{t - 3}{t^2 + 1}$

13. $\frac{x - 4}{x - 3} + \frac{x - 1}{3 - x}$

14. $\frac{x - 4}{x - 3} - \frac{x - 1}{3 - x}$

15. $\frac{5}{t - 1} + \frac{3}{t}$

16. $\frac{1}{x^2 - 16} - \frac{x + 4}{x^2 - 3x - 4}$

17. $\frac{1}{x - 1} + \frac{4}{x^2 - 1} - \frac{2}{x^2 - 2x + 1}$

ANSWERS

1. ______

2. ______

3. ______

4. ______

5. ______

6. ______

7. ______

8. ______

9. ______

10. ______

11. ______

12. ______

13. ______

14. ______

15. ______

16. ______

17. ______

ANSWERS

18. ____________

19. ____________

20. ____________

21. ____________

22. ____________

23. ____________

24. ____________

25. ____________

26. ____________

27. ____________

28. ____________

29. ____________

30. ____________

18. Simplify: $\dfrac{9 - \dfrac{1}{y^2}}{3 - \dfrac{1}{y}}$.

Solve.

19. $\dfrac{7}{y} - \dfrac{1}{3} = \dfrac{1}{4}$

20. $\dfrac{15}{x} - \dfrac{15}{x-2} = -2$

21. The reciprocal of 3 less than a number is four times the reciprocal of the number itself. What is the number?

22. A sample of 125 spark plugs contained 4 defective spark plugs. How many defective spark plugs would you expect to find in a sample of 500?

23. One car travels 20 mph faster than another on a freeway. While one goes 225 mi, the other goes 325 mi. Find the speed of each car.

24. Solve $L = \dfrac{Mt - g}{t}$ for t.

SKILL MAINTENANCE

25. Factor: $16a^2 - 49$.

26. Simplify: $\left(\dfrac{3x^2}{y^3}\right)^{-4}$.

27. Subtract:
$(5x^2 - 19x + 34) - (-8x^2 + 10x - 42)$.

28. The product of two consecutive integers is 462. Find the integers.

SYNTHESIS

29. Team A and team B work together and complete a job in $2\frac{6}{7}$ hr. It would take team B 6 hr longer, working alone, to do the job than it would team A. How long would it take each of them to do the job working alone?

30. Simplify: $1 + \dfrac{1}{1 + \dfrac{1}{1 + \dfrac{1}{a}}}$.

Cumulative Review: Chapters 1–5

Evaluate.

1. $\dfrac{2x+5}{y-10}$ when $x = 2$ and $y = 5$

2. $4 - x^3$ when $x = -2$

Simplify.

3. $x - [x - 2(x + 3)]$

4. $(2x^{-2})^{-2}(3x)^3$

5. $\dfrac{24x^8}{18x^{-2}}$

6. $\dfrac{2t^2 + 8t - 42}{2t^2 + 13t - 7}$

7. $\dfrac{\dfrac{2}{x} + 1}{\dfrac{x}{x+2}}$

8. $\dfrac{a^2 - 16}{a^2 - 8a + 16}$

Add. Simplify, if possible.

9. $\dfrac{9}{14} + \left(-\dfrac{5}{21}\right)$

10. $\dfrac{2x + y}{x^2y} + \dfrac{x + 2y}{xy^2}$

11. $\dfrac{z}{z^2 - 1} + \dfrac{2}{z + 1}$

12. $(2x^4 + 5x^3 + 4) + (3x^3 - 2x + 5)$

Subtract. Simplify, if possible.

13. $1.53 - (-0.8)$

14. $(x^2 - xy - y^2) - (x^2 - y^2)$

15. $\dfrac{3}{x^2 - 9} - \dfrac{x}{9 - x^2}$

16. $\dfrac{2x}{x^2 - x - 20} - \dfrac{4}{x^2 - 10x + 25}$

Multiply. Simplify, if possible.

17. $(1.3)(-0.5)(2)$

18. $3x^2(2x^2 + 4x - 5)$

19. $\left(3t + \frac{1}{2}\right)\left(3t - \frac{1}{2}\right)$

20. $(2p - q)^2$

21. $(3x + 5)(x - 4)$

22. $(2x^2 + 1)(2x^2 - 1)$

23. $\dfrac{6t + 6}{t^3 - 2t^2} \cdot \dfrac{t^3 - 3t^2 + 2t}{3t + 3}$

24. $\dfrac{a^2 - 1}{a^2} \cdot \dfrac{2a}{1 - a}$

Divide. Simplify, if possible.

25. $(3x^3 - 7x^2 + 9x - 5) \div (x - 1)$

26. $-\dfrac{21}{25} \div \dfrac{28}{15}$

27. $\dfrac{x^2 - x - 2}{4x^3 + 8x^2} \div \dfrac{x^2 - 2x - 3}{2x^2 + 4x}$

28. $\dfrac{3 - 3x}{x^2} \div \dfrac{x - 1}{4x}$

Factor completely.

29. $4x^3 + 12x^2 - 9x - 27$

30. $x^2 + 7x - 8$

31. $3x^2 - 14x - 5$

32. $16y^2 + 40xy + 25x^2$

33. $3x^3 + 24x^2 + 45x$

34. $2x^2 - 2$

35. $x^2 - 28x + 196$

36. $4y^3 + 10y^2 + 12y + 30$

Solve.

37. $2(x - 3) = 5(x + 3)$

38. $2x(3x + 4) = 0$

39. $x^2 = 8x$

40. $x^2 + 16 = 8x$

41. $x - 5 \le 2x + 4$

42. $3x^2 = 27$

43. $\frac{1}{3}x - \frac{2}{5} = \frac{4}{5}x + \frac{1}{3}$

44. $\frac{x}{3} = \frac{3}{x}$

45. $\frac{x + 5}{2x + 1} = \frac{x - 7}{2x - 1}$

46. $\frac{1}{3}x\left(2x - \frac{1}{5}\right) = 0$

47. $\frac{3 - x}{x - 1} = \frac{2}{x - 1}$

48. $\frac{3}{2x + 5} = \frac{2}{5 - x}$

49. $\frac{1}{x} + \frac{1}{y} = \frac{1}{z}$, for z

50. $\frac{3N}{T} = D$, for N

Solve.

51. The sum of three consecutive integers is 99. What are the integers?

52. The speed of one bicyclist is 2 km/h faster than the speed of another bicyclist. The first bicyclist travels 60 km in the same time that it takes the second to travel 50 km. Find the speed of each bicyclist.

53. A swimming pool can be filled in 5 hr by hose A alone and in 6 hr by hose B alone. How long would it take to fill the tank if both hoses were working?

54. The sum of the page numbers on the facing pages of a book is 69. What are the page numbers?

55. The product of the page numbers on two facing pages of a book is 272. Find the page numbers.

56. In a cake recipe, the ratio of flour to sugar is $\frac{6}{5}$. If 2 cups of flour are used, how many cups of sugar are used?

57. The area of a circle is 35π more than the circumference. Find the length of the radius.

58. The sum of the squares of two consecutive odd positive integers is 202. Find the integers.

SYNTHESIS

Solve.

59. $(2x - 1)^2 = (x + 3)^2$

60. $\frac{x + 2}{3x + 2} = \frac{1}{x}$

61. $\dfrac{2 + \dfrac{2}{x}}{x + 2 + \dfrac{1}{x}} = \dfrac{x + 2}{3}$

62. $\frac{x^6 x^4}{x^9 x^{-1}} = \frac{5^{14}}{25^6}$

63. Find the reciprocal of $\frac{1 - x}{x + 3} + \frac{x + 1}{2 - x}$.

64. Find the reciprocal of 2.0×10^{-8} and express in scientific notation.

6

Graphs of Equations and Inequalities

INTRODUCTION

We now study the graphs of equations in two variables. These variables will be raised only to the first power, and there will be no products or quotients involving variables. We also learn to graph such equations, called *linear*. The graphs of linear equations are straight lines. Later, we assign numbers called *slopes* to describe the way the lines slant. We will also consider variation and the graphing of inequalities in two variables.

AN APPLICATION

The number N of computer instructions per second varies directly as the speed S of its internal processor. A processor with a speed of 25 megahertz can perform 2,000,000 instructions per second. How many instructions will the same processor perform if it is running at a speed of 40 megahertz?

THE MATHEMATICS

We translate the problem to this equation:

$$N = 80{,}000S.$$

We can then substitute 40 for S and compute N. This is an example of a linear equation.

Objectives for Review

The review objectives to be tested in addition to the material in this chapter are as follows.

[1.8d] Simplify expressions using rules for order of operations.
[4.7b] Solve quadratic equations by factoring and then using the principle of zero products.
[5.6a] Solve rational equations.
[5.7a, b] Solve applied problems and proportion problems using rational equations.

Pretest: Chapter 6

Graph on a plane.

1. $y = -x$

2. $x = -4$

3. $4x - 5y = 20$

4. $y = \frac{2}{3}x - 1$

5. In which quadrant is the point $(-4, -1)$ located?

6. Determine whether the ordered pair $(-2, 4)$ is a solution of the equation $2x - y = 0$.

Find the slope, if it exists, of the line.

7. $-4x + y = 6$

8. $y = 3$

9. Find the slope and the y-intercept of the line $x - 3y = 7$.

10. Find the slope, if it exists, of the line containing the points $(3, 0)$ and $(3, 6)$.

11. Find an equation of the line containing the points $(3, -1)$ and $(1, -3)$.

12. Find an equation of the line containing the point $(-1, 3)$ and having slope 4.

13. Find an equation of variation where y varies directly as x and $y = 10$ when $x = 4$.

14. Find an equation of variation where y varies inversely as x and $y = 10$ when $x = 4$.

Graph on a plane.

15. $y < x + 2$

16. $2y - 3x \geq 6$

Determine whether the graphs of the equations are parallel, perpendicular, or neither.

17. $y - 3x = 9,$
$y - 3x = 7$

18. $-x + 2y = 7,$
$2x + y = 4$

19. $y = \frac{2}{3}x - 5,$
$y = -\frac{3}{2}x + 4$

20. Determine whether the ordered pair $(-3, 4)$ is a solution of $2x + 5y < 17$.

6.1 Graphs

We often see graphs of various kinds in newspapers and magazines and on television. Graphs are now even more common because they are easy to prepare using a computer. We first solve some problems with commonly used graphs.

a Problem Solving with Graphs

Circle Graphs

Circle graphs, or pie charts, are often used to show the percent of a quantity for each particular item in a related group.

Example 1 This circle graph shows expenses as a percent of income for a family of four, according to the Bureau of Labor Statistics.

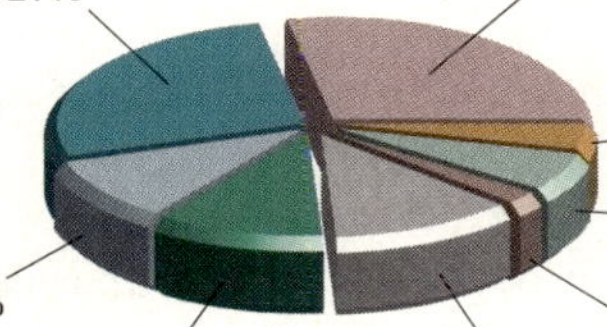

A family with a \$2000 monthly income would typically spend how much on housing per month?

1. ***Familiarize.*** The graph tells us that housing is 27% of income. We let $y =$ the amount spent on housing.
2. ***Translate.*** We restate and translate the problem as follows:

What	is	27%	of	income?	Restate
↓	↓	↓	↓	↓	
y	$=$	27%	$\cdot$	\$2000	Translate

3. ***Solve.*** We solve by carrying out the computation:

$$y = 0.27 \cdot \$2000 = \$540.$$

4. ***Check.*** We leave the check to the student.
5. ***State.*** The family would spend \$540 on housing.

Do Exercise 1.

Bar Graphs

A *bar graph* is convenient for showing comparisons. The bars may be vertical or horizontal. In Example 2, the number of years of school completed are located on the horizontal axis. Then corresponding income levels are located on the vertical axis.

Objectives

After finishing Section 6.1, you should be able to:

- **a** Solve problems related to circle, bar, and line graphs.
- **b** Plot points associated with ordered pairs of numbers.
- **c** Determine the quadrant in which a point lies.
- **d** Find the coordinates of a point on a graph.

For Extra Help

TAPE 14 | TAPE 12B | MAC: 6 IBM: 6

1. Consider the circle graph in Example 1.
   a) A family with a \$2000 monthly income would typically spend how much on transportation per month?
   b) What percent of the income is spent on food?

Answers on page A-5

2. Consider the bar graph in Example 2.

a) What is the average annual income for a person who has completed only high school?

b) What is the greatest average annual income for a person who doesn't finish high school?

Answers on page A-5

EXAMPLE 2 A recent survey of 2000 individuals produced the following information on average income based on years of schooling.

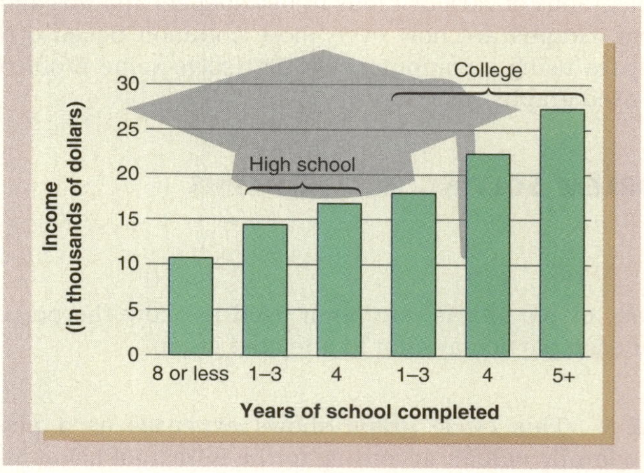

a) What is the average annual income for a person who has completed two years of high school?

b) How many years of schooling does it take to expect an average annual income of at least $20,000?

a) We go to the right, across the bottom, to the bar representing income for a person with 1–3 yr of high school. We then go up to the top of the bar and, from there, back to the left to read approximately $14,000 on the income scale.

b) We go up the left-hand scale of the graph to the $20,000 mark and read to the right, until we come to a bar crossing our path. Moving down on that bar, we find that at least 4 yr of college are needed.

Do Exercise 2.

LINE GRAPHS

We often use *line graphs* to show change over time. Certain horizontal units are associated with vertical units. Each association determines a point. When the points are connected by line segments, we get a line graph.

EXAMPLE 3 This line graph shows the closing Dow Jones Industrial Average (DJIA) for each of six weeks.

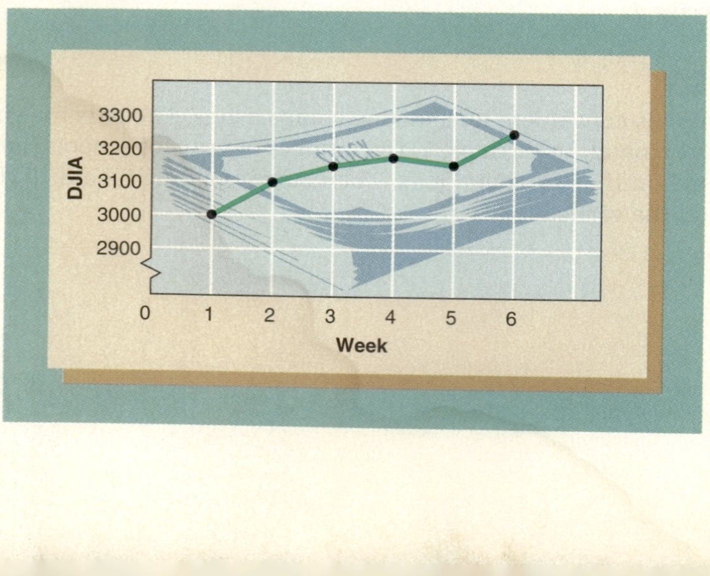

a) For which week was the DJIA closing the lowest?

b) Between which two weeks did the DJIA closing decrease?

c) For which week was the DJIA closing about 3200?

a) During the six weeks, the lowest closing was 3000 at the end of the first week.

b) Reading the graph from left to right, we see that the line went down only between the fourth and fifth weeks.

c) We locate 3200 on the DJIA scale and then move to the right until we reach the line. At that point, we move down to the "Week" scale and read the information we are seeking. The DJIA closing was about 3200 for the fourth week.

Do Exercise 3.

3. Consider the line graph in Example 3.

a) For which week was the DJIA closing about 3000?

b) For which week was the DJIA closing about 3150?

b POINTS AND ORDERED PAIRS

We have graphed numbers on a line. To enable us to graph an equation that contains two variables, we now learn to graph number pairs on a plane.

On a number line, each point is the graph of a number. On a plane, each point is the graph of a number pair. We use two perpendicular number lines called **axes**. They cross at a point called the **origin**. The arrows show the positive directions. Consider the ordered pair (3, 4). The numbers in an ordered pair are called **coordinates**. In (3, 4), the **first coordinate** is 3 and the **second coordinate** is 4. To plot (3, 4), we start at the origin and move horizontally to the 3. Then we move up vertically 4 units and make a "dot."

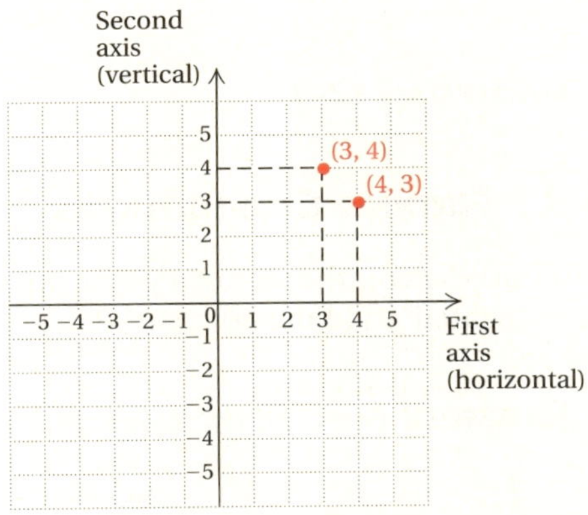

The point (4, 3) is also plotted. Note that (3, 4) and (4, 3) give different points. The order of the numbers in the pair is indeed important. They are called **ordered pairs** because it makes a difference which number comes first. The coordinates of the origin are (0, 0) even though it is usually labeled either with the number 0 or not at all.

EXAMPLE 4 Plot the point (−3, 4).

The first number, −3, is negative. Starting at the origin, we move −3 units in the horizontal direction (3 units to the left). The second number, 4, is positive. We move 4 units in the vertical direction (up).

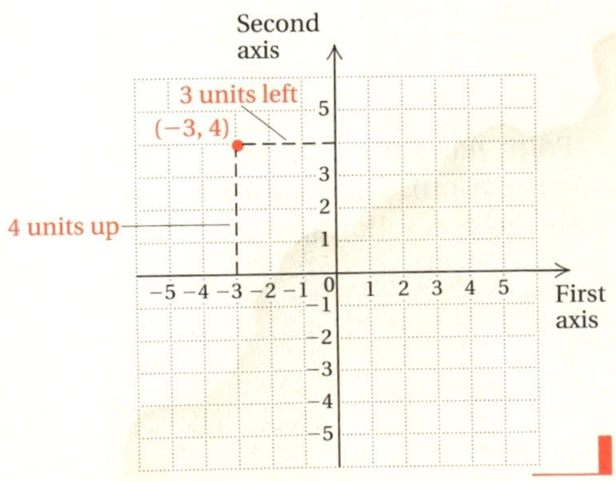

Do Exercises 4–11.

Plot these points on the graph below.

4. (4, 5)

5. (5, 4)

6. (−2, 5)

7. (−3, −4)

8. (5, −3)

9. (−2, −1)

10. (0, −3)

11. (2, 0)

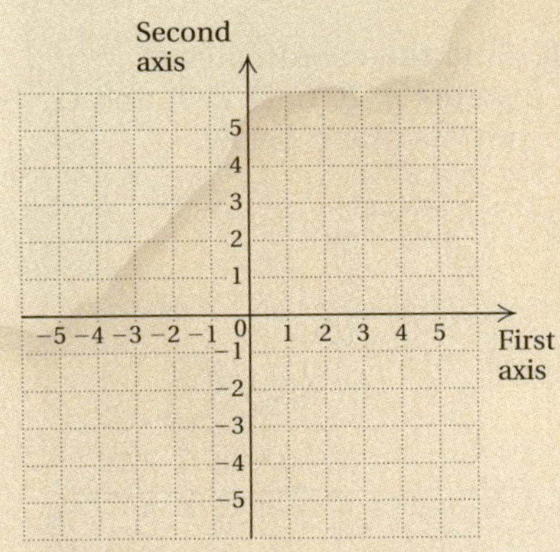

Answers on page A-5

c QUADRANTS

This figure shows some points and their coordinates. In region I (the *first quadrant*), both coordinates of any point are positive. In region II (the *second quadrant*), the first coordinate is negative and the second positive. In region III (the *third quadrant*), both coordinates are negative. In region IV (the *fourth quadrant*), the first coordinate is positive and the second is negative.

EXAMPLE 5 In which quadrant are the points (−4, 5), (5, −5), (2, 4), and (−2, −5) located?

The point (−4, 5) is in the second quadrant. The point (5, −5) is in the fourth quadrant. The point (2, 4) is in the first quadrant. The point (−2, −5) is in the third quadrant.

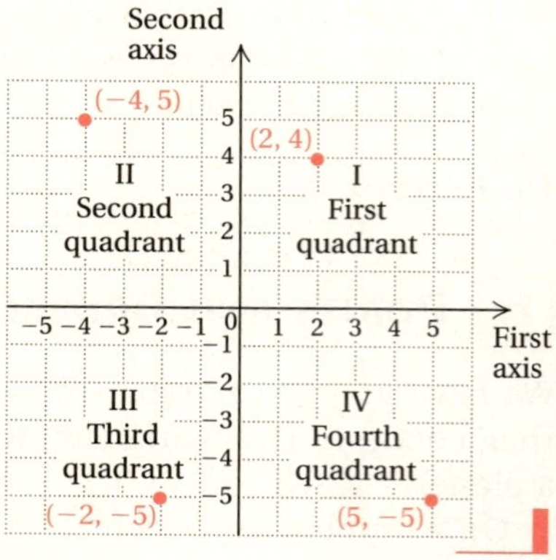

Do Exercises 12–17.

d FINDING COORDINATES

To find the coordinates of a point, we see how far to the right or left of zero it is located and how far up or down.

EXAMPLE 6 Find the coordinates of points *A*, *B*, *C*, *D*, *E*, *F*, and *G*.

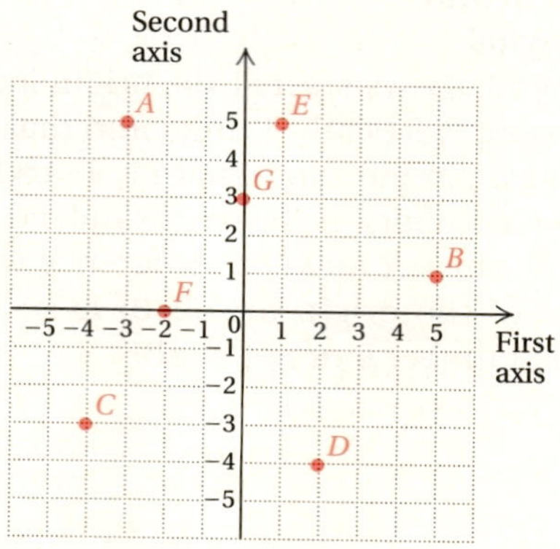

Point *A* is 3 units to the left (horizontal direction) and 5 units up (vertical direction). Its coordinates are (−3, 5). The coordinates of the other points are as follows:

B: (5, 1); *C*: (−4, −3); *D*: (2, −4);
E: (1, 5); *F*: (−2, 0); *G*: (0, 3).

Do Exercise 18.

12. What can you say about the coordinates of a point in the third quadrant?

13. What can you say about the coordinates of a point in the fourth quadrant?

In which quadrant is the point located?

14. (5, 3)

15. (−6, −4)

16. (10, −14)

17. (−13, 9)

18. Find the coordinates of points *A*, *B*, *C*, *D*, *E*, *F*, and *G* on the graph below.

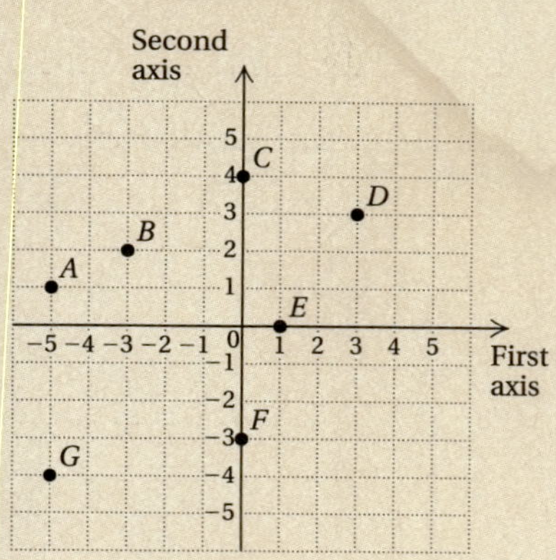

Answers on page A-5

NAME SECTION DATE

Exercise Set 6.1

a Use the circle graph in Example 1 to answer Exercises 1 and 2.

1. What percent of the monthly income is spent on medical expenses?

2. A family with a $2000 monthly income would typically spend how much on clothing per month?

Use the following circle graph for Exercises 3–8. The graph shows music preferences.

Music Preferences According to Music Store Sales

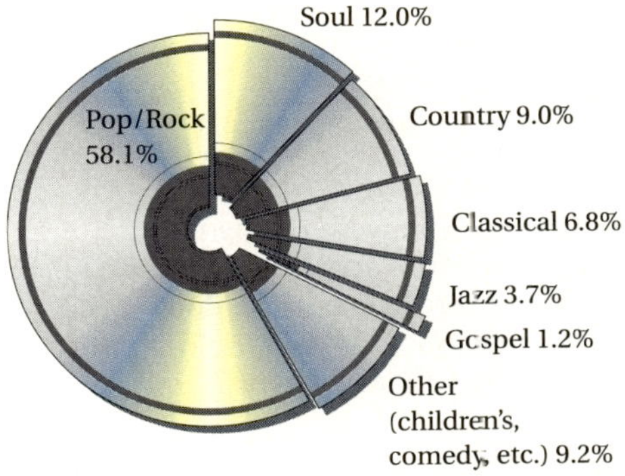

Source: National Association of Recording Merchandisers

3. What percent of all recordings sold are jazz?

4. What percent of all recordings sold are country?

5. Together, what percent of all recordings sold are either soul or pop/rock?

6. Together, what percent of all recordings sold are either classical or jazz?

7. A music store sells 3000 recordings a month. How many are pop/rock? soul? country?

8. A music store sells 2500 recordings a month. How many are pop/rock? classical? gospel?

Use the bar graph in Example 2 to answer Exercises 9–11.

9. How much more income can be expected if 5 yr, rather than 4 yr, of college are completed?

10. How much more can you expect to earn annually if you complete high school than if you complete only the eighth grade?

11. How much more income can you expect if you complete 4 yr of college than if you complete from 1–3 yr?

Use the line graph in Example 3 to answer Exercises 12–14.

12. For which week was the DJIA closing the highest?

13. For which week was the DJIA closing about 3100?

14. About how many points was the increase in the DJIA between weeks 1 and 6?

ANSWERS

1. __________
2. __________
3. __________
4. __________
5. __________
6. __________
7. __________
8. __________
9. __________
10. __________
11. __________
12. __________
13. __________
14. __________

ANSWERS

15. ______

16. ______

17. ______

18. ______

19. ______

20. ______

21. ______

22. ______

23. ______

24. ______

25. ______

26. ______

Use the following bar graph for Exercises 15–20. The graph shows the number of calories burned per hour during each activity by a person weighing 152 lb.

15. About how many calories are burned hiking?

16. About how many calories are burned jogging?

17. About how many more calories are burned jogging than hiking?

18. About how many more calories are burned playing tennis than sleeping?

19. Suppose you were trying to lose weight by exercising and had to choose one of these exercises, but the doctor told you not to jog. What would be the most beneficial exercise?

20. What is the average number of calories burned per hour in these exercises?

Use the following line graph for Exercises 21–26. The graph shows the estimated sales (in millions of dollars) for a company for several years.

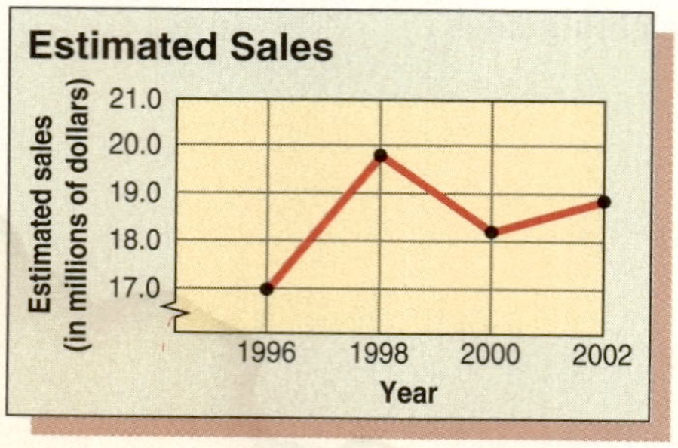

21. In what year are estimated sales the greatest?

22. In what year are estimated sales the least?

23. What are estimated sales in 1996?

24. How much greater are estimated sales in 1998 than in 2000?

25. What are estimated sales in 2002?

26. How much greater are estimated sales in 2002 than in 1996?

b

27. Plot these points.

(2, 5) (−1, 3) (3, −2) (−2, −4)
(0, 4) (0, −5) (5, 0) (−5, 0)

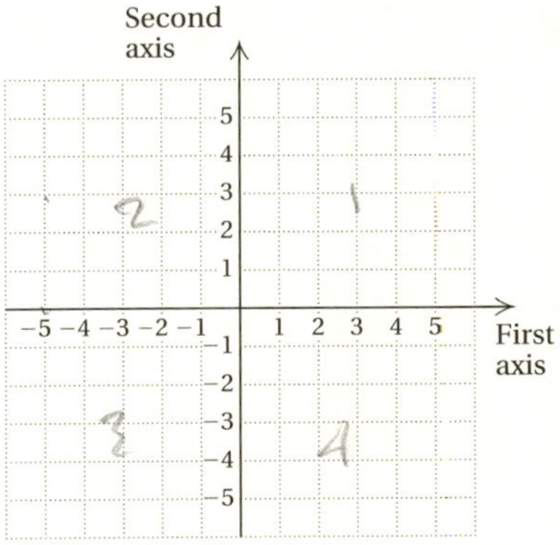

28. Plot these points.

(4, 4) (−2, 4) (5, −3) (−5, −5)
(0, 4) (0, −4) (3, 0) (−4, 0)

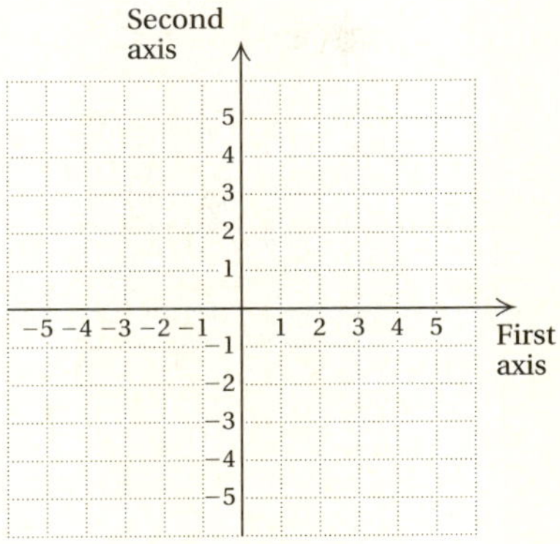

c In which quadrant is the point located?

29. (−5, 3) **30.** (1, −12) **31.** (100, −1) **32.** (−2.5, 35.6)

33. (−6, −29) **34.** (3.6, 105.9) **35.** (3.8, 9.2) **36.** (−895, −492)

37. In quadrant III, first coordinates are always ________ and second coordinates are always ________.

38. In quadrant II, ________ coordinates are always positive and ________ coordinates are always negative.

Answers

27. See graph.
28. See graph.
29. ________
30. ________
31. ________
32. ________
33. ________
34. ________
35. ________
36. ________
37. ________
38. ________

ANSWERS

39. ______

40. ______

41. ______

42. ______

43. ______

44. ______

45. ______

46. ______

47. ______

48. ______

49. ______

50. ______

51. ______

52. ______

53. ______

54. ______

55. ______

56. ______

d

39. Find the coordinates of points A, B, C, D, and E.

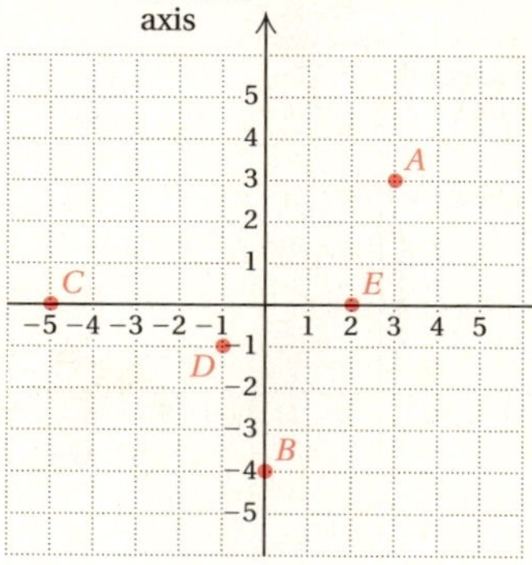

40. Find the coordinates of points A, B, C, D, and E.

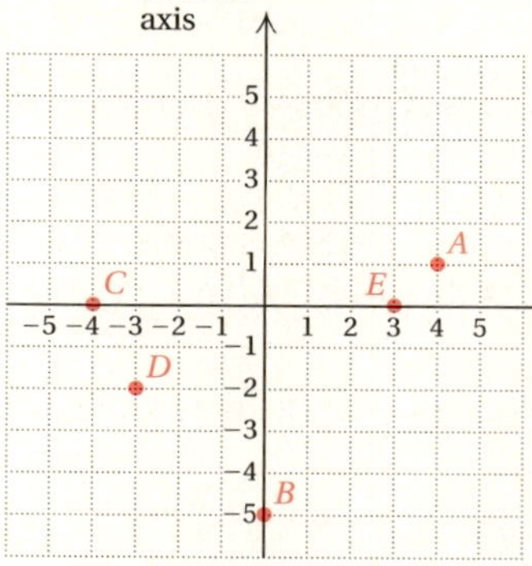

SKILL MAINTENANCE

41. A family making \$19,600 a year will spend \$5096 on food. At this rate, how much would a family making \$20,500 spend on food?

42. Two cars leave town at the same time and drive in opposite directions. One travels at 54 mph and the other travels at 43 mph. In how many hours will they be 291 mi apart?

Compute.

43. $256 \div 4 \cdot 16 - 37$

44. $5^2 - 3^2 - 2^2 \div 4$

45. $1000 \div 100 \div 10$

46. $10^3 \div 5^2 - 4 \cdot 8$

SYNTHESIS

In Exercises 47–50, tell in which quadrant(s) the point can be located.

47. The first coordinate is positive.

48. The second coordinate is negative.

49. The first and second coordinates are equal.

50. The first coordinate is the additive inverse of the second coordinate.

51. The points $(-1, 1)$, $(4, 1)$, and $(4, -5)$ are three vertices of a rectangle. Find the coordinates of the fourth vertex.

52. Three parallelograms share the vertices $(-2, -3)$, $(-1, 2)$, and $(4, -3)$. Find the fourth vertex of each parallelogram.

53. Graph eight points such that the sum of the coordinates in each pair is 6.

54. Graph eight points such that the first coordinate minus the second coordinate is 1.

55. Find the perimeter of a rectangle whose vertices have coordinates $(5, 3)$, $(5, -2)$, $(-3, -2)$, and $(-3, 3)$.

56. Find the area of a triangle whose vertices have coordinates $(0, 9)$, $(0, -4)$, and $(5, -4)$.

6.2 Graphing Linear Equations

A **linear equation** is an equation whose graph is a straight line. We now learn how to graph certain types of linear equations.

a Solutions of Equations

An equation with two variables has *ordered pairs* of numbers for solutions. If not directed otherwise, alphabetical *order* of the variables corresponds to the *order* in the pair.

Example 1 Determine whether $(-3, -5)$ is a solution of the equation $y = 2x + 1$.

We substitute:

y	$= 2x + 1$	
-5	$2(-3) + 1$	We substitute -3 for x and -5 for y (alphabetical order of variables).
	$-6 + 1$	
	-5	TRUE

Since the equation becomes true, $(-3, -5)$ is a solution.

Similarly, in Example 1, we can show that $(3, 7)$ and $(0, 1)$ are also solutions. In fact, there are more solutions than we can list: an infinite number of solutions.

Example 2 Determine whether $(-2, 3)$ is a solution of $2t = 4s - 8$.

We substitute:

$2t$	$= 4s - 8$	
$2 \cdot 3$	$4(-2) - 8$	We substitute -2 for s and 3 for t.
6	$-8 - 8$	
	-16	FALSE

Since the equation becomes false, $(-2, 3)$ is not a solution.

Do Exercises 1 and 2.

b Graphing Equations of the Type $y = mx$ and $y = mx + b$

The equations considered in Examples 1 and 2 have an infinite number of solutions, meaning that we cannot list them all. Because of this, it is convenient to make a drawing, or **graph**, that represents the solutions.

> To *graph* an equation means to make a drawing that represents its solutions.

The graphs of linear equations of the type $y = mx$ and $y = mx + b$ are straight lines. If an equation has a graph that is a straight line, we can graph it by plotting two or more points and then drawing a line through them.

Objectives

After finishing Section 6.2, you should be able to:

a Determine whether an ordered pair of numbers is a solution of an equation with two variables.

b Graph equations of the type $y = mx$ and $y = mx + b$.

For Extra Help

TAPE 15 TAPE 12B MAC: 6 IBM: 6

1. Determine whether $(2, 3)$ is a solution of $y = 2x + 3$.

2. Determine whether $(-2, 4)$ is a solution of $4q - 3p = 22$.

Answers on page A-5

Graph.

3. $y = 3x$

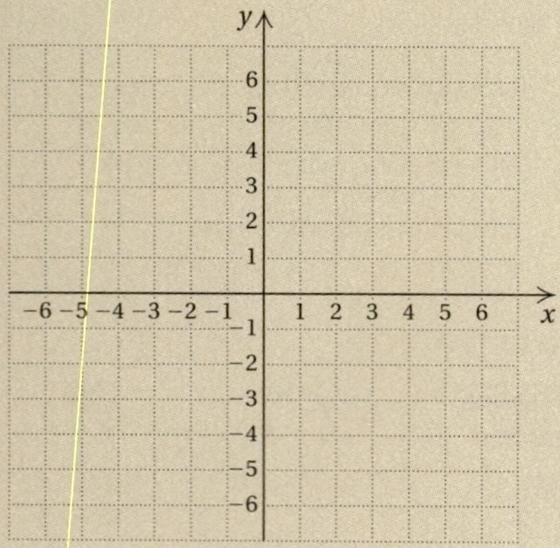

4. $y = \frac{1}{2}x$

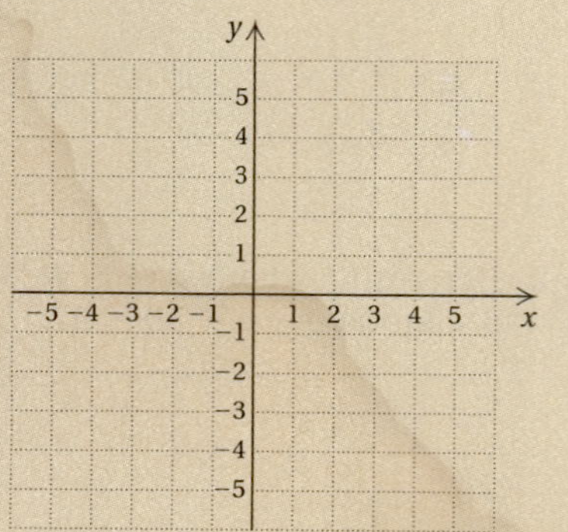

Answers on page A-5

EXAMPLE 3 Graph: $y = x$.

We will use alphabetical order. Thus the first (horizontal) axis will be the x-axis and the second (vertical) axis will be the y-axis. Next, we find some ordered pairs that are solutions of the equation, keeping the results in a table. We choose *any* number for x and then find y by substitution. In this case, it is easy. Here are a few:

Let $x = 0$. Then $y = x = 0$. We get a solution: the ordered pair (0, 0).

Let $x = 1$. Then $y = x = 1$. We get a solution: the ordered pair (1, 1).

Let $x = 5$. Then $y = x = 5$. We get a solution: (5, 5).

Let $x = -2$. Then $y = x = -2$. We get a solution: $(-2, -2)$.

Let $x = -4$. Then $y = x = -4$. We get a solution: $(-4, -4)$.

We gather our results in a table like the one shown below. Then we plot the points. We look for a pattern in the points plotted. It looks as if the points resemble a straight line. We draw the line with a ruler. Since the line is the graph of the equation $y = x$, we label the line $y = x$ on the graph paper.

x	y $y = x$	(x, y)
0	0	(0, 0)
1	1	(1, 1)
5	5	(5, 5)
−2	−2	(−2, −2)
−4	−4	(−4, −4)

(1) Choose x.
(2) Compute y.
(3) Form the pair (x, y).
(4) Plot the points.

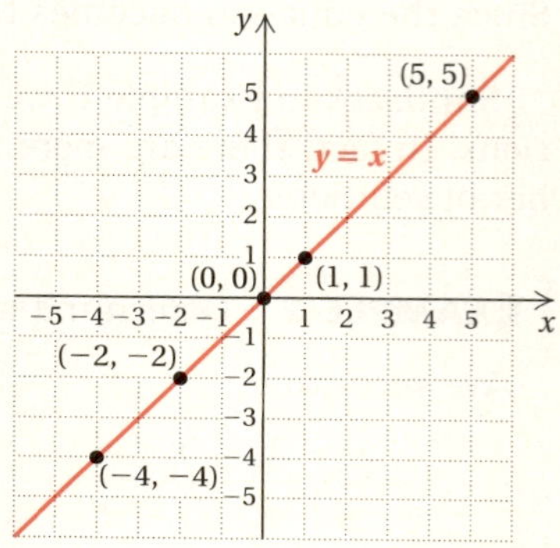

EXAMPLE 4 Graph: $y = 2x$.

We find some ordered pairs that are solutions. Since the graph is a line, we really need to find only two, but we will usually plot a third point as a check. We keep the results in a table. We choose *any* number for x and then determine y by substitution. Since $y = 2x$, we find y by doubling x. Suppose we choose 3 for x. Then

$$y = 2x = 2 \cdot 3 = 6.$$

We get a solution: the ordered pair (3, 6). Suppose we choose 0 for x. Then

$$y = 2x = 2 \cdot 0 = 0.$$

We get a solution: the ordered pair (0, 0). For a third point, we make a negative choice for x. We now have enough points to plot the line, but if we wish we can compute more. If a number takes us off the graph paper, we generally do not use it. Continuing in this manner, we get a table like the one shown below.

Now we plot these points. We draw the line, or graph, with a ruler and label it $y = 2x$.

x	y $y = 2x$	(x, y)
3	6	(3, 6)
1	2	(1, 2)
0	0	(0, 0)
−2	−4	(−2, −4)
−3	−6	(−3, −6)

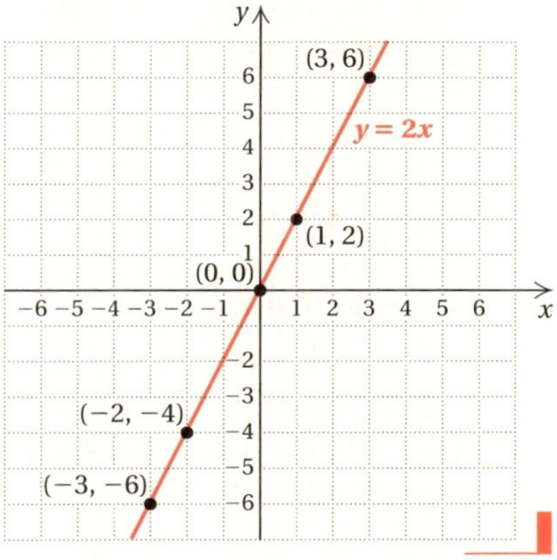

Do Exercises 3 and 4 on the preceding page.

EXAMPLE 5 Graph: $y = -3x$.

We make a table of solutions. Then we plot points. We draw the line with a ruler and label it $y = -3x$.

x	y $y = -3x$	(x, y)
0	0	(0, 0)
1	−3	(1, −3)
−1	3	(−1, 3)
2	−6	(2, −6)
−2	6	(−2, 6)

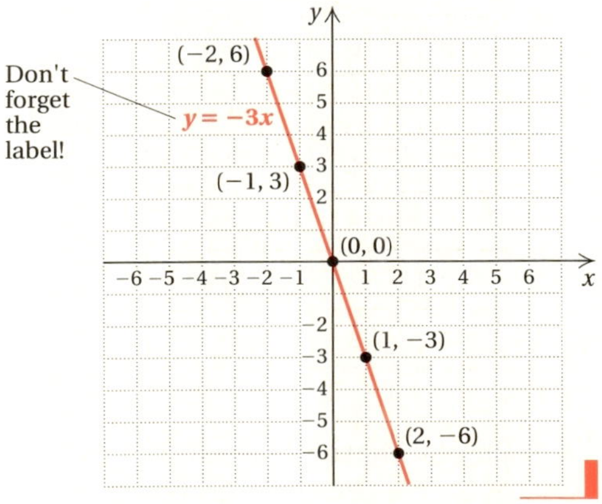

Do Exercises 5 and 6.

EXAMPLE 6 Graph: $y = -\frac{5}{3}x$.

We make a table of solutions.

When $x = 0$, $y = -\frac{5}{3} \cdot 0 = 0$.
When $x = 3$, $y = -\frac{5}{3} \cdot 3 = -5$.
When $x = -3$, $y = -\frac{5}{3} \cdot (-3) = 5$.
When $x = 1$, $y = -\frac{5}{3} \cdot 1 = -\frac{5}{3}$.

Note that if we substitute multiples of 3, we can avoid fractions.

Next we plot the points and complete the graph by drawing a line through them.

Graph.

5. $y = -x$ (or $-1 \cdot x$)

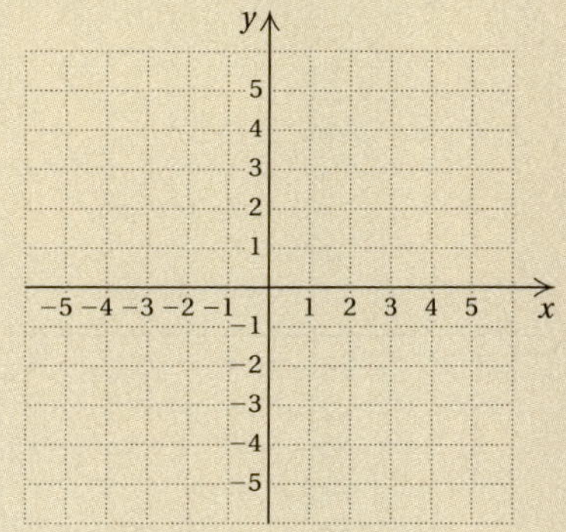

6. $y = -2x$

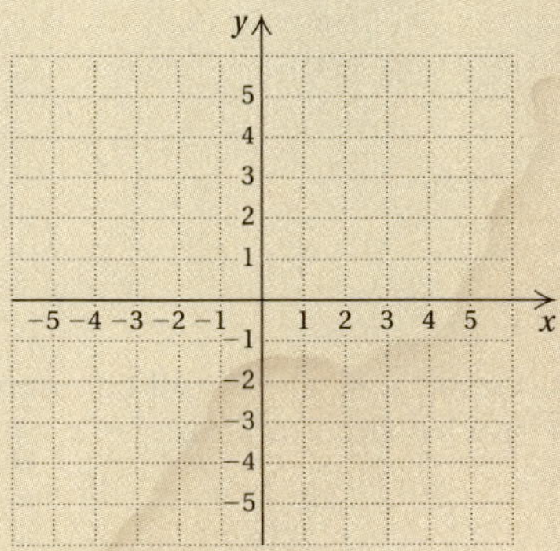

Answers on page A-5

Graph.

7. $y = \frac{3}{4}x$

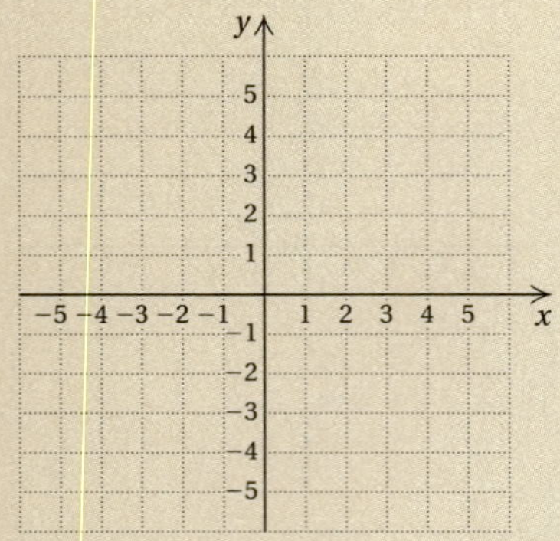

8. $y = -\frac{4}{5}x$

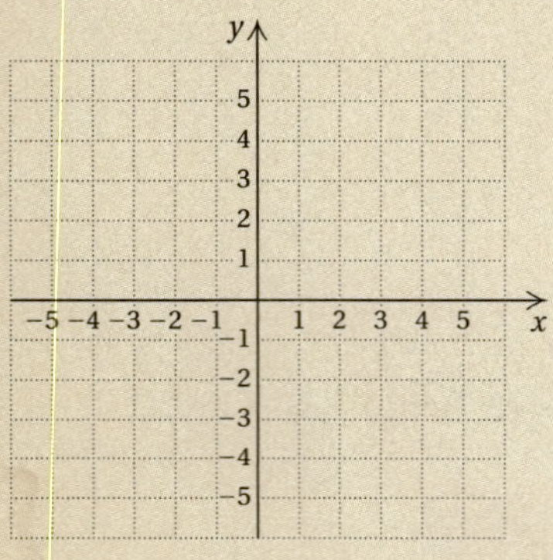

9. Graph $y = x + 3$ and compare it with the graph of $y = x$.

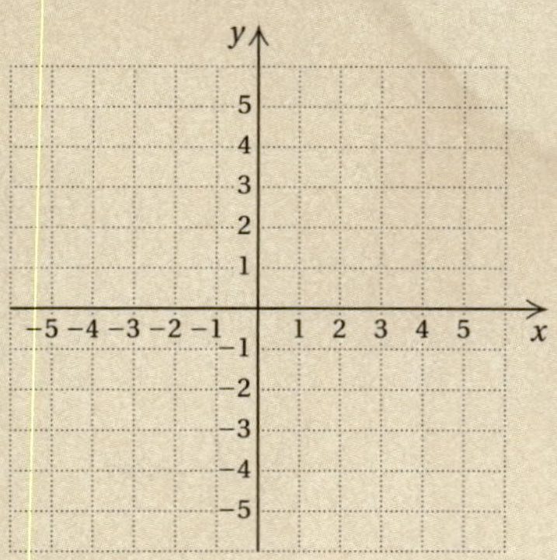

Answers on page A-5

	y
x	$y = -\frac{5}{3}x$
0	0
3	−5
−3	5
1	$-\frac{5}{3}$

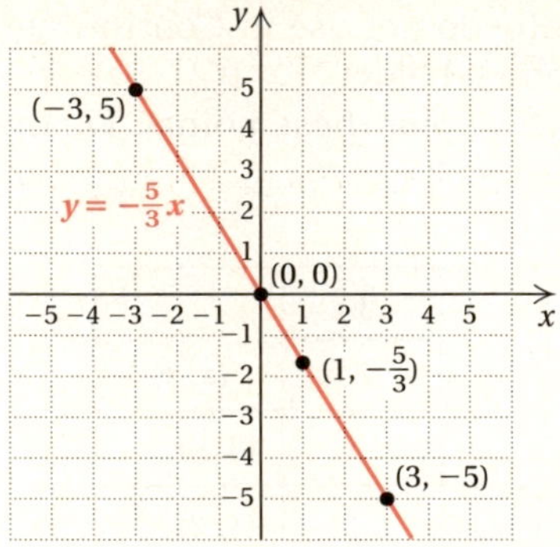

Do Exercises 7 and 8.

Every equation $y = mx$ has a graph that is a straight line. It contains the origin, (0, 0). What will happen if we add a number b on the right side to get an equation $y = mx + b$?

EXAMPLE 7 Graph $y = x$ and $y = x + 2$ using the same set of axes. Compare.

We first make a table containing values for both equations.

	y	y
x	$y = x$	$y = x + 2$
0	0	2
1	1	3
−1	−1	1
2	2	4
−2	−2	0
3	3	5

We then plot these points and draw a blue line for $y = x$ and a red line for $y = x + 2$. We see that the graph of $y = x + 2$ can be obtained from the graph of $y = x$ by moving, or translating, the graph of $y = x$ up 2 units.

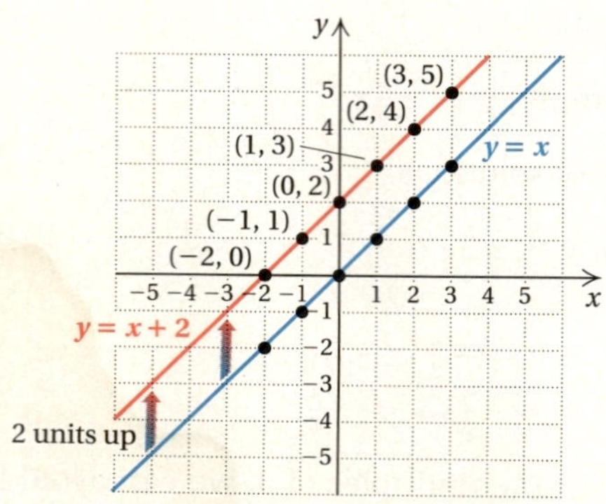

Do Exercises 9 and 10. (Exercise 10 is on the following page.)

EXAMPLE 8 Graph $y = 2x$ and $y = 2x - 3$ using the same set of axes. Compare.

We first make a table containing values for both equations.

x	y $y = 2x$	y $y = 2x - 3$
0	0	-3
1	2	-1
2	4	1
-1	-2	-5

The graph of $y = 2x - 3$ looks just like the graph of $y = 2x$, but $y = 2x$ is moved, or translated, down 3 units.

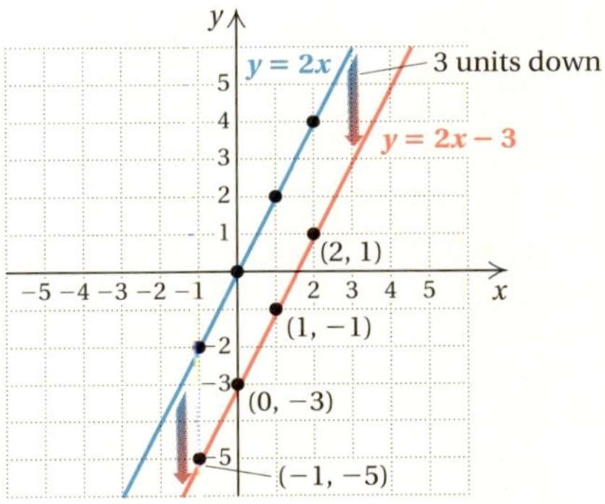

Do Exercise 11.

EXAMPLE 9 Graph: $y = \frac{2}{5}x + 4$.

We make a table of values. Using multiples of 5 avoids fractions.

When $x = 0$, $y = \frac{2}{5} \cdot 0 + 4 = 0 + 4 = 4$.
When $x = 5$, $y = \frac{2}{5} \cdot 5 + 4 = 2 + 4 = 6$.
When $x = -5$, $y = \frac{2}{5} \cdot (-5) + 4 = -2 + 4 = 2$.

Since two points determine a line, that is all we really need to graph a line, but we will usually plot a third point as a check.

x	y
0	4
5	6
-5	2

10. Graph $y = x - 1$ and compare it with the graph of $y = x$.

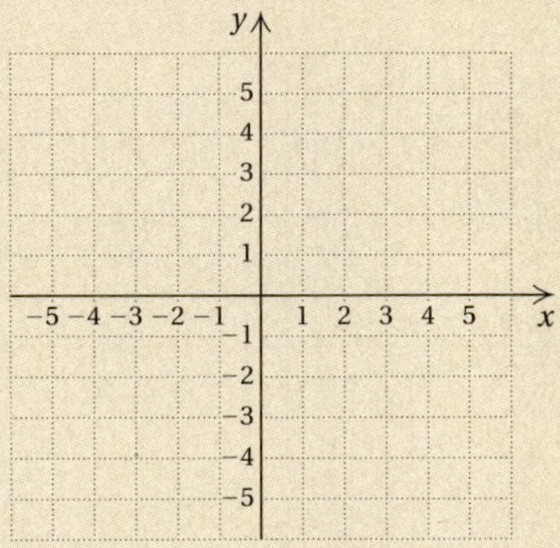

11. Graph $y = 2x + 3$ and compare it with the graph of $y = 2x$.

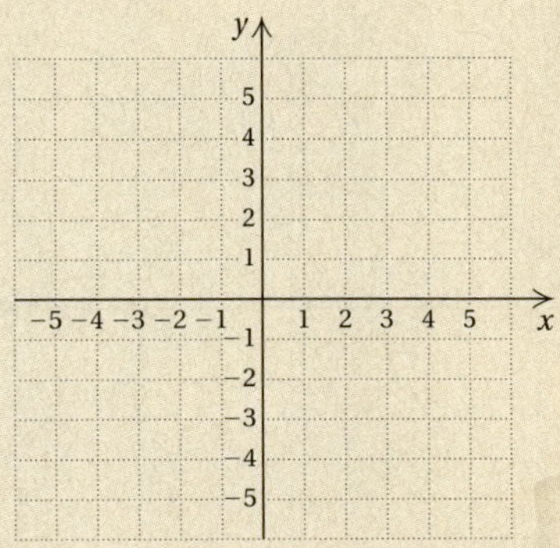

12. Graph: $y = \dfrac{3}{5}x + 2$.

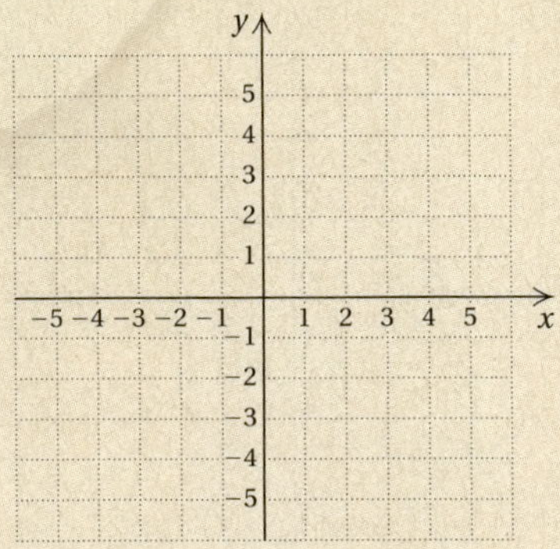

Answers on page A-5

Graph.

13. $y = \frac{3}{5}x - 2$

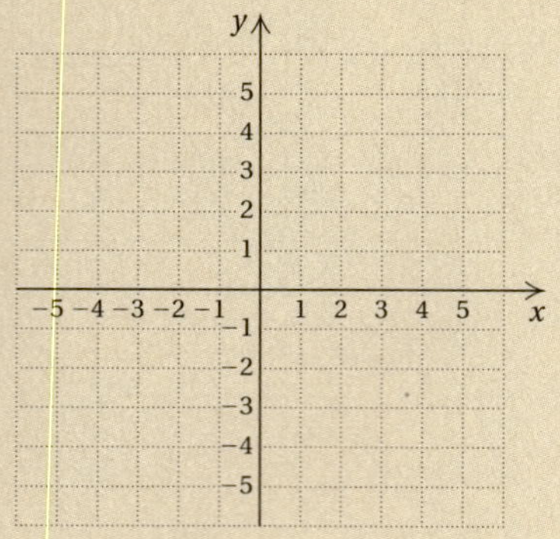

14. $y = -\frac{3}{5}x - 1$

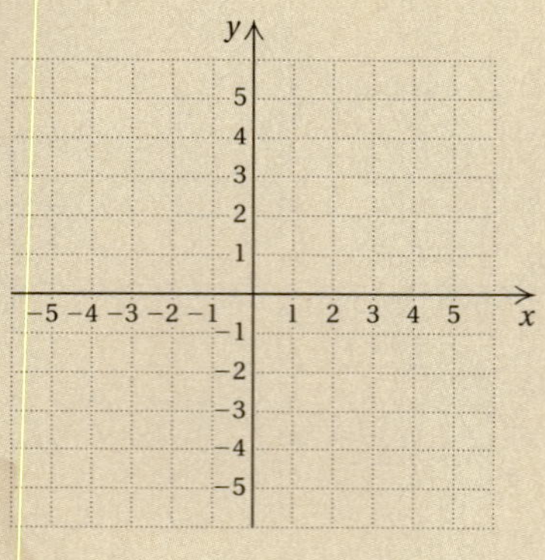

15. $y = -\frac{3}{5}x + 4$

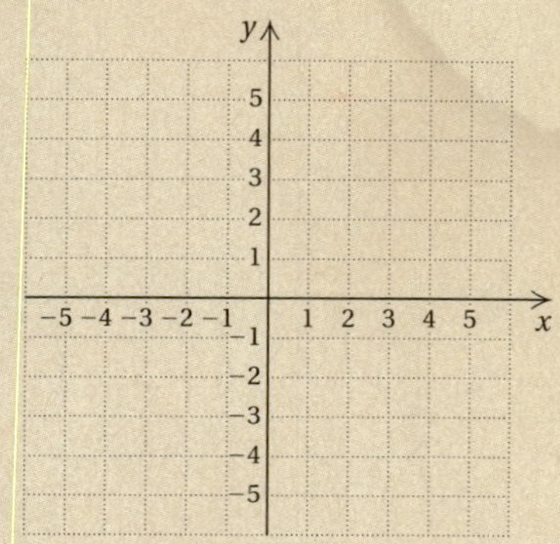

Answers on page A-5

We draw the graph of $y = \frac{2}{5}x + 4$.

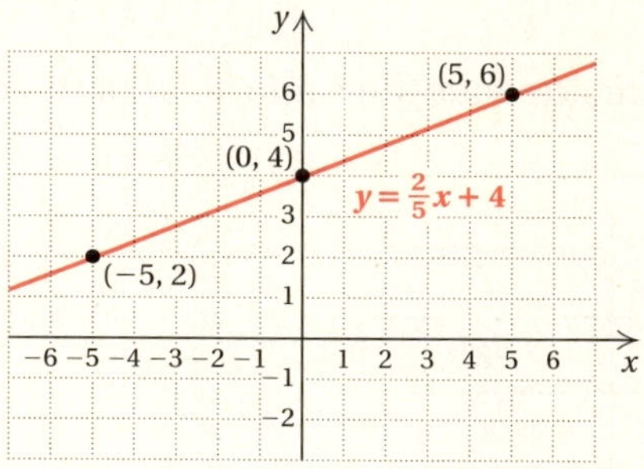

EXAMPLE 10 Graph: $y = -\frac{3}{4}x - 2$.

We first make a table of values.

When $x = 0$, $y = -\frac{3}{4} \cdot 0 - 2 = 0 - 2 = -2$.
When $x = 4$, $y = -\frac{3}{4} \cdot 4 - 2 = -3 - 2 = -5$.
When $x = -4$, $y = -\frac{3}{4}(-4) - 2 = 3 - 2 = 1$.

x	y
0	−2
4	−5
−4	1

We plot these points and draw a line through them. This line is the graph of the equation. We label the graph $y = -\frac{3}{4}x - 2$. (for the rows 0, −2 and 4, −5)

We plot this point for a check to see whether it is on the line. (for the row −4, 1)

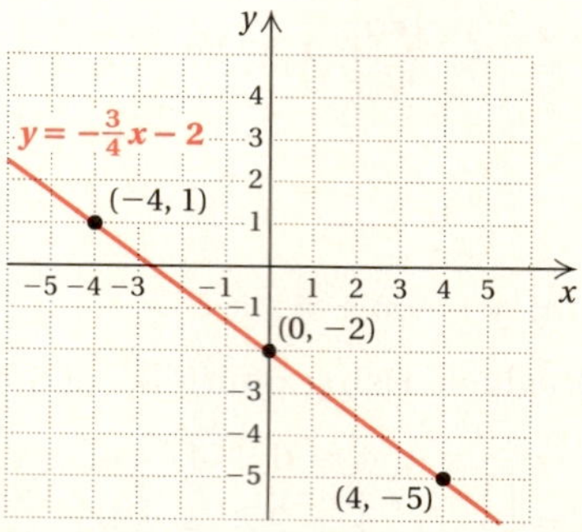

Do Exercises 12–15. (Exercise 12 is on the preceding page.)

NAME SECTION DATE

Exercise Set 6.2

a Determine whether the given point is a solution of the equation.

1. (2, 5); $y = 3x - 1$

2. (4, 11); $y = 2x + 3$

3. (2, −3); $3x - y = 4$

4. (4, −1); $8x + y = 10$

5. (0, −4); $4p + 2q = -9$

6. (−2, −1); $2a + 2b = -7$

b Graph.

7. $y = 4x$

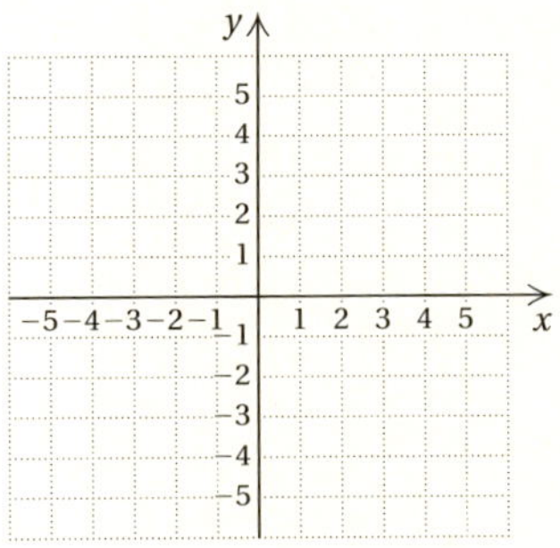

8. $y = 2x$

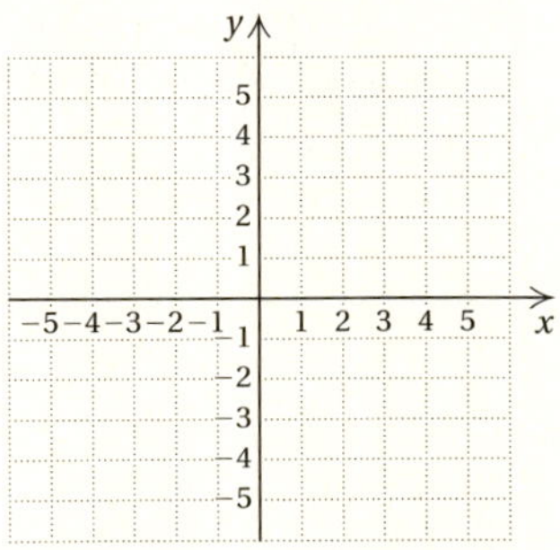

9. $y = -2x$

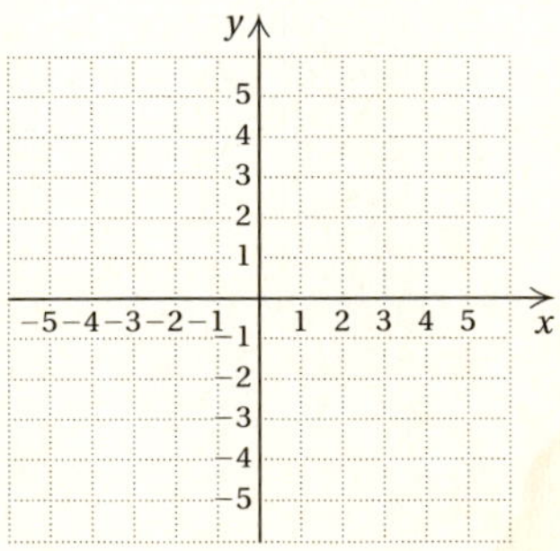

10. $y = -4x$

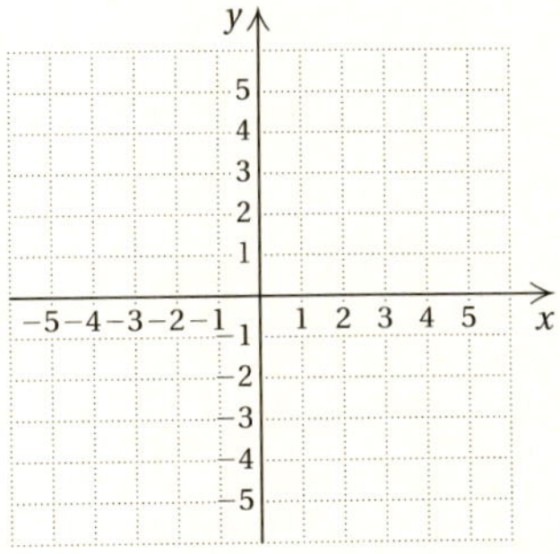

11. $y = \frac{1}{3}x$

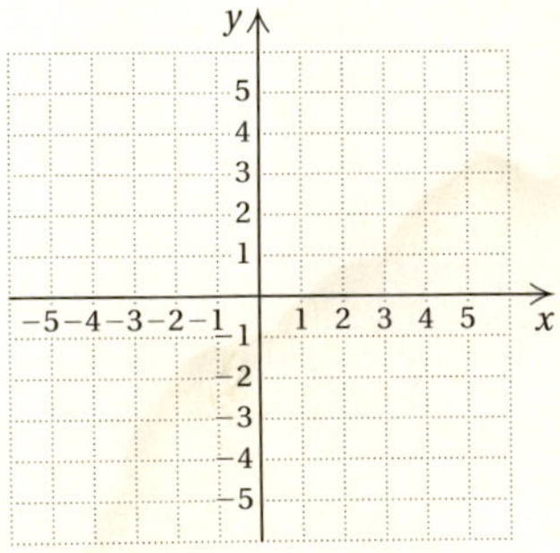

12. $y = \frac{1}{4}x$

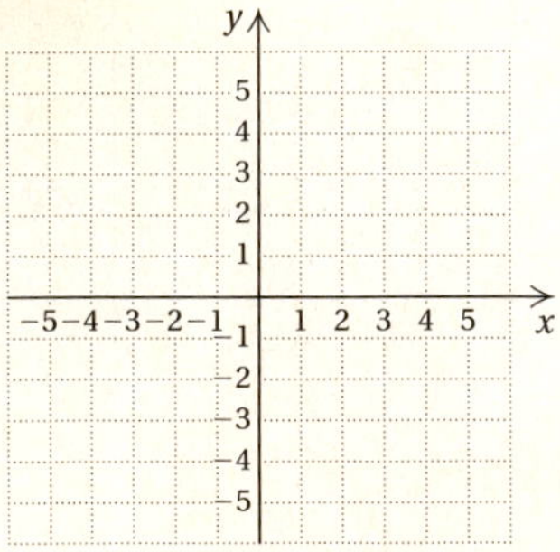

ANSWERS

1. ______
2. ______
3. ______
4. ______
5. ______
6. ______

13. $y = -\frac{3}{2}x$

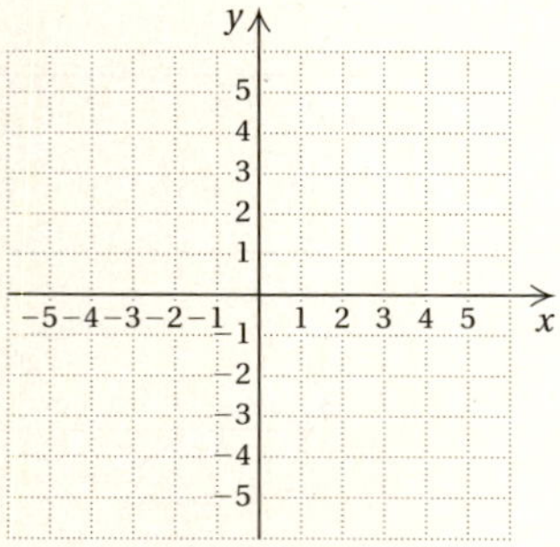

14. $y = -\frac{5}{4}x$

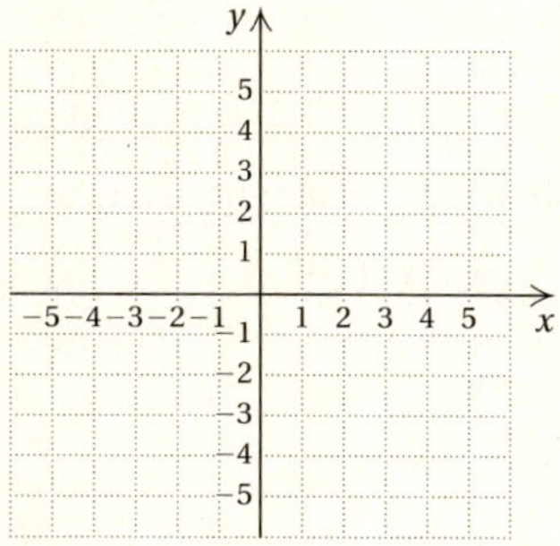

15. $y = x + 1$

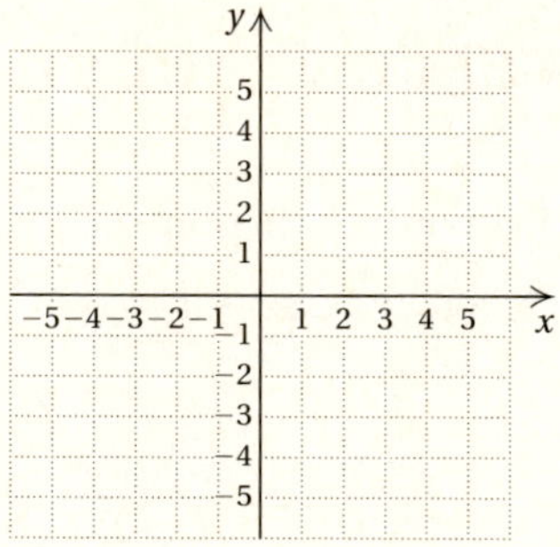

16. $y = -x + 1$

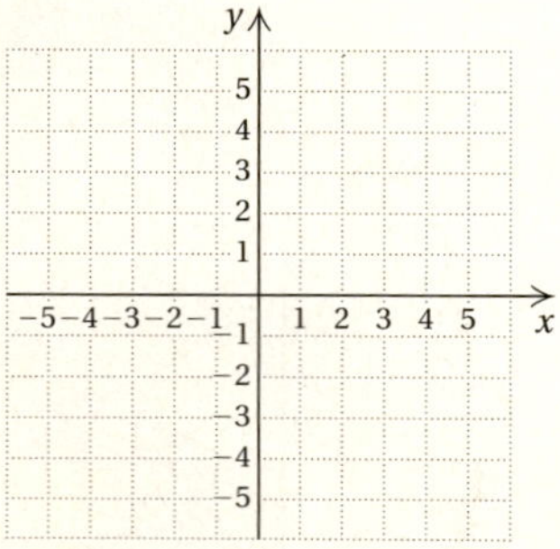

17. $y = 2x + 2$

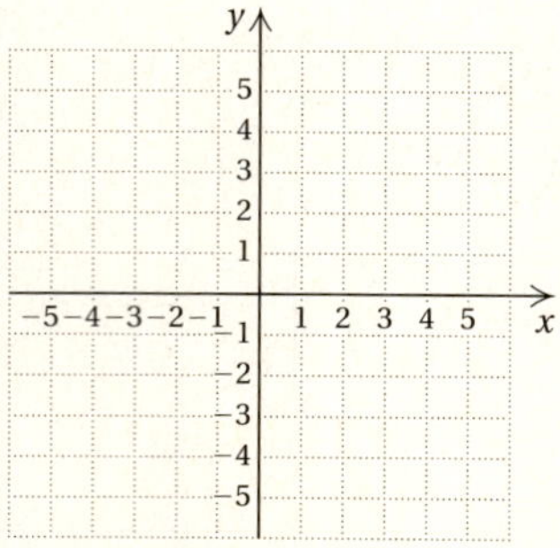

18. $y = 3x - 2$

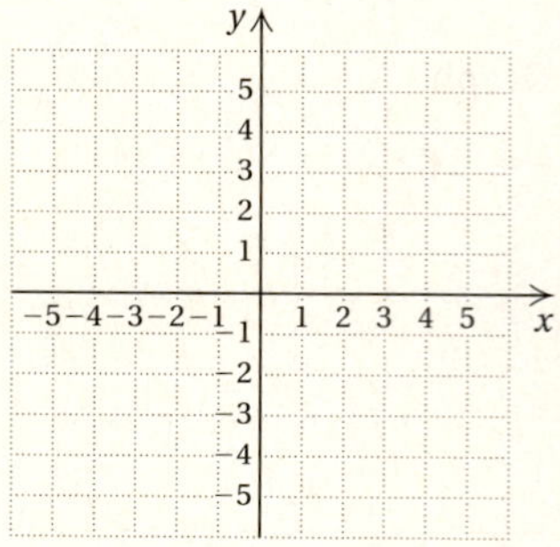

19. $y = \frac{1}{3}x - 1$

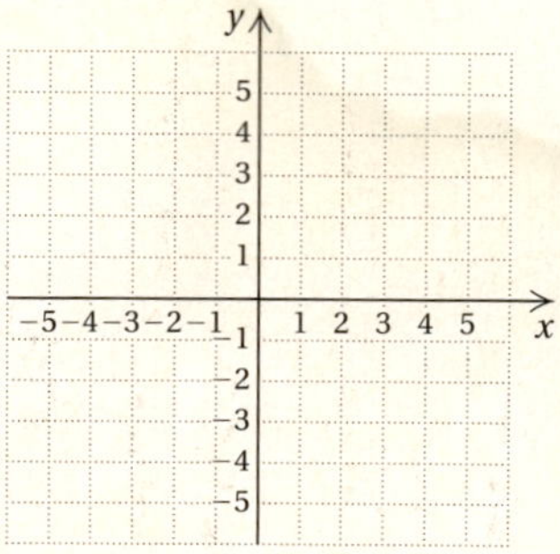

20. $y = \frac{1}{2}x + 1$

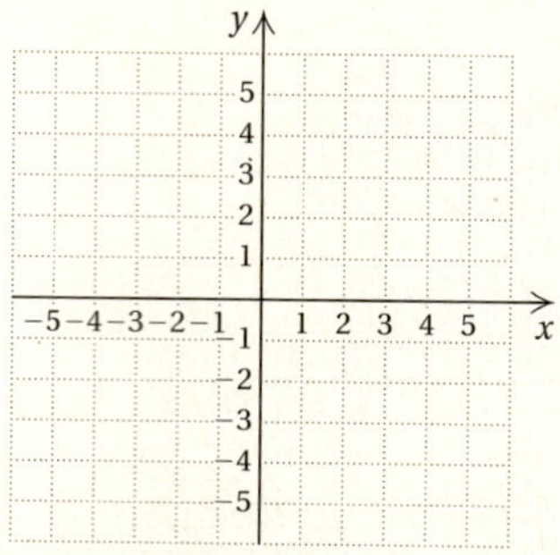

21. $y = -x - 3$

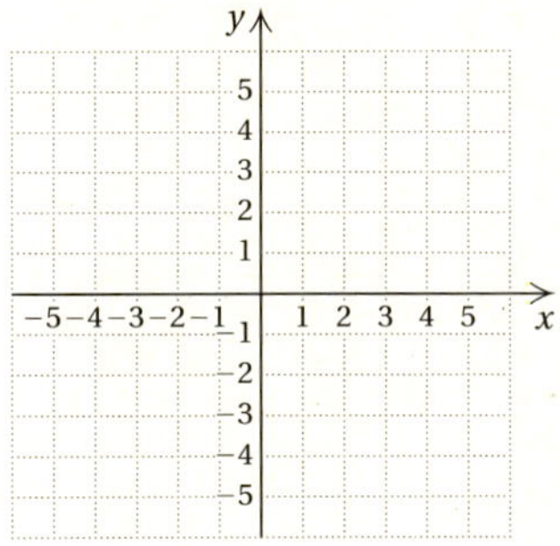

22. $y = -x - 2$

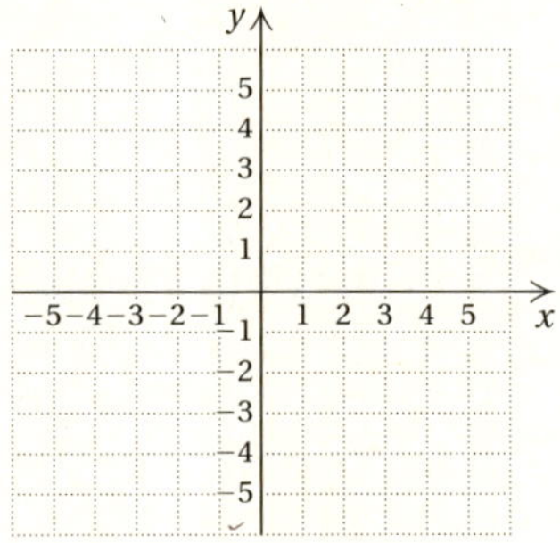

23. $y = \frac{5}{2}x + 3$

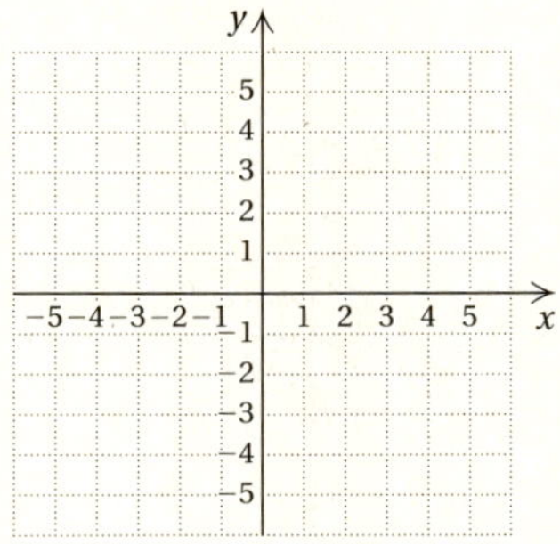

24. $y = \frac{5}{3}x - 2$

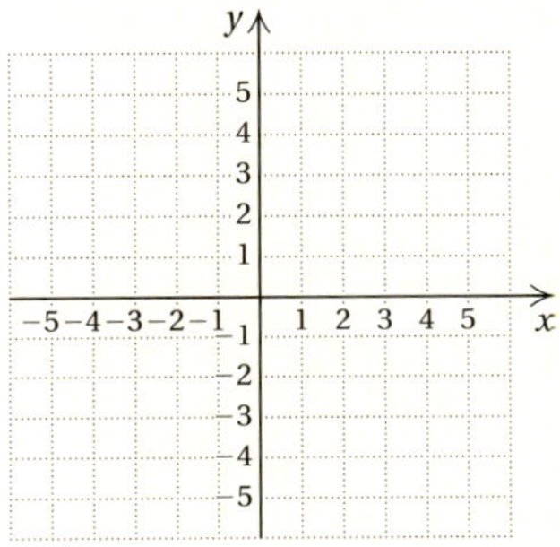

25. $y = -\frac{5}{3}x - 2$

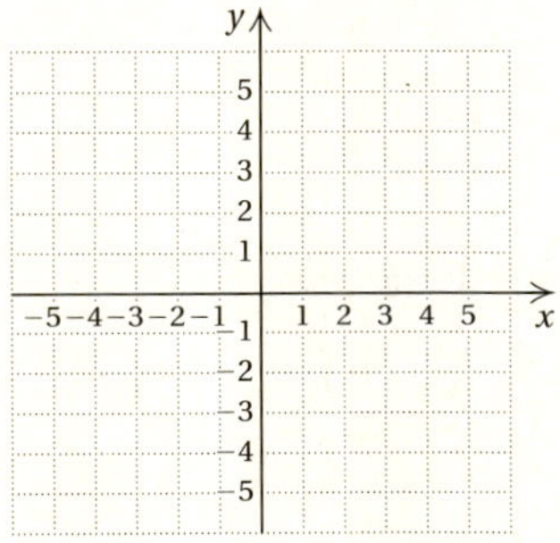

26. $y = -\frac{2}{3}x + 1$

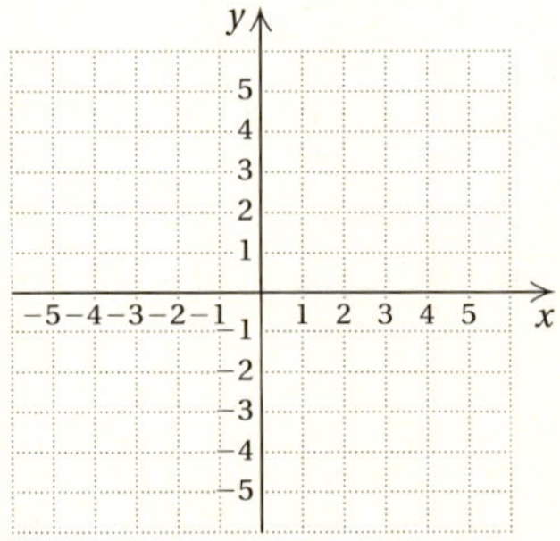

27. $y = x$

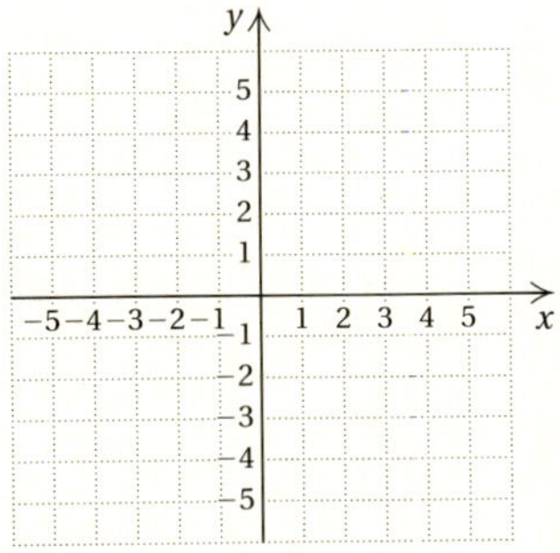

28. $y = -x$

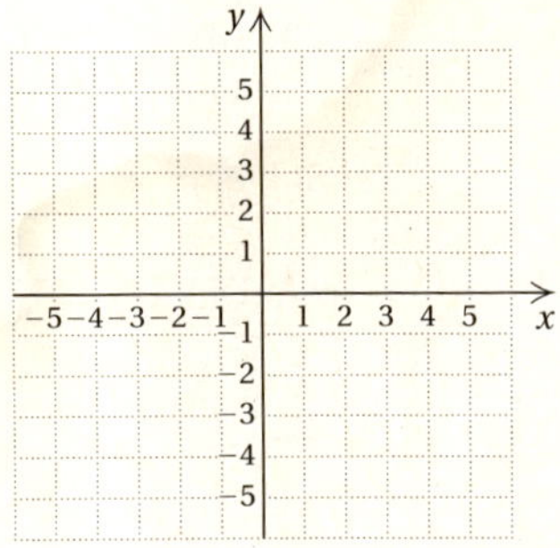

Answers

29. See graph.

30. See graph.

31. See graph.

32. See graph.

33. ______

34. ______

35. ______

36. ______

37. ______

38. ______

39. ______

40. ______

29. $y = 3 - 2x$

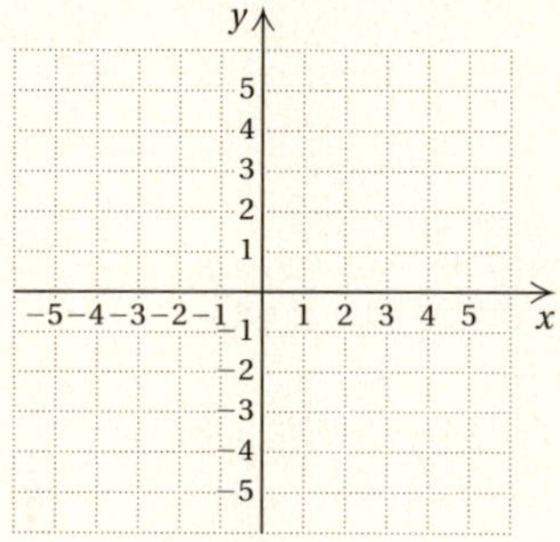

30. $y = 7 - 5x$

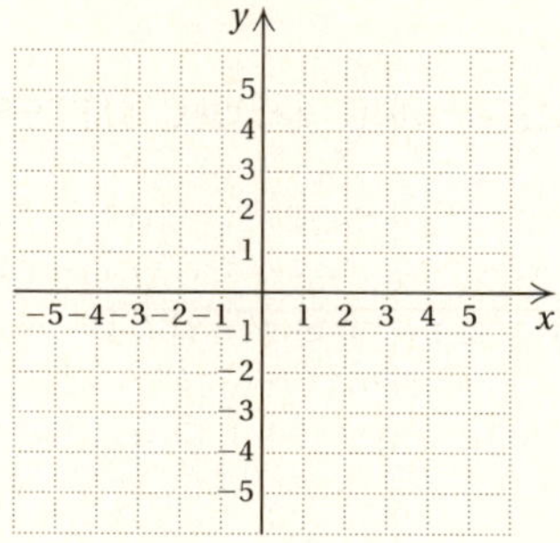

31. $y = \frac{4}{3} - \frac{1}{3}x$

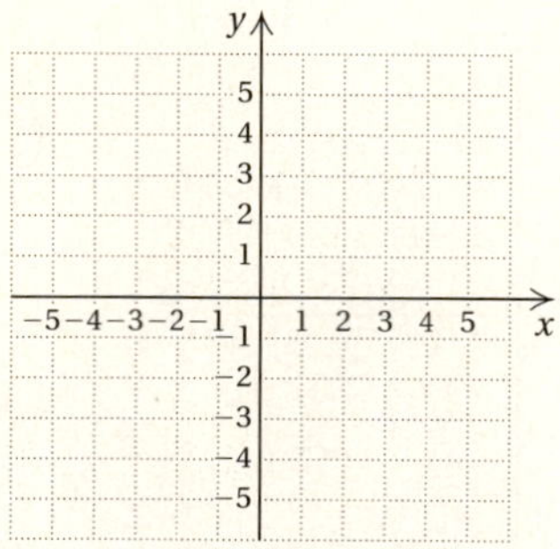

32. $y = -\frac{1}{4}x - \frac{1}{2}$

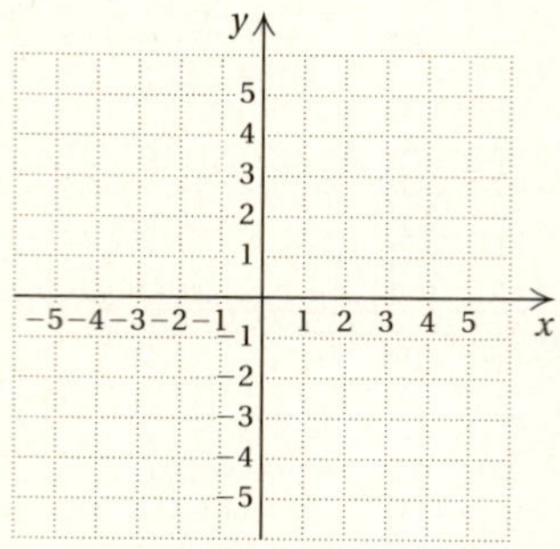

SKILL MAINTENANCE

33. An airplane flew for 7 hr with a 5-km/h tailwind. The return flight against the wind took 8 hr. Find the speed of the plane in still air.

Solve.

34. $\frac{3}{x - 5} + \frac{1}{x + 5} = \frac{2}{x^2 - 25}$

35. $25t^2 - 49 = 0$

36. $x^2 - 4x = 0$

37. $\frac{x^2}{x - 1} = \frac{1}{x - 1}$

38. $\frac{4}{x + 2} = \frac{x^2}{x + 2}$

SYNTHESIS

39. Find all the whole-number solutions of $x + y = 6$.

40. Find three solutions of $y = |x|$.

6.3 More on Graphing Linear Equations

a GRAPHING USING INTERCEPTS

We graphed equations of the type $y = mx$ and $y = mx + b$ in Section 6.2. We now consider equations of the type $Ax + By = C$, where $A \neq 0$ and $B \neq 0$. These equations can be graphed conveniently using **intercepts**. Look at the graph of $-2x + y = 4$ shown below. We could graph this equation by solving for y to get $y = 2x + 4$ and proceed as before, but we want to develop a faster method.

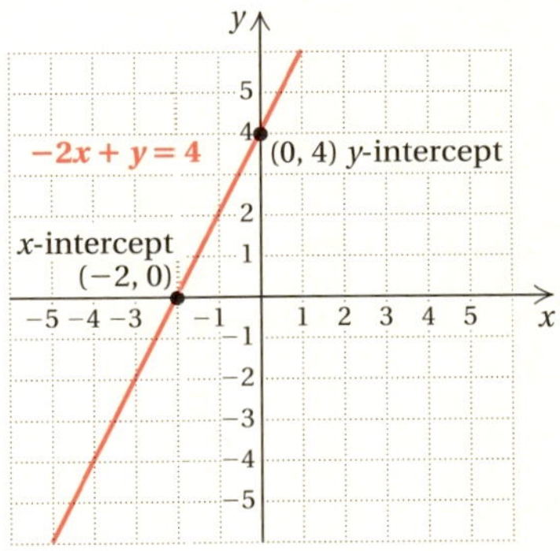

The y-intercept is (0, 4). It occurs where the line crosses the y-axis and thus will always have 0 as the first coordinate. The x-intercept is (−2, 0). It occurs where the line crosses the x-axis and thus will always have 0 as the second coordinate.

Do Exercise 1.

We find intercepts as follows.

> The y-intercept is $(0, b)$. To find b, let $x = 0$ and solve the original equation for y.
>
> The x-intercept is $(a, 0)$. To find a, let $y = 0$ and solve the original equation for x.

Now let us draw a graph using intercepts.

EXAMPLE 1 Graph: $4x + 3y = 12$.

To find the y-intercept, we let $x = 0$. Then we solve for y:

$$4 \cdot 0 + 3y = 12$$
$$3y = 12$$
$$y = 4.$$

Thus, (0, 4) is the y-intercept. Note that finding this intercept amounts to covering up the x-term and solving the rest of the equation.

To find the x-intercept, we let $y = 0$. Then we solve for x:

$$4x + 3 \cdot 0 = 12$$
$$4x = 12$$
$$x = 3.$$

OBJECTIVES

After finishing Section 6.3, you should be able to:

a Find the intercepts of a linear equation, and graph using intercepts.

b Graph equations of the type $x = a$ or $y = b$.

FOR EXTRA HELP

TAPE 15 TAPE 13A MAC: 6 IBM: 6

1. Look at the graph shown below.

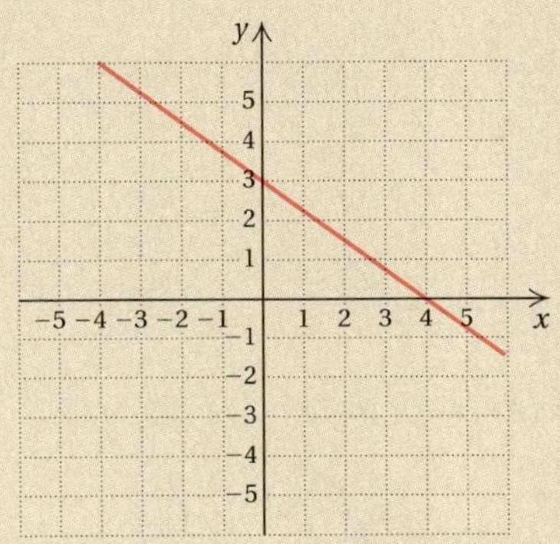

a) Find the coordinates of the x-intercept.

b) Find the coordinates of the y-intercept.

Answers on page A-6

Graph using intercepts.

2. $2x + 3y = 6$

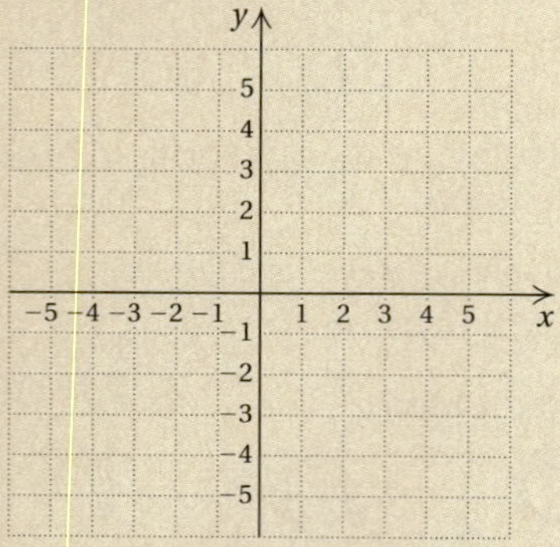

3. $3y - 4x = 12$

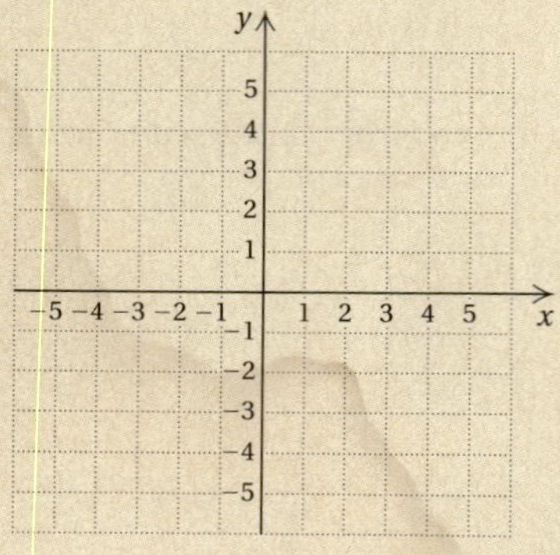

Answers on page A-6

Thus, (3, 0) is the x-intercept. Note that finding this intercept amounts to covering up the y-term and solving the rest of the equation.

We plot these points and draw the line, or graph.

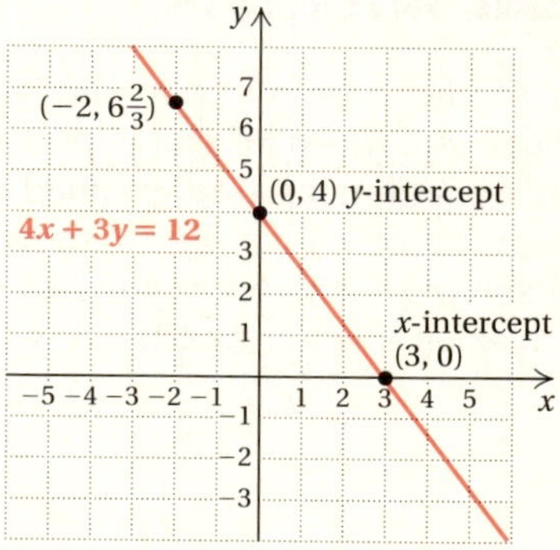

A third point should be used as a check. We substitute any convenient value for x and solve for y. In this case, we choose $x = -2$. Then

$$4(-2) + 3y = 12 \qquad \text{Substituting } -2 \text{ for } x$$
$$-8 + 3y = 12$$
$$3y = 12 + 8 = 20$$
$$y = \tfrac{20}{3}, \text{ or } 6\tfrac{2}{3}. \qquad \text{Solving for } y$$

It appears that the point $\left(-2, 6\frac{2}{3}\right)$ is on the graph, though graphing fractional values can be inexact. The graph is probably correct.

Graphs of equations of the type $y = mx$ pass through the origin. Thus the x-intercept and the y-intercept are the same, (0, 0). In such cases, we must calculate another point in order to complete the graph. Another point would also have to be calculated if a check is desired.

Do Exercises 2 and 3.

b Equations Whose Graphs Are Horizontal or Vertical Lines

Example 2 Graph: $y = 3$.

Consider $y = 3$. We can also think of this equation as $0 \cdot x + y = 3$. No matter what number we choose for x, we find that y is 3. We make up a table with all 3's in the y-column.

x	y
	3
	3
	3

y must be 3.

Choose any number for *x*. →

x	y
−2	3
0	3
4	3

← y-intercept (row 0, 3)

Now when we plot the ordered pairs $(-2, 3)$, $(0, 3)$, and $(4, 3)$ and connect the points, we will obtain a horizontal line. Any ordered pair $(x, 3)$ is a solution. So the line is parallel to the x-axis with y-intercept $(0, 3)$.

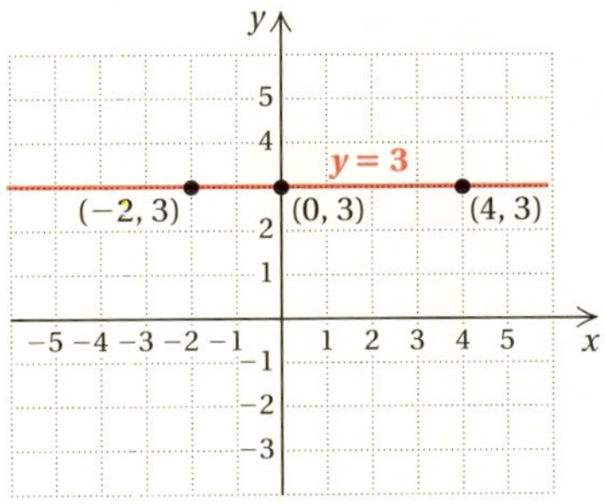

EXAMPLE 3 Graph: $x = -4$.

Consider $x = -4$. We can also think of this equation as $x + 0 \cdot y = -4$. We make up a table with all -4's in the x-column.

x	y
−4	
−4	
−4	
−4	

x must be −4.

Choose any number for *y*. →

x	y
−4	−5
−4	1
−4	3
−4	0

x-intercept → (last row)

When we plot the ordered pairs $(-4, -5)$, $(-4, 1)$, $(-4, 3)$, and $(-4, 0)$ and connect the points, we will obtain a vertical line. Any ordered pair $(-4, y)$ is a solution. So the line is parallel to the y-axis with x-intercept $(-4, 0)$.

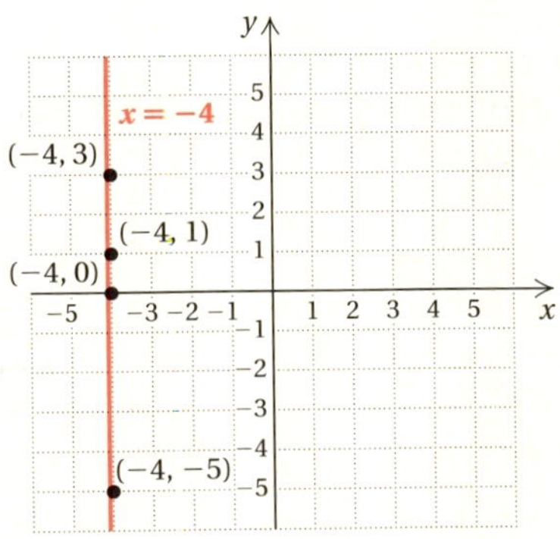

The graph of $y = b$ is a horizontal line. The y-intercept is $(0, b)$.
The graph of $x = a$ is a vertical line. The x-intercept is $(a, 0)$.

Do Exercises 4–7. (Exercise 7 is on the following page.)

Graph.

4. $x = 5$

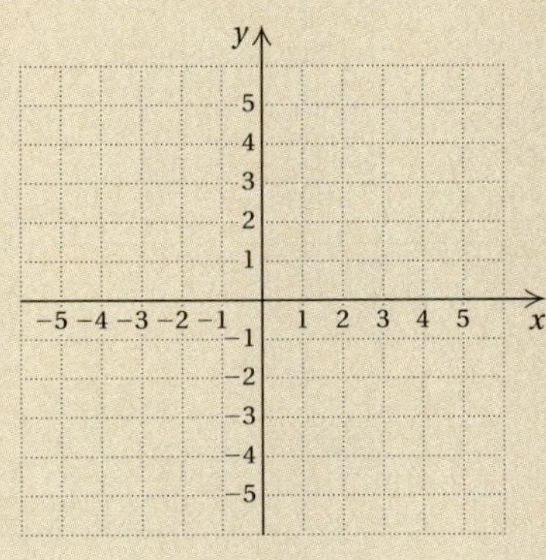

5. $y = -2$

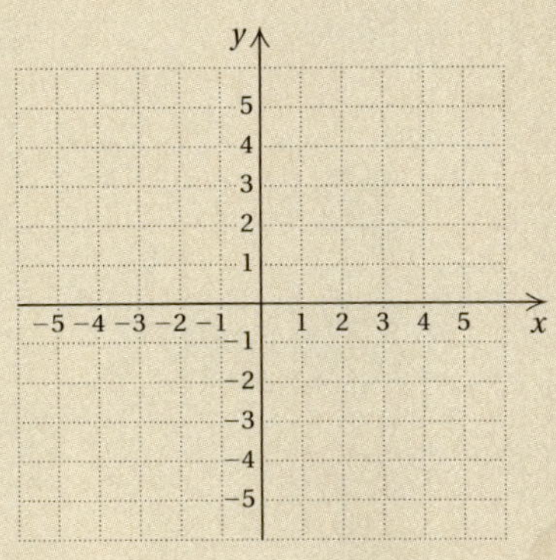

6. $x = 0$

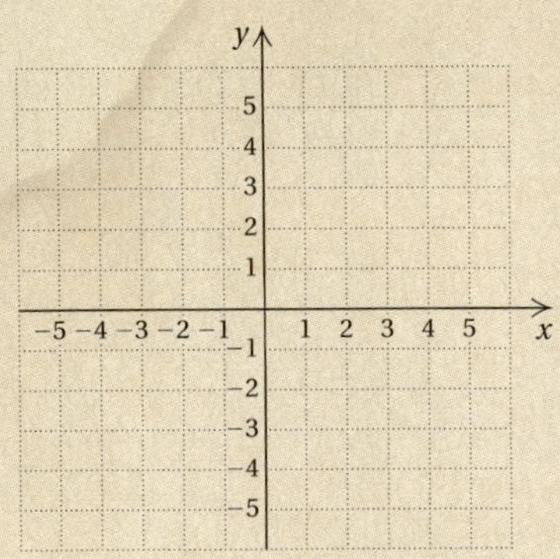

Answers on page A-6

7. Graph: $x = -3$.

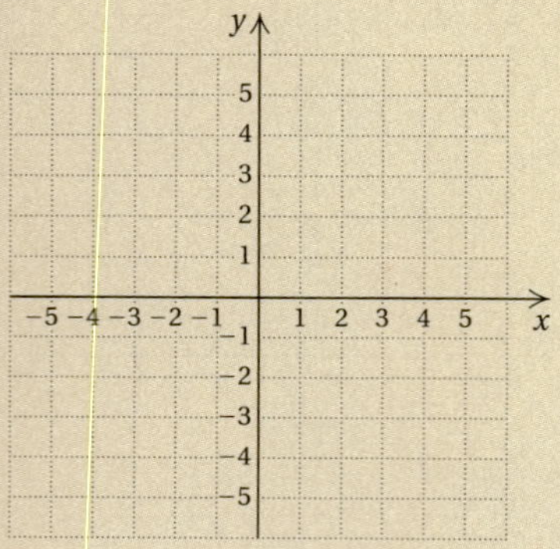

Answer on page A-6

The following is a general procedure for graphing linear equations.

TO GRAPH LINEAR EQUATIONS

1. Is the equation of the type $x = a$ or $y = b$? If so, the graph will be a line parallel to an axis; $x = a$ is vertical and $y = b$ is horizontal.

   Examples.

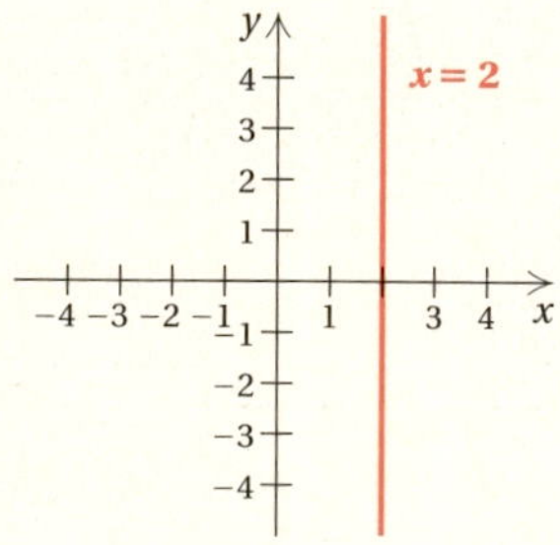

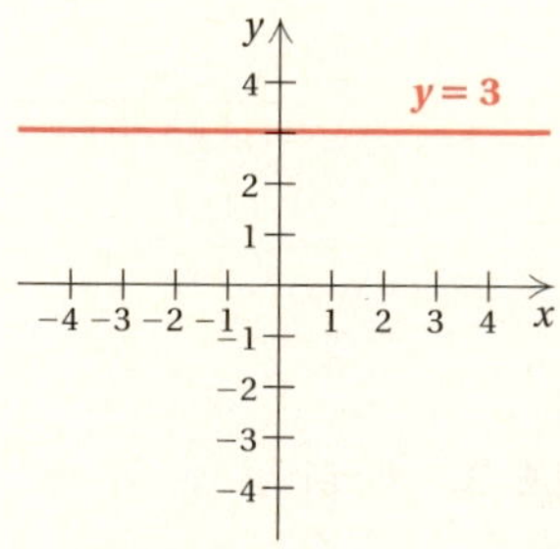

2. If the line is of the type $y = mx$, both intercepts are the origin, (0, 0). Plot (0, 0) and one other point.

   Example.

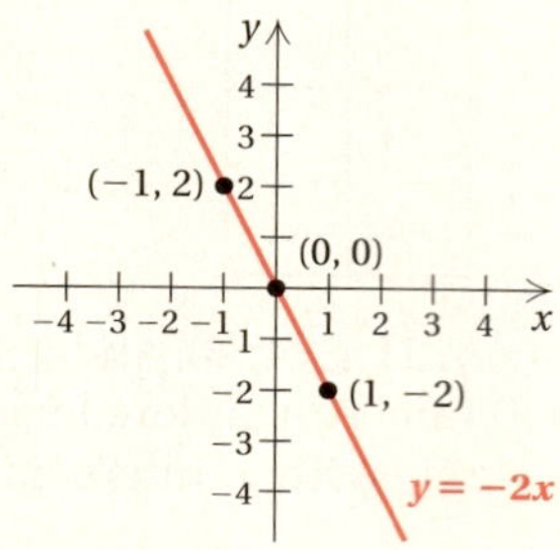

3. If the line is of the type $y = mx + b$, plot the y-intercept $(0, b)$ and one other point.

   Example.

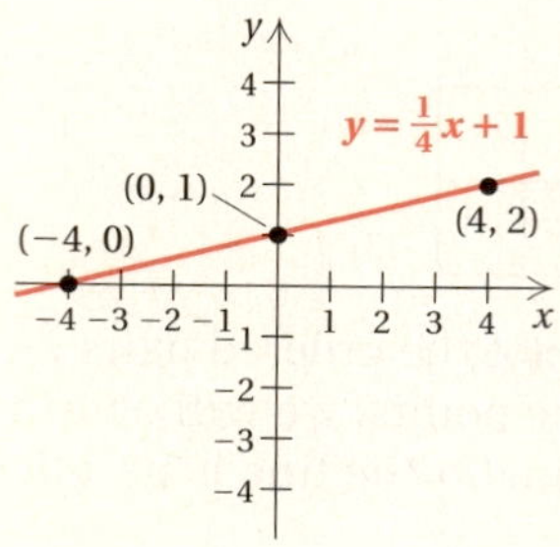

4. If the equation is of the type $Ax + By = C$, but not of the type $x = a$, $y = b$, $y = mx$, or $y = mx + b$, graph using intercepts. If the intercepts are too close together, choose another point farther from the origin.

   Examples.

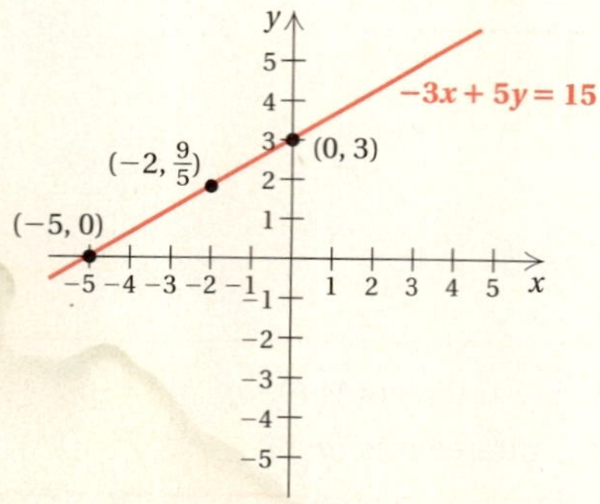

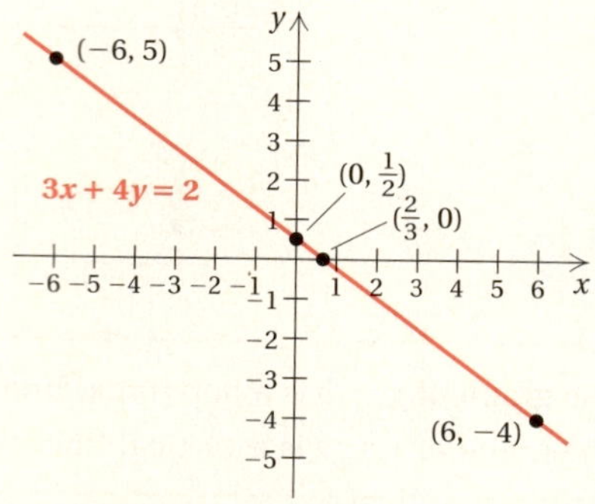

5. In any case, use a third point as a check. Calculate more points if needed.

NAME SECTION DATE

Exercise Set 6.3

a Find the intercepts. Then graph.

1. $x + 3y = 6$

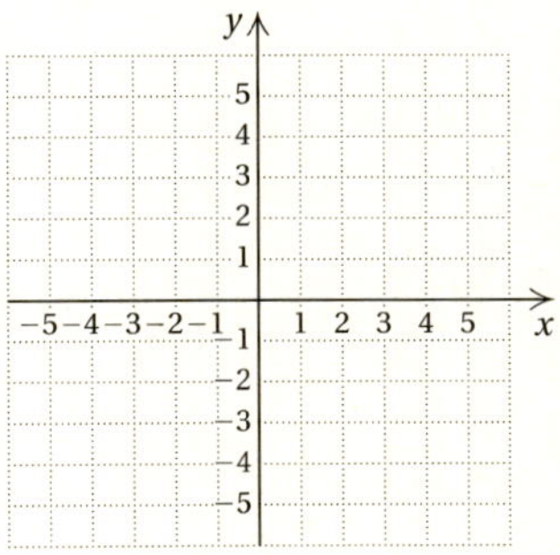

2. $x + 2y = 2$

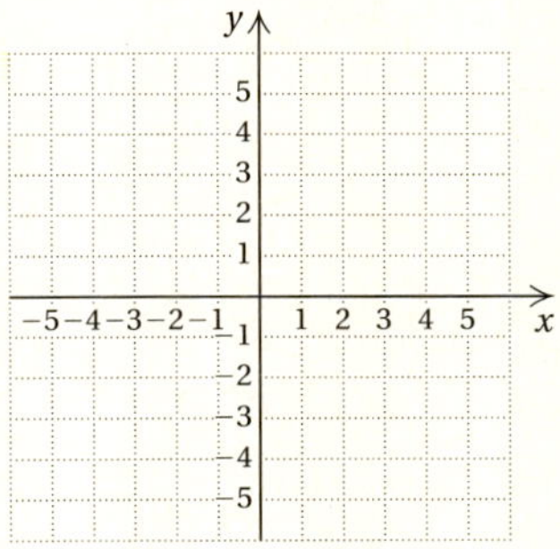

3. $-x + 2y = 4$

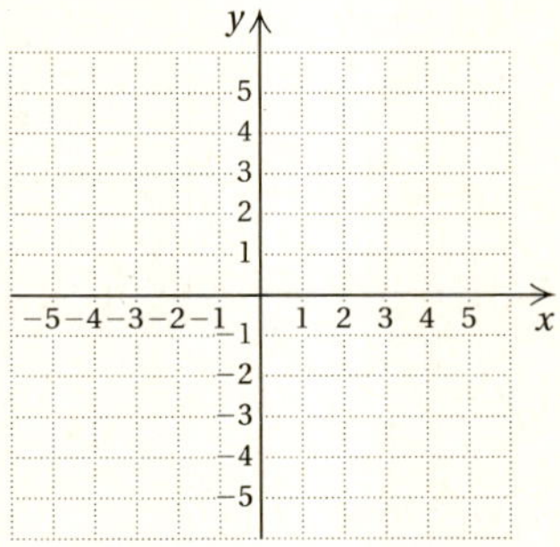

4. $-x + y = 5$

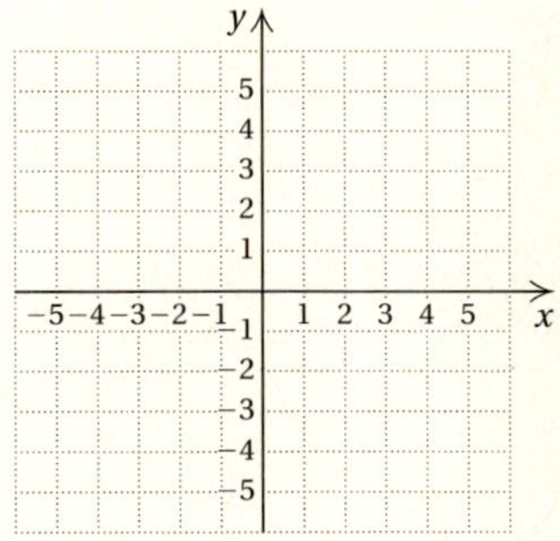

5. $3x + y = 6$

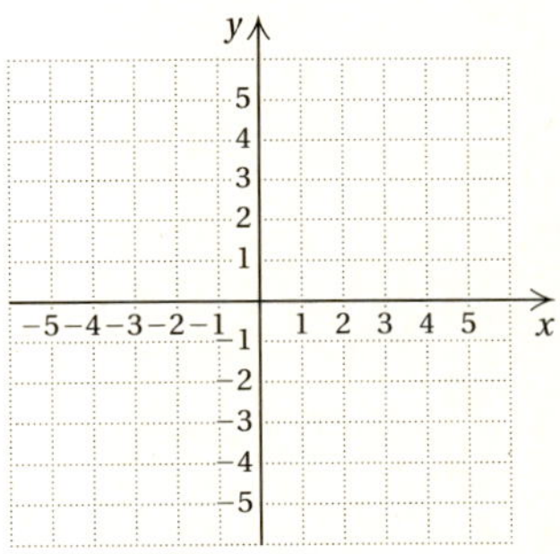

6. $2x + y = 6$

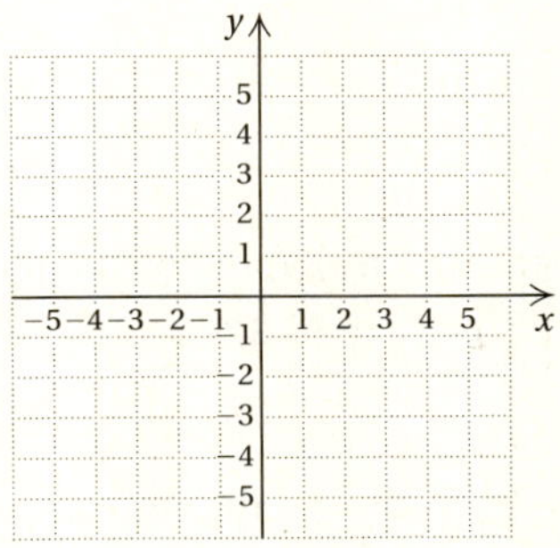

7. $2y - 2 = 6x$

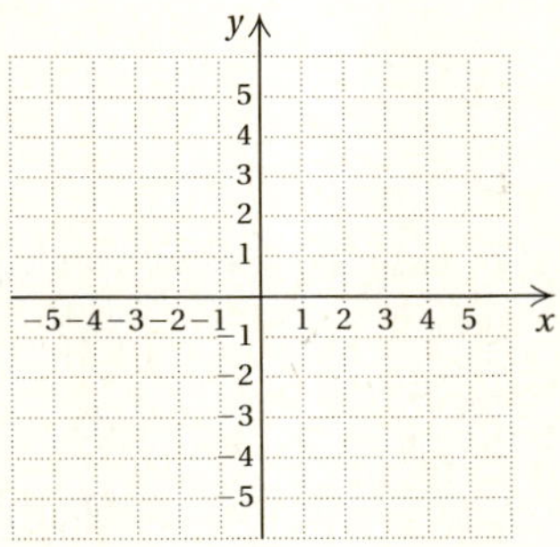

8. $3y - 6 = 9x$

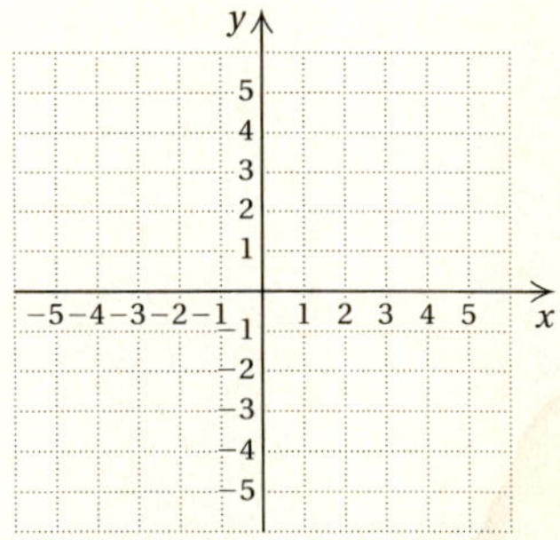

Find the intercepts of the equation. Then graph using your own graph paper. Attach graphs to this page.

9. $3x - 9 = 3y$

10. $5x - 10 = 5y$

11. $2x - 3y = 6$

12. $2x - 5y = 10$

13. $4x + 5y = 20$

14. $2x + 6y = 12$

15. $2x + 3y = 8$

16. $x - 1 = y$

17. $x - 3 = y$

18. $2x - 1 = y$

19. $3x - 2 = y$

20. $4x - 3y = 12$

21. $6x - 2y = 18$

22. $7x + 2y = 6$

23. $3x + 4y = 5$

24. $y = -4 - 4x$

25. $y = -3 - 3x$

ANSWERS

26.–46. See graph.

47. ______

48. ______

49. ______

50. ______

51. ______

52. ______

53. ______

54. ______

55. ______

56. ______

26. $-3x = 6y - 2$ **27.** $-4x = 8y - 5$ **28.** $3 = 2x - 5y$

29. $y - 3x = 0$ **30.** $x + 2y = 0$

b Graph.

31. $x = -2$

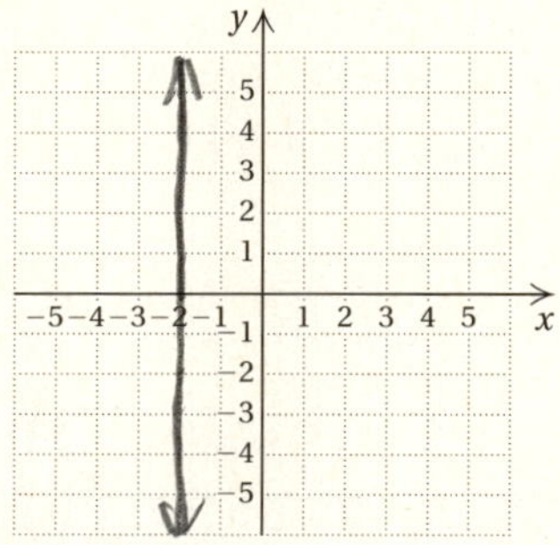

32. $x = 1$

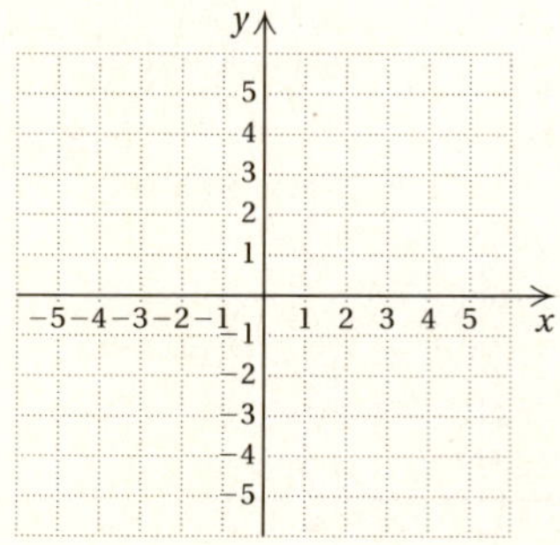

33. $y = 2$

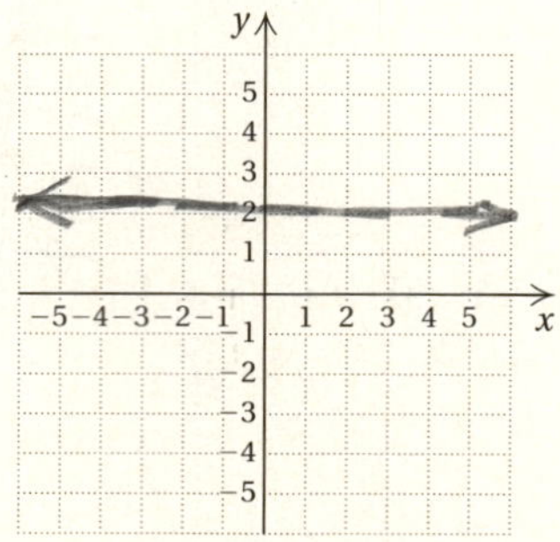

34. $y = -4$

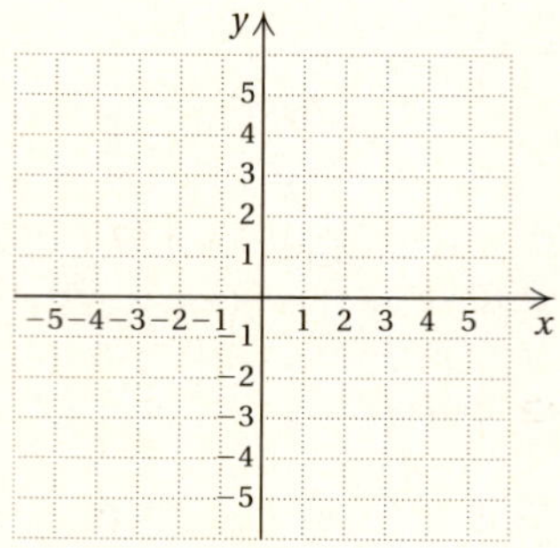

Graph using your own graph paper. Attach graphs to this page.

35. $x = 2$ **36.** $x = 3$ **37.** $y = 0$ **38.** $y = -1$

39. $x = \frac{3}{2}$ **40.** $x = -\frac{5}{2}$ **41.** $3y = -5$ **42.** $12y = 45$

43. $4x + 3 = 0$ **44.** $-3x + 12 = 0$ **45.** $48 - 3y = 0$ **46.** $63 + 7y = 0$

SKILL MAINTENANCE

47. Crew A can paint a house in 10 hr. Crew B can paint the same house in 12 hr. How many hours would it take both crews to paint the house if they worked together?

48. One car travels 20 km/h faster than another. While one car travels 250 km, the other travels 140 km. Find the speed of each car.

Solve.

49. $\frac{7}{1 + x} - 1 = \frac{5x}{x^2 + 3x + 2}$

50. $\frac{x}{x + 1} = 3 - \frac{33}{x^2 - 1}$

51. $\frac{1}{t} = \frac{1}{8} - \frac{1}{3}$

52. $x + 2 = \frac{15}{x}$

SYNTHESIS

53. Write an equation for the y-axis.

54. Write an equation for the x-axis.

55. Find the coordinates of the point of intersection of the graphs of the equations $x = -3$ and $y = 6$.

56. Find the value of m in $y = mx + 3$ so that the x-intercept of its graph will be $(2, 0)$.

6.4 *Slope and Equations of Lines*

a SLOPE

The graphs of some linear equations slant upward from left to right. Others slant downward. Some are vertical and some are horizontal. Some slant more steeply than others. We now look for a way to describe such possibilities with numbers.

Consider a line with two points marked P and Q. As we move from P to Q, the y-coordinate changes from 1 to 3 and the x-coordinate changes from 2 to 6. The change in y is $3 - 1$, or 2. The change in x is $6 - 2$, or 4.

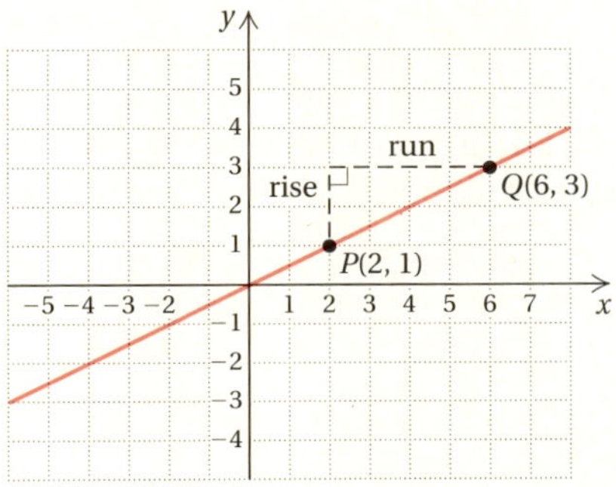

We call the change in y the **rise** and the change in x the **run**. The ratio rise/run is the same for any two points on a line. We call this ratio the **slope**. Slope describes the slant of a line. The slope of the line in the graph above is given by

$$\frac{\text{rise}}{\text{run}} = \frac{\text{the change in } y}{\text{the change in } x}, \text{ or } \frac{2}{4}, \text{ or } \frac{1}{2}.$$

> The *slope* of a line containing points (x_1, y_1) and (x_2, y_2) is given by
>
> $$m = \frac{\text{rise}}{\text{run}} = \frac{\text{the change in } y}{\text{the change in } x} = \frac{y_2 - y_1}{x_2 - x_1}.$$

EXAMPLE 1 Graph the line containing the points $(-4, 3)$ and $(2, -6)$ and find the slope.

The graph is shown below. From $(-4, 3)$ and $(2, -6)$, we see that the change in y, or the rise, is $-6 - 3$, or -9. The change in x, or the run, is $2 - (-4)$, or 6. We consider (x_1, y_1) to be $(-4, 3)$ and (x_2, y_2) to be $(2, -6)$:

$$\text{Slope} = \frac{\text{rise}}{\text{run}} = \frac{\text{change in } y}{\text{change in } x}$$

$$= \frac{-6 - 3}{2 - (-4)}$$

$$= \frac{-9}{6} = -\frac{9}{6}, \text{ or } -\frac{3}{2}.$$

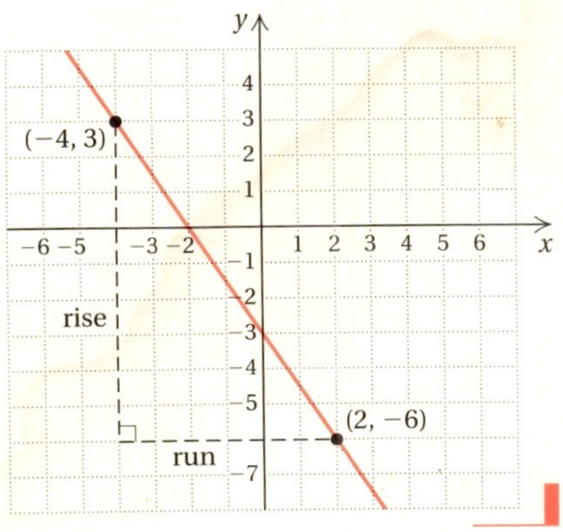

OBJECTIVES

After finishing Section 6.4, you should be able to:

a Given two points of a line, find the slope.

b Find the slope of a line from an equation.

c Given any equation, derive the equivalent slope–intercept equation and determine the slope and the y-intercept.

d Find an equation of a line given a point on the line and the slope, or given two points on the line.

FOR EXTRA HELP

TAPE 15 TAPE 13A MAC: 6 IBM: 6

Graph the line containing the points and find the slope in two different ways.

1. (−2, 3) and (3, 5)

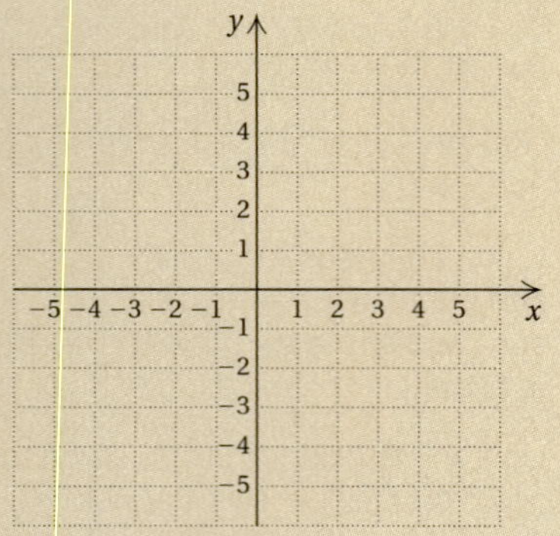

2. (0, −3) and (−3, 2)

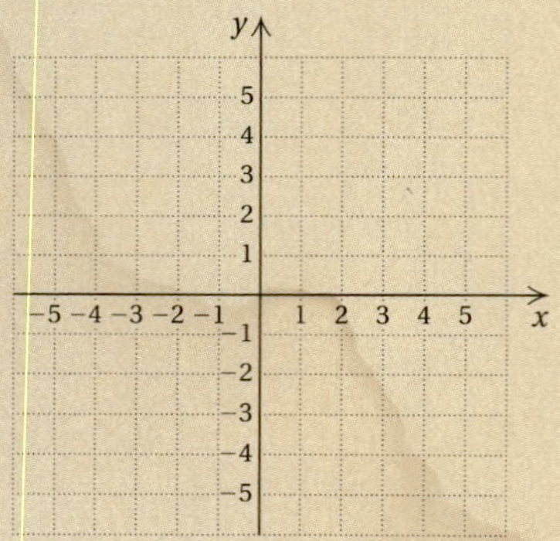

Answers on page A-6

When we use the formula

$$m = \frac{y_2 - y_1}{x_2 - x_1},$$

we can subtract in two ways. We must remember, however, to subtract the y-coordinates in the same order that we subtract the x-coordinates. Let us do Example 1 again, where we consider (x_1, y_1) to be $(2, -6)$ and (x_2, y_2) to be $(-4, 3)$:

$$\text{Slope} = \frac{\text{change in } y}{\text{change in } x} = \frac{3 - (-6)}{-4 - 2} = \frac{9}{-6} = -\frac{3}{2}.$$

The slope of a line tells how it slants. A line with positive slope slants up from left to right. The larger the slope, the steeper the slant. A line with negative slope slants downward from left to right.

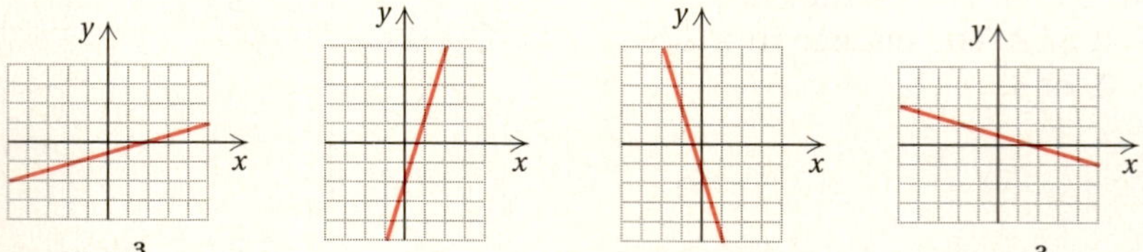

$m = \frac{3}{10}$ $\quad m = \frac{10}{3}$ $\quad m = -\frac{10}{3}$ $\quad m = -\frac{3}{10}$

Do Exercises 1 and 2.

b FINDING THE SLOPE FROM AN EQUATION

It is possible to find the slope of a line from its equation. Let us consider the equation $y = 2x + 3$, which is in the form $y = mx + b$. We can find two points by choosing convenient values for x—say, 0 and 1—and substituting to find the corresponding y-values. We find the two points on the line to be (0, 3) and (1, 5). The slope of the line is found using the definition of slope:

$$m = \frac{\text{change in } y}{\text{change in } x} = \frac{5 - 3}{1 - 0} = \frac{2}{1} = 2.$$

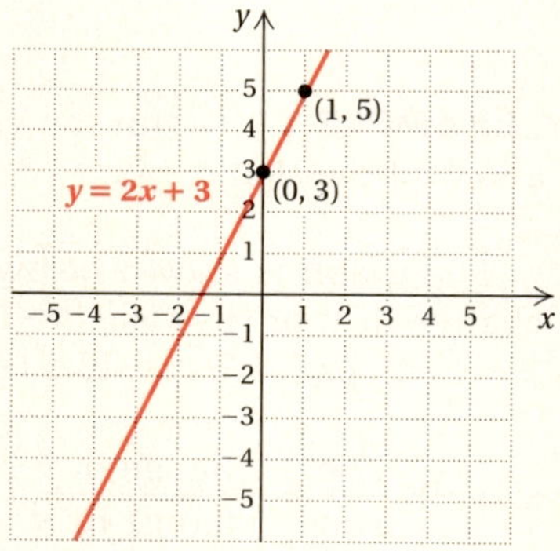

The slope is 2. Note that this is also the coefficient of the x-term in the equation $y = 2x + 3$.

> The slope of the line $y = mx + b$ is m. To find the slope of a nonvertical line, solve the linear equation in x and y for y and get the resulting equation in the form $y = mx + b$. The coefficient of the x-term, m, is the slope of the line.

EXAMPLE 2 Find the slope of the line $2x + 3y = 7$.

We solve for y in two ways. We first subtract $2x$. Then we either multiply by $\frac{1}{3}$ or divide by 3:

$$2x + 3y = 7 \qquad\qquad 2x + 3y = 7$$

$$3y = -2x + 7 \qquad\qquad 3y = -2x + 7$$

$$y = \frac{1}{3}(-2x + 7) \qquad\qquad y = \frac{-2x + 7}{3}$$

$$y = -\frac{2}{3}x + \frac{7}{3}; \qquad\qquad y = -\frac{2}{3}x + \frac{7}{3}.$$

The slope is $-\frac{2}{3}$.

Do Exercises 3 and 4.

What about the slope of a horizontal or a vertical line?

EXAMPLE 3 Find the slope of the line $y = 5$.

We can think of $y = 5$ as $y = 0x + 5$. Then from this equation, we see that $m = 0$. Consider the points $(-3, 5)$ and $(4, 5)$, which are on the line. The change in $y = 5 - 5$, or 0. The change in $x = -3 - 4$, or -7. We have

$$m = \frac{5 - 5}{-3 - 4} = \frac{0}{-7} = 0.$$

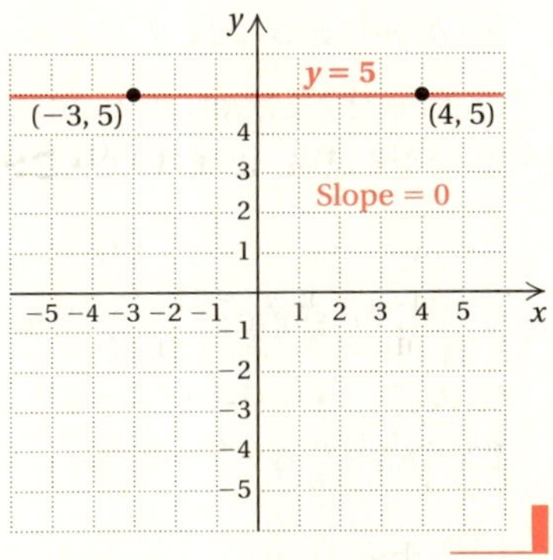

Any two points on a horizontal line have the same y-coordinate. Thus the change in y is 0.

EXAMPLE 4 Find the slope of the line $x = -4$.

Consider the points $(-4, 3)$ and $(-4, -2)$, which are on the line. The change in $y = 3 - (-2)$, or 5. The change in $x = -4 - (-4)$, or 0. We have

$$m = \frac{3 - (-2)}{-4 - (-4)} = \frac{5}{0}.$$ **Undefined**

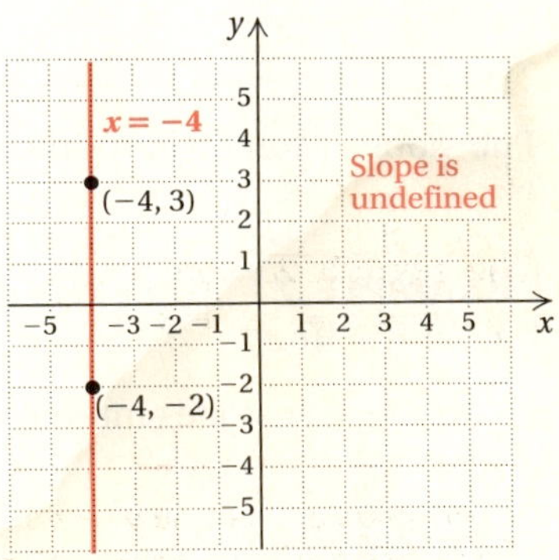

Since division by 0 is undefined, the slope of this line is undefined. The answer in this example is "The slope of this line is undefined."

Find the slope of the line.

3. $4x + 4y = 7$

4. $5x - 4y = 8$

Answers on page A-6

Find the slope, if it exists, of the line.

5. $x = 7$

6. $y = -5$

Find the slope and the y-intercept.

7. $y = 5x$

8. $y = -\frac{3}{2}x - 6$

9. $3x + 4y = 15$

10. $2y = 4x - 17$

11. $-7x - 5y = 22$

12. A line has slope 3.5 and y-intercept $(0, -23)$. Find an equation of the line.

Answers on page A-6

A horizontal line has slope 0. The slope of a vertical line is undefined.

Do Exercises 5 and 6.

C THE SLOPE–INTERCEPT EQUATION OF A LINE

In the equation $y = mx + b$, we know that m is the slope. What is the y-intercept? To find out, we let $x = 0$ and solve for y:

$$\begin{aligned} y &= mx + b \\ &= m(0) + b \\ &= b. \end{aligned}$$

Thus the y-intercept is $(0, b)$.

THE SLOPE–INTERCEPT EQUATION: $y = mx + b$

The equation $y = mx + b$ is called the *slope–intercept equation*. The slope is m and the y-intercept is $(0, b)$.

EXAMPLE 5 Find the slope and the y-intercept of $y = 3x - 4$.

Since the equation is already in the form $y = mx + b$, we simply read the slope and the y-intercept from the equation.

$$y = 3x - 4$$

The slope is 3. The y-intercept is $(0, -4)$.

EXAMPLE 6 Find the slope and the y-intercept of $2x - 3y = 8$.

We first solve for y:

$$\begin{aligned} 2x - 3y &= 8 \\ -3y &= -2x + 8 && \text{This equation is not yet solved for } y. \\ y &= -\tfrac{1}{3}(-2x + 8) && \text{Multiplying by } -\tfrac{1}{3} \\ y &= \tfrac{2}{3}x - \tfrac{8}{3}. \end{aligned}$$

Caution! Only the coefficient of x is the slope.

The slope is $\frac{2}{3}$ and the y-intercept is $\left(0, -\frac{8}{3}\right)$.

EXAMPLE 7 A line has slope -2.4 and y-intercept 11. Find an equation of the line.

We use the slope–intercept equation and substitute -2.4 for m and 11 for b:

$$\begin{aligned} y &= mx + b \\ y &= -2.4x + 11 \end{aligned}$$

Do Exercises 7–12.

d THE POINT–SLOPE EQUATION OF A LINE

Suppose we know the slope of a line and a certain point on that line. We can use the slope–intercept equation to find an equation of the line.

EXAMPLE 8 Find the equation of the line with slope 3 that contains the point (4, 1).

Using the point (4, 1), we substitute 4 for x and 1 for y in $y = mx + b$. We also substitute 3 for m, the slope. Then we solve for b:

$$y = mx + b$$
$$1 = 3 \cdot 4 + b \qquad \textbf{Substituting}$$
$$-11 = b. \qquad \textbf{Solving for } b\textbf{, the } y\textbf{-intercept}$$

We use the equation $y = mx + b$ and substitute 3 for m and -11 for b:

$$y = 3x - 11.$$

Do Exercises 13 and 14.

Let us redo Example 8 using another method. The new method leads us to another equation for a line: the *point–slope equation.* Consider a line with slope 3 that contains the point (4, 1), as shown. Suppose (x, y) is any point on this line. Using the definition of slope and the two points (4, 1) and (x, y), we get

$$m = \frac{\text{change in } y}{\text{change in } x} = \frac{y - 1}{x - 4}.$$

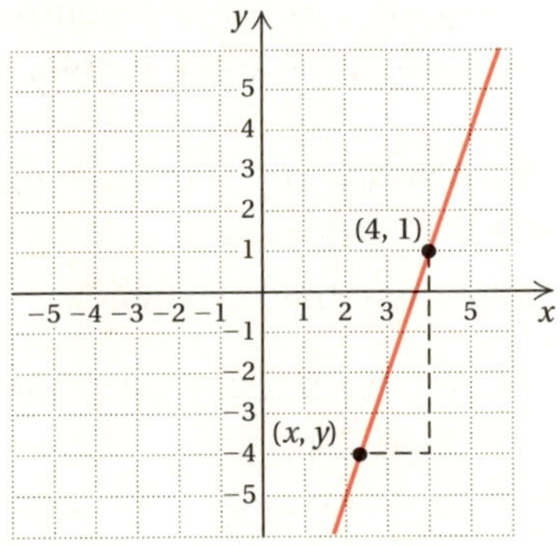

We know that the slope is 3, so

$$3 = \frac{y - 1}{x - 4}, \quad \text{or} \quad \frac{y - 1}{x - 4} = 3.$$

We solve for y, multiplying by $x - 4$ to obtain

$$y - 1 = 3(x - 4)$$
$$y - 1 = 3x - 12$$
$$y = 3x - 11.$$

We can find a general equation for a line with slope m containing the point (x_1, y_1) as follows. Consider any other point (x, y) of the line.

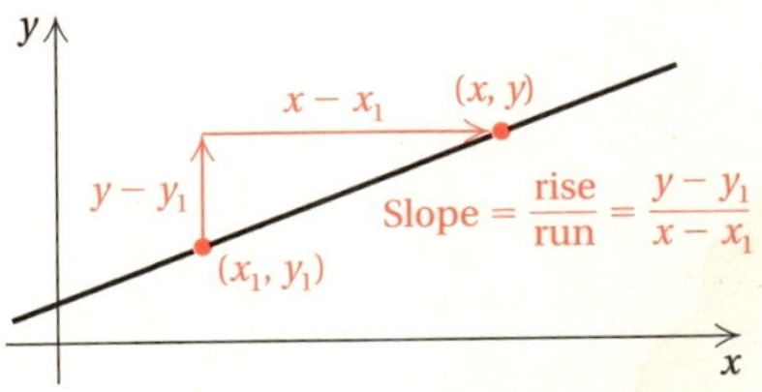

From this drawing, we see that the slope of the line is

$$\frac{y - y_1}{x - x_1}.$$

Find an equation of the line that contains the given point and has the given slope.

13. (4, 2), $m = 5$

14. (−2, 1), $m = -3$

Answers on page A-6

Find an equation of the line containing the given point and having the given slope.

15. (3, 5), $m = 6$

16. (1, 4), $m = -\frac{2}{3}$

Find an equation of the line containing the given points.

17. (2, 4) and (3, 5)

18. (−1, 2) and (−3, −2)

Slope has many real-world applications. For example, numbers like 2%, 3%, and 6% are often used to represent the grade of a road. Such a number is meant to tell how steep a road up a hill or mountain is. For example, a 3% grade means that for every horizontal distance of 100 ft, the road rises 3 ft. The concept of grade also occurs in cardiology when a person runs on a treadmill. A physician may change the steepness of the treadmill to measure its effect on heartbeat.

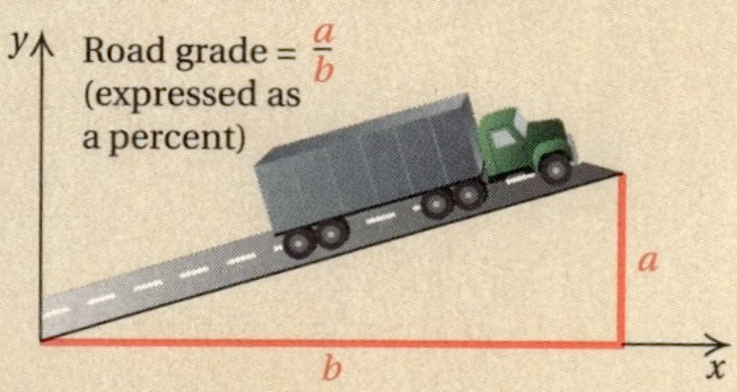

Answers on page A-6

The slope is also m. Thus we have the equation

$$\frac{y - y_1}{x - x_1} = m.$$

Multiplying by $x - x_1$ gives us the following equation.

> **THE POINT–SLOPE EQUATION: $y - y_1 = m(x - x_1)$**
>
> A nonvertical line that contains a point (x_1, y_1) with slope m has an equation $y - y_1 = m(x - x_1)$.

EXAMPLE 9 Find an equation of the line with slope 5 that contains the point (−2, −3).

We substitute 5 for m, −2 for x_1, and −3 for y_1:

$$\begin{aligned} y - y_1 &= m(x - x_1) && \text{Using the point–slope equation} \\ y - (-3) &= 5[x - (-2)] \\ y + 3 &= 5(x + 2) \\ y + 3 &= 5x + 10 \\ y &= 5x + 7. \end{aligned}$$

Do Exercises 15 and 16.

We can also use the point–slope equation to find an equation of a line containing two given points.

EXAMPLE 10 Find an equation of the line containing the points (2, 3) and (−6, 1).

First we find the slope:

$$m = \frac{3 - 1}{2 - (-6)} = \frac{2}{8}, \text{ or } \frac{1}{4}.$$

Either point can be used for (x_1, y_1), but we use (2, 3):

$$\begin{aligned} y - y_1 &= m(x - x_1) && \text{Using the point–slope equation} \\ y - 3 &= \tfrac{1}{4}(x - 2) && \text{Substituting 2 for } x_1\text{, 3 for } y_1\text{, and } \tfrac{1}{4} \text{ for } m \\ y - 3 &= \tfrac{1}{4}x - \tfrac{1}{2} \\ y &= \tfrac{1}{4}x + \tfrac{5}{2}. \end{aligned}$$

Try substituting the point (−6, 1) for (x_1, y_1). You will find that you will obtain the same equation.

Do Exercises 17 and 18.

NAME SECTION DATE

Exercise Set 6.4

a Find the slope, if it exists, of the line containing the given pair of points.

1. (3, 2) and (−1, 2)

2. (5, 1) and (−3, 6)

3. (−2, 4) and (3, 0)

4. (2, −4) and (−3, 2)

5. (3, 0) and (6, 2)

6. (4, 0) and (5, 7)

7. (−3, −2) and (−5, −6)

8. (−7, −6) and (−4, −2)

9. $\left(-2, \frac{1}{2}\right)$ and $\left(-5, \frac{1}{2}\right)$

10. (8, −3) and (10, −3)

11. (9, −4) and (9, −7)

12. (−8, 5) and (−8, 2)

b Find the slope, if it exists, of the line.

13. $3x + 2y = 6$

14. $3x - y = 4$

15. $x + 4y = 8$

16. $x + 5y = 10$

17. $-2x + y = 4$

18. $-5x + y = 5$

19. $x = -8$

20. $x = -2$

21. $y = 2$

22. $y = 17$

23. $x = 9$

24. $y = -9$

25. $y = -6$

26. $x = 4$

c Find the slope and the y-intercept of the line.

27. $y = -4x - 9$

28. $y = -2x + 3$

29. $y = 1.8x$

ANSWERS

1.

2.

3.

4.

5.

6.

7.

8.

9.

10.

11.

12.

13.

14.

15.

16.

17.

18.

19.

20.

21.

22.

23.

24.

25.

26.

27.

28.

29.

Answers

30. ______
31. ______
32. ______
33. ______
34. ______
35. ______
36. ______
37. ______
38. ______
39. ______
40. ______
41. ______
42. ______
43. ______
44. ______
45. ______
46. ______
47. ______
48. ______
49. ______
50. ______
51. ______
52. ______
53. ______
54. ______
55. ______
56. ______
57. ______
58. ______
59. ______
60. ______
61. ______
62. ______
63. ______
64. ______

30. $y = -27.4x$

31. $-8x - 7y = 21$

32. $-2x - 8y = 16$

33. $9x = 3y + 5$

34. $4x = 9y + 7$

35. $5x + 4y = 12$

36. $-6x = 4y + 2$

37. $y = -17$

38. $y = 28$

d Find an equation of the line containing the given point and having the given slope.

39. $(-3, 0)$, $m = -2$

40. $(2, 5)$, $m = 5$

41. $(2, 4)$, $m = \frac{3}{4}$

42. $\left(\frac{1}{2}, 2\right)$, $m = -1$

43. $(2, -6)$, $m = 1$

44. $(4, -2)$, $m = 6$

45. $(0, 3)$, $m = -3$

46. $(-3, 0)$, $m = -3$

Find an equation of the line that contains the given pair of points.

47. $(12, 16)$ and $(1, 5)$

48. $(-6, 1)$ and $(2, 3)$

49. $(0, 4)$ and $(4, 2)$

50. $(0, 0)$ and $(4, 2)$

51. $(3, 2)$ and $(1, 5)$

52. $(-4, 1)$ and $(-1, 4)$

53. $(-4, 5)$ and $(-2, -3)$

54. $(-2, -4)$ and $(2, -1)$

SKILL MAINTENANCE

Simplify.

55. $11 \cdot 6 \div 3 \cdot 2 \div 7$

56. $2^4 - 2^4 \div 2^2 - 2$

57. $[10 - 3(7 - 2)]$

58. $5^3 - 4^2 + 6(5 \cdot 7 + 4 \cdot 3)$

59. $\dfrac{4^3 + 2^2}{5^3 - 4^2}$

60. $(4^3 + 2^2) \cdot (5^3 - 4^2)$

SYNTHESIS

61. Find an equation of the line that contains the point $(2, -3)$ and has the same slope as the line $3x - y + 4 = 0$.

62. Find an equation of the line that has the same y-intercept as the line $x - 3y = 6$ and contains the point $(5, -1)$.

63. Find an equation of the line with the same slope as $3x - 2y = 8$ and the same y-intercept as $2y + 3x = -4$.

64. Graph several equations that have the same slope. How are they related?

6.5 *Parallel and Perpendicular Lines*

When we graph a pair of linear equations, there are three possibilities:

1. The graphs are the same.
2. The graphs intersect at exactly one point.
3. The graphs are parallel (they do not intersect).

OBJECTIVES

After finishing Section 6.5, you should be able to:

a Determine whether the graphs of two linear equations are parallel.

b Determine whether the graphs of two linear equations are perpendicular.

FOR EXTRA HELP

TAPE 16 | TAPE 13B | MAC: 6 IBM: 6

a PARALLEL LINES

The graphs shown at the right are of the linear equations

$$y = 2x + 5$$

and $\quad y = 2x - 3.$

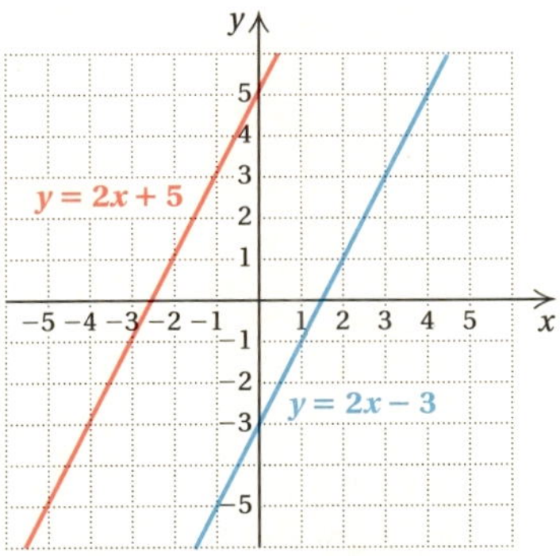

The slope of each line is 2. The y-intercepts are $(0, 5)$ and $(0, -3)$ and are different. The lines do not intersect and are parallel.

PARALLEL LINES

Parallel nonvertical lines have the same slope and different y-intercepts.
Parallel horizontal lines have equations $y = p$ and $y = q$, where $p \neq q$.
Parallel vertical lines have equations $x = p$ and $x = q$, where $p \neq q$.

EXAMPLE 1 Determine whether the graphs of $y = -3x + 4$ and $6x + 2y = -10$ are parallel.

The graphs of these equations are shown below. By simply graphing, we may find it difficult to determine whether lines are parallel. Sometimes they may intersect only very far from the origin. We can use the preceding results about slopes, y-intercepts, and parallel lines to determine for certain whether lines are parallel.

We first solve each equation for y. In this case, the first equation is already solved for y.

a) $y = -3x + 4$

b) $6x + 2y = -10$

$$2y = -6x - 10$$

$$y = \frac{1}{2}(-6x - 10)$$

$$y = -3x - 5$$

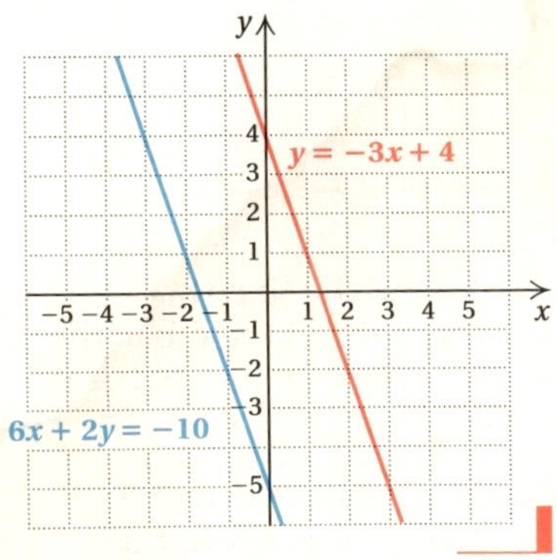

The slope of each line is -3. The y-intercepts are $(0, 4)$ and $(0, -5)$ and are different. The lines are parallel.

Determine whether the graphs of the pair of equations are parallel.

1. $y - 3x = 1,$
   $-2y = 3x + 2$

2. $3x - y = -5,$
   $y - 3x = -2$

Determine whether the graphs of the pair of equations are perpendicular.

3. $y = -\frac{3}{4}x + 7,$
   $y = \frac{4}{3}x - 9$

4. $4x - 5y = 8,$
   $6x + 9y = -12$

Answers on page A-6

Do Exercises 1 and 2.

b PERPENDICULAR LINES

Perpendicular lines in a plane are lines that intersect at a right angle. The measure of a right angle is 90°. The lines whose graphs are shown below are perpendicular. You can check this approximately by using a protractor or placing a rectangular piece of paper at the intersection.

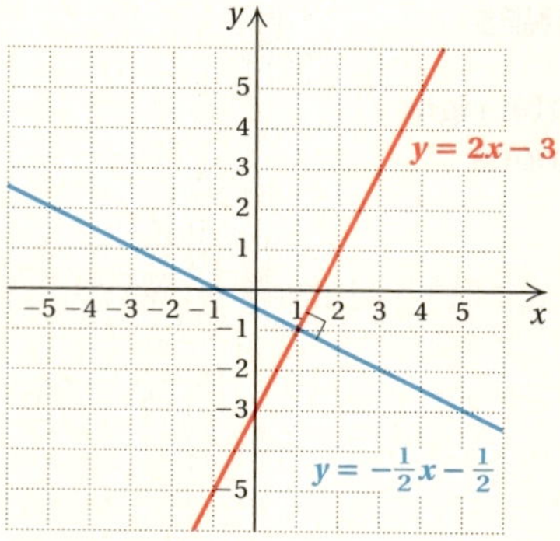

The slopes of the lines are 2 and $-\frac{1}{2}$. Note that $2\left(-\frac{1}{2}\right) = -1$. That is, the product of the slopes is -1.

> Two nonvertical lines are perpendicular if the product of their slopes is -1. (If one line has slope m, the slope of the line perpendicular to it is $-1/m$.)
>
> If one equation in a pair of perpendicular lines is vertical, then the other is horizontal. These equations are of the form $x = a$ and $y = b$.

EXAMPLE 2 Determine whether the graphs of $3y = 9x + 3$ and $6y + 2x = 6$ are perpendicular.

The graphs are shown below, but they are not necessary in order to determine whether the lines are perpendicular.

We first solve each equation for y in order to determine the slopes:

a) $3y = 9x + 3$
$y = \frac{1}{3}(9x + 3)$
$y = 3x + 1;$

b) $6y + 2x = 6$
$6y = -2x + 6$
$y = \frac{1}{6}(-2x + 6)$
$y = -\frac{1}{3}x + 1.$

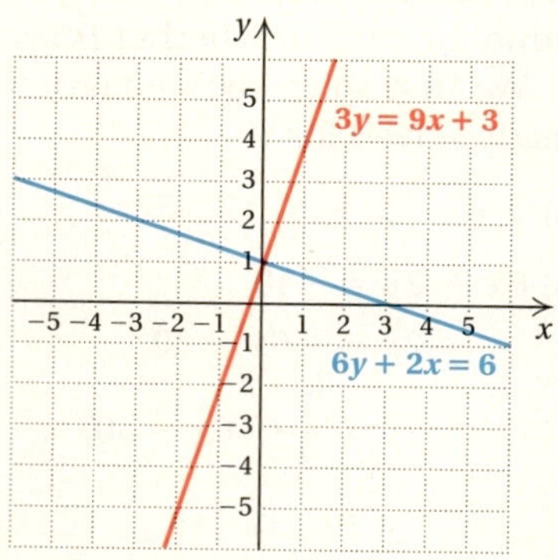

The slopes are 3 and $-\frac{1}{3}$. The product of the slopes is $3\left(-\frac{1}{3}\right) = -1$. The lines are perpendicular.

Do Exercises 3 and 4.

NAME SECTION DATE

Exercise Set 6.5

a Determine whether the graphs of the equations are parallel lines.

1. $x + 4 = y,$
$y - x = -3$

2. $3x - 4 = y,$
$y - 3x = 8$

3. $y + 3 = 6x,$
$-6x - y = 2$

4. $y = -4x + 2,$
$-5 = -2y + 8x$

5. $10y + 32x = 16.4,$
$y + 3.5 = 0.3125x$

6. $y = 6.4x + 8.9,$
$5y - 32x = 5$

7. $y = 2x + 7,$
$5y + 10x = 20$

8. $y + 5x = -6,$
$3y + 5x = -15$

9. $3x - y = -9,$
$2y - 6x = -2$

10. $y - 6 = -6x,$
$-2x + y = 5$

11. $x = 3,$
$x = 4$

12. $y = 1,$
$y = -2$

b Determine whether the graphs of the equations are perpendicular lines.

13. $y = -4x + 3,$
$4y + x = -1$

14. $y = -\frac{2}{3}x + 4,$
$3x + 2y = 1$

15. $x + y = 6,$
$4y - 4x = 12$

16. $2x - 5y = -3,$
$5x + 2y = 6$

ANSWERS

1. ______
2. ______
3. ______
4. ______
5. ______
6. ______
7. ______
8. ______
9. ______
10. ______
11. ______
12. ______
13. ______
14. ______
15. ______
16. ______

ANSWERS

17. ______
18. ______
19. ______
20. ______
21. ______
22. ______
23. ______
24. ______
25. ______
26. ______
27. ______
28. ______
29. ______
30. ______
31. ______
32. ______
33. ______
34. ______
35. ______
36. ______
37. ______

17. $y = -0.3125x + 11,$
$y - 3.2x = -14$

18. $y = -6.4x - 7,$
$64y - 5x = 32$

19. $y = -x + 8,$
$x - y = -1$

20. $2x + 6y = -3,$
$12y = 4x + 20$

21. $\frac{3}{8}x - \frac{y}{2} = 1,$
$\frac{4}{3}x - y + 1 = 0$

22. $\frac{1}{2}x + \frac{3}{4}y = 6,$
$-\frac{3}{2}x + y = 4$

SKILL MAINTENANCE

23. A train leaves a station and travels west at 70 km/h. Two hours later, a second train leaves on a parallel track and travels west at 90 km/h. When will it overtake the first train?

24. One car travels 10 km/h faster than another. While one car travels 130 km, the other travels 140 km. What is the speed of each car?

Solve.

25. $x^2 - 10x + 25 = 0$

26. $\frac{x^2}{x + 4} = \frac{16}{x + 4}$

27. $\frac{2}{3} - \frac{5}{6} = \frac{1}{x}$

28. $\frac{x + 1}{2x - 3} = 8$

SYNTHESIS

29. Find an equation of a line that contains the point (0, 6) and is parallel to $y - 3x = 4$.

30. Find an equation of the line that contains the point $(-2, 4)$ and is parallel to $y = 2x - 3$.

31. Find an equation of the line that contains the point (0, 2) and is perpendicular to the line $3y - x = 0$.

32. Find an equation of the line that contains the point (1, 0) and is perpendicular to $2x + y = -4$.

33. Find an equation of the line that has x-intercept $(-2, 0)$ and is parallel to $4x - 8y = 12$.

34. Find the value of k so that $4y = kx - 6$ and $5x + 20y = 12$ are parallel.

35. Find the value of k so that $4y = kx - 6$ and $5x + 20y = 12$ are perpendicular.

The lines in each graph are perpendicular. Find an equation of each line.

36.

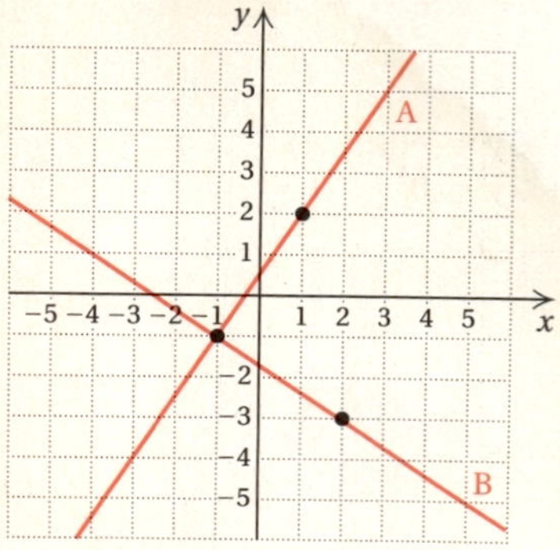

37.

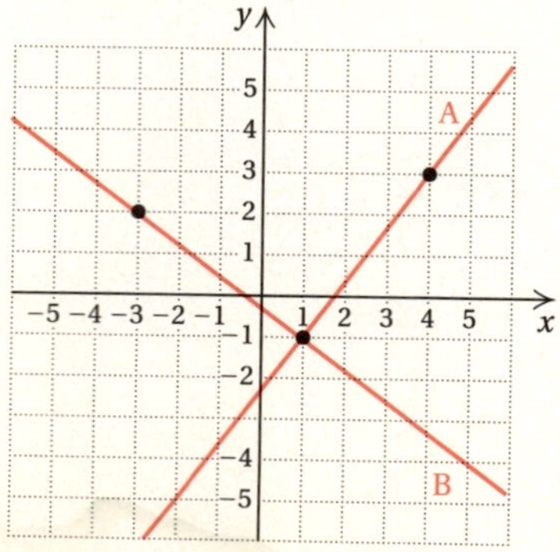

6.6 Graphing Inequalities in Two Variables

OBJECTIVES

After finishing Section 6.6, you should be able to:

a Determine whether an ordered pair of numbers is a solution of an inequality in two variables.

b Graph linear inequalities.

FOR EXTRA HELP

TAPE 16 TAPE 13B MAC: 6 IBM: 6

A graph of an inequality is a drawing that represents its solutions. An inequality in one variable can be graphed on a number line. An inequality in two variables can be graphed on a coordinate plane.

a SOLUTIONS OF INEQUALITIES IN TWO VARIABLES

The solutions of inequalities in two variables are ordered pairs.

EXAMPLE 1 Determine whether $(-3, 2)$ is a solution of $5x + 4y < 13$.

We use alphabetical order to replace x by -3 and y by 2.

$5x + 4y < 13$	
$5(-3) + 4 \cdot 2$	13
$-15 + 8$	
-7	TRUE

Since $-7 < 13$ is true, $(-3, 2)$ is a solution.

EXAMPLE 2 Determine whether $(6, 8)$ is a solution of $5x + 4y < 13$.

We use alphabetical order to replace x by 6 and y by 8.

$5x + 4y < 13$	
$5(6) + 4(8)$	13
$30 + 32$	
62	FALSE

Since $62 < 13$ is false, $(6, 8)$ is not a solution.

Do Exercises 1 and 2.

1. Determine whether $(4, 3)$ is a solution of $3x - 2y < 1$.

2. Determine whether $(2, -5)$ is a solution of $4x + 7y \geq 12$.

b GRAPHING INEQUALITIES IN TWO VARIABLES

EXAMPLE 3 Graph: $y > x$.

We first graph the line $y = x$ for comparison. Every solution of $y = x$ is an ordered pair like $(3, 3)$. The first and second coordinates are the same. The graph of $y = x$ is shown on the left below. We draw it dashed because these points are *not* solutions of $y > x$.

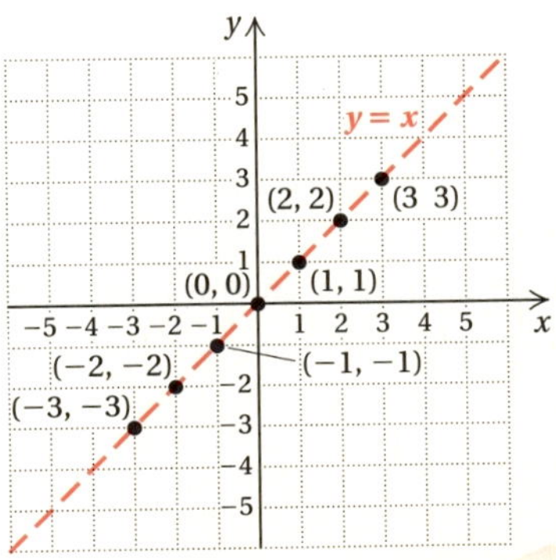

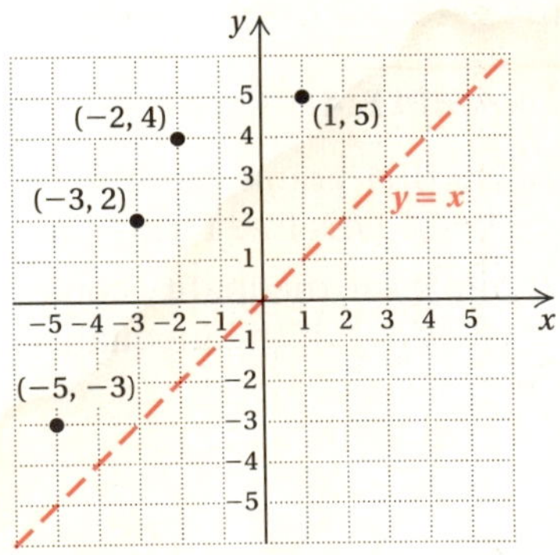

Answers on page A-6

3. Graph: $y < x$.

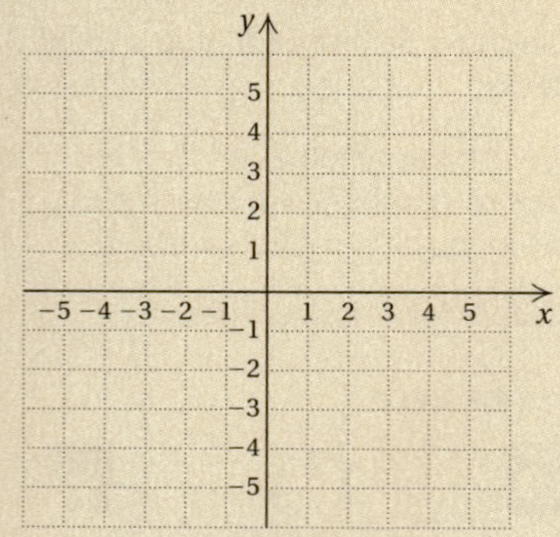

Answer on page A-6

Now look at the graph on the right on the preceding page. Several ordered pairs are plotted in the half-plane above the line $y = x$. Each is a solution of $y > x$.

We can check a pair such as $(-2, 4)$ as follows:

$y > x$		
4	-2	TRUE

It turns out that any point on the same side of $y = x$ as $(-2, 4)$ is also a solution. *If we know that one point in a half-plane is a solution, then all points in that half-plane are solutions.* The graph of $y > x$ is shown below. (Solutions will be indicated by color shading throughout.) We shade the half-plane above $y = x$.

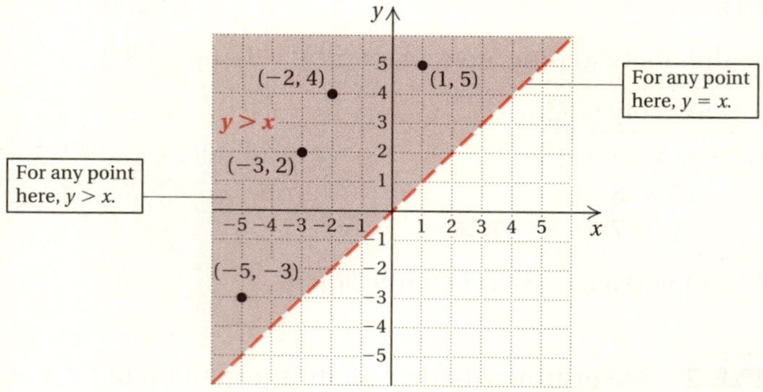

Do Exercise 3.

A **linear inequality** is one that we can get from a linear equation by changing the equals symbol to an inequality symbol. Every linear equation has a graph that is a straight line. The graph of a linear inequality is a half-plane, sometimes including the line along the edge.

To graph an inequality in two variables:

1. Replace the inequality symbol with an equals sign and graph this related equation.
2. If the inequality symbol is $<$ or $>$, draw the line dashed. If the inequality symbol is $\leq$ or $\geq$, draw the line solid.
3. The graph consists of a half-plane, either above or below or left or right of the line, and, if the line is solid, the line as well. To determine which half-plane to shade, choose a point not on the line as a test point. Substitute to find whether that point is a solution of the inequality. If so, shade the half-plane containing that point. If not, shade the half-plane on the opposite side of the line.

EXAMPLE 4 Graph: $5x - 2y < 10$.

1. We first graph the line $5x - 2y = 10$. The intercepts are $(0, -5)$ and $(2, 0)$. This line forms the boundary of the solutions of the inequality.
2. Since the inequality contains the $<$ symbol, points on the line are not solutions of the inequality, so we draw a dashed line.

3. To determine which half-plane to shade, we consider a test point *not* on the line. We try $(3, -2)$ and substitute:

$5x - 2y < 10$	
$5(3) - 2(-2)$	10
$15 + 4$	
19	

FALSE

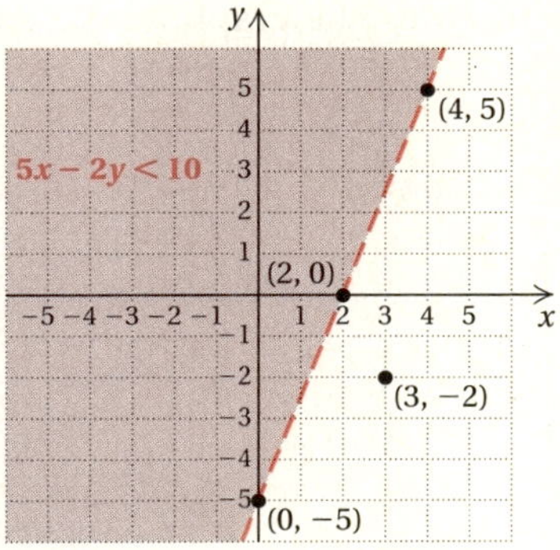

Since this inequality is false, the point $(3, -2)$ is *not* a solution; no point in the half-plane containing $(3, -2)$ is a solution. Thus the points in the opposite half-plane are solutions. The graph is shown above.

Do Exercise 4.

4. Graph: $2x + 4y < 8$.

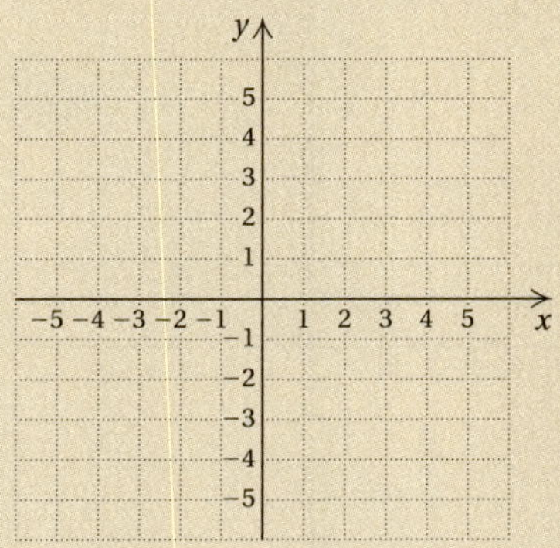

Graph.

5. $3x - 5y < 15$

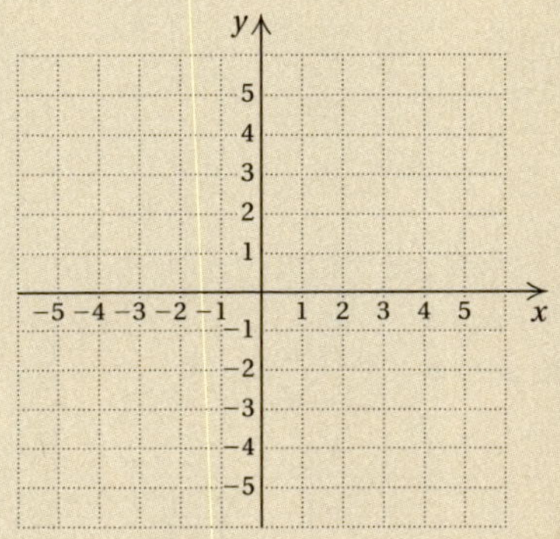

6. $2x + 3y \geq 12$

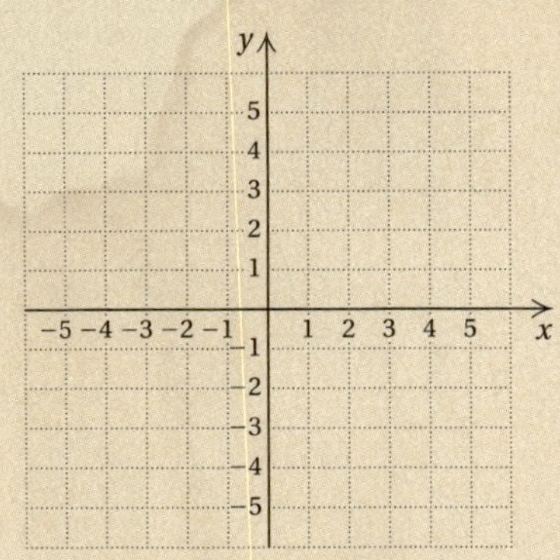

EXAMPLE 5 Graph: $2x + 3y \leq 6$.

First we graph the line $2x + 3y = 6$. The intercepts are $(0, 2)$ and $(3, 0)$. Since the inequality contains the $\leq$ symbol, we draw the line solid to indicate that any pair on the line is a solution. Next, we choose a test point that does not belong to the line. We substitute to determine whether this point is a solution. The origin $(0, 0)$ is generally an easy one to use:

$2x + 3y \leq 6$	
$2 \cdot 0 + 3 \cdot 0$	6
0	

TRUE

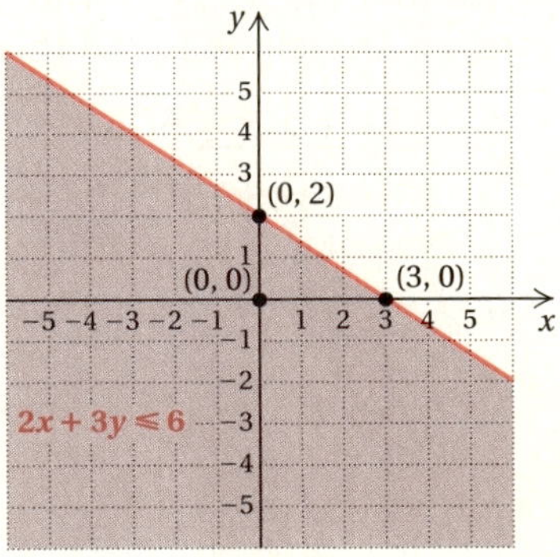

We see that $(0, 0)$ is a solution, so we shade the lower half-plane. Had the substitution given us a false inequality, we would have shaded the other half-plane.

Do Exercises 5 and 6.

EXAMPLE 6 Graph $x < 3$ on a plane.

There is a missing variable in this inequality. If we graph it on a line, its graph is as follows:

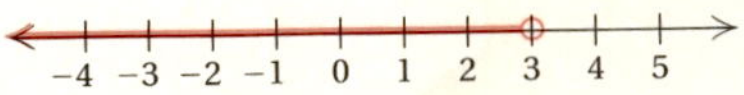

Answers on page A-6

Graph.

7. $x > -3$

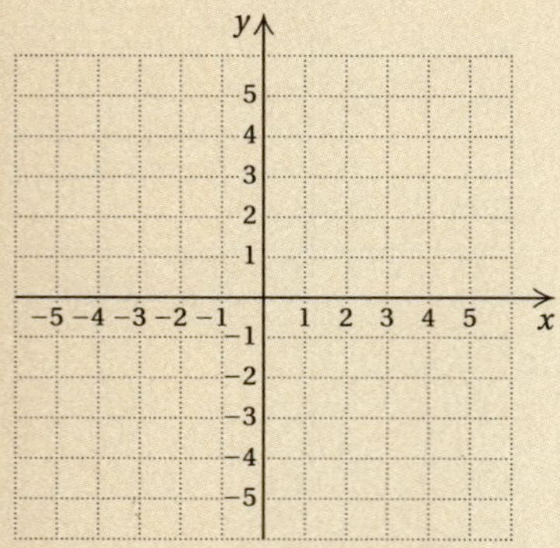

8. $y \leq 4$

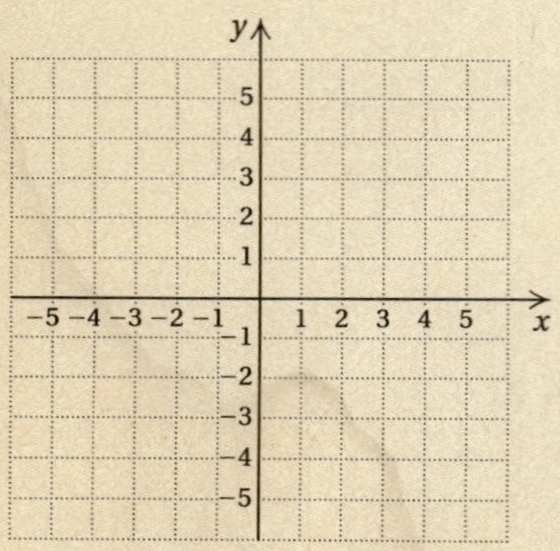

Answers on page A-6

However, we can also write this inequality as $x + 0y < 3$ and consider graphing it on a plane. We use the same technique that we have used with the other examples. We first graph the related equation $x = 3$ on the plane and draw the graph with a dashed line since the inequality symbol is $<$.

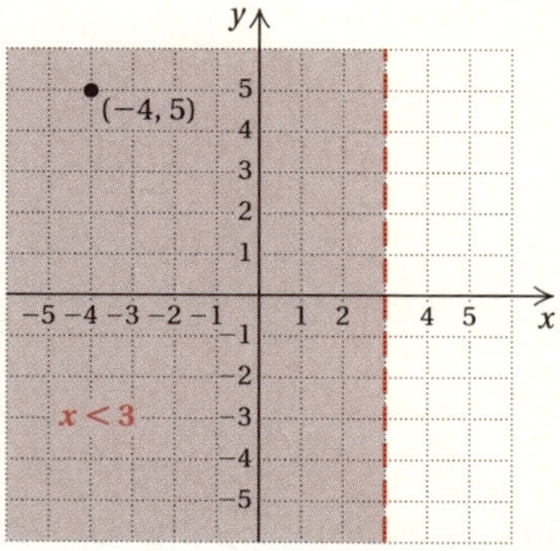

The rest of the graph is a half-plane either to the right or to the left of the line $x = 3$. To determine which, we consider a test point, $(-4, 5)$:

$$\begin{array}{r|l} \multicolumn{2}{c}{x + 0y < 3} \\ \hline -4 + 0(5) & 3 \\ -4 & \end{array} \quad \text{TRUE}$$

We see that $(-4, 5)$ is a solution, so all the pairs in the half-plane containing $(-4, 5)$ are solutions. We shade that half-plane.

We see from the graph that the solutions of $x < 3$ are all those ordered pairs whose first coordinates are less than 3.

EXAMPLE 7 Graph $y \geq -4$ on a plane.

We first graph $y = -4$ using a solid line to indicate that all points on the line are solutions. We then use $(2, 3)$ as a test point and substitute:

$$\begin{array}{r|l} \multicolumn{2}{c}{0x + y \geq -4} \\ \hline 0(2) + 3 & -4 \\ 3 & \end{array} \quad \text{TRUE}$$

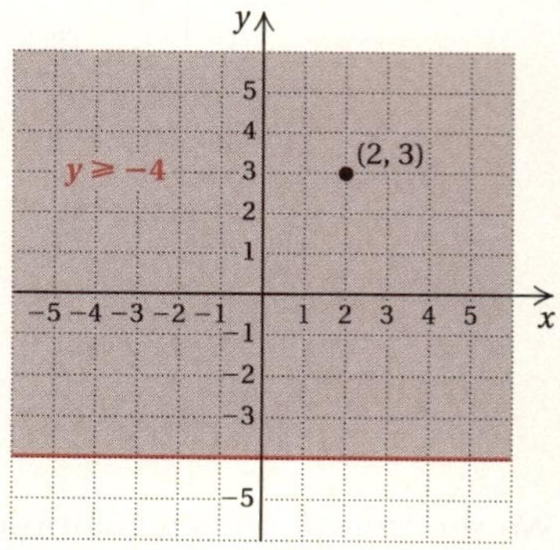

Since $(2, 3)$ is a solution, all points in the half-plane containing $(2, 3)$ are solutions. Note that this half-plane consists of all ordered pairs whose second coordinate is greater than or equal to -4.

Do Exercises 7 and 8.

NAME SECTION DATE

Exercise Set 6.6

a

1. Determine whether $(-3, -5)$ is a solution of

$$-x - 3y < 18.$$

2. Determine whether $(2, -3)$ is a solution of

$$5x - 4y \geq 1.$$

3. Determine whether $\left(\frac{1}{2}, -\frac{1}{4}\right)$ is a solution of

$$7y - 9x \leq -3.$$

4. Determine whether $(-8, 5)$ is a solution of

$$x + 0 \cdot y > 4.$$

ANSWERS

1. ________

2. ________

3. ________

4. ________

b Graph on a plane.

5. $x > 2y$

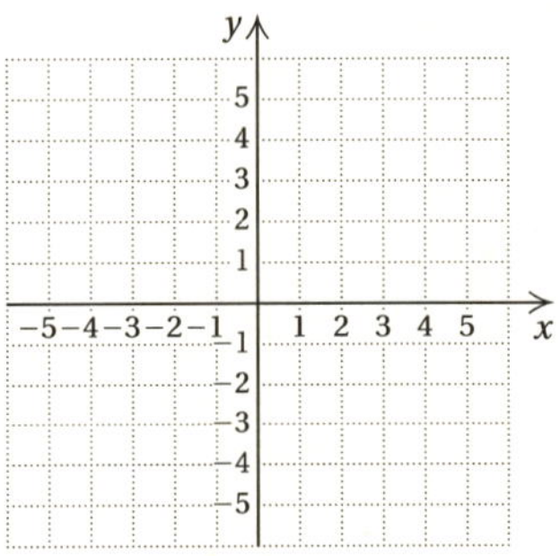

6. $x > 3y$

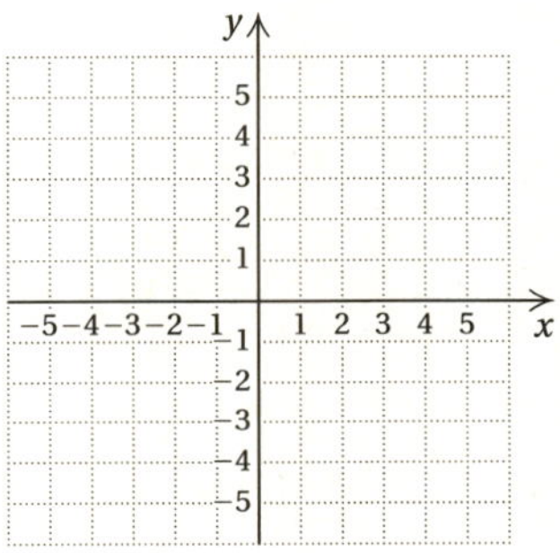

7. $y \leq x - 3$

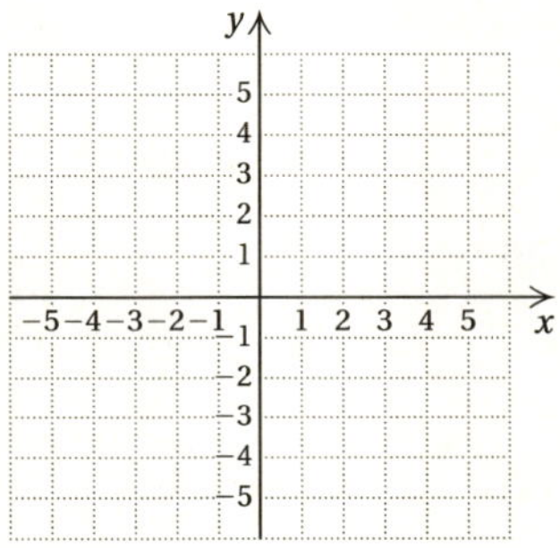

8. $y \leq x - 5$

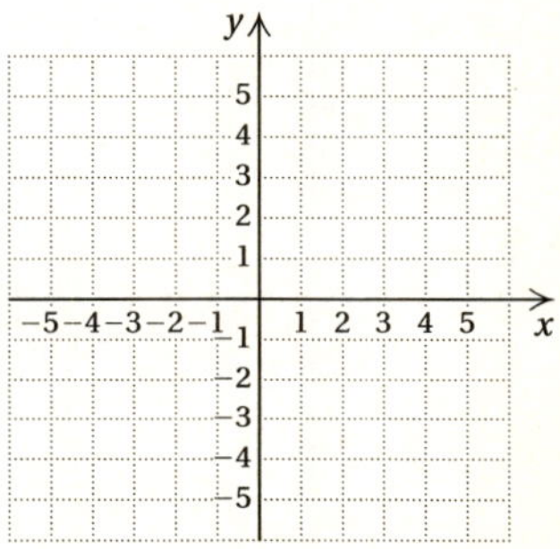

9. $y < x + 1$

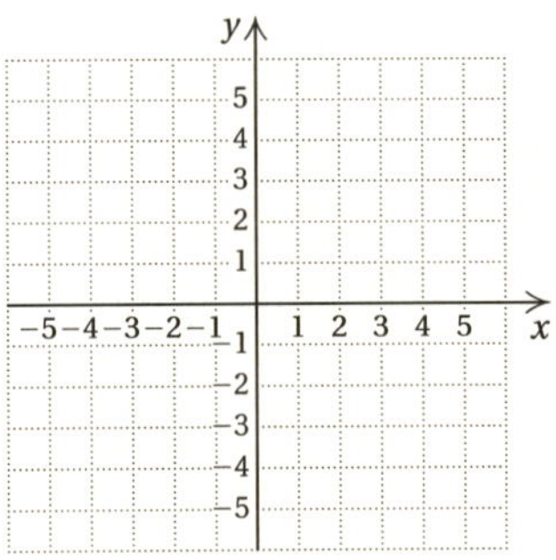

10. $y < x + 4$

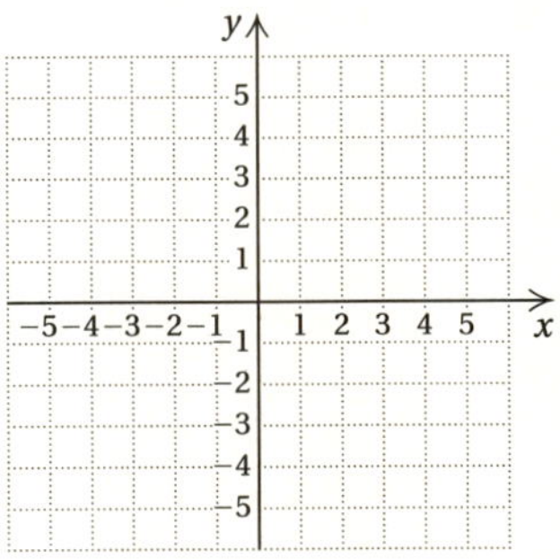

11. $y \geq x - 2$

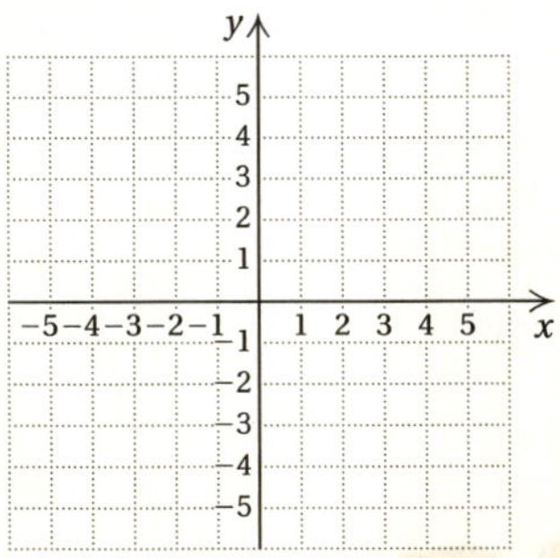

12. $y \geq x - 1$

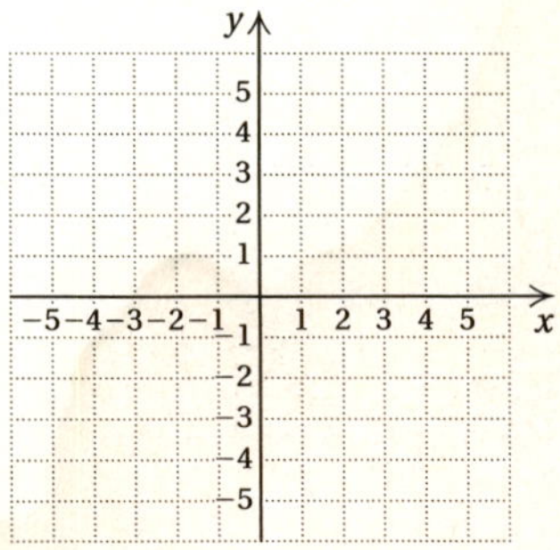

13. $y \leq 2x - 1$

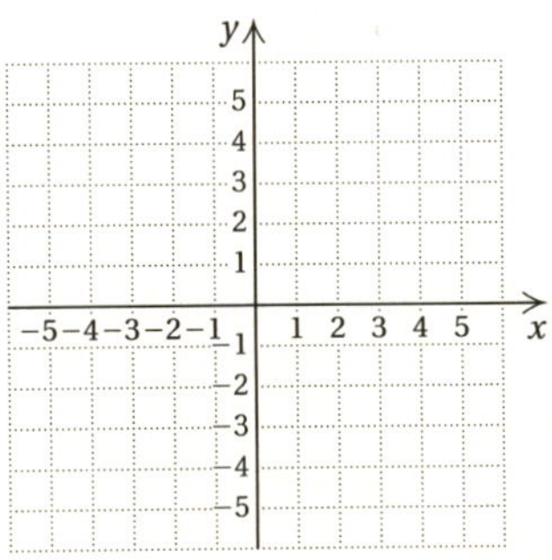

14. $y \leq 3x + 2$

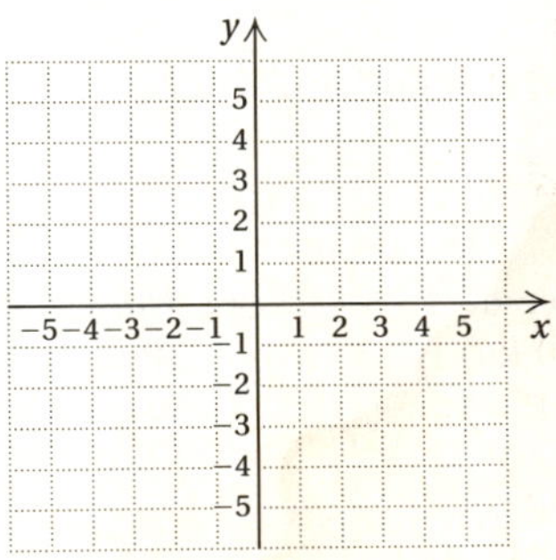

15. $x + y \leq 3$

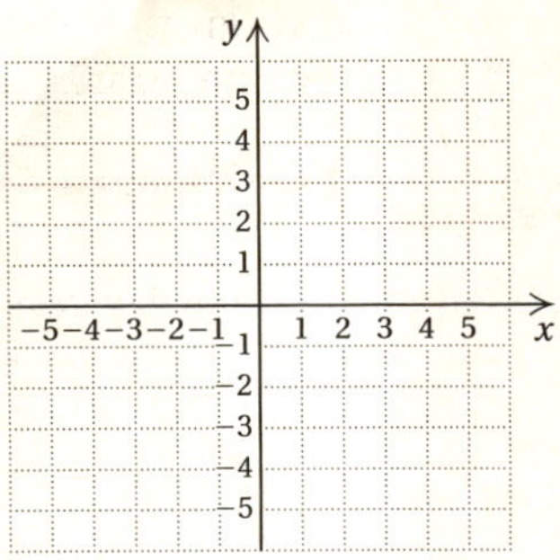

16. $x + y \leq 4$

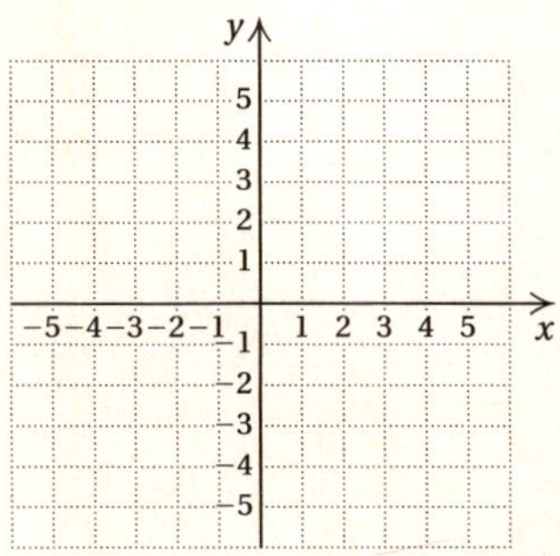

17. $x - y > 7$

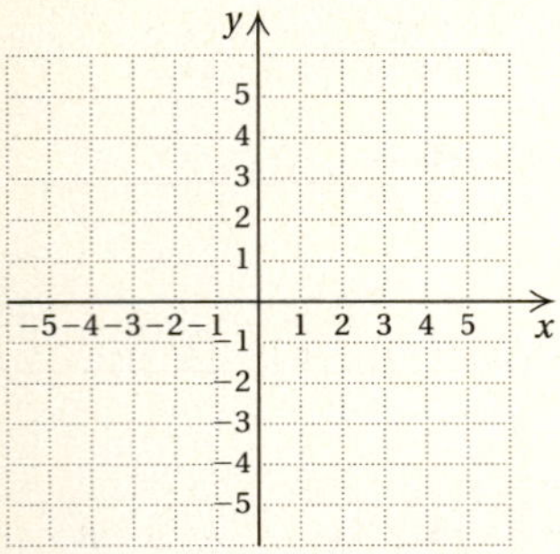

18. $x - y > -2$

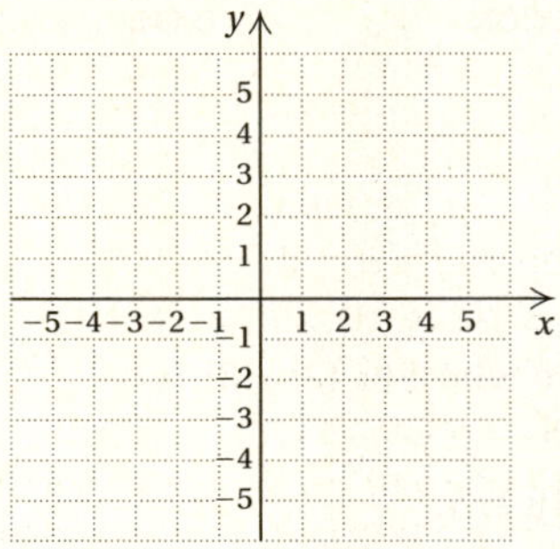

19. $2x + 3y \leq 12$

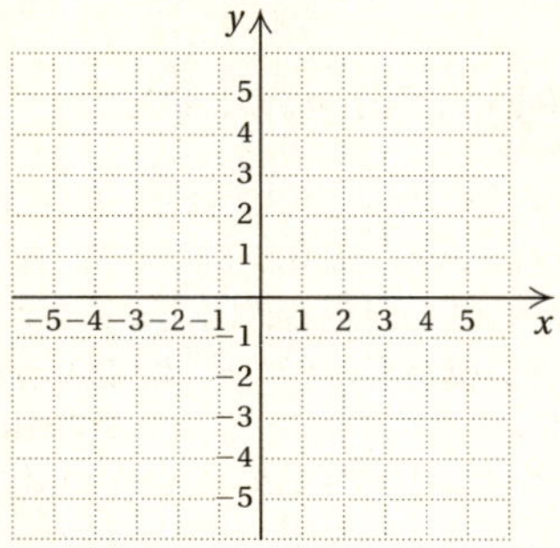

20. $5x + 4y \geq 20$

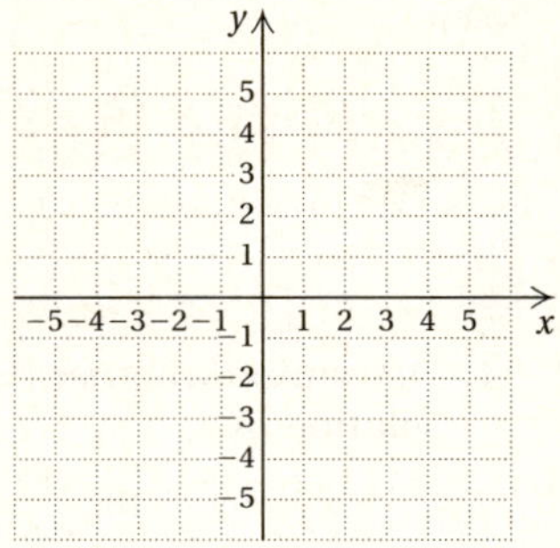

21. $y \geq 1 - 2x$

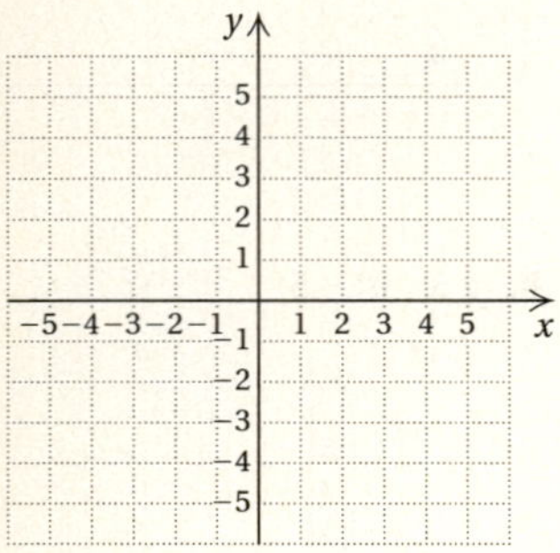

22. $y - 2x \leq -1$

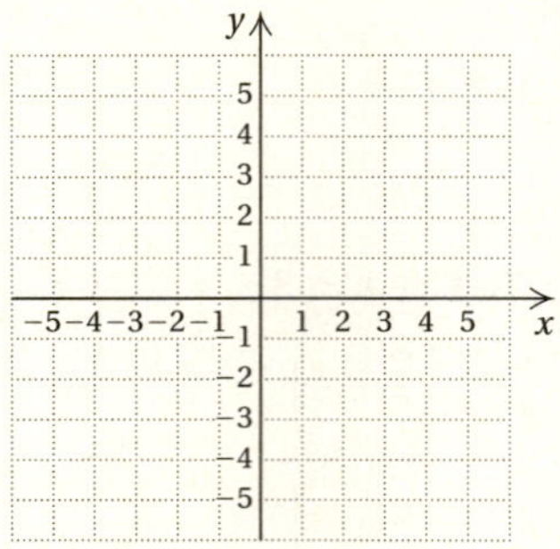

23. $y + 4x > 0$

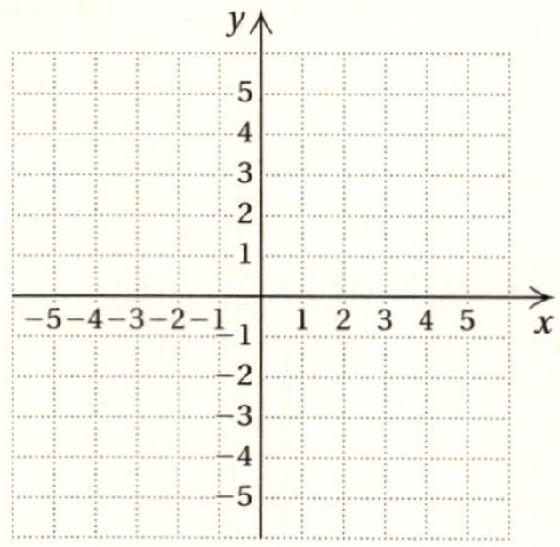

24. $y - x < 0$

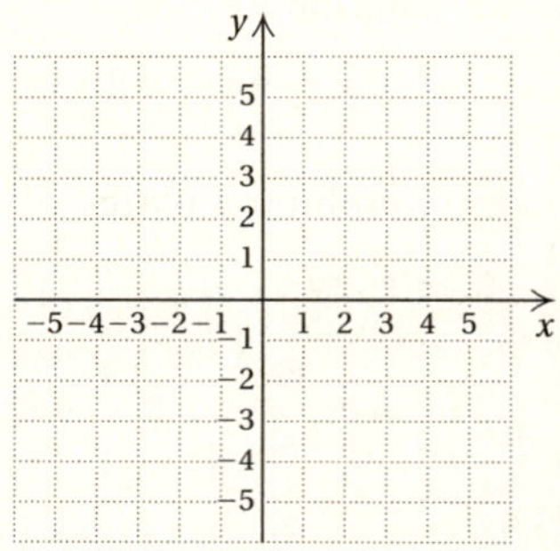

25. $y \leq 3$

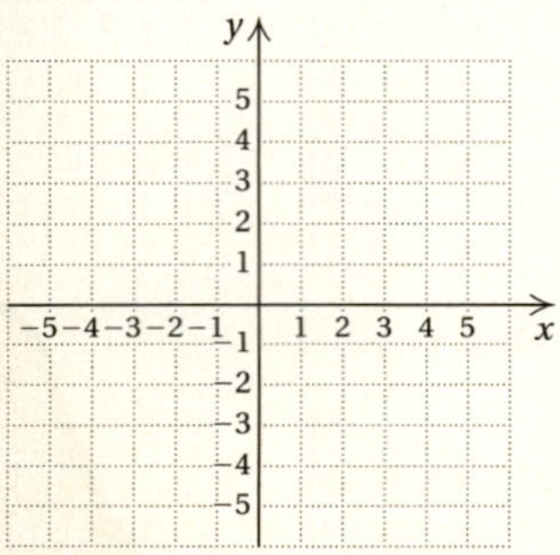

26. $y > -1$

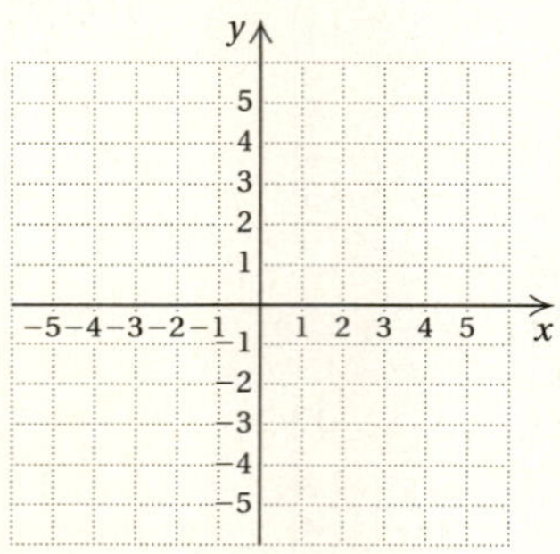

27. $x \geq -1$

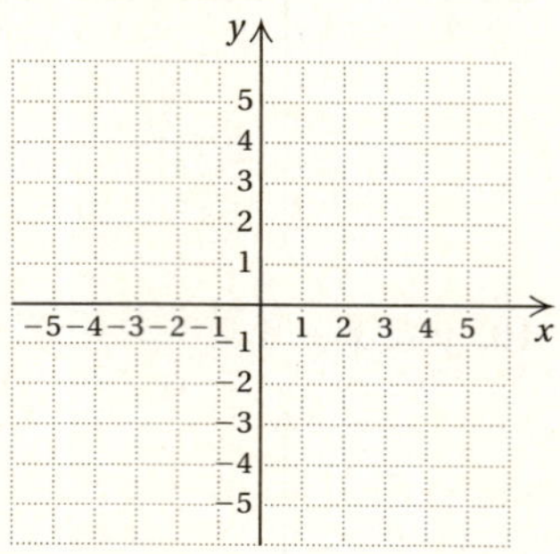

28. $x < 0$

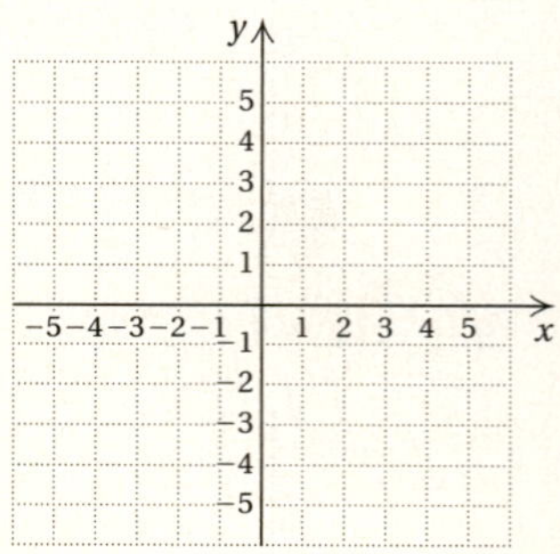

SKILL MAINTENANCE

Calculate.

29. $3^7 \div 3^4 \div 3^3 \div 3$

30. $\dfrac{37 - 5(4 - 6)}{2 \cdot 6 + 8}$

Solve.

31. $\dfrac{12}{x} = \dfrac{48}{x + 9}$

32. $x + 5 = -\dfrac{6}{x}$

33. $x^2 + 16 = 8x$

34. $12x^2 + 17x = 5$

SYNTHESIS

35. *Elevators.* Many elevators have a capacity of 1 metric ton (1000 kg). Suppose c children, each weighing 35 kg, and a adults, each weighing 75 kg, are on an elevator. Find and graph an inequality that asserts that the elevator is overloaded.

36. *Hockey wins and losses.* A hockey team figures that it needs at least 60 points for the season in order to make the playoffs. A win w is worth 2 points and a tie t is worth 1 point. Find and graph an inequality that describes the situation.

ANSWERS

29. ______

30. ______

31. ______

32. ______

33. ______

34. ______

35. ______

36. ______

6.7 *Direct and Inverse Variation*

a EQUATIONS OF DIRECT VARIATION

A bicycle is traveling at 10 km/h. In 1 hr, it goes 10 km. In 2 hr, it goes 20 km. In 3 hr, it goes 30 km, and so on. We will use the number of hours as the first coordinate and the number of kilometers traveled as the second coordinate: (1, 10), (2, 20), (3, 30), (4, 40), and so on. Note that as the first number gets larger, so does the second. Note also that the ratio of distance to time for each of these ordered pairs is $\frac{10}{1}$, or 10.

Whenever a situation produces pairs of numbers in which the *ratio is constant*, we say that there is **direct variation.** Here the distance varies directly as the time:

$$\frac{d}{t} = 10 \text{ (a constant)}, \quad \text{or} \quad d = 10t.$$

> **DIRECT VARIATION**
>
> If a situation translates to an equation described by $y = kx$, where k is a positive constant, $y = kx$ is called an *equation of direct variation,* and k is called the *variation constant.* We say that y varies directly as x.

The terminologies

"y varies as x,"

"y is directly proportional to x,"

and "y is proportional to x"

also imply direct variation and are used in many situations. The constant k is often referred to as a **constant of proportionality.**

When there is direct variation $y = kx$, the variation constant can be found if one pair of values of x and y is known. Then other values can be found.

EXAMPLE 1 Find an equation of variation where y varies directly as x and $y = 7$ when $x = 25$.

We substitute to find k:

$$y = kx$$
$$7 = k \cdot 25$$
$$\frac{7}{25} = k, \quad \text{or } k = 0.28.$$

Then the equation of variation is

$$y = 0.28x.$$

Note that the answer is an *equation.*

Do Exercises 1 and 2.

OBJECTIVES

After finishing Section 6.7, you should be able to:

a Find an equation of direct variation given a pair of values of the variables.

b Solve problems involving direct variation.

c Find an equation of inverse variation given a pair of values of the variables.

d Solve problems involving inverse variation.

FOR EXTRA HELP

TAPE 16 | TAPE 14A | MAC: 6 IBM: 6

Find an equation of variation where y varies directly as x and the following is true.

1. $y = 84$ when $x = 12$

2. $y = 50$ when $x = 80$

Answers on page A-6

3. The cost C of operating a television varies directly as the number n of hours that it is in operation. It costs \$14.00 to operate a standard-size color TV continuously for 30 days. At this rate, how much would it cost to operate the TV for 1 day? for 1 hour?

4. The weight V of an object on Venus varies directly as its weight E on earth. A person weighing 165 lb on earth would weigh 145.2 lb on Venus. How much would a person weighing 198 lb on earth weigh on Venus?

Answers on page A-6

b SOLVING PROBLEMS WITH DIRECT VARIATION

EXAMPLE 2 It is known that the karat rating K of a gold object varies directly as the actual percentage P of gold in the object. A 14-karat gold ring is 58.25% gold. What is the percentage of gold in a 24-karat ring?

1., 2. ***Familiarize*** and ***Translate.*** The problem states that we have direct variation between the variables K and P. Thus an equation $K = kP$, $k > 0$, applies. As the percentage of gold increases, the karat rating increases. The letters K and k represent different quantities.

3. ***Solve.*** The mathematical manipulation has two steps. First, we find the equation of variation by substituting known values for K and P to find k. Second, we compute the percentage of gold in a 24-karat ring.

a) First, we find an equation of variation:

$$K = kP$$

$$14 = k(0.5825) \quad \text{Substituting 14 for } K \text{ and 58.25\%, or 0.5825, for } P$$

$$\frac{14}{0.5825} = k$$

$$24.03 \approx k. \quad \text{Dividing and rounding to the nearest hundredth}$$

The equation of variation is $K = 24.03P$.

b) We then use the equation to find the percentage of gold in a 24-karat ring:

$$K = 24.03P$$

$$24 = 24.03P \quad \text{Substituting 24 for } K$$

$$\frac{24}{24.03} = P$$

$$0.999 \approx P$$

$$99.9\% \approx P.$$

4. ***Check.*** The check might be done by repeating the computations. You might also do some reasoning about the answer. The karat rating increased from 14 to 24. Similarly, the percentage increased from 58.25% to 99.9%.

5. ***State.*** A 24-karat ring is 99.9% gold.

Do Exercises 3 and 4.

Let us consider direct variation from the standpoint of a graph. The graph of $y = kx$, $k > 0$, always goes through the origin and rises from left to right. Note that as x increases, y increases; and as x decreases, y decreases. This is why the terminology "direct" is used. What one variable does, the other does as well.

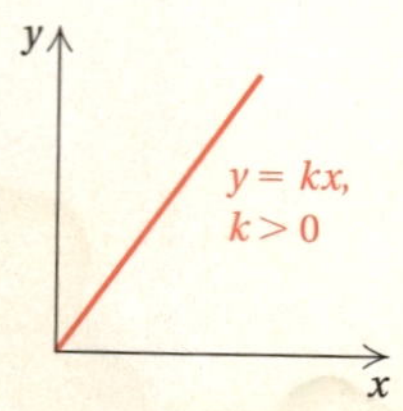

C EQUATIONS OF INVERSE VARIATION

A car is traveling a distance of 10 km. At a speed of 10 km/h, it will take 1 hr. At 20 km/h, it will take $\frac{1}{2}$ hr. At 30 km/h, it will take $\frac{1}{3}$ hr, and so on. This determines a set of pairs of numbers, all having the same product:

$$(10, 1), \quad \left(20, \tfrac{1}{2}\right), \quad \left(30, \tfrac{1}{3}\right), \quad \left(40, \tfrac{1}{4}\right), \quad \text{and so on.}$$

Note that as the first number gets larger, the second number gets smaller. Whenever a situation produces pairs of numbers whose *product is constant,* we say that there is **inverse variation.** Here the time varies inversely as the speed:

$$rt = 10 \text{ (a constant)}, \quad \text{or} \quad t = \frac{10}{r}.$$

INVERSE VARIATION

If a situation translates to an equation described by $y = k/x$, where k is a positive constant, $y = k/x$ is called an *equation of inverse variation.* We say that y varies inversely as x.

The terminology "y is inversely proportional to x" also implies inverse variation and is used in some situations.

EXAMPLE 3 Find an equation of variation where y varies inversely as x and $y = 145$ when $x = 0.8$.

We substitute to find k:

$$y = \frac{k}{x}$$

$$145 = \frac{k}{0.8}$$

$$(0.8)145 = k$$

$$116 = k.$$

The equation of variation is

$$y = \frac{116}{x}.$$

Do Exercises 5 and 6.

The graph of $y = k/x$, $k > 0$, is shaped like the following figure for positive values of x. (You need not know how to graph such equations at this time.) Note that as x increases, y decreases; and as x decreases, y increases. This is why the terminology "inverse" is used. One variable does the opposite of what the other does.

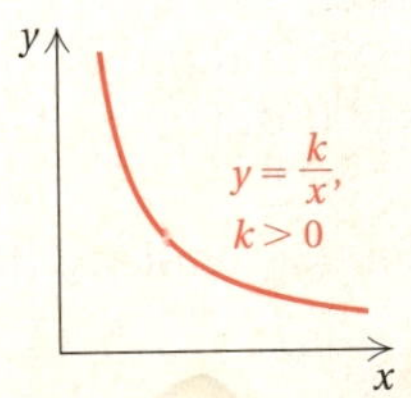

Find an equation of variation where y varies inversely as x and the following is true.

5. $y = 105$ when $x = 0.6$

6. $y = 45$ when $x = 20$

Answers on page A-6

7. In Example 4, how long would it take 10 people to do the job?

8. The time required to drive a fixed distance varies inversely as the speed r. It takes 5 hr at 60 km/h to drive a fixed distance. How long would it take at 40 km/h?

Answers on page A-6

d SOLVING PROBLEMS WITH INVERSE VARIATION

Often in an applied situation we must decide which kind of variation, if any, might apply to the problem.

EXAMPLE 4 A maintenance supervisor notes that it takes 4 hr for 20 people to wash and wax the floors in a building. How long would it then take 25 people to do the job?

1. ***Familiarize.*** Think about the problem situation. What kind of variation would be used? It seems reasonable that the more people there are working on the job, the less time it will take to finish. (One might argue that too many people in a crowded area would be counterproductive, but we will disregard that possibility.) Thus inverse variation might apply. We let $T =$ the time to do the job, in hours, and $N =$ the number of people. Assuming inverse variation, we know that an equation $T = k/N$, $k > 0$, applies. As the number of people increases, the time it takes to do the job decreases.

2. ***Translate.*** We write an equation of variation:

$$T = \frac{k}{N}.$$

Time varies inversely as the number of people.

3. ***Solve.*** The mathematical manipulation has two steps. First, we find the equation of variation by substituting known values to find k. Second, we compute the amount of time it would take 25 people to do the job.

a) First, we find an equation of variation:

$$T = \frac{k}{N}$$

$$4 = \frac{k}{20} \quad \text{Substituting 4 for } T \text{ and 20 for } N$$

$$20 \cdot 4 = k$$

$$80 = k.$$

The equation of variation is $T = \dfrac{80}{N}$.

b) We then use the equation to find the amount of time that it takes 25 people to do the job:

$$T = \frac{80}{N}$$

$$T = \frac{80}{25} \quad \text{Substituting 25 for } N$$

$$T = 3.2.$$

4. ***Check.*** The check might be done by repeating the computations. We might also analyze the results. The number of people increased from 20 to 25. Did the time decrease? It did, and this confirms what we expect with inverse variation.

5. ***State.*** It should take 3.2 hr for 25 people to do the job.

Do Exercises 7 and 8.

NAME SECTION DATE

Exercise Set 6.7

a Find an equation of variation where y varies directly as x and the following are true.

1. $y = 36$ when $x = 9$

2. $y = 60$ when $x = 16$

3. $y = 0.8$ when $x = 0.5$

4. $y = 0.7$ when $x = 0.4$

5. $y = 630$ when $x = 175$

6. $y = 400$ when $x = 125$

7. $y = 500$ when $x = 60$

8. $y = 200$ when $x = 300$

b Solve.

9. A person's paycheck P varies directly as the number H of hours worked. For working 15 hr, the pay is \$78.75. Find the pay for 35 hr of work.

10. The interest I earned in 1 yr on a fixed principal varies directly as the interest rate r. An investment earns \$53.55 at an interest rate of 4.25%. How much will the investment earn at a rate of 5.75%?

11. A chef is planning meals in a refreshment tent at a golf tournament. The number of servings S of meat that can be obtained from a turkey varies directly as its weight W. From a turkey weighing 14 kg, one can get 40 servings of meat. How many servings can be obtained from an 8-kg turkey?

12. The number of servings S of meat that can be obtained from round steak varies directly as the weight W. From 9 kg of round steak, one can get 70 servings of meat. How many servings can one get from 12 kg of round steak?

ANSWERS

1. ________

2. ________

3. ________

4. ________

5. ________

6. ________

7. ________

8. ________

9. ________

10. ________

11. ________

12. ________

ANSWERS

13. ____________

14. ____________

15. ____________

16. ____________

17. ____________

18. ____________

19. ____________

20. ____________

21. ____________

22. ____________

23. ____________

24. ____________

25. ____________

26. ____________

13. The weight M of an object on the moon varies directly as its weight E on earth. A person who weighs 171.6 lb on earth weighs 28.6 lb on the moon. How much would a 220-lb person weigh on the moon?

14. The weight M of an object on Mars varies directly as its weight E on earth. A person who weighs 209 lb on earth weighs 79.42 lb on Mars. How much would a 176-lb person weigh on Mars?

15. The number of computer instructions N per second varies directly as the speed S of its internal processor. A processor with a speed of 25 megahertz can perform 2,000,000 instructions per second. How many instructions will the same processor perform if it is running at a speed of 40 megahertz?

16. The number of kilograms W of water in a human body varies directly as the total body weight B. A person weighing 75 kg contains 54 kg of water. How many kilograms of water are in a person weighing 95 kg?

c Find an equation of variation where y varies inversely as x and the following are true.

17. $y = 3$ when $x = 25$

18. $y = 2$ when $x = 45$

19. $y = 10$ when $x = 8$

20. $y = 10$ when $x = 7$

21. $y = 6.25$ when $x = 0.16$

22. $y = 0.125$ when $x = 8$

23. $y = 50$ when $x = 42$

24. $y = 25$ when $x = 42$

25. $y = 0.2$ when $x = 0.3$

26. $y = 0.4$ when $x = 0.6$

b **d** Solve.

27. A production line produces 15 compact disc players every 8 hr. How many players can it produce in 37 hr?
   a) What kind of variation might apply to this situation?
   b) Solve the problem.

28. A person works for 15 hr and makes \$93.75. How much will the person make by working 35 hr?
   a) What kind of variation might apply to this situation?
   b) Solve the problem.

29. It takes 4 hr for 9 cooks to prepare the food for a wedding rehearsal dinner. How long will it take 8 cooks to prepare the dinner?
   a) What kind of variation might apply to this situation?
   b) Solve the problem.

30. It takes 16 hr for 2 people to resurface a tennis court. How long will it take 6 people to do the job?
   a) What kind of variation might apply to this situation?
   b) Solve the problem.

31. The current I in an electrical conductor varies inversely as the resistance R of the conductor. The current is 2 amperes when the resistance is 960 ohms. What is the current when the resistance is 540 ohms?

32. The volume V of a gas varies inversely as the pressure P on it. The volume of a gas is 200 cm^3 under a pressure of 32 kg/cm^2. What will be its volume under a pressure of 20 kg/cm^2?

33. The number of files N of the same size that a computer's hard disk will hold varies inversely as the size of the files. Loretta's hard disk will hold 1600 files if each is 50,000 bytes long. How many files will the disk hold if each is 125,000 bytes long?

34. The time t required to empty a tank varies inversely as the rate r of pumping. A pump can empty a tank in 90 min at a rate of 1200 L/min. How long will it take the pump to empty the tank at a rate of 2000 L/min?

35. The apparent size A of an object varies inversely as the distance d of the object from the eye. A flagpole 30 ft from an observer appears to be 27.5 ft tall. How tall will the same flagpole appear to be if it is 100 ft from the eye?

36. The time t required to drive a fixed distance varies inversely as the speed r. It takes 5 hr at 55 mph to drive a fixed distance. How long would it take at 40 mph?

ANSWERS

27. a) ______
b) ______
28. a) ______
b) ______
29. a) ______
b) ______
30. a) ______
b) ______
31. ______
32. ______
33. ______
34. ______
35. ______
36. ______

ANSWERS

37. ______
38. ______
39. ______
40. ______
41. ______
42. ______
43. ______
44. ______
45. ______
46. ______
47. ______
48. ______
49. ______
50. ______
51. ______
52. ______

SKILL MAINTENANCE

Solve.

37. $\dfrac{x+2}{x+5} = \dfrac{x-4}{x-6}$

38. $\dfrac{x-3}{x-5} = \dfrac{x+5}{x+1}$

39. $x^2 - 25x + 144 = 0$

40. $t^2 + 21t + 108 = 0$

41. $35x^2 + 8 = 34x$

42. $14x^2 - 19x - 3 = 0$

SYNTHESIS

Write an equation of direct variation for the situation. If possible, give a value for k.

43. The perimeter P of an equilateral polygon varies directly as the length s of a side.

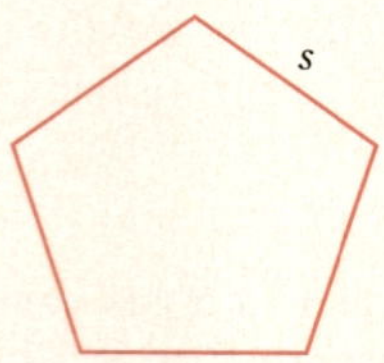

44. The circumference C of a circle varies directly as the radius r.

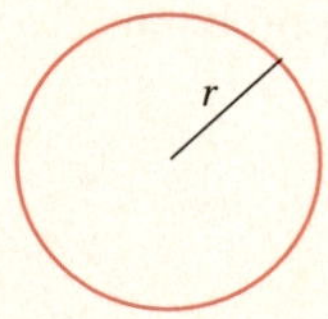

45. The cost C of building a new house varies directly as the area A of the floor space of the house.

46. The number of bags B of peanuts sold at a baseball game varies directly as the number of people N in attendance.

Write an equation of variation for the situation.

47. The square of the pitch P of a vibrating string varies directly as the tension t on the string.

48. In a stream, the amount S of salt carried varies directly as the sixth power of the speed V of the stream.

49. The power P in a windmill varies directly as the cube of the wind speed V.

50. The volume V of a sphere varies directly as the cube of the radius r.

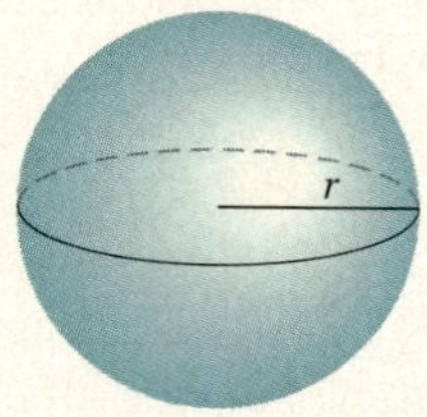

51. The number of revolutions N of a tire rolling over a given distance varies inversely as the circumference C of the tire.

52. The intensity of illumination I from a light source varies inversely as the square of the distance d from the source.

CRITICAL THINKING

CALCULATOR CONNECTION

Graphing calculators and computer graphing software are often referred to as *graphers*. The following exercises are designed to make use of a grapher. We assume that you have studied an instructional manual and are familiar with some of the related terminology. Many of these exercises can also be done by using a scientific calculator to compute y-values.

1. Graph each equation using a grapher and a $[-10, 10, -10, 10]$ viewing box.

a) $y = 3x - 1$

b) $y = |x - 1|$

c) $y = x^2 - 4$

d) $3x - 4y = 12$

e) $y = \frac{1}{x}$

f) $y = x^3 - 3x + 2$

Which have graphs that are straight lines?

2. Graph each equation in a $[-10, 10, -10, 10]$ viewing box. Then use the Trace feature to find the coordinates of at least three points on the graph.

a) $y = 4x - 3$

b) $y = x^2 + 4x + 3$

c) $y = (x - 1)^2$

3. To solve equations and find the x-intercepts of a graph, we use the Zoom feature to zoom in on the regions in which the graph crosses the x-axis. The closer we zoom in on the numbers, the better our approximation. Find the x-intercepts of each graph to three decimal places.

a) $y = 0.8x - 2.3$

b) $y = x^2 - 5.3$

c) $y = |x| - 2$

4. *Average price of a movie ticket.* The average price of a movie ticket can be estimated by the equation

$$P = 0.1522Y - 298.592,$$

where $Y =$ the year and $P =$ the average price, in dollars. The price is lower than what might be expected due to senior-citizen discounts, children's prices, and special volume discounts.

a) Use the equation to predict the average price of a ticket in 1996, 2000, and 2005.

b) Graph the equation. Let Y be the first coordinate and P be the second.

EXTENDED SYNTHESIS EXERCISES

1. Find three different ordered pairs such that the product of the coordinates is -8.

2. Dolores is paid \$4 per hour on her new job. When she works on Saturdays, she is paid \$6 per hour. Suppose one week Dolores works x hours on Friday and y hours on Saturday.

a) Find an algebraic expression for the total earnings E for the two days.

b) Suppose Dolores earned a total of \$38 for the two days and that she worked a whole number of hours each day. Find all possible solutions.

3. Show that if p varies directly as q, then q varies directly as p.

4. The area of a circle varies directly as the square of the length of the radius. What is the variation constant?

5. In chess, the queen can move on a line either diagonally, horizontally, or vertically, as shown in the figure below. What slopes are possible for these lines?

6. In chess, the knight can move to any of the eight squares shown by a, b, c, d, e, f, g, and h. If lines are drawn from the beginning to the end of the move, what slopes are possible for these lines?

(continued)

CRITICAL THINKING

Determine whether inverse variation seems to apply to each situation.

7. The number of hours that a student watches television per week and the student's grade point average

8. The cost of mailing a large package in the United States and the distance that it travels

9. The number of years that a car is owned and its resale value

10. A runner's speed in a race and the time that it takes to run it

11. The weight of a turkey and the cooking time

12. The number of plays that it takes to go 80 yd for a touchdown and the average gain per play

EXERCISES FOR THINKING AND WRITING

1. Briefly describe the concept of slope.

2. Graph $x < 1$ on both a number line and a plane, and explain the difference between the graphs.

3. Graph the following equations using the same set of axes.

$$y = x, \quad y = 2x, \quad y = 5x,$$

$$y = -3x, \quad y = \frac{1}{2}x, \quad y = -\frac{1}{2}x.$$

Look for patterns and then write a paragraph explaining how the slopes of these lines change as the coefficient of x changes.

4. Do all graphs of linear equations have y-intercepts? Explain.

5. Do all graphs of linear equations have x-intercepts? Explain.

6. Are the equations

$$y = -\tfrac{2}{5}x + 3 \quad \text{and} \quad 2x + 5y = 15$$

equivalent? Why or why not?

7. The graph of the line $y = 3x - 4$ is shown below.

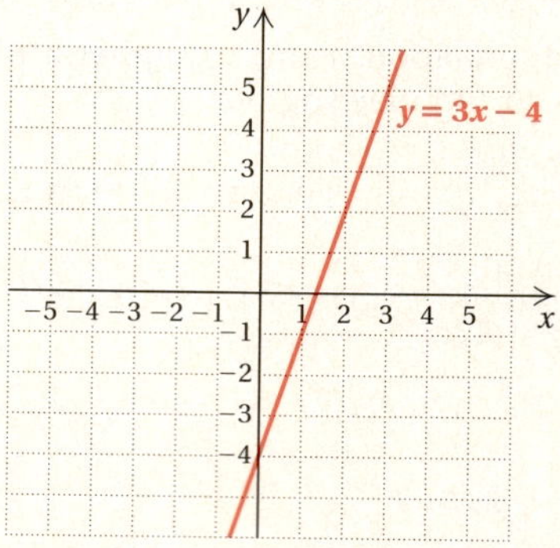

Suppose you graphed each of the following equations on the same set of axes. How would the graph of each equation be related to the one already graphed?

a) $y = 3x + 6$

b) $y = 6x - 4$

c) $2y = 6x - 8$

d) $4 = 3x - y$

e) $-y = -3x + 4$

f) $y = \dfrac{-x}{3} - 4$

Summary and Review: Chapter 6

Important Properties and Formulas

Slope $= m = \dfrac{y_2 - y_1}{x_2 - x_1}$

Slope–Intercept Equation:	$y = mx + b$
Point–Slope Equation:	$y - y_1 = m(x - x_1)$
Parallel Lines:	**Slopes equal, y-intercepts different**
Perpendicular Lines:	**Product of slopes $= -1$**

Review Exercises

The review objectives to be tested in addition to the material in this chapter are [1.8d], [4.7b], [5.6a], and [5.7a, b].

1. This line graph shows the prime rate (the interest rate charged by banks to their best customers) in June for several years.

a) What was the highest prime rate?

b) Between what two consecutive years did the prime rate decrease the most?

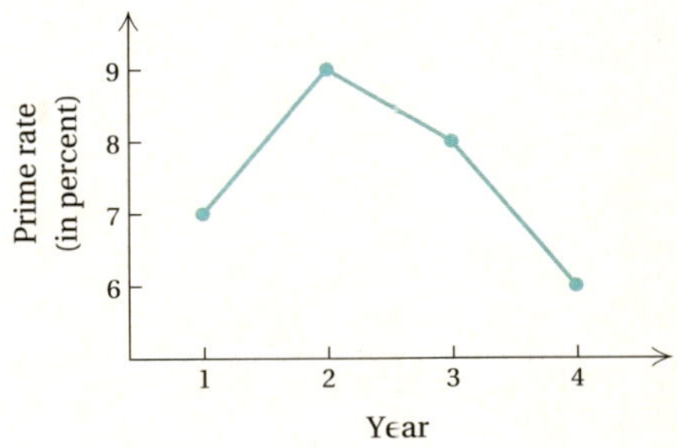

Find the coordinates of each point.

2. A **3.** B **4.** C

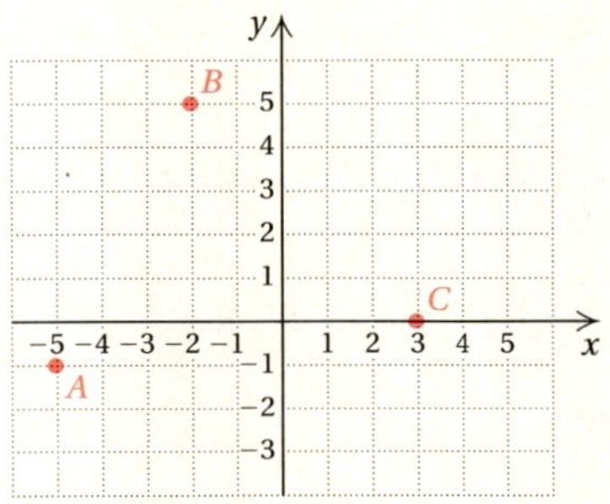

Plot these points using graph paper.

5. $(2, 5)$ **6.** $(0, -3)$ **7.** $(-4, -2)$

In which quadrant is the point located?

8. $(3, -8)$ **9.** $(-20, -14)$ **10.** $(4.9, 1.3)$

Determine whether the given point is a solution of the equation $2y - x = 10$.

11. $(2, -6)$ **12.** $(0, 5)$

Graph on a plane.

13. $y = 2x - 5$ **14.** $y = -\frac{3}{4}x$ **15.** $y = -x + 4$ **16.** $y = 3 - 4x$

17. $5x - 2y = 10$ **18.** $y = 3$ **19.** $4x + 3 = 0$ **20.** $x - 2y = 6$

Find the slope, if it exists, of the line containing the given pair of points.

21. $(6, 8)$ and $(-2, -4)$ **22.** $(5, 1)$ and $(-1, 1)$

23. $(-3, 0)$ and $(-3, 5)$ **24.** $(-8.3, 4.6)$ and $(-9.9, 1.4)$

Find the slope, if it exists, of the line.

25. $y = -6$

26. $x = 90$

27. $4x + 3y = -12$

Find the slope and the y-intercept of the line.

28. $y = -9x + 46$

29. $x + y = 9$

30. $3x - 5y = 4$

Find an equation of the line containing the given point and with the given slope.

31. $(1, 2), \quad m = 3$

32. $(-2, -5), \quad m = \frac{2}{3}$

33. $(0, -4), \quad m = -2$

Find an equation of the line that contains the given pair of points.

34. $(5, 7)$ and $(-1, 1)$

35. $(2, 0)$ and $(-4, -3)$

Determine whether the graphs of the equations are parallel, perpendicular, or neither.

36. $4x + y = 6,$
$4x + y = 8$

37. $2x + y = 10,$
$y = \frac{1}{2}x - 4$

38. $x + 4y = 8,$
$x = -4y - 10$

39. $3x - y = 6,$
$3x + y = 8$

Determine whether the given point is a solution of the inequality $x - 2y > 1$.

40. $(0, 0)$

41. $(1, 3)$

42. $(4, -1)$

Graph on a plane.

43. $x < y$

44. $x + 2y \geq 4$

45. $x > -2$

Find an equation of variation where y varies directly as x and the following are true.

46. $y = 12$ when $x = 4$

47. $y = 4$ when $x = 8$

48. $y = 0.4$ when $x = 0.5$

Find an equation of variation where y varies inversely as x and the following are true.

49. $y = 5$ when $x = 6$

50. $y = 0.5$ when $x = 2$

51. $y = 1.3$ when $x = 0.5$

Solve.

52. A person's paycheck P varies directly as the number of hours H worked. The pay is \$165.00 for working 20 hr. Find the pay for 30 hr of work.

53. It takes 5 hr for 2 washing machines to wash a fixed amount. How long would it take 10 washing machines? (The number of hours varies inversely as the number of washing machines.)

SKILL MAINTENANCE

54. Judd can paint a shed alone in 5 hr. Mo can paint the same shed alone in 10 hr. How long would it take both of them working together to paint the shed?

55. Compute: $13 \cdot 6 \div 3 \cdot 26 \div 13$.

Solve.

56. $\dfrac{x^2}{x - 4} = \dfrac{16}{x - 4}$

57. $a^2 + 6a - 55 = 0$

NAME SECTION DATE

Test: Chapter 6

Consider the bar graph shown here for Exercises 1–4.

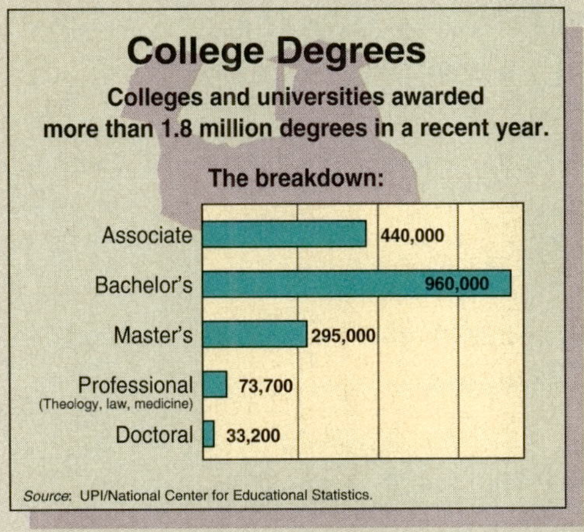

1. What kind of degree was awarded most?

2. How many more bachelor's degrees than associate degrees were awarded?

3. How many more master's degrees than doctoral degrees were awarded?

4. In all, how many graduate degrees were awarded; that is, how many master's, doctoral, and professional degrees were awarded?

In which quadrant is the given point located?

5. $\left(-\frac{1}{2}, 7\right)$

6. $(-5, -6)$

Find the coordinates of the point.

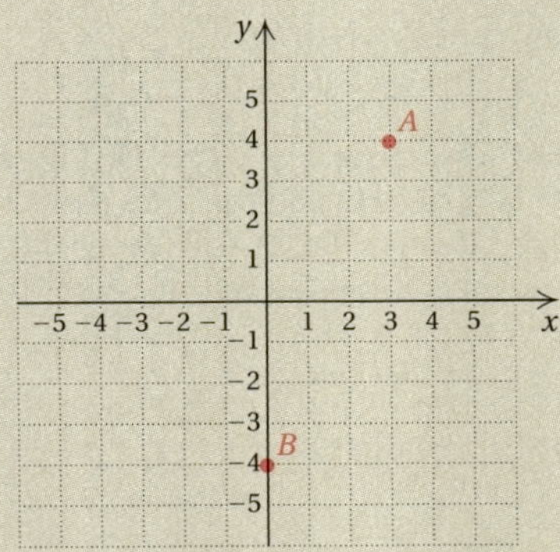

7. A

8. B

9. Determine whether the ordered pair $(2, -4)$ is a solution of the equation $y - 3x = -10$.

Graph.

10. $y = 2x - 1$

11. $2x - 4y = -8$

12. $y = 5$

13. $y = -\frac{3}{2}x$

14. $2x + 8 = 0$

Find the slope, if it exists, of the line containing the given pair of points.

15. $(4, 7)$ and $(4, -1)$

16. $(9, 2)$ and $(-3, -5)$

Find the slope, if it exists, of the given line.

17. $y = -7$

18. $x = 6$

Answers

1. ______
2. ______
3. ______
4. ______
5. ______
6. ______
7. ______
8. ______
9. ______
10. ______
11. ______
12. ______
13. ______
14. ______
15. ______
16. ______
17. ______
18. ______

ANSWERS

19. ______

20. ______

21. ______

22. ______

23. ______

24. ______

25. ______

26. ______

27. ______

28. ______

29. ______

30. ______

31. ______

32. ______

33. ______

34. ______

35. ______

36. ______

37. ______

38. ______

39. ______

40. ______

41. ______

42. ______

43. ______

Find the slope and the y-intercept.

19. $y = 2x - \frac{1}{4}$

20. $-4x + 3y = -6$

Find an equation of the line that contains the given point and has the given slope.

21. $(3, 5)$, $m = 1$

22. $(-2, 0)$, $m = -3$

Find an equation of the line that contains the given pair of points.

23. $(1, 1)$ and $(2, -2)$

24. $(4, -1)$ and $(-4, -3)$

Determine whether the graphs of the equations are parallel, perpendicular, or neither.

25. $2x + y = 8$,
$2x + y = 4$

26. $2x + 5y = 2$,
$y = 2x + 4$

27. $x + 2y = 8$,
$-2x + y = 8$

Determine whether the given point is a solution of the inequality $3y - 2x < -2$.

28. $(0, 0)$

29. $(-4, -10)$

Graph on a plane.

30. $y > x - 1$

31. $2x - y \leq 4$

Find an equation of variation where y varies directly as x and the following are true.

32. $y = 6$ when $x = 3$

33. $y = 1.5$ when $x = 3$

Find an equation of variation where y varies inversely as x and the following are true.

34. $y = 6$ when $x = 2$

35. $y = \frac{1}{3}$ when $x = 3$

36. The distance d traveled by a train varies directly as the time t that it travels. The train travels 60 km in $\frac{1}{2}$ hr. How far will it travel in 2 hr?

37. It takes 3 hr for 2 cement mixers to mix a certain amount. The number of hours varies inversely as the number of cement mixers. How long would it take 5 cement mixers to do the job?

SKILL MAINTENANCE

38. The speed of a freight train is 15 mph slower than the speed of a passenger train. The freight train travels 360 mi in the same time that it takes the passenger train to travel 420 mi. Find the speed of each train.

39. Compute: $\dfrac{3^2 - 2^3}{2^2 + 3 - 12 \div 2}$.

Solve.

40. $3x^2 + 14x - 5 = 0$

41. $\dfrac{x+1}{x+3} = \dfrac{x+2}{x+4}$

SYNTHESIS

42. Find the area and the perimeter of a rectangle whose vertices are $(-3, 1)$, $(5, 1)$, $(5, 8)$, and $(-3, 8)$.

43. Find the slope–intercept equation of the line that contains the point $(-4, 1)$ and has the same slope as the line $2x - 3y = -6$.

Cumulative Review: Chapters 1–6

1. Find the absolute value: $|3.5|$.

2. Identify the coefficient of each term of the polynomial

$$x^3 - 2x^2 + x - 1.$$

3. Identify the degree of each term and the degree of the polynomial

$$x^3 - 2x^2 + x - 1.$$

4. Classify this polynomial as a monomial, binomial, trinomial, or none of these:

$$x^3 - 2x^2 + x - 1.$$

5. Collect like terms: $x^2 - 3x^3 - 4x^2 + 5x^3 - 2$.

Simplify.

6. $\frac{1}{2}x - \left[\frac{3}{8}x - \left(\frac{2}{3} + \frac{1}{4}x\right) - \frac{1}{3}\right]$

7. $\left(\frac{2x^3}{3x^{-1}}\right)^{-2}$

8. $\dfrac{\frac{4}{x} - \frac{6}{x^2}}{\frac{5}{x} + \frac{7}{2x}}$

Perform the indicated operations. Simplify, if possible.

9. $(5xy^2 - 6x^2y^2 - 3xy^3) - (-4xy^3 + 7xy^2 - 2x^2y^2)$

10. $(4x^4 + 6x^3 - 6x^2 - 4) + (2x^5 + 2x^4 - 4x^3 - 4x^2 + 3x - 5)$

11. $\frac{2y + 4}{21} \cdot \frac{7}{y^2 + 4y + 4}$

12. $\frac{x^2 - 9}{x^2 + 8x + 15} \div \frac{x - 3}{2x + 10}$

13. $\frac{x^2}{x - 4} + \frac{16}{4 - x}$

14. $\frac{5x}{x^2 - 4} - \frac{-3}{2 - x}$

Multiply.

15. $(2.5a + 7.5)(0.4a - 1.2)$

16. $(2x^2 - 1)(x^3 + x - 3)$

17. $(2x^3 + 1)(2x^3 - 1)$

18. $(6x - 5)^2$

19. $4x(3x^3 + 4x^2 + x)$

20. $(2x^5 + 3)(3x^2 - 6)$

Solve.

21. $x - [x - (x - 1)] = 2$

22. $2x^2 + 7x = 4$

23. $x^2 + x - 20 = 0$

24. $3(x - 2) \leq 4(x + 5)$

25. $x(x - 4) = 0$

26. $x^2 = 10x$

27. $2x^2 = 800$

28. $t = ax + ay$, for a

29. $\frac{5x - 2}{4} - \frac{4x - 5}{3} = 1$

30. $\frac{2x}{x - 3} - \frac{6}{x} = \frac{18}{x^2 - 3x}$

Factor.

31. $-6 - 2x - 12y$

32. $x^2 - 10x + 24$

33. $2x^2 - 18$

34. $m^4 + 2m^3 - 3m - 6$

35. $16x^2 + 40x + 25$

36. $8x^2 + 10x + 3$

Solve.

37. The product of a number and 1 more than the number is 20. Find the number.

38. A person's salary varies directly as the number of hours worked. For working 9 hr, the salary is \$117. Find the salary for working 6 hr.

39. Money is borrowed at 6% simple interest. After 1 yr, \$2650 pays off the loan. How much was originally borrowed?

40. One car travels 105 mi in the same time that a car traveling 10 mph slower travels 75 mi. Find the speed of each car.

41. If the sides of a square are increased by 2 ft, the sum of the areas of the two squares is 452 ft^2. Find the length of a side of the original square.

42. One number is 7 more than another number. The quotient of the larger divided by the smaller is $\frac{5}{4}$. Find the numbers.

Graph on a plane.

43. $y = \dfrac{1}{2}x$

44. $3x - 5y = 15$

45. $y = 1$

46. $y < -x - 2$

47. $x \leq -3$

48. Find an equation of variation where y varies directly as x and $y = 8$ when $x = 12$.

49. Find an equation of variation where y varies inversely as x and $y = 20$ when $x = 0.5$.

Find the slope, if it exists, of the line containing the given pair of points.

50. $(-2, 6)$ and $(-2, -1)$

51. $(-4, 1)$ and $(3, -2)$

52. Find the slope and the y-intercept of $4x - 3y = 6$.

53. Find an equation for the line containing the point $(2, -3)$ and having slope $m = -4$.

54. Find an equation of the line containing the points $(-1, -3)$ and $(5, -2)$.

Determine whether the graphs of the equations are parallel, perpendicular, or neither.

55. $2x = 7 - 3y,$
$7 + 2x = 3y$

56. $x - y = 4,$
$y = x + 5$

SYNTHESIS

57. Compute: $(x + 7)(x - 4) - (x + 8)(x - 5)$.

58. Multiply: $[4y^3 - (y^2 - 3)][4y^3 + (y^2 - 3)]$.

59. Factor: $2a^{32} - 13{,}122b^{40}$.

60. Solve: $(x - 4)(x + 7)(x - 12) = 0$.

61. Find an equation of the line that contains the point $(-3, -2)$ and is parallel to the line $2x - 3y = -12$.

62. Find all numbers for which the complex rational expression is undefined.

$$\frac{\dfrac{1}{x} + x}{2 + \dfrac{1}{x - 3}}$$

7

Systems of Equations

INTRODUCTION

We now consider how two graphs of linear equations might intersect. Such a point of intersection is a solution of what is called a *system of equations.* Many problems involve two facts about two quantities and are easier to solve by translating to a system of two equations in two variables. Systems of equations have extensive applications in many fields such as sociology, psychology, business, education, engineering, and science.

AN APPLICATION

The state of Wyoming is roughly in the shape of a rectangle whose perimeter is 1280 mi. The length is 90 mi more than the width. Find the length and the width.

THE MATHEMATICS

We let l = the length and w = the width. The problem then translates to this pair of equations:

$$\left.\begin{aligned} 2l + 2w &= 1280, \\ l &= 90 + w. \end{aligned}\right\}$$

This is a *system of equations.*

Objectives for Review

The review objectives to be tested in addition to the material in this chapter are as follows.

[3.1d, e] **Use the product and the quotient rules to multiply and divide exponential expressions with like bases.**

[5.1c] **Simplify rational expressions.**

[5.5a] **Subtract rational expressions.**

[6.3a] **Find the intercepts of a linear equation, and graph using intercepts.**

Pretest: Chapter 7

1. Determine whether the ordered pair $(-1, 1)$ is a solution of the system of equations

$$2x + y = -1,$$
$$3x - 2y = -5.$$

2. Solve this system by graphing.

$$2x = y + 1,$$
$$2x - y = 5$$

Solve by the substitution method.

3. $x + y = 7,$
$x = 2y + 1$

4. $2x - 3y = 7,$
$x + y = 1$

Solve by the elimination method.

5. $2x - y = 1,$
$2x + y = 2$

6. $2x - 3y = -4,$
$3x - 4y = -7$

7. $\frac{3}{5}x - \frac{1}{4}y = 4,$
$\frac{1}{5}x + \frac{3}{4}y = 8$

8. Find two numbers whose sum is 74 and whose difference is 26.

9. Two angles are complementary. One angle is 15° more than twice the other. Find the angles. (Complementary angles are angles whose sum is 90°.)

10. A train leaves a station and travels north at 96 mph. Two hours later, a second train leaves on a parallel track and travels north at 120 mph. When will it overtake the first train?

7.1 *Systems of Equations in Two Variables*

OBJECTIVES

After finishing Section 7.1, you should be able to:

a Determine whether an ordered pair is a solution of a system of equations.

b Solve systems of two linear equations in two variables by graphing.

FOR EXTRA HELP

TAPE 17 | TAPE 14A | MAC: 7 IBM: 7

a SYSTEMS OF EQUATIONS AND SOLUTIONS

Many problems can be solved more easily by translating to two equations in two variables. The following is such a **system of equations:**

$$x + y = 8,$$
$$2x - y = 1.$$

> A *solution* of a system of two equations is an ordered pair that makes both equations true.

Consider the system shown above. Look at the graph below. **Recall that a graph of an equation is a drawing that represents its solution set. Each point on the graph corresponds to a solution of that equation.** Which points (ordered pairs) are solutions of *both* equations?

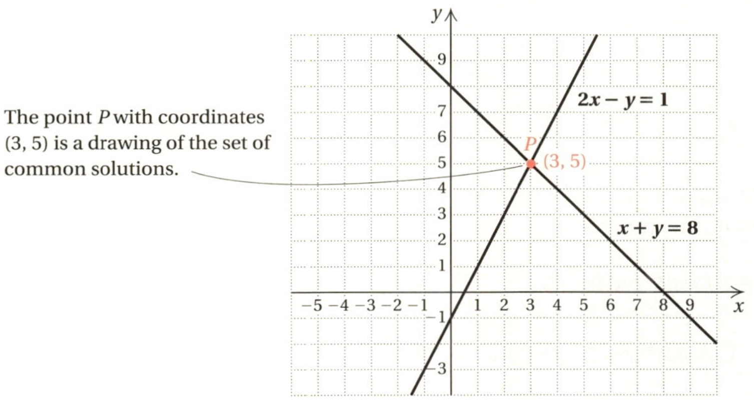

The graph shows that there is only one. It is the point P where the graphs cross. This point looks as if its coordinates are (3, 5). We check to see if (3, 5) is a solution of *both* equations, substituting 3 for x and 5 for y.

CHECK:

$$\begin{array}{c|c} x + y & 8 \\ \hline 3 + 5 & 8 \\ 8 & \end{array} \quad \text{TRUE}$$

$$\begin{array}{c|c} 2x - y & 1 \\ \hline 2 \cdot 3 - 5 & 1 \\ 6 - 5 & \\ 1 & \end{array} \quad \text{TRUE}$$

There is just one solution of the system of equations. It is (3, 5). In other words, $x = 3$ and $y = 5$.

Determine whether the given ordered pair is a solution of the system of equations.

1. (2, −3); $x = 2y + 8$, $2x + y = 1$

2. (20, 40); $a = \frac{1}{2}b$, $b - a = 60$

Answers on page A-6

EXAMPLE 1 Determine whether (1, 2) is a solution of the system

$$y = x + 1,$$
$$2x + y = 4.$$

We check by substituting alphabetically 1 for x and 2 for y.

CHECK:

$$\begin{array}{c|c} y & x + 1 \\ \hline 2 & 1 + 1 \\ & 2 \end{array} \quad \text{TRUE} \qquad \begin{array}{c|c} 2x + y & 4 \\ \hline 2 \cdot 1 + 2 & 4 \\ 2 + 2 & \\ 4 & \end{array} \quad \text{TRUE}$$

This checks, so (1, 2) is a solution of the system.

EXAMPLE 2 Determine whether (−3, 2) is a solution of the system

$$p + q = -1,$$
$$q + 3p = 4.$$

We check by substituting alphabetically −3 for p and 2 for q.

CHECK:

$$\begin{array}{c|c} p + q & -1 \\ \hline -3 + 2 & -1 \\ -1 & \end{array} \quad \text{TRUE} \qquad \begin{array}{c|c} q + 3p & 4 \\ \hline 2 + 3(-3) & 4 \\ 2 - 9 & \\ -7 & \end{array} \quad \text{FALSE}$$

The point (−3, 2) is not a solution of $q + 3p = 4$. Thus it is not a solution of the system.

Example 2 illustrates that an ordered pair may be a solution of one equation but not *both*. If that is the case, it is *not* a solution of the system.

Do Exercises 1 and 2.

b GRAPHING SYSTEMS OF EQUATIONS

Recall that the **graph** of an equation is a drawing that represents its solution set. If the graph of an equation is a line, then every point on the line corresponds to an ordered pair that is a solution of the equation. If we graph a **system** of two linear equations, we graph both equations and find the coordinates of the points of intersection, if any exist.

EXAMPLE 3 Solve this system of equations by graphing:

$$x + y = 6,$$
$$x = y + 2.$$

We graph the equations using any of the methods studied in Chapter 6. Point P with coordinates (4, 2) looks as if it is the solution.

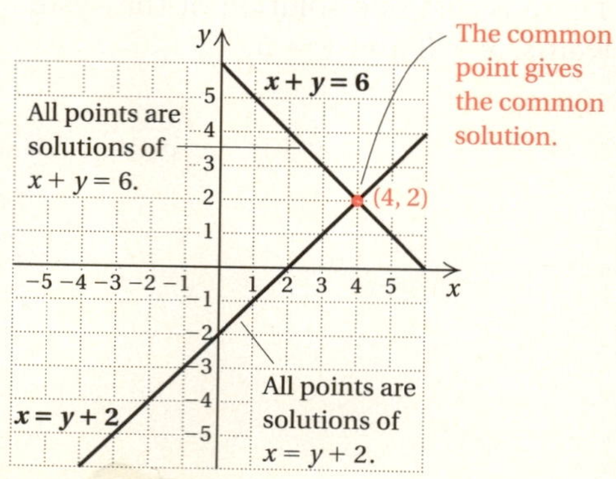

We check the pair as follows.

CHECK:

$x + y = 6$		
$4 + 2$	6	
6		TRUE

$x = y + 2$		
4	$2 + 2$	
	4	TRUE

The solution is (4, 2).

Graphing is not perfectly accurate, so solving by graphing may give only approximate solutions.

Do Exercise 3.

Sometimes the equations in a system have graphs that are parallel lines.

EXAMPLE 4 Solve this system of equations by graphing:

$$y = 3x + 4,$$
$$y = 3x - 3.$$

We graph the equations, again using any of the methods studied in Chapter 6. The lines have the same slope, 3, and different y-intercepts, (0, 4) and (0, −3), so they are parallel.

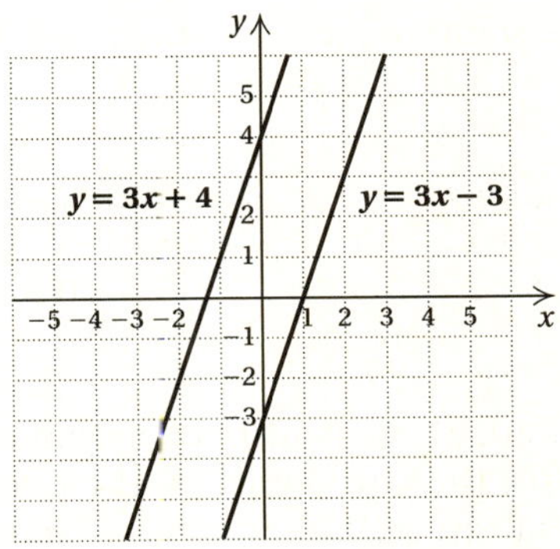

There is no point at which they cross, so the system has no solution. The solution set is the empty set, denoted ∅, or { }.

Do Exercise 4.

3. Solve this system by graphing:

$$2x + y = 1,$$
$$x = 2y + 8.$$

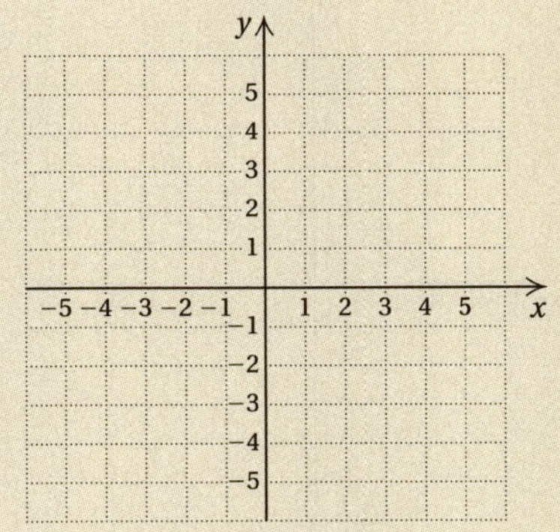

4. Solve this system by graphing:

$$y + 4 = x,$$
$$x - y = -2.$$

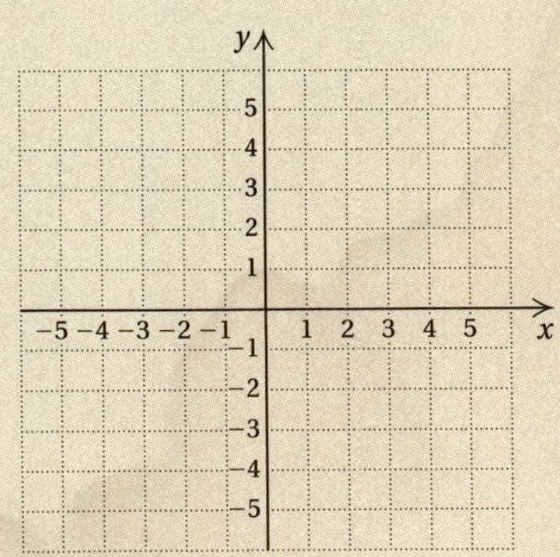

Answers on page A-6

5. Solve this system by graphing:

$$2x + y = 4,$$
$$-6x - 3y = -12.$$

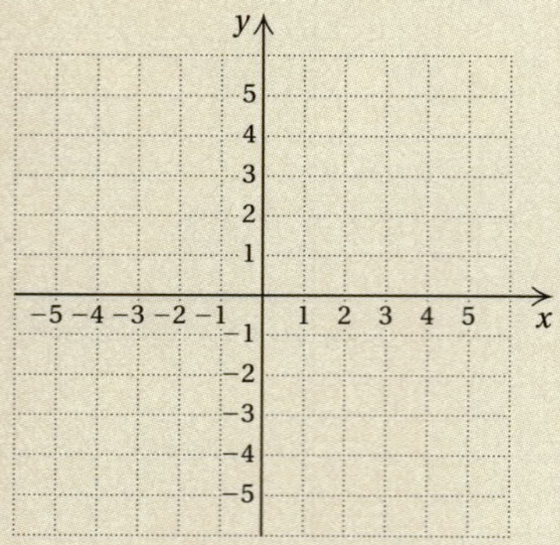

Answer on page A-6

Sometimes the equations in a system have the same graph.

EXAMPLE 5 Solve this system by graphing:

$$2x + 3y = 6,$$
$$-8x - 12y = -24.$$

We graph the equations and see that the graphs are the same. Thus any solution of one of the equations is a solution of the other. Each equation has an infinite number of solutions, some of which are indicated on the graph.

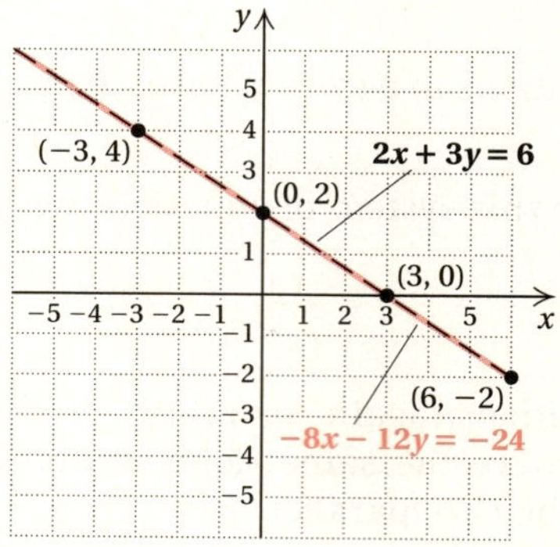

We check one such solution, (0, 2), which is the y-intercept of each equation.

CHECK:

$2x + 3y$	$= 6$	
$2(0) + 3(2)$	6	
$0 + 6$		
6		TRUE

$-8x - 12y$	$= -24$	
$-8(0) - 12(2)$	-24	
$0 - 24$		
-24		TRUE

On your own, check that (−3, 4) is also a solution of the system. If (0, 2) and (−3, 4) are solutions, then all points on the line containing them are solutions. The system has an infinite number of solutions.

Do Exercise 5.

When we graph a system of two equations in two variables, one of the following three things can happen.

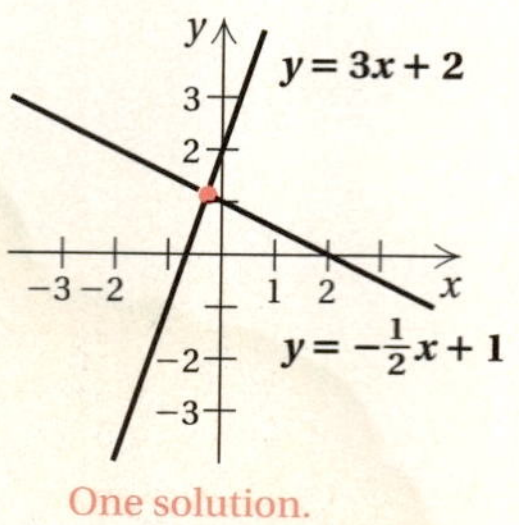

One solution.
Graphs intersect.

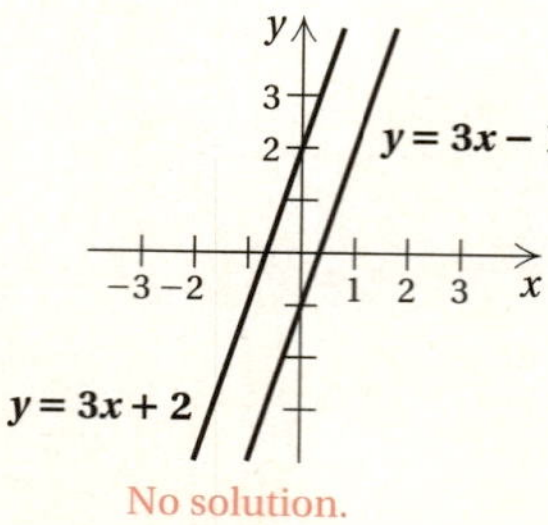

No solution.
Graphs are parallel.

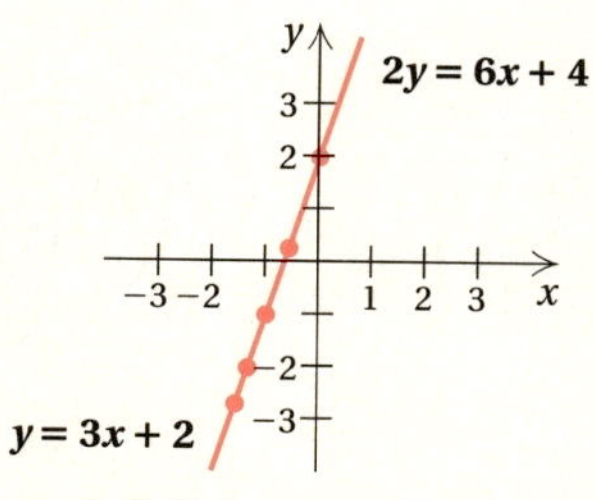

Infinitely many solutions.
Equations have the same graph.

NAME SECTION DATE

Exercise Set 7.1

a Determine whether the given ordered pair is a solution of the system of equations. Use alphabetical order of the variables.

1. (1, 5); $5x - 2y = -5,$
$3x - 7y = -32$

2. (3, 2); $2x + 3y = 12,$
$x - 4y = -5$

3. (4, 2); $3b - 2a = -2,$
$b + 2a = 8$

4. (6, −6); $t + 2s = 6,$
$t - s = -12$

5. (15, 20); $3x - 2y = 5,$
$6x - 5y = -10$

6. (−1, −5); $4r + s = -9,$
$3r = 2 + s$

7. (−1, 1); $x = -1,$
$x - y = -2$

8. (−3, 4); $2x = -y - 2,$
$y = -4$

9. (18, 3); $y = \frac{1}{6}x,$
$2x - y = 33$

10. (−3, 1); $y = -\frac{1}{3}x,$
$3y = -5x - 12$

b Solve the system of equations by graphing.

11. $x - y = 2,$
$x + y = 6$

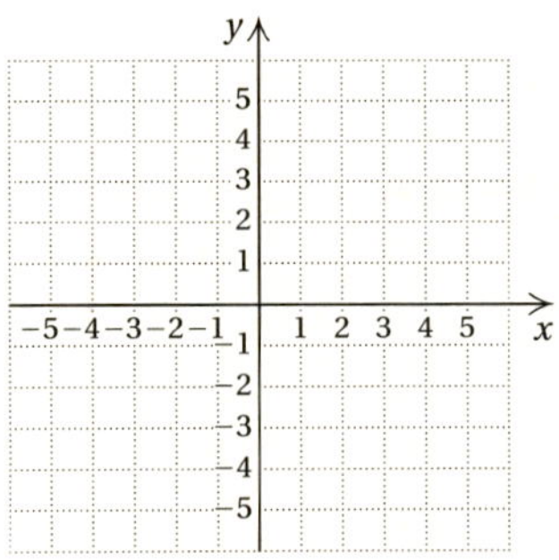

12. $x + y = 3,$
$x - y = 1$

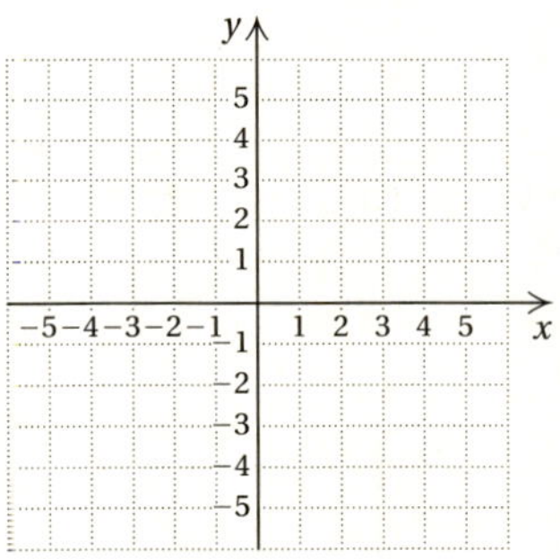

13. $8x - y = 29,$
$2x + y = 11$

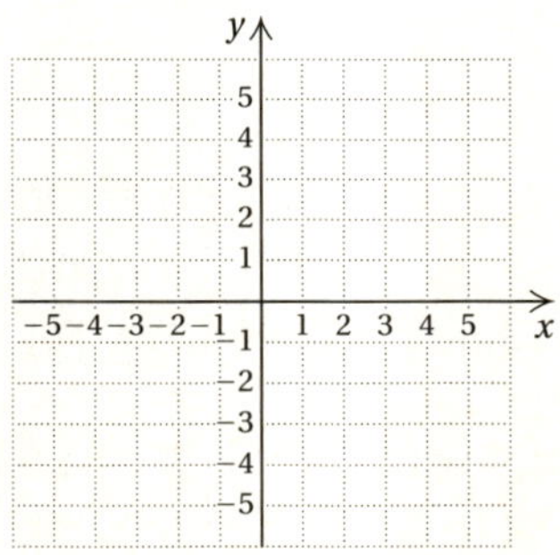

14. $4x - y = 10,$
$3x + 5y = 19$

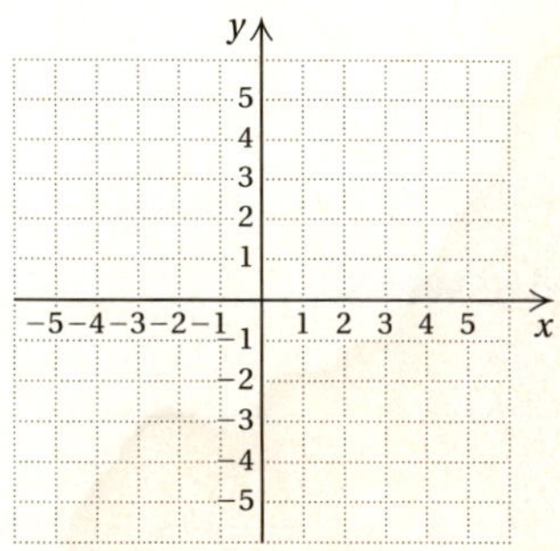

15. $u = v,$
$4u = 2v - 6$

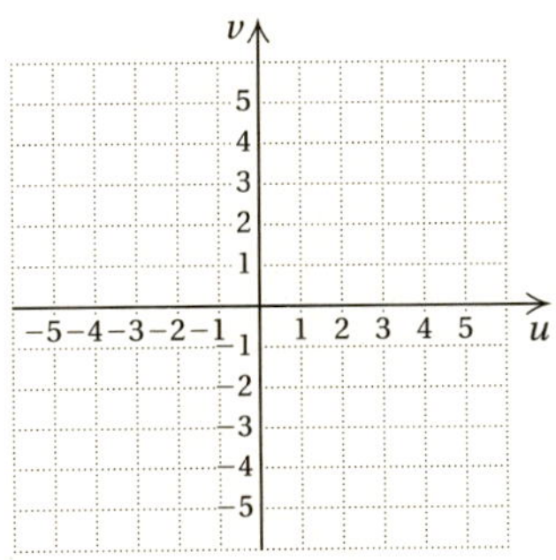

16. $x = 3y,$
$3y - 6 = 2x$

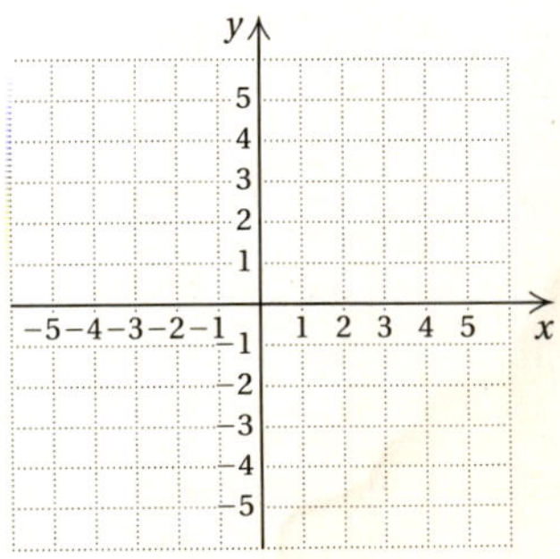

17. $x = -y,$
$x + y = 4$

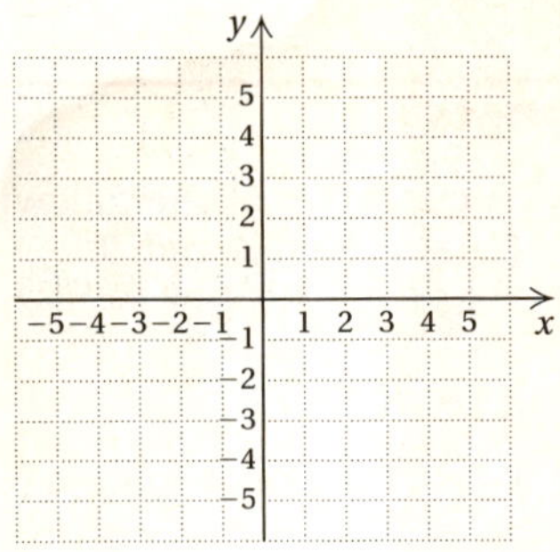

18. $-3x = 5 - y,$
$2y = 6x + 10$

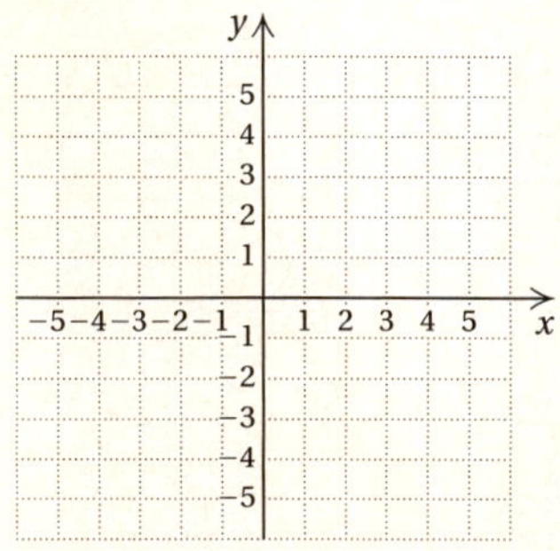

ANSWERS

1. ______
2. ______
3. ______
4. ______
5. ______
6. ______
7. ______
8. ______
9. ______
10. ______

19. $a = \frac{1}{2}b + 1,$
$a - 2b = -2$

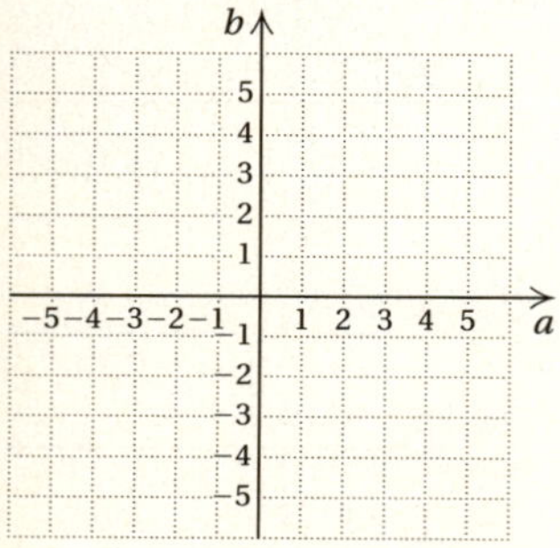

20. $x = \frac{1}{3}y + 2,$
$-2x - y = 1$

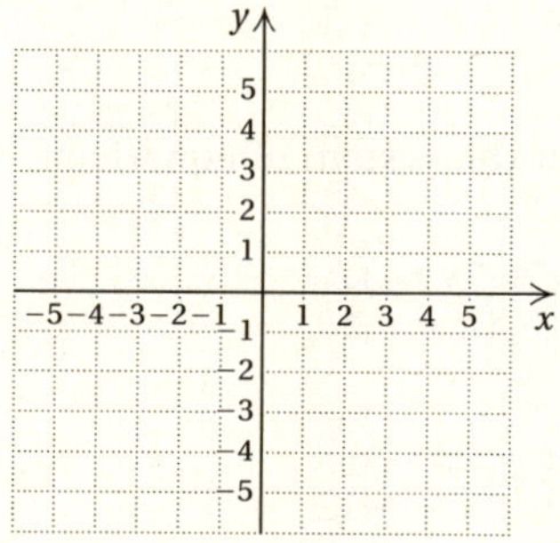

21. $y - 2x = 0,$
$y = 6x - 2$

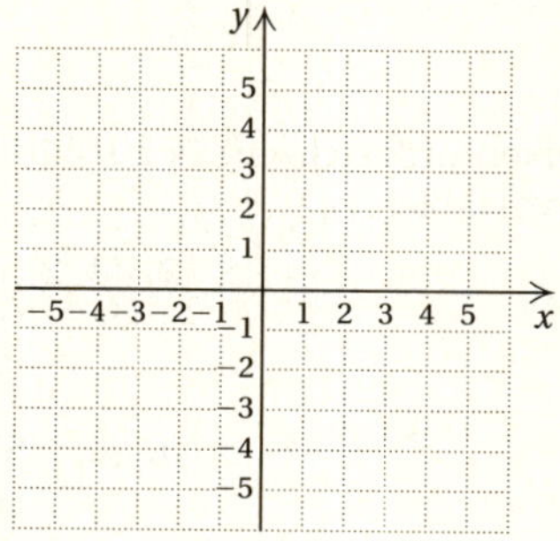

22. $y = 3x,$
$y = -3x + 2$

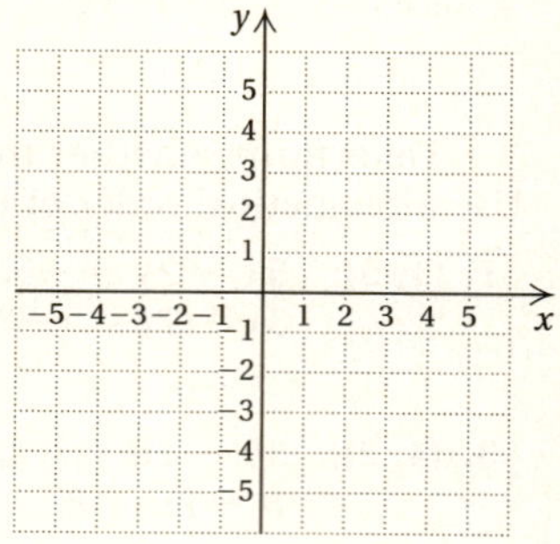

23. $x + y = 9,$
$3x + 3y = 27$

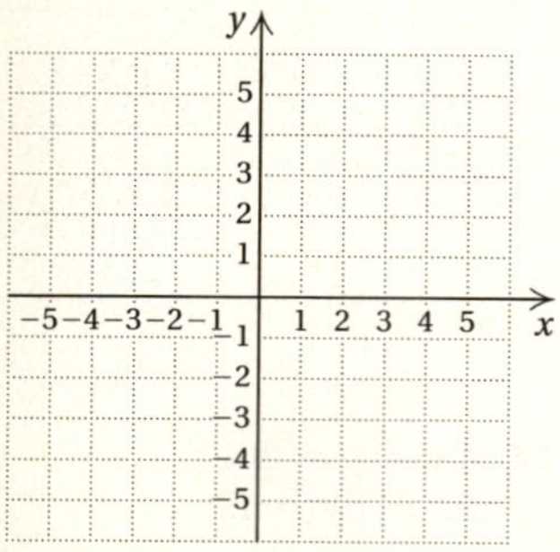

24. $x + y = 4,$
$x + y = -4$

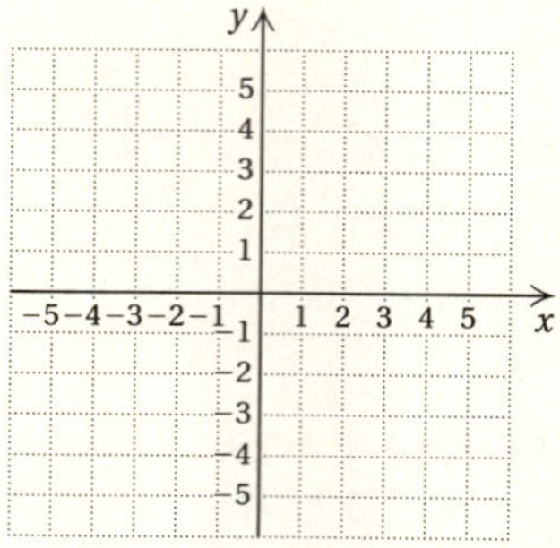

25. $x = 5,$
$y = -3$

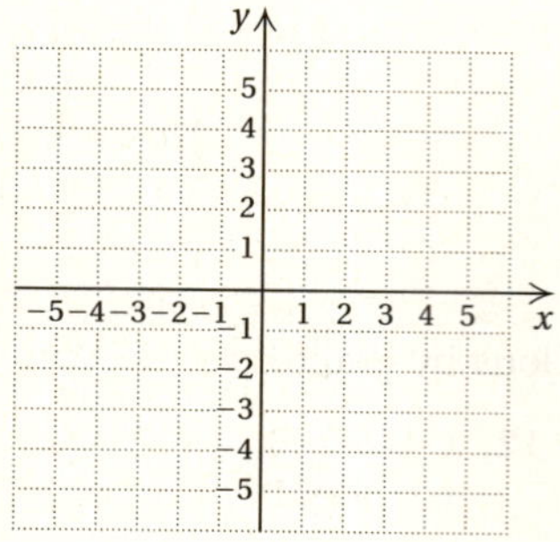

26. $y = 2,$
$x = -4$

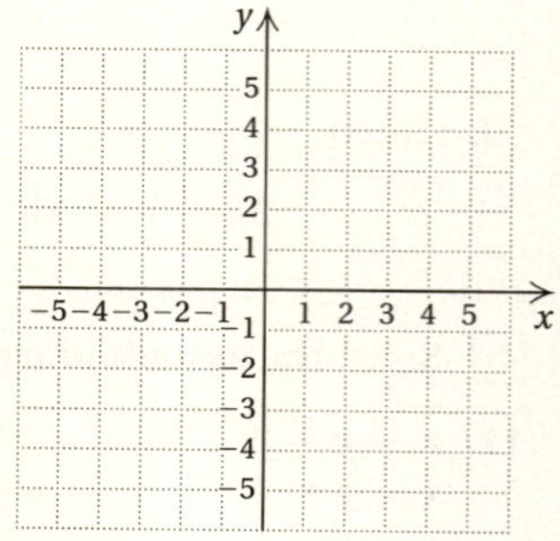

SKILL MAINTENANCE

27. Multiply: $(9x^{-5})(12x^{-8})$.

28. Divide: $\frac{9x^{-5}}{3x^{-8}}$.

Simplify.

29. $\frac{1}{x} - \frac{1}{x^2} + \frac{1}{x+1}$

30. $\frac{3-x}{x-2} - \frac{x-7}{2-x}$

31. $\frac{x+2}{x-4} - \frac{x+1}{x+4}$

32. $\frac{2x^2 - x - 15}{x^2 - 9}$

SYNTHESIS

33. The solution of the following system is $(2, -3)$. Find A and B.

$$Ax - 3y = 13,$$
$$x - By = 8$$

34. Graph this system. What happens when you check your possible solution?

$$x - 2y = 6,$$
$$3x + 2y = 4$$

35. Find a system of equations with $(6, -2)$ as a solution. Answers may vary.

36. Find an equation to go with $5x + 2y = 11$ so that the solution of the system is $(3, -2)$. Answers may vary.

ANSWERS

27. ______

28. ______

29. ______

30. ______

31. ______

32. ______

33. ______

34. ______

35. ______

36. ______

7.2 The Substitution Method

Consider the following system of equations:

$$3x + 7y = 5,$$
$$6x - 7y = 1.$$

Suppose we try to solve this system graphically. We obtain this graph.

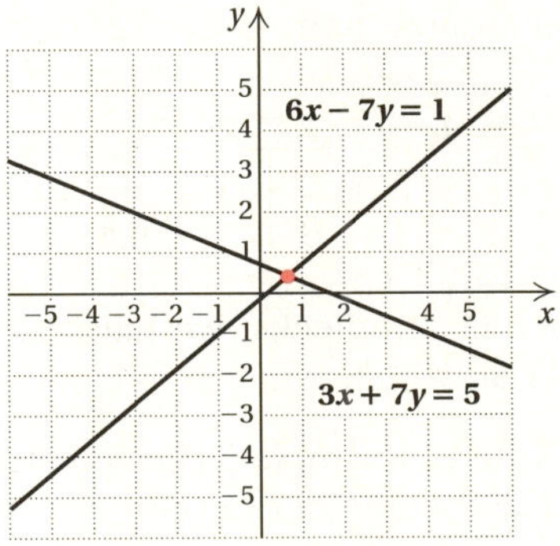

What is the solution? It is rather difficult to tell exactly. It would appear that fractions are involved. It turns out that the solution is $\left(\frac{2}{3}, \frac{3}{7}\right)$. We need techniques involving algebra to determine the solution exactly. Graphing helps us picture the solution of a system of equations, but solving by graphing, though practical in many applied situations, is not always fast or accurate in cases where solutions are not integers. We now learn other methods using algebra. Because they use algebra, they are called **algebraic**.

OBJECTIVES

After finishing Section 7.2, you should be able to:

a Solve a system of two equations in two variables by the substitution method when one of the equations has a variable alone on one side.

b Solve a system of two equations in two variables by the substitution method when neither equation has a variable alone on one side.

c Solve problems by translating to a system of two equations and then solving using the substitution method.

FOR EXTRA HELP

TAPE 17 | TAPE 14B | MAC: 7 IBM: 7

a SOLVING BY THE SUBSTITUTION METHOD

One nongraphical method for solving systems is known as the **substitution method.** In Example 1, we use the substitution method to solve a system we graphed in Example 3 of Section 7.1.

EXAMPLE 1 Solve the system

$$x + y = 6, \quad (1)$$
$$x = y + 2. \quad (2)$$

Equation (2) says that x and $y + 2$ name the same thing. Thus in Equation (1), we can substitute $y + 2$ for x:

$$x + y = 6 \quad \text{Equation (1)}$$
$$(y + 2) + y = 6. \quad \text{Substituting } y + 2 \text{ for } x$$

This last equation has only one variable. We solve it:

$$y + 2 + y = 6 \quad \text{Removing parentheses}$$
$$2y + 2 = 6 \quad \text{Collecting like terms}$$
$$2y + 2 - 2 = 6 - 2 \quad \text{Subtracting 2 on both sides}$$
$$2y = 4 \quad \text{Simplifying}$$
$$\frac{2y}{2} = \frac{4}{2} \quad \text{Dividing by 2}$$
$$y = 2. \quad \text{Simplifying}$$

1. Solve by the substitution method. Do not graph.

$$x + y = 5,$$
$$x = y + 1$$

2. Solve by the substitution method:

$$a - b = 4,$$
$$b = 2 - a.$$

Answers on page A-6

We have found the y-value of the solution. To find the x-value, we return to the original pair of equations. Substituting into either equation will give us the x-value. We choose Equation (2) because it has x alone on one side:

$$\begin{aligned} x &= y + 2 && \text{Equation (2)} \\ &= 2 + 2 && \text{Substituting 2 for } y \\ &= 4. \end{aligned}$$

The ordered pair (4, 2) may be a solution. We check.

CHECK:

$x + y = 6$		
$4 + 2$	6	
6		TRUE

$x = y + 2$		
4	$2 + 2$	
	4	TRUE

Since (4, 2) checks, we have the solution. We could also express the answer as $x = 4$, $y = 2$.

Note in Example 1 that substituting 2 for y in Equation (1) will also give us the x-value of the solution:

$$\begin{aligned} x + y &= 6 \\ x + 2 &= 6 \\ x &= 4. \end{aligned}$$

Note also that we are using alphabetical order in listing the coordinates in an ordered pair. That is, since x precedes y, we have 4 before 2 in the pair (4, 2).

Do Exercise 1.

EXAMPLE 2 Solve the system

$$s = 13 - 3t, \quad (1)$$
$$s + t = 5. \quad (2)$$

We substitute $13 - 3t$ for s in Equation (2):

$$\begin{aligned} s + t &= 5 && \text{Equation (2)} \\ (13 - 3t) + t &= 5. && \text{Substituting } 13 - 3t \text{ for } s \end{aligned}$$

Remember to use parentheses when you substitute. Then remove them carefully.

Now we solve for t:

$$\begin{aligned} 13 - 2t &= 5 && \text{Collecting like terms} \\ -2t &= -8 && \text{Subtracting 13} \\ t &= \frac{-8}{-2}, \text{ or } 4. && \text{Dividing by } -2 \end{aligned}$$

Next we substitute 4 for t in Equation (1) of the original system:

$$\begin{aligned} s &= 13 - 3t && \text{Equation (1)} \\ &= 13 - 3 \cdot 4 && \text{Substituting 4 for } t \\ &= 1. \end{aligned}$$

The pair (1, 4) checks and is the solution. Remember: We list the answer in alphabetical order (s, t)—that is, 1 first and 4 second.

Do Exercise 2.

b SOLVING FOR THE VARIABLE FIRST

Sometimes neither equation of a pair has a variable alone on one side. Then we solve one equation for one of the variables and proceed as before, substituting into the *other* equation. If possible, we solve in either equation for a variable that has a coefficient of 1.

EXAMPLE 3 Solve the system

$$x - 2y = 6, \quad (1)$$
$$3x + 2y = 4. \quad (2)$$

We solve one equation for one variable. Since the coefficient of x is 1 in Equation (1), it is easier to solve that equation for x:

$$x - 2y = 6 \quad \text{Equation (1)}$$
$$x = 6 + 2y. \quad \text{Adding } 2y \quad (3)$$

We substitute $6 + 2y$ for x in Equation (2) of the original pair and solve for y:

$$3x + 2y = 4 \quad \text{Equation (2)}$$
$$3(6 + 2y) + 2y = 4 \quad \text{Substituting } 6 + 2y \text{ for } x$$
$$18 + 6y + 2y = 4 \quad \text{Removing parentheses}$$
$$18 + 8y = 4 \quad \text{Collecting like terms}$$
$$8y = -14 \quad \text{Subtracting 18}$$
$$y = \frac{-14}{8} \text{ or } -\frac{7}{4}. \quad \text{Dividing by 8}$$

To find x, we go back to either of the original Equations (1) or (2) or to Equation (3), which we solved for x. It is generally easier to use an equation like Equation (3) where we have solved for a specific variable. We substitute $-\frac{7}{4}$ for y in Equation (3) and compute x:

$$x = 6 + 2y \quad \text{Equation (3)}$$
$$= 6 + 2\left(-\frac{7}{4}\right) \quad \text{Substituting } -\frac{7}{4} \text{ for } y$$
$$= 6 - \frac{7}{2} = \frac{5}{2}.$$

We check the ordered pair $\left(\frac{5}{2}, -\frac{7}{4}\right)$.

CHECK:

$x - 2y = 6$	
$\frac{5}{2} - 2\left(-\frac{7}{4}\right)$	6
$\frac{5}{2} + \frac{7}{2}$	
$\frac{12}{2}$	
6	TRUE

$3x + 2y = 4$	
$3 \cdot \frac{5}{2} + 2\left(-\frac{7}{4}\right)$	4
$\frac{15}{2} - \frac{7}{2}$	
$\frac{8}{2}$	
4	TRUE

Since $\left(\frac{5}{2}, -\frac{7}{4}\right)$ checks, it is the solution.

This solution would have been difficult to find graphically because it involves fractions.

Do Exercise 3.

3. Solve:

$$x - 2y = 8,$$
$$2x + y = 8.$$

Answer on page A-6

4. The perimeter of a rectangle is 76 cm. The length is 17 cm more than the width. Find the length and the width.

C Solving Problems

Now let us use the substitution method to solve a problem.

EXAMPLE 4 The state of Wyoming is roughly in the shape of a rectangle whose perimeter is 1280 mi. The length is 90 mi more than the width. Find the length and the width.

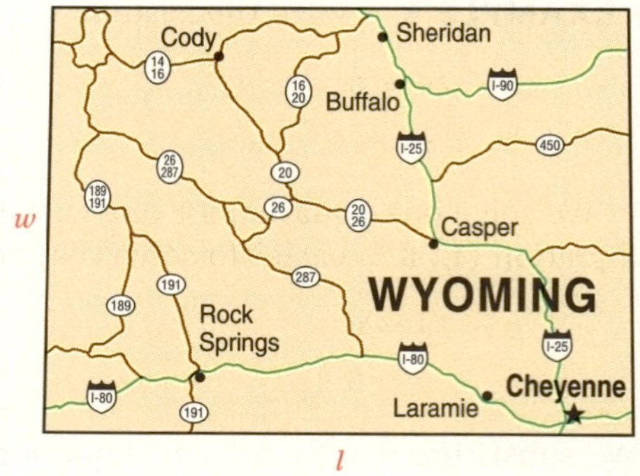

1. ***Familiarize.*** We make a drawing and label it, calling the length l and the width w.

2. ***Translate.*** The perimeter of the rectangle is $2l + 2w$. We translate the first statement.

The perimeter is 1280 mi.

$$2l + 2w = 1280$$

We translate the second statement.

The length is 90 mi greater than the width.

$$l = 90 + w$$

We have translated to a system of equations:

$$2l + 2w = 1280, \quad (1)$$
$$l = 90 + w. \quad (2)$$

3. ***Solve.*** We solve the system. We substitute $90 + w$ for l in the first equation and solve:

$$2(90 + w) + 2w = 1280 \quad \text{Substituting } 90 + w \text{ for } l \text{ in Equation (1)}$$
$$180 + 2w + 2w = 1280 \quad \text{Removing parentheses}$$
$$180 + 4w = 1280 \quad \text{Collecting like terms}$$
$$4w = 1100 \quad \text{Subtracting 180}$$
$$w = \frac{1100}{4}, \text{ or } 275. \quad \text{Dividing by 4}$$

We go back to the original equations and substitute 275 for w:

$$l = 90 + w \quad \text{Equation (2)}$$
$$= 90 + 275 \quad \text{Substituting 275 for } w$$
$$= 365.$$

4. ***Check.*** A possible solution is a length of 365 mi and a width of 275 mi. The perimeter would then be $2(365) + 2(275)$, or $730 + 550$, or 1280. Also, the length is 90 mi greater than the width. These check.

5. ***State.*** The length is 365 mi, and the width is 275 mi.

This problem illustrates that many problems that can be solved by translating to *one* equation in *one* variable are actually easier to solve by translating to *two* equations in *two* variables.

Do Exercise 4.

Answer on page A-6

NAME SECTION DATE

Exercise Set 7.2

a Solve by the substitution method.

1. $x + y = 10,$
   $y = x + 8$

2. $x + y = 4,$
   $y = 2x + 1$

3. $y = x - 6,$
   $x + y = -2$

4. $y = x + 1,$
   $2x + y = 4$

5. $y = 2x - 5,$
   $3y - x = 5$

6. $y = 2x + 1,$
   $x + y = -2$

7. $x = -2y,$
   $x + 4y = 2$

8. $r = -3s,$
   $r + 4s = 10$

b Solve by the substitution method. First, solve one equation for one variable.

9. $x - y = 6,$
   $x + y = -2$

10. $s + t = -4,$
    $s - t = 2$

11. $y - 2x = -6,$
    $2y - x = 5$

12. $x - y = 5,$
    $x + 2y = 7$

13. $2x + 3y = -2,$
    $2x - y = 9$

14. $x + 2y = 10,$
    $3x + 4y = 8$

15. $x - y = -3,$
    $2x + 3y = -6$

16. $3b + 2a = 2,$
    $-2b + a = 8$

17. $r - 2s = 0,$
    $4r - 3s = 15$

18. $y - 2x = 0,$
    $3x + 7y = 17$

c Solve.

19. The sum of two numbers is 37. One number is 5 more than the other. Find the numbers.

20. The sum of two numbers is 26. One number is 12 more than the other. Find the numbers.

21. Find two numbers whose sum is 52 and whose difference is 28.

22. Find two numbers whose sum is 63 and whose difference is 5.

ANSWERS

1. ______
2. ______
3. ______
4. ______
5. ______
6. ______
7. ______
8. ______
9. ______
10. ______
11. ______
12. ______
13. ______
14. ______
15. ______
16. ______
17. ______
18. ______
19. ______
20. ______
21. ______
22. ______

ANSWERS

23. ______

24. ______

25. ______

26. ______

27. ______

28. ______

29. See graph.

30. See graph.

31. See graph.

32. See graph.

33. See graph.

34. See graph.

35. ______

36. ______

37. ______

38. ______

39. ______

40. ______

41. ______

23. The difference between two numbers is 12. Two times the larger is five times the smaller. What are the numbers?

24. The difference between two numbers is 18. Twice the smaller number plus three times the larger is 74. What are the numbers?

25. The state of Colorado is roughly in the shape of a rectangle whose perimeter is 1300 mi. The width is 110 mi less than the length. Find the length and the width.

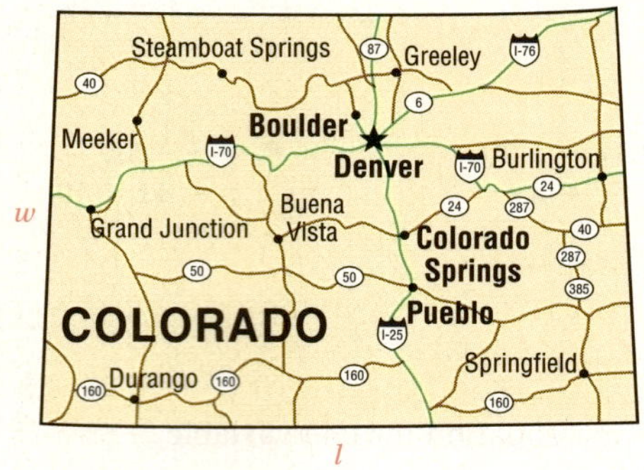

26. The perimeter of a standard-sized rectangular rug is 42 ft. The length is 3 ft more than the width. Find the length and the width.

27. The perimeter of a rectangular rose garden is 400 m. The length is 3 m more than twice the width. Find the length and the width.

28. The perimeter of a rectangular advertisement is 876 cm. The length is 1 cm less than three times the width. Find the length and the width.

SKILL MAINTENANCE

Graph.

29. $2x - 3y = 6$

30. $2x + 3y = 6$

31. $2x - 3 = 0$

32. $y = 2x - 5$

33. $x = -3$

34. $y = 4$

SYNTHESIS

Solve by the substitution method.

35. ▦ $y - 2.35x = -5.97,$
$2.14y - x = 4.88$

36. $\frac{1}{4}(a - b) = 2,$
$\frac{1}{6}(a + b) = 1$

37. $\frac{x}{2} + \frac{3y}{2} = 2,$
$\frac{x}{5} - \frac{y}{2} = 3$

38. The perimeter of a football field (excluding the end zones) is $306\frac{2}{3}$ yd. The length is $46\frac{2}{3}$ yd longer than the width. Find the length and the width.

39. The perimeter of an isosceles triangle is 324 cm. The base is 33 cm longer than each of the other two sides. Find the lengths of the sides of the triangle.

40. Consider this system of equations:

$3y + 3x = 14,$
$y = -x + 4.$

Try to solve by the substitution method. Can you explain your results?

41. Consider this system of equations:

$y = x + 5,$
$-3x + 3y = 15.$

Try to solve by the substitution method. Can you explain your results?

7.3 *The Elimination Method*

a SOLVING BY THE ELIMINATION METHOD

The **elimination method** for solving systems of equations makes use of the *addition principle.* Some systems are much easier to solve using this method. Trying to solve the system in Example 1 by substitution would necessitate the use of fractions and extra steps. Instead we use the elimination method.

EXAMPLE 1 Solve the system

$$2x + 3y = 13, \quad (1)$$
$$4x - 3y = 17. \quad (2)$$

The key to the advantage of the elimination method for solving this system involves the $3y$ in one equation and the $-3y$ in the other. The terms are opposites. If we add the terms on the sides of the equations, these terms will add to 0, and in effect, the variable y will be eliminated.

We will use the addition principle for equations. According to Equation (2), $4x - 3y$ and 17 are the same number. Thus we can use a vertical form and add $4x - 3y$ to the left side of Equation (1) and 17 to the right side—in effect, adding the same number on both sides of Equation (1):

$$\begin{aligned} 2x + 3y &= 13 && (1)\\ 4x - 3y &= 17 && (2)\\ \hline 6x + 0y &= 30. && \text{Adding} \end{aligned}$$

We have "eliminated" one variable. This is why we call this the **elimination method.** We now have an equation with just one variable that can be solved for x:

$$\begin{aligned} 6x &= 30\\ x &= 5. \end{aligned}$$

Next we substitute 5 for x in either of the original equations:

$$\begin{aligned} 2x + 3y &= 13 && \text{Equation (1)}\\ 2(5) + 3y &= 13 && \text{Substituting 5 for } x\\ 10 + 3y &= 13\\ 3y &= 3\\ y &= 1. && \text{Solving for } y \end{aligned}$$

We check the ordered pair (5, 1).

CHECK:

$2x + 3y = 13$	
$2(5) + 3(1)$	13
$10 + 3$	
13	TRUE

$4x - 3y = 17$	
$4(5) - 3(1)$	17
$20 - 3$	
17	TRUE

Since (5, 1) checks, it is the solution.

Do Exercises 1 and 2.

OBJECTIVES

After finishing Section 7.3, you should be able to:

a Solve a system of two equations in two variables using the elimination method when no multiplication is necessary.

b Solve a system of two equations in two variables using the elimination method when multiplication is necessary.

c Solve problems by translating to a system of two equations and then solving using the elimination method.

FOR EXTRA HELP

TAPE 17 TAPE 14B MAC: 7 IBM: 7

Solve using the elimination method.

1. $x + y = 5,$
$2x - y = 4$

2. $-2x + y = -4,$
$2x - 5y = 12$

Answers on page A-6

3. Solve. Multiply one equation by -1 first.

$$5x + 3y = 17,$$
$$5x - 2y = -3$$

4. Solve the system

$$3x - 2y = -30,$$
$$5x - 2y = -46.$$

Answers on page A-6

b USING THE MULTIPLICATION PRINCIPLE FIRST

The elimination method allows us to eliminate a variable. We may need to multiply by certain numbers first, however, so that terms become opposites.

EXAMPLE 2 Solve the system

$$2x + 3y = 8, \quad (1)$$
$$x + 3y = 7. \quad (2)$$

If we add, we will not eliminate a variable. However, if the $3y$ were $-3y$ in one equation, we could eliminate y. Thus we multiply on both sides of Equation (2) by -1 and then add, using a vertical form:

$$\begin{aligned} 2x + 3y &= 8 && \text{Equation (1)} \\ -x - 3y &= -7 && \text{Multiplying Equation (2) by } -1 \\ \hline x \quad &= 1. && \text{Adding} \end{aligned}$$

Now we substitute 1 for x in one of the original equations:

$$\begin{aligned} x + 3y &= 7 && \text{Equation (2)} \\ 1 + 3y &= 7 && \text{Substituting 1 for } x \\ 3y &= 6 \\ y &= 2. && \text{Solving for } y \end{aligned}$$

We can check the ordered pair (1, 2).

CHECK:

$2x + 3y = 8$		
$2 \cdot 1 + 3 \cdot 2$	8	
$2 + 6$		
8		TRUE

$x + 3y = 7$		
$1 + 3 \cdot 2$	7	
$1 + 6$		
7		TRUE

Since (1, 2) checks, it is the solution.

Do Exercises 3 and 4.

In Example 2, we used the multiplication principle, multiplying by -1. However, we often need to multiply by something other than -1.

EXAMPLE 3 Solve the system

$$3x + 6y = -6, \quad (1)$$
$$5x - 2y = 14. \quad (2)$$

Looking at the terms with variables, we see that if $-2y$ were $-6y$, we would have terms that are opposites. We can achieve this by multiplying on both sides of Equation (2) by 3. Then we add:

$$\begin{aligned} 3x + 6y &= -6 && \text{Equation (1)} \\ 15x - 6y &= 42 && \text{Multiplying Equation (2) by 3} \\ \hline 18x \quad &= 36 && \text{Adding} \\ x &= 2. && \text{Solving for } x \end{aligned}$$

We go back to Equation (1) and substitute 2 for x:

$$
\begin{aligned}
3 \cdot 2 + 6y &= -6 && \text{Substituting} \\
6 + 6y &= -6 \\
6y &= -12 \\
y &= -2. && \text{Solving for } y
\end{aligned}
$$

We check the ordered pair $(2, -2)$.

CHECK:

$3x + 6y = -6$		
$3 \cdot 2 + 6 \cdot (-2)$	-6	
$6 + (-12)$		
-6		TRUE

$5x - 2y = 14$		
$5 \cdot 2 - 2 \cdot (-2)$	14	
$10 - (-4)$		
14		TRUE

Since $(2, -2)$ checks, it is the solution.

Do Exercises 5 and 6.

We have used the elimination method in Examples 1–3. Part of the strategy in doing so is making a decision about which variable to eliminate. So long as the algebra has been carried out correctly, the solution can be found by eliminating *either* variable. We multiply so that terms involving the variable to be eliminated are opposites. It is helpful to first get each equation in a form equivalent to $Ax + By = C$.

EXAMPLE 4 Solve the system

$$
\begin{aligned}
3y + 1 + 2x &= 0, && (1) \\
5x &= 7 - 4y. && (2)
\end{aligned}
$$

We first rewrite each equation in a form equivalent to $Ax + By = C$:

$$
\begin{aligned}
2x + 3y &= -1, && (1) \quad \text{Subtracting 1 on both sides and rearranging terms} \\
5x + 4y &= 7. && (2) \quad \text{Adding } 4y \text{ on both sides}
\end{aligned}
$$

We decide to eliminate the x-term. We do this by multiplying on both sides of Equation (1) by 5 and on both sides of Equation (2) by -2. Then we add:

$$
\begin{aligned}
10x + 15y &= -5 && \text{Multiplying on both sides of Equation (1) by 5} \\
-10x - 8y &= -14 && \text{Multiplying on both sides of Equation (2) by } -2 \\
\hline
7y &= -19 && \text{Adding} \\
y &= \frac{-19}{7}, \text{ or } -\frac{19}{7}. && \text{Dividing by 7}
\end{aligned}
$$

We substitute $-\frac{19}{7}$ for y in one of the original equations:

$$
\begin{aligned}
2x + 3y &= -1 && \text{Equation (1)} \\
2x + 3\left(-\tfrac{19}{7}\right) &= -1 && \text{Substituting } -\tfrac{19}{7} \text{ for } y \\
2x - \tfrac{57}{7} &= -1 \\
2x &= -1 + \tfrac{57}{7} \\
2x &= -\tfrac{7}{7} + \tfrac{57}{7} \\
2x &= \tfrac{50}{7} \\
x &= \tfrac{50}{7} \cdot \tfrac{1}{2}, \text{ or } \tfrac{25}{7}. && \text{Solving for } x
\end{aligned}
$$

We check the ordered pair $\left(\frac{25}{7}, -\frac{19}{7}\right)$.

Solve the system.

5. $4a + 7b = 11,$
$2a + 3b = 5$

6. $3x - 8y = 2,$
$5x + 2y = -12$

Caution! Solving a *system* of equations in two variables requires finding an ordered *pair* of numbers. Once you have solved for one variable, don't forget the other.

Answers on page A-6

7. Solve the system

$$3x = 5 + 2y,$$
$$2x + 3y - 1 = 0.$$

8. Solve the system

$$2x + y = 15,$$
$$4x + 2y = 23.$$

9. Solve the system

$$5x - 2y = 3,$$
$$-15x + 6y = -9.$$

Answers on page A-6

CHECK:

$2x + 3y = -1$		
$2\left(\frac{25}{7}\right) + 3\left(-\frac{19}{7}\right)$	-1	
$\frac{50}{7} - \frac{57}{7}$		
$-\frac{7}{7}$		
-1		TRUE

$5x + 4y = 7$		
$5\left(\frac{25}{7}\right) + 4\left(-\frac{19}{7}\right)$	7	
$\frac{125}{7} - \frac{76}{7}$		
$\frac{49}{7}$		
7		TRUE

The solution is $\left(\frac{25}{7}, -\frac{19}{7}\right)$.

Do Exercise 7.

Let us consider a system with no solution and see what happens when we apply the elimination method.

EXAMPLE 5 Solve the system

$$y - 3x = 2, \quad (1)$$
$$y - 3x = 1. \quad (2)$$

We multiply by -1 on both sides of Equation (2) and then add:

$$\begin{aligned} y - 3x &= 2 \\ -y + 3x &= -1 \quad \text{Multiplying by } -1 \\ \hline 0 &= 1. \quad \text{Adding} \end{aligned}$$

We obtain a false equation, $0 = 1$, so there is *no solution*. The slope–intercept forms of these equations are

$$y = 3x + 2,$$
$$y = 3x + 1.$$

The slopes are the same and the y-intercepts are different. Thus the lines are parallel. They do not intersect.

Do Exercise 8.

Sometimes there is an infinite number of solutions. Let us look at a system that we graphed in Example 5 of Section 7.1.

EXAMPLE 6 Solve the system

$$2x + 3y = 6, \quad (1)$$
$$-8x - 12y = -24. \quad (2)$$

We multiply on both sides of Equation (1) by 4 and then add the two equations:

$$\begin{aligned} 8x + 12y &= 24 \quad \text{Multiplying by 4} \\ -8x - 12y &= -24 \\ \hline 0 &= 0. \quad \text{Adding} \end{aligned}$$

We have eliminated both variables, and what remains, $0 = 0$, is an equation easily seen to be true. If this happens when we use the elimination method, we have an infinite number of solutions.

Do Exercise 9.

When decimals or fractions appear, we first multiply to clear them. Then we proceed as before.

EXAMPLE 7 Solve the system

$$\frac{1}{3}x + \frac{1}{2}y = -\frac{1}{6}, \quad (1)$$

$$\frac{1}{2}x + \frac{2}{5}y = \frac{7}{10}. \quad (2)$$

The number 6 is a multiple of all the denominators of Equation (1). The number 10 is a multiple of all the denominators of Equation (2). We multiply on both sides of Equation (1) by 6 and on both sides of Equation (2) by 10:

$$6\left(\frac{1}{3}x + \frac{1}{2}y\right) = 6\left(-\frac{1}{6}\right) \qquad 10\left(\frac{1}{2}x + \frac{2}{5}y\right) = 10\left(\frac{7}{10}\right)$$

$$6 \cdot \frac{1}{3}x + 6 \cdot \frac{1}{2}y = -1 \qquad 10 \cdot \frac{1}{2}x + 10 \cdot \frac{2}{5}y = 7$$

$$2x + 3y = -1; \qquad 5x + 4y = 7.$$

The resulting system is

$$2x + 3y = -1,$$

$$5x + 4y = 7.$$

As we saw in Example 4, the solution of this system is $\left(\frac{25}{7}, -\frac{19}{7}\right)$.

Do Exercise 10.

The following is a summary that compares the graphical, substitution, and elimination methods for solving systems of equations.

METHOD	STRENGTHS	WEAKNESSES
Graphical	Can "see" solution.	Inexact when solution involves numbers that are not integers or are very large and off the graph.
Substitution	Works well when solutions are not integers. Easy to use when a variable is alone on one side.	Introduces extensive computations with fractions for more complicated systems where coefficients are not 1 or −1. Cannot "see" solution.
Elimination	Works well when solutions are not integers, when coefficients are not 1 or −1, and when coefficients involve decimals or fractions.	Cannot "see" solution.

When deciding which method to use, consider the preceding chart and directions from your instructor. The situation is like having a piece of wood to cut and three saws with which to cut it. The saw you use depends on the type of wood, the type of cut you are making, and how you want the wood to turn out.

10. Solve the system

$$\frac{1}{2}x + \frac{3}{10}y = \frac{1}{5},$$

$$\frac{3}{5}x + y = -\frac{2}{5}.$$

Answer on page A-6

11. Budget Rent-A-Car rents a car at a daily rate of \$41.95 plus 43 cents per mile. Speedo Rentzit rents a car for \$44.95 plus 39 cents per mile. For what mileage are the costs the same?

Answer on page A-6

C Solving Problems

We now use the elimination method to solve a problem.

EXAMPLE 8 At one time, Value Rent-A-Car rented compact cars at a daily rate of \$43.95 plus 40 cents per mile. Thrifty Rent-A-Car rented compact cars at a daily rate of \$42.95 plus 42 cents per mile. For what mileage are the costs the same?

1. ***Familiarize.*** To become familiar with the problem, we make a guess. Suppose a person rents a compact car from each rental agency and drives it 100 mi. The total cost at Value is \$43.95 + \$0.40(100) = \$43.95 + \$40.00, or \$83.95. The total cost at Thrifty is \$42.95 + \$0.42(100) = \$42.95 + \$42.00, or \$84.95. Note that we converted all of our money units to dollars. The resulting costs are very nearly the same, so our guess is close. We can, of course, refine our guess. Instead, we will use algebra to solve the problem. We let M = the number of miles driven and C = the total cost of the car rental.

2. ***Translate.*** We translate the first statement, using \$0.40 for 40 cents. It helps to reword the problem before translating.

\$43.95	plus	40 cents	times	Number of miles driven	is	Cost.	Rewording
↓	↓	↓	↓	↓	↓	↓	
\$43.95	+	\$0.40	·	M	=	C	Translating

We translate the second statement, but again it helps to reword it first.

\$42.95	plus	42 cents	times	Number of miles driven	is	Cost.	Rewording
↓	↓	↓	↓	↓	↓	↓	
\$42.95	+	\$0.42	·	M	=	C	Translating

We have now translated to a system of equations:

$$43.95 + 0.40M = C,$$
$$42.95 + 0.42M = C.$$

3. ***Solve.*** We solve the system of equations. We clear the system of decimals by multiplying on both sides by 100. Then we multiply the second equation by -1 and add:

$$\begin{aligned} 4395 + 40M &= 100C \\ -4295 - 42M &= -100C \\ \hline 100 - 2M &= 0 \\ 100 &= 2M \\ 50 &= M. \end{aligned}$$

4. ***Check.*** For 50 mi, the cost of the Value car is \$43.95 + \$0.40(50), or \$43.95 + \$20, or \$63.95, and the cost of the Thrifty car is \$42.95 + \$0.42(50), or \$42.95 + \$21, or \$63.95. Thus the costs are the same when the mileage is 50.

5. ***State.*** When the cars are driven 50 mi, the costs will be the same.

Do Exercise 11.

NAME SECTION DATE

Exercise Set 7.3

a Solve using the elimination method.

1. $x - y = 7,$
$x + y = 5$

2. $x + y = 11,$
$x - y = 7$

3. $x + y = 8,$
$-x + 2y = 7$

4. $x + y = 6,$
$-x + 3y = -2$

5. $5x - y = 5,$
$3x + y = 11$

6. $2x - y = 8,$
$3x + y = 12$

7. $4a + 3b = 7,$
$-4a + b = 5$

8. $7c + 5d = 18,$
$c - 5d = -2$

9. $8x - 5y = -9,$
$3x + 5y = -2$

10. $3a - 3b = -15,$
$-3a - 3b = -3$

11. $4x - 5y = 7,$
$-4x + 5y = 7$

12. $2x + 3y = 4,$
$-2x - 3y = -4$

ANSWERS

1. ________

2. ________

3. ________

4. ________

5. ________

6. ________

7. ________

8. ________

9. ________

10. ________

11. ________

12. ________

ANSWERS

13. ______
14. ______
15. ______
16. ______
17. ______
18. ______
19. ______
20. ______
21. ______
22. ______
23. ______
24. ______
25. ______
26. ______
27. ______
28. ______
29. ______
30. ______

b Solve using the multiplication principle first. Then add.

13. $x + y = -7,$
$3x + y = -9$

14. $-x - y = 8,$
$2x - y = -1$

15. $3x - y = 8,$
$x + 2y = 5$

16. $x + 3y = 19,$
$x - y = -1$

17. $x - y = 5,$
$4x - 5y = 17$

18. $x + y = 4,$
$5x - 3y = 12$

19. $2w - 3z = -1,$
$3w + 4z = 24$

20. $7p + 5q = 2,$
$8p - 9q = 17$

21. $2a + 3b = -1,$
$3a + 5b = -2$

22. $3x - 4y = 16,$
$5x + 6y = 14$

23. $x = 3y,$
$5x + 14 = y$

24. $5a = 2b,$
$2a + 11 = 3b$

25. $2x + 5y = 16,$
$3x - 2y = 5$

26. $3p - 2q = 8,$
$5p + 3q = 7$

27. $p = 32 + q,$
$3p = 8q + 6$

28. $3x = 8y + 11,$
$x + 6y - 8 = 0$

29. $3x - 2y = 10,$
$-6x + 4y = -20$

30. $2x + y = 13,$
$4x + 2y = 23$

31. $0.06x + 0.05y = 0.07,$
$0.4x - 0.3y = 1.1$

32. $1.8x - 2y = 0.9,$
$0.04x + 0.18y = 0.15$

33. $\frac{1}{3}x + \frac{3}{2}y = \frac{5}{4},$
$\frac{3}{4}x - \frac{5}{6}y = \frac{3}{8}$

34. $x - \frac{3}{2}y = 13,$
$\frac{3}{2}x - y = 17$

c Solve.

35. A family plans to rent a van to move a child to college. Quick-Haul rents a 10-ft moving van at a daily rate of \$19.95 plus 39 cents per mile. Another company rents the same size van for \$39.95 plus 29 cents per mile. For what mileage are the costs the same?

36. Elite Rent-A-Car rents a basic car at a daily rate of \$45.95 plus 40 cents per mile. Another company rents a basic car for \$46.95 plus 20 cents per mile. For what mileage are the costs the same?

37. Two angles are supplementary. One is 30° more than two times the other. Find the angles. (**Supplementary angles** are angles whose sum is 180°.)

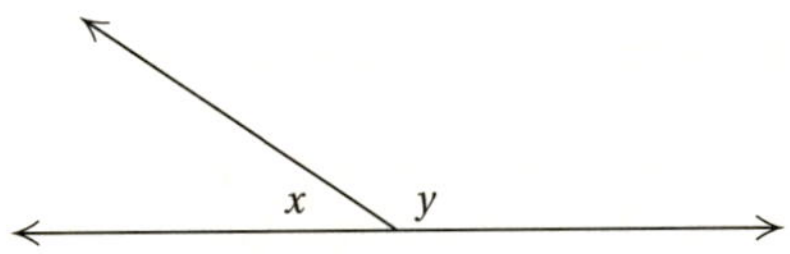

Supplementary angles

38. Two angles are supplementary. One is 8° less than three times the other. Find the angles.

39. Two angles are complementary. Their difference is 34°. Find the angles. (**Complementary angles** are angles whose sum is 90°.)

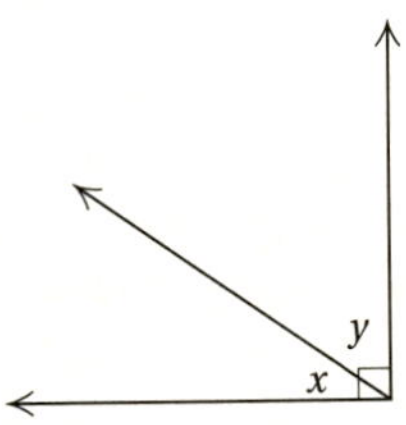

Complementary angles

40. Two angles are complementary. One angle is 42° more than one-half the other. Find the angles.

ANSWERS

31. ______

32. ______

33. ______

34. ______

35. ______

36. ______

37. ______

38. ______

39. ______

40. ______

ANSWERS

41. ______

42. ______

43. ______

44. ______

45. ______

46. ______

47. ______

48. ______

49. ______

50. ______

51. ______

52. ______

53. ______

54. ______

55. ______

56. ______

41. The Rolling Velvet Horse Farm allots 650 hectares to plant hay and oats. The owners know that their needs are best met if they plant 180 hectares more of hay than of oats. How many hectares of each should they plant?

42. In a vineyard, a vintner uses 820 hectares to plant Chardonnay and Riesling grapes. The vintner knows that the profits will be greatest by planting 140 hectares more of Chardonnay than of Riesling. How many hectares of each grape should be planted?

SKILL MAINTENANCE

Simplify.

43. $(a^2b^{-3})(a^5b^{-6})$

44. $\dfrac{a^2b^{-3}}{a^5b^{-6}}$

45. $\dfrac{x^2 - 5x + 6}{x^2 - 4}$

46. $\dfrac{x^2 - 25}{x^2 - 10x + 25}$

Subtract.

47. $\dfrac{x - 2}{x + 3} - \dfrac{2x - 5}{x - 4}$

48. $\dfrac{x + 7}{x^2 - 1} - \dfrac{3}{x + 1}$

SYNTHESIS

49. Will's age is 20% of his father's age. Twenty years from now, Will's age will be 52% of his father's age. How old are Will and his father now?

50. If 5 is added to a woman's age and the total is divided by 5, the result will be her daughter's age. Five years ago, the woman's age was eight times her daughter's age. Find their present ages.

51. When the base of a triangle is increased by 2 ft and the height is decreased by 1 ft, the height becomes one-third of the base, and the area becomes 24 ft^2. Find the original dimensions of the triangle.

52. Several ancient Chinese books included problems that can be solved by translating to systems of equations. *Arithmetical Rules in Nine Sections* is a book of 246 problems compiled by a Chinese mathematician, Chang Tsang, who died in 152 B.C. One of the problems is: Suppose there are a number of rabbits and pheasants confined in a cage. In all, there are 35 heads and 94 feet. How many rabbits and how many pheasants are there? Solve the problem.

Solve.

53. $3(x - y) = 9,$
$x + y = 7$

54. $2(x - y) = 3 + x,$
$x = 3y + 4$

55. $2(5a - 5b) = 10,$
$-5(6a + 2b) = 10$

56. $\dfrac{x}{3} + \dfrac{y}{2} = 1\dfrac{1}{3},$
$x + 0.05y = 4$

7.4 More on Solving Problems

OBJECTIVE

After finishing Section 7.4, you should be able to:

a Solve problems by translating them to systems of two equations in two variables.

FOR EXTRA HELP

TAPE 18 | TAPE 15A | MAC: 7 IBM: 7

a We continue solving problems using the five steps for problem solving and our methods for solving systems of equations.

EXAMPLE 1 Denny's® is a national restaurant firm. The ad shown here once appeared on the tables as a special.

Determine the price of one item from the A side of the menu and the price of one item from the B side of the menu.

1, 2. ***Familiarize*** and ***Translate.*** The ad gives us the system of equations at the outset:

$$a + b = \$5.49, \quad (1)$$
$$a + 2b = \$6.99, \quad (2)$$

where a = the price of one item from the A side of the menu and b = the price of one item from the B side of the menu.

3. ***Solve.*** We solve the system of equations. Which method should we use? As we discussed in Section 7.3, any method can be used. Each has its advantages and disadvantages. We decide to proceed with the elimination method, because we see that if we multiply Equation (1) by -1 and then add, the a-terms can be eliminated:

$$\begin{aligned} -a - b &= -5.49 \qquad \text{Multiplying by } -1 \\ a + 2b &= 6.99 \\ \hline b &= 1.50. \qquad \text{Adding} \end{aligned}$$

We substitute 1.50 for b in Equation (1) and solve for a:

$$\begin{aligned} a + b &= 5.49 \\ a + 1.50 &= 5.49 \\ a &= 3.99. \end{aligned}$$

1. A fast-food restaurant is running a promotion in which a hamburger and two pieces of chicken costs $2.39 and a hamburger and one piece of chicken costs $1.69. Determine the cost of one hamburger and the cost of one piece of chicken.

Answer on page A-7

4. ***Check.*** The sum of the two prices is $3.99 + $1.50, or $5.49. The A price plus twice the B price is $3.99 + 2($1.50) = $3.99 + $3.00, or $6.99. The prices check. Sometimes a "common sense" check is appropriate. If you look at the foods on the B side of the menu, it does not seem reasonable that such items would sell for $1.50 each. Note that the menu does not say that you can buy a B item alone for $1.50. That price is a bonus for buying an A item. Mathematically, the prices given do stand, though it is doubtful that you can buy a B item by itself.
5. ***State.*** The price of one A item is $3.99, and the price of one B item is $1.50.

Do Exercise 1.

EXAMPLE 2 Caleb is 21 years older than Tanya. In 6 yr, Caleb will be twice as old as Tanya. How old are they now?

1. ***Familiarize.*** Let us consider some conditions of the problem. We let C = Caleb's age now and T = Tanya's age now. How will the ages relate in 6 yr? In 6 yr, Tanya will be $T + 6$ and Caleb will be $C + 6$. We make a table to organize our information.

	CALEB	TANYA	
AGE NOW	C	T	→ $C = 21 + T$
AGE IN 6 YR	$C + 6$	$T + 6$	→ $C + 6 = 2(T + 6)$

2. ***Translate.*** From the present ages, we get the following rewording and translation.

Caleb's age	is	21	more than	Tanya's age.	Rewording
↓	↓	↓	↓	↓	
C	$=$	21	$+$	T	Translating

From their ages in 6 yr, we get the following rewording and translation.

Caleb's age in 6 yr	will be	twice	Tanya's age in 6 yr	Rewording
↓	↓	↓	↓	
$C + 6$	$=$	$2 \cdot$	$(T + 6)$	Translating

The problem has been translated to the following system of equations:

$$C = 21 + T, \quad (1)$$
$$C + 6 = 2(T + 6). \quad (2)$$

3. ***Solve.*** We solve the system of equations. This time we use the substitution method since there is a variable alone on one side. We substitute $21 + T$ for C in Equation (2):

$$C + 6 = 2(T + 6)$$
$$(21 + T) + 6 = 2(T + 6)$$
$$T + 27 = 2T + 12$$
$$15 = T.$$

We find C by substituting 15 for T in the first equation:

$$\begin{aligned} C &= 21 + T \\ &= 21 + 15 \\ &= 36. \end{aligned}$$

4. ***Check.*** Caleb's age is 36, which is 21 more than 15, Tanya's age. In 6 yr, when Caleb will be 42 and Tanya 21, Caleb's age will be twice Tanya's age.
5. ***State.*** Caleb is now 36 and Tanya is 15.

Do Exercise 2.

EXAMPLE 3 There were 411 people at a movie. Admission was \$7.00 for adults and \$3.75 for children, and receipts totaled \$2678.75. The box office manager lost the records of how many adults and how many children attended. Can you use algebra to help her? How many adults and how many children did attend?

1. ***Familiarize.*** There are many ways in which to familiarize ourselves with a problem situation. This time, let us make a guess and do some calculations. The total number of people at the movie was 411, so we choose numbers that total 411. Let us try

240 adults and
171 children.

How much money was taken in? The problem says that adults paid \$7.00 each, so the total amount of money collected from the adults was

240(\$7), or \$1680.

The children paid \$3.75 each, so the total amount of money collected from the children was

171(\$3.75), or \$641.25.

This makes the total receipts \$1680 + \$641.25, or \$2321.25.

Our guess is not the answer to the problem because the total taken in, according to the problem, was \$2678.75. If we were to continue guessing, we would need to add more adults and fewer children, since our first guess was too low. The steps we have used to see if our guesses are correct help us to understand the actual steps involved in solving the problem.

Let us list the information in a table. That usually helps in the familiarization process. We let a = the number of adults and c = the number of children.

	ADULTS	CHILDREN	TOTAL	
ADMISSION	\$7.00	\$3.75		
NUMBER ATTENDING	a	c	411	→ $a + c = 411$
MONEY TAKEN IN	$7.00a$	$3.75c$	\$2678.75	→ $7.00a + 3.75c = 2678.75$

2. ***Translate.*** The total number of people attending was 411, so

$$a + c = 411.$$

2. Marvella is 26 yr older than Malcolm. In 5 yr, Marvella will be twice as old as Malcolm. How old are they now?

Complete the following table to aid with the familiarization.

	MARVELLA	MALCOLM	
AGE NOW	M	K	→ $M = (\quad) + K$
AGE IN 5 YR			→ $M + 5 = (\quad)(K + 5)$

Answer on page A-7

3. There were 166 paid admissions to a game. The price was \$2.10 each for adults and \$0.75 each for children. The amount taken in was \$293.25. How many adults and how many children attended?

Complete the following table to aid with the familiarization.

	ADULTS	CHILDREN	TOTAL	
PAID		\$0.75		
NUMBER ATTENDING	x	y		→ $x + y = ($ $)$
MONEY TAKEN IN			\$293.25	→ $2.10x + ($ $)$ $= 293.25$

Answer on page A-7

The amount taken in from the adults was $7.00a$, and the amount taken in from the children was $3.75c$. These amounts are in dollars. The total was \$2678.75, so we have

$$7.00a + 3.75c = 2678.75.$$

We can multiply on both sides by 100 to clear decimals. Thus we have a translation to a system of equations:

$$a + c = 411, \quad (1)$$
$$700a + 375c = 267{,}875. \quad (2)$$

3. *Solve.* We solve the system of equations. We use the elimination method since the equations are both in the form $Ax + By = C$. (A case can certainly be made for using the substitution method since we can solve for one of the variables quite easily in the first equation. Very often a decision is just a matter of choice.) We multiply on both sides of Equation (1) by -375 and then add:

$$-375a - 375c = -154{,}125 \quad \text{Multiplying by } -375$$
$$700a + 375c = 267{,}875$$
$$325a = 113{,}750 \quad \text{Adding}$$
$$a = \frac{113{,}750}{325} \quad \text{Dividing by 325}$$
$$a = 350.$$

We go back to Equation (1) and substitute 350 for a:

$$a + c = 411$$
$$350 + c = 411$$
$$c = 61.$$

4. *Check.* We leave the check to the student. It is similar to what we did in the familiarization step.

5. *State.* 350 adults and 61 children attended.

Do Exercise 3.

EXAMPLE 4 A chemist has one solution that is 80% acid (that is, 8 parts are acid and 2 parts are water) and another solution that is 30% acid. What is needed is 200 L of a solution that is 62% acid. The chemist will prepare it by mixing the two solutions. How much of each should be used?

1. *Familiarize.* We can draw a picture of the situation. The chemist uses x liters of the first solution and y liters of the second solution.

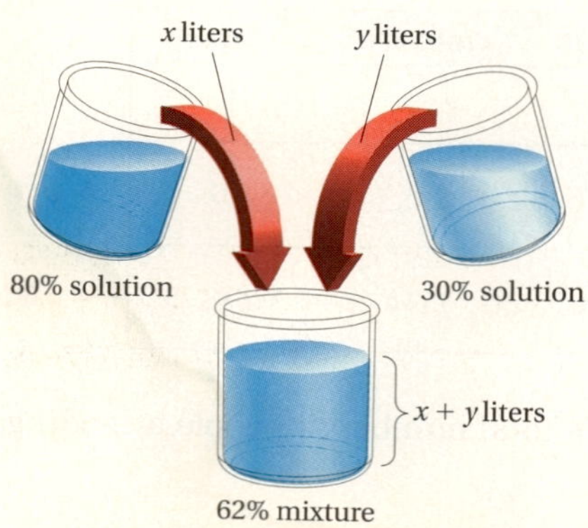

We can also arrange the information in a table.

TYPE OF SOLUTION	FIRST	SECOND	MIXTURE	
AMOUNT OF SOLUTION	x	y	200 L	→ $x + y = 200$
PERCENT OF ACID	80%	30%	62%	
AMOUNT OF ACID IN SOLUTION	80%x	30%y	62% × 200, or 124 L	→ 80%x + 30%y = 124

2. ***Translate.*** The chemist uses x liters of the first solution and y liters of the second. Since the total is to be 200 L, we have

Total amount of solution: $x + y = 200.$

The amount of acid in the new mixture is to be 62% of 200 L, or 124 L. The amounts of acid from the two solutions are 80%x and 30%y. Thus,

Total amount of acid: $80\%x + 30\%y = 124$

or $0.8x + 0.3y = 124.$

We clear decimals by multiplying on both sides by 10:

$$10(0.8x + 0.3y) = 10 \cdot 124$$
$$8x + 3y = 1240.$$

Thus we have a translation to a system of equations:

$$x + y = 200, \quad (1)$$
$$8x + 3y = 1240. \quad (2)$$

3. ***Solve.*** We solve the system. We use the elimination method, again because equations are in the form $Ax + By = C$ and a multiplication in one equation will allow us to eliminate a variable, but substitution would also work. We multiply on both sides of Equation (1) by -3 and then add:

$$-3x - 3y = -600 \qquad \textbf{Multiplying by } -3$$
$$8x + 3y = 1240$$
$$5x \quad = 640 \qquad \textbf{Adding}$$
$$x = \frac{640}{5} \qquad \textbf{Dividing by 5}$$
$$x = 128.$$

We go back to Equation (1) and substitute 128 for x:

$$x + y = 200$$
$$128 + y = 200$$
$$y = 72.$$

The solution is $x = 128$ and $y = 72$.

4. ***Check.*** The sum of 128 and 72 is 200. Also, 80% of 128 is 102.4 and 30% of 72 is 21.6. These add up to 124.

5. ***State.*** The chemist should use 128 L of the 80%-acid solution and 72 L of the 30%-acid solution.

Do Exercise 4.

4. One solution is 50% alcohol and a second is 70% alcohol. How much of each should be mixed to make 30 L of a solution that is 55% alcohol?

Complete the following table to aid in the familiarization.

TYPE OF SOLUTION	FIRST	SECOND	MIXTURE	
AMOUNT OF SOLUTION	x	y		→ $x + y = (\ )$
PERCENT OF ALCOHOL		70%	55%	
AMOUNT OF ALCOHOL IN SOLUTION				→ $(\ ) + 70\%y = (\ )$

Answer on page A-7

EXAMPLE 5 A bulk wholesaler wishes to mix some candy worth 45 cents per pound and some worth 80 cents per pound to make 350 lb of a mixture worth 65 cents per pound. How much of each should be used?

1. ***Familiarize.*** Arranging the information in a table will help. We let $x =$ the amount of 45-cents candy and $y =$ the amount of 80-cents candy.

TYPE OF CANDY	INEXPENSIVE CANDY	EXPENSIVE CANDY	MIXTURE	
COST OF CANDY	45 cents	80 cents	65 cents	
AMOUNT (IN POUNDS)	x	y	350	→ $x + y = 350$
TOTAL COST	$45x$	$80y$	65 cents · (350), or 22,750 cents	→ $45x + 80y = 22{,}750$

Note the similarity of this problem to Example 3. Here we consider candy instead of tickets.

2. ***Translate.*** We translate as follows. From the second row of the table, we find that

Total amount of candy: $x + y = 350$.

Our second equation will come from the costs. The value of the inexpensive candy, in cents, is $45x$ (x pounds at 45 cents per pound). The value of the expensive candy is $80y$, and the value of the mixture is 65×350, or 22,750 cents. Thus we have

Total cost of mixture: $45x + 80y = 22{,}750$.

Remember the problem-solving tip about dimension symbols. In this last equation, all expressions stand for cents. We could have expressed them all in dollars, but we do not want some in cents and some in dollars. Thus we have a translation to a system of equations:

$$\begin{aligned} x + \quad y &= 350, \\ 45x + 80y &= 22{,}750. \end{aligned}$$

3. ***Solve.*** We solve the system using the elimination method again. We multiply on both sides of Equation (1) by -45 and then add:

$$\begin{aligned} -45x - 45y &= -15{,}750 && \text{Multiplying by } -45 \\ 45x + 80y &= \quad 22{,}750 \\ \hline 35y &= \quad 7{,}000 && \text{Adding} \\ y &= \frac{7{,}000}{35} \\ y &= 200. \end{aligned}$$

We go back to Equation (1) and substitute 200 for y:

$$\begin{aligned} x + y &= 350 \\ x + 200 &= 350 \\ x &= 150. \end{aligned}$$

4. ***Check.*** We consider $x = 150$ lb and $y = 200$ lb. The sum is 350 lb. The value of the candy is $45(150) + 80(200)$, or 22,750 cents. These values check.

5. ***State.*** The grocer should mix 150 lb of the 45-cents candy with 200 lb of the 80-cents candy.

Do Exercise 5.

EXAMPLE 6 A student assistant at the university copy center has some nickels and dimes to use for change when students make copies. The value of the coins is \$3.70. There are 13 more dimes than nickels. How many of each kind of coin are there?

1. ***Familiarize.*** We let $d =$ the number of dimes and $n =$ the number of nickels.

2. ***Translate.*** We have one equation at once:

$$d = n + 13.$$

The value of the nickels, in cents, is $5n$, since each is worth 5 cents. The value of the dimes, in cents, is $10d$, since each coin is worth 10 cents. The total value is given as \$3.70. Since we have the values of the nickels and dimes *in cents*, we must use cents for the total value. This is 370. This gives us another equation:

$$10d + 5n = 370.$$

We now have a system of equations:

$$d = n + 13, \quad (1)$$
$$10d + 5n = 370. \quad (2)$$

3. ***Solve.*** Since we have d alone on one side of one equation, we use the substitution method. We substitute $n + 13$ for d in Equation (2):

$$10d + 5n = 370$$
$$10(n + 13) + 5n = 370 \quad \text{Substituting } n + 13 \text{ for } d$$
$$10n + 130 + 5n = 370 \quad \text{Removing parentheses}$$
$$15n + 130 = 370 \quad \text{Collecting like terms}$$
$$15n = 240 \quad \text{Subtracting 130}$$
$$n = \frac{240}{15}, \text{ or } 16. \quad \text{Dividing by 15}$$

We substitute 16 for n in either of the original equations to find d. We use Equation (1):

$$d = n + 13$$
$$= 16 + 13$$
$$= 29.$$

4. ***Check.*** We have 29 dimes and 16 nickels. There are 13 more dimes than nickels. The value of the coins is 29(\$0.10) + 16(\$0.05), which is \$3.70. This checks.

5. ***State.*** The student has 29 dimes and 16 nickels.

Do Exercise 6.

5. Grass seed A is worth \$1.40 per pound and seed B is worth \$1.75 per pound. How much of each should be mixed to make 50 lb of a mixture worth \$1.54 per pound?

Complete the following table to aid in the familiarization.

TYPE OF SEED	A	B	MIXTURE	
COST OF SEED	\$1.40		\$1.54	
AMOUNT (IN POUNDS)	x	y		$\rightarrow x + y = (\ \ )$
MIXTURE		$1.75y$		$\rightarrow 1.40x + 1.75y = (\ \ )$

6. On a table are 20 coins, quarters and dimes. Their value is \$3.05. How many of each kind of coin are there?

Answers on page A-7

You should look back over Examples 3–6. The problems are quite similar in their structure. Compare them and try to see the similarities. The problems in Examples 3–6 are often called *mixture problems.* In each case, a situation is considered in two different ways. These problems provide a pattern, or model, for many related problems.

PROBLEM-SOLVING TIP

When solving problems, see if they are patterned or modeled after other problems that you have studied.

SIDELIGHTS

STUDY TIPS: EXTRA TIPS ON PROBLEM SOLVING

We have often presented some tips and guidelines to enhance your learning abilities. The following tips, which are focused on problem solving, summarize some points already considered and propose some new ones.

- *The following are the five steps for problem solving:*

1. ***Familiarize* yourself with the problem situation.**
2. ***Translate* the problem to an equation.** As you study more mathematics, you will find that the translation may be to some other kind of mathematical language, such as an inequality.
3. ***Solve* the equation.** If the translation is to some other kind of mathematical language, you would carry out some kind of mathematical manipulation—in the case of an inequality, you would solve it.
4. ***Check* the answer in the original problem.** This does not mean to check in the translated equation. It means to go back to the original worded problem.
5. ***State* the answer to the problem clearly.**

For Step 4 on checking, some further comment is appropriate. *You may be able to translate to an equation and to solve the equation, but you may find that none of the solutions of the equation is the solution of the original problem.* To see how this can happen, consider this example.

EXAMPLE The sum of two consecutive even integers is 537. Find the integers.

1. ***Familiarize.*** Suppose we let $x =$ the first number. Then $x + 2 =$ the second number.
2. ***Translate.*** The problem can be translated to the following equation: $x + (x + 2) = 537$.
3. ***Solve.*** We solve the equation as follows:

$$2x + 2 = 537$$
$$2x = 535$$
$$x = \frac{535}{2}, \text{ or } 267.5.$$

4. ***Check.*** Then $x + 2 = 269.5$. However, the numbers are not only not even, but they are not integers.
5. ***State.*** The problem has no solution.

The following are some other tips.

- *To be good at problem solving, do lots of problems.* The situation is similar to learning a skill, such as playing golf. At first you may not be successful, but the more you practice and work at improving your skills, the more successful you will become. For problem solving, do more than just two or three odd-numbered assigned problems. Do them all, and if you have time, do the even-numbered problems as well. Then find another book on the same subject and do problems in that book.
- *Look for patterns when solving problems.* You will eventually see patterns in similar kinds of problems. For example, there is a pattern in the way that you solve problems involving consecutive integers.
- *When translating to an equation, or some other mathematical language, consider the dimensions of the variables and the constants in the equation.* The variables that represent length should all be in the same unit, those that represent money should all be in dollars or in cents, and so on.

NAME SECTION DATE

Exercise Set 7.4

a Solve.

1. The Huxtables generate twice as much trash as their neighbors, the Simpsons. Together, the two households produce 14 bags of trash each month. How much trash does each household produce?

2. In a summer-league basketball game, the Shooting Stars scored 46 points on a combination of 20 two- and three-point baskets. How many two-point shots were made and how many three-point shots were made?

3. The Kuyatts' house is twice as old as the Marconis' house. Eight years ago, the Kuyatts' house was three times as old as the Marconis' house. How old is each house?

4. David is twice as old as his daughter. In 4 yr, David's age will be three times what his daughter's age was 6 yr ago. How old are they now?

5. Randy is four times as old as Mandy. In 12 yr, Mandy's age will be half of Randy's. How old are they now?

6. Madonna is twice as old as Ramon. The sum of their ages 7 yr ago was 13. How old are they now?

7. A collection of dimes and quarters is worth \$15.25. There are 103 coins in all. How many of each are there?

8. A collection of nickels and dimes is worth \$2.90. There are 19 more nickels than dimes. How many of each are there?

9. Mr. Cholesterol's Pizza Parlor charges \$1.99 for a slice of pizza and a soda and \$5.48 for three slices of pizza and two sodas. Determine the cost of one soda and the cost of one slice of pizza.

10. Cassandra has a number of \$50 and \$100 savings bonds to use for part of her college expenses. The total value of the bonds is \$1250. There are 7 more \$50 bonds than \$100 bonds. How many of each type of bond does she have?

ANSWERS

1. ____
2. ____
3. ____
4. ____
5. ____
6. ____
7. ____
8. ____
9. ____
10. ____

ANSWERS

11. ____________

12. ____________

13. ____________

14. ____________

15. ____________

16. ____________

17. ____________

18. ____________

11. There were 203 tickets sold for a volleyball game. For activity-card holders, the price was \$1.25, and for non-cardholders, the price was \$2. The total amount of money collected was \$310. How many of each type of ticket were sold?

12. There were 429 people at a play. Admission was \$8 each for adults and \$4.50 each for children. The total receipts were \$2641. How many adults and how many children attended?

13. Following the baseball season, the players on a junior college team decided to go to a major-league baseball game. Ticket prices for the game are shown in the table below. They bought 29 tickets of two types, Upper Box and Lower Reserved. The cost of all the tickets was \$284. How many of each kind of ticket did they buy?

TICKET INFORMATION	
Lower Box	\$12.50
Upper Box	\$10.00
Lower Reserved	\$ 9.50
Upper Reserved	\$ 8.00
General Admission	\$ 6.50

14. A faculty group bought tickets for the game in Exercise 13, but they bought 54 tickets of two types, Lower Box and Upper Box. The cost of all their tickets was \$577.50. How many of each kind of ticket did they buy?

15. Solution A is 50% acid and solution B is 80% acid. How many liters of each should be used to make 100 L of a solution that is 68% acid? (*Hint:* 68% of what is acid?) Complete the following to aid in the familiarization.

TYPE OF SOLUTION	A	B	MIXTURE	
AMOUNT OF SOLUTION	x	y	L	$\rightarrow x + y = (\quad)$
PERCENT OF ACID	50%		68%	
AMOUNT OF ACID IN SOLUTION		80%y	68% × 100, or L	$\rightarrow$ 50%x + ( ) = ( )

16. Solution A is 30% alcohol and solution B is 75% alcohol. How much of each should be used to make 100 L of a solution that is 50% alcohol?

17. At a local "paint swap," Gayle found large supplies of Skylite Pink (12.5% red pigment) and MacIntosh Red (20% red pigment). How many gallons of each color should Gayle pick up in order to mix a gallon of Summer Rose (17% red pigment)?

18. A solution containing 30% insecticide is to be mixed with a solution containing 50% insecticide to make 200 L of a solution containing 42% insecticide. How much of each solution should be used?

19. A coffee shop mixes Brazilian coffee worth \$5 per pound with Turkish coffee worth \$8 per pound. The mixture is to sell for \$7 per pound. How much of each type of coffee should be used to make a 300-lb mixture? Complete the following table to aid in the familiarization.

20. The Java Joint wishes to mix Kenyan coffee beans that sell for \$7.25 per pound with Venezuelan beans that sell for \$8.50 per pound to form a 50-lb batch of Morning Blend that sells for \$8.00 per pound. How many pounds of Kenyan beans and how many pounds of Venezuelan beans should be used to make the blend?

TYPE OF COFFEE	BRAZILIAN	TURKISH	MIXTURE	
COST OF COFFEE	\$5		\$7	
AMOUNT (IN POUNDS)	x	y	300	$\rightarrow x + y = (\quad)$
MIXTURE		$8y$	\$7(300), or \$2100	$\rightarrow 5x + (\quad) = 2100$

21. Grass seed A is worth \$2.50 per pound and seed B is worth \$1.75 per pound. How much of each would you use to make 75 lb of a mixture worth \$2.14 per pound?

22. A customer has asked a caterer to provide 60 lb of nuts, 60% of which are to be cashews. The caterer has available mixtures of 70% cashews and 45% cashews. How many pounds of each mixture should be used?

23. You are taking a test in which items of type A are worth 10 points and items of type B are worth 15 points. It takes 3 min to complete each item of type A and 6 min to complete each item of type B. The total time allowed is 60 min and you can do exactly 16 questions. Your score is 180 points by using the entire 60 min. How many questions of each type did you answer correctly?

24. A goldsmith has two alloys that are different purities of gold. The first is three-fourths pure gold and the second is five-twelfths pure gold. How many ounces of each should be melted and mixed in order to obtain a 6-oz mixture that is two-thirds pure gold?

25. A printer knows that a page of print contains 1300 words if large type is used and 1850 words if small type is used. A document containing 18,526 words fills exactly 12 pages. How many pages are in the large type? in the small type?

26. A merchant has two kinds of paint. If 9 gal of the inexpensive paint is mixed with 7 gal of the expensive paint, the mixture will be worth \$19.70 per gallon. If 3 gal of the inexpensive paint is mixed with 5 gal of the expensive paint, the mixture will be worth \$19.825 per gallon. What is the price per gallon of each type of paint?

ANSWERS

19. ______

20. ______

21. ______

22. ______

23. ______

24. ______

25. ______

26. ______

ANSWERS

27. ______

28. ______

29. ______

30. ______

31. ______

32. ______

33. ______

34. ______

35. ______

36. ______

37. ______

38. ______

SYNTHESIS

27. An automobile radiator contains 16 L of antifreeze and water. This mixture is 30% antifreeze. How much of this mixture should be drained and replaced with pure antifreeze so that the mixture will be 50% antifreeze?

28. An employer has a daily payroll of \$1225 when employing some workers at \$80 per day and others at \$85 per day. When the number of \$80 workers is increased by 50% and the number of \$85 workers is decreased by $\frac{1}{5}$, the new daily payroll is \$1540. How many were originally employed at each rate?

29. One day, 1315 people visited a zoo. Admission was \$8.50 per person, but some used coupons to save 20% off the regular admission price. The total receipts for the day were \$10,242.50. How many used a coupon to visit the zoo?

30. A two-digit number is six times the sum of its digits. The tens digit is 1 more than the units digit. Find the number.

31. A farmer has 100 L of milk that is 4.6% butterfat. How much skim milk (no butterfat) should be mixed with it to make milk that is 3.2% butterfat?

32. A tank contains 8000 L of a solution that is 40% acid. How much water should be added to make a solution that is 30% acid?

33. A total of \$27,000 is invested, part of it at 12% and part of it at 13%. The total interest after one year is \$3385. How much was invested at each rate?

34. A student earned \$288 on investments. If \$1100 was invested at one yearly rate and \$1800 at a rate that was 1.5% higher, find the two rates of interest.

35. A flavored drink manufacturer mixes flavoring worth \$1.45 per ounce with sugar worth \$0.05 per ounce. The mixture sells for \$0.106 per ounce. How much of each should be mixed to fill a 20-oz can?

36. A framing shop charges \$0.40 per inch for a certain kind of frame. A customer is looking for a frame whose length is 3 in. longer than the width. The clerk recommends using a frame that is 2 in. longer and 1 in. wider. The second frame will cost \$22.40. What are the dimensions of the first frame?

37. Together, a baseball bat, ball, and glove cost \$204.85. The glove costs \$75 more than the ball, and the bat costs \$40 more than the glove. How much does each cost?

38. In Lewis Carroll's "Through the Looking Glass," Tweedledum says to Tweedledee, "The sum of your weight and twice mine is 361 pounds." Then Tweedledee says to Tweedledum, "Contrariwise, the sum of your weight and twice mine is 362 pounds." Find the weight of Tweedledum and Tweedledee.

7.5 Motion Problems

a We studied problems involving motion in Chapter 5. Here we solve certain motion problems whose solutions can be found using systems of equations. Recall the motion formula.

> **THE MOTION FORMULA**
>
> Distance = Rate (or speed) · Time
>
> $d = rt$

We have five steps for problem solving. The following tips are also helpful when solving motion problems.

> **TIPS FOR SOLVING MOTION PROBLEMS**
>
> 1. Draw a diagram using an arrow or arrows to represent distance and the direction of each object in motion.
> 2. Organize the information in a chart.
> 3. Look for as many things as you can that are the same so that you can write equations.

EXAMPLE 1 A train leaves Stanton traveling east at 35 miles per hour (mph). An hour later, another train leaves Stanton on a parallel track at 40 mph. How far from Stanton will the second (or faster) train catch up with the first (or slower) train?

1. *Familiarize.* We first make a drawing.

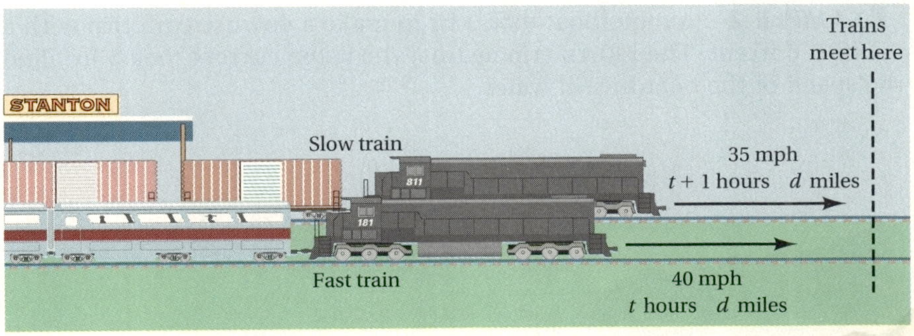

From the drawing, we see that the distances are the same. Let us call the distance d. We don't know the times. Let t = the time for the faster train. Then the time for the slower train = $t + 1$, since it left 1 hr earlier. We can organize the information in a chart.

$$d \quad = \quad r \quad \cdot \quad t$$

	DISTANCE	SPEED	TIME	
SLOW TRAIN	d	35	$t + 1$	→ $d = 35(t + 1)$
FAST TRAIN	d	40	t	→ $d = 40t$

OBJECTIVE

After finishing Section 7.5, you should be able to:

a Solve motion problems using the formula $d = rt$.

FOR EXTRA HELP

TAPE 18 | TAPE 15A | MAC: 7 IBM: 7

1. A car leaves Spokane traveling north at 56 km/h. Another car leaves Spokane one hour later traveling north at 84 km/h. How far from Spokane will the second car catch up with the first? (*Hint:* The cars travel the same distance.)

2. *Translate.* In motion problems, we look for things that are the same so that we can write equations. From each row of the chart, we get an equation, $d = rt$. Thus we have two equations:

$$d = 35(t + 1), \qquad (1)$$
$$d = 40t. \qquad (2)$$

3. *Solve.* Since we have a variable alone on one side, we solve the system using the substitution method:

$35(t + 1) = 40t$ Using the substitution method (substituting $35(t + 1)$ for d in Equation 2)

$35t + 35 = 40t$ Removing parentheses

$35 = 5t$ Subtracting $35t$

$\frac{35}{5} = t$ Dividing by 5

$7 = t.$

The problem asks us to find how far from Stanton the fast train catches up with the other. Thus we need to find d. We can do this by substituting 7 for t in the equation $d = 40t$:

$$d = 40(7)$$
$$= 280.$$

4. *Check.* If the time is 7 hr, then the distance that the slow train travels is $35(7 + 1)$, or 280 mi. The fast train travels $40(7)$, or 280 mi. Since the distances are the same, we know how far from Stanton the trains will be when the fast train catches up with the other.

5. *State.* The fast train will catch up with the slow train 280 mi from Stanton.

Do Exercise 1.

EXAMPLE 2 A motorboat took 3 hr to make a downstream trip with a 6-km/h current. The return trip against the same current took 5 hr. Find the speed of the boat in still water.

1. *Familiarize.* We first make a drawing. From the drawing, we see that the distances are the same. Let's call the distance d. Let r = the speed of the boat in still water. Then, when the boat is traveling downstream,

Answer on page A-7

its speed is $r + 6$ (the current helps the boat along). When it is traveling upstream, its speed is $r - 6$ (the current holds the boat back). We can organize the information in a chart. In this case, the distances are the same, so we use the formula $d = rt$.

	DISTANCE (d)	SPEED ($= r$)	TIME ($\cdot\ t$)	
DOWNSTREAM	d	$r + 6$	3	$\rightarrow d = (r + 6)3$
UPSTREAM	d	$r - 6$	5	$\rightarrow d = (r - 6)5$

2. ***Translate.*** From each row of the chart, we get an equation, $d = rt$:

$d = (r + 6)3,$ **(1)**

$d = (r - 6)5.$ **(2)**

3. ***Solve.*** Since there is a variable alone on one side of an equation, we solve the system using substitution:

$(r + 6)3 = (r - 6)5$ **Substituting $(r + 6)3$ for d in Equation (2)**

$3r + 18 = 5r - 30$ **Removing parentheses**

$-2r + 18 = -30$ **Subtracting $5r$**

$-2r = -48$ **Subtracting 18**

$r = \frac{-48}{-2}$, or 24. **Dividing by -2**

4. ***Check.*** When $r = 24$, $r + 6 = 30$, and $30 \cdot 3 = 90$, the distance downstream. When $r = 24$, $r - 6 = 18$, and $18 \cdot 5 = 90$, the distance upstream. In both cases, we get the same distance.

In this type of problem, a problem-solving tip to keep in mind is "Have I found what the problem asked for?" We could solve for a certain variable but still have not answered the question of the original problem. For example, we might have found speed, r, when the problem wanted distance, d. In this problem, we want the speed of the boat in still water, and that is r.

5. ***State.*** The speed in still water is 24 km/h.

MORE TIPS FOR SOLVING MOTION PROBLEMS

1. Translating to a system of equations eases the solution of many motion problems.
2. When checking, be sure that you have solved for what the original problem asked for.

Do Exercise 2.

2. An airplane flew for 5 hr with a 25-km/h tail wind. The return flight against the same wind took 6 hr. Find the speed of the airplane in still air. (*Hint:* The distance is the same both ways. The speeds are $r + 25$ and $r - 25$, where r is the speed in still air.)

Answer on page A-7

3. Two cars leave town at the same time traveling in opposite directions. One travels at 48 mph and the other at 60 mph. How far apart will they be 3 hr later? (*Hint:* The times are the same. Be *sure* to make a drawing.)

4. Two cars leave town at the same time traveling in the same direction. One travels at 35 mph and the other at 40 mph. In how many hours will they be 15 mi apart? (*Hint:* The times are the same. Be *sure* to make a drawing.)

Answers on page A-7

As we saw in Chapter 5, there are motion problems that can be solved with just one equation. The following is another such problem.

EXAMPLE 3 Two cars leave York at the same time traveling in opposite directions. One travels at 60 mph and the other at 30 mph. In how many hours will they be 150 mi apart?

1. *Familiarize.* We first make a drawing.

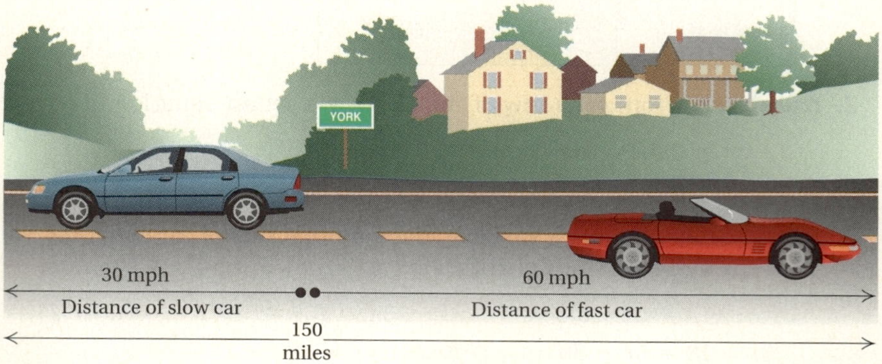

From the wording of the problem and the drawing, we see that the distances may *not* be the same. But the times the cars travel are the same, so we can use just t for time. We can organize the information in a chart.

	d DISTANCE	$=$ r SPEED	$\cdot$ t TIME
FAST CAR	Distance of fast car	60	t
SLOW CAR	Distance of slow car	30	t
TOTAL	150		

2. *Translate.* From the drawing we see that

(Distance of fast car) + (Distance of slow car) = 150.

Then using $d = rt$ in each row of the table, we get $60t + 30t = 150$.

3. *Solve.* We solve the equation:

$$60t + 30t = 150$$

$$90t = 150 \quad \text{Collecting like terms}$$

$$t = \frac{150}{90}, \text{ or } \frac{5}{3}, \text{ or } 1\frac{2}{3} \text{ hours.} \quad \text{Dividing by 90}$$

4. *Check.* When $t = \frac{5}{3}$ hr,

$$(\text{Distance of fast car}) + (\text{Distance of slow car}) = 60\left(\frac{5}{3}\right) + 30\left(\frac{5}{3}\right)$$

$$= 100 + 50, \text{ or } 150 \text{ mi.}$$

Thus the time of $\frac{5}{3}$ hr, or $1\frac{2}{3}$ hr, checks.

5. *State.* In $1\frac{2}{3}$ hr, the cars will be 150 mi apart.

Do Exercises 3 and 4.

NAME SECTION DATE

Exercise Set 7.5

a Solve.

1. Two cars leave town at the same time going in the same direction. One travels at 30 mph and the other travels at 46 mph. In how many hours will they be 72 mi apart? Complete the following table to aid the translation.

$d = r \cdot t$

	DISTANCE	SPEED	TIME
SLOW CAR	Distance of slow car		t
FAST CAR	Distance of fast car	46	

2. A truck and a car leave a service station at the same time and travel in the same direction. The truck travels at 55 mph and the car at 40 mph. They can maintain CB radio contact within a range of 10 mi. When will they lose contact? Complete the following table to aid the translation.

$d = r \cdot t$

	DISTANCE	SPEED	TIME
TRUCK	Distance of truck	55	
CAR	Distance of car		t

3. A train leaves a station and travels east at 72 mph. Three hours later, a second train leaves on a parallel track and travels east at 120 mph. When will it overtake the first train? Complete the following table to aid the translation.

$d = r \cdot t$

	DISTANCE	SPEED	TIME	
SLOW TRAIN	d		$t + 3$	$\rightarrow d = 72(\quad)$
FAST TRAIN	d	120		$\rightarrow d = (\quad)t$

4. A private airplane leaves an airport and flies due south at 192 mph. Two hours later, a jet leaves the same airport and flies due south at 960 mph. When will the jet overtake the plane?

5. A canoeist paddled for 4 hr with a 6-km/h current to reach a campsite. The return trip against the same current took 10 hr. Find the speed of the canoe in still water. Complete the following table to aid the translation.

$d = r \cdot t$

	DISTANCE	SPEED	TIME	
DOWNSTREAM	d	$r + 6$		$\rightarrow d = (\quad)4$
UPSTREAM	d		10	$\rightarrow \quad = (r - 6)10$

6. An airplane flew for 4 hr with a 20-km/h tail wind. The return flight against the same wind took 5 hr. Find the speed of the plane in still air.

ANSWERS

1. ______

2. ______

3. ______

4. ______

5. ______

6. ______

ANSWERS

7. ____________

8. ____________

9. ____________

10. ____________

11. ____________

12. ____________

13. ____________

14. ____________

15. ____________

16. ____________

17. ____________

18. ____________

19. ____________

20. ____________

21. ____________

22. ____________

23. ____________

7. It takes a passenger train 2 hr less time than it takes a freight train to make the trip from Central City to Clear Creek. The passenger train averages 96 km/h, while the freight train averages 64 km/h. How far is it from Central City to Clear Creek?

8. It takes a small jet 4 hr less time than it takes a propeller-driven plane to travel from Glen Rock to Oakville. The jet averages 637 km/h, while the propeller plane averages 273 km/h. How far is it from Glen Rock to Oakville?

9. An airplane took 2 hr to fly 600 mi against a head wind. The return trip with the wind took $1\frac{2}{3}$ hr. Find the speed of the plane in still air.

10. It took 3 hr to row a boat 18 mi against the current. The return trip with the current took $1\frac{1}{2}$ hr. Find the speed of the rowboat in still water.

11. A toddler takes off running down the sidewalk at 230 ft/min. One minute later, a worried mother runs after the child at 660 ft/min. When will the mother overtake the toddler?

12. Two airplanes start at the same time and fly toward each other from points 1000 km apart at rates of 420 km/h and 330 km/h. When will they meet?

13. A motorcycle breaks down and the rider must walk the rest of the way to work. The motorcycle was being driven at 45 mph, and the rider walks at a speed of 6 mph. The distance from home to work is 25 mi, and the total time for the trip was 2 hr. How far did the motorcycle go before it broke down?

14. A student walks and jogs to college each day. The student averages 5 km/h walking and 9 km/h jogging. The distance from home to college is 8 km, and the student makes the trip in 1 hr. How far does the student jog?

SKILL MAINTENANCE

15. $\dfrac{x^2 - 3x - 10}{x^2 - 2x - 15}$

16. $\dfrac{2x^2 - 50}{x^2 - 25}$

17. $\dfrac{6x^2 + 15x - 36}{2x^2 - 5x + 3}$

18. $\dfrac{x^3 + 3x^2 + 2x + 6}{2x^3 + 6x^2 + x + 3}$

19. $\dfrac{5x^8y^4}{10x^3y}$

20. $\dfrac{x^2 - 1}{x^4 - 1}$

SYNTHESIS

21. Charles Lindbergh flew the Spirit of St. Louis in 1927 from New York to Paris at an average speed of 107.4 mph. Eleven years later, Howard Hughes flew the same route, averaged 217.1 mph, and took 16 hr and 57 min less time. Find the length of their route.

22. A car travels from one town to another at a speed of 32 mph. If it had gone 4 mph faster, it could have made the trip in $\frac{1}{2}$ hr less time. How far apart are the towns?

23. An afternoon sightseeing cruise up river and back down river is scheduled to last 1 hr. The speed of the current is 4 mph, and the speed of the riverboat in still water is 12 mph. How far upstream should the pilot travel before turning around?

CRITICAL THINKING

CALCULATOR CONNECTION

1. Solve each system using a grapher to find points of intersection. Solve each equation for y before graphing. Use changes of viewing box and the Zoom and Trace features to help find the solutions to three decimal places.

   a) $3x + 7y = 5,$
   $6x - 7y = 1$

   b) $14x - 10y = 2600,$
   $24x + 20y = 520$

   c) $y - 2.35x = -5.97,$
   $2.14y - x = 4.88$

   d) $4.05x + 2.53y = 1.23,$
   $1.12x + 1.43y = 2.81$

   e) $8y = 3x + 9.6,$
   $-19.2 = 6x - 16y$

   f) $4y - 10.4 = 1.5x,$
   $9.6 = 8y - 3x$

2. Solve each system in Exercise 1 using a scientific calculator to do the computations. Find the solutions to three decimal places.
3. An airplane flew for 4.23 hr with a 25.5-mph tail wind. The return flight against the same wind took 4.97 hr. Find the speed of the plane in still air.

EXTENDED SYNTHESIS EXERCISES

1. Find an equation to go with $2x - 13y = 9$ so that the resulting system has $(-2, -1)$ as its solution.
2. During a publicity campaign, a sports book publisher gave away a combined total of 5000 miniature baseball bats and sports cards. The bats cost 21¢ each and the cards cost 14¢ each. The publisher spent a total of $826 on the gifts. How many of each gift did they buy?
3. Against a head wind, Amelia computes her flight time for a trip of 2900 mi at 5 hr. If the head wind were half as much, the flight would take 4 hr and 50 min. Find the speed of the head wind and the speed of the plane in still air.
4. If $2x + 3y = 27$ and $x + 2y = 7$, find $\dfrac{3x + 5y}{2}$.
5. If $x + 4y = 9$ and $4x + 3y = 7$, find $5x + 7y$.
6. If $4x - 3y = 12$ and $x = \dfrac{7y}{4}$, find y.
7. If $x + y = 3$ and $x - y = 2$, find $4x$.
8. To deliver a package, a messenger must travel at a speed of 60 mph on land and then use a motorboat whose speed is 20 mph in still water. While delivering the package, the messenger first travels on land to a dock and then travels on a river against a current of 4 mph. The messenger reaches the destination in 4.5 hr and then returns to the starting point in 3.5 hr. How far did the messenger travel on land and how far on water?
9. An airplane took $2\frac{1}{2}$ hr to fly 625 mi with the wind. It took 4 hr and 10 min to make the return trip against the same wind. Find the speed of the wind and the speed of the plane in still air.

10. A rectangle has a perimeter of P ft. The width is 5 ft less than the length. Find the length and the width in terms of P.

Solve each system.

11. $3(7 - a) - 2(1 + 2b) + 5 = 0,$
    $3a + 2b - 18 = 0$
12. $3(x - y) + 4(x + y) = 10,$
    $2(x + y) - 5(x - y) = 16$

Each of the following shows the graph of a system of equations. Find the equations.

13. 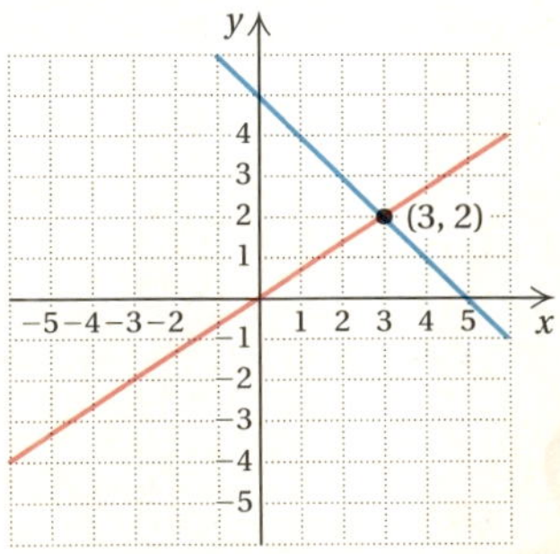

14. 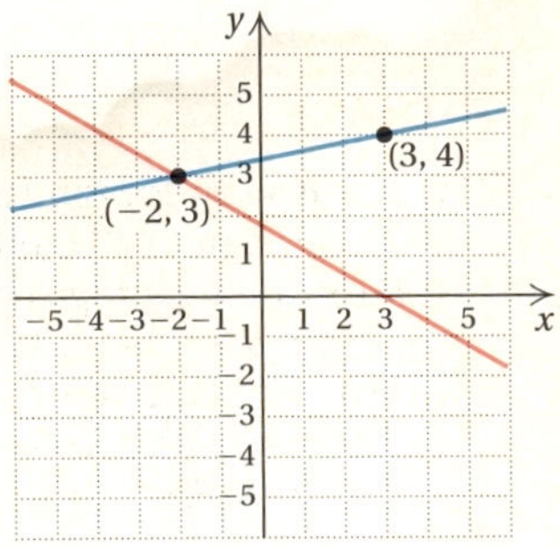

(continued)

15.

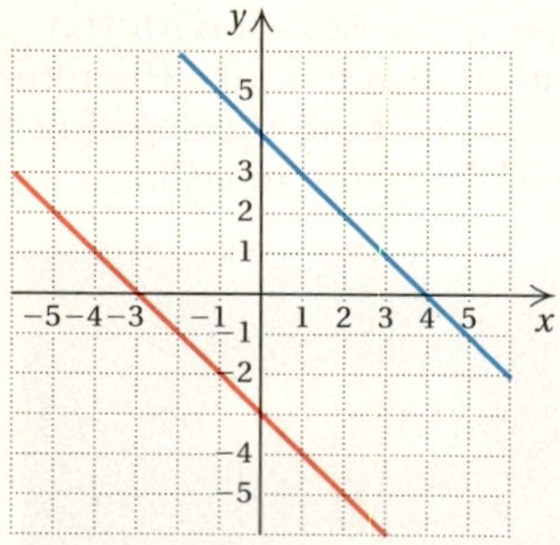

16.

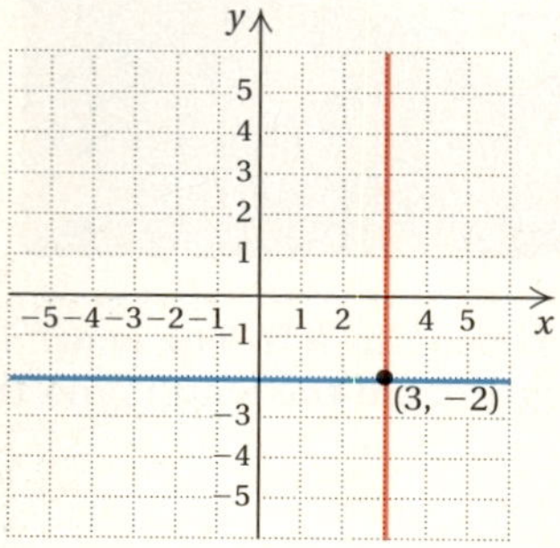

Exercises for Thinking and Writing

1. Briefly compare the strengths and the weaknesses of the graphical, substitution, and elimination methods.
2. Write a system of equations with no solution. Explain why the system has no solution.
3. Write a system of equations with infinitely many solutions. Explain why the system has infinitely many solutions.
4. Explain the advantages of using a system of equations to solve certain kinds of problems.

Each of the following lists the steps in solving a certain system of equations, but an error occurs. Find and describe the error and correct the answer.

5.
$$\begin{aligned} 3x - y &= 4 \\ 2x + y &= 16 \\ \hline 5x \quad &= 20 \\ x &= 4 \end{aligned}$$

$$\begin{aligned} 3x - y &= 4 \\ 3(4) - y &= 4 \\ y &= 4 - 12 \\ y &= -8 \end{aligned}$$

The solution is $(4, -8)$.

6.
$$\begin{aligned} 4x - y &= 17 \\ -x - y &= 7 \\ \hline 3x &= 24 \\ x &= 8 \end{aligned}$$

$$\begin{aligned} 4(8) - y &= 17 \\ 32 - y &= 17 \\ -y &= -15 \\ y &= 15 \end{aligned}$$

The solution is $(8, 15)$.

Summary and Review: Chapter 7

Important Properties and Formulas

Motion Formula: $d = rt$

Review Exercises

The review objectives to be tested in addition to the material in this chapter are [3.1d, e], [5.1c], [5.5a], and [6.3a].

Determine whether the given ordered pair is a solution of the system of equations.

1. $(6, -1)$; $x - y = 3$, $2x + 5y = 6$

2. $(2, -3)$; $2x + y = 1$, $x - y = 5$

3. $(-2, 1)$; $x + 3y = 1$, $2x - y = -5$

4. $(-4, -1)$; $x - y = 3$, $x + y = -5$

Solve the system of equations by graphing.

5. $x + y = 4$, $x - y = 8$

6. $x + 3y = 12$, $2x - 4y = 4$

7. $y = 5 - x$, $3x - 4y = -20$

8. $3x - 2y = -4$, $2y - 3x = -2$

Solve using the substitution method.

9. $y = 5 - x$, $3x - 4y = -20$

10. $x + 2y = 6$, $2x + 3y = 8$

11. $3x + y = 1$, $x - 2y = 5$

12. $x + y = 6$, $y = 3 - 2x$

13. $s + t = 5$, $s = 13 - 3t$

14. $x - y = 4$, $y = 2 - x$

Solve using the elimination method.

15. $x + y = 4$, $2x - y = 5$

16. $x + 2y = 9$, $3x - 2y = -5$

17. $x - y = 8$, $2x + y = 7$

18. $\frac{2}{3}x + y = -\frac{5}{3}$, $x - \frac{1}{3}y = -\frac{13}{3}$

19. $2x + 3y = 8$, $5x + 2y = -2$

20. $5x - 2y = 2$, $3x - 7y = 36$

21. $-x - y = -5$, $2x - y = 4$

22. $6x + 2y = 4$, $10x + 7y = -8$

23. $-6x - 2y = 5$, $12x + 4y = -10$

Solve.

24. The sum of two numbers is 8. Their difference is 12. Find the numbers.

25. The sum of two numbers is 27. One-half of the first number plus one-third of the second number is 11. Find the numbers.

26. The perimeter of a rectangle is 96 cm. The length is 27 cm more than the width. Find the length and the width.

27. An airplane flew for 4 hr with a 15-km/h tail wind. The return flight against the wind took 5 hr. Find the speed of the airplane in still air.

28. There were 508 people at a rock concert. Orchestra seats cost $25 per person and balcony seats cost $18. The total receipts were $11,223. Find the number of orchestra seats and the number of balcony seats sold.

29. Solution A is 30% alcohol, and solution B is 60% alcohol. How much of each is needed to make 80 L of a solution that is 45% alcohol?

30. Jeff is three times as old as his son. In 9 yr, Jeff will be twice as old as his son. How old is each now?

31. A zoo has an Asian and an African elephant. The African elephant weighs 2400 kg more than the Asian elephant. Together, they weigh 12,000 kg. How much does each elephant weigh?

SKILL MAINTENANCE

Simplify.

32. $t^{-5} \cdot t^{13}$

33. $\dfrac{t^{-5}}{t^{13}}$

34. Subtract: $\dfrac{x}{x^2 - 9} - \dfrac{x - 1}{x^2 - 5x + 6}$.

35. Graph: $2y - x = 6$.

SYNTHESIS

36. The solution of the following system is (6, 2). Find C and D.

$$2x - Dy = 6,$$
$$Cx + 4y = 14$$

37. Solve:

$$3(x - y) = 4 + x,$$
$$x = 5y + 2.$$

38. A stablehand agreed to work for 1 year. At the end of that time, she was to receive $240 and one horse. After 7 months she quit the job, but still received the horse and $100. What was the value of the horse?

39. The sum of the digits of a two-digit number is 6. When the digits are reversed, the new number is 18 more than the original number. Find the original number.

NAME SECTION DATE

Test: Chapter 7

1. Determine whether the given ordered pair is a solution of the system of equations.

$(-2, -1);\quad x = 4 + 2y,$
$\qquad 2y - 3x = 4$

2. Solve this system by graphing:

$x - y = 3,$
$x - 2y = 4.$

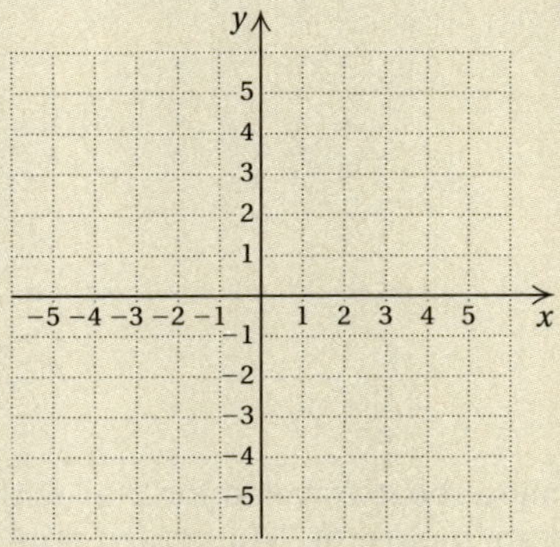

Solve using the substitution method.

3. $y = 6 - x,$
$2x - 3y = 22$

4. $x + 2y = 5,$
$x + y = 2$

5. $y = 5x - 2,$
$y - 2 = 5x$

Solve using the elimination method.

6. $x - y = 6,$
$3x + y = -2$

7. $\frac{1}{2}x - \frac{1}{3}y = 8,$
$\frac{2}{3}x + \frac{1}{2}y = 5$

8. $4x + 5y = 5,$
$6x + 7y = 7$

9. $2x + 3y = 13,$
$3x - 5y = 10$

ANSWERS

1. ______

2. ______

3. ______

4. ______

5. ______

6. ______

7. ______

8. ______

9. ______

ANSWERS

10. ______

11. ______

12. ______

13. ______

14. ______

15. ______

16. ______

17. ______

18. ______

19. ______

10. The perimeter of a rectangular field is 8266 yd. The length is 84 yd more than the width. Find the length and the width.

11. The difference of two numbers is 12. One-fourth of the larger number plus one-half of the smaller is 9. Find the numbers.

12. A motorboat traveled for 2 hr with an 8-km/h current. The return trip against the same current took 3 hr. Find the speed of the motorboat in still water.

13. Solution A is 25% acid, and solution B is 40% acid. How much of each is needed to make 60 L of a solution that is 30% acid?

SKILL MAINTENANCE

14. Subtract: $\frac{1}{x^2 - 16} - \frac{x - 4}{x^2 - 3x - 4}$.

15. Graph: $3x - 4y = -12$.

Simplify.

16. $(2x^{-2}y^7)(5x^6y^{-9})$

17. $\frac{a^4b^2}{a^{-6}b^8}$

SYNTHESIS

18. Find the numbers C and D such that $(-2, 3)$ is a solution of the system

$$Cx - 4y = 7,$$
$$3x + Dy = 8.$$

19. You are in line at a ticket window. There are two more people ahead of you than there are behind you. In the entire line, there are three times as many people as there are behind you. How many are ahead of you in line?

Cumulative Review: Chapters 1–7

Compute and simplify.

1. $-2[1.4 - (-0.8 - 1.2)]$

2. $(1.3 \times 10^8)(2.4 \times 10^{-10})$

3. $\left(-\frac{1}{6}\right) \div \left(\frac{2}{9}\right)$

4. $\frac{2^{12}2^{-7}}{2^8}$

Simplify.

5. $\frac{x^2 - 9}{2x^2 - 7x + 3}$

6. $\frac{t^2 - 16}{(t + 4)^2}$

7. $\frac{x - \frac{x}{x+2}}{\frac{2}{x} - \frac{1}{x+2}}$

Perform the indicated operations and simplify.

8. $(1 - 3x^2)(2 - 4x^2)$

9. $(2a^2b - 5ab^2)^2$

10. $(3x^2 + 4y)(3x^2 - 4y)$

11. $-2x^2(x - 2x^2 + 3x^3)$

12. $(1 + 2x)(4x^2 - 2x + 1)$

13. $\left(8 - \frac{1}{3}x\right)\left(8 + \frac{1}{3}x\right)$

14. $(-8y^2 - y + 2) - (y^3 - 6y^2 + y - 5)$

15. $(2x^3 - 3x^2 - x - 1) \div (2x - 1)$

16. $\frac{7}{5x - 25} + \frac{x + 7}{5 - x}$

17. $\frac{2x - 1}{x - 2} - \frac{2x}{2 - x}$

18. $\frac{y^2 + y}{y^2 + y - 2} \cdot \frac{y + 2}{y^2 - 1}$

19. $\frac{7x + 7}{x^2 - 2x} \div \frac{14}{3x - 6}$

Factor completely.

20. $6x^5 - 36x^3 + 9x^2$

21. $16y^4 - 81$

22. $3x^2 + 10x - 8$

23. $4x^4 - 12x^2y + 9y^2$

24. $3m^3 + 6m^2 - 45m$

25. $x^3 + x^2 - x - 1$

Solve.

26. $3x - 4(x + 1) = 5$

27. $x(2x - 5) = 0$

28. $5x + 3 \geq 6(x - 4) + 7$

29. $1.5x - 2.3x = 0.4(x - 0.9)$

30. $2x^2 = 338$

31. $3x^2 + 15 = 14x$

32. $\frac{2}{x} - \frac{3}{x - 2} = \frac{1}{x}$

33. $1 + \frac{3}{x} + \frac{x}{x + 1} = \frac{1}{x^2 + x}$

34. $y = 2x - 9,$
$2x + 3y = -3$

35. $6x + 3y = -6,$
$-2x + 5y = 14$

36. $2x = y - 2,$
$3y - 6x = 6$

37. $\frac{1}{x} - \frac{1}{y} = \frac{1}{xy}$, for x

Solve.

38. The vice-president of a sorority has \$100 to spend on promotional buttons. There is a set-up fee of \$18 and a cost of 35¢ per button. How many buttons can she purchase?

39. It takes David 15 hr to put a roof on a house. It takes Loren 9 hr to put a roof on the same type of house. How long would it take if they worked together?

40. The length of one leg of a right triangle is 8 m. The length of the hypotenuse is 4 m longer than the length of the other leg. Find the lengths of the hypotenuse and the other leg.

41. To determine the number of fish in a lake, a conservationist catches 85 fish, tags them, and throws them back into the lake. Later, 60 fish are caught, 25 of which are tagged. How many fish are in the lake?

42. The height of a triangle is 3 cm less than the base. The area is 27 cm^2. Find the height and the base.

43. The height h of a parallelogram of fixed area varies inversely as the base b. Suppose that the height is 24 ft when the base is 15 ft. Find the height when the base is 5 ft. What is the variation constant?

44. Two cars leave town at the same time going in the same direction. One travels 50 mph and the other travels 55 mph. In how many hours will they be 50 mi apart?

45. Solution A is 20% alcohol, and solution B is 60% alcohol. How much of each should be used to make 10 L of a solution that is 50% alcohol?

46. Find an equation of variation where y varies directly as x and $y = 2.4$ when $x = 12$.

47. Find the slope of the line containing the points $(2, 3)$ and $(-1, 3)$.

48. Find the slope and the y-intercept of the line $2x + 3y = 6$.

49. Find an equation of the line that contains the points $(-5, 6)$ and $(2, -4)$.

50. Find an equation of the line containing the point $(0, -3)$ and having the slope $m = 6$.

Graph on a plane.

51. $y = -2$

52. $2x + 5y = 10$

53. $y \leq 5x$

54. $5x - 1 < 24$

Solve by graphing.

55. $x = 5 + y,$
$x - y = 1$

56. $3x - y = 4,$
$x + 3y = -2$

SYNTHESIS

57. The solution of the following system is $(-5, 2)$. Find A and B.

$$3x - Ay = -7,$$
$$Bx + 4y = 15$$

58. Solve: $x^2 + 2 < 0$.

59. Simplify:

$$\frac{x - 5}{x + 3} - \frac{x^2 - 6x + 5}{x^2 + x - 2} \div \frac{x^2 + 4x + 3}{x^2 + 3x + 2}.$$

60. Find the value of k so that $y - kx = 4$ and $10x - 3y = -12$ are perpendicular.

8

Radical Expressions and Equations

INTRODUCTION

The formula on this page illustrates the use of another type of expression in problem solving. It is called a *radical expression* and involves a square root. We say that 3 is a square root of 9 because $3^2 = 9$. In this chapter, we study manipulations of radical expressions in addition, subtraction, multiplication, division, and simplifying. Finally, we consider another equation-solving principle and apply it to problem solving.

AN APPLICATION

How far to the horizon can you see through an airplane window at a certain height?

THE MATHEMATICS

At a height of h meters, you can see V kilometers to the horizon, where V is given by the formula

$$V = \underbrace{3.5\sqrt{h}}.$$

This is a radical expression.

Objectives for Review

The review objectives to be tested in addition to the material in this chapter are as follows.

[5.2b] Divide rational expressions.

[6.7b] Solve problems involving direct variation.

[7.3a, b] Solve a system of two equations in two variables using the elimination method.

[7.3c] Solve problems by translating to a system of two equations and then solving using the elimination method.

Pretest: Chapter 8

1. Find the square roots of 49.

2. Identify the radicand in $\sqrt{3t}$.

Determine whether the expression is meaningful as a real number. Write "yes" or "no."

3. $\sqrt{-47}$

4. $\sqrt{81}$

5. Approximate $\sqrt{47}$ to three decimal places.

6. Solve: $\sqrt{2x+1} = 3$.

Assume henceforth that *all* expressions under radicals represent positive numbers.

Simplify.

7. $\sqrt{4x^2}$

8. $4\sqrt{18} - 2\sqrt{8} + \sqrt{32}$

9. $(2 - \sqrt{3})^2$

10. $(2 - \sqrt{3})(2 + \sqrt{3})$

Multiply and simplify.

11. $\sqrt{6}\sqrt{10}$

12. $(2\sqrt{6} - 1)^2$

Divide and simplify.

13. $\dfrac{\sqrt{15}}{\sqrt{3}}$

14. $\sqrt{\dfrac{24a^7}{3a^3}}$

15. In a right triangle, $a = 5$ and $b = 8$. Find c, the length of the hypotenuse. Give an exact answer and an approximation to three decimal places.

16. How long is a guy wire reaching from the top of a 12-m pole to a point 7 m from the base of the pole?

17. Rationalize the denominator:

$$\frac{\sqrt{5}}{\sqrt{x}}.$$

18. Rationalize the denominator:

$$\frac{8}{6 + \sqrt{5}}.$$

8.1 *Introduction to Square Roots and Radical Expressions*

a SQUARE ROOTS

When we raise a number to the second power, we have squared the number. Sometimes we may need to find the number that was squared. We call this process finding a square root of a number.

> The number c is a *square root* of a if $c^2 = a$.

Every positive number has two square roots. For example, the square roots of 25 are 5 and -5 because $5^2 = 25$ and $(-5)^2 = 25$. The positive square root is also called the **principal square root.** The symbol $\sqrt{}$ is called a **radical** symbol. The radical symbol represents only the principal square root. Thus, $\sqrt{25} = 5$. To name the negative square root of a number, we use $-\sqrt{}$. The number 0 has only one square root, 0.

EXAMPLE 1 Find the square roots of 81.

The square roots are 9 and -9.

EXAMPLE 2 Find $\sqrt{225}$.

The symbol $\sqrt{225}$ represents the principal square root. There are two square roots, 15 and -15. We want the positive square root since this is what $\sqrt{}$ represents. Thus, $\sqrt{225} = 15$.

Table 2 at the back of the book contains a list of squares and square roots. It would be helpful to memorize the squares of whole numbers from 0 to 25.

EXAMPLE 3 Find $-\sqrt{64}$.

The symbol $\sqrt{64}$ represents the positive square root. Then $-\sqrt{64}$ represents the negative square root. That is, $\sqrt{64} = 8$, so $-\sqrt{64} = -8$.

Do Exercises 1–10.

b APPROXIMATING SQUARE ROOTS

We often need to use rational numbers to approximate square roots that are irrational. Such approximations can be found using a calculator with a square root key $\boxed{\sqrt{}}$. They can also be found using Table 2 at the back of the book.

OBJECTIVES

After finishing Section 8.1, you should be able to:

- **a** Find the principal square roots and their opposites of the whole numbers from 0^2 to 25^2.
- **b** Approximate square roots of real numbers using a calculator or a table.
- **c** Solve problems involving applications of square roots.
- **d** Identify radicands of radical expressions.
- **e** Identify whether a radical expression represents a real number.
- **f** Simplify a radical expression with a perfect-square radicand.

FOR EXTRA HELP

TAPE 18	TAPE 15B	MAC: 8A IBM: 8A

Find the square roots.

1. 36 **2.** 64

3. 121 **4.** 144

Find the following.

5. $\sqrt{16}$ **6.** $\sqrt{49}$

7. $\sqrt{100}$ **8.** $\sqrt{441}$

9. $-\sqrt{49}$ **10.** $-\sqrt{169}$

Answers on page A-7

Use a calculator or Table 2. Approximate to three decimal places.

11. $\sqrt{30}$

12. $-\sqrt{98}$

13. In the situation of Example 5, find the number of spaces needed when the average number of arrivals in peak hours is **(a)** 64; **(b)** 83.

Identify the radicand.

14. $\sqrt{45 + x}$

15. $\sqrt{\dfrac{x}{x + 2}}$

16. $8\sqrt{x^2 + 4}$

Answers on page A-7

EXAMPLE 4 Use a calculator or Table 2 to approximate $\sqrt{10}$. Round to three decimal places.

Before finding square roots on a calculator, you will need to consult the instruction manual. Calculators vary in their methods of operation.

$$\sqrt{10} \approx 3.162277660 \quad \text{Using a calculator with a 10-digit readout}$$

Different calculators give different numbers of digits in their readouts. This may cause some variance in answers. We round to the third decimal place. Then $\sqrt{10} \approx 3.162$. This can also be found in Table 2.

Do Exercises 11 and 12.

c APPLICATIONS OF SQUARE ROOTS

We now consider an application involving a formula with a radical expression.

EXAMPLE 5 *Parking lot arrival spaces.* A parking lot has attendants to park cars, and it uses spaces in which cars are left before they are taken to permanent parking stalls. The number N of such spaces needed is approximated by the formula

$$N = 2.5\sqrt{A},$$

where A is the average number of arrivals in peak hours. Find the number of spaces needed when the average number of arrivals in peak hours is 77.

We substitute 77 into the formula and find an approximation:

$$\begin{aligned} N &= 2.5\sqrt{77} \\ &\approx 2.5(8.774964387) \\ &\approx 21.93741097 \\ &\approx 22. \end{aligned}$$

Note that we round up to 22 because a fractional part of a parking space cannot be used. To ensure that we have enough, we need 22.

Do Exercise 13.

d RADICANDS AND RADICAL EXPRESSIONS

When an expression is written under a radical, we have a **radical expression.** Here are some examples:

$$\sqrt{14}, \quad \sqrt{x}, \quad 8\sqrt{x^2 + 4}, \quad \sqrt{\frac{x^2 - 5}{2}}.$$

The expression written under the radical is called the **radicand.**

EXAMPLES Identify the radicand in each expression.

6. $\sqrt{x}$ — The radicand is x.

7. $6\sqrt{y^2 - 5}$ — The radicand is $y^2 - 5$.

8. $\sqrt{\dfrac{a - b}{a + b}}$ — The radicand is $\dfrac{a - b}{a + b}$.

Do Exercises 14–16 on the preceding page.

e EXPRESSIONS THAT ARE MEANINGFUL AS REAL NUMBERS

The square of any nonzero number is always positive. For example, $8^2 = 64$ and $(-11)^2 = 121$. There are no real numbers that can be squared to get negative numbers.

Radical expressions with negative radicands do not represent real numbers.

Thus the following expressions do not represent real numbers (they are meaningless as real numbers):

$$\sqrt{-100}, \qquad \sqrt{-49}, \qquad -\sqrt{-3}.$$

Later in your study of mathematics, you may encounter a number system called the **complex numbers** in which negative numbers have square roots.

Do Exercises 17–20.

f PERFECT-SQUARE RADICANDS

The expression $\sqrt{x^2}$, with a perfect-square radicand, can be troublesome. If x represents a nonnegative number, $\sqrt{x^2}$ simplifies to x. If x represents a negative number, $\sqrt{x^2}$ simplifies to $-x$ (the opposite of x). That is because $\sqrt{a}$ denotes the *principal* square root of a.

Suppose $x = 3$. Then $\sqrt{x^2} = \sqrt{3^2}$, which is $\sqrt{9}$, or 3. Suppose $x = -3$. Then $\sqrt{x^2} = \sqrt{(-3)^2}$, which is $\sqrt{9}$, or 3, the *absolute value* of -3. In either case, when replacements for x are considered to be any real number, it follows that $\sqrt{x^2} = |x|$.

For any real number A,

$$\sqrt{A^2} = |A|.$$

(That is, for any real number A, the principal square root of A^2 is the absolute value of A.)

EXAMPLES Simplify. Assume that expressions under radicals represent any real number.

9. $\sqrt{(3x)^2} = |3x|$ Absolute-value notation is necessary.

10. $\sqrt{a^2b^2} = \sqrt{(ab)^2} = |ab|$

11. $\sqrt{x^2 + 2x + 1} = \sqrt{(x + 1)^2} = |x + 1|$

Do Exercises 21–24.

Determine whether the expression is meaningful as a real number. Write "yes" or "no."

17. $-\sqrt{25}$

18. $\sqrt{-25}$

19. $-\sqrt{-36}$

20. $-\sqrt{36}$

Simplify. Assume that expressions under radicals represent any real number.

21. $\sqrt{(xy)^2}$

22. $\sqrt{x^2y^2}$

23. $\sqrt{(x - 11)^2}$

24. $\sqrt{x^2 + 8x + 16}$

Answers on page A-7

Simplify. Assume that expressions under radicals represent nonnegative real numbers.

25. $\sqrt{(xy)^2}$

26. $\sqrt{x^2y^2}$

27. $\sqrt{(x-11)^2}$

28. $\sqrt{x^2+8x+16}$

29. $\sqrt{25y^2}$

30. $\sqrt{\frac{1}{4}t^2}$

Answers on page A-7

Fortunately, in most uses of radicals, it can be assumed that expressions under radicals are nonnegative or positive. Indeed, many computers are programmed to consider only nonnegative radicands. Suppose that $x \geq 0$. Then

$$\sqrt{x^2} = |x| = x,$$

since x is nonnegative.

For any nonnegative real number A,

$$\sqrt{A^2} = A.$$

(That is, for any nonnegative real number A, the principal square root of A^2 is A.)

EXAMPLES Simplify. Assume that expressions under radicals represent nonnegative real numbers.

12. $\sqrt{(3x)^2} = 3x$ — Since $3x$ is assumed to be nonnegative

13. $\sqrt{a^2b^2} = \sqrt{(ab)^2} = ab$ — Since ab is assumed to be nonnegative

14. $\sqrt{x^2+2x+1} = \sqrt{(x+1)^2} = x+1$ — Since $x+1$ is assumed to be nonnegative

Do Exercises 25–30.

Henceforth, in this text we will assume that all expressions under radicals represent nonnegative real numbers.

We make this assumption in order to eliminate some confusion and because it is valid in many applications. As you study further in mathematics, however, you will frequently have to make a determination about expressions under radicals being nonnegative or positive. This will often be necessary in calculus.

NAME SECTION DATE

Exercise Set 8.1

a Find the square roots.

1. 4 **2.** 1 **3.** 9 **4.** 16

5. 100 **6.** 121 **7.** 169 **8.** 144

Simplify.

9. $\sqrt{4}$ **10.** $\sqrt{1}$ **11.** $-\sqrt{9}$ **12.** $-\sqrt{25}$ **13.** $-\sqrt{36}$

14. $-\sqrt{81}$ **15.** $-\sqrt{225}$ **16.** $\sqrt{400}$ **17.** $\sqrt{361}$ **18.** $\sqrt{441}$

b Use a calculator or Table 2 to approximate these square roots. Round to three decimal places.

19. $\sqrt{5}$ **20.** $\sqrt{8}$ **21.** $\sqrt{17}$

22. $\sqrt{19}$ **23.** $\sqrt{91}$ **24.** $\sqrt{43}$

c Solve. Use the formula $N = 2.5\sqrt{A}$ of Example 5.

25. Find the number of spaces needed when the average number of arrivals is **(a)** 25; **(b)** 89.

26. Find the number of spaces needed when the average number of arrivals is **(a)** 62; **(b)** 100.

d Identify the radicand.

27. $\sqrt{a - 4}$ **28.** $\sqrt{t + 10}$ **29.** $5\sqrt{t^2 + 1}$

30. $9\sqrt{x^2 + 16}$ **31.** $x^2y\sqrt{\frac{3}{x + 2}}$ **32.** $ab^2\sqrt{\frac{a}{a - b}}$

ANSWERS

1. ____
2. ____
3. ____
4. ____
5. ____
6. ____
7. ____
8. ____
9. ____
10. ____
11. ____
12. ____
13. ____
14. ____
15. ____
16. ____
17. ____
18. ____
19. ____
20. ____
21. ____
22. ____
23. ____
24. ____
25. a) ____
b) ____
26. a) ____
b) ____
27. ____
28. ____
29. ____
30. ____
31. ____
32. ____

ANSWERS

33. ______
34. ______
35. ______
36. ______
37. ______
38. ______
39. ______
40. ______
41. ______
42. ______
43. ______
44. ______
45. ______
46. ______
47. ______
48. ______
49. ______
50. ______
51. ______
52. ______
53. ______
54. ______
55. ______
56. ______
57. ______
58. ______
59. ______
60. ______
61. ______
62. ______
63. ______
64. ______
65. ______

e Determine whether the expression is meaningful as a real number. Write "yes" or "no."

33. $\sqrt{-16}$ **34.** $\sqrt{-81}$ **35.** $-\sqrt{81}$ **36.** $-\sqrt{64}$

f Simplify. Remember that we have assumed that expressions under radicals represent nonnegative real numbers.

37. $\sqrt{c^2}$ **38.** $\sqrt{x^2}$ **39.** $\sqrt{9x^2}$

40. $\sqrt{16y^2}$ **41.** $\sqrt{(ab)^2}$ **42.** $\sqrt{(6y)^2}$

43. $\sqrt{(34d)^2}$ **44.** $\sqrt{(53b)^2}$ **45.** $\sqrt{(x + 3)^2}$

46. $\sqrt{(d - 3)^2}$ **47.** $\sqrt{a^2 - 10a + 25}$ **48.** $\sqrt{x^2 + 2x + 1}$

49. $\sqrt{4a^2 - 20a + 25}$ **50.** $\sqrt{9p^2 + 12p + 4}$

SKILL MAINTENANCE

51. The amount F that a family spends on food varies directly as its income I. A family making \$19,600 a year will spend \$5096 on food. At this rate, how much would a family making \$20,500 spend on food?

Divide and simplify.

52. $\dfrac{x - 3}{x + 4} \div \dfrac{x^2 - 9}{x + 4}$ **53.** $\dfrac{x^2 - x - 2}{x - 1} \div \dfrac{x - 2}{x^2 - 1}$

54. $\dfrac{x^2 + 10x - 11}{x^2 - 1} \div \dfrac{x + 11}{x + 1}$ **55.** $\dfrac{x^4 - 16}{x^4 - 1} \div \dfrac{x^2 + 4}{x^2 + 1}$

SYNTHESIS

Simplify.

56. $\sqrt{\sqrt{625}}$ **57.** $\sqrt{3^2 + 4^2}$

58. Between what two consecutive integers is $-\sqrt{33}$?

Use a calculator to approximate the square root. Round to three decimal places.

59. $\sqrt{12.8}$ **60.** $\sqrt{930}$ **61.** $\sqrt{1043.89}$

Solve.

62. $\sqrt{x^2} = 6$ **63.** $\sqrt{y^2} = -7$ **64.** $t^2 = 49$

65. Suppose the area of a square is 3. Find the length of a side.

8.2 Multiplying and Simplifying with Radical Expressions

OBJECTIVES

After finishing Section 8.2, you should be able to:

- a Simplify radical expressions.
- b Simplify radical expressions where radicands are powers.
- c Multiply radical expressions and simplify, if possible.
- d Approximate square roots not in Table 2 using factoring and the square root table.

FOR EXTRA HELP

TAPE 19 | TAPE 15B | MAC: 8A IBM: 8A

a SIMPLIFYING BY FACTORING

To see how to multiply with radical notation, consider the following.

a) $\sqrt{9} \cdot \sqrt{4} = 3 \cdot 2 = 6$ **This is a product of square roots.**

b) $\sqrt{9 \cdot 4} = \sqrt{36} = 6$ **This is the square root of a product.**

Note that

$$\sqrt{9} \cdot \sqrt{4} = \sqrt{9 \cdot 4}.$$

Do Exercise 1.

We can multiply radical expressions by multiplying the radicands.

> **THE PRODUCT RULE FOR RADICALS**
>
> For any nonnegative radicands A and B,
>
> $$\sqrt{A} \cdot \sqrt{B} = \sqrt{A \cdot B}.$$
>
> (The product of square roots is the square root of the product of the radicands.)

EXAMPLES Multiply.

1. $\sqrt{5}\sqrt{7} = \sqrt{5 \cdot 7} = \sqrt{35}$

2. $\sqrt{8}\sqrt{8} = \sqrt{8 \cdot 8} = \sqrt{64} = 8$

3. $\sqrt{\frac{2}{3}}\sqrt{\frac{4}{5}} = \sqrt{\frac{2}{3} \cdot \frac{4}{5}} = \sqrt{\frac{8}{15}}$

4. $\sqrt{2x}\sqrt{3x - 1} = \sqrt{2x(3x - 1)}$
$= \sqrt{6x^2 - 2x}$

Do Exercises 2–5.

To factor radical expressions, we can use the product rule for radicals in reverse. That is,

> $$\sqrt{AB} = \sqrt{A}\sqrt{B}.$$

In some cases, we can simplify after factoring.

> A radical expression is simplified when its radicand has no factors that are perfect squares.

When simplifying a radical expression, we first determine whether a radicand is a perfect square. Then we determine whether it has perfect-square factors. The radicand is then factored and the radical expression simplified using the preceding rule.

---

1. Simplify.
   a) $\sqrt{25} \cdot \sqrt{16}$

   b) $\sqrt{25 \cdot 16}$

Multiply.

2. $\sqrt{3}\sqrt{11}$

3. $\sqrt{5}\sqrt{5}$

4. $\sqrt{x}\sqrt{x + 1}$

5. $\sqrt{x + 2}\sqrt{x - 2}$

Answers on page A-7

Simplify by factoring.

6. $\sqrt{32}$

7. $\sqrt{x^2 + 14x + 49}$

8. $\sqrt{25x^2}$

9. $\sqrt{36m^2}$

10. $\sqrt{92}$

11. $\sqrt{x^2 - 20x + 100}$

12. $\sqrt{64t^2}$

13. $\sqrt{100a^2}$

Answers on page A-7

Compare the following:

$$\sqrt{50} = \sqrt{10 \cdot 5} = \sqrt{10}\sqrt{5};$$
$$\sqrt{50} = \sqrt{25 \cdot 2} = \sqrt{25}\sqrt{2} = 5\sqrt{2}.$$

In the second case, the radicand has the perfect-square factor 25. If you do not recognize perfect-square factors, try factoring the radicand into its prime factors. For example,

$$\sqrt{50} = \sqrt{2 \cdot \underbrace{5 \cdot 5}} = 5\sqrt{2}.$$

Perfect square (a pair of the same numbers)

Radical expressions in which the radicand has no perfect-square factors, such as $5\sqrt{2}$, are considered to be in simplest form.

Examples

5. $\sqrt{18} = \sqrt{9 \cdot 2}$ — Identifying a perfect-square factor and factoring the radicand. The factor 9 is a perfect square.

$= \sqrt{9} \cdot \sqrt{2}$ — Factoring into a product of radicals

$= 3\sqrt{2}$ — $3\sqrt{2}$ means $3 \cdot \sqrt{2}$

The radicand has no factors that are perfect squares.

6. $\sqrt{48t} = \sqrt{16 \cdot 3t}$ — Identifying a perfect-square factor and factoring the radicand. The factor 16 is a perfect square.

$= \sqrt{16}\sqrt{3t}$ — Factoring into a product of radicals

$= 4\sqrt{3t}$ — Taking a square root

7. $\sqrt{20t^2} = \sqrt{4 \cdot t^2 \cdot 5}$ — Identifying perfect-square factors and factoring the radicand. The factors 4 and t^2 are perfect squares.

$= \sqrt{4}\sqrt{t^2}\sqrt{5}$ — Factoring into a product of several radicals

$= 2t\sqrt{5}$ — Taking square roots. No absolute-value signs are necessary since we have assumed that expressions under radicals are nonnegative.

8. $\sqrt{x^2 - 6x + 9} = \sqrt{(x - 3)^2} = x - 3$ — No absolute-value signs are necessary since we have assumed that expressions under radicals are nonnegative.

9. $\sqrt{36x^2} = \sqrt{36}\sqrt{x^2} = 6x$, or $\sqrt{36x^2} = \sqrt{(6x)^2} = 6x$

10. $\sqrt{3x^2 + 6x + 3} = \sqrt{3(x^2 + 2x + 1)}$ — Factoring the radicand

$= \sqrt{3}\sqrt{x^2 + 2x + 1}$ — Factoring into a product of radicals

$= \sqrt{3}\sqrt{(x + 1)^2}$

$= \sqrt{3}\,(x + 1)$ — Taking the square root

Do Exercises 6–13.

b Simplifying Square Roots of Powers

To take the square root of an even power such as x^{10}, we note that $x^{10} = (x^5)^2$. Then

$$\sqrt{x^{10}} = \sqrt{(x^5)^2} = x^5.$$

We can find the answer by taking half the exponent. That is,

$$\sqrt{x^{10}} = x^5. \leftarrow \tfrac{1}{2}(10) = 5$$

EXAMPLES Simplify.

11. $\sqrt{x^6} = \sqrt{(x^3)^2} = x^3$ ← $\frac{1}{2}(6) = 3$

12. $\sqrt{x^8} = x^4$

13. $\sqrt{t^{22}} = t^{11}$

Do Exercises 14–16.

If an odd power occurs, we express the power in terms of the largest even power. Then we simplify the even power as in Examples 11–13.

EXAMPLE 14 Simplify by factoring: $\sqrt{x^9}$.

$$\begin{aligned}\sqrt{x^9} &= \sqrt{x^8 \cdot x}\\ &= \sqrt{x^8}\sqrt{x}\\ &= x^4\sqrt{x}\end{aligned}$$

Note in Example 14 that $\sqrt{x^9} \neq x^3$.

EXAMPLE 15 Simplify by factoring: $\sqrt{32x^{15}}$.

$\sqrt{32x^{15}} = \sqrt{16x^{14}(2x)}$ — The largest even power is 14. Then we factor the radicand.

$= \sqrt{16}\sqrt{x^{14}}\sqrt{2x}$ — Factoring into a product of radicals

$= 4x^7\sqrt{2x}$ — Simplifying

Do Exercises 17 and 18.

c MULTIPLYING AND SIMPLIFYING

Sometimes we can simplify after multiplying. We leave the radicand in factored form and factor further to determine perfect-square factors. Then we simplify the perfect-square factors.

EXAMPLE 16 Multiply and then simplify by factoring: $\sqrt{2}\sqrt{14}$.

$\sqrt{2}\sqrt{14} = \sqrt{2 \cdot 14}$ — Multiplying

$= \sqrt{2 \cdot 2 \cdot 7}$ — Factoring

$= \sqrt{2 \cdot 2}\sqrt{7}$ — Looking for perfect-square factors; pairs of factors

$= 2\sqrt{7}$

Do Exercises 19 and 20.

EXAMPLE 17 Multiply and then simplify by factoring: $\sqrt{3x^2}\sqrt{9x^3}$.

$\sqrt{3x^2}\sqrt{9x^3} = \sqrt{3x^2 \cdot 9x^3}$ — Multiplying

$= \sqrt{9 \cdot x^2 \cdot x^2 \cdot 3 \cdot x}$ — Looking for perfect-square factors or largest even powers

Perfect squares are listed first.

$= \sqrt{9}\sqrt{x^2}\sqrt{x^2}\sqrt{3x}$

$= 3 \cdot x \cdot x \cdot \sqrt{3x}$

$= 3x^2\sqrt{3x}$

Simplify.

14. $\sqrt{t^4}$

15. $\sqrt{t^{20}}$

16. $\sqrt{h^{46}}$

Simplify by factoring.

17. $\sqrt{x^7}$

18. $\sqrt{24x^{11}}$

Multiply and simplify.

19. $\sqrt{3}\sqrt{6}$

20. $\sqrt{2}\sqrt{50}$

Answers on page A-7

Multiply and simplify.

21. $\sqrt{2x^3}\sqrt{8x^3y^4}$

22. $\sqrt{10xy^2}\sqrt{5x^2y^3}$

Approximate the square roots using Table 2. Round to three decimal places.

23. $\sqrt{275}$

24. $\sqrt{102}$

Answers on page A-7

In doing an example like the preceding one, it might be helpful to do more factoring, as follows:

$$\sqrt{3x^2} \cdot \sqrt{9x^3} = \sqrt{3 \cdot \underline{x \cdot x} \cdot \underline{3 \cdot 3} \cdot \underline{x \cdot x} \cdot x}.$$

Then we look for pairs of factors, as shown, and simplify perfect-square factors:

$$= 3 \cdot x \cdot x\sqrt{3x}$$
$$= 3x^2\sqrt{3x}.$$

Do Exercises 21 and 22.

We know that $\sqrt{AB} = \sqrt{A}\sqrt{B}$. That is, the square root of a product is the product of the square roots. What about the square root of a sum? That is, is the square root of a sum equal to the sum of the square roots? To check, consider $\sqrt{A + B}$ and $\sqrt{A} + \sqrt{B}$ when $A = 16$ and $B = 9$:

$$\sqrt{A + B} = \sqrt{16 + 9} = \sqrt{25} = 5;$$

and

$$\sqrt{A} + \sqrt{B} = \sqrt{16} + \sqrt{9} = 4 + 3 = 7.$$

Thus we see the following.

Caution! The square root of a sum is not the sum of the square roots.

$$\sqrt{A + B} \neq \sqrt{A} + \sqrt{B}$$

d Approximating Square Roots

If you are using a table to approximate square roots, you may find that some numbers are too large to be listed in the table. For example, Table 2 goes only to 100. We may still be able to find approximate square roots for other numbers, however. We do so by first looking for the largest perfect-square factor, if there is one. If there is none, we use any factorization for which all factors appear in Table 2.

EXAMPLES Approximate the square roots using Table 2.

18. $\sqrt{160} = \sqrt{16 \cdot 10}$ — Factoring the radicand. (Make one factor a perfect square, if you can.)

$= \sqrt{16}\sqrt{10}$ — Factoring the radical expression

$= 4\sqrt{10} \approx 4(3.162) \approx 12.648$ — From Table 2, $\sqrt{10} \approx 3.162$.

19. $\sqrt{341} = \sqrt{11 \cdot 31}$ — Factoring into a product where each factor is in Table 2. (There is no perfect-square factor.)

$= \sqrt{11}\sqrt{31}$

$\approx 3.317 \times 5.568$ — From Table 2

≈ 18.469 — Rounded to three decimal places

If the approximations in Examples 18 and 19 were done on a calculator, the simplifying would not be necessary and the square roots would be found directly, but there is variance in the answers. For example, on a calculator with a 10-digit readout, we would get

$$\sqrt{160} \approx 12.64911064 \quad \text{and} \quad \sqrt{341} \approx 18.46618531.$$

Do Exercises 23 and 24.

NAME SECTION DATE

Exercise Set 8.2

a Simplify by factoring.

1. $\sqrt{12}$ **2.** $\sqrt{8}$ **3.** $\sqrt{75}$ **4.** $\sqrt{50}$ **5.** $\sqrt{20}$

6. $\sqrt{45}$ **7.** $\sqrt{600}$ **8.** $\sqrt{300}$ **9.** $\sqrt{9x}$ **10.** $\sqrt{4y}$

11. $\sqrt{48x}$ **12.** $\sqrt{40m}$ **13.** $\sqrt{16a}$ **14.** $\sqrt{49b}$ **15.** $\sqrt{64y^2}$

16. $\sqrt{9x^2}$ **17.** $\sqrt{13x^2}$ **18.** $\sqrt{23s^2}$ **19.** $\sqrt{8t^2}$ **20.** $\sqrt{125a^2}$

21. $\sqrt{180}$ **22.** $\sqrt{320}$ **23.** $\sqrt{288y}$ **24.** $\sqrt{363p}$ **25.** $\sqrt{28x^2}$

26. $\sqrt{20x^2}$ **27.** $\sqrt{8x^2 + 8x + 2}$ **28.** $\sqrt{20x^2 - 20x + 5}$

29. $\sqrt{36y + 12y^2 + y^3}$ **30.** $\sqrt{x - 2x^2 + x^3}$

ANSWERS

1. ______
2. ______
3. ______
4. ______
5. ______
6. ______
7. ______
8. ______
9. ______
10. ______
11. ______
12. ______
13. ______
14. ______
15. ______
16. ______
17. ______
18. ______
19. ______
20. ______
21. ______
22. ______
23. ______
24. ______
25. ______
26. ______
27. ______
28. ______
29. ______
30. ______

ANSWERS

31. ____

32. ____

33. ____

34. ____

35. ____

36. ____

37. ____

38. ____

39. ____

40. ____

41. ____

42. ____

43. ____

44. ____

45. ____

46. ____

47. ____

48. ____

49. ____

50. ____

51. ____

52. ____

53. ____

54. ____

55. ____

56. ____

b Simplify by factoring.

31. $\sqrt{x^6}$ **32.** $\sqrt{x^{18}}$ **33.** $\sqrt{x^{12}}$ **34.** $\sqrt{x^{16}}$

35. $\sqrt{x^5}$ **36.** $\sqrt{x^3}$ **37.** $\sqrt{t^{19}}$ **38.** $\sqrt{p^{17}}$

39. $\sqrt{(y-2)^8}$ **40.** $\sqrt{(x+3)^6}$ **41.** $\sqrt{4(x+5)^{10}}$ **42.** $\sqrt{16(a-7)^4}$

43. $\sqrt{36m^3}$ **44.** $\sqrt{250y^3}$ **45.** $\sqrt{8a^5}$ **46.** $\sqrt{12b^7}$

47. $\sqrt{104p^{17}}$ **48.** $\sqrt{284m^{23}}$ **49.** $\sqrt{448x^6y^3}$ **50.** $\sqrt{243x^5y^4}$

c Multiply and then simplify by factoring, if possible.

51. $\sqrt{3}\sqrt{18}$ **52.** $\sqrt{5}\sqrt{10}$ **53.** $\sqrt{15}\sqrt{6}$

54. $\sqrt{3}\sqrt{27}$ **55.** $\sqrt{18}\sqrt{14x}$ **56.** $\sqrt{12}\sqrt{18x}$

57. $\sqrt{3x}\sqrt{12y}$

58. $\sqrt{7x}\sqrt{21y}$

59. $\sqrt{13}\sqrt{13}$

60. $\sqrt{11}\sqrt{11x}$

61. $\sqrt{5b}\sqrt{15b}$

62. $\sqrt{6a}\sqrt{18a}$

63. $\sqrt{2t}\sqrt{2t}$

64. $\sqrt{7a}\sqrt{7a}$

65. $\sqrt{ab}\sqrt{ac}$

66. $\sqrt{xy}\sqrt{xz}$

67. $\sqrt{2x^2y}\sqrt{4xy^2}$

68. $\sqrt{15mn^2}\sqrt{5m^2n}$

69. $\sqrt{18}\sqrt{18}$

70. $\sqrt{16}\sqrt{16}$

71. $\sqrt{5}\sqrt{2x-1}$

72. $\sqrt{3}\sqrt{4x+2}$

73. $\sqrt{x+2}\sqrt{x+2}$

74. $\sqrt{x-9}\sqrt{x-9}$

75. $\sqrt{18x^2y^3}\sqrt{6xy^4}$

76. $\sqrt{12x^3y^2}\sqrt{8xy}$

77. $\sqrt{50x^4y^6}\sqrt{10xy}$

78. $\sqrt{10xy^2}\sqrt{5x^2y^3}$

ANSWERS

57. ______
58. ______
59. ______
60. ______
61. ______
62. ______
63. ______
64. ______
65. ______
66. ______
67. ______
68. ______
69. ______
70. ______
71. ______
72. ______
73. ______
74. ______
75. ______
76. ______
77. ______
78. ______

ANSWERS

79. ______
80. ______
81. ______
82. ______
83. ______
84. ______
85. ______
86. ______
87. ______
88. ______
89. ______
90. ______
91. ______
92. ______
93. ______
94. ______
95. ______
96. ______
97. ______
98. ______
99. ______
100. ______
101. ______
102. ______
103. ______
104. ______
105. ______
106. ______
107. ______
108. ______
109. ______

d Approximate the square roots using Table 2. Round to three decimal places.

79. $\sqrt{125}$ **80.** $\sqrt{288}$ **81.** $\sqrt{360}$ **82.** $\sqrt{105}$

83. $\sqrt{700}$ **84.** $\sqrt{143}$ **85.** $\sqrt{122}$ **86.** $\sqrt{2000}$

Speed of a skidding car. How do police determine the speed of a car after an accident? The formula

$$r = 2\sqrt{5L}$$

can be used to approximate the speed r, in miles per hour, of a car that has left a skid mark of length L, in feet.

87. What was the speed of a car that left skid marks of 20 ft? of 150 ft?

88. What was the speed of a car that left skid marks of 30 ft? of 70 ft?

SKILL MAINTENANCE

Solve.

89. $x - y = -6,$
$x + y = 2$

90. $3x + 5y = 6,$
$5x + 3y = 4$

91. $3x - 2y = 4,$
$2x + 5y = 9$

92. $4a - 5b = 25,$
$a - b = 7$

93. The perimeter of a rectangle is 642 ft. The length is 15 ft greater than the width. Find the area of the rectangle.

SYNTHESIS

Factor.

94. $\sqrt{5x - 5}$ **95.** $\sqrt{x^2 - x - 2}$ **96.** $\sqrt{x^2 - 36}$

97. $\sqrt{2x^2 - 5x - 12}$ **98.** $\sqrt{x^3 - 2x^2}$ **99.** $\sqrt{a^2 - b^2}$

Simplify.

100. $\sqrt{0.01}$ **101.** $\sqrt{0.25}$ **102.** $\sqrt{x^8}$ **103.** $\sqrt{9a^6}$

Multiply and then simplify by factoring.

104. $\sqrt{a}(\sqrt{a^3} - 5)$

105. $(\sqrt{2y})(\sqrt{3})(\sqrt{8y})$

106. $\sqrt{18(x - 2)}\sqrt{20(x - 2)^3}$

107. $\sqrt{27(x + 1)}\sqrt{12y(x + 1)^2}$

108. $\sqrt{2^{109}}\sqrt{x^{306}}\sqrt{x^{11}}$

109. $\sqrt{x}\sqrt{2x}\sqrt{10x^5}$

8.3 Quotients Involving Square Roots

OBJECTIVES

After finishing Section 8.3, you should be able to:

a Divide radical expressions.

b Simplify square roots of quotients.

c Rationalize the denominator of a radical expression.

d Approximate radical expressions involving division.

FOR EXTRA HELP

TAPE 19 | TAPE 16A | MAC: 8A IBM: 8A

a DIVIDING RADICAL EXPRESSIONS

Consider the expressions

$$\frac{\sqrt{25}}{\sqrt{16}} \quad \text{and} \quad \sqrt{\frac{25}{16}}.$$

Let us evaluate them separately:

a) $\dfrac{\sqrt{25}}{\sqrt{16}} = \dfrac{5}{4}$ because $\sqrt{25} = 5$ and $\sqrt{16} = 4$;

b) $\sqrt{\dfrac{25}{16}} = \dfrac{5}{4}$ because $\dfrac{5}{4} \cdot \dfrac{5}{4} = \dfrac{25}{16}$.

We see that both expressions represent the same number. This suggests that the quotient of two square roots is the square root of the quotient of the radicands.

THE QUOTIENT RULE FOR RADICALS

For any nonnegative number A and any positive number B,

$$\frac{\sqrt{A}}{\sqrt{B}} = \sqrt{\frac{A}{B}}.$$

(The quotient of two square roots is the square root of the quotient of the radicands.)

EXAMPLES Divide and simplify.

1. $\dfrac{\sqrt{27}}{\sqrt{3}} = \sqrt{\dfrac{27}{3}} = \sqrt{9} = 3$

2. $\dfrac{\sqrt{30a^5}}{\sqrt{6a^2}} = \sqrt{\dfrac{30a^5}{6a^2}} = \sqrt{5a^3} = \sqrt{a^2 \cdot 5a} = \sqrt{a^2} \cdot \sqrt{5a} = a\sqrt{5a}$

Do Exercises 1–3.

Divide and simplify.

1. $\dfrac{\sqrt{96}}{\sqrt{6}}$

2. $\dfrac{\sqrt{75}}{\sqrt{3}}$

3. $\dfrac{\sqrt{42x^5}}{\sqrt{7x^2}}$

b ROOTS OF QUOTIENTS

To find the square root of a quotient, we can reverse the quotient rule for radicals. We can take the square root of a quotient by taking the square roots of the numerator and the denominator separately.

For any nonnegative number A and any positive number B,

$$\sqrt{\frac{A}{B}} = \frac{\sqrt{A}}{\sqrt{B}}.$$

(We can take the square roots of the numerator and the denominator separately.)

Answers on page A-7

Simplify.

4. $\sqrt{\frac{16}{9}}$

5. $\sqrt{\frac{1}{25}}$

6. $\sqrt{\frac{36}{x^2}}$

Simplify.

7. $\sqrt{\frac{18}{32}}$

8. $\sqrt{\frac{2250}{2560}}$

9. $\sqrt{\frac{98y}{2y^{11}}}$

Answers on page A-7

EXAMPLES Simplify by taking the square roots of the numerator and the denominator separately.

3. $\sqrt{\frac{25}{9}} = \frac{\sqrt{25}}{\sqrt{9}} = \frac{5}{3}$ Taking the square roots of the numerator and the denominator

4. $\sqrt{\frac{1}{16}} = \frac{\sqrt{1}}{\sqrt{16}} = \frac{1}{4}$ Taking the square roots of the numerator and the denominator

5. $\sqrt{\frac{49}{t^2}} = \frac{\sqrt{49}}{\sqrt{t^2}} = \frac{7}{t}$

Do Exercises 4–6.

We are assuming that expressions for numerators are nonnegative and expressions for denominators are positive. Thus we need not be concerned about absolute-value signs or zero denominators.

Sometimes a rational expression can be simplified to one that has a perfect-square numerator and a perfect-square denominator.

EXAMPLES Simplify.

6. $\sqrt{\frac{18}{50}} = \sqrt{\frac{9 \cdot 2}{25 \cdot 2}} = \sqrt{\frac{9}{25} \cdot \frac{2}{2}} = \sqrt{\frac{9}{25} \cdot 1}$

$= \sqrt{\frac{9}{25}} = \frac{\sqrt{9}}{\sqrt{25}} = \frac{3}{5}$

7. $\sqrt{\frac{2560}{2890}} = \sqrt{\frac{256 \cdot 10}{289 \cdot 10}} = \sqrt{\frac{256}{289} \cdot \frac{10}{10}} = \sqrt{\frac{256}{289} \cdot 1}$

$= \sqrt{\frac{256}{289}} = \frac{\sqrt{256}}{\sqrt{289}} = \frac{16}{17}$

8. $\frac{\sqrt{48x^3}}{\sqrt{3x^7}} = \sqrt{\frac{48x^3}{3x^7}} = \sqrt{\frac{16}{x^4}} = \frac{\sqrt{16}}{\sqrt{x^4}} = \frac{4}{x^2}$

Do Exercises 7–9.

c RATIONALIZING DENOMINATORS

Sometimes in mathematics it is useful to find an equivalent expression without a radical in the denominator. This provides a standard notation for expressing results. The procedure for finding such an expression is called **rationalizing the denominator.** We carry this out by multiplying by 1 in either of two ways. One way is to multiply by 1 under the radical to make the denominator a perfect square. Another way is to multiply by 1 outside the radical to make the denominator a perfect square.

EXAMPLE 9 Rationalize the denominator: $\sqrt{\frac{2}{3}}$.

METHOD 1 We multiply by 1, choosing $\frac{3}{3}$ for 1. This makes the denominator a perfect square:

$$\begin{aligned} \sqrt{\frac{2}{3}} &= \sqrt{\frac{2}{3}\cdot\frac{3}{3}} && \text{Multiplying by 1} \\ &= \sqrt{\frac{6}{9}} \\ &= \frac{\sqrt{6}}{\sqrt{9}} \\ &= \frac{\sqrt{6}}{3}. \end{aligned}$$

METHOD 2 We can also rationalize by first taking the square roots of the numerator and the denominator. Then we multiply by 1, using $\sqrt{3}/\sqrt{3}$:

$$\begin{aligned} \sqrt{\frac{2}{3}} &= \frac{\sqrt{2}}{\sqrt{3}} \\ &= \frac{\sqrt{2}}{\sqrt{3}}\cdot\frac{\sqrt{3}}{\sqrt{3}} && \text{Multiplying by 1} \\ &= \frac{\sqrt{2}\cdot\sqrt{3}}{\sqrt{3}\cdot\sqrt{3}} \\ &= \frac{\sqrt{6}}{\sqrt{9}} \\ &= \frac{\sqrt{6}}{3}. \end{aligned}$$

Do Exercise 10.

We can always multiply by 1 to make a denominator a perfect square. Then we can take the square root of the denominator.

EXAMPLE 10 Rationalize the denominator: $\sqrt{\frac{5}{18}}$.

The denominator 18 is not a perfect square. Factoring, we get $18 = 3\cdot 3\cdot 2$. If we had another factor of 2, however, we would have a perfect square, 36. Thus we multiply by 1, choosing $\frac{2}{2}$. This makes the denominator a perfect square.

$$\sqrt{\frac{5}{18}} = \sqrt{\frac{5}{18}\cdot\frac{2}{2}} = \sqrt{\frac{10}{36}} = \frac{\sqrt{10}}{\sqrt{36}} = \frac{\sqrt{10}}{6}$$

10. Rationalize the denominator:

$$\sqrt{\frac{3}{5}}.$$

Answer on page A-7

Rationalize the denominator.

11. $\sqrt{\frac{5}{8}}$

(*Hint:* Multiply the radicand by $\frac{2}{2}$.)

12. $\frac{10}{\sqrt{3}}$

Rationalize the denominator.

13. $\frac{\sqrt{3}}{\sqrt{7}}$

14. $\frac{\sqrt{5}}{\sqrt{r}}$

15. $\frac{\sqrt{64y^2}}{\sqrt{7}}$

Approximate to three decimal places.

16. $\sqrt{\frac{2}{3}}$

17. $\frac{\sqrt{10}}{\sqrt{13}}$

Answers on page A-7

EXAMPLE 11 Rationalize the denominator: $\frac{8}{\sqrt{7}}$.

This time we obtain an expression without a radical in the denominator by multiplying by 1, choosing $\sqrt{7}/\sqrt{7}$:

$$\frac{8}{\sqrt{7}} = \frac{8}{\sqrt{7}} \cdot \frac{\sqrt{7}}{\sqrt{7}} = \frac{8\sqrt{7}}{\sqrt{49}} = \frac{8\sqrt{7}}{7}.$$

Do Exercises 11 and 12.

EXAMPLE 12 Rationalize the denominator: $\frac{\sqrt{3}}{\sqrt{2}}$.

We look at the denominator. It is $\sqrt{2}$. We multiply by 1, choosing $\sqrt{2}/\sqrt{2}$:

$$\frac{\sqrt{3}}{\sqrt{2}} = \frac{\sqrt{3}}{\sqrt{2}} \cdot \frac{\sqrt{2}}{\sqrt{2}} = \frac{\sqrt{3}\cdot\sqrt{2}}{\sqrt{2}\cdot\sqrt{2}} = \frac{\sqrt{6}}{\sqrt{4}} = \frac{\sqrt{6}}{2}, \text{ or } \frac{1}{2}\sqrt{6}.$$

EXAMPLES Rationalize the denominator.

13. $\frac{\sqrt{5}}{\sqrt{x}} = \frac{\sqrt{5}}{\sqrt{x}} \cdot \frac{\sqrt{x}}{\sqrt{x}}$ Multiplying by 1

$$= \frac{\sqrt{5}\sqrt{x}}{\sqrt{x}\sqrt{x}} = \frac{\sqrt{5x}}{x}$$

14. $\frac{\sqrt{49a^5}}{\sqrt{12}} = \frac{\sqrt{49a^5}}{\sqrt{12}} \cdot \frac{\sqrt{3}}{\sqrt{3}}$ Multiplying by 1 using $\sqrt{3}/\sqrt{3}$ because $\sqrt{3}\cdot\sqrt{12}$ gives a perfect-square radicand in $\sqrt{36}$.

$$= \frac{\sqrt{49a^5}\sqrt{3}}{\sqrt{12}\sqrt{3}} = \frac{\sqrt{49a^4\cdot 3a}}{\sqrt{36}} = \frac{7a^2\sqrt{3a}}{6}$$

Do Exercises 13–15.

d APPROXIMATING EXPRESSIONS WITH SQUARE ROOTS

We can use a calculator to approximate square roots of quotients.

EXAMPLE 15 Approximate $\sqrt{\frac{3}{5}}$ to three decimal places.

We divide and then approximate the square root:

$$\sqrt{\frac{3}{5}} = \sqrt{0.6} \approx 0.774596669 \approx 0.775.$$

Do Exercises 16 and 17.

NAME SECTION DATE

Exercise Set 8.3

a Divide and simplify.

1. $\dfrac{\sqrt{18}}{\sqrt{2}}$

2. $\dfrac{\sqrt{20}}{\sqrt{5}}$

3. $\dfrac{\sqrt{108}}{\sqrt{3}}$

4. $\dfrac{\sqrt{60}}{\sqrt{15}}$

5. $\dfrac{\sqrt{65}}{\sqrt{13}}$

6. $\dfrac{\sqrt{45}}{\sqrt{15}}$

7. $\dfrac{\sqrt{3}}{\sqrt{75}}$

8. $\dfrac{\sqrt{3}}{\sqrt{48}}$

9. $\dfrac{\sqrt{12}}{\sqrt{75}}$

10. $\dfrac{\sqrt{18}}{\sqrt{32}}$

11. $\dfrac{\sqrt{8x}}{\sqrt{2x}}$

12. $\dfrac{\sqrt{18b}}{\sqrt{2b}}$

13. $\dfrac{\sqrt{63y^3}}{\sqrt{7y}}$

14. $\dfrac{\sqrt{48x^3}}{\sqrt{3x}}$

b Simplify.

15. $\sqrt{\dfrac{16}{49}}$

16. $\sqrt{\dfrac{9}{49}}$

17. $\sqrt{\dfrac{1}{36}}$

18. $\sqrt{\dfrac{1}{4}}$

19. $-\sqrt{\dfrac{16}{81}}$

ANSWERS

1. ______
2. ______
3. ______
4. ______
5. ______
6. ______
7. ______
8. ______
9. ______
10. ______
11. ______
12. ______
13. ______
14. ______
15. ______
16. ______
17. ______
18. ______
19. ______

ANSWERS

20. ______
21. ______
22. ______
23. ______
24. ______
25. ______
26. ______
27. ______
28. ______
29. ______
30. ______
31. ______
32. ______
33. ______
34. ______
35. ______
36. ______
37. ______
38. ______
39. ______
40. ______
41. ______
42. ______
43. ______
44. ______
45. ______
46. ______
47. ______
48. ______

20. $-\sqrt{\frac{25}{49}}$ **21.** $\sqrt{\frac{64}{289}}$ **22.** $\sqrt{\frac{81}{361}}$ **23.** $\sqrt{\frac{1690}{1960}}$ **24.** $\sqrt{\frac{1210}{6250}}$

25. $\sqrt{\frac{25}{x^2}}$ **26.** $\sqrt{\frac{36}{a^2}}$ **27.** $\sqrt{\frac{9a^2}{625}}$ **28.** $\sqrt{\frac{x^2y^2}{256}}$

c Rationalize the denominator.

29. $\sqrt{\frac{2}{5}}$ **30.** $\sqrt{\frac{2}{7}}$ **31.** $\sqrt{\frac{7}{8}}$ **32.** $\sqrt{\frac{3}{8}}$ **33.** $\sqrt{\frac{1}{12}}$

34. $\sqrt{\frac{7}{12}}$ **35.** $\sqrt{\frac{5}{18}}$ **36.** $\sqrt{\frac{1}{18}}$ **37.** $\frac{3}{\sqrt{5}}$ **38.** $\frac{4}{\sqrt{3}}$

39. $\sqrt{\frac{8}{3}}$ **40.** $\sqrt{\frac{12}{5}}$ **41.** $\sqrt{\frac{3}{x}}$ **42.** $\sqrt{\frac{2}{x}}$ **43.** $\sqrt{\frac{x}{y}}$

44. $\sqrt{\frac{a}{b}}$ **45.** $\sqrt{\frac{x^2}{20}}$ **46.** $\sqrt{\frac{x^2}{18}}$ **47.** $\frac{\sqrt{7}}{\sqrt{2}}$ **48.** $\frac{\sqrt{3}}{\sqrt{5}}$

49. $\frac{\sqrt{9}}{\sqrt{8}}$ 50. $\frac{\sqrt{4}}{\sqrt{27}}$ 51. $\frac{\sqrt{3}}{\sqrt{2}}$ 52. $\frac{\sqrt{2}}{\sqrt{5}}$ 53. $\frac{2}{\sqrt{2}}$

54. $\frac{3}{\sqrt{3}}$ 55. $\frac{\sqrt{5}}{\sqrt{11}}$ 56. $\frac{\sqrt{7}}{\sqrt{27}}$ 57. $\frac{\sqrt{7}}{\sqrt{12}}$ 58. $\frac{\sqrt{5}}{\sqrt{18}}$

59. $\frac{\sqrt{48}}{\sqrt{32}}$ 60. $\frac{\sqrt{56}}{\sqrt{40}}$ 61. $\frac{\sqrt{450}}{\sqrt{18}}$ 62. $\frac{\sqrt{224}}{\sqrt{14}}$ 63. $\frac{\sqrt{3}}{\sqrt{x}}$

64. $\frac{\sqrt{2}}{\sqrt{y}}$ 65. $\frac{4y}{\sqrt{5}}$ 66. $\frac{8x}{\sqrt{3}}$ 67. $\frac{\sqrt{a^3}}{\sqrt{8}}$ 68. $\frac{\sqrt{x^3}}{\sqrt{27}}$

69. $\frac{\sqrt{56}}{\sqrt{12x}}$ 70. $\frac{\sqrt{45}}{\sqrt{8a}}$ 71. $\frac{\sqrt{27c}}{\sqrt{32c^3}}$ 72. $\frac{\sqrt{7x^3}}{\sqrt{12x}}$ 73. $\frac{\sqrt{y^5}}{\sqrt{xy^2}}$

74. $\frac{\sqrt{x^3}}{\sqrt{xy}}$ 75. $\frac{\sqrt{45mn^2}}{\sqrt{32m}}$ 76. $\frac{\sqrt{16a^4b^6}}{\sqrt{128a^6b^6}}$

ANSWERS

49. ____
50. ____
51. ____
52. ____
53. ____
54. ____
55. ____
56. ____
57. ____
58. ____
59. ____
60. ____
61. ____
62. ____
63. ____
64. ____
65. ____
66. ____
67. ____
68. ____
69. ____
70. ____
71. ____
72. ____
73. ____
74. ____
75. ____
76. ____

ANSWERS

77.
78.
79.
80.
81.
82.
83.
84.
85.
86.
87.
88.
89.
90.
91.
92.
93.
94.
95.
96.
97.
98.
99.
100.
101.
102.
103.
104.
105.
106.
107.
108.

d Approximate to three decimal places.

77. $\sqrt{\frac{1}{3}}$ **78.** $\sqrt{\frac{3}{2}}$ **79.** $\sqrt{\frac{7}{8}}$ **80.** $\sqrt{\frac{3}{8}}$ **81.** $\sqrt{\frac{1}{12}}$ **82.** $\sqrt{\frac{5}{12}}$

83. $\sqrt{\frac{1}{2}}$ **84.** $\sqrt{\frac{1}{7}}$ **85.** $\frac{17}{\sqrt{20}}$ **86.** $\frac{24}{\sqrt{13}}$ **87.** $\frac{\sqrt{13}}{\sqrt{18}}$ **88.** $\frac{\sqrt{7}}{\sqrt{18}}$

SKILL MAINTENANCE

Solve.

89. $x = y + 2,$ $x + y = 6$

90. $4x - y = 10,$ $4x + y = 70$

91. $2x - 3y = 7,$ $2x - 3y = 9$

92. $2x - 3y = 7,$ $-4x + 6y = -14$

93. $x + y = -7,$ $x - y = 2$

94. $2x + 3y = 8,$ $5x - 4y = -2$

SYNTHESIS

The period T of a pendulum is the time it takes to move from one side to the other and back. A formula for the period is

$$T = 2\pi\sqrt{\frac{L}{32}},$$

where T is in seconds and L is in feet. Use 3.14 for π.

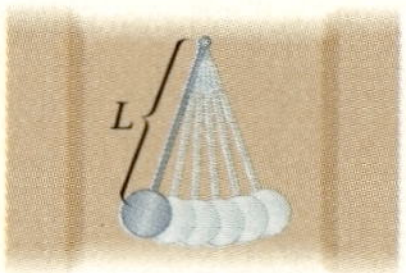

95. Find the periods of pendulums of lengths 2 ft, 8 ft, 64 ft, and 100 ft.

96. Find the period of a pendulum of length $\frac{2}{3}$ in.

97. The pendulum of a grandfather clock is $(32/\pi^2)$ ft long. How long does it take to swing from one side to the other?

98. The pendulum of a grandfather clock is $(45/\pi^2)$ ft long. How long does it take to swing from one side to the other?

Rationalize the denominator.

99. $\sqrt{\frac{5}{1600}}$ **100.** $\sqrt{\frac{3}{1000}}$ **101.** $\sqrt{\frac{1}{5x^3}}$ **102.** $\sqrt{\frac{3x^2y}{a^2x^5}}$

103. $\sqrt{\frac{3a}{b}}$ **104.** $\sqrt{\frac{1}{5zw^2}}$ **105.** $\sqrt{0.009}$ **106.** $\sqrt{0.012}$

Simplify.

107. $\sqrt{\frac{1}{x^2} - \frac{2}{xy} + \frac{1}{y^2}}$ **108.** $\sqrt{2 - \frac{4}{z^2} + \frac{2}{z^4}}$

8.4 Addition, Subtraction, and More Multiplication

a Addition and Subtraction

We can add any two real numbers. The sum of 5 and $\sqrt{2}$ can be expressed as

$$5 + \sqrt{2}.$$

We cannot simplify this unless we use rational approximations. However, when we have *like radicals*, a sum can be simplified using the distributive laws and collecting like terms. **Like radicals** have the same radicands.

Example 1 Add: $3\sqrt{5} + 4\sqrt{5}$.

Suppose we were considering $3x + 4x$. Recall that to add, we use the distributive laws as follows:

$$3x + 4x = (3 + 4)x = 7x.$$

The situation is similar in this example, but we let $x = \sqrt{5}$:

$$3\sqrt{5} + 4\sqrt{5} = (3 + 4)\sqrt{5} \quad \text{Using the distributive law to factor out } \sqrt{5}$$
$$= 7\sqrt{5}.$$

To add or subtract as we did in Example 1, the radicands must be the same. Sometimes after simplifying the radical terms, we discover that we have like radicals.

Examples Add or subtract. Simplify, if possible, by collecting like radical terms.

2. $5\sqrt{2} - \sqrt{18} = 5\sqrt{2} - \sqrt{9 \cdot 2}$ Factoring 18
$= 5\sqrt{2} - \sqrt{9}\sqrt{2}$
$= 5\sqrt{2} - 3\sqrt{2}$
$= (5 - 3)\sqrt{2}$ Using the distributive law to factor out the common factor, $\sqrt{2}$
$= 2\sqrt{2}$

3. $\sqrt{4x^3} + 7\sqrt{x} = \sqrt{4x^2 \cdot x} + 7\sqrt{x}$
$= 2x\sqrt{x} + 7\sqrt{x}$
$= (2x + 7)\sqrt{x}$ Using the distributive law to factor out $\sqrt{x}$

Don't forget the parentheses!

4. $\sqrt{x^3 - x^2} + \sqrt{4x - 4} = \sqrt{x^2(x - 1)} + \sqrt{4(x - 1)}$ Factoring radicands
$= \sqrt{x^2}\sqrt{x - 1} + \sqrt{4}\sqrt{x - 1}$
$= x\sqrt{x - 1} + 2\sqrt{x - 1}$
$= (x + 2)\sqrt{x - 1}$ Using the distributive law to factor out the common factor, $\sqrt{x - 1}$

Don't forget the parentheses!

Do Exercises 1–5.

Objectives

After finishing Section 8.4, you should be able to:

- **a** Add or subtract with radical notation, using the distributive law to simplify.
- **b** Multiply expressions involving radicals, where some of the expressions contain more than one term.
- **c** Rationalize denominators having two terms.

For Extra Help

TAPE 19 | TAPE 16A | MAC: 8A IBM: 8A

Add or subtract and simplify by collecting like radical terms, if possible.

1. $3\sqrt{2} + 9\sqrt{2}$

2. $8\sqrt{5} - 3\sqrt{5}$

3. $2\sqrt{10} - 7\sqrt{40}$

4. $\sqrt{24} + \sqrt{54}$

5. $\sqrt{9x + 9} - \sqrt{4x + 4}$

Answers on page A-7

Add or subtract.

6. $\sqrt{2} + \sqrt{\frac{1}{2}}$

7. $\sqrt{\frac{5}{3}} + \sqrt{\frac{3}{5}}$

Answers on page A-7

Sometimes rationalizing denominators enables us to combine like radicals.

EXAMPLE 5 Add: $\sqrt{3} + \sqrt{\frac{1}{3}}$.

$$\begin{aligned}\sqrt{3} + \sqrt{\frac{1}{3}} &= \sqrt{3} + \sqrt{\frac{1}{3} \cdot \frac{3}{3}} && \text{Multiplying by 1 in order to rationalize the denominator}\\ &= \sqrt{3} + \sqrt{\frac{3}{9}}\\ &= \sqrt{3} + \frac{\sqrt{3}}{\sqrt{9}}\\ &= \sqrt{3} + \frac{\sqrt{3}}{3}\\ &= 1 \cdot \sqrt{3} + \frac{1}{3}\sqrt{3}\\ &= \left(1 + \frac{1}{3}\right)\sqrt{3} && \text{Factoring out the common factor, } \sqrt{3}\\ &= \frac{4}{3}\sqrt{3}\end{aligned}$$

Do Exercises 6 and 7.

b MULTIPLICATION

Now let us multiply where some of the expressions may contain more than one term. To do this, we use procedures already studied in this chapter as well as the distributive law and special products for multiplying with polynomials.

EXAMPLE 6 Multiply: $\sqrt{2}(\sqrt{3} + \sqrt{7})$.

$$\begin{aligned}\sqrt{2}(\sqrt{3} + \sqrt{7}) &= \sqrt{2}\sqrt{3} + \sqrt{2}\sqrt{7} && \text{Multiplying using a distributive law}\\ &= \sqrt{6} + \sqrt{14} && \text{Using the rule for multiplying with radicals}\end{aligned}$$

EXAMPLE 7 Multiply: $(2 + \sqrt{3})(5 - 4\sqrt{3})$.

$$\begin{aligned}(2 + \sqrt{3})(5 - 4\sqrt{3}) &= 2 \cdot 5 - 2 \cdot 4\sqrt{3} + \sqrt{3} \cdot 5 - \sqrt{3} \cdot 4\sqrt{3} && \text{Using FOIL}\\ &= 10 - 8\sqrt{3} + 5\sqrt{3} - 4 \cdot 3\\ &= 10 - 12 - 3\sqrt{3}\\ &= -2 - 3\sqrt{3}\end{aligned}$$

EXAMPLE 8 Multiply: $(\sqrt{3} - \sqrt{x})(\sqrt{3} + \sqrt{x})$.

$$(\sqrt{3} - \sqrt{x})(\sqrt{3} + \sqrt{x}) = (\sqrt{3})^2 - (\sqrt{x})^2 \qquad \text{Using } (A - B)(A + B) = A^2 - B^2$$
$$= 3 - x$$

EXAMPLE 9 Multiply: $(3 - \sqrt{p})^2$.

$$(3 - \sqrt{p})^2 = 3^2 - 2 \cdot 3 \cdot \sqrt{p} + (\sqrt{p})^2 \qquad \text{Using } (A - B)^2 = A^2 - 2AB + B^2$$
$$= 9 - 6\sqrt{p} + p$$

EXAMPLE 10 Multiply: $(2 - \sqrt{5})(2 + \sqrt{5})$.

$$(2 - \sqrt{5})(2 + \sqrt{5}) = 2^2 - (\sqrt{5})^2 \qquad \text{Using } (A - B)(A + B) = A^2 - B^2$$
$$= 4 - 5$$
$$= -1$$

Do Exercises 8–12.

c MORE ON RATIONALIZING DENOMINATORS

Note in Examples 8 and 10 that the results have no radicals. This will happen whenever we multiply expressions such as $\sqrt{a} - \sqrt{b}$ and $\sqrt{a} + \sqrt{b}$, where a and b are rational numbers. We see this in the following:

$$(\sqrt{a} + \sqrt{b})(\sqrt{a} - \sqrt{b}) = (\sqrt{a})^2 - (\sqrt{b})^2 = a - b.$$

Expressions such as $\sqrt{3} - \sqrt{x}$ and $\sqrt{3} + \sqrt{x}$ are known as **conjugates**; so too are $2 + \sqrt{5}$ and $2 - \sqrt{5}$. We can use conjugates to rationalize a denominator that involves a sum or difference of two terms, where one or both are radicals. To do so, we multiply by 1 using the conjugate in the numerator and the denominator.

Do Exercises 13–15.

EXAMPLE 11 Rationalize the denominator: $\dfrac{3}{2 + \sqrt{5}}$.

We multiply by 1 using the conjugate of $2 + \sqrt{5}$, which is $2 - \sqrt{5}$, as the numerator and the denominator:

$$\frac{3}{2 + \sqrt{5}} = \frac{3}{2 + \sqrt{5}} \cdot \frac{2 - \sqrt{5}}{2 - \sqrt{5}} \qquad \text{Multiplying by 1}$$
$$= \frac{3(2 - \sqrt{5})}{(2 + \sqrt{5})(2 - \sqrt{5})} \qquad \text{Multiplying}$$
$$= \frac{6 - 3\sqrt{5}}{2^2 - (\sqrt{5})^2}$$
$$= \frac{6 - 3\sqrt{5}}{4 - 5}$$
$$= \frac{6 - 3\sqrt{5}}{-1}$$
$$= -6 + 3\sqrt{5}, \text{ or } 3\sqrt{5} - 6.$$

Multiply.

8. $\sqrt{3}(\sqrt{5} + \sqrt{2})$

9. $(1 - \sqrt{2})(4 + 3\sqrt{5})$

10. $(\sqrt{2} + \sqrt{a})(\sqrt{2} - \sqrt{a})$

11. $(5 + \sqrt{x})^2$

12. $(3 - \sqrt{7})(3 + \sqrt{7})$

Find the conjugate of the expression.

13. $7 + \sqrt{5}$

14. $\sqrt{5} - \sqrt{2}$

15. $1 - \sqrt{x}$

Answers on page A-7

Rationalize the denominator.

16. $\dfrac{6}{7 + \sqrt{5}}$

17. $\dfrac{\sqrt{5} + \sqrt{2}}{\sqrt{5} - \sqrt{2}}$

18. Rationalize the denominator:

$$\frac{7}{1 - \sqrt{x}}.$$

Answers on page A-7

EXAMPLE 12 Rationalize the denominator: $\dfrac{\sqrt{3} + \sqrt{5}}{\sqrt{3} - \sqrt{5}}$.

We multiply by 1 using the conjugate of $\sqrt{3} - \sqrt{5}$, which is $\sqrt{3} + \sqrt{5}$, as the numerator and the denominator:

$$\frac{\sqrt{3} + \sqrt{5}}{\sqrt{3} - \sqrt{5}} = \frac{\sqrt{3} + \sqrt{5}}{\sqrt{3} - \sqrt{5}} \cdot \frac{\sqrt{3} + \sqrt{5}}{\sqrt{3} + \sqrt{5}} \quad \text{Multiplying by 1}$$

$$= \frac{(\sqrt{3} + \sqrt{5})^2}{(\sqrt{3} - \sqrt{5})(\sqrt{3} + \sqrt{5})}$$

$$= \frac{(\sqrt{3})^2 + 2\sqrt{3}\sqrt{5} + (\sqrt{5})^2}{(\sqrt{3})^2 - (\sqrt{5})^2}$$

$$= \frac{3 + 2\sqrt{15} + 5}{3 - 5}$$

$$= \frac{8 + 2\sqrt{15}}{-2}$$

$$= \frac{2(4 + \sqrt{15})}{2(-1)} \quad \text{Factoring in order to simplify}$$

$$= \frac{2}{2} \cdot \frac{4 + \sqrt{15}}{-1}$$

$$= \frac{4 + \sqrt{15}}{-1} = -4 - \sqrt{15}.$$

Do Exercises 16 and 17.

EXAMPLE 13 Rationalize the denominator: $\dfrac{5}{2 + \sqrt{x}}$.

We multiply by 1 using the conjugate of $2 + \sqrt{x}$, which is $2 - \sqrt{x}$, as the numerator and the denominator:

$$\frac{5}{2 + \sqrt{x}} = \frac{5}{2 + \sqrt{x}} \cdot \frac{2 - \sqrt{x}}{2 - \sqrt{x}} \quad \text{Multiplying by 1}$$

$$= \frac{5(2 - \sqrt{x})}{(2 + \sqrt{x})(2 - \sqrt{x})}$$

$$= \frac{5 \cdot 2 - 5 \cdot \sqrt{x}}{2^2 - (\sqrt{x})^2}$$

$$= \frac{10 - 5\sqrt{x}}{4 - x}.$$

Do Exercise 18.

NAME SECTION DATE

Exercise Set 8.4

a Add or subtract. Simplify by collecting like radical terms, if possible.

1. $7\sqrt{3} + 9\sqrt{3}$

2. $6\sqrt{2} + 8\sqrt{2}$

3. $7\sqrt{5} - 3\sqrt{5}$

4. $8\sqrt{2} - 5\sqrt{2}$

5. $6\sqrt{x} + 7\sqrt{x}$

6. $9\sqrt{y} + 3\sqrt{y}$

7. $4\sqrt{d} - 13\sqrt{d}$

8. $2\sqrt{a} - 17\sqrt{a}$

9. $5\sqrt{8} + 15\sqrt{2}$

10. $3\sqrt{12} + 2\sqrt{3}$

11. $\sqrt{27} - 2\sqrt{3}$

12. $7\sqrt{50} - 3\sqrt{2}$

13. $\sqrt{45} - \sqrt{20}$

14. $\sqrt{27} - \sqrt{12}$

15. $\sqrt{72} + \sqrt{98}$

16. $\sqrt{45} + \sqrt{80}$

ANSWERS

1. ________
2. ________
3. ________
4. ________
5. ________
6. ________
7. ________
8. ________
9. ________
10. ________
11. ________
12. ________
13. ________
14. ________
15. ________
16. ________

ANSWERS

17. ______

18. ______

19. ______

20. ______

21. ______

22. ______

23. ______

24. ______

25. ______

26. ______

27. ______

28. ______

29. ______

30. ______

31. ______

32. ______

17. $2\sqrt{12} + \sqrt{27} - \sqrt{48}$

18. $9\sqrt{8} - \sqrt{72} + \sqrt{98}$

19. $\sqrt{18} - 3\sqrt{8} + \sqrt{50}$

20. $3\sqrt{18} - 2\sqrt{32} - 5\sqrt{50}$

21. $2\sqrt{27} - 3\sqrt{48} + 3\sqrt{12}$

22. $3\sqrt{48} - 2\sqrt{27} - 3\sqrt{12}$

23. $\sqrt{4x} + \sqrt{81x^3}$

24. $\sqrt{12x^2} + \sqrt{27}$

25. $\sqrt{27} - \sqrt{12x^2}$

26. $\sqrt{81x^3} - \sqrt{4x}$

27. $\sqrt{8x + 8} + \sqrt{2x + 2}$

28. $\sqrt{12x + 12} + \sqrt{3x + 3}$

29. $\sqrt{x^5 - x^2} + \sqrt{9x^3 - 9}$

30. $\sqrt{16x - 16} + \sqrt{25x^3 - 25x^2}$

31. $4a\sqrt{a^2b} + a\sqrt{a^2b^3} - 5\sqrt{b^3}$

32. $3x\sqrt{y^3x} - x\sqrt{yx^3} + y\sqrt{y^3x}$

33. $\sqrt{3} - \sqrt{\frac{1}{3}}$

34. $\sqrt{2} - \sqrt{\frac{1}{2}}$

35. $5\sqrt{2} + 3\sqrt{\frac{1}{2}}$

36. $4\sqrt{3} + 2\sqrt{\frac{1}{3}}$

37. $\sqrt{\frac{2}{3}} - \sqrt{\frac{1}{6}}$

38. $\sqrt{\frac{1}{2}} - \sqrt{\frac{1}{8}}$

b Multiply.

39. $\sqrt{3}(\sqrt{5} - 1)$

40. $\sqrt{2}(\sqrt{2} + \sqrt{3})$

41. $(\sqrt{2} + 8)(\sqrt{2} - 8)$

42. $(1 + \sqrt{7})(1 - \sqrt{7})$

43. $(\sqrt{6} - \sqrt{5})(\sqrt{6} + \sqrt{5})$

44. $(\sqrt{3} + \sqrt{10})(\sqrt{3} - \sqrt{10})$

45. $(3\sqrt{5} - 2)(\sqrt{5} + 1)$

46. $(\sqrt{5} - 2\sqrt{2})(\sqrt{10} - 1)$

47. $(\sqrt{x} - \sqrt{y})^2$

48. $(\sqrt{w} + 11)^2$

ANSWERS

33. ______

34. ______

35. ______

36. ______

37. ______

38. ______

39. ______

40. ______

41. ______

42. ______

43. ______

44. ______

45. ______

46. ______

47. ______

48. ______

ANSWERS

49. ______
50. ______
51. ______
52. ______
53. ______
54. ______
55. ______
56. ______
57. ______
58. ______
59. ______
60. ______
61. ______
62. ______
63. ______
64. ______
65. ______
66. ______
67. ______
68. ______
69. ______

c Rationalize the denominator.

49. $\dfrac{2}{\sqrt{3} - \sqrt{5}}$ **50.** $\dfrac{5}{3 + \sqrt{7}}$ **51.** $\dfrac{\sqrt{3} - \sqrt{2}}{\sqrt{3} + \sqrt{2}}$ **52.** $\dfrac{2 - \sqrt{7}}{\sqrt{3} - \sqrt{2}}$

53. $\dfrac{4}{\sqrt{10} + 1}$ **54.** $\dfrac{6}{\sqrt{11} - 3}$ **55.** $\dfrac{1 - \sqrt{7}}{3 + \sqrt{7}}$ **56.** $\dfrac{2 + \sqrt{8}}{1 - \sqrt{5}}$

57. $\dfrac{3}{4 + \sqrt{x}}$ **58.** $\dfrac{8}{2 - \sqrt{x}}$ **59.** $\dfrac{3 + \sqrt{2}}{8 - \sqrt{x}}$ **60.** $\dfrac{4 - \sqrt{3}}{6 + \sqrt{y}}$

SKILL MAINTENANCE

61. The time t that it takes a bus to travel a fixed distance varies inversely as its speed r. At a speed of 40 mph, it takes $\frac{1}{2}$ hr to travel a fixed distance. How long will it take to travel the same distance at 60 mph? Describe the variation constant.

62. Solution A is 3% alcohol, and solution B is 6% alcohol. A service station attendant wants to mix the two to get 80 gal of a solution that is 5.4% alcohol. How many gallons of each should the attendant use?

SYNTHESIS

63. Three students were asked to simplify $\sqrt{10} + \sqrt{50}$. Their answers were $\sqrt{10}(1 + \sqrt{5})$, $\sqrt{10} + 5\sqrt{2}$, and $\sqrt{2}(5 + \sqrt{5})$. Which, if any, are correct?

Add or subtract.

64. $\frac{3}{5}\sqrt{24} + \frac{2}{5}\sqrt{150} - \sqrt{96}$

65. $\frac{1}{3}\sqrt{27} + \sqrt{8} + \sqrt{300} - \sqrt{18} - \sqrt{162}$

66. Evaluate $\sqrt{a^2 + b^2}$ and $\sqrt{a^2} + \sqrt{b^2}$ when $a = 2$ and $b = 3$.

67. On the basis of Exercise 66, determine whether $\sqrt{a^2 + b^2}$ and $\sqrt{a^2} + \sqrt{b^2}$ are equivalent.

Determine whether each of the following is true. Show why or why not.

68. $(\sqrt{x + 2})^2 = x + 2$

69. $(3\sqrt{x + 2})^2 = 9(x + 2)$

8.5 *Radical Equations*

a Solving Radical Equations

The following are examples of *radical equations:*

$$\sqrt{2x} - 4 = 7, \qquad \sqrt{x+1} = \sqrt{2x-5}.$$

A **radical equation** has variables in one or more radicands. To solve radical equations, we first convert them to equations without radicals. We do this by squaring both sides of the equation, using the following principle.

The Principle of Squaring

If an equation $a = b$ is true, then the equation $a^2 = b^2$ is true.

To solve radical equations, we first try to get a radical by itself. That is, we try to isolate the radical. Then we use the principle of squaring. This allows us to eliminate one radical.

Example 1 Solve: $\sqrt{2x} - 4 = 7$.

$$\begin{aligned} \sqrt{2x} - 4 &= 7 \\ \sqrt{2x} &= 11 && \text{Adding 4 to isolate the radical} \\ (\sqrt{2x})^2 &= 11^2 && \text{Squaring both sides} \\ 2x &= 121 \\ x &= \frac{121}{2} && \text{Dividing by 2} \end{aligned}$$

Check:

$$\begin{array}{r|l} \sqrt{2x} - 4 & = 7 \\ \hline \sqrt{2 \cdot \frac{121}{2}} - 4 & 7 \\ \sqrt{121} - 4 & \\ 11 - 4 & \\ 7 & \end{array} \quad \text{TRUE}$$

The solution is $\frac{121}{2}$.

Do Exercise 1.

Example 2 Solve: $2\sqrt{x+2} = \sqrt{x+10}$.

Each radical is already isolated. We proceed with the principle of squaring.

$$\begin{aligned} (2\sqrt{x+2})^2 &= (\sqrt{x+10})^2 && \text{Squaring both sides} \\ 2^2(\sqrt{x+2})^2 &= (\sqrt{x+10})^2 && \text{Raising the product to the power 2 on the left} \\ 4(x+2) &= x + 10 && \text{Simplifying} \\ 4x + 8 &= x + 10 && \text{Removing parentheses} \\ 3x &= 2 && \text{Subtracting } x \text{ and } 8 \\ x &= \frac{2}{3} && \text{Dividing by 3} \end{aligned}$$

Objectives

After finishing Section 8.5, you should be able to:

- **a** Solve radical equations with one or more radical terms isolated, using the principle of squaring once.
- **b** Solve radical expressions with two radical terms, using the principle of squaring twice.
- **c** Solve applied problems using radical equations.

For Extra Help

TAPE 20 | TAPE 16B | MAC: 8B IBM: 8B

1. Solve: $\sqrt{3x} - 5 = 3$.

Answer on page A-7

Solve.

2. $\sqrt{3x+1} = \sqrt{2x+3}$

3. $3\sqrt{x+1} = \sqrt{x+12}$

4. Solve: $x - 1 = \sqrt{x+5}$.

Answers on page A-7

CHECK:

$2\sqrt{x+2} = \sqrt{x+10}$	
$2\sqrt{\frac{2}{3}+2}$	$\sqrt{\frac{2}{3}+10}$
$2\sqrt{\frac{8}{3}}$	$\sqrt{\frac{32}{3}}$
$4\sqrt{\frac{2}{3}}$	$4\sqrt{\frac{2}{3}}$ TRUE

The number $\frac{2}{3}$ checks. The solution is $\frac{2}{3}$.

Do Exercises 2 and 3.

It is important to check when using the principle of squaring. This principle may not produce equivalent equations. When we square both sides of an equation, the new equation may have solutions that the first one does not. For example, the equation

$$x = 1 \qquad \textbf{(1)}$$

has just one solution, the number 1. When we square both sides, we get

$$x^2 = 1, \qquad \textbf{(2)}$$

which has two solutions, 1 and -1. Thus the equations $x = 1$ and $x^2 = 1$ do not have the same solutions and thus are not equivalent. Whereas it is true that any solution of Equation (1) is a solution of Equation (2), it is *not* true that any solution of Equation (2) is a solution of Equation (1).

> When the principle of squaring is used to solve an equation, solutions of an equation found by squaring *must* be checked in the original equation!

Sometimes we may need to apply the principle of zero products after squaring. (See Section 4.7.)

EXAMPLE 3 Solve: $x - 5 = \sqrt{x+7}$.

$$\begin{aligned} x - 5 &= \sqrt{x+7} \\ (x-5)^2 &= (\sqrt{x+7})^2 && \text{Using the principle of squaring} \\ x^2 - 10x + 25 &= x + 7 \\ x^2 - 11x + 18 &= 0 \\ (x-9)(x-2) &= 0 && \text{Factoring} \\ x - 9 = 0 \quad &\text{or} \quad x - 2 = 0 && \text{Using the principle of zero products} \\ x = 9 \quad &\text{or} \quad x = 2 \end{aligned}$$

CHECK:

$x - 5 = \sqrt{x+7}$	
$9 - 5$	$\sqrt{9+7}$
4	4 TRUE

$x - 5 = \sqrt{x+7}$	
$2 - 5$	$\sqrt{2+7}$
-3	3 FALSE

The number 9 checks, but 2 does not. Thus the solution is 9.

Do Exercise 4.

EXAMPLE 4 Solve: $3 + \sqrt{27 - 3x} = x$.

In this case, we must first isolate the radical.

$$
\begin{aligned}
3 + \sqrt{27 - 3x} &= x \\
\sqrt{27 - 3x} &= x - 3 && \text{Subtracting 3 to isolate the radical} \\
(\sqrt{27 - 3x})^2 &= (x - 3)^2 && \text{Using the principle of squaring} \\
27 - 3x &= x^2 - 6x + 9 \\
0 &= x^2 - 3x - 18 && \text{We can have 0 on the left.} \\
0 &= (x - 6)(x + 3) && \text{Factoring} \\
x - 6 = 0 \quad &\text{or} \quad x + 3 = 0 && \text{Using the principle of zero products} \\
x = 6 \quad &\text{or} \quad x = -3
\end{aligned}
$$

CHECK:

$3 + \sqrt{27 - 3x} = x$	
$3 + \sqrt{27 - 3 \cdot 6}$	6
$3 + \sqrt{9}$	
$3 + 3$	
6	TRUE

$3 + \sqrt{27 - 3x} = x$	
$3 + \sqrt{27 - 3 \cdot (-3)}$	-3
$3 + \sqrt{27 + 9}$	
$3 + \sqrt{36}$	
$3 + 6$	
9	FALSE

The number 6 checks, but -3 does not. The solution is 6.

Do Exercise 5.

Suppose that in Example 4 we do not isolate the radical before squaring. Then we get an expression on the left side of the equation in which we have *not* eliminated the radical:

$$
\begin{aligned}
(3 + \sqrt{27 - 3x})^2 &= (x)^2 \\
3^2 + 2 \cdot 3 \cdot \sqrt{27 - 3x} + (\sqrt{27 - 3x})^2 &= x^2 \\
9 + 6\sqrt{27 - 3x} + (27 - 3x) &= x^2.
\end{aligned}
$$

In fact, we have ended up with a more complicated expression than the one we squared.

b USING THE PRINCIPLE OF SQUARING MORE THAN ONCE

Sometimes when we have two radical terms, we may need to apply the principle of squaring a second time.

EXAMPLE 5 Solve: $\sqrt{x} - 1 = \sqrt{x - 5}$.

$$
\begin{aligned}
\sqrt{x} - 1 &= \sqrt{x - 5} \\
(\sqrt{x} - 1)^2 &= (\sqrt{x - 5})^2 && \text{Using the principle of squaring} \\
(\sqrt{x})^2 - 2 \cdot \sqrt{x} \cdot 1 + 1^2 &= x - 5 && \text{Using } (A - B)^2 = A^2 - 2AB + B^2 \text{ on the left side} \\
x - 2\sqrt{x} + 1 &= x - 5 && \text{Simplifying} \\
-2\sqrt{x} &= -6 && \text{Isolating the radical} \\
\sqrt{x} &= 3 \\
(\sqrt{x})^2 &= 3^2 && \text{Using the principle of squaring} \\
x &= 9
\end{aligned}
$$

The check is left to the student. The number 9 checks and is the solution.

5. Solve: $1 + \sqrt{1 - x} = x$.

Answer on page A-7

6. Solve: $\sqrt{x} - 1 = \sqrt{x - 3}$.

7. How far can you see to the horizon through an airplane window at a height of 8000 m?

8. How far can a sailor see to the horizon from the top of a 20-m mast?

9. A technician can see 49 km to the horizon from the top of a radio tower. How high is the tower?

Answers on page A-7

The following is a procedure for solving radical equations.

> To solve radical equations:
>
> 1. Isolate one of the radical terms.
> 2. Use the principle of squaring.
> 3. If a radical term remains, perform steps (1) and (2) again.
> 4. Solve the equation and check possible solutions.

Do Exercise 6.

c APPLICATIONS

How far can you see from a given height? There is a formula for this. At a height of h meters, you can see V kilometers to the horizon. These numbers are related as follows:

$$V = 3.5\sqrt{h}. \qquad \textbf{(1)}$$

EXAMPLE 6 How far to the horizon can you see through an airplane window at a height, or altitude, of 9000 m?

We substitute 9000 for h in Equation (1) and find an approximation.

METHOD 1 We use a calculator and approximate $\sqrt{9000}$ directly:

$$V = 3.5\sqrt{9000} \approx 3.5(94.86832981) \approx 332.039 \text{ km}.$$

METHOD 2 We simplify and then approximate:

$$V = 3.5\sqrt{9000} = 3.5\sqrt{900 \cdot 10} = 3.5 \times 30 \times \sqrt{10}$$
$$\approx 3.5 \times 30 \times 3.162277660 \approx 332.039 \text{ km}.$$

You can see about 332 km at a height of 9000 m.

Do Exercises 7 and 8.

EXAMPLE 7 A person can see 50.4 km to the horizon from the top of a cliff. What is the altitude of the person's eyes?

We substitute 50.4 for V in Equation (1) and solve:

$$50.4 = 3.5\sqrt{h}$$
$$\frac{50.4}{3.5} = \sqrt{h}$$
$$14.4 = \sqrt{h}$$
$$(14.4)^2 = (\sqrt{h})^2$$
$$207.36 = h.$$

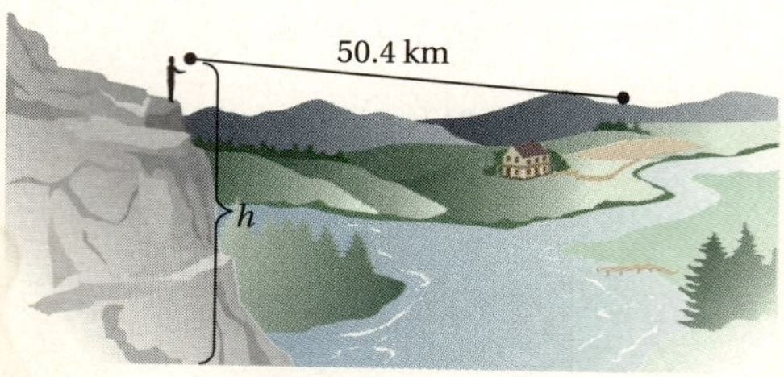

The altitude of the person's eyes is about 207 m.

Do Exercise 9.

NAME SECTION DATE

Exercise Set 8.5

a Solve.

1. $\sqrt{x} = 6$
2. $\sqrt{x} = 1$
3. $\sqrt{x} = 4.3$
4. $\sqrt{x} = 6.2$
5. $\sqrt{y + 4} = 13$
6. $\sqrt{y - 5} = 21$
7. $\sqrt{2x + 4} = 25$
8. $\sqrt{2x + 1} = 13$
9. $3 + \sqrt{x - 1} = 5$
10. $4 + \sqrt{y - 3} = 11$
11. $6 - 2\sqrt{3n} = 0$
12. $8 - 4\sqrt{5n} = 0$
13. $\sqrt{5x - 7} = \sqrt{x + 10}$
14. $\sqrt{4x - 5} = \sqrt{x + 9}$

ANSWERS

1. ________
2. ________
3. ________
4. ________
5. ________
6. ________
7. ________
8. ________
9. ________
10. ________
11. ________
12. ________
13. ________
14. ________

ANSWERS

15. ______

16. ______

17. ______

18. ______

19. ______

20. ______

21. ______

22. ______

23. ______

24. ______

25. ______

26. ______

27. ______

28. ______

15. $\sqrt{x} = -7$

16. $\sqrt{x} = -5$

17. $\sqrt{2y + 6} = \sqrt{2y - 5}$

18. $2\sqrt{3x - 2} = \sqrt{2x - 3}$

19. $x - 7 = \sqrt{x - 5}$

20. $\sqrt{x + 7} = x - 5$

21. $x - 9 = \sqrt{x - 3}$

22. $\sqrt{x + 18} = x - 2$

23. $2\sqrt{x - 1} = x - 1$

24. $x + 4 = 4\sqrt{x + 1}$

25. $\sqrt{5x + 21} = x + 3$

26. $\sqrt{27 - 3x} = x - 3$

27. $\sqrt{2x - 1} + 2 = x$

28. $x = 1 + 6\sqrt{x - 9}$

29. $\sqrt{x^2 + 6} - x + 3 = 0$

30. $\sqrt{x^2 + 5} - x + 2 = 0$

31. $\sqrt{(p + 6)(p + 1)} - 2 = p + 1$

32. $\sqrt{(4x + 5)(x + 4)} = 2x + 5$

33. $\sqrt{4x - 10} = \sqrt{2 - x}$

34. $\sqrt{2 - x} = \sqrt{3x - 7}$

b Solve. Use the principle of squaring twice.

35. $\sqrt{x - 5} = 5 - \sqrt{x}$

36. $\sqrt{x + 9} = 1 + \sqrt{x}$

37. $\sqrt{y + 8} - \sqrt{y} = 2$

38. $\sqrt{3x + 1} = 1 - \sqrt{x + 4}$

c Solve.

Use $V = 3.5\sqrt{h}$ for Exercises 39–42.

39. How far can a sailor see to the horizon from the top of a 24-m mast?

40. How far can you see to the horizon through an airplane window at a height of 9800 m?

41. A sailor can see 21.4 km to the horizon from the top of a mast. How high is the mast?

42. A person can see 371 km to the horizon from an airplane window. How high is the airplane?

ANSWERS

29. ______

30. ______

31. ______

32. ______

33. ______

34. ______

35. ______

36. ______

37. ______

38. ______

39. ______

40. ______

41. ______

42. ______

ANSWERS

43. ____________

44. ____________

45. ____________

46. ____________

47. ____________

48. ____________

49. ____________

50. ____________

51. ____________

52. ____________

53. ____________

54. ____________

55. ____________

56. ____________

57. ____________

58. ____________

The formula $r = 2\sqrt{5L}$ can be used to approximate the speed r, in miles per hour, of a car that has left a skid mark of length L, in feet.

43. How far will a car skid at 60 mph? at 100 mph?

44. How far will a car skid at 50 mph? at 70 mph?

45. Find a number such that the square root of 4 more than five times the number is 8.

46. Find the number such that twice its square root is 14.

SKILL MAINTENANCE

Divide and simplify.

47. $\dfrac{x^2 - 49}{x + 8} \div \dfrac{x^2 - 14x + 49}{x^2 + 15x + 56}$

48. $\dfrac{x - 2}{x - 3} \div \dfrac{x - 4}{x - 5}$

49. $\dfrac{a^2 - 25}{6} \div \dfrac{a + 5}{3}$

50. $\dfrac{x - 2}{x + 3} \div \dfrac{x^2 - 4x + 4}{x^2 - 9}$

51. Two angles are supplementary. One angle is 3° less than twice the other. Find the measures of the angles.

52. Two angles are complementary. The sum of the measure of the first angle and half the second is 64°. Find the measures of the angles.

SYNTHESIS

Solve.

53. $\sqrt{5x^2 + 5} = 5$

54. $\sqrt{x} = -x$

55. $4 + \sqrt{19 - x} = 6 + \sqrt{4 - x}$

56. $x = (x - 2)\sqrt{x}$

57. $\sqrt{x + 3} = \dfrac{8}{\sqrt{x - 9}}$

58. $\dfrac{12}{\sqrt{5x + 6}} = \sqrt{2x + 5}$

8.6 *Right Triangles and Applications*

OBJECTIVES

After finishing Section 8.6, you should be able to:

a Given the lengths of any two sides of a right triangle, find the length of the third side.

b Solve applied problems involving right triangles.

FOR EXTRA HELP

TAPE 20 | TAPE 16B | MAC: 8B IBM: 8B

a RIGHT TRIANGLES

A **right triangle** is a triangle with a 90° angle, as shown in the figure below. The small square in the corner indicates the 90° angle.

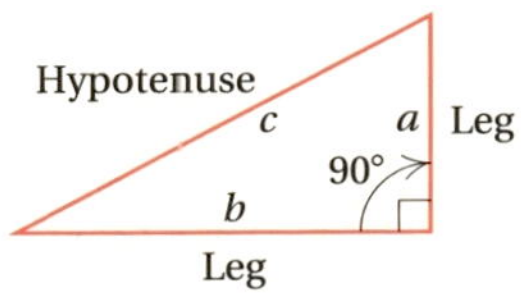

In a right triangle, the longest side is called the **hypotenuse**. It is also the side opposite the right angle. The other two sides are called **legs**. We generally use the letters a and b for the lengths of the legs and c for the length of the hypotenuse. They are related as follows.

> **THE PYTHAGOREAN THEOREM**
>
> In any right triangle, if a and b are the lengths of the legs and c is the length of the hypotenuse, then
>
> $$a^2 + b^2 = c^2.$$
>
> The equation $a^2 + b^2 = c^2$ is called the *Pythagorean equation.*

The Pythagorean theorem is named after the ancient Greek mathematician Pythagoras (569?–500? B.C.). It is uncertain who actually proved this result the first time. The proof can be found in most geometry books.

If we know the lengths of any two sides of a right triangle, we can find the length of the third side.

EXAMPLE 1 Find the length of the hypotenuse of this right triangle. Give an exact answer and an approximation to three decimal places.

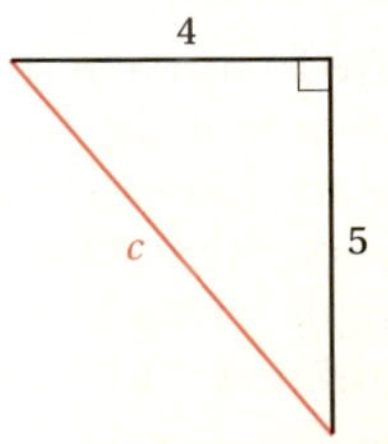

$$
\begin{aligned}
4^2 + 5^2 &= c^2 && \text{Substituting in the Pythagorean equation}\\
16 + 25 &= c^2\\
41 &= c^2\\
c &= \sqrt{41}\\
&\approx 6.403 && \text{Using a calculator or Table 2}
\end{aligned}
$$

EXAMPLE 2 Find the length of the leg of this right triangle. Give an exact answer and an approximation to three decimal places.

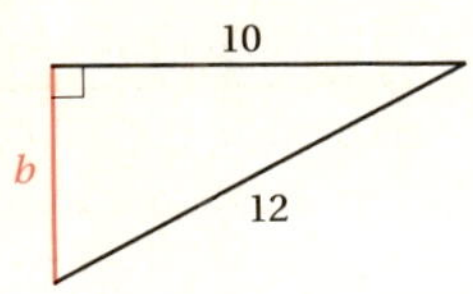

$$
\begin{aligned}
10^2 + b^2 &= 12^2 && \text{Substituting in the Pythagorean equation}\\
100 + b^2 &= 144\\
b^2 &= 144 - 100\\
b^2 &= 44\\
b &= \sqrt{44}\\
&\approx 6.633 && \text{Using a calculator or Table 2}
\end{aligned}
$$

Do Exercises 1 and 2.

1. Find the length of the hypotenuse of this right triangle. Give an exact answer and an approximation to three decimal places.

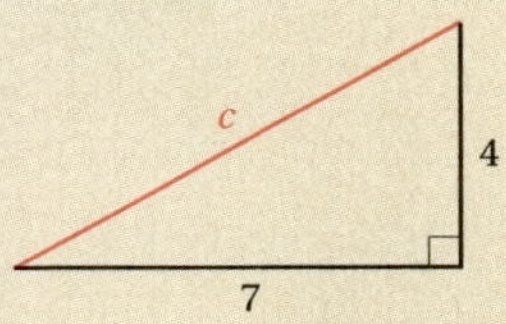

2. Find the length of the leg of this right triangle. Give an exact answer and an approximation to three decimal places.

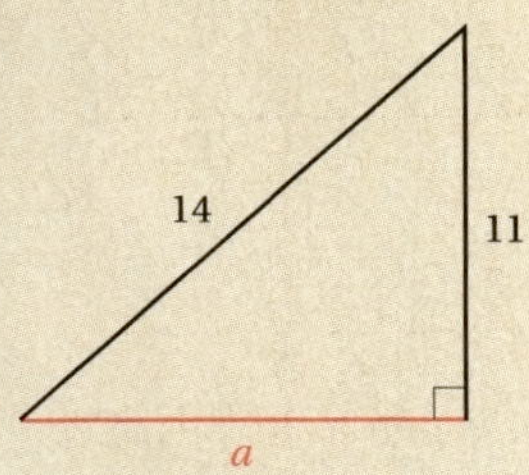

Answers on page A-7

Find the length of the leg of the right triangle. Give an exact answer and an approximation to three decimal places.

3.

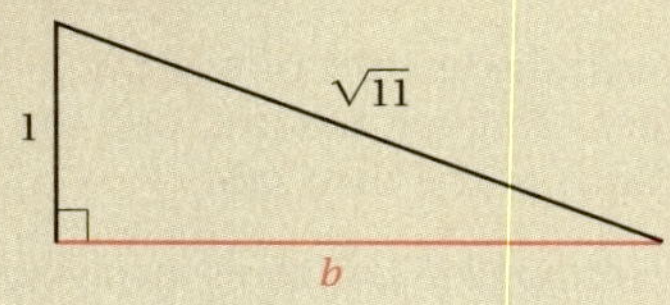

4.

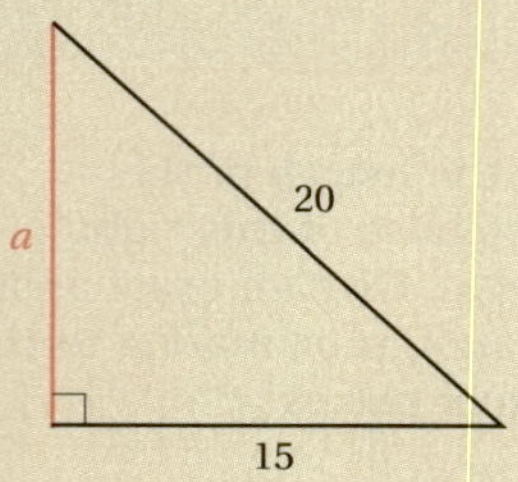

5. How long is a guy wire reaching from the top of a 15-ft pole to a point on the ground 10 ft from the pole? Give an exact answer and an approximation to three decimal places.

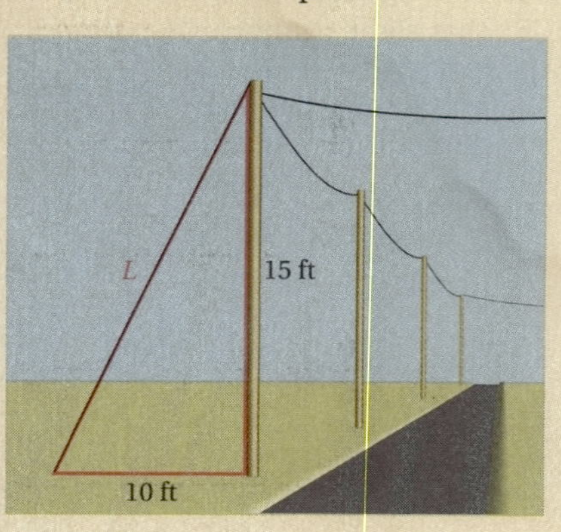

Answers on page A-7

EXAMPLE 3 Find the length of the leg of this right triangle. Give an exact answer and an approximation to three decimal places.

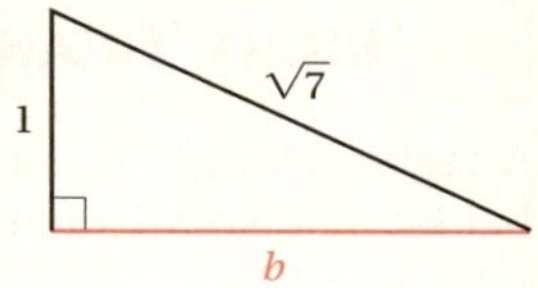

$$1^2 + b^2 = (\sqrt{7})^2 \quad \text{Substituting in the Pythagorean equation}$$
$$1 + b^2 = 7$$
$$b^2 = 7 - 1 = 6$$
$$b = \sqrt{6} \approx 2.449 \quad \text{Using a calculator or Table 2}$$

EXAMPLE 4 Find the length of the leg of this right triangle. Give an exact answer and an approximation to three decimal places.

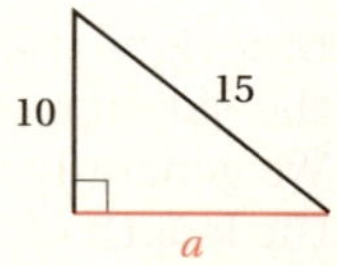

$$a^2 + 10^2 = 15^2$$
$$a^2 + 100 = 225$$
$$a^2 = 225 - 100$$
$$a^2 = 125$$
$$a = \sqrt{125} \approx 11.180 \quad \text{Using a calculator}$$

In Example 4, if you use Table 2 to find an approximation, you will need to simplify before finding an approximation:

$$\sqrt{125} = \sqrt{25 \cdot 5} = 5\sqrt{5} \approx 5(2.236) \approx 11.180.$$

A variance in answers can occur depending on the procedure used.

Do Exercises 3 and 4.

b APPLICATIONS

EXAMPLE 5 A slow-pitch softball diamond is actually a square 65 ft on a side. How far is it from home plate to second base? (This can be helpful information when lining up the bases.) Give an exact answer and an approximation to three decimal places.

a) We first make a drawing. We note that the first and second base lines, together with a line from home to second, form a right triangle. We label the unknown distance d.

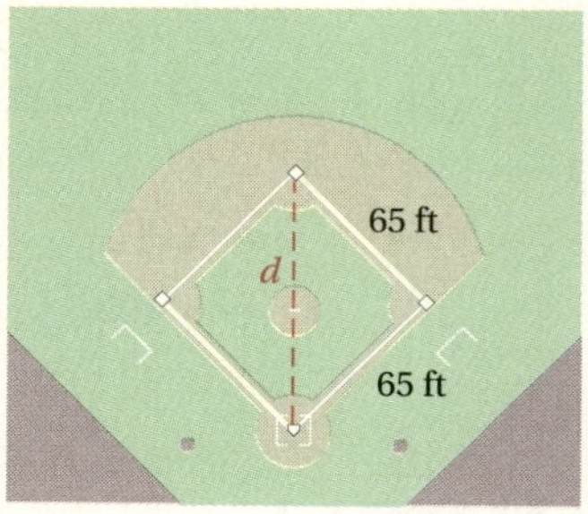

b) We know that $65^2 + 65^2 = d^2$. We solve this equation:

$$4225 + 4225 = d^2$$
$$8450 = d^2.$$

Exact answer: $\sqrt{8450}$ ft $= d$ *Approximation:* 91.924 ft $\approx d$

If you use Table 2 to find an approximation, you will need to simplify before finding an approximation in the table:

$$d = \sqrt{8450} = \sqrt{25 \cdot 169 \cdot 2} = \sqrt{25}\sqrt{169}\sqrt{2}$$
$$\approx 5(13)(1.414) \approx 91.910.$$

Note that we get a variance in the last two decimal places.

Do Exercise 5.

NAME SECTION DATE

Exercise Set 8.6

a Find the length of the third side of the right triangle. Give an exact answer and an approximation to three decimal places.

1.

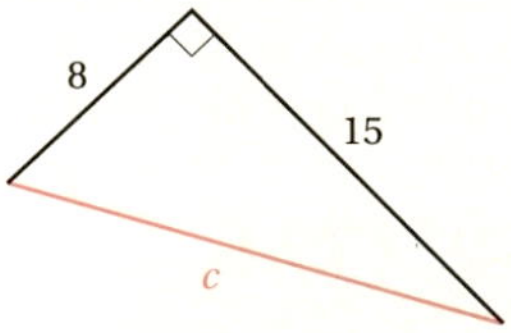

2.

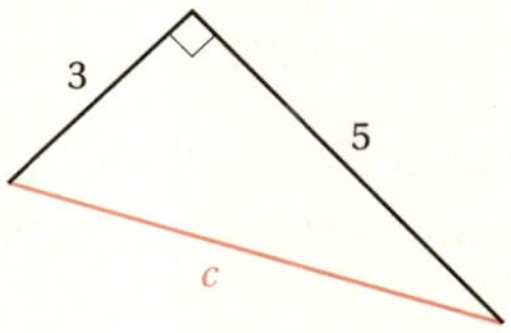

3.

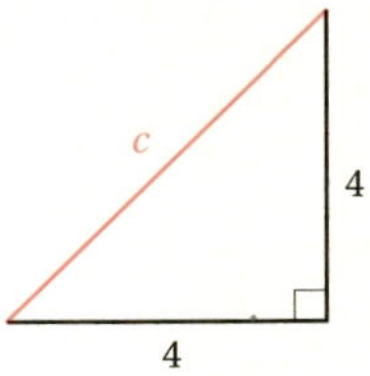

4.

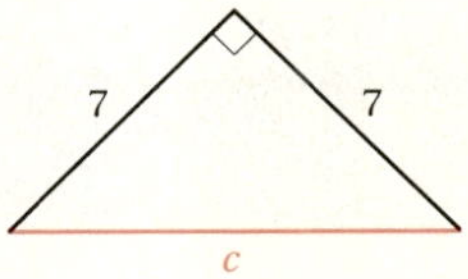

5.

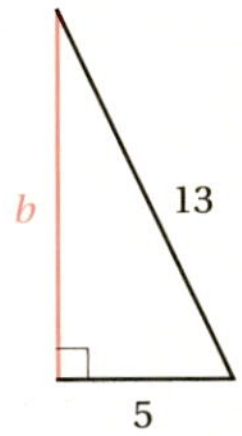

6.

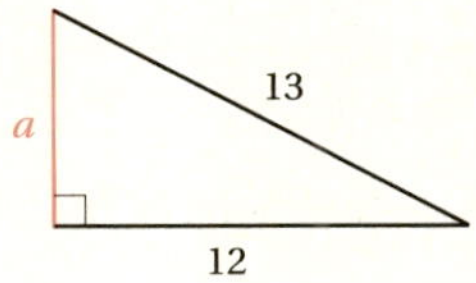

7.

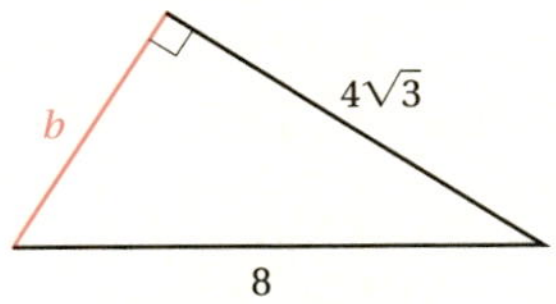

8.

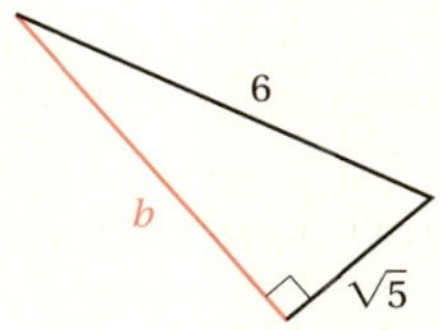

In a right triangle, find the length of the side not given. Give an exact answer and an approximation to three decimal places.

9. $a = 10, \quad b = 24$

10. $a = 5, \quad b = 12$

11. $a = 9, \quad c = 15$

12. $a = 18, \quad c = 30$

13. $b = 1, \quad c = \sqrt{5}$

14. $b = 1, \quad c = \sqrt{2}$

15. $a = 1, \quad c = \sqrt{3}$

16. $a = \sqrt{3}, \quad b = \sqrt{5}$

17. $c = 10, \quad b = 5\sqrt{3}$

18. $a = 5, \quad b = 5$

b Solve. Don't forget to make a drawing. Give an exact answer and an approximation to three decimal places.

19. Find the length of a diagonal of a square whose sides are 3 cm long.

20. A 10-m ladder is leaning against a building. The bottom of the ladder is 5 m from the building. How high is the top of the ladder?

ANSWERS

1. ________
2. ________
3. ________
4. ________
5. ________
6. ________
7. ________
8. ________
9. ________
10. ________
11. ________
12. ________
13. ________
14. ________
15. ________
16. ________
17. ________
18. ________
19. ________
20. ________

Answers

21. ______

22. ______

23. ______

24. ______

25. ______

26. ______

27. ______

28. ______

29. ______

30. ______

31. ______

32. ______

33. ______

21. The largest regulation soccer field is 100 yd wide and 130 yd long. Find the length of a diagonal of such a field.

22. How long is a guy wire reaching from the top of a 12-ft pole to a point 8 ft from the pole?

23. An airplane is flying at an altitude of 4100 ft. The slanted distance directly to the airport is 15,100 ft. How far is the airplane horizontally from the airport?

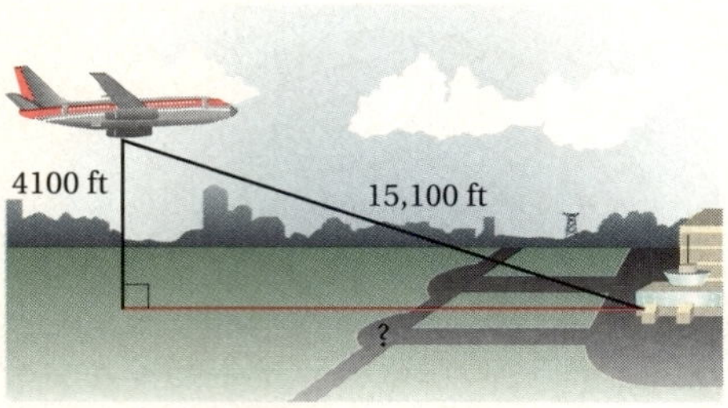

24. A surveyor had poles located at points P, Q, and R. The distances that the surveyor was able to measure are marked on the drawing. What is the approximate distance from P to R?

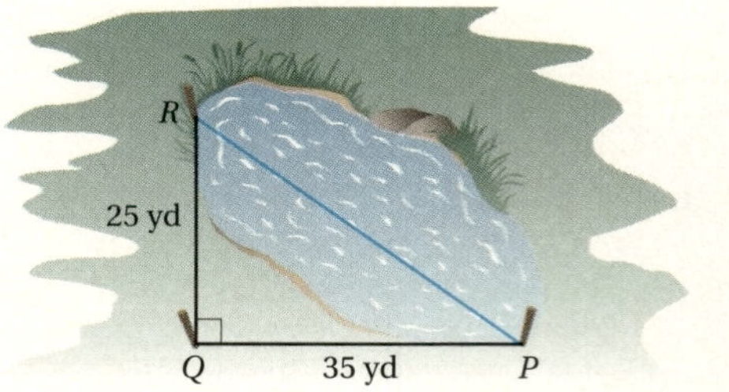

SKILL MAINTENANCE

Solve.

25. $5x + 7 = 8y,$
$3x = 8y - 4$

26. $5x + y = 17,$
$-5x + 2y = 10$

27. $3x - 4y = -11,$
$5x + 6y = 12$

28. $x + y = -9,$
$x - y = -11$

SYNTHESIS

29. Two cars leave a service station at the same time. One car travels east at a speed of 50 mph, and the other travels south at a speed of 60 mph. After one-half hour, how far apart are they?

30. The length and the width of a rectangle are given by consecutive integers. The area of the rectangle is 90 cm^2. Find the length of a diagonal of the rectangle.

Find x.

31.

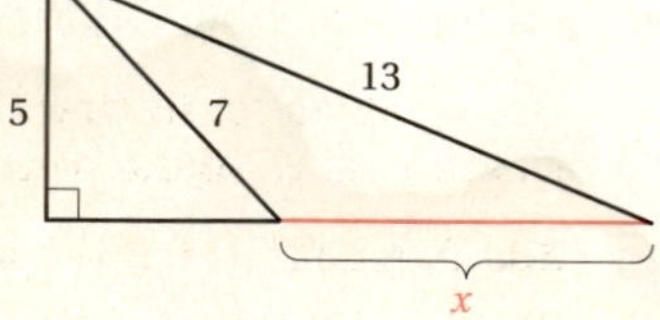

32.

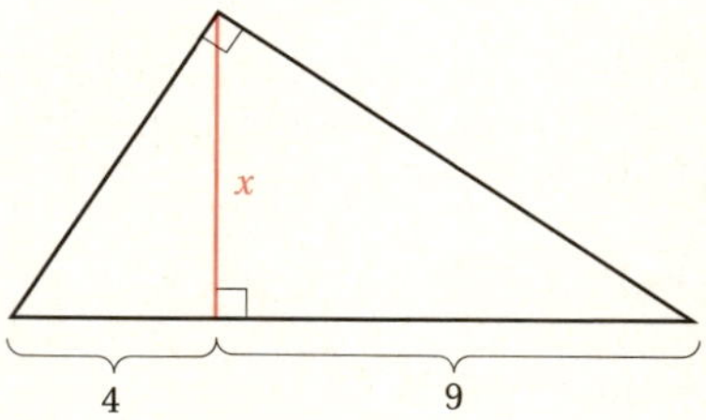

33.

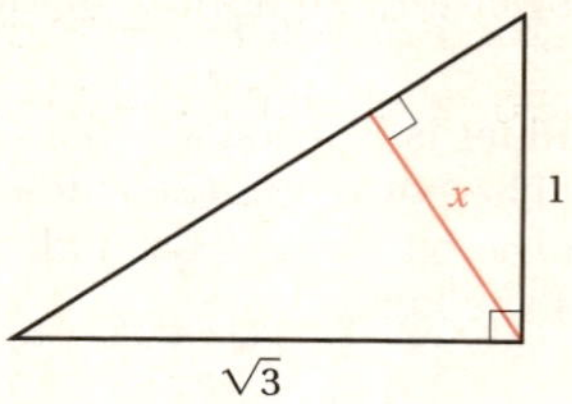

CRITICAL THINKING

CALCULATOR CONNECTION

1. a) Approximate each of the following and look for patterns:

$\sqrt{49}$, $\sqrt{490}$, $\sqrt{4900}$, $\sqrt{49{,}000}$, $\sqrt{490{,}000}$.

b) On the basis of the patterns you found in part (a), approximate $\sqrt{4{,}900{,}000}$ and $\sqrt{49{,}000{,}000}$ without using your calculator.

Find the missing lengths.

2.

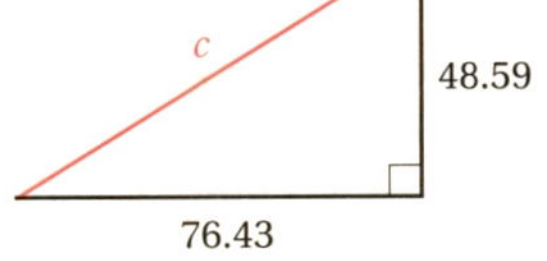

3.

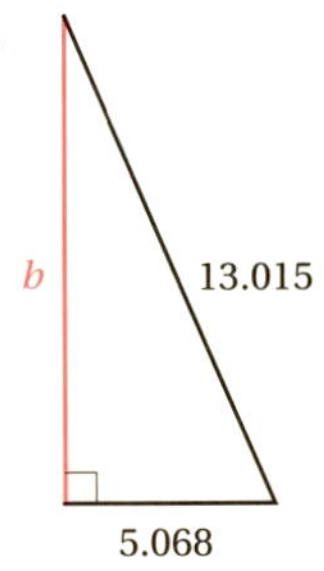

4.

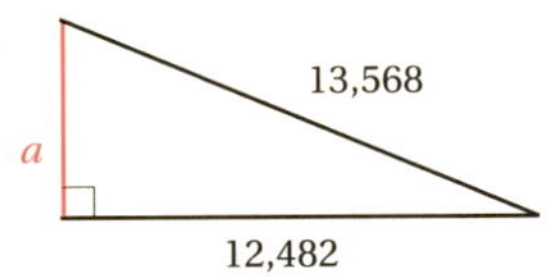

5.

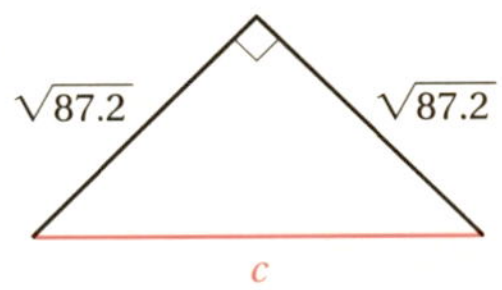

Assume that x is a positive number. Place one of $<$, $=$, or $>$ in each blank to make a true sentence.

6. $14 \;\square\; \sqrt{195}$

7. $\sqrt{450} \;\square\; 15\sqrt{2}$

8. $\sqrt{15}\sqrt{17} \;\square\; 16$

9. $25 \;\square\; \sqrt{625}$

10. $7\sqrt{2} \;\square\; 3\sqrt{11}$

11. $5\sqrt{0.64x} \;\square\; 4\sqrt{x}$

12. $100\sqrt{90x} \;\square\; 90\sqrt{100x}$

13. $\sqrt{12x} + 4\sqrt{2x} \;\square\; 4\sqrt{5x}$

14. $5\sqrt{7} \;\square\; 4\sqrt{11}$

15. $\sqrt{12x} - 4\sqrt{2x} \;\square\; 4\sqrt{3x}$

Use a graphing calculator to find the points of intersection of each pair of graphs.

16. $y = \sqrt{x}$, $y = x^2 - 2x$

17. $y = \sqrt{4 - x}$, $y = x^2 - 5$

18. $y = \sqrt{x + 7}$, $y = x - 5$
Relate this to Example 3 in Section 8.5.

19. $y = \sqrt{x} - 1$, $y = \sqrt{x - 5}$
Relate this to Example 5 in Section 8.5.

Wind Chill Temperature We can use square roots to consider an application involving the effect of wind on the feeling of cold in the winter. Because wind speed enhances the loss of heat from the skin, we feel colder when there is wind than when there is not. The *wind chill temperature* is what the temperature would have to be with no wind in order to give the same chilling effect. A formula for finding the wind chill temperature, T_W, is

$$T_W = 91.4 - \frac{(10.45 + 6.68\sqrt{v} - 0.447v)(457 - 5T)}{110},$$

where T is the actual temperature given by a thermometer, in degrees Fahrenheit, and v is the wind speed, in miles per hour.

Use a calculator to find the wind chill temperature in each case. Round to the nearest degree.

20. $T = 30°\text{F}$, $v = 25$ mph

21. $T = 10°\text{F}$, $v = 25$ mph

22. $T = 20°\text{F}$, $v = 20$ mph

23. $T = 20°\text{F}$, $v = 40$ mph

24. $T = -10°\text{F}$, $v = 30$ mph

25. $T = -30°\text{F}$, $v = 30$ mph

(*continued*)

Divide and simplify.

40. $\dfrac{\sqrt{27}}{\sqrt{45}}$

41. $\dfrac{\sqrt{45x^2y}}{\sqrt{54y}}$

42. Rationalize the denominator: $\dfrac{4}{2 + \sqrt{3}}$.

Simplify.

43. $10\sqrt{5} + 3\sqrt{5}$

44. $\sqrt{80} - \sqrt{45}$

45. $3\sqrt{2} - 5\sqrt{\dfrac{1}{2}}$

46. $(2 + \sqrt{3})^2$

47. $(2 + \sqrt{3})(2 - \sqrt{3})$

In a right triangle, find the length of the side not given.

48. $a = 15, \quad c = 25$

49. $a = 1, \quad b = \sqrt{2}$

50. Find the length of the diagonal of a square whose sides are 7 m long.

Solve.

51. $\sqrt{x - 3} = 7$

52. $\sqrt{5x + 3} = \sqrt{2x - 1}$

53. $\sqrt{x} = \sqrt{x - 5} + 1$

54. $1 + x = \sqrt{1 + 5x}$

Solve.

55. The formula $r = 2\sqrt{5L}$ can be used to approximate the speed r, in miles per hour, of a car that has left a skid mark of length L, in feet. How far will a car skid at 90 mph?

SKILL MAINTENANCE

Solve.

56. There were 14,000 people at a farm-aid concert. Tickets were $12.00 at the door and $10.00 if purchased in advance. Total receipts were $159,400. How many people bought their tickets in advance?

57. A person's paycheck P varies directly as the number of hours H worked. For working 15 hr, the pay is $168.75. Find the pay for 40 hr of work.

58. Solve:

$$2x - 3y = 4,$$
$$3x + 4y = 2.$$

59. Divide and simplify:

$$\frac{x^2 - 10x + 25}{x^2 + 14x + 49} \div \frac{x^2 - 25}{x^2 - 49}.$$

SYNTHESIS

60. Simplify: $\sqrt{\sqrt{\sqrt{256}}}$.

61. Solve $A = \sqrt{a^2 + b^2}$ for b.

NAME SECTION DATE

Test: Chapter 8

1. Find the square roots of 81.

Simplify.

2. $\sqrt{64}$

3. $-\sqrt{25}$

Approximate the expression involving square roots to three decimal places.

4. $\sqrt{116}$

5. $\sqrt{87}$

6. $\dfrac{3}{\sqrt{3}}$

7. Identify the radicand in $\sqrt{4 - y^3}$.

Determine whether the expression is meaningful as a real number. Write "yes" or "no."

8. $\sqrt{24}$

9. $\sqrt{-23}$

Simplify.

10. $\sqrt{a^2}$

11. $\sqrt{36y^2}$

Multiply.

12. $\sqrt{5}\sqrt{6}$

13. $\sqrt{x - 8}\sqrt{x + 8}$

Simplify by factoring.

14. $\sqrt{27}$

15. $\sqrt{25x - 25}$

16. $\sqrt{t^5}$

Multiply and simplify.

17. $\sqrt{5}\sqrt{10}$

18. $\sqrt{3ab}\sqrt{6ab^3}$

ANSWERS

1. ________

2. ________

3. ________

4. ________

5. ________

6. ________

7. ________

8. ________

9. ________

10. ________

11. ________

12. ________

13. ________

14. ________

15. ________

16. ________

17. ________

18. ________

ANSWERS

19. ______

20. ______

21. ______

22. ______

23. ______

24. ______

25. ______

26. ______

27. ______

28. ______

29. ______

30. ______

31. ______

32. ______

33. ______

34. ______

35. ______

36. ______

37. ______

38. ______

39. ______

40. ______

Simplify.

19. $\sqrt{\frac{27}{12}}$

20. $\sqrt{\frac{144}{a^2}}$

Rationalize the denominator.

21. $\sqrt{\frac{2}{5}}$

22. $\sqrt{\frac{2x}{y}}$

Divide and simplify.

23. $\frac{\sqrt{27}}{\sqrt{32}}$

24. $\frac{\sqrt{35x}}{\sqrt{80xy^2}}$

Add or subtract.

25. $3\sqrt{18} - 5\sqrt{18}$

26. $\sqrt{5} + \sqrt{\frac{1}{5}}$

Simplify.

27. $(4 - \sqrt{5})^2$

28. $(4 - \sqrt{5})(4 + \sqrt{5})$

29. Rationalize the denominator: $\frac{10}{4 - \sqrt{5}}$.

30. In a right triangle, $a = 8$ and $b = 4$. Find c.

Solve.

31. $\sqrt{3x} + 2 = 14$

32. $\sqrt{6x + 13} = x + 3$

33. $\sqrt{1 - x} + 1 = \sqrt{6 - x}$

34. A person can see 247.49 km to the horizon from an airplane window. How high is the airplane? Use the formula $V = 3.5\sqrt{h}$.

SKILL MAINTENANCE

35. The perimeter of a rectangle is 118 yd. The width is 18 yd less than the length. Find the area of the rectangle.

36. The number of switches N that a production line can make varies directly as the time it operates. It can make 7240 switches in 6 hr. How many can it make in 13 hr?

37. Solve:

$$-6x + 5y = 10,$$
$$5x + 6y = 12.$$

38. Divide and simplify:

$$\frac{x^2 - 11x + 30}{x^2 - 12x + 35} \div \frac{x^2 - 36}{x^2 - 14x + 49}.$$

SYNTHESIS

Simplify.

39. $\sqrt{\sqrt{\sqrt{625}}}$

40. $\sqrt{y^{16n}}$

Cumulative Review: Chapters 1–8

1. Evaluate $x^3 - x^2 + x - 1$ when $x = -2$.

2. Collect like terms:

$$2x^3 - 7 + \frac{3}{7}x^2 - 6x^3 - \frac{4}{7}x^2 + 5.$$

3. Find all numbers for which the expression is undefined:

$$\frac{x - 6}{2x + 1}.$$

4. Determine whether the expression is meaningful as a real number. Write "yes" or "no."

$$\sqrt{-24}$$

Simplify.

5. $(2 + \sqrt{3})(2 - \sqrt{3})$

6. $-\sqrt{196}$

7. $\sqrt{3}\sqrt{75}$

8. $(1 - \sqrt{2})^2$

9. $\dfrac{\sqrt{162}}{\sqrt{125}}$

10. $2\sqrt{45} + 3\sqrt{20}$

Perform the indicated operations and simplify.

11. $(3x^4 - 2y^5)(3x^4 + 2y^5)$

12. $(x^2 + 4)^2$

13. $\left(2x + \dfrac{1}{4}\right)\left(4x - \dfrac{1}{2}\right)$

14. $\dfrac{x}{2x - 1} - \dfrac{3x + 2}{1 - 2x}$

15. $(3x^2 - 2x^3) - (x^3 - 2x^2 + 5) + (3x^2 - 5x + 5)$

16. $\dfrac{2x + 2}{3x - 9} \cdot \dfrac{x^2 - 8x + 15}{x^2 - 1}$

17. $\dfrac{2x^2 - 2}{2x^2 + 7x + 3} \div \dfrac{4x - 4}{2x^2 - 5x - 3}$

18. $(3x^3 - 2x^2 + x - 5) \div (x - 2)$

Simplify.

19. $\sqrt{2x^2 - 4x + 2}$

20. $x^{-9} \cdot x^{-3}$

21. $\sqrt{\dfrac{50}{2x^8}}$

22. $\dfrac{x - \dfrac{1}{x}}{1 - \dfrac{x - 1}{2x}}$

Factor completely.

23. $3 - 12x^8$

24. $12t - 4t^2 - 48t^4$

25. $6x^2 - 28x + 16$

26. $4x^3 + 4x^2 - x - 1$

27. $16x^4 - 56x^2 + 49$

28. $x^2 + 3x - 180$

Solve.

29. $x^2 = -17x$

30. $-3x < 30 + 2x$

31. $\dfrac{1}{x} + \dfrac{2}{3} = \dfrac{1}{4}$

32. $x^2 - 30 = x$

33. $-4(x + 5) \geq 2(x + 5) - 3$

34. $2x^2 = 162$

35. $\sqrt{2x - 1} + 5 = 14$

36. $\sqrt{4x} + 1 = \sqrt{x} + 4$

37. $\dfrac{1}{4}x + \dfrac{2}{3}x = \dfrac{2}{3} - \dfrac{3}{4}x$

38. $\dfrac{x}{x - 1} - \dfrac{x}{x + 1} = \dfrac{1}{2x - 2}$

39. $x = y + 3,$
$3y - 4x = -13$

40. $2x - 3y = 30,$
$5y - 2x = -46$

41. $\dfrac{E}{r} = \dfrac{R + r}{R}$, for R

42. $4A = pr + pq$, for p

Graph on a plane.

43. $3y - 3x > -6$

44. $x = 5$

45. $2x - 6y = 12$

46. Solve by graphing:

$$x + 2y = 9,$$
$$2x + y = 6.$$

47. Find an equation of the line containing the points $(1, -2)$ and $(5, 9)$.

48. Find the slope and the y-intercept of the line $5x - 3y = 9$.

Determine whether the lines are parallel, perpendicular, or neither.

49. $2x - 3y = 5,$
$3x + 2y = 5$

50. $x - y = 3,$
$2y - 5 = 2x$

Solve.

51. The second angle of a triangle is twice as large as the first. The third angle is 48° less than the sum of the other two angles. Find the measures of the angles.

52. The cost of 6 hamburgers and 4 milkshakes is \$11.40. Three hamburgers and 1 milkshake cost \$4.80. Find the cost of a hamburger and the cost of a milkshake.

53. An 8-m ladder is leaning against a building. The bottom of the ladder is 4 m from the building. How high is the top of the ladder?

54. The amount C that a family spends on housing varies directly as its income I. A family making \$25,000 a year will spend \$6250 a year for housing. How much will a family making \$30,000 a year spend for housing?

55. A sample of 150 resistors contained 12 defective resistors. How many defective resistors would you expect in 250 resistors?

56. The length of a rectangle is 3 m greater than the width. The area of the rectangle is 180 m^2. Find the length and the width.

57. A collection of dimes and quarters is worth \$19.00. There are 115 coins in all. How many of each are there?

58. The winner of an election won by a margin of 2 to 1, with 238 votes. How many voted in the election?

59. Money is invested in an account at 10.5% simple interest. At the end of 1 yr, there is \$2873 in the account. How much was originally invested?

60. A person traveled 600 mi in one direction. The return trip took 2 hr longer at a speed that was 10 mph less. Find the speed going.

SYNTHESIS

Write a true sentence using < or >.

61. $-4 \ \blacksquare \ -3$

62. $|-4| \ \blacksquare \ |-3|$

63. A tank contains 200 L of a 30%-salt solution. How much pure water should be added to make a solution that is 12% salt?

64. Solve: $\sqrt{x} + 1 = y,$
$\sqrt{x} + \sqrt{y} = 5.$

9 Quadratic Equations

Introduction

A *quadratic equation* contains a polynomial of second degree. In this chapter, we first learn to solve quadratic equations by factoring. Because certain quadratic equations are difficult to solve by factoring, however, we also learn to use the *quadratic formula*, which is a "recipe" for finding solutions of quadratic equations. We also apply these equation-solving skills to problem solving. Then we graph quadratic equations.

An Application

The World Trade Center in New York is 1368 ft tall. How long would it take an object to fall to the ground from the top?

The Mathematics

We let t = the time, in seconds, required for the object to fall to the ground from the top. We determine t by solving the equation

$$\underbrace{1368 = 16t^2}.$$

↑

This is a *quadratic equation.*

Objectives for Review

The review objectives to be tested in addition to the material in this chapter are as follows.

[6.7c] Find an equation of inverse variation given a pair of values of the variables.

[8.2c] Multiply radical expressions and simplify, if possible.

[8.4a] Add or subtract with radical notation, using the distributive law to simplify.

[8.6b] Solve applied problems involving right triangles.

Pretest: Chapter 9

Solve.

1. $x^2 + 9 = 6x$

2. $x^2 - 7 = 0$

3. $3x^2 + 3x - 1 = 0$

4. $5y^2 - 3y = 0$

5. $\dfrac{3}{3x + 2} - \dfrac{2}{3x + 4} = 1$

6. $(x + 4)^2 = 5$

7. Solve $x^2 - 2x - 5 = 0$ by completing the square. Show your work.

8. Solve $A = n^2 - pn$ for n.

9. The length of a rectangle is three times the width. The area is 48 cm^2. Find the length and the width.

10. Find the x-intercepts: $y = 2x^2 + x - 4$.

11. The current in a stream moves at a speed of 2 km/h. A boat travels 24 km upstream and 24 km downstream in a total time of 5 hr. What is the speed of the boat in still water?

12. Graph: $y = 4 - x^2$.

9.1 *Introduction to Quadratic Equations*

Objectives

After finishing Section 9.1, you should be able to:

a. Write a quadratic equation in standard form $ax^2 + bx + c = 0$, $a > 0$, and determine the coefficients a, b, and c.

b. Solve quadratic equations of the type $ax^2 + bx = 0$, where $b \neq 0$, by factoring.

c. Solve quadratic equations of the type $ax^2 + bx + c = 0$, where $b \neq 0$ and $c \neq 0$, by factoring.

d. Solve problems involving quadratic equations.

FOR EXTRA HELP

TAPE 20 TAPE 17A MAC: 9 IBM: 9

a STANDARD FORM

The following are **quadratic equations.** They contain polynomials of second degree.

$$x^2 + 7x - 5 = 0, \quad 3t^2 - \tfrac{1}{2}t = 9, \quad 5y^2 = -6y, \quad 5m^2 = 15$$

The quadratic equation

$$4x^2 + 7x - 5 = 0$$

is said to be in **standard form.** The quadratic equation

$$4x^2 = 5 - 7x$$

is equivalent to the preceding equation, but it is *not* in standard form.

> A quadratic equation of the type $ax^2 + bx + c = 0$, where a, b, and c are real-number constants and $a > 0$, is called the *standard form of a quadratic equation.*

Often a quadratic equation is defined so that $a \neq 0$. We use $a > 0$ to ease the proof of the quadratic formula, which we consider later, and to ease solving by factoring, which we review in this section. Suppose we are studying an equation like $-3x^2 + 8x - 2 = 0$. It is not in standard form. We can find an equivalent equation that is in standard form by multiplying on both sides by -1:

$$-1(-3x^2 + 8x - 2) = -1(0)$$
$$3x^2 - 8x + 2 = 0.$$

EXAMPLES Write in standard form and determine a, b, and c.

1. $4x^2 + 7x - 5 = 0$ The equation is already in standard form.

$a = 4$; $b = 7$; $c = -5$

2. $3x^2 - 0.5x = 9$

$3x^2 - 0.5x - 9 = 0$ Subtracting 9. This is standard form.

$a = 3$; $b = -0.5$; $c = -9$

3. $-4y^2 = 5y$

$-4y^2 - 5y = 0$ Subtracting $5y$

Not positive!

$4y^2 + 5y = 0$ Multiplying by -1. This is standard form.

$a = 4$; $b = 5$; $c = 0$

Do Exercises 1–4.

Write in standard form and determine a, b, and c.

1. $y^2 = 8y$

2. $3 - x^2 = 9x$

3. $3x + 5x^2 = x^2 - 4 + 2x$

4. $5x^2 = 21$

Answers on page A-7

Solve.

5. $2x^2 + 9x = 0$

6. $10x^2 - 6x = 0$

Answers on page A-7

b SOLVING QUADRATIC EQUATIONS OF THE TYPE $ax^2 + bx = 0$

Sometimes we can use factoring and the principle of zero products to solve quadratic equations. We are actually reviewing methods that we introduced in Section 4.7.

When $c = 0$ and $b \neq 0$, we can always factor and use the principle of zero products (see Section 4.7 for a review).

EXAMPLE 4 Solve: $7x^2 + 2x = 0$.

$$
\begin{aligned}
7x^2 + 2x &= 0 \\
x(7x + 2) &= 0 && \text{Factoring} \\
x = 0 \quad \text{or} \quad 7x + 2 &= 0 && \text{Using the principle of zero products} \\
x = 0 \quad \text{or} \quad 7x &= -2 \\
x = 0 \quad \text{or} \quad x &= -\tfrac{2}{7}
\end{aligned}
$$

CHECK: For 0:

$$
\begin{array}{c|c}
\multicolumn{2}{c}{7x^2 + 2x = 0} \\
\hline
7 \cdot 0^2 + 2 \cdot 0 & 0 \\
0 & \text{TRUE}
\end{array}
$$

For $-\frac{2}{7}$:

$$
\begin{array}{c|c}
\multicolumn{2}{c}{7x^2 + 2x = 0} \\
\hline
7\left(-\frac{2}{7}\right)^2 + 2\left(-\frac{2}{7}\right) & 0 \\
7\left(\frac{4}{49}\right) - \frac{4}{7} & \\
\frac{4}{7} - \frac{4}{7} & \\
0 & \text{TRUE}
\end{array}
$$

The solutions are 0 and $-\frac{2}{7}$.

You may be tempted to divide each term in an equation like the one in Example 4 by x. This method would yield the equation

$$7x + 2 = 0,$$

whose only solution is $-\frac{2}{7}$. In effect, since 0 is also a solution of the original equation, we have divided by 0. The error of such division means the loss of one of the solutions.

EXAMPLE 5 Solve: $20x^2 - 15x = 0$.

$$
\begin{aligned}
20x^2 - 15x &= 0 \\
5x(4x - 3) &= 0 && \text{Factoring} \\
5x = 0 \quad \text{or} \quad 4x - 3 &= 0 && \text{Using the principle of zero products} \\
x = 0 \quad \text{or} \quad 4x &= 3 \\
x = 0 \quad \text{or} \quad x &= \tfrac{3}{4}
\end{aligned}
$$

The solutions are 0 and $\frac{3}{4}$.

A quadratic equation of the type $ax^2 + bx = 0$, where $c = 0$ and $b \neq 0$, will always have 0 as one solution and a nonzero number as the other solution.

Do Exercises 5 and 6.

c SOLVING QUADRATIC EQUATIONS OF THE TYPE $ax^2 + bx + c = 0$

When neither b nor c is 0, we can sometimes solve by factoring.

EXAMPLE 6 Solve: $5x^2 - 8x + 3 = 0$.

$$\begin{aligned} 5x^2 - 8x + 3 &= 0 \\ (5x - 3)(x - 1) &= 0 && \text{Factoring} \\ 5x - 3 = 0 \quad &\text{or} \quad x - 1 = 0 && \text{Using the principle of zero products} \\ 5x = 3 \quad &\text{or} \quad x = 1 \\ x = \tfrac{3}{5} \quad &\text{or} \quad x = 1 \end{aligned}$$

The solutions are $\frac{3}{5}$ and 1.

EXAMPLE 7 Solve: $(y - 3)(y - 2) = 6(y - 3)$.

We write the equation in standard form and then try to factor:

$$\begin{aligned} y^2 - 5y + 6 &= 6y - 18 && \text{Multiplying} \\ y^2 - 11y + 24 &= 0 && \text{Standard form} \\ (y - 8)(y - 3) &= 0 \\ y - 8 = 0 \quad &\text{or} \quad y - 3 = 0 \\ y = 8 \quad &\text{or} \quad y = 3. \end{aligned}$$

The solutions are 8 and 3.

Do Exercises 7 and 8.

Recall that to solve a rational equation, we multiply on both sides by the LCM of all the denominators. We may obtain a quadratic equation after a few steps. When that happens, we know how to finish solving, but we must remember to check possible solutions because a replacement may result in division by 0.

EXAMPLE 8 Solve: $\dfrac{3}{x - 1} + \dfrac{5}{x + 1} = 2$.

We multiply by the LCM, which is $(x - 1)(x + 1)$:

$$(x - 1)(x + 1) \cdot \left(\frac{3}{x - 1} + \frac{5}{x + 1}\right) = 2 \cdot (x - 1)(x + 1).$$

We use the distributive law on the left:

$$\begin{aligned} (x - 1)(x + 1) \cdot \frac{3}{x - 1} + (x - 1)(x + 1) \cdot \frac{5}{x + 1} &= 2(x - 1)(x + 1) \\ 3(x + 1) + 5(x - 1) &= 2(x - 1)(x + 1) \\ 3x + 3 + 5x - 5 &= 2(x^2 - 1) \\ 8x - 2 &= 2x^2 - 2 \\ 0 &= 2x^2 - 8x \\ 0 &= 2x(x - 4) && \text{Factoring} \\ 2x = 0 \quad \text{or} \quad x - 4 &= 0 \\ x = 0 \quad \text{or} \quad x &= 4. \end{aligned}$$

Solve.

7. $4x^2 + 5x - 6 = 0$

8. $(x - 1)(x + 1) = 5(x - 1)$

Answers on page A-7

9. Solve:

$$\frac{20}{x+5} - \frac{1}{x-4} = 1.$$

10. Use $d = \dfrac{n^2 - 3n}{2}$.

a) A heptagon has 7 sides. How many diagonals does it have?

b) A polygon has 44 diagonals. How many sides does it have?

Answers on page A-7

CHECK:

For 0:

$$\begin{array}{c|c} \frac{3}{x-1} + \frac{5}{x+1} & = 2 \\ \hline \frac{3}{0-1} + \frac{5}{0+1} & 2 \\ \frac{3}{-1} + \frac{5}{1} & \\ -3 + 5 & \\ 2 & \quad \text{TRUE} \end{array}$$

For 4:

$$\begin{array}{c|c} \frac{3}{x-1} + \frac{5}{x+1} & = 2 \\ \hline \frac{3}{4-1} + \frac{5}{4+1} & 2 \\ \frac{3}{3} + \frac{5}{5} & \\ 1 + 1 & \\ 2 & \quad \text{TRUE} \end{array}$$

The solutions are 0 and 4.

Do Exercise 9.

d SOLVING PROBLEMS

EXAMPLE 9 The number of diagonals d of a polygon of n sides is given by the formula

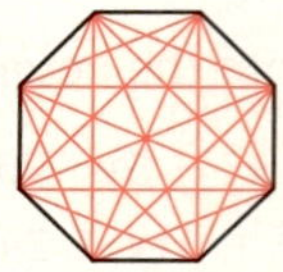

$$d = \frac{n^2 - 3n}{2}.$$

If a polygon has 27 diagonals, how many sides does it have?

1. ***Familiarize.*** We can make a drawing to familiarize ourselves with the problem. We draw an octagon (8 sides) and count the diagonals and see that there are 20. Let us check this in the formula. We evaluate the formula for $n = 8$:

$$d = \frac{8^2 - 3(8)}{2} = \frac{64 - 24}{2} = \frac{40}{2} = 20.$$

2. ***Translate.*** We know that the number of diagonals is 27. We substitute 27 for d:

$$27 = \frac{n^2 - 3n}{2}.$$

This gives us a translation.

3. ***Solve.*** We solve the equation for n, reversing the equation first for convenience:

$$\frac{n^2 - 3n}{2} = 27$$

$$n^2 - 3n = 54 \qquad \text{Multiplying by 2 to clear fractions}$$

$$n^2 - 3n - 54 = 0$$

$$(n - 9)(n + 6) = 0$$

$$n - 9 = 0 \quad \text{or} \quad n + 6 = 0$$

$$n = 9 \quad \text{or} \quad n = -6.$$

4. ***Check.*** Since the number of sides cannot be negative, -6 cannot be a solution. We leave it to the student to show by substitution that 9 checks.

5. ***State.*** The polygon has 9 sides (it is a nonagon).

Do Exercise 10.

NAME SECTION DATE

Exercise Set 9.1

a Write standard form and determine a, b, and c.

1. $x^2 - 3x + 2 = 0$

2. $x^2 - 8x - 5 = 0$

3. $7x^2 = 4x - 3$

4. $9x^2 = x + 5$

5. $5 = -2x^2 + 3x$

6. $3x - 1 = 5x^2 + 9$

b Solve.

7. $x^2 + 5x = 0$

8. $x^2 + 7x = 0$

9. $3x^2 + 6x = 0$

10. $4x^2 + 8x = 0$

11. $5x^2 = 2x$

12. $11x = 3x^2$

13. $4x^2 + 4x = 0$

14. $8x^2 - 8x = 0$

15. $0 = 10x^2 - 30x$

16. $0 = 10x^2 - 50x$

17. $11x = 55x^2$

18. $33x^2 = -11x$

19. $14t^2 = 3t$

20. $6m = 19m^2$

21. $5y^2 - 3y^2 = 72y + 9y$

22. $63p - 16p^2 = 17p + 58p^2$

c Solve.

23. $x^2 + 8x - 48 = 0$

24. $x^2 - 16x + 48 = 0$

25. $5 + 6x + x^2 = 0$

26. $x^2 + 10 + 11x = 0$

27. $18 = 7p + p^2$

28. $t^2 + 14t = -24$

29. $-15 = -8y + y^2$

30. $q^2 + 14 = 9q$

31. $x^2 + 10x + 25 = 0$

32. $x^2 + 6x + 9 = 0$

33. $r^2 = 8r - 16$

34. $x^2 + 1 = 2x$

35. $6x^2 + x - 2 = 0$

36. $2x^2 - 11x + 15 = 0$

37. $3a^2 = 10a + 8$

38. $15b - 9b^2 = 4$

ANSWERS

1. ______
2. ______
3. ______
4. ______
5. ______
6. ______
7. ______
8. ______
9. ______
10. ______
11. ______
12. ______
13. ______
14. ______
15. ______
16. ______
17. ______
18. ______
19. ______
20. ______
21. ______
22. ______
23. ______
24. ______
25. ______
26. ______
27. ______
28. ______
29. ______
30. ______
31. ______
32. ______
33. ______
34. ______
35. ______
36. ______
37. ______
38. ______

ANSWERS

39. ______
40. ______
41. ______
42. ______
43. ______
44. ______
45. ______
46. ______
47. ______
48. ______
49. ______
50. ______
51. ______
52. ______
53. ______
54. ______
55. ______
56. ______
57. ______
58. ______
59. ______
60. ______
61. ______
62. ______
63. ______
64. ______
65. ______
66. ______
67. ______
68. ______
69. ______
70. ______
71. ______
72. ______
73. ______
74. ______
75. ______
76. ______
77. ______
78. ______
79. ______
80. ______

39. $6x^2 - 4x = 10$

40. $3x^2 - 7x = 20$

41. $2t^2 + 12t = -10$

42. $12w^2 - 5w = 2$

43. $t(t - 5) = 14$

44. $6z^2 + z - 1 = 0$

45. $t(9 + t) = 4(2t + 5)$

46. $3y^2 + 8y = 12y + 15$

47. $16(p - 1) = p(p + 8)$

48. $(2x - 3)(x + 1) = 4(2x - 3)$

49. $(t - 1)(t + 3) = t - 1$

50. $(x - 2)(x + 2) = x + 2$

Solve.

51. $\dfrac{24}{x - 2} + \dfrac{24}{x + 2} = 5$

52. $\dfrac{8}{x + 2} + \dfrac{8}{x - 2} = 3$

53. $\dfrac{1}{x} + \dfrac{1}{x + 6} = \dfrac{1}{4}$

54. $\dfrac{1}{x} + \dfrac{1}{x + 9} = \dfrac{1}{20}$

55. $1 + \dfrac{12}{x^2 - 4} = \dfrac{3}{x - 2}$

56. $\dfrac{5}{t - 3} - \dfrac{30}{t^2 - 9} = 1$

57. $\dfrac{r}{r - 1} + \dfrac{2}{r^2 - 1} = \dfrac{8}{r + 1}$

58. $\dfrac{x + 2}{x^2 - 2} = \dfrac{2}{3 - x}$

59. $\dfrac{x - 1}{1 - x} = -\dfrac{x + 8}{x - 8}$

60. $\dfrac{4 - x}{x - 4} + \dfrac{x + 3}{x - 3} = 0$

d Solve.

61. A decagon is a figure with 10 sides. How many diagonals does a decagon have?

62. A hexagon is a figure with 6 sides. How many diagonals does a hexagon have?

63. A polygon has 14 diagonals. How many sides does it have?

64. A polygon has 9 diagonals. How many sides does it have?

SKILL MAINTENANCE

Simplify.

65. $\sqrt{20}$

66. $\sqrt{\dfrac{2890}{2560}}$

67. $\sqrt{\dfrac{3240}{2560}}$

68. $\sqrt{88}$

69. $\sqrt{405}$

70. $\sqrt{1020}$

SYNTHESIS

Solve.

71. $4m^2 - (m + 1)^2 = 0$

72. $x^2 + \sqrt{22}\,x = 0$

73. $\sqrt{5}\,x^2 - x = 0$

74. $\sqrt{7}\,x^2 + \sqrt{3}\,x = 0$

75. $\dfrac{5}{y + 4} - \dfrac{3}{y - 2} = 4$

76. $\dfrac{2z + 11}{2z + 8} = \dfrac{3z - 1}{z - 1}$

77. Solve $ax^2 + bx = 0$ for x.

78. 🖩 Solve: $0.0025x^2 + 70{,}400x = 0$.

Solve.

79. $z - 10\sqrt{z} + 9 = 0$ (Let $x = \sqrt{z}$.)

80. $(x - 2)^2 + 3(x - 2) = 4$

9.2 *Solving Quadratic Equations by Completing the Square*

OBJECTIVES

After finishing Section 9.2, you should be able to:

- **a** Solve quadratic equations of the type $ax^2 = p$.
- **b** Solve quadratic equations of the type $(x + c)^2 = d$.
- **c** Solve quadratic equations by completing the square.
- **d** Solve certain problems involving quadratic equations of the type $ax^2 = p$.

FOR EXTRA HELP

TAPE 21   TAPE 17A   MAC: 9 IBM: 9

a SOLVING QUADRATIC EQUATIONS OF THE TYPE $ax^2 = p$

For equations of the type $ax^2 = p$, we first solve for x^2 and then apply the *principle of square roots,* which states that a positive number has two square roots. The number 0 has one square root, 0.

> **THE PRINCIPLE OF SQUARE ROOTS**
>
> The equation $x^2 = d$ has two real solutions when $d > 0$. The solutions are $\sqrt{d}$ and $-\sqrt{d}$.
>
> The equation $x^2 = 0$ has 0 as its only solution.
>
> The equation $x^2 = d$ has no real-number solution when $d < 0$.

EXAMPLE 1 Solve: $x^2 = 3$.

$$x^2 = 3$$

$$x = \sqrt{3} \quad \text{or} \quad x = -\sqrt{3} \qquad \text{Using the principle of square roots}$$

CHECK:

For $\sqrt{3}$:

$x^2 = 3$	
$(\sqrt{3})^2$	3
3	TRUE

For $-\sqrt{3}$:

$x^2 = 3$	
$(-\sqrt{3})^2$	3
3	TRUE

The solutions are $\sqrt{3}$ and $-\sqrt{3}$.

Do Exercise 1.

EXAMPLE 2 Solve: $\frac{1}{3}x^2 = 0$.

$$\frac{1}{3}x^2 = 0$$

$$x^2 = 0 \qquad \text{Multiplying by 3}$$

$$x = 0 \qquad \text{Using the principle of square roots}$$

The solution is 0.

Do Exercise 2.

EXAMPLE 3 Solve: $-3x^2 + 7 = 0$.

$$-3x^2 + 7 = 0$$

$$-3x^2 = -7 \qquad \text{Subtracting 7}$$

$$x^2 = \frac{-7}{-3} \qquad \text{Dividing by } -3$$

$$x^2 = \frac{7}{3}$$

$$x = \sqrt{\frac{7}{3}} \quad \text{or} \quad x = -\sqrt{\frac{7}{3}} \qquad \text{Using the principle of square roots}$$

$$x = \sqrt{\frac{7}{3} \cdot \frac{3}{3}} \quad \text{or} \quad x = -\sqrt{\frac{7}{3} \cdot \frac{3}{3}} \qquad \text{Rationalizing the denominators}$$

$$x = \frac{\sqrt{21}}{3} \quad \text{or} \quad x = -\frac{\sqrt{21}}{3}$$

1. Solve: $x^2 = 10$.

2. Solve: $6x^2 = 0$.

Answers on page A-7

3. Solve: $2x^2 - 3 = 0$.

Solve.

4. $(x - 3)^2 = 16$

5. $(x + 4)^2 = 11$

Solve.

6. $x^2 - 6x + 9 = 64$

7. $x^2 - 2x + 1 = 5$

Answers on page A-7

CHECK: For $\frac{\sqrt{21}}{3}$:

$$\begin{array}{r|l} -3x^2 + 7 = 0 & \\ \hline -3\left(\frac{\sqrt{21}}{3}\right)^2 + 7 & 0 \\ -3 \cdot \frac{21}{9} + 7 & \\ -7 + 7 & \\ 0 & \quad \text{TRUE} \end{array}$$

For $-\frac{\sqrt{21}}{3}$:

$$\begin{array}{r|l} -3x^2 + 7 = 0 & \\ \hline -3\left(-\frac{\sqrt{21}}{3}\right)^2 + 7 & 0 \\ -3 \cdot \frac{21}{9} + 7 & \\ -7 + 7 & \\ 0 & \quad \text{TRUE} \end{array}$$

The solutions are $\frac{\sqrt{21}}{3}$ and $-\frac{\sqrt{21}}{3}$.

Do Exercise 3.

b SOLVING QUADRATIC EQUATIONS OF THE TYPE $(x + c)^2 = d$

The equation $(x - 5)^2 = 9$ can be solved by using the principle of square roots. We will see that other equations can be made to look like this one.

In an equation of the type $(x + c)^2 = d$, we have the square of a binomial equal to a constant. We can use the principle of square roots to solve such an equation.

EXAMPLE 4 Solve: $(x - 5)^2 = 9$.

$$(x - 5)^2 = 9$$
$$x - 5 = 3 \quad \text{or} \quad x - 5 = -3 \qquad \text{Using the principle of square roots}$$
$$x = 8 \quad \text{or} \quad x = 2$$

The solutions are 8 and 2.

EXAMPLE 5 Solve: $(x + 2)^2 = 7$.

$$(x + 2)^2 = 7$$
$$x + 2 = \sqrt{7} \quad \text{or} \quad x + 2 = -\sqrt{7} \qquad \text{Using the principle of square roots}$$
$$x = -2 + \sqrt{7} \quad \text{or} \quad x = -2 - \sqrt{7}$$

The solutions are $-2 + \sqrt{7}$ and $-2 - \sqrt{7}$, or simply $-2 \pm \sqrt{7}$ (read "-2 plus or minus $\sqrt{7}$").

Do Exercises 4 and 5.

In Examples 4 and 5, the left sides of the equations are squares of binomials. If we can express an equation in such a form, we can proceed as we did in those examples.

EXAMPLE 6 Solve: $x^2 + 8x + 16 = 49$.

$$x^2 + 8x + 16 = 49 \qquad \text{The left side is the square of a binomial.}$$
$$(x + 4)^2 = 49$$
$$x + 4 = 7 \quad \text{or} \quad x + 4 = -7 \qquad \text{Using the principle of square roots}$$
$$x = 3 \quad \text{or} \quad x = -11$$

The solutions are 3 and -11.

Do Exercises 6 and 7.

C Completing the Square

We have seen that a quadratic equation like $(x - 5)^2 = 9$ can be solved by using the principle of square roots. We also noted that an equation like $x^2 + 8x + 16 = 49$ can be solved in the same manner because the expression on the left side is the square of a binomial, $(x + 4)^2$. This second procedure is the basis for a method of solving quadratic equations called **completing the square.** *It can be used to solve any quadratic equation.*

Suppose we have the following quadratic equation:

$$x^2 + 10x = 4.$$

If we could add to both sides of the equation a constant that would make the expression on the left the square of a binomial, we could then solve the equation using the principle of square roots.

How can we determine what to add to $x^2 + 10x$ in order to construct the square of a binomial? We want to find a number a such that the following equation is satisfied:

$$(x + a)(x + a) = x^2 + 10x + a^2$$

$$ax + ax = 2ax$$

Thus, a is such that $2ax = 10x$. Solving for a, we get

$$a = \frac{10x}{2x} = \frac{10}{2} = 5;$$

that is, a is half of the coefficient of x in $x^2 + 10x$. Since $a^2 = \left(\frac{10}{2}\right)^2 = 5^2 = 25$, we add 25 to our original expression:

$$x^2 + 10x + 25 \text{ is the square of } x + 5;$$

that is,

$$x^2 + 10x + 25 = (x + 5)^2.$$

> To *complete the square* of an expression like $x^2 + bx$, we take half of the coefficient of x and square. Then we add that number, which is $(b/2)^2$.

Returning to solve our original equation, we first add 25 on both sides to complete the square. Then we solve as follows:

$$x^2 + 10x = 4 \quad \text{Original equation}$$

$$x^2 + 10x + 25 = 4 + 25 \quad \text{Adding 25: } \left(\tfrac{10}{2}\right)^2 = 5^2 = 25$$

$$(x + 5)^2 = 29$$

$$x + 5 = \sqrt{29} \quad \text{or} \quad x + 5 = -\sqrt{29} \quad \text{Using the principle of square roots}$$

$$x = -5 + \sqrt{29} \quad \text{or} \quad x = -5 - \sqrt{29}.$$

The solutions are $-5 \pm \sqrt{29}$.

We have seen that a quadratic equation $(x + c)^2 = d$ can be solved by using the principle of square roots. Any quadratic equation can be put in this form by completing the square. Then we can solve as before.

Solve.

8. $x^2 - 6x + 8 = 0$

9. $x^2 + 8x - 20 = 0$

10. Solve: $x^2 - 12x + 23 = 0$.

11. Solve: $x^2 - 3x - 10 = 0$.

Answers on page A-7

EXAMPLE 7 Solve: $x^2 + 6x + 8 = 0$.

We have

$$x^2 + 6x + 8 = 0$$
$$x^2 + 6x = -8. \quad \text{Subtracting 8}$$

We take half of 6 and square it, to get 9. Then we add 9 on *both* sides of the equation. This makes the left side the square of a binomial. We have now completed the square.

$$x^2 + 6x + 9 = -8 + 9 \quad \text{Adding 9}$$
$$(x + 3)^2 = 1$$
$$x + 3 = 1 \quad \text{or} \quad x + 3 = -1 \quad \text{Using the principle of square roots}$$
$$x = -2 \quad \text{or} \quad x = -4$$

The solutions are -2 and -4.

Do Exercises 8 and 9.

EXAMPLE 8 Solve $x^2 - 4x - 7 = 0$ by completing the square.

$$x^2 - 4x - 7 = 0$$
$$x^2 - 4x = 7 \quad \text{Adding 7}$$
$$x^2 - 4x + 4 = 7 + 4 \quad \text{Adding 4: } \left(\frac{-4}{2}\right)^2 = (-2)^2 = 4$$
$$(x - 2)^2 = 11$$
$$x - 2 = \sqrt{11} \quad \text{or} \quad x - 2 = -\sqrt{11} \quad \text{Using the principle of square roots}$$
$$x = 2 + \sqrt{11} \quad \text{or} \quad x = 2 - \sqrt{11}$$

The solutions are $2 \pm \sqrt{11}$.

Do Exercise 10.

Example 7, as well as the following example, can be solved more easily by factoring. We solved it by completing the square only to illustrate that completing the square can be used to solve any quadratic equation.

EXAMPLE 9 Solve $x^2 + 3x - 10 = 0$ by completing the square.

We have

$$x^2 + 3x - 10 = 0$$
$$x^2 + 3x = 10$$
$$x^2 + 3x + \frac{9}{4} = 10 + \frac{9}{4} \quad \text{Adding } \tfrac{9}{4}\text{: } \left(\frac{3}{2}\right)^2 = \frac{9}{4}$$
$$\left(x + \frac{3}{2}\right)^2 = \frac{40}{4} + \frac{9}{4} = \frac{49}{4}$$
$$x + \frac{3}{2} = \frac{7}{2} \quad \text{or} \quad x + \frac{3}{2} = -\frac{7}{2} \quad \text{Using the principle of square roots}$$
$$x = \frac{4}{2} \quad \text{or} \quad x = -\frac{10}{2}$$
$$x = 2 \quad \text{or} \quad x = -5.$$

The solutions are 2 and -5.

Do Exercise 11.

When the coefficient of x^2 is not 1, we can make it 1, as shown in the following example.

EXAMPLE 10 Solve $2x^2 = 3x + 1$ by completing the square.

We first obtain standard form. Then we multiply on both sides by $\frac{1}{2}$ to make the x^2-coefficient 1.

$$2x^2 = 3x + 1$$

$$2x^2 - 3x - 1 = 0 \quad \text{Finding standard form}$$

$$\frac{1}{2}(2x^2 - 3x - 1) = \frac{1}{2} \cdot 0 \quad \text{Multiplying by } \frac{1}{2} \text{ to make the } x^2\text{-coefficient 1}$$

$$x^2 - \frac{3}{2}x - \frac{1}{2} = 0$$

$$x^2 - \frac{3}{2}x = \frac{1}{2} \quad \text{Adding } \frac{1}{2}$$

$$x^2 - \frac{3}{2}x + \frac{9}{16} = \frac{1}{2} + \frac{9}{16} \quad \text{Adding } \frac{9}{16}\text{: } \left[\frac{1}{2}\left(-\frac{3}{2}\right)\right]^2 = \left[-\frac{3}{4}\right]^2 = \frac{9}{16}$$

$$\left(x - \frac{3}{4}\right)^2 = \frac{8}{16} + \frac{9}{16} \quad \text{Finding a common denominator}$$

$$\left(x - \frac{3}{4}\right)^2 = \frac{17}{16}$$

$$x - \frac{3}{4} = \frac{\sqrt{17}}{4} \quad \text{or} \quad x - \frac{3}{4} = -\frac{\sqrt{17}}{4} \quad \text{Using the principle of square roots}$$

$$x = \frac{3}{4} + \frac{\sqrt{17}}{4} \quad \text{or} \quad x = \frac{3}{4} - \frac{\sqrt{17}}{4}$$

The solutions are $\dfrac{3 \pm \sqrt{17}}{4}$.

SOLVING BY COMPLETING THE SQUARE

To solve a quadratic equation $ax^2 + bx + c = 0$ by completing the square:

1. If $a \neq 1$, multiply by $1/a$ so that the x^2-coefficient is 1.
2. If the x^2-coefficient is 1, add so that the equation is in the form

$$x^2 + bx = -c, \quad \text{or} \quad x^2 + \frac{b}{a}x = -\frac{c}{a} \quad \text{if step (1) has been applied.}$$

3. Take half of the x-coefficient and square it. Add the result on both sides of the equation.
4. Express the side with the variables as the square of a binomial.
5. Use the principle of square roots and complete the solution.

Do Exercise 12.

12. Solve: $2x^2 + 3x - 3 = 0$.

Answer on page A-7

13. The Texas Building in Houston is 1002 ft tall. How long would it take an object to fall to the ground from the top?

Answer on page A-7

d APPLICATIONS

EXAMPLE 11 The World Trade Center in New York is 1368 ft tall. How long would it take an object to fall to the ground from the top?

1. ***Familiarize.*** If we did not know anything about this problem, we might consider looking up a formula in a mathematics or physics book. A formula that fits this situation is

$$s = 16t^2,$$

where s is the distance, in feet, traveled by a body falling freely from rest in t seconds. This formula is actually an approximation in that it does not account for air resistance. In this problem, we know the distance s to be 1368 ft. We want to determine the time t for the object to reach the ground.

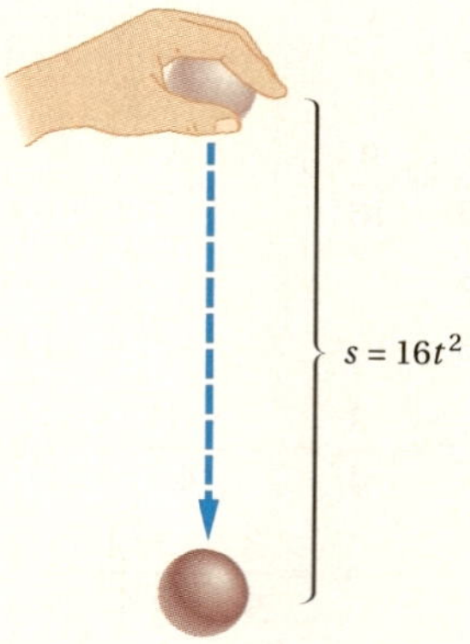

2. ***Translate.*** We know that the distance is 1368 and that we need to solve for t. We substitute 1368 for s:

$$1368 = 16t^2.$$

This gives us a translation.

3. ***Solve.*** We solve the equation:

$$1368 = 16t^2$$

$$\frac{1368}{16} = t^2 \quad \text{Solving for } t^2$$

$$85.5 = t^2 \quad \text{Dividing}$$

$$\sqrt{85.5} = t \quad \text{or} \quad -\sqrt{85.5} = t \quad \text{Using the principle of square roots}$$

$$9.2 \approx t \quad \text{or} \quad -9.2 \approx t. \quad \text{Using a calculator to find the square root and rounding to the nearest tenth}$$

4. ***Check.*** The number -9.2 cannot be a solution because time cannot be negative in this situation. We substitute 9.2 in the original equation:

$$s = 16(9.2)^2 = 16(84.64) = 1354.24.$$

This is close. Remember that we approximated a solution, $t \approx 9.2$. Thus we have a check.

5. ***State.*** It takes about 9.2 sec for an object to fall to the ground from the top of the World Trade Center.

Do Exercise 13.

NAME SECTION DATE

Exercise Set 9.2

a Solve.

1. $x^2 = 121$
2. $x^2 = 100$
3. $5x^2 = 35$
4. $5x^2 = 45$
5. $5x^2 = 3$
6. $2x^2 = 9$
7. $4x^2 - 25 = 0$
8. $9x^2 - 4 = 0$
9. $3x^2 - 49 = 0$
10. $5x^2 - 16 = 0$
11. $4y^2 - 3 = 9$
12. $36y^2 - 25 = 0$
13. $49y^2 - 64 = 0$
14. $8x^2 - 400 = 0$

b Solve.

15. $(x + 3)^2 = 16$
16. $(x - 4)^2 = 25$
17. $(x + 3)^2 = 21$
18. $(x - 3)^2 = 6$
19. $(x + 13)^2 = 8$
20. $(x - 13)^2 = 64$
21. $(x - 7)^2 = 12$
22. $(x + 1)^2 = 14$
23. $(x + 9)^2 = 34$
24. $(t + 5)^2 = 49$
25. $\left(x + \frac{3}{2}\right)^2 = \frac{7}{2}$
26. $\left(y - \frac{3}{4}\right)^2 = \frac{17}{16}$
27. $x^2 - 6x + 9 = 64$
28. $p^2 - 10p + 25 = 100$
29. $x^2 + 14x + 49 = 64$
30. $t^2 + 8t + 16 = 36$

c Solve by completing the square. Show your work.

31. $x^2 - 6x - 16 = 0$
32. $x^2 + 8x + 15 = 0$
33. $x^2 + 22x + 21 = 0$
34. $x^2 + 14x - 15 = 0$
35. $x^2 - 2x - 5 = 0$
36. $x^2 - 4x - 11 = 0$

ANSWERS

1. ______
2. ______
3. ______
4. ______
5. ______
6. ______
7. ______
8. ______
9. ______
10. ______
11. ______
12. ______
13. ______
14. ______
15. ______
16. ______
17. ______
18. ______
19. ______
20. ______
21. ______
22. ______
23. ______
24. ______
25. ______
26. ______
27. ______
28. ______
29. ______
30. ______
31. ______
32. ______
33. ______
34. ______
35. ______
36. ______

ANSWERS

37. ______
38. ______
39. ______
40. ______
41. ______
42. ______
43. ______
44. ______
45. ______
46. ______
47. ______
48. ______
49. ______
50. ______
51. ______
52. ______
53. ______
54. ______
55. ______
56. ______
57. ______
58. ______
59. ______
60. ______
61. ______
62. ______
63. ______
64. ______
65. ______
66. ______
67. ______
68. ______
69. ______
70. ______
71. ______
72. ______
73. ______
74. ______
75. ______

37. $x^2 - 22x + 102 = 0$ **38.** $x^2 - 18x + 74 = 0$ **39.** $x^2 + 10x - 4 = 0$

40. $x^2 - 10x - 4 = 0$ **41.** $x^2 - 7x - 2 = 0$ **42.** $x^2 + 7x - 2 = 0$

43. $x^2 + 3x - 28 = 0$ **44.** $x^2 - 3x - 28 = 0$ **45.** $x^2 + \frac{3}{2}x - \frac{1}{2} = 0$

46. $x^2 - \frac{3}{2}x - 2 = 0$ **47.** $2x^2 + 3x - 17 = 0$ **48.** $2x^2 - 3x - 1 = 0$

49. $3x^2 + 4x - 1 = 0$ **50.** $3x^2 - 4x - 3 = 0$ **51.** $2x^2 = 9x + 5$

52. $2x^2 = 5x + 12$ **53.** $6x^2 + 11x = 10$ **54.** $4x^2 + 12x = 7$

d Solve.

55. The height of the Sears Tower in Chicago is 1451 ft (excluding TV towers and antennas). How long would it take an object to fall to the ground from the top?

56. Library Square Tower in Los Angeles is 1012 ft tall. How long would it take an object to fall from the top?

57. The world record for free-fall to the ground, by a man without a parachute, is 311 ft and is held by Dar Robinson. Approximately how long did the fall take?

58. The world record for free-fall by a woman to the ground, without a parachute, into a cushioned landing area is 175 ft and is held by Kitty O'Neill. Approximately how long did the fall take?

SKILL MAINTENANCE

59. Find an equation of variation where y varies inversely as x, and $y = 235$ when $x = 0.6$.

60. The time T to do a certain job varies inversely as the number N of people working. It takes 5 hr for 24 people to wash and wax the floors in a building. How long would it take 36 people to do the job?

Multiply and simplify.

61. $\sqrt{3x} \cdot \sqrt{6x}$ **62.** $\sqrt{8x^2} \cdot \sqrt{24x^3}$ **63.** $3\sqrt{t} \cdot \sqrt{t}$ **64.** $\sqrt{x^2} \cdot \sqrt{x^5}$

SYNTHESIS

Find b such that the trinomial is a square.

65. $x^2 + bx + 36$ **66.** $x^2 + bx + 55$ **67.** $x^2 + bx + 128$

68. $4x^2 + bx + 16$ **69.** $x^2 + bx + c$ **70.** $ax^2 + bx + c$

Solve.

71. 🖩 $4.82x^2 = 12{,}000$ **72.** $1 = \frac{1}{7}x^2$ **73.** $\frac{x}{2} = \frac{32}{x}$

74. $\frac{x}{9} = \frac{36}{4x}$ **75.** $\frac{4}{m^2 - 7} = 1$

9.3 *The Quadratic Formula*

We learn to complete the square to prove a general formula that can be used to solve quadratic equations even when they cannot be solved by factoring.

OBJECTIVES

After finishing Section 9.3, you should be able to:

a Solve quadratic equations using the quadratic formula.

b Find approximate solutions of quadratic equations using a calculator or a square-root table.

FOR EXTRA HELP

TAPE 21 | TAPE 17B | MAC: 9 IBM: 9

a SOLVING USING THE QUADRATIC FORMULA

Each time you solve by completing the square, you continually do nearly the same thing. When we repeat the same kind of computation many times, we look for a formula so we can speed up our work. Consider

$$ax^2 + bx + c = 0, \quad a > 0.$$

Let us solve by completing the square. As we carry out the steps, compare them with Example 10 in the preceding section.

$$x^2 + \frac{b}{a}x + \frac{c}{a} = 0 \qquad \text{Multiplying by } \frac{1}{a}$$

$$x^2 + \frac{b}{a}x = -\frac{c}{a} \qquad \text{Adding } -\frac{c}{a}$$

Half of $\frac{b}{a}$ is $\frac{b}{2a}$. The square is $\frac{b^2}{4a^2}$. Thus we add $\frac{b^2}{4a^2}$ on both sides.

$$x^2 + \frac{b}{a}x + \frac{b^2}{4a^2} = -\frac{c}{a} + \frac{b^2}{4a^2} \qquad \text{Adding } \frac{b^2}{4a^2}$$

$$\left(x + \frac{b}{2a}\right)^2 = -\frac{4ac}{4a^2} + \frac{b^2}{4a^2} \qquad \text{Factoring the left side and finding a common denominator on the right}$$

$$\left(x + \frac{b}{2a}\right)^2 = \frac{b^2 - 4ac}{4a^2}$$

$$x + \frac{b}{2a} = \sqrt{\frac{b^2 - 4ac}{4a^2}} \quad \text{or} \quad x + \frac{b}{2a} = -\sqrt{\frac{b^2 - 4ac}{4a^2}} \qquad \text{Using the principle of square roots}$$

Since $a > 0$, $\sqrt{4a^2} = 2a$, so we can simplify as follows:

$$x + \frac{b}{2a} = \frac{\sqrt{b^2 - 4ac}}{2a} \quad \text{or} \quad x + \frac{b}{2a} = -\frac{\sqrt{b^2 - 4ac}}{2a}.$$

Thus,

$$x = -\frac{b}{2a} + \frac{\sqrt{b^2 - 4ac}}{2a} \quad \text{or} \quad x = -\frac{b}{2a} - \frac{\sqrt{b^2 - 4ac}}{2a},$$

so $$x = -\frac{b}{2a} \pm \frac{\sqrt{b^2 - 4ac}}{2a},$$

or $$x = \frac{-b \pm \sqrt{b^2 - 4ac}}{2a}.$$

We now have the following.

THE QUADRATIC FORMULA

The solutions of $ax^2 + bx + c = 0$ are given by

$$x = \frac{-b \pm \sqrt{b^2 - 4ac}}{2a}.$$

1. Solve using the quadratic formula:

$$2x^2 = 4 - 7x.$$

Answer on page A-8

Note that the formula also holds when $a < 0$. A similar proof would show this, but we will not consider it here.

EXAMPLE 1 Solve $5x^2 - 8x = -3$ using the quadratic formula.

We first find standard form and determine a, b, and c:

$$5x^2 - 8x + 3 = 0,$$

$$a = 5, \quad b = -8, \quad c = 3.$$

We then use the quadratic formula:

$$x = \frac{-b \pm \sqrt{b^2 - 4ac}}{2a}$$

$$x = \frac{-(-8) \pm \sqrt{(-8)^2 - 4 \cdot 5 \cdot 3}}{2 \cdot 5} \qquad \text{Substituting}$$

Caution! Be sure to write the fraction bar all the way across.

$$x = \frac{8 \pm \sqrt{64 - 60}}{10}$$

$$x = \frac{8 \pm \sqrt{4}}{10}$$

$$x = \frac{8 \pm 2}{10}$$

$$x = \frac{8 + 2}{10} \quad \text{or} \quad x = \frac{8 - 2}{10}$$

$$x = \frac{10}{10} \quad \text{or} \quad x = \frac{6}{10}$$

$$x = 1 \quad \text{or} \quad x = \frac{3}{5}.$$

The solutions are 1 and $\frac{3}{5}$.

Do Exercise 1.

It would have been easier to solve the equation in Example 1 by factoring. We used the quadratic formula only to illustrate that it can be used to solve any quadratic equation. The following is a general procedure for solving a quadratic equation.

SOLVING USING THE QUADRATIC FORMULA

To solve a quadratic equation:

1. Check to see if it is in the form $ax^2 = p$ or $(x + c)^2 = d$. If it is, use the principle of square roots as in Section 9.2.
2. If it is not in the form of (1), write it in standard form, $ax^2 + bx + c = 0$ with a and b nonzero.
3. Then try factoring.
4. If it is not possible to factor or if factoring seems difficult, use the quadratic formula.

The solutions of a quadratic equation can always be found using the quadratic formula. They cannot always be found by factoring. (When $b^2 - 4ac \geq 0$, the equation has real-number solutions. When $b^2 - 4ac < 0$, the equation has no real-number solutions.)

The expression $b^2 - 4ac$ is called the **discriminant**. The square root of the discriminant is part of the quadratic formula.

When using the quadratic formula, it is wise to compute the discriminant first. If it is negative, there are no real-number solutions because we are taking the square root of a negative number. If it is a perfect square, you can solve by factoring if you wish.

EXAMPLE 2 Solve $x^2 + 3x - 10 = 0$ using the quadratic formula.

The equation is in standard form. So we determine a, b, and c:

$$x^2 + 3x - 10 = 0,$$
$$a = 1, \quad b = 3, \quad c = -10.$$

We compute the discriminant:

$$b^2 - 4ac = 3^2 - 4 \cdot 1 \cdot (-10) = 9 + 40 = 49.$$

The discriminant is positive and is also a perfect square, so we could use factoring to solve. But for purposes of illustration, we will use the quadratic formula (try factoring on your own):

$$x = \frac{-3 \pm \sqrt{49}}{2(1)} = \frac{-3 \pm 7}{2}.$$

Thus,

$$x = \frac{-3 + 7}{2} = \frac{4}{2} = 2 \quad \text{or} \quad x = \frac{-3 - 7}{2} = \frac{-10}{2} = -5.$$

The solutions are 2 and -5.

Do Exercise 2.

EXAMPLE 3 Solve $x^2 = 4x + 7$ using the quadratic formula. Compare with Example 8 in Section 9.2.

We first find standard form and determine a, b, and c:

$$x^2 - 4x - 7 = 0,$$
$$a = 1, \quad b = -4, \quad c = -7.$$

We then compute the discriminant:

$$b^2 - 4ac = (-4)^2 - 4 \cdot (1) \cdot (-7) = 16 + 28 = 44.$$

The discriminant is positive, so there are real-number solutions. They are given by

$$x = \frac{-(-4) \pm \sqrt{44}}{2(1)} \qquad \text{Substituting into the quadratic formula}$$
$$= \frac{4 \pm \sqrt{44}}{2} = \frac{4 \pm \sqrt{4 \cdot 11}}{2} = \frac{4 \pm \sqrt{4}\sqrt{11}}{2}$$
$$= \frac{4 \pm 2\sqrt{11}}{2} = \frac{2 \cdot 2 \pm 2\sqrt{11}}{2 \cdot 1} \qquad \text{Factoring out 2 in the numerator and the denominator}$$
$$= \frac{2(2 \pm \sqrt{11})}{2 \cdot 1} = \frac{2}{2} \cdot \frac{2 \pm \sqrt{11}}{1} = 2 \pm \sqrt{11}.$$

The solutions are $2 + \sqrt{11}$ and $2 - \sqrt{11}$, or $2 \pm \sqrt{11}$.

Do Exercise 3.

2. Solve using the quadratic formula:

$$x^2 - 3x - 10 = 0.$$

3. Solve using the quadratic formula:

$$x^2 + 4x = 7.$$

Answers on page A-8

4. Solve using the quadratic formula:

$$x^2 = x - 1.$$

5. Solve using the quadratic formula:

$$5x^2 - 8x = 3.$$

6. Approximate the solutions to the equation in Margin Exercise 5. Round to the nearest tenth.

Answers on page A-8

EXAMPLE 4 Solve $x^2 + x = -1$ using the quadratic formula.

We first find standard form and determine a, b, and c:

$$x^2 + x + 1 = 0, \qquad a = 1, \quad b = 1, \quad c = 1.$$

We then compute the discriminant:

$$b^2 - 4ac = 1^2 - 4 \cdot 1 \cdot 1 = 1 - 4 = -3.$$

Since the discriminant is negative, there are no real-number solutions because square roots of negative numbers do not exist as real numbers.

Do Exercise 4.

EXAMPLE 5 Solve $3x^2 = 7 - 2x$ using the quadratic formula.

We first find standard form and determine a, b, and c:

$$3x^2 + 2x - 7 = 0, \qquad a = 3, \quad b = 2, \quad c = -7.$$

We then compute the discriminant:

$$b^2 - 4ac = 2^2 - 4 \cdot 3 \cdot (-7) = 4 + 84 = 88.$$

This is positive, so there are real-number solutions. They are given by

$$x = \frac{-2 \pm \sqrt{88}}{2(3)} \qquad \text{Substituting into the quadratic formula}$$

$$= \frac{-2 \pm \sqrt{4 \cdot 22}}{6} = \frac{-2 \pm 2\sqrt{22}}{6}$$

$$= \frac{2(-1 \pm \sqrt{22})}{2 \cdot 3} \qquad \text{Factoring out 2 in the numerator and the denominator}$$

$$= \frac{-1 \pm \sqrt{22}}{3}.$$

The solutions are $\dfrac{-1 + \sqrt{22}}{3}$ and $\dfrac{-1 - \sqrt{22}}{3}$, or $\dfrac{-1 \pm \sqrt{22}}{3}$.

Do Exercise 5.

b APPROXIMATE SOLUTIONS

A calculator or Table 2 can be used to approximate solutions.

EXAMPLE 6 Use a calculator or Table 2 to approximate to the nearest tenth the solutions to the equation in Example 5.

Using a calculator or Table 2, we see that $\sqrt{22} \approx 4.690$. Thus we have

$$\frac{-1 + \sqrt{22}}{3} \approx \frac{-1 + 4.690}{3} \quad \text{or} \quad \frac{-1 - \sqrt{22}}{3} \approx \frac{-1 - 4.690}{3}$$

$$\approx \frac{3.69}{3} \quad \text{or} \quad \approx \frac{-5.69}{3}$$

$$\approx 1.2 \text{ to the nearest tenth} \quad \text{or} \quad \approx -1.9 \text{ to the nearest tenth.}$$

The approximate solutions are 1.2 and -1.9.

Do Exercise 6.

NAME SECTION DATE

Exercise Set 9.3

a Solve. Try factoring first. If factoring is not possible or is difficult, use the quadratic formula.

1. $x^2 - 4x = 21$

2. $x^2 + 8x = 9$

3. $x^2 = 6x - 9$

4. $x^2 = 24x - 144$

5. $3y^2 - 2y - 8 = 0$

6. $3y^2 - 7y + 4 = 0$

7. $4x^2 + 4x = 15$

8. $4x^2 + 12x = 7$

9. $x^2 - 9 = 0$

10. $x^2 - 16 = 0$

11. $x^2 - 2x - 2 = 0$

12. $x^2 - 2x - 11 = 0$

13. $y^2 - 10y + 22 = 0$

14. $y^2 + 6y - 1 = 0$

15. $x^2 + 4x + 4 = 7$

16. $x^2 - 2x + 1 = 5$

17. $3x^2 + 8x + 2 = 0$

18. $3x^2 - 4x - 2 = 0$

19. $2x^2 - 5x = 1$

20. $4x^2 + 4x = 5$

21. $2y^2 - 2y - 1 = 0$

22. $4y^2 + 4y - 1 = 0$

23. $2t^2 + 6t + 5 = 0$

24. $4y^2 + 3y + 2 = 0$

25. $3x^2 = 5x + 4$

26. $2x^2 + 3x = 1$

27. $2y^2 - 6y = 10$

ANSWERS

1. ______
2. ______
3. ______
4. ______
5. ______
6. ______
7. ______
8. ______
9. ______
10. ______
11. ______
12. ______
13. ______
14. ______
15. ______
16. ______
17. ______
18. ______
19. ______
20. ______
21. ______
22. ______
23. ______
24. ______
25. ______
26. ______
27. ______

ANSWERS

28. ______
29. ______
30. ______
31. ______
32. ______
33. ______
34. ______
35. ______
36. ______
37. ______
38. ______
39. ______
40. ______
41. ______
42. ______
43. ______
44. ______
45. ______
46. ______
47. ______
48. ______
49. ______
50. ______
51. ______
52. ______
53. ______
54. ______
55. ______
56. ______

28. $5m^2 = 3 + 11m$

29. $\frac{x^2}{x+3} - \frac{5}{x+3} = 0$

30. $\frac{x^2}{x-4} - \frac{7}{x-4} = 0$

31. $x + 2 = \frac{3}{x+2}$

32. $x - 3 = \frac{5}{x-3}$

33. $\frac{1}{x} + \frac{1}{x+1} = \frac{1}{3}$

34. $\frac{1}{x} + \frac{1}{x+6} = \frac{1}{5}$

b Solve using the quadratic formula. Use a calculator or Table 2 to approximate the solutions to the nearest tenth.

35. $x^2 - 4x - 7 = 0$

36. $x^2 + 2x - 2 = 0$

37. $y^2 - 6y - 1 = 0$

38. $y^2 + 10y + 22 = 0$

39. $4x^2 + 4x = 1$

40. $4x^2 = 4x + 1$

41. $3x^2 - 8x + 2 = 0$

42. $3x^2 + 4x - 2 = 0$

SKILL MAINTENANCE

43. Multiply and simplify: $\sqrt{3x^2}\sqrt{9x^3}$.

44. Subtract: $\sqrt{54} - \sqrt{24}$.

45. Simplify: $\sqrt{80}$.

46. Rationalize the denominator: $\sqrt{\frac{7}{3}}$.

Add or subtract.

47. $\sqrt{18} + \sqrt{50} - 3\sqrt{8}$

48. $\sqrt{81x^3} - \sqrt{4x}$

SYNTHESIS

Solve.

49. $5x + x(x - 7) = 0$

50. $x(3x + 7) - 3x = 0$

51. $3 - x(x - 3) = 4$

52. $x(5x - 7) = 1$

53. $(y + 4)(y + 3) = 15$

54. $(y + 5)(y - 1) = 27$

55. $x^2 + (x + 2)^2 = 7$

56. $x^2 + (x + 1)^2 = 5$

9.4 Formulas

OBJECTIVE

After finishing Section 9.4, you should be able to:

a Solve a formula for a given letter.

FOR EXTRA HELP

TAPE 21 | TAPE 17B | MAC: 9 IBM: 9

a To solve a formula for a given letter, we try to get the letter alone on one side.

EXAMPLE 1 Solve for h: $V = 3.5\sqrt{h}$ (the distance to the horizon).

This is a radical equation. Recall that we first isolate the radical. Then we use the principle of squaring.

$$\frac{V}{3.5} = \sqrt{h} \quad \text{Isolating the radical}$$

$$\left(\frac{V}{3.5}\right)^2 = (\sqrt{h})^2 \quad \text{Using the principle of squaring (Section 8.5)}$$

$$\frac{V^2}{12.25} = h \quad \text{Simplifying}$$

EXAMPLE 2 Solve for g: $T = 2\pi\sqrt{\frac{L}{g}}$ (the period of a pendulum).

$$\frac{T}{2\pi} = \sqrt{\frac{L}{g}} \quad \text{Isolating the radical}$$

$$\left(\frac{T}{2\pi}\right)^2 = \left(\sqrt{\frac{L}{g}}\right)^2 \quad \text{Using the principle of squaring}$$

$$\frac{T^2}{4\pi^2} = \frac{L}{g}$$

$$gT^2 = 4\pi^2 L \quad \text{Multiplying by } 4\pi^2 g \text{ to clear fractions}$$

$$g = \frac{4\pi^2 L}{T^2} \quad \text{Dividing by } T^2 \text{ to get } g \text{ alone}$$

Do Exercises 1–3.

In most formulas, the letters represent nonnegative numbers, so we need not use absolute values when taking square roots.

EXAMPLE 3 *Torricelli's theorem.* The speed v of a liquid leaving a bucket from an opening is related to the height h of the top of the liquid above the opening by the formula

$$h = \frac{v^2}{2g}.$$

Solve for v.

Since v^2 appears by itself and there is no expression involving v, we first solve for v^2. Then we use the principle of square roots, taking only the nonnegative square root because v is nonnegative.

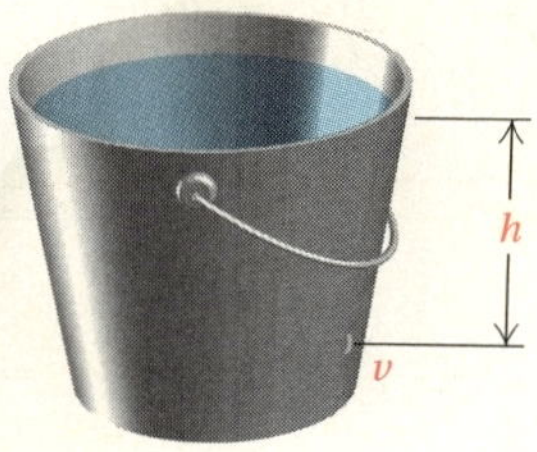

$$2gh = v^2 \quad \text{Multiplying by } 2g \text{ to clear fractions}$$

$$\sqrt{2gh} = v \quad \text{Using the principle of square roots. Assume that } v \text{ is nonnegative.}$$

Do Exercise 4.

1. Solve for L: $r = 2\sqrt{5L}$ (the speed of a skidding car).

2. Solve for L: $T = 2\pi\sqrt{\frac{L}{g}}$.

3. Solve for m: $c = \sqrt{\frac{E}{m}}$.

4. Solve for r: $A = \pi r^2$ (the area of a circle).

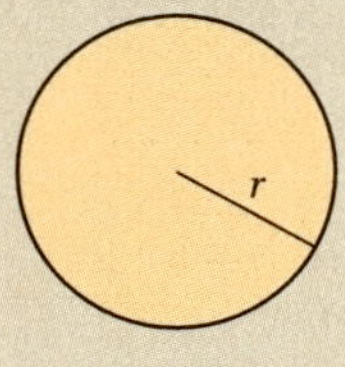

Answers on page A-8

5. Solve for d: $C = P(d - 1)^2$.

6. Solve for n: $N = n^2 - n$.

7. Solve for t: $h = vt + 8t^2$.

Answers on page A-8

EXAMPLE 4 Solve for r: $A = P(1 + r)^2$ (a compound-interest formula).

$$A = P(1 + r)^2$$

$$\frac{A}{P} = (1 + r)^2 \quad \text{Dividing by } P$$

$$\sqrt{\frac{A}{P}} = 1 + r \quad \text{Using the principle of square roots. Assume that } 1 + r \text{ is nonnegative.}$$

$$-1 + \sqrt{\frac{A}{P}} = r \quad \text{Subtracting 1 to get } r \text{ alone}$$

Do Exercise 5.

Sometimes we must use the quadratic formula to solve a formula for a certain letter.

EXAMPLE 5 Solve for n: $d = \dfrac{n^2 - 3n}{2}$, where d is the number of diagonals of an n-sided polygon.

This time there is a term involving n as well as an n^2-term. Thus we must use the quadratic formula.

$$d = \frac{n^2 - 3n}{2}$$

$$n^2 - 3n = 2d \quad \text{Multiplying by 2 to clear fractions}$$

$$n^2 - 3n - 2d = 0 \quad \text{Finding standard form}$$

$$a = 1, \quad b = -3, \quad c = -2d \quad \text{The letter } d \text{ represents a constant.}$$

$$n = \frac{-b \pm \sqrt{b^2 - 4ac}}{2a} \quad \text{Quadratic formula}$$

$$= \frac{-(-3) \pm \sqrt{(-3)^2 - 4 \cdot 1 \cdot (-2d)}}{2 \cdot 1} \quad \text{Substituting into the quadratic formula}$$

$$= \frac{3 + \sqrt{9 + 8d}}{2} \quad \text{Using the positive root}$$

Do Exercise 6.

EXAMPLE 6 Solve for t: $S = gt + 16t^2$.

$$S = gt + 16t^2$$

$$16t^2 + gt - S = 0 \quad \text{Finding standard form}$$

$$a = 16, \quad b = g, \quad c = -S$$

$$t = \frac{-b \pm \sqrt{b^2 - 4ac}}{2a}$$

$$= \frac{-g \pm \sqrt{g^2 - 4 \cdot 16 \cdot (-S)}}{2 \cdot 16} \quad \text{Substituting into the quadratic formula}$$

$$= \frac{-g + \sqrt{g^2 + 64S}}{32} \quad \text{Using the positive root}$$

Do Exercise 7.

NAME SECTION DATE

Exercise Set 9.4

a Solve for the indicated letter.

1. $P = 17\sqrt{Q}$, for Q

2. $A = 1.4\sqrt{t}$, for t

3. $v = \sqrt{\frac{2gE}{m}}$, for E

4. $Q = \sqrt{\frac{aT}{c}}$, for T

5. $S = 4\pi r^2$, for r

6. $E = mc^2$, for c

7. $P = kA^2 + mA$, for A

8. $Q = ad^2 - cd$, for d

9. $c^2 = a^2 + b^2$, for a

10. $c = \sqrt{a^2 + b^2}$, for b

11. $s = 16t^2$, for t

12. $V = \pi r^2 h$, for r

13. $A = \pi r^2 + 2\pi rh$, for r

14. $A = 2\pi r^2 + 2\pi rh$, for r

15. $F = \frac{Av^2}{400}$, for v

16. $A = \frac{\pi r^2 S}{360}$, for r

17. $c = \sqrt{a^2 + b^2}$, for a

18. $c^2 = a^2 + b^2$, for b

19. $h = \frac{a}{2}\sqrt{3}$, for a
(The height of an equilateral triangle with sides of length a)

20. $d = s\sqrt{2}$, for s
(The hypotenuse of an isosceles right triangle with s the length of the legs)

ANSWERS

1. ________
2. ________
3. ________
4. ________
5. ________
6. ________
7. ________
8. ________
9. ________
10. ________
11. ________
12. ________
13. ________
14. ________
15. ________
16. ________
17. ________
18. ________
19. ________
20. ________

ERS

21. ______

22. ______

23. ______

24. ______

25. ______

26. ______

27. ______

28. ______

29. ______

30. ______

31. ______

32. ______

33. ______

34. ______

35. a) ______

b) ______

36. ______

37. ______

38. ______

21. $n = aT^2 - 4T + m$, for T

22. $y = ax^2 + bx + c$, for x

23. $v = 2\sqrt{\dfrac{2kT}{\pi m}}$, for T

24. $E = \dfrac{1}{2}mv^2 + mgy$, for v

25. $3x^2 = d^2$, for x

26. $c = \sqrt{\dfrac{E}{m}}$, for E

27. $N = \dfrac{n^2 - n}{2}$, for n

28. $M = \dfrac{m}{\sqrt{1 - \left(\dfrac{v}{c}\right)^2}}$, for c

SKILL MAINTENANCE

In a right triangle, find the length of the side not given. Give an exact answer and an approximation to three decimal places.

29. $a = 4$, $b = 7$

30. $b = 11$, $c = 14$

31. $a = 4$, $b = 5$

32. $a = 10$, $c = 12$

33. $c = 8\sqrt{17}$, $a = 2$

34. $a = \sqrt{2}$, $b = \sqrt{3}$

SYNTHESIS

35. The circumference C of a circle is given by $C = 2\pi r$.

a) Solve $C = 2\pi r$ for r.

b) The area is given by $A = \pi r^2$. Express the area in terms of the circumference C.

36. In reference to Exercise 35, express the circumference C in terms of the area A.

37. Solve $3ax^2 - x - 3ax + 1 = 0$ for x.

38. Solve $h = 16t^2 + vt + s$ for t.

9.5 *Solving Problems*

a USING QUADRATIC EQUATIONS TO SOLVE PROBLEMS

OBJECTIVE

After finishing Section 9.5, you should be able to:

a Solve problems using quadratic equations.

FOR EXTRA HELP

TAPE 22 | TAPE 18A | MAC: 9 IBM: 9

EXAMPLE 1 The area of a rectangle is 76 in^2. The length is 7 in. longer than three times the width. Find the dimensions of the rectangle.

1. ***Familiarize.*** We first make a drawing and label it with both known and unknown information. We let w = the width of the rectangle. The length of the rectangle is 7 in. longer than three times the width. Thus the length is $3w + 7$.

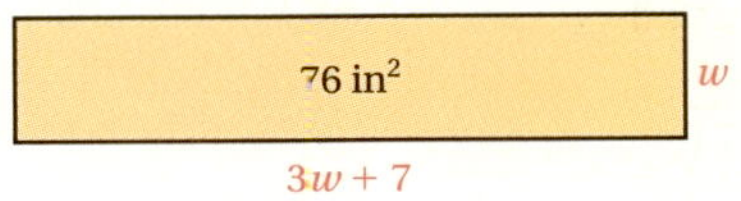

2. ***Translate.*** Recall that area is length × width. Thus we have two expressions for the area of the rectangle: $(3w + 7)(w)$ and 76. This gives us a translation:

$$(3w + 7)(w) = 76.$$

3. ***Solve.*** We solve the equation:

$$3w^2 + 7w = 76$$
$$3w^2 + 7w - 76 = 0$$
$$(3w + 19)(w - 4) = 0 \quad \text{Factoring (the quadratic formula could also be used)}$$
$$3w + 19 = 0 \quad \text{or} \quad w - 4 = 0 \quad \text{Using the principle of zero products}$$
$$3w = -19 \quad \text{or} \quad w = 4$$
$$w = -\tfrac{19}{3} \quad \text{or} \quad w = 4.$$

4. ***Check.*** We check in the original problem. We know that $-\frac{19}{3}$ is not a solution because width cannot be negative. When $w = 4$, $3w + 7 = 19$, and the area is 4(19), or 76. This checks.

5. ***State.*** The width of the rectangle is 4 in., and the length is 19 in.

Do Exercise 1.

1. The area of a rectangle is 68 in^2. The length is 1 in. longer than three times the width. Find the dimensions of the rectangle.

EXAMPLE 2 The hypotenuse of a right triangle is 6 m long. One leg is 1 m longer than the other. Find the lengths of the legs. Round to the nearest tenth.

1. ***Familiarize.*** We first make a drawing, letting s = the length of one leg. Then $s + 1$ = the length of the other leg.

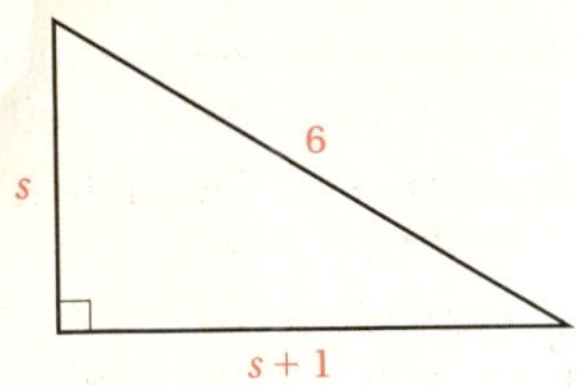

2. ***Translate.*** To translate, we use the Pythagorean equation:

$$s^2 + (s + 1)^2 = 6^2.$$

Answer on page A-8

2. The hypotenuse of a right triangle is 4 cm long. One leg is 1 cm longer than the other. Find the lengths of the legs. Round to the nearest tenth.

3. ***Solve.*** We solve the equation:

$$s^2 + (s+1)^2 = 6^2$$
$$s^2 + s^2 + 2s + 1 = 36$$
$$2s^2 + 2s - 35 = 0.$$

Since we cannot factor, we use the quadratic formula:

$$a = 2, \quad b = 2, \quad c = -35$$
$$s = \frac{-b \pm \sqrt{b^2 - 4ac}}{2a}$$
$$= \frac{-2 \pm \sqrt{2^2 - 4 \cdot 2(-35)}}{2 \cdot 2}$$
$$= \frac{-2 \pm \sqrt{4 + 280}}{4} = \frac{-2 \pm \sqrt{284}}{4}$$
$$= \frac{-2 \pm \sqrt{4 \cdot 71}}{4} = \frac{-2 \pm 2 \cdot \sqrt{71}}{2 \cdot 2}$$
$$= \frac{2(-1 \pm \sqrt{71})}{2 \cdot 2} = \frac{2}{2} \cdot \frac{-1 \pm \sqrt{71}}{2}$$
$$= \frac{-1 \pm \sqrt{71}}{2}.$$

Using a calculator or Table 2, we get an approximation: $\sqrt{71} \approx 8.426$. Thus,

$$\frac{-1 + 8.426}{2} \approx 3.7 \quad \text{or} \quad \frac{-1 - 8.426}{2} \approx -4.7.$$

4. ***Check.*** Since the length of a leg cannot be negative, -4.7 does not check. But 3.7 does check. If the smaller leg is 3.7, the other leg is 4.7. Then

$$(3.7)^2 + (4.7)^2 = 13.69 + 22.09 = 35.78.$$

Using a calculator, we get $\sqrt{35.78} \approx 5.98 \approx 6$. Note that our check is not exact because we are using an approximation for $\sqrt{71}$.

5. ***State.*** One leg is about 3.7 m long, and the other is about 4.7 m long.

Do Exercise 2.

EXAMPLE 3 The current in a stream moves at a speed of 2 km/h. A boat travels 24 km upstream and 24 km downstream in a total time of 5 hr. What is the speed of the boat in still water?

1. ***Familiarize.*** We first make a drawing. The distances are the same. We let r = the speed of the boat in still water. Then when the boat is traveling upstream, its speed is $r - 2$. When it is traveling downstream, its speed is $r + 2$. We let t_1 represent the time it takes the boat to go upstream and t_2 the time it takes to go downstream. We summarize in a table.

Answer on page A-8

	d	r	t
UPSTREAM	24	$r - 2$	t_1
DOWNSTREAM	24	$r + 2$	t_2
TOTAL TIME			5

$\longrightarrow t_1 = \dfrac{24}{r-2}$

$\longrightarrow t_2 = \dfrac{24}{r+2}$

2. ***Translate.*** Recall the basic formula for motion: $d = rt$. From it we can obtain an equation for time: $t = d/r$. Total time consists of the time to go upstream, t_1, plus the time to go downstream, t_2. Using $t = d/r$ and the rows of the table, we have

$$t_1 = \frac{24}{r-2} \quad \text{and} \quad t_2 = \frac{24}{r+2}.$$

Since the total time is 5 hr, $t_1 + t_2 = 5$, and we have

$$\frac{24}{r-2} + \frac{24}{r+2} = 5.$$

3. ***Solve.*** We solve the equation. We multiply on both sides by the LCM, which is $(r-2)(r+2)$:

$$(r-2)(r+2) \cdot \left[\frac{24}{r-2} + \frac{24}{r+2}\right] = (r-2)(r+2)5 \qquad \textbf{Multiplying by the LCM}$$

$$(r-2)(r+2) \cdot \frac{24}{r-2} + (r-2)(r+2) \cdot \frac{24}{r+2} = (r^2 - 4)5$$

$$24(r+2) + 24(r-2) = 5r^2 - 20$$

$$24r + 48 + 24r - 48 = 5r^2 - 20$$

$$-5r^2 + 48r + 20 = 0$$

$$5r^2 - 48r - 20 = 0 \qquad \textbf{Multiplying by } \mathbf{-1}$$

$$(5r + 2)(r - 10) = 0 \qquad \textbf{Factoring}$$

$$5r + 2 = 0 \quad \text{or} \quad r - 10 = 0 \qquad \textbf{Using the principle of zero products}$$

$$5r = -2 \quad \text{or} \quad r = 10$$

$$r = -\tfrac{2}{5} \quad \text{or} \quad r = 10.$$

4. ***Check.*** Since speed cannot be negative, $-\frac{2}{5}$ cannot be a solution. But suppose the speed of the boat in still water is 10 km/h. The speed upstream is then $10 - 2$, or 8 km/h. The speed downstream is $10 + 2$, or 12 km/h. The time upstream, using $t = d/r$, is 24/8, or 3 hr. The time downstream is 24/12, or 2 hr. The total time is 5 hr. This checks.

5. ***State.*** The speed of the boat in still water is 10 km/h.

Do Exercise 3.

3. The speed of a boat in still water is 12 km/h. The boat travels 45 km upstream and 45 km downstream in a total time of 8 hr. What is the speed of the stream? (*Hint:* Let $s =$ the speed of the stream. Then $12 - s$ is the speed upstream and $12 + s$ is the speed downstream. Note also that $12 - s$ cannot be negative, because the boat must be going faster than the current if it is moving forward.)

Answer on page A-8

SIDELIGHTS

HANDLING DIMENSION SYMBOLS

In many applications, we add, subtract, multiply, and divide quantities having units, or dimensions, such as ft, km, sec, hr, and so on. For example, to find average speed, we divide total distance by total time. What results is notation very much like a rational expression.

EXAMPLE 1 A car travels 150 km in 2 hr. What is its average speed?

$$\text{Speed} = \frac{150 \text{ km}}{2 \text{ hr}}, \text{ or } 75 \frac{\text{km}}{\text{hr}}$$

(The standard abbreviation for km/hr is km/h, but it does not suit our present discussion well.)

The symbol km/hr makes it look as if we are dividing kilometers by hours. It may be argued that we can divide only numbers. Nevertheless, we treat dimension symbols, such as km, ft, and hr, as if they were numerals or variables, obtaining correct results mechanically.

EXAMPLE 2 Compare

$$\frac{150x}{2y} = \frac{150}{2} \cdot \frac{x}{y} = 75\,\frac{x}{y}$$

with

$$\frac{150 \text{ km}}{2 \text{ hr}} = \frac{150}{2}\,\frac{\text{km}}{\text{hr}} = 75\,\frac{\text{km}}{\text{hr}}.$$

EXAMPLE 3 Compare

$$3x + 2x = (3 + 2)x = 5x$$

with

$$3 \text{ ft} + 2 \text{ ft} = (3 + 2) \text{ ft} = 5 \text{ ft}.$$

EXAMPLE 4 Compare

$$5x \cdot 3x = 15x^2$$

with

$$5 \text{ ft} \cdot 3 \text{ ft} = 15 \text{ ft}^2 \text{ (square feet)}.$$

EXAMPLE 5 Compare

$$5x \cdot 8y = 40xy$$

with

$$5 \text{ men} \cdot 8 \text{ hours} = 40 \text{ man-hours}.$$

If 5 men work 8 hours, the total amount of labor is 40 man-hours, which is the same as 4 men working 10 hours.

EXAMPLE 6 Compare

$$\frac{300x \cdot 240y}{15t} = 4800\frac{xy}{t}$$

with

$$\frac{300 \text{ kW} \cdot 240 \text{ hr}}{15 \text{ da}} = 4800\,\frac{\text{kW-hr}}{\text{da}}.$$

If an electrical device uses 300 kW (kilowatts) for 240 hr over a period of 15 days, its rate of usage of energy is 4800 kilowatt-hours per day. The standard abbreviation for kilowatt-hours is kWh.

These "multiplications" and "divisions" can have humorous interpretations. For example,

2 barns · 4 dances = 8 barn-dances,

2 dances · 4 dances = 8 dances^2 (8 square dances),

and

Ice · Ice · Ice = Ice^3 (Ice cubed).

However, the fact that such amusing examples exist causes us no trouble, since they do not come up in practice.

EXERCISES

Find average speeds, given total distance and total time.

1. 90 mi, 6 hr
2. 640 km, 20 hr
3. 9.9 m, 3 sec
4. 76 ft, 4 min

Perform these calculations.

5. 45 ft + 23 ft
6. 55 km/hr − 27 km/hr
7. 28 g − 17 g
8. 3.4 lb + 5.2 lb
9. $\frac{3 \text{ in.} \cdot 8 \text{ lb}}{6 \text{ sec}}$
10. $\frac{60 \text{ men} \cdot 8 \text{ hr}}{20 \text{ da}}$
11. $36 \text{ ft} \cdot \frac{1 \text{ yd}}{3 \text{ ft}}$
12. $55\frac{\text{mi}}{\text{hr}} \cdot 4 \text{ hr}$
13. $5 \text{ ft}^3 + 11 \text{ ft}^3$
14. $\frac{3 \text{ lb}}{14 \text{ ft}} \cdot \frac{7 \text{ lb}}{6 \text{ ft}}$
15. Divide \$4850 by 5 days.
16. Divide \$25.60 by 8 hr.

NAME SECTION DATE

Exercise Set 9.5

a Solve.

1. The length of a rectangle is 3 cm greater than the width. The area is 70 cm^2. Find the length and the width.

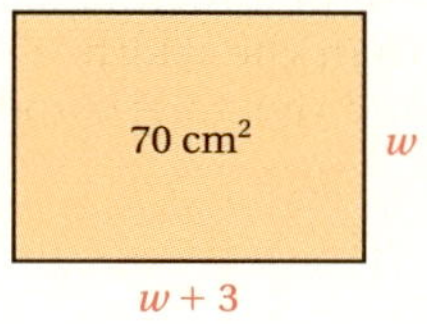

2. The length of a rectangle is 2 cm greater than the width. The area is 80 cm^2. Find the length and the width.

3. The hypotenuse of a right triangle is 26 yd long. One leg is 14 yd longer than the other. Find the lengths of the legs.

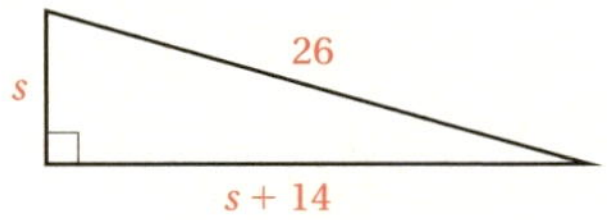

4. The hypotenuse of a right triangle is 25 ft long. One leg is 17 ft longer than the other. Find the lengths of the legs.

5. The width of a rectangle is 4 cm less than the length. The area is 320 cm^2. Find the length and the width.

6. The width of a rectangle is 3 cm less than the length. The area is 340 cm^2. Find the length and the width.

7. The length of a rectangle is twice the width. The area is 50 m^2. Find the length and the width.

8. A *square* is a carpenter's tool in the shape of a right triangle. One side, or leg, of a square is 8 in. longer than the other. The length of the hypotenuse is $8\sqrt{13}$ in. Find the lengths of the legs of the square.

Answers

1. ______

2. ______

3. ______

4. ______

5. ______

6. ______

7. ______

8. ______

ANSWERS

9. ______

10. ______

11. ______

12. ______

13. ______

14. ______

15. ______

16. ______

Find the approximate answers for Exercises 9–14. Round to the nearest tenth.

9. The hypotenuse of a right triangle is 8 m long. One leg is 2 m longer than the other. Find the lengths of the legs.

10. The hypotenuse of a right triangle is 5 cm long. One leg is 2 cm longer than the other. Find the lengths of the legs.

11. The length of a rectangle is 2 in. greater than the width. The area is 20 in^2. Find the length and the width.

12. The length of a rectangle is 3 ft greater than the width. The area is 15 ft^2. Find the length and the width.

13. The length of a rectangle is twice the width. The area is 20 cm^2. Find the length and the width.

14. The length of a rectangle is twice the width. The area is 10 m^2. Find the length and the width.

15. A picture frame measures 25 cm by 20 cm. There is 266 cm^2 of picture showing. The frame is of uniform thickness. Find the thickness of the frame.

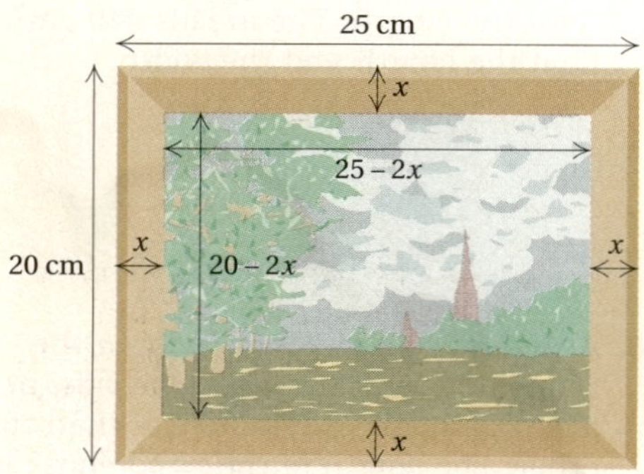

16. A tablecloth measures 96 in. by 72 in. It is laid on a tabletop with an area of 5040 in^2, and hangs over the edge by the same amount on all sides. By how many inches does the cloth hang over the edge?

17. The current in a stream moves at a speed of 3 km/h. A boat travels 40 km upstream and 40 km downstream in a total time of 14 hr. What is the speed of the boat in still water? Complete the following table to help with the familiarization.

	d	r	t
UPSTREAM		$r - 3$	t_1
DOWNSTREAM	40		t_2

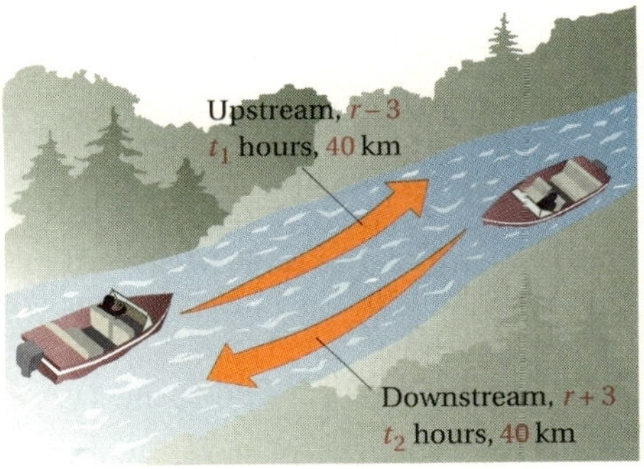

18. The current in a stream moves at a speed of 3 km/h. A boat travels 45 km upstream and 45 km downstream in a total time of 8 hr. What is the speed of the boat in still water?

19. The current in a stream moves at a speed of 4 mph. A boat travels 4 mi upstream and 12 mi downstream in a total time of 2 hr. What is the speed of the boat in still water?

20. The current in a stream moves at a speed of 4 mph. A boat travels 5 mi upstream and 13 mi downstream in a total time of 2 hr. What is the speed of the boat in still water?

21. The speed of a boat in still water is 10 km/h. The boat travels 12 km upstream and 28 km downstream in a total time of 4 hr. What is the speed of the stream?

22. The speed of a boat in still water is 8 km/h. The boat travels 60 km upstream and 60 km downstream in a total time of 16 hr. What is the speed of the stream?

ANSWERS

17. ______

18. ______

19. ______

20. ______

21. ______

22. ______

ANSWERS

23. ______

24. ______

25. ______

26. ______

27. ______

28. ______

29. ______

30. ______

31. ______

32. ______

33. ______

34. ______

35. ______

36. ______

37. ______

38. ______

23. An airplane flies 738 mi against the wind and 1062 mi with the wind in a total time of 9 hr. The speed of the airplane in still air is 200 mph. What is the speed of the wind?

24. An airplane flies 520 km against the wind and 680 km with the wind in a total time of 4 hr. The speed of the airplane in still air is 300 km/h. What is the speed of the wind?

25. The speed of a boat in still water is 9 km/h. The boat travels 80 km upstream and 80 km downstream in a total time of 18 hr. What is the speed of the stream?

26. The speed of a boat in still water is 10 km/h. The boat travels 48 km upstream and 48 km downstream in a total time of 10 hr. What is the speed of the stream?

SKILL MAINTENANCE

Add or subtract.

27. $5\sqrt{2} + \sqrt{18}$

28. $7\sqrt{40} - 2\sqrt{10}$

29. $\sqrt{4x^3} - 7\sqrt{x}$

30. $\sqrt{24} - \sqrt{54}$

31. $\sqrt{2} + \sqrt{\frac{1}{2}}$

32. $\sqrt{3} - \sqrt{\frac{1}{3}}$

SYNTHESIS

33. Two consecutive integers have squares that differ by 25. Find the integers.

34. Find the area of a square for which the diagonal is one unit longer than the length of the sides.

35. Find r in this figure. Round to the nearest hundredth.

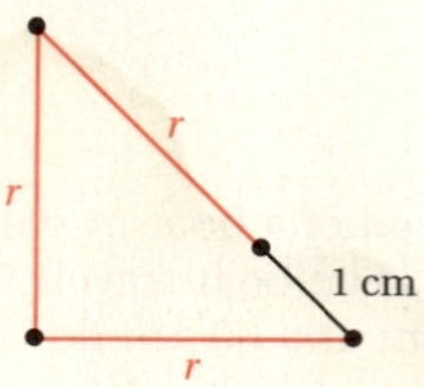

36. A 20-ft pole is struck by lightning and, while not completely broken, falls over and touches the ground 10 ft from the bottom of the pole. How high up did the pole break?

37. What should the diameter d of a pizza be so that it has the same area as two 10-in. pizzas? Do you get more to eat with a 13-in. pizza or with two 10-in. pizzas?

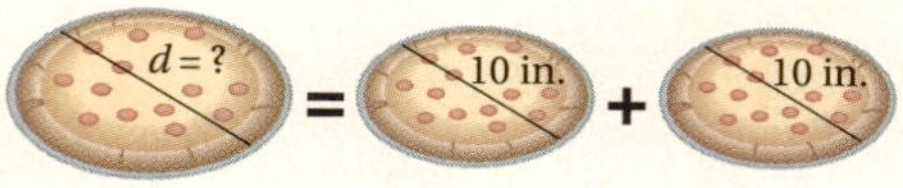

38. The width of a dollar bill is 9 cm less than the length. The area is 102.96 cm^2. Find the length and the width.

9.6 Graphs of Quadratic Equations

Objectives

After finishing Section 9.6, you should be able to:

a Graph quadratic equations.

b Find the x-intercepts of a quadratic equation.

FOR EXTRA HELP

TAPE 22 | TAPE 18A | MAC: 9 IBM: 9

In this section, we will graph equations of the form

$$y = ax^2 + bx + c, \quad a \neq 0.$$

The polynomial on the right side of the equation is of second degree, or **quadratic**. Examples of the types of equations we are going to graph are

$$y = x^2, \qquad y = x^2 + 2x - 3, \qquad y = -2x^2 + 3.$$

a Graphing Quadratic Equations of the Type $y = ax^2 + bx + c$

Graphs of quadratic equations of the type $y = ax^2 + bx + c$ (where $a \neq 0$) are always cup-shaped. They have a **line of symmetry** like the dashed lines shown in the figures below. If we fold on this line, the two halves will match exactly. The curve goes on forever. The top or bottom point where the curve changes is called the **vertex**. The second coordinate is either the largest value of y or the smallest value of y. The vertex is also thought of as a turning point. Graphs of quadratic equations are called **parabolas**.

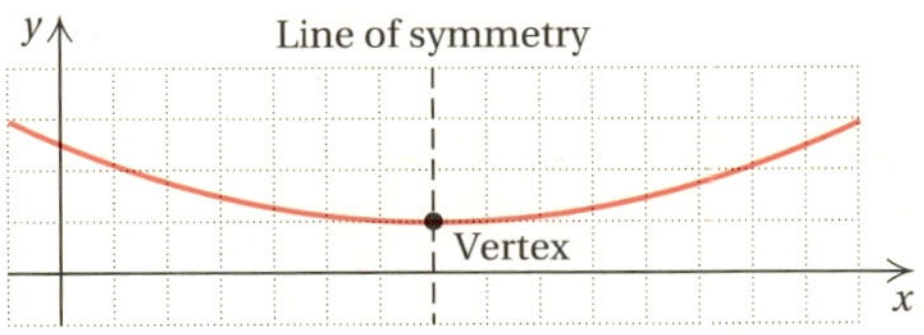

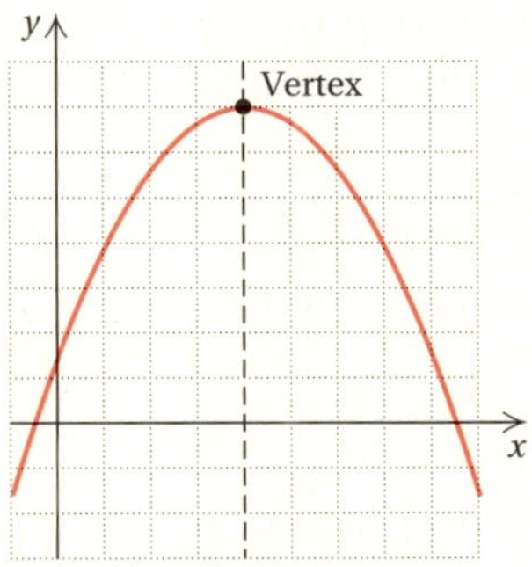

To graph a quadratic equation, we begin by choosing some numbers for x and computing the corresponding values of y.

EXAMPLE 1 Graph: $y = x^2$.

We choose numbers for x and find the corresponding values for y. Then we plot the ordered pairs (x, y) resulting from the computations and connect them with a smooth curve.

For $x = -3$, $y = x^2 = (-3)^2 = 9$.
For $x = -2$, $y = x^2 = (-2)^2 = 4$.
For $x = -1$, $y = x^2 = (-1)^2 = 1$.
For $x = 0$, $y = x^2 = (0)^2 = 0$.
For $x = 1$, $y = x^2 = (1)^2 = 1$.
For $x = 2$, $y = x^2 = (2)^2 = 4$.
For $x = 3$, $y = x^2 = (3)^2 = 9$.

x	y	(x, y)
−3	9	(−3, 9)
−2	4	(−2, 4)
−1	1	(−1, 1)
0	0	(0, 0)
1	1	(1, 1)
2	4	(2, 4)
3	9	(3, 9)

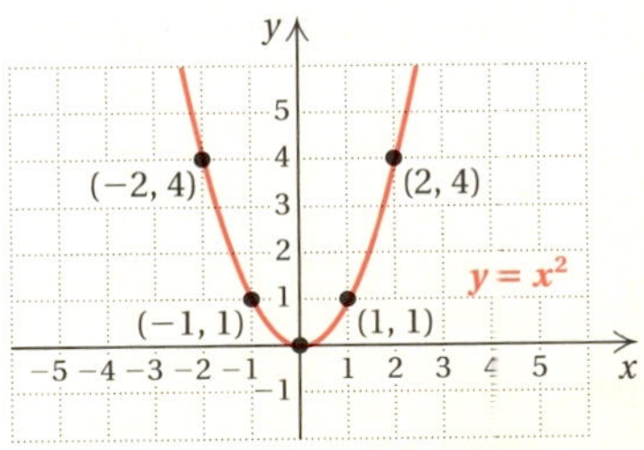

In Example 1, the vertex is the point (0, 0). The second coordinate of the vertex, 0, is the smallest y-value. The y-axis is the line of symmetry. Parabolas whose equations are $y = ax^2$ always have the origin (0, 0) as the vertex and the y-axis as the line of symmetry.

How do we graph a general equation? There are many methods, some of which you will study in your next mathematics course. Our goal here is to give you a basic graphing technique that is fairly easy to apply. A key in the graphing is knowing the vertex. By graphing it and then choosing x-values on both sides of the vertex, we can compute more points and complete the graph.

FINDING THE VERTEX

For a parabola given by the quadratic equation $y = ax^2 + bx + c$:

1. The x-coordinate of the vertex is $-\dfrac{b}{2a}$.
2. The second coordinate of the vertex is found by substituting the x-coordinate into the equation and computing y.

The proof that the vertex can be found in this way can be shown by completing the square in a manner similar to the proof of the quadratic formula, but it will not be considered here.

EXAMPLE 2 Graph: $y = -2x^2 + 3$.

We first find the vertex. The x-coordinate of the vertex is

$$-\frac{b}{2a} = -\frac{0}{2(-2)} = 0.$$

We substitute 0 for x into the equation to find the second coordinate of the vertex:

$$y = -2x^2 + 3 = -2(0)^2 + 3 = 3.$$

The vertex is (0, 3). The line of symmetry is $x = 0$, which is the y-axis. We choose some x-values on both sides of the vertex and graph the parabola.

For $x = 1$, $y = -2x^2 + 3 = -2(1)^2 + 3 = -2 + 3 = 1$.
For $x = -1$, $y = -2x^2 + 3 = -2(-1)^2 + 3 = -2 + 3 = 1$.
For $x = 2$, $y = -2x^2 + 3 = -2(2)^2 + 3 = -8 + 3 = -5$.
For $x = -2$, $y = -2x^2 + 3 = -2(-2)^2 + 3 = -8 + 3 = -5$.

x	y	
0	3	← This is the vertex.
1	1	
−1	1	
2	−5	
−2	−5	

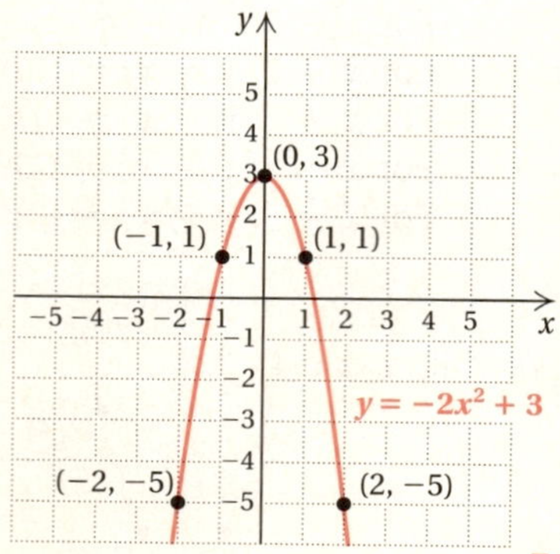

There are two other tips you might use when graphing quadratic equations. The first involves the coefficient of x^2. Note that a in $y = ax^2 + bx + c$ tells us whether the graph opens up or down. When a is positive, as in Example 1, the graph opens up; when a is negative, as in Example 2, the graph opens down. It is also helpful to plot the y-intercept. It occurs when $x = 0$.

TIPS FOR GRAPHING QUADRATIC EQUATIONS

1. Graphs of quadratic equations $y = ax^2 + bx + c$ are all parabolas. They are *smooth* cup-shaped symmetric curves, with no sharp points or kinks in them.
2. The graph of $y = ax^2 + bx + c$ opens up if $a > 0$. It opens down if $a < 0$.
3. Find the y-intercept. It occurs when $x = 0$, and it is easy to compute.

EXAMPLE 3 Graph: $y = x^2 + 2x - 3$.

We first find the vertex. The x-coordinate of the vertex is

$$-\frac{b}{2a} = -\frac{2}{2(1)} = -1.$$

We substitute -1 for x into the equation to find the second coordinate of the vertex:

$$\begin{aligned} y &= x^2 + 2x - 3 \\ &= (-1)^2 + 2(-1) - 3 \\ &= 1 - 2 - 3 \\ &= -4. \end{aligned}$$

The vertex is $(-1, -4)$. The line of symmetry is $x = -1$.

We choose some x-values on both sides of $x = -1$—say, -2, -3, -4 and 0, 1, 2—and graph the parabola. Since the coefficient of x^2 is 1, which is positive, we know that the graph opens up. Be sure to find y when $x = 0$. This gives the y-intercept.

x	y	
-1	-4	← Vertex
0	-3	← y-intercept
-2	-3	
1	0	
-3	0	
2	5	
-4	5	

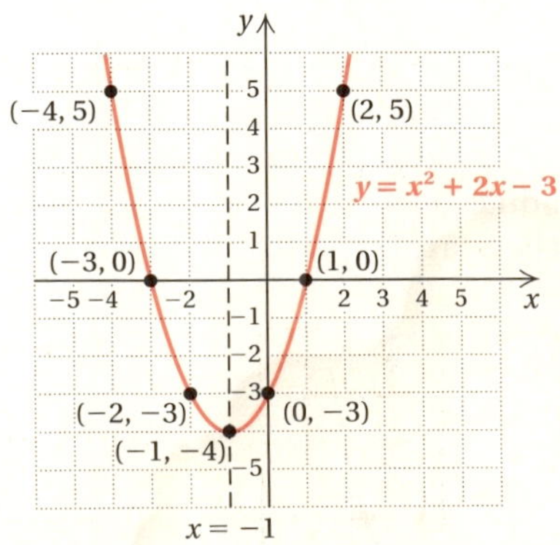

Do Exercises 1–3.

Graph. List the ordered pair for the vertex.

1. $y = x^2 - 3$

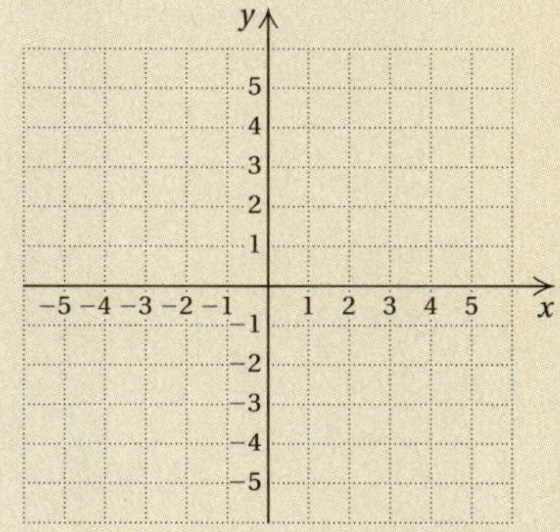

2. $y = -3x^2 + 6x$

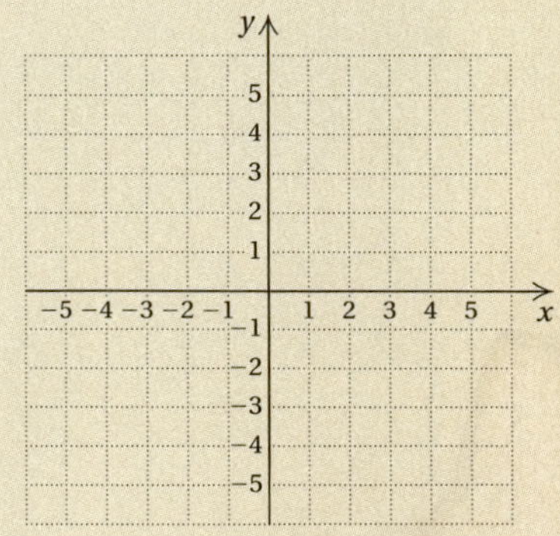

3. $y = x^2 - 4x + 4$

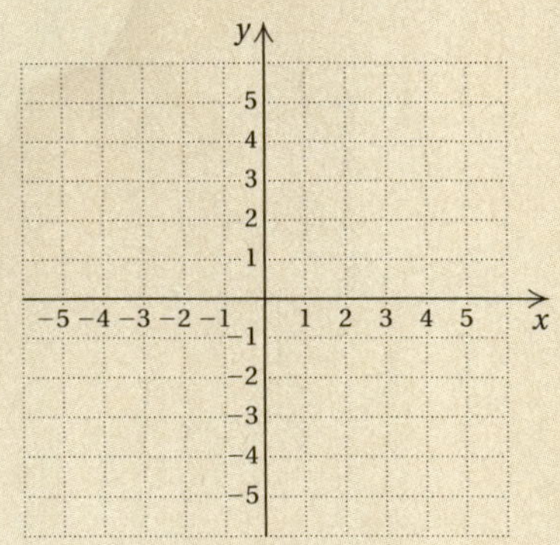

Answers on page A-8

Find the x-intercepts.

4. $y = x^2 - 3$

5. $y = x^2 + 6x + 8$

6. $y = -2x^2 - 4x + 1$

7. $y = x^2 + 3$

Answers on page A-8

b FINDING THE X-INTERCEPTS OF A QUADRATIC EQUATION

The x-intercepts of $y = ax^2 + bx + c$ occur at those values of x for which $y = 0$. Thus the first coordinates of the x-intercepts are solutions of the equation

$$0 = ax^2 + bx + c.$$

We have been studying how to find such numbers in Sections 9.1–9.3.

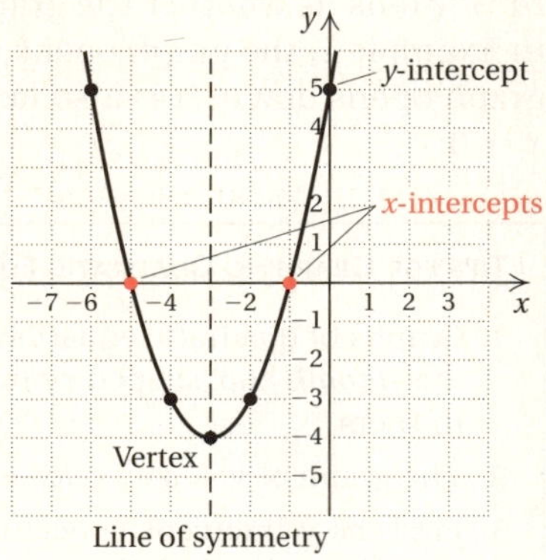

EXAMPLE 4 Find the x-intercepts of $y = x^2 - 4x + 1$.

We solve the equation

$$x^2 - 4x + 1 = 0.$$

Factoring is not convenient, so we use the quadratic formula:

$$a = 1, \quad b = -4, \quad c = 1$$

$$\begin{aligned} x &= \frac{-b \pm \sqrt{b^2 - 4ac}}{2a} \\ &= \frac{-(-4) \pm \sqrt{(-4)^2 - 4(1)(1)}}{2(1)} \\ &= \frac{4 \pm \sqrt{16 - 4}}{2} \\ &= \frac{4 \pm \sqrt{12}}{2} = \frac{4 \pm \sqrt{4 \cdot 3}}{2} \\ &= \frac{4 \pm 2\sqrt{3}}{2} = \frac{2 \cdot 2 \pm 2\sqrt{3}}{2 \cdot 1} \\ &= \frac{2}{2} \cdot \frac{2 \pm \sqrt{3}}{1} = 2 \pm \sqrt{3}. \end{aligned}$$

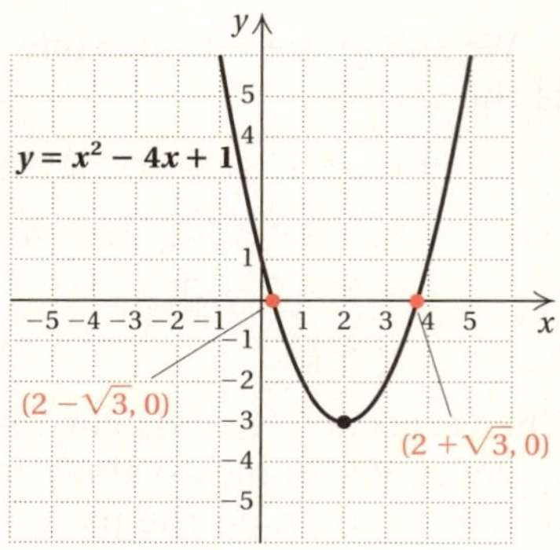

The x-intercepts are $(2 - \sqrt{3}, 0)$ and $(2 + \sqrt{3}, 0)$.

The discriminant, $b^2 - 4ac$, tells how many real-number solutions the equation $0 = ax^2 + bx + c$ has, so it also tells how many x-intercepts there are.

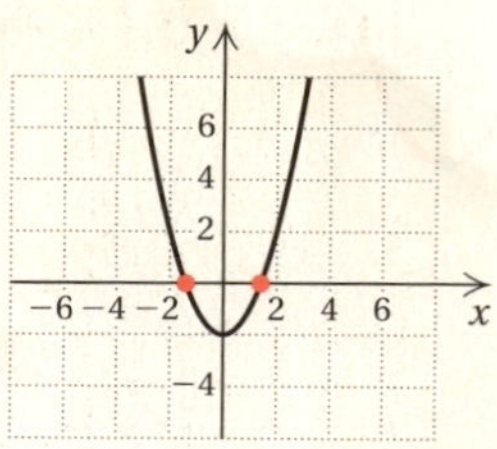

$y = x^2 - 2$

$b^2 - 4ac = 8 > 0$

Two real solutions

Two x-intercepts

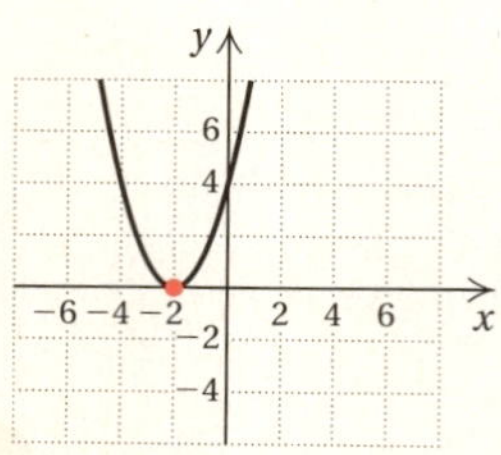

$y = x^2 + 4x + 4$

$b^2 - 4ac = 0$

One real solution

One x-intercept

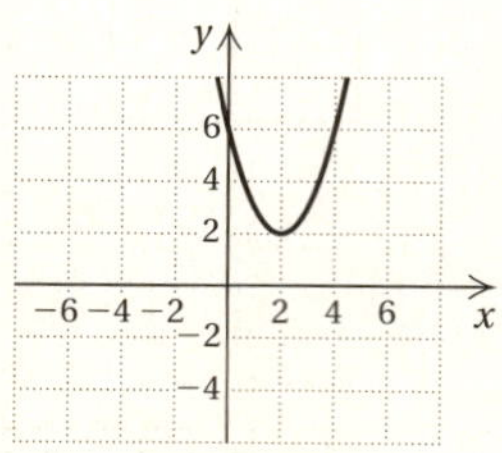

$y = x^2 - 4x + 6$

$b^2 - 4ac = -8 < 0$

No real solutions

No x-intercepts

Do Exercises 4–7.

NAME SECTION DATE

Exercise Set 9.6

a Graph the quadratic equation. List the ordered pair for the vertex.

1. $y = x^2 + 1$

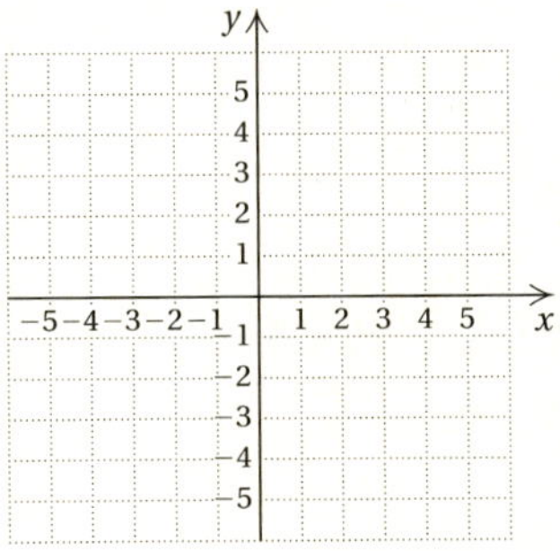

2. $y = 2x^2$

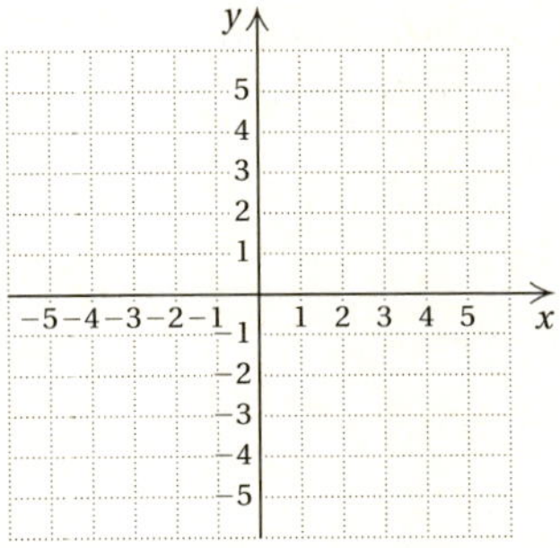

3. $y = -1 \cdot x^2$

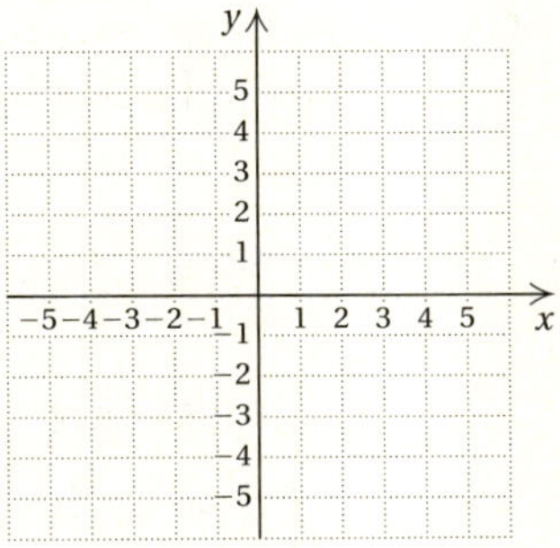

4. $y = x^2 - 1$

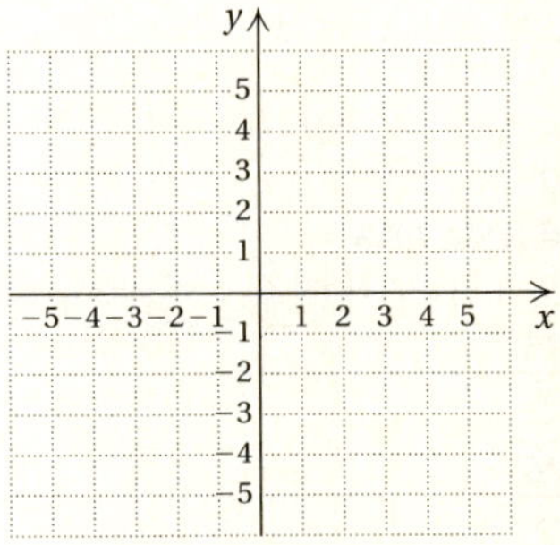

5. $y = -x^2 + 2x$

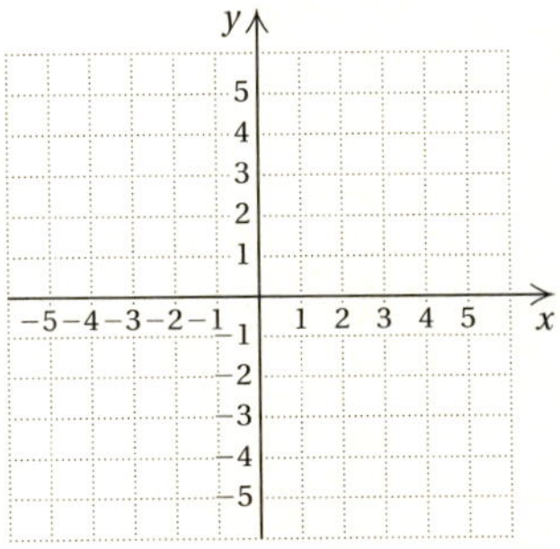

6. $y = x^2 + x - 2$

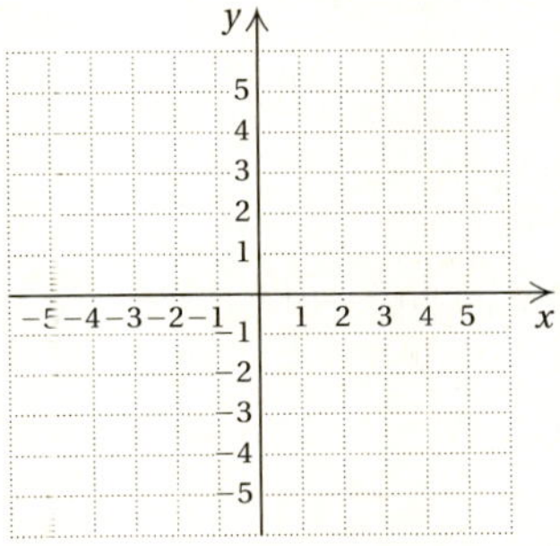

7. $y = 5 - x - x^2$

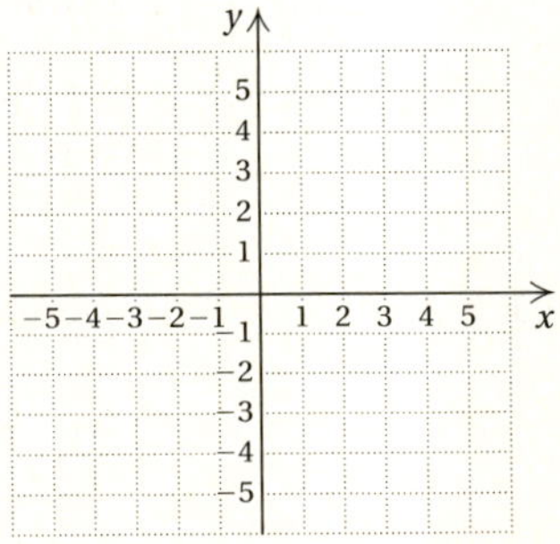

8. $y = x^2 + 2x + 1$

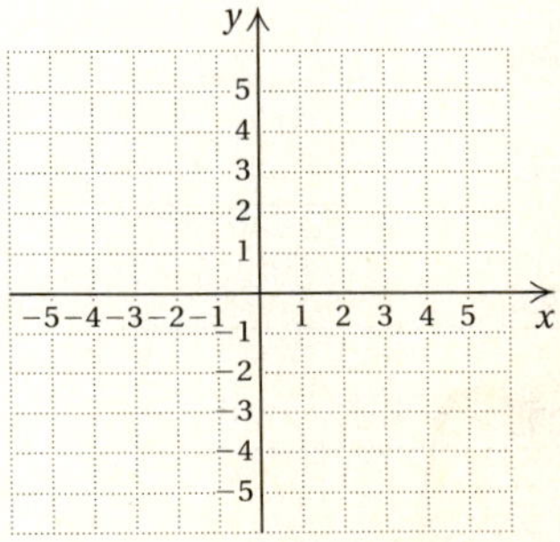

9. $y = x^2 - 2x + 1$

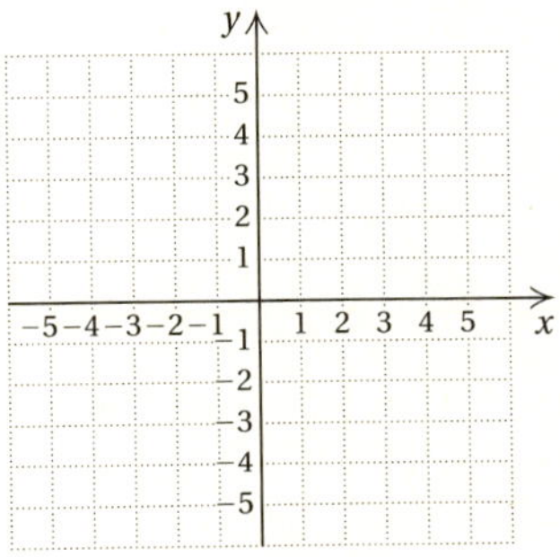

10. $y = -\frac{1}{2}x^2$

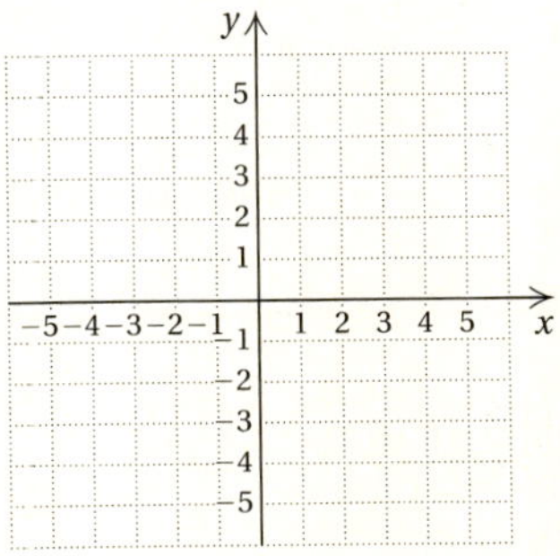

11. $y = -x^2 + 2x + 3$

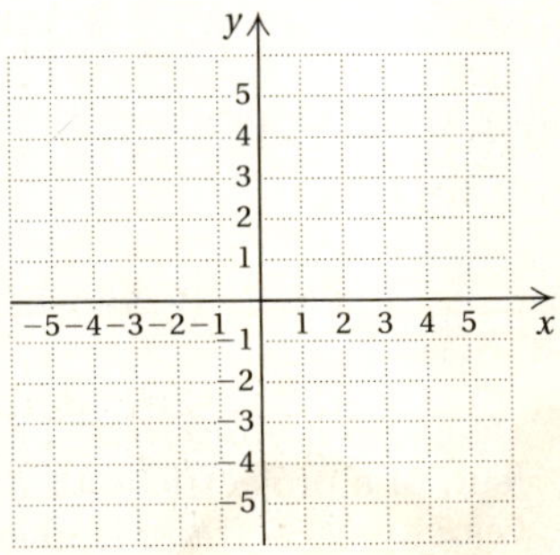

12. $y = -x^2 - 2x + 3$

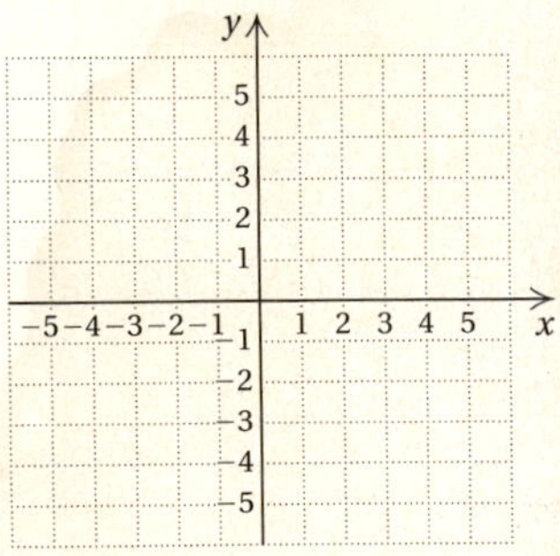

13. $y = -2x^2 - 4x + 1$

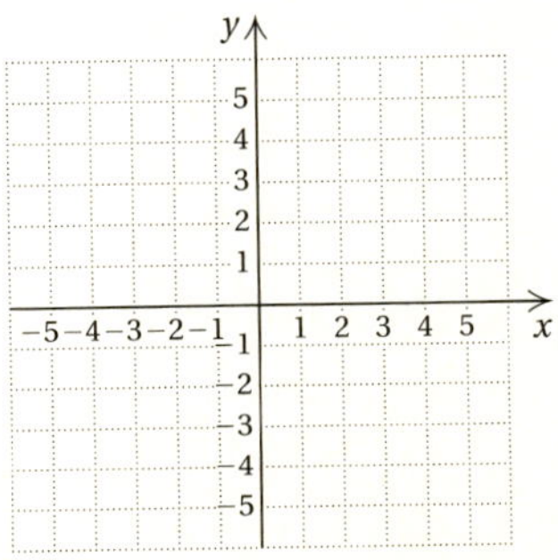

14. $y = 2x^2 + 4x - 1$

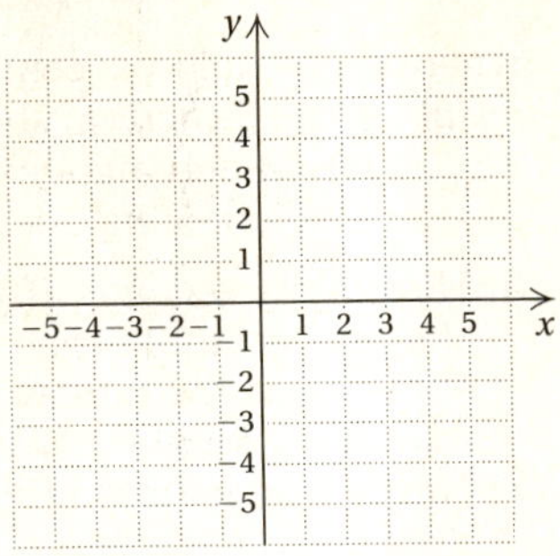

Graph the quadratic equation. Use your own graph paper.

15. $y = \frac{1}{4}x^2$

16. $y = -0.1x^2$

17. $y = 3 - x^2$

18. $y = x^2 + 3$

19. $y = -x^2 + x - 1$

20. $y = x^2 + 2x$

21. $y = -2x^2$

22. $y = -x^2 - 1$

23. $y = x^2 - x - 6$

24. $y = 6 + x - x^2$

ANSWERS

25. ______

26. ______

27. ______

28. ______

29. ______

30. ______

31. ______

32. ______

33. ______

34. ______

35. ______

36. ______

37. ______

38. ______

39. ______

40. ______

41. a) ______

b) ______

c) ______

b Find the x-intercepts.

25. $y = x^2 - 2$

26. $y = x^2 - 7$

27. $y = x^2 + 5x$

28. $y = x^2 - 4x$

29. $y = 8 - x - x^2$

30. $y = 8 + x - x^2$

31. $y = x^2 - 6x + 9$

32. $y = x^2 + 10x + 25$

33. $y = -x^2 - 4x + 1$

34. $y = x^2 + 4x - 1$

35. $y = x^2 + 9$

36. $y = x^2 + 1$

SKILL MAINTENANCE

Add.

37. $\sqrt{x^3 - x^2} + \sqrt{4x - 4}$

38. $\sqrt{8} + \sqrt{50} + \sqrt{98} + \sqrt{128}$

Multiply and simplify.

39. $\sqrt{2}\sqrt{14}$

40. $\sqrt{5y^4}\sqrt{125y}$

SYNTHESIS

41. *Height of a projectile.* The height H, in feet, of a projectile with an initial velocity of 96 ft/sec is given by the equation

$$H = -16t^2 + 96t,$$

where t = time, in seconds. Use the graph of this function, shown here, or any equation-solving technique to answer the following questions.

a) How many seconds after launch is the projectile 128 ft above ground?
b) When does the projectile reach its maximum height?
c) How many seconds after launch does the projectile return to the ground?

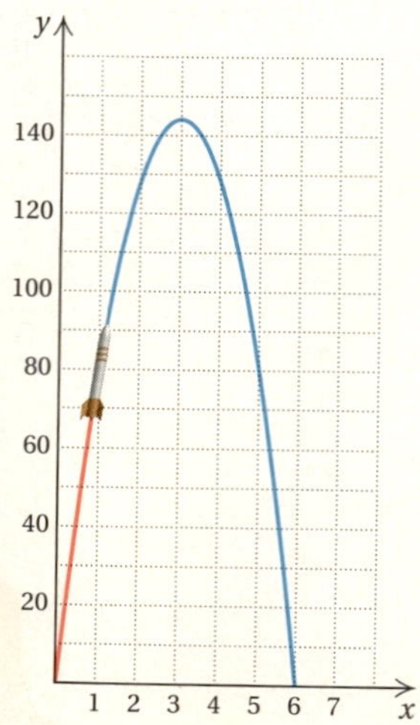

CRITICAL THINKING

CALCULATOR CONNECTION

Use a scientific calculator or a graphing calculator to solve each of the following quadratic equations. Round answers to three decimal places.

1. $4.82x^2 = 12{,}000$
2. $0.0045x^2 + 68{,}535x = 0$
3. $x^2 + 6.2x - 1.1 = 0$
4. $3.2x^2 = 7.8 - 2.3x$
5. $5.24x^2 - 8.35x + 3.88 = 0$
6. $56.9x^2 - 77.4x = 33.2$
7. What must be added to $x^2 - 75.836x$ in order to make it the square of a binomial?

Use a graphing calculator to graph each of the following equations.

8. $y = -2.3x^2 + 4.1x + 1.8$
9. $y = -2.3x^2 - 4.1x + 1.8$
10. $y = 2.3x^2 + 4.1x - 1.8$
11. On the basis of Exercises 9 and 10 and by creating more graphs, if necessary, compare the graphs of $y = ax^2 + bx + c$ and $y = -ax^2 - bx - c$.

Solve the following systems of equations graphically.

12. $y = x^2 - 4x + 1,$ $y = 1 - x$
13. $y = -2x^2 + 4x + 1,$ $y = x^2 - 2x + 1$
14. $y = -2x^2 - 4x + 1,$ $y = x^2 + 2x$
15. $y = 0.1x^2 + 6.2x + 8.3,$ $y = -2.2x^2 - 4.6x + 1.9$

EXTENDED SYNTHESIS EXERCISES

1. Explain how you might use the quadratic formula to factor $x^2 - 4x - 7$.

Find a quadratic equation whose solutions are the given numbers.

2. $0, -5$
3. $-2, -3$
4. $\frac{2}{3}, -6$
5. $-\frac{3}{8}, \frac{4}{5}$
6. Use the two solutions, as given by the quadratic formula, to find a formula for the sum of the solutions of any quadratic equation. Then find a formula for the product. Next, without solving, determine the sum and the product of the solutions of the equation $2x^2 - 5x + 3 = 0$.
7. One solution of the equation $3x^2 + bx - 4 = 0$ is known to be -7. Use the formulas from Exercise 6 to find the other solution and b.
8. In downtown San Francisco, the Transamerica building is 853 ft tall and the Bank of America building is 776 ft tall. How much longer would it take an object dropped from the top of the Transamerica building to reach the ground than an object dropped from the top of the Bank of America building?

(continued)

9. In the right triangle below, find the missing length x.

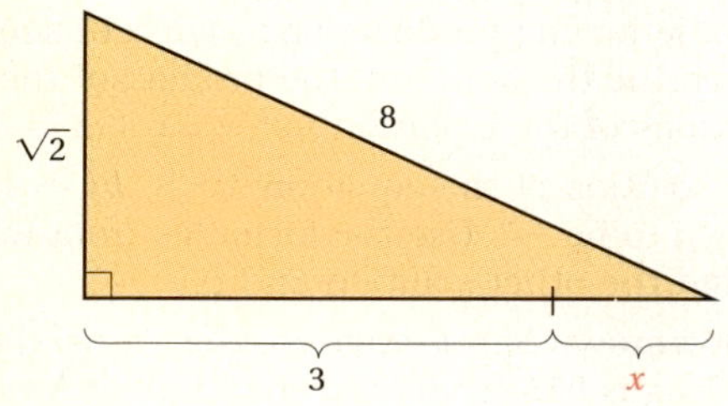

Solve.

10. $\dfrac{1}{x^2 + 3x - 4} + \dfrac{1}{x^2 + 2x - 8} = \dfrac{1}{x^2 - 8x + 12}$

11. The solutions of a quadratic equation $ax^2 + bx + c = 0$ are given by

$$\frac{-11 \pm \sqrt{481}}{36}.$$

   a) Find a possible value of a.

   b) Find a possible value of b.

   c) Find a possible value of c.

   d) Find a quadratic equation that has these numbers as solutions.

EXERCISES FOR THINKING AND WRITING

1. Briefly explain the connection between the real-number solutions of a quadratic equation and its x-intercepts.

2. Find a quadratic equation $ax^2 + bx + c = 0$ with exactly one real-number solution and explain the connection with the graph of $y = ax^2 + bx + c$.

3. List the names and give an example of as many types of equations as you can that you have learned to solve in this text.

4. List a quadratic equation with no real-number solutions. How can you use that equation to find an equation in the form $y = ax^2 + bx + c$ that does not cross the x-axis?

Find the errors in each of the following solutions of quadratic equations.

5. $$\begin{aligned} (x + 6)^2 &= 16 \\ x + 6 &= \sqrt{16} \\ x + 6 &= 4 \\ x &= -2 \end{aligned}$$

6. $$\begin{aligned} x^2 + 2x - 8 &= 0 \\ (x + 4)(x - 2) &= 0 \\ x = 4 \quad &\text{or} \quad x = -2 \end{aligned}$$

7. $$\begin{aligned} x^2 + 20x &= 0 \\ x(x + 20) &= 0 \\ x + 20 &= 0 \\ x &= 20 \end{aligned}$$

8. $$\begin{aligned} x^2 + x &= 6 \\ x(x + 1) &= 6 \\ x = 6 \quad &\text{or} \quad x + 1 = 6 \\ x = 6 \quad &\text{or} \quad x = 5 \end{aligned}$$

Summary and Review: Chapter 9

Important Properties and Formulas

Standard Form: $ax^2 + bx + c = 0,\ a > 0$

Principle of Square Roots: The equation $x^2 = d$, where $d > 0$, has two solutions, $\sqrt{d}$ and $-\sqrt{d}$. The solution of $x^2 = 0$ is 0.

Quadratic Formula: $x = \dfrac{-b \pm \sqrt{b^2 - 4ac}}{2a}$

Discriminant: $b^2 - 4ac$

The x-coordinate of the vertex of a parabola $= -\dfrac{b}{2a}$.

Review Exercises

The review objectives to be tested in addition to the material in this chapter are [6.7c], [8.2c], [8.4a], and [8.6b].

Solve.

1. $8x^2 = 24$

2. $5x^2 - 8x + 3 = 0$

3. $x^2 - 2x - 10 = 0$

4. $3y^2 + 5y = 2$

5. $(x + 8)^2 = 13$

6. $9x^2 = 0$

7. $5t^2 - 7t = 0$

8. $9x^2 - 6x - 9 = 0$

9. $x^2 + 6x = 9$

10. $1 + 4x^2 = 8x$

11. $6 + 3y = y^2$

12. $3m = 4 + 5m^2$

13. $3x^2 = 4x$

14. $40 = 5y^2$

15. $\dfrac{15}{x} - \dfrac{15}{x + 2} = 2$

16. $x + \dfrac{1}{x} = 2$

Solve by completing the square. Show your work.

17. $3x^2 - 2x - 5 = 0$

18. $x^2 - 5x + 2 = 0$

Approximate the solutions to the nearest tenth.

19. $x^2 - 5x + 2 = 0$

20. $4y^2 + 8y + 1 = 0$

21. Solve for T: $V = \dfrac{1}{2}\sqrt{1 + \dfrac{T}{L}}$.

Graph the quadratic equation.

22. $y = 2 - x^2$

23. $y = x^2 - 4x - 2$

Find the x-intercepts.

24. $y = 2 - x^2$

25. $y = x^2 - 4x - 2$

Solve.

26. The hypotenuse of a right triangle is 5 m long. One leg is 3 m longer than the other. Find the lengths of the legs. Round to the nearest tenth.

27. The length of a rectangle is 1 m greater than the width. The area is 132 m^2. Find the length and the width.

28. The current in a stream moves at a speed of 2 km/h. A boat travels 56 km upstream and 64 km downstream in a total time of 4 hr. What is the speed of the boat in still water?

SKILL MAINTENANCE

Multiply and simplify.

29. $\sqrt{18a}\sqrt{2}$

30. $\sqrt{12xy^2}\sqrt{5xy}$

31. Find an equation of variation where y varies inversely as x and $y = 16$ when $x = 0.0625$.

32. The sides of a rectangle are 1 and $\sqrt{2}$. Find the length of a diagonal.

Add or subtract.

33. $5\sqrt{11} + 7\sqrt{11}$

34. $2\sqrt{90} - \sqrt{40}$

SYNTHESIS

35. Two consecutive integers have squares that differ by 63. Find the integers.

36. Find b such that the trinomial $x^2 + bx + 49$ is a square.

37. Solve: $x - 4\sqrt{x} - 5 = 0$.

38. A square with sides of length s has the same area as a circle with radius of 5 in. Find s.

NAME SECTION DATE

Test: Chapter 9

Solve.

1. $7x^2 = 35$

2. $7x^2 + 8x = 0$

3. $48 = t^2 + 2t$

4. $3y^2 - 5y = 2$

5. $(x - 8)^2 = 13$

6. $x^2 = x + 3$

7. $m^2 - 3m = 7$

8. $10 = 4x + x^2$

9. $3x^2 - 7x + 1 = 0$

10. $x - \frac{2}{x} = 1$

11. $\frac{4}{x} - \frac{4}{x+2} = 1$

12. Solve $x^2 - 4x - 10 = 0$ by completing the square. Show your work.

13. Approximate the solutions to $x^2 - 4x - 10 = 0$ to the nearest tenth.

14. Solve for n: $d = an^2 + bn$.

ANSWERS

1. ______
2. ______
3. ______
4. ______
5. ______
6. ______
7. ______
8. ______
9. ______
10. ______
11. ______
12. ______
13. ______
14. ______

ANSWERS

15. ______

16. ______

17. ______

18. ______

19. ______

20. ______

21. ______

22. ______

23. ______

24. ______

25. ______

Graph.

15. $y = 4 - x^2$

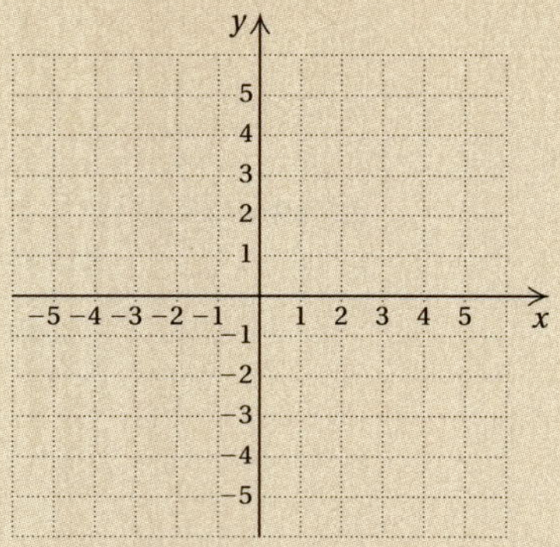

16. $y = -x^2 + x + 5$

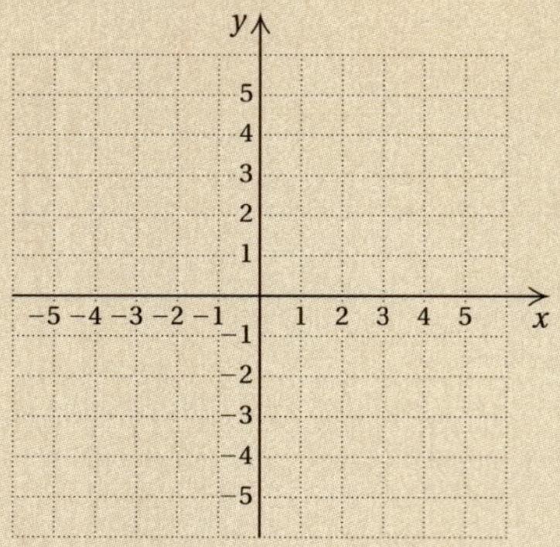

17. Find the x-intercepts: $y = -x^2 + x + 5$.

Solve.

18. The width of a rectangle is 4 m less than the length. The area is 16.25 m^2. Find the length and the width.

19. The current in a stream moves at a speed of 2 km/h. A boat travels 44 km upstream and 52 km downstream in a total of 4 hr. What is the speed of the boat in still water?

SKILL MAINTENANCE

20. Subtract: $\sqrt{240} - \sqrt{60}$.

21. Multiply and simplify: $\sqrt{7xy}\,\sqrt{14x^2y}$.

22. Find an equation of variation where y varies inversely as x and $y = 32$ when $x = 0.125$.

23. The sides of a rectangle are $\sqrt{2}$ and $\sqrt{3}$. Find the length of a diagonal.

SYNTHESIS

24. Find the side of a square whose diagonal is 5 ft longer than a side.

25. Solve this system for x. Use the substitution method.

$$x - y = 2,$$
$$xy = 4$$

Cumulative Review: Chapters 1–9

1. What is the meaning of x^3?

2. Evaluate $(x - 3)^2 + 5$ when $x = 10$.

3. Find decimal notation: $-\frac{3}{11}$.

4. Find the LCM of 15 and 48.

5. Find the absolute value: $|-7|$

Compute and simplify.

6. $-6 + 12 + (-4) + 7$

7. $2.8 - (-12.2)$

8. $-\frac{3}{8} \div \frac{5}{2}$

9. $13 \cdot 6 \div 3 \cdot 2 \div 13$

10. Remove parentheses and simplify: $4m + 9 - (6m + 13)$.

Solve.

11. $3x = -24$

12. $3x + 7 = 2x - 5$

13. $3(y - 1) - 2(y + 2) = 0$

14. $x^2 - 8x + 15 = 0$

15. $y - x = 1,$
$y = 3 - x$

16. $x + y = 17,$
$x - y = 7$

17. $4x - 3y = 3,$
$3x - 2y = 4$

18. $x^2 - x - 6 = 0$

19. $x^2 + 3x = 5$

20. $3 - x = \sqrt{x^2 - 3}$

21. $5 - 9x \leq 19 + 5x$

22. $-\frac{7}{8}x + 7 = \frac{3}{8}x - 3$

23. $0.6x - 1.8 = 1.2x$

24. $-3x > 24$

25. $23 - 19y - 3y \geq -12$

26. $3y^2 = 30$

27. $(x - 3)^2 = 6$

28. $\frac{6x - 2}{2x - 1} = \frac{9x}{3x + 1}$

29. $\frac{2x}{x + 1} = 2 - \frac{5}{2x}$

30. $\frac{2x}{x + 3} + \frac{6}{x} + 7 = \frac{18}{x^2 + 3x}$

31. $\sqrt{x + 9} = \sqrt{2x - 3}$

Solve the formula for the given letter.

32. $A = \frac{4b}{t}$, for b

33. $\frac{1}{t} = \frac{1}{m} - \frac{1}{n}$, for m

34. $r = \sqrt{\dfrac{A}{\pi}}$, for A

35. $y = ax^2 - bx$, for x

Simplify.

36. $x^{-6} \cdot x^2$

37. $\dfrac{y^3}{y^{-4}}$

38. $(2y^6)^2$

39. Collect like terms and arrange in descending order: $2x - 3 + 5x^3 - 2x^3 + 7x^3 + x$.

Compute and simplify.

40. $(4x^3 + 3x^2 - 5) + (3x^3 - 5x^2 + 4x - 12)$

41. $(6x^2 - 4x + 1) - (-2x^2 + 7)$

42. $-2y^2(4y^2 - 3y + 1)$

43. $(2t - 3)(3t^2 - 4t + 2)$

44. $\left(t - \dfrac{1}{4}\right)\left(t + \dfrac{1}{4}\right)$

45. $(3m - 2)^2$

46. $(15x^2y^3 + 10xy^2 + 5) - (5xy^2 - x^2y^2 - 2)$

47. $(x^2 - 0.2y)(x^2 + 0.2y)$

48. $(3p + 4q^2)^2$

49. $\dfrac{4}{2x - 6} \cdot \dfrac{x - 3}{x + 3}$

50. $\dfrac{3a^4}{a^2 - 1} \div \dfrac{2a^3}{a^2 - 2a + 1}$

51. $\dfrac{3}{3x - 1} + \dfrac{4}{5x}$

52. $\dfrac{2}{x^2 - 16} - \dfrac{x - 3}{x^2 - 9x + 20}$

Factor.

53. $8x^2 - 4x$

54. $25x^2 - 4$

55. $6y^2 - 5y - 6$

56. $m^2 - 8m + 16$

57. $x^3 - 8x^2 - 5x + 40$

58. $3a^4 + 6a^2 - 72$

59. $16x^4 - 1$

60. $49a^2b^2 - 4$

61. $9x^2 + 30xy + 25y^2$

62. $2ac - 6ab - 3db + dc$

63. $15x^2 + 14xy - 8y^2$

Simplify.

64. $\dfrac{\dfrac{3}{x} + \dfrac{1}{2x}}{\dfrac{1}{3x} - \dfrac{3}{4x}}$

65. $\sqrt{49}$

66. $-\sqrt{625}$

67. $\sqrt{64x^2}$

68. Multiply: $\sqrt{a + b}\sqrt{a - b}$.

69. Multiply and simplify: $\sqrt{32ab}\sqrt{6a^4b^2}$.

Simplify.

70. $\sqrt{150}$

71. $\sqrt{243x^3y^2}$

72. $\sqrt{\dfrac{100}{81}}$

73. $\sqrt{\dfrac{64}{x^2}}$

74. $4\sqrt{12} + 2\sqrt{12}$

75. Divide and simplify: $\dfrac{\sqrt{72}}{\sqrt{45}}$.

76. In a right triangle, $a = 9$ and $c = 41$. Find b.

Graph on a plane.

77. $y = \dfrac{1}{3}x - 2$

78. $2x + 3y = -6$

79. $y = -3$

80. $4x - 3y > 12$

81. $y = x^2 + 2x + 1$

82. $x \geq -3$

83. Solve $9x^2 - 12x - 2 = 0$ by completing the square. Show your work.

84. Approximate the solutions of $4x^2 = 4x + 1$ to the nearest tenth.

Solve.

85. What percent of 52 is 13?

86. 12 is 20% of what?

87. The speed of a boat in still water is 8 km/h. It travels 60 km upstream and 60 km downstream in a total time of 16 hr. What is the speed of the stream?

88. The length of a rectangle is 7 m more than the width. The length of a diagonal is 13 m. Find the length.

89. Three-fifths of the automobiles entering the city each morning will be parked in city parking lots. There are 3654 such parking spaces filled each morning. How many cars enter the city each morning?

90. A candy shop mixes nuts worth $1.10 per pound with another variety worth $0.80 per pound to make 42 lb of a mixture worth $0.90 per pound. How many pounds of each kind of nuts should be used?

91. In checking records, a contractor finds that crew A can resurface a tennis court in 8 hr. Crew B can do the same job in 10 hr. How long would they take if they worked together?

92. A student's paycheck varies directly as the number of hours worked. The pay was $242.52 for 43 hr of work. What would the pay be for 80 hr of work? Explain the meaning of the variation constant.

93. Determine whether the graphs of the following equations are parallel, perpendicular, or neither.

$$y - x = 4,$$
$$3y + x = 8$$

94. Find the slope and the y-intercept:

$$-6x + 3y = -24.$$

95. Find the slope of the line containing the points $(-5, -6)$ and $(-4, 9)$.

Find an equation of variation where:

96. y varies directly as x and $y = 100$ when $x = 10$.

97. y varies inversely as x and $y = 100$ when $x = 10$.

SYNTHESIS

98. Solve: $|x| = 12$.

99. Simplify: $\sqrt{\sqrt{\sqrt{81}}}$.

100. Find b such that the trinomial $x^2 - bx + 225$ is a square.

101. Find x.

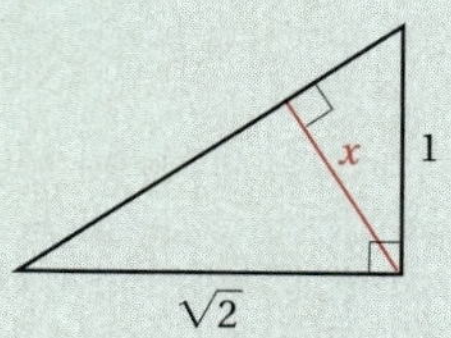

Determine whether the pair of expressions is equivalent.

102. $x^2 - 9, \quad (x - 3)(x + 3)$

103. $\dfrac{x + 3}{3}, \quad x$

104. $(x + 5)^2, \quad x^2 + 25$

105. $\sqrt{x^2 + 16}, \quad x + 4$

106. $\sqrt{x^2}, \quad |x|$

NAME SECTION DATE

Final Examination

1. Evaluate $x^3 + 5$ when $x = -10$.

2. Find the LCM of 16 and 24.

3. Find the absolute value of $|-9|$.

Compute and simplify.

4. $-6.3 + (-8.4) + 5$

5. $-8 - (-3)$

6. $\frac{3}{11} \cdot \left(-\frac{22}{7}\right)$

7. Remove parentheses and simplify:

$4y - 5(9 - 3y)$.

8. Simplify:

$2^3 - 14 \cdot 10 + (3 + 4)^3$.

Solve.

9. $x + 8 = 13.6$

10. $4x = -28$

11. $5x + 3 = 2x - 27$

12. $5(x - 3) - 2(x + 3) = 0$

13. $x^2 - 2x - 24 = 0$

14. $y = x - 7,$
$2x + y = 5$

15. $5x - 3y = -1,$
$4x + 2y = 30$

16. $\frac{1}{x} - 2 = 8x$

17. $\sqrt{x^2 - 11} = x - 1$

ANSWERS

1. ______
2. ______
3. ______
4. ______
5. ______
6. ______
7. ______
8. ______
9. ______
10. ______
11. ______
12. ______
13. ______
14. ______
15. ______
16. ______
17. ______

ANSWERS

18. ______

19. ______

20. ______

21. ______

22. ______

23. ______

24. ______

25. ______

26. ______

27. ______

28. ______

29. ______

30. ______

31. ______

32. ______

33. ______

34. ______

35. ______

36. ______

18. $x^2 = 7 - 3x$

19. $2 - 3x \leq 12 - 7x$

Solve the formula for the given letter.

20. $A = \dfrac{Bw + 1}{w}$, for w

21. $K = MT + 2$, for M

Simplify.

22. $\dfrac{x^8}{x^{-2}}$

23. $(x^{-5})^2$

24. $x^{-5} \cdot x^{-7}$

25. Collect like terms and arrange in descending order:

$2y^3 - 3 + 4y^3 - 3y^2 + 12 - y.$

Compute and simplify.

26. $(2x^2 - 6x + 3) - (4x^2 + 2x - 4)$

27. $-3t^2(2t^4 + 4t^2 + 1)$

28. $(4x - 1)(x^2 - 5x + 2)$

29. $(x - 8)(x + 8)$

30. $(2m - 7)^2$

31. $(3ab^2 + 2c)^2$

32. $(3x^2 - 2y)(3x^2 + 4y)$

33. $\dfrac{x}{x^2 - 9} \cdot \dfrac{x - 3}{x^3}$

34. $\dfrac{3x^5}{4x - 4} \div \dfrac{x}{x^2 - 2x + 1}$

35. $\dfrac{2}{3x - 1} + \dfrac{1}{4x}$

36. $\dfrac{3}{x - 3} - \dfrac{x - 1}{x^2 - 2x - 3}$

Factor.

37. $3x^3 - 15x$

38. $16x^2 - 25$

39. $6x^2 - 13x + 6$

40. $x^2 - 10x + 25$

41. $2ax + 6bx - ay - 3by$

42. $x^8 - 81y^4$

Simplify.

43. $\sqrt{72}$

44. $\dfrac{\sqrt{54}}{\sqrt{45}}$

45. $2\sqrt{8} + 3\sqrt{18}$

46. $\sqrt{24a^2b}\sqrt{a^3b^2}$

Graph on a plane.

47. $3x + 2y = -4$

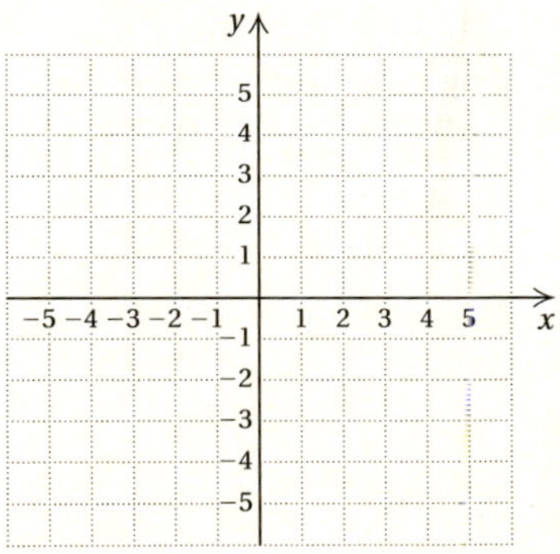

48. $x = -2$

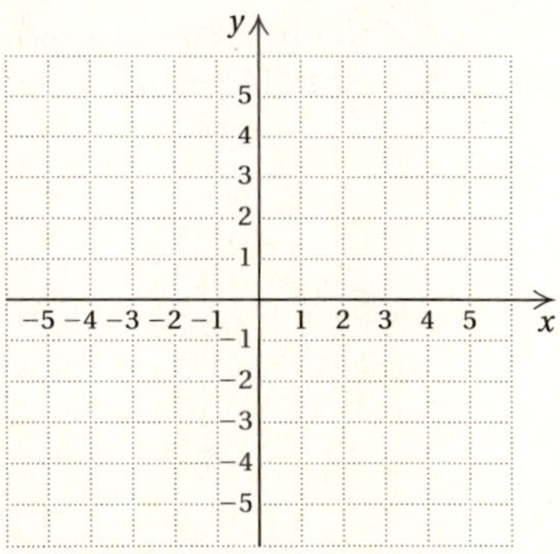

49. $3x - 2y < 6$

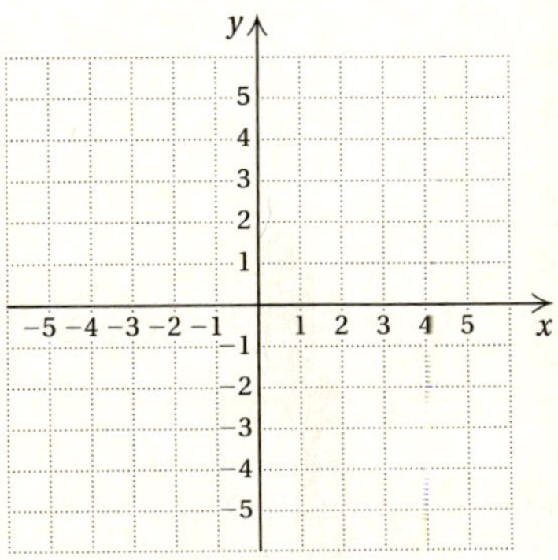

50. $y = x^2 - 2x + 1$

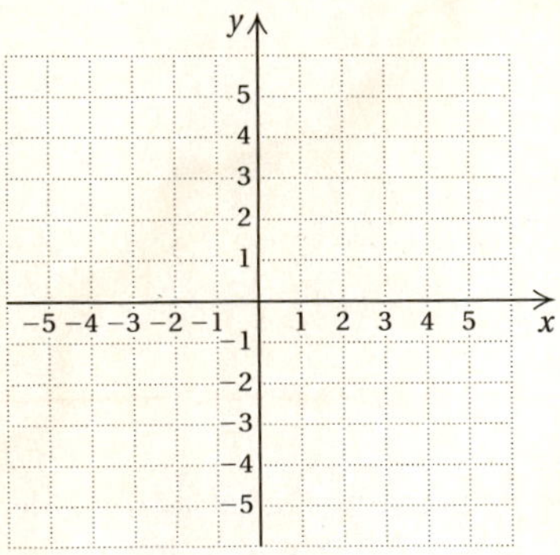

ANSWERS

37. ______________

38. ______________

39. ______________

40. ______________

41. ______________

42. ______________

43. ______________

44. ______________

45. ______________

46. ______________

47. ______________

48. ______________

49. ______________

50. ______________

ANSWERS

51. ______

52. ______

53. ______

54. ______

55. ______

56. ______

57. ______

58. ______

59. ______

60. ______

Solve.

51. The sum of the squares of two consecutive odd integers is 74. Find the integers.

52. Solution A is 75% alcohol and solution B is 50% alcohol. How much of each is needed to make 60 L of a solution that is $66\frac{2}{3}\%$ alcohol?

53. An airplane flew for 6 hr with a 10-km/h tail wind. The return flight against the same wind took 8 hr. Find the speed of the plane in still air.

54. The width of a rectangle is 3 m less than the length. The area is 88 m^2. Find the length and the width.

55. Find the slope of the line containing the points $(-2, 3)$ and $(4, -5)$.

56. Determine whether the graphs of the following equations are parallel, perpendicular, or neither.

$$y = 2x + 7,$$
$$2y + x = 6$$

Find an equation of variation where:

57. y varies directly as x and $y = 200$ when $x = 25$.

58. y varies inversely as x and $y = 200$ when $x = 25$.

SYNTHESIS

59. A side of a square is 5 less than a side of an equilateral triangle. The perimeter of the square is the same as the perimeter of the triangle. Find the length of a side of the square and the length of a side of the triangle.

60. Find c such that the trinomial $x^2 - 24x + c$ is a square.

TABLE 1
Fractional and Decimal Equivalents

Fractional Notation	Decimal Notation	Percent Notation
$\frac{1}{10}$	0.1	10%
$\frac{1}{8}$	0.125	12.5%, or $12\frac{1}{2}$%
$\frac{1}{6}$	$0.16\overline{6}$	$16.\overline{6}$%, or $16\frac{2}{3}$%
$\frac{1}{5}$	0.2	20%
$\frac{1}{4}$	0.25	25%
$\frac{3}{10}$	0.3	30%
$\frac{1}{3}$	$0.33\overline{3}$	$33.\overline{3}$%, or $33\frac{1}{3}$%
$\frac{3}{8}$	0.375	37.5%, or $37\frac{1}{2}$%
$\frac{2}{5}$	0.4	40%
$\frac{1}{2}$	0.5	50%
$\frac{3}{5}$	0.6	60%
$\frac{5}{8}$	0.625	62.5%, or $62\frac{1}{2}$%
$\frac{2}{3}$	$0.66\overline{6}$	$66.\overline{6}$%, or $66\frac{2}{3}$%
$\frac{7}{10}$	0.7	70%
$\frac{3}{4}$	0.75	75%
$\frac{4}{5}$	0.8	80%
$\frac{5}{6}$	$0.83\overline{3}$	$83.\overline{3}$%, or $83\frac{1}{3}$%
$\frac{7}{8}$	0.875	87.5%, or $87\frac{1}{2}$%
$\frac{9}{10}$	0.9	90%
$\frac{1}{1}$	1	100%

TABLE 2

Square Roots

N	√N	N	√N	N	√N	N	√N
2	1.414	27	5.196	52	7.211	77	8.775
3	1.732	28	5.292	53	7.280	78	8.832
4	2	29	5.385	54	7.348	79	8.888
5	2.236	30	5.477	55	7.416	80	8.944
6	2.449	31	5.568	56	7.483	81	9
7	2.646	32	5.657	57	7.550	82	9.055
8	2.828	33	5.745	58	7.616	83	9.110
9	3	34	5.831	59	7.681	84	9.165
10	3.162	35	5.916	60	7.746	85	9.220
11	3.317	36	6	61	7.810	86	9.274
12	3.464	37	6.083	62	7.874	87	9.327
13	3.606	38	6.164	63	7.937	88	9.381
14	3.742	39	6.245	64	8	89	9.434
15	3.873	40	6.325	65	8.062	90	9.487
16	4	41	6.403	66	8.124	91	9.539
17	4.123	42	6.481	67	8.185	92	9.592
18	4.243	43	6.557	68	8.246	93	9.644
19	4.359	44	6.633	69	8.307	94	9.695
20	4.472	45	6.708	70	8.367	95	9.747
21	4.583	46	6.782	71	8.426	96	9.798
22	4.690	47	6.856	72	8.485	97	9.849
23	4.796	48	6.928	73	8.544	98	9.899
24	4.899	49	7	74	8.602	99	9.950
25	5	50	7.071	75	8.660	100	10
26	5.099	51	7.141	76	8.718		

VIDEOTAPE INDEX

The following is a list of text exercises that appear on the "Math Makes a Difference" videotapes.

Text/Video Section	Exercise Numbers
Section R.1	1, 17, 19, 35, 47
Section R.2	3, 7, 11, 23, 25, 31, 33, 35, 43, 47
Section R.3	1, 7, 13, 15, 17, 27, 31, 35, 45, 51
Section R.4	7, 15, 21, 31, 33, 41
Section R.5	1, 3, 7, 47
Section 1.1	7, 9, 15
Section 1.2	5, 29, 39, 51, 53, 55, 59
Section 1.3	8, 11, 15, 31, 39, 44, 63
Section 1.4	1, 9, 21, 93, 97
Section 1.5	9, 15, 23, 31, 67
Section 1.6	9, 11, 15, 17, 21, 25, 49, 61
Section 1.7	3, 7, 49, 51, 65, 79, 111
Section 1.8	1, 11, 31, 39, 51, 81
Section 2.1	3, 15, 21, 35
Section 2.2	19, 23, 25, 33
Section 2.3	5, 11, 15, 19, 25, 37, 39, 47, 57, 65, 75
Section 2.4	9, 19
Section 2.5	9, 30
Section 2.6	9, 11, 23, 25
Section 2.7	3, 11, 19, 23, 51, 73
Section 2.8	1, 7, 9, 15, 17
Section 3.1	17, 23, 25, 47, 49, 53, 59, 63, 67
Section 3.2	2, 3, 11, 45, 57
Section 3.3	15, 25, 33, 35, 45, 51, 67, 73
Section 3.4	7, 13, 17, 23, 29, 33, 43
Section 3.5	5, 21, 27, 35, 45, 53
Section 3.6	13, 55, 81
Section 3.7	1, 9, 11, 23, 31, 39, 57
Section 3.8	21, 25
Section 4.1	1, 13, 17, 19, 31
Section 4.2	1, 13, 47, 53
Section 4.3	31
Section 4.4	5, 15, 19
Section 4.5	1, 3, 5, 29, 33, 41, 45, 47, 51, 55, 67, 75
Section 4.6	5, 15, 17, 39, 48, 76
Section 4.7	1, 17, 25, 43, 55
Section 4.8	7, 17
Section 5.1	5, 7, 15, 25, 31, 51, 55, 69
Section 5.2	3, 5, 7, 13, 27, 29
Section 5.3	5, 15, 19, 21, 23, 25, 29, 39
Section 5.4	3, 7, 9, 13, 31, 35, 53
Section 5.5	5, 13, 38
Section 5.6	1, 11, 23, 33
Section 5.7	3, 9, 13, 17, 21
Section 5.8	1, 7, 11, 19
Section 5.9	1, 9, 13
Section 6.1	none
Section 6.2	1, 6, 9, 11, 21
Section 6.3	5
Section 6.4	3, 9, 11, 13, 15, 19, 21, 31, 37, 51
Section 6.5	3, 11, 15, 21
Section 6.6	7, 19
Section 6.7	7, 9, 15, 17, 21, 27
Section 7.1	3, 12, 21
Section 7.2	5, 11, 25
Section 7.3	19
Section 7.4	21
Section 7.5	5, 7, 13
Section 8.1	19, 21, 25, 27, 29, 31, 39, 45, 47
Section 8.2	7, 13, 15, 27, 33, 37, 53, 67, 81
Section 8.3	1, 9, 13, 17, 19, 29, 37, 41, 57, 73, 79, 85
Section 8.4	5, 17, 49
Section 8.5	11, 23, 29
Section 8.6	none
Section 9.1	3, 9, 19, 23, 33, 53, 63
Section 9.2	3, 5, 25, 27, 35, 47, 57
Section 9.3	3, 5, 9, 11, 23
Section 9.4	14, 23, 27
Section 9.5	17
Section 9.6	9

APPENDIX

Using a Scientific or Graphing Calculator

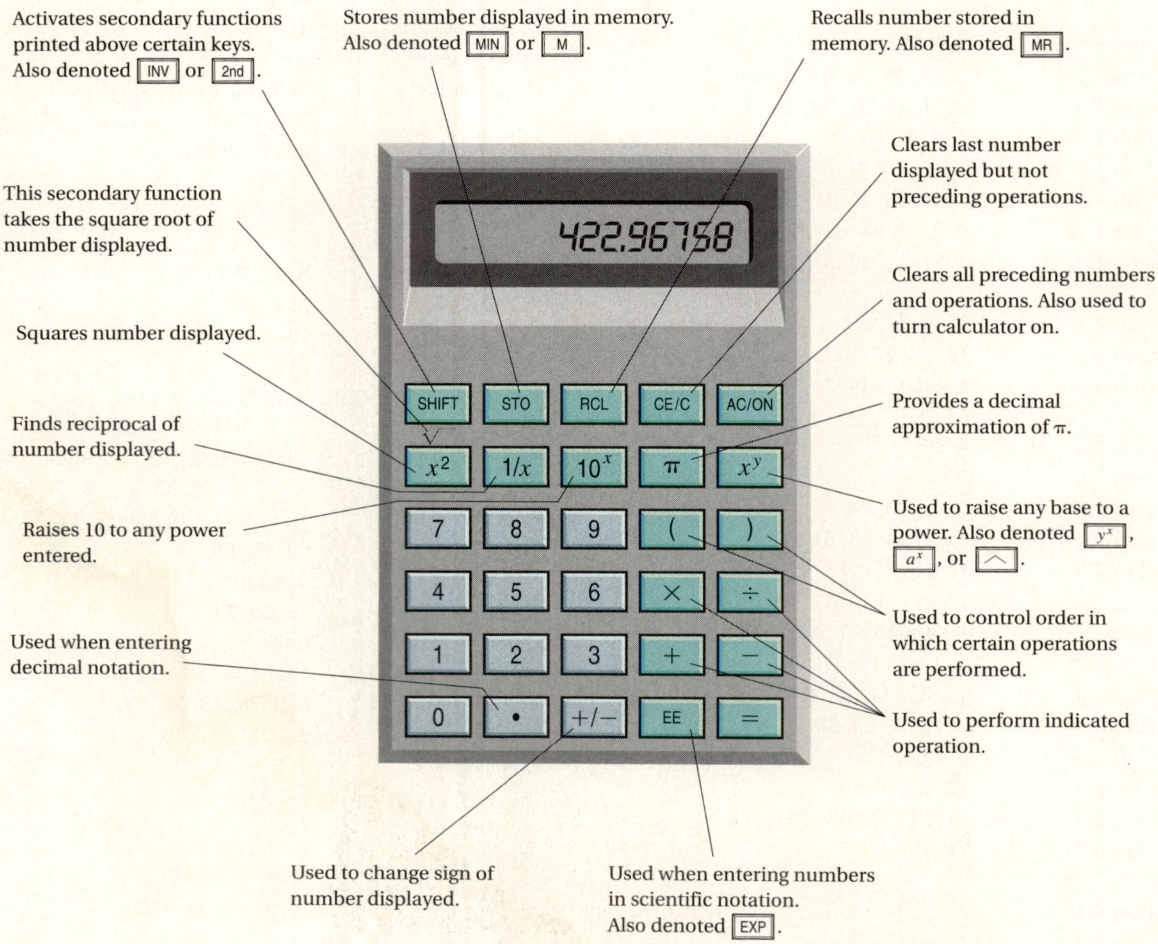

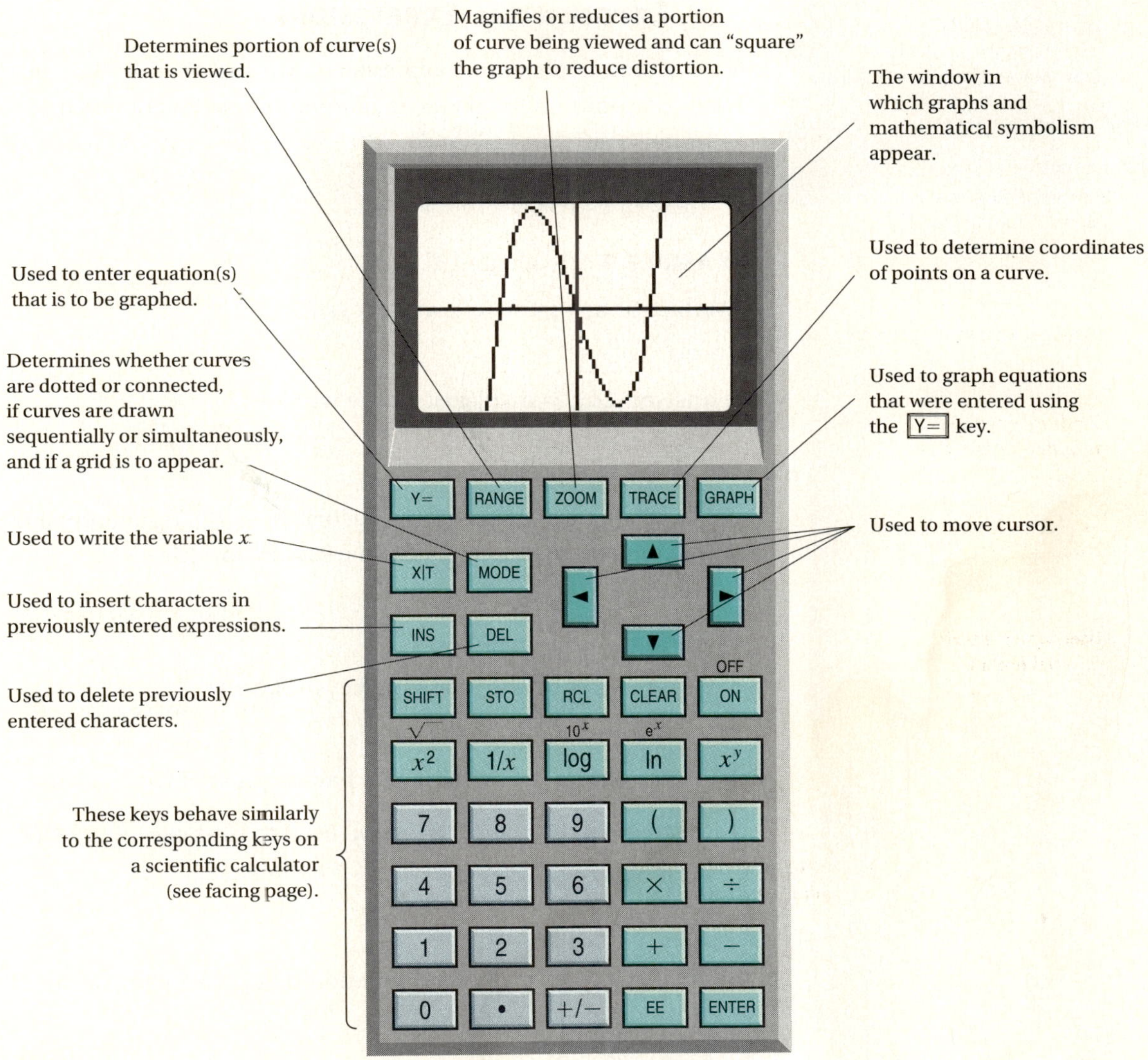
Magnifies or reduces a portion of curve being viewed and can "square" the graph to reduce distortion.
Determines portion of curve(s) that is viewed.
The window in which graphs and mathematical symbolism appear.
Used to determine coordinates of points on a curve.
Used to enter equation(s) that is to be graphed.
Determines whether curves are dotted or connected, if curves are drawn sequentially or simultaneously, and if a grid is to appear.
Used to graph equations that were entered using the Y= key.
Used to write the variable x.
Used to move cursor.
Used to insert characters in previously entered expressions.
Used to delete previously entered characters.
These keys behave similarly to the corresponding keys on a scientific calculator (see facing page).
Y=
RANGE
ZOOM
TRACE
GRAPH
X|T
MODE
INS
DEL
OFF
SHIFT
STO
RCL
CLEAR
ON
√
10^x
e^x
x²
1/x
log
ln
x^y
7
8
9
(
)
4
5
6
×
÷
1
2
3
+
−
0
•
+/−
EE
ENTER

OBJECTIVES

After finishing this appendix, you should be able to use a calculator to:

- **a** Evaluate exponential expressions.
- **b** Evaluate expressions using the built-in order of operations.
- **c** Calculate with fractions and mixed numerals.
- **d** Find square roots.
- **e** Calculate with real numbers.
- **f** Check solutions of equations.
- **g** Evaluate polynomials.
- **h** Calculate with scientific notation.
- **i** Graph linear equations.
- **j** Solve systems of equations.

Using a Scientific or Graphing Calculator

Calculator usage has become commonplace in education, business, and daily living. For this reason, we have included a brief section on the basic operation of a scientific calculator and a graphing calculator.

Features and required procedures vary widely among calculators. If you are unfamiliar with the use of your particular calculator, consult its manual or your instructor.

The calculator is a remarkable tool for both students and teachers, but remember, it is not a substitute for learning the concepts.

a EXPONENTIAL EXPRESSIONS

To evaluate exponential expressions, we use the [x^y] key to raise any base to a power. On some calculators, this key is denoted [y^x], [a^x], or [^]. The keystrokes will make more sense if we say "raised to the" as we press the [x^y] key.

EXAMPLE 1 Evaluate: 3^5.

We find 3^5 using the following keystrokes:

[3] [x^y] [5] [=].

The answer, 243, is displayed in the window:

243 .

Since raising a number to the second power is frequently encountered in mathematics, most calculators have a key for this, [x^2].

EXAMPLE 2 Evaluate: 15^2.

We find 15^2 using the following keystrokes:

[1] [5] [x^2].

The display shows 225 without our having pressed the [=] key.

In Example 2, the [x^y] key can be used to evaluate 15^2, but the [x^2] key requires fewer keystrokes.

Do Exercises 1–5.

We can easily raise decimal notation to a power using a calculator.

EXAMPLE 3 Find: $(1.2)^3$.

To find $(1.2)^3$, we press the following keys:

[1] [.] [2] [x^y] [3] [=].

The answer, 1.728, appears in the window.

Do Exercises 6 and 7.

Evaluate.

1. 3^8

2. 15^3

3. 23^2

4. 2^{10}

5. 7^4

Evaluate.

6. $(1.8)^4$

7. $(0.3)^5$

Answers on page A-34

b ORDER OF OPERATIONS

The order of operations is built into most scientific calculators.

EXAMPLE 4 Calculate: $36 \div 2 \cdot 3 - 4 \cdot 4$.

To calculate $36 \div 2 \cdot 3 - 4 \cdot 4$, we press the following keys:

[3] [6] [÷] [2] [×] [3] [−] [4] [×] [4] [=].

The answer, 38, is displayed in the window:

38 .

When parentheses appear in the problem, we must enter the operation preceding the parentheses.

EXAMPLE 5 Calculate: $36 \div (2 \cdot 3 - 4) \cdot 4$.

We press the following keys:

[3] [6] [÷] [(] [2] [×] [3] [−] [4] [)] [×] [4] [=].

The answer, 72, is displayed in the window.

EXAMPLE 6 Calculate: $(15 + 3)^3 + 4(12 - 7)^2$.

We calculate using the following keystrokes:

[(] [1] [5] [+] [3] [)] [x^y] [3] [+] [4] [×] [(] [1] [2] [−] [7] [)] [x^2] [=].

The answer is 5932.

Do Exercises 8–12.

Even when the order of operations is built in, parentheses must be inserted at times.

EXAMPLE 7 Calculate: $\frac{80}{8 - 6}$.

We press

[8] [0] [÷] [(] [8] [−] [6] [)] [=].

The display reads 40, the correct answer:

40 .

In Example 7, if we had not inserted the parentheses, 80 would have been divided by 8 *before* the 6 had been subtracted, given an incorrect answer of 4.

[8] [0] [÷] [8] [−] [6] [=] 4

Wrong!

Do Exercises 13 and 14.

Calculate.

8. $68 - 8 \div 4 + 3 \cdot 5$

9. $50 - 8 \cdot 3 + 4(5^2 - 2)$

10. $3 + 4[15 - 2(6 - 3)]$

11. $[3 + 2(10 - 4)^2] - 20 \div 5$

12. $\{(150 \cdot 5) \div [(3 \cdot 16) \div (8 \cdot 3)]\} + 25 \cdot (12 \div 4)$

Calculate.

13. $\frac{1200}{30 - 18}$

14. $\frac{50 - 5}{5 + 10}$

Answers on page A-34

Calculate.

15. $\frac{7}{8} - \frac{1}{3}$

16. $\frac{2}{3} + \frac{1}{4} - \frac{2}{5}$

17. $3\frac{1}{4} + 9\frac{5}{6}$

18. $\frac{2}{5} \cdot \frac{14}{11} - \frac{1}{2}$

19. $2\frac{1}{2} \div 1\frac{1}{4}$

Evaluate and express the answer in decimal notation.

20. $\frac{3}{5}\left(\frac{19}{4} - \frac{1}{8}\right)$

21. $\frac{3}{8} + \frac{3}{4} + \frac{1}{2}$

22. $\frac{1}{8}\left(\frac{1}{2} \div \frac{2}{3}\right) + \frac{7}{32}$

Answers on page A-34

c CALCULATING WITH FRACTIONAL NOTATION

The [a b/c] key allows us to compute with fractional notation and mixed numerals.

EXAMPLE 8 Calculate: $\frac{1}{2} + 3\frac{1}{5} - \frac{4}{9}$.

To enter the problem, we use the following keystrokes:

[1] [a b/c] [2] [+] [3] [a b/c] [1] [a b/c] [5] [−] [4] [a b/c] [9] [=].

The display in the window is in the form

3 ⌋ 23 ⌋ 90 .

This means that the answer is $3\frac{23}{90}$.

If we press the [Shift] key and the [d/c] key, the answer is converted from a mixed numeral to fractional notation:

293 ⌋ 90 ,

which means $\frac{293}{90}$.

Do Exercises 15–19.

The [a b/c] key can also convert an answer in fractional notation to decimal notation.

EXAMPLE 9 Calculate $\frac{1}{2}\left(\frac{3}{5} - \frac{1}{8}\right)$ and express the answer in decimal notation.

We press the following keys:

[1] [a b/c] [2] [×] [(] [3] [a b/c] [5] [−] [1] [a b/c] [8] [)] [=].

The display reads

19 ⌋ 80 ,

which means $\frac{19}{80}$. We now press [a b/c] to convert $\frac{19}{80}$ to decimal notation:

0.2375 .

The answer is 0.2375.

Do Exercises 20–22.

d FINDING SQUARE ROOTS

To find square roots, we use the [√‾] key.

EXAMPLES Simplify.

10. $\sqrt{64}$

We press the following keys:

[6] [4] [√‾].

The display shows 8 without our having used the [=] key.

11. $\sqrt{0.000625}$

We enter

[.] [0] [0] [0] [6] [2] [5] [√].

The answer is 0.025.

Do Exercises 23–26.

We can approximate some square roots as follows.

EXAMPLE 12 Approximate $\sqrt{31}$ to three decimal places.

To find $\sqrt{31}$, we press

[3] [1] [√].

The display shows

5.5677644 .

Thus, $\sqrt{31} \approx 5.568$, if we round to three decimal places. The answer seems reasonable since $5^2 < 31 < 6^2$.

Do Exercises 27–30.

e OPERATIONS WITH REAL NUMBERS

To enter a negative number, we use the [+/−] key. To enter −5, we press [5] and then [+/−]. The display then reads

−5 .

EXAMPLE 13 Evaluate: $-8 - (-2.3)$.

We press the following keys:

[8] [+/−] [−] [2] [.] [3] [+/−] [=].

The answer is −5.7.

Do Exercises 31–34.

EXAMPLE 14 Evaluate: $-7(2 - 9) - 20$.

We enter

[7] [+/−] [×] [(] [2] [−] [9] [)] [−] [2] [0] [=].

The answer is 29.

Do Exercises 35–37.

Simplify.

23. $\sqrt{0.0001}$

24. $\sqrt{169}$

25. $\sqrt{62{,}500}$

26. $\sqrt{0.001764}$

Approximate the square root to three decimal places.

27. $\sqrt{11}$

28. $\sqrt{270}$

29. $\sqrt{6.4}$

30. $\sqrt{0.009}$

Evaluate.

31. $-5 - (-13)$

32. $-13 + 72 + (-20) + (-23)$

33. $(-6)^2$

34. $35 - (-16) - (-21) + 9^2$

Evaluate.

35. $-8 + 4(7 - 9) + 5$

36. $-3[2 + (-5)]$

37. $7[4 - (-3)] + 5[3^2 - (-4)]$

Answers on page A-34

Check to see if the given number is a solution of the equation using a calculator.

38. $3(5x - 9) + 2(x + 3) = -5(3x - 7) + 8$; 2

39. $2(3x - 5) + 7x = 3x - (x + 8) + 20$; 3

40. $20(x - 39) = 5x - 432$; $23\frac{1}{5}$

41. $-\frac{1}{2}x + 8 = 1\frac{1}{2}x - 6$; -28

42. $-(x - 3) - (x - 4) = 2x + 3$; 1

Evaluate the polynomial.

43. $3x^2 + 5x - 8$; when $x = -1$

44. $4x^2 + 12x + 19$; when $x = 3$

45. $-17x^2 - (-10x^4) + 3x - 8$; when $x = 2$

46. $6 - 7x - 2x^2$; when $x = -5$

47. $-2x^2 - 16x + 6$; when $x = -2$

Answers on page A-34

f Checking Solutions of Equations

Calculators can be used to check solutions of equations. We can replace the variable with the solution and evaluate each side of the equation separately.

EXAMPLE 15 Check to see if 5 is a solution of the equation:

$$3(x - 5) + 4(x + 3) = 2(x - 6) + 34.$$

To check using a calculator, we first evaluate the left side:

[3] [×] [(] [5] [−] [5] [)] [+] [4] [×] [(] [5] [+] [3] [)] [=].

The display reads 32:

32 .

We then evaluate the right side:

[2] [×] [(] [5] [−] [6] [)] [+] [3] [4] [=].

Again, the display reads 32:

32 .

Since the left side is the same as the right side, the solution checks.

Do Exercises 38–42.

g Evaluating Polynomials

To evaluate a polynomial for a specific value of the variable, we replace the variable with the given value.

EXAMPLE 16 Evaluate the polynomial $2x^2 + 5x - 4$ when $x = 6$.

We replace x with 6:

$$2(6)^2 + 5(6) - 4.$$

Then we press the following keys:

[2] [×] [6] [x^2] [+] [5] [×] [6] [−] [4] [=].

The answer, 98, is displayed in the window:

98 .

EXAMPLE 17 Evaluate $-12x^2 - (-5x^3) - 15$ when $x = -2$.

We first replace x with -2:

$$-12(-2)^2 - (-5)(-2)^3 - 15.$$

We then press the following keys:

[1] [2] [+/−] [x] [2] [+/−] [x^2] [−] [5] [+/−] [x] [2] [+/−] [x^y] [3] [−] [1] [5] [=].

The display reads -103.

Do Exercises 43–47.

h CALCULATING WITH SCIENTIFIC NOTATION

Most scientific calculators use either [EE] or [EXP] to enter numbers written in scientific notation.

EXAMPLE 18 Enter the number 3.265×10^8 into a calculator.

We use the following keystrokes:

[3] [·] [2] [6] [5] [EE] [8] or [3] [·] [2] [6] [5] [EXP] [8].

The display will read

3.265 08 or 3.265^{08}.

Do Exercises 48–50.

When an answer to a calculation contains more digits than the display window allows, the calculator will give the answer in scientific notation.

EXAMPLE 19 Multiply: $6{,}000{,}000 \times 5{,}000{,}000$.

We calculate using the following keystrokes:

[6] [0] [0] [0] [0] [0] [0] [×] [5] [0] [0] [0] [0] [0] [0] [=].

The display in the window is in the form

3 13 or 3^{13}.

This means that the answer is 30,000,000,000,000.

Do Exercises 51–55.

i GRAPHING LINEAR EQUATIONS

The primary use for a graphing calculator is to graph equations. One feature common to all graphers is the window. This refers to the rectangular portion of the screen in which a graph appears. Windows are described by four numbers [L, R, B, T], which represent the Left and Right endpoints of the x-axis and the Bottom and Top endpoints of the y-axis. A Range key is sometimes used to set these dimensions.

EXAMPLE 20 Graph the equation $2x + y = 3$ using a standard $[-10, 10, -10, 10]$ window.

Although each graphing calculator has its own set of procedures for drawing graphs, almost all of them require that the equation be written in the form $y = \ldots$.

Solving $2x + y = 3$ for y, we get

$$y = -2x + 3.$$

We enter this equation into a calculator with keystrokes as follows:

[y=] [(−)] [2] [X|T] [+] [3] [ENTER].

Enter the number into a calculator.

48. 8.3×10^9

49. 2.591×10^{-4}

50. 4.77×10^{23}

Perform the indicated operations.

51. $8{,}000{,}000 \times 7{,}000{,}000{,}000$

52. $0.04 \div 2{,}000{,}000{,}000$

53. $(2.6 \times 10^5) \div 10^9$

54. $(4.2 \times 10^{14})(9.05 \times 10^{-6})$

55. $\dfrac{(7.5 \times 10^{-9})(4.0 \times 10^{-3})}{(2.5 \times 10^{12})(8.0 \times 10^{20})}$

Answers on page A-34

Then we press GRAPH .

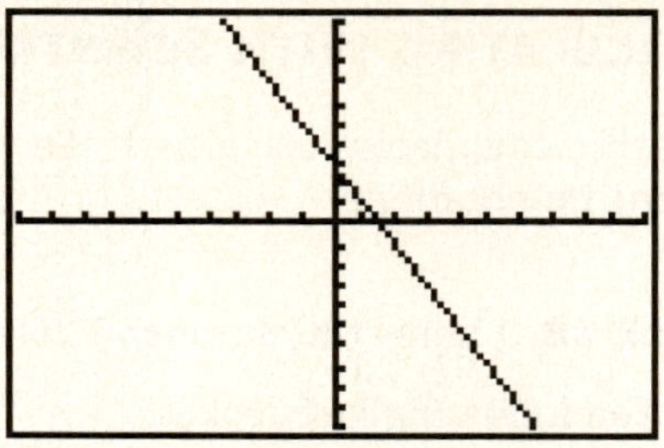

Do Exercise 56.

Once you have graphed an equation on a graphing calculator, you can investigate some of its points by using the Trace feature. Usually a Trace key is pressed to access this feature. Once it has been pressed, a cursor (often blinking) appears somewhere on the graph while its x- and y-coordinates are shown elsewhere on the screen. These coordinates will change as the cursor moves along the graph.

Do Exercises 57–60.

j SOLVING SYSTEMS OF EQUATIONS

A graphing calculator can also be used to solve a system of equations, especially if it has a Zoom feature that can be used to magnify the intersection. If your calculator does not have this feature, you can shrink the window's dimensions. By doing so repeatedly and using the Trace feature, you can determine the coordinates of the point of intersection to the desired accuracy.

EXAMPLE 21 Solve the system

$$7x - y = 9,$$
$$6x + 3y = 1.$$

Round each coordinate to the nearest tenth.

We first solve each equation for y:

$$y = 7x - 9,$$
$$y = -2x + \tfrac{1}{3}.$$

We then enter each equation of the system into a calculator. Next we activate the Trace feature so that the coordinates of any point of intersection are displayed. Then we zoom in as many times as necessary to determine the coordinates to the nearest tenth.

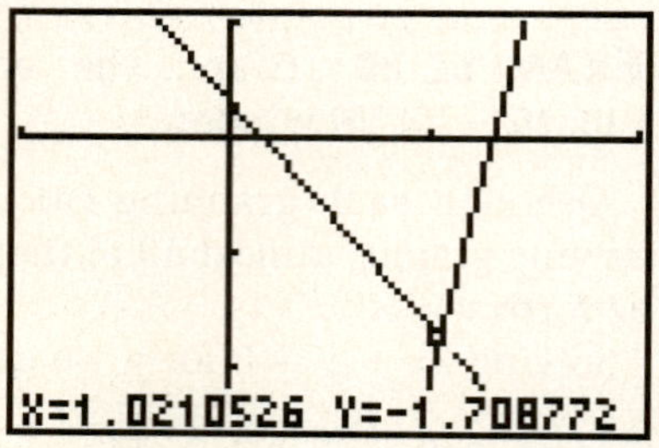

From the graph, we see that the solution of the system is the ordered pair $(1.0, -1.7)$.

The technique described above can be used to check possible solutions of any system of equations. Using the calculator is especially advantageous for systems in which the equations have noninteger coefficients.

Do Exercises 61–64.

56. Graph: $4x + y = 3$.

Graph the equation. Then use the Trace feature to find the coordinates of at least three points on the graph.

57. $y = 2x - 1$

58. $4x - y = 5$

59. $x + 2y = 9$

60. $2x - 4y = 1$

Solve the system of equations. Round each coordinate to the nearest tenth.

61. $y = x - 6,\ y = 3x + 1$

62. $2x - y = 3,\ y = x + 4$

63. $y = -4.56x + 12.95,$
$y = 7.88x - 6.77$

64. $5x - 2y = 4,\ x + 3y = 2$

Answers on page A-34

Answers

Margin Exercise Answers

CHAPTER R

Margin Exercises, Section R.1, pp. 3–6

1. 1, 3, 9 **2.** 1, 2, 4, 8, 16 **3.** 1, 2, 3, 6, 9, 18 **4.** 1, 2, 3, 4, 6, 8, 12, 24 **5.** 13 **6.** $2 \cdot 2 \cdot 2 \cdot 2 \cdot 3$ **7.** $2 \cdot 5 \cdot 5$ **8.** $2 \cdot 5 \cdot 7 \cdot 11$ **9.** 15, 30, 45, 60, . . . **10.** 40 **11.** 54 **12.** 360 **13.** 18 **14.** 24 **15.** 36 **16.** 210

Margin Exercises, Section R.2, pp. 10–15

1. $\frac{8}{12}$ **2.** $\frac{21}{28}$ **3.** $\frac{14}{16}, \frac{21}{24}, \frac{28}{32}$; answers may vary **4.** $\frac{2}{5}$ **5.** $\frac{19}{9}$ **6.** $\frac{8}{3}$ **7.** $\frac{1}{2}$ **8.** 4 **9.** $\frac{5}{2}$ **10.** $\frac{35}{16}$ **11.** $\frac{7}{5}$ **12.** 2 **13.** $\frac{23}{15}$ **14.** $\frac{3}{4}$ **15.** $\frac{19}{40}$ **16.** $\frac{7}{36}$ **17.** $\frac{11}{4}$ **18.** $\frac{7}{15}$ **19.** $\frac{1}{5}$ **20.** 3 **21.** $\frac{21}{20}$ **22.** $\frac{8}{21}$ **23.** $\frac{5}{6}$ **24.** $\frac{8}{15}$ **25.** $\frac{1}{64}$ **26.** 81

Margin Exercises, Section R.3, pp. 19–24

1. $\frac{568}{1000}$ **2.** $\frac{23}{10}$ **3.** $\frac{8904}{100}$ **4.** 4.131 **5.** 0.4131 **6.** 5.73 **7.** 284.455 **8.** 268.63 **9.** 27.676 **10.** 64.683 **11.** 99.59 **12.** 239.883 **13.** 5.868 **14.** 0.5868 **15.** 51.53808 **16.** 48.9 **17.** 15.82 **18.** 1.28 **19.** 17.95 **20.** 856 **21.** 0.85 **22.** 0.625 **23.** $0.\overline{6}$ **24.** $7.\overline{63}$ **25.** 2.8 **26.** 13.9 **27.** 7.0 **28.** 7.83 **29.** 34.68 **30.** 0.03 **31.** 0.943 **32.** 8.004 **33.** 43.112 **34.** 37.401 **35.** 7459.355 **36.** 7459.35 **37.** 7459.4 **38.** 7459 **39.** 7460

Margin Exercises, Section R.4, pp. 27–28

1. 0.873 **2.** 1 **3.** $\frac{53}{100}$ **4.** $\frac{456}{1000}$ **5.** $\frac{23}{10,000}$ **6.** 677% **7.** 99.44% **8.** 25% **9.** 87.5% **10.** $66.\overline{6}\%$, or $66\frac{2}{3}\%$

Margin Exercises, Section R.5, pp. 31–33

1. 4^3 **2.** 6^5 **3.** $(1.08)^2$ **4.** 10,000 **5.** 512 **6.** 1.331 **7.** 13 **8.** 1000 **9.** 250 **10.** 178 **11.** 2 **12.** 8000 **13.** 48 **14.** 3.55625

CHAPTER 1

Margin Exercises, Section 1.1, pp. 43–46

1. $2128 + x = 2866$ **2.** 64 **3.** 28 **4.** 60 **5.** 192 ft^2 **6.** 3.375 hr **7.** 25 **8.** 16 **9.** $x - 8$ **10.** $y + 8$, or $8 + y$ **11.** $m - 4$ **12.** $\frac{1}{2}p$ **13.** $6 + 8x$, or $8x + 6$ **14.** $a - b$ **15.** $59\%x$, or $0.59x$ **16.** $xy - 200$ **17.** $p + q$

Margin Exercises, Section 1.2, pp. 51–56

1. 8; −5 **2.** 134; −76 **3.** −10; 156 **4.** −147; 289

5.

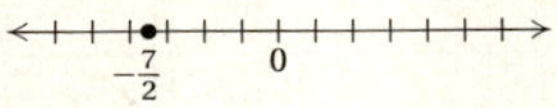

6.

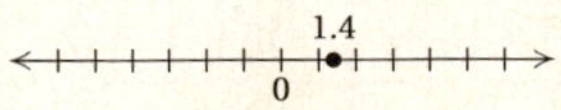

7.

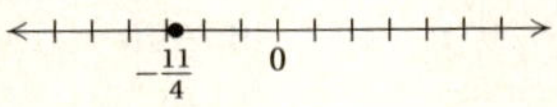

8. −0.375 **9.** $-0.\overline{54}$ **10.** $1.\overline{3}$ **11.** < **12.** < **13.** > **14.** > **15.** > **16.** < **17.** < **18.** > **19.** $7 > -5$ **20.** $4 < x$ **21.** False **22.** True **23.** True **24.** 8 **25.** 0 **26.** 9 **27.** $\frac{2}{3}$ **28.** 5.6

Margin Exercises, Section 1.3, pp. 59–62

1. −6 **2.** −3 **3.** −8 **4.** 4 **5.** 0 **6.** −2 **7.** −11 **8.** −12 **9.** 2 **10.** −4 **11.** −2 **12.** 0 **13.** −22 **14.** 3 **15.** 0.53 **16.** 2.3 **17.** −7.7 **18.** −6.2 **19.** $-\frac{2}{9}$ **20.** $-\frac{19}{20}$ **21.** −58 **22.** −56 **23.** −14 **24.** −12 **25.** 4 **26.** −8.7 **27.** 7.74 **28.** $\frac{8}{9}$ **29.** 0 **30.** −12 **31.** −14; 14 **32.** −1; 1 **33.** 19; −19 **34.** 1.6; −1.6 **35.** $-\frac{2}{3}; \frac{2}{3}$ **36.** $\frac{9}{8}; -\frac{9}{8}$ **37.** 4 **38.** 13.4 **39.** 0 **40.** $-\frac{1}{4}$

Margin Exercises, Section 1.4, pp. 65–67

1. −10 **2.** 3 **3.** −5 **4.** −2 **5.** −11 **6.** 4 **7.** −2 **8.** −6 **9.** −16 **10.** 7.1 **11.** 3 **12.** 0 **13.** $\frac{3}{2}$ **14.** −8 **15.** 7 **16.** −3 **17.** −23.3 **18.** 0 **19.** −9 **20.** 17 **21.** 12.7 **22.** $17 profit **23.** 50°C

Margin Exercises, Section 1.5, pp. 73–75

1. 20; 10; 0; −10; −20; −30 **2.** −18 **3.** −100 **4.** −80 **5.** $-\frac{5}{9}$ **6.** −30.033 **7.** $-\frac{7}{10}$ **8.** −10; 0; 10; 20; 30 **9.** 12 **10.** 32 **11.** 35 **12.** $\frac{20}{63}$ **13.** $\frac{2}{3}$ **14.** 13.455 **15.** −30 **16.** 30 **17.** 0 **18.** $-\frac{8}{3}$ **19.** −30 **20.** −30.75 **21.** $-\frac{5}{3}$ **22.** 120 **23.** −120 **24.** 6 **25.** 4; −4 **26.** 9; −9 **27.** 48; 48

Margin Exercises, Section 1.6, pp. 79–82

1. −2 **2.** 5 **3.** −3 **4.** 8 **5.** −6 **6.** $-\frac{30}{7}$ **7.** Undefined **8.** 0 **9.** $\frac{3}{2}$ **10.** $-\frac{4}{5}$ **11.** $-\frac{1}{3}$ **12.** −5 **13.** $\frac{1}{1.6}$ **14.** $\frac{2}{3}$ **15.** First row: $\frac{2}{3}, -\frac{2}{3}, \frac{3}{2}$; second row: $-\frac{5}{4}, \frac{5}{4}, -\frac{4}{5}$; third row: 0, 0, undefined; fourth row: 1, −1, 1; fifth row: −8, 8, $-\frac{1}{8}$; sixth row: −4.5, 4.5, $-\frac{1}{4.5}$ **16.** $\frac{4}{7} \cdot \left(-\frac{5}{3}\right)$ **17.** $5 \cdot \left(-\frac{1}{8}\right)$ **18.** $(a - b) \cdot \left(\frac{1}{7}\right)$ **19.** $-23 \cdot a$ **20.** $-5 \cdot \left(\frac{1}{7}\right)$ **21.** $-\frac{20}{21}$ **22.** $-\frac{12}{5}$ **23.** $\frac{16}{7}$ **24.** −7 **25.** $\frac{5}{-6}, -\frac{5}{6}$ **26.** $\frac{-8}{7}, \frac{8}{-7}$ **27.** $\frac{-10}{3}, -\frac{10}{3}$

Margin Exercises, Section 1.7, pp. 85–92

1.

	x + x	2x
x = 3	6	6
x = −6	−12	−12
x = 4.8	9.6	9.6

2.

	x + 3x	5x
x = 2	8	10
x = −6	−24	−30
x = 4.8	19.2	24

3. $\frac{6}{8}$ **4.** $\frac{3t}{4t}$ **5.** $\frac{3}{4}$ **6.** $-\frac{4}{3}$ **7.** 1; 1 **8.** −10; −10 **9.** $9 + x$ **10.** qp **11.** $t + xy$, or $yx + t$, or $t + yx$ **12.** 19; 19 **13.** 150; 150 **14.** $(r + s) + 7$ **15.** $(9a)b$ **16.** $(4t)u$, $(tu)4$, $t(4u)$; answers may vary **17.** $(2 + r) + s$, $(r + s) + 2$, $s + (r + 2)$; answers may vary **18.** (a) 63; (b) 63 **19.** (a) 80; (b) 80 **20.** (a) 28; (b) 28 **21.** (a) 8; (b) 8 **22.** (a) −4; (b) −4 **23.** (a) −25; (b) −25 **24.** $5x, -8y, 3$ **25.** $-4y, -2x, 3z$ **26.** $3x - 15$ **27.** $5x + 5$ **28.** $\frac{2}{3}p + \frac{2}{3}q - \frac{2}{3}t$ **29.** $-2x + 6$ **30.** $5x - 10y + 20z$ **31.** $-5x + 10y - 20z$ **32.** $6(x - 2)$ **33.** $3(x - 2y + 3)$ **34.** $b(x + y - z)$ **35.** $2(8a - 18b + 21)$ **36.** $\frac{1}{8}(3x - 5y + 7)$ **37.** $-4(3x - 8y + 4z)$ **38.** $3x$ **39.** $6x$ **40.** $-8x$ **41.** $0.59x$ **42.** $3x + 3y$ **43.** $-4x - 5y - 7$

Margin Exercises, Section 1.8, pp. 97–100

1. $-x - 2$ **2.** $-5x - 2y - 8$ **3.** $-6 + t$ **4.** $-x + y$ **5.** $4a - 3t + 10$ **6.** $-18 + m + 2n - 4z$ **7.** $2x - 9$ **8.** $3y + 2$ **9.** $2x - 7$ **10.** $3y + 3$ **11.** $-2a + 8b - 3c$ **12.** $-9x - 8y$ **13.** $-16a + 18$ **14.** $-26a + 41b - 48c$ **15.** $3x - 7$ **16.** 2 **17.** 18 **18.** 6 **19.** 17 **20.** $5x - y - 8$ **21.** −1237 **22.** 8 **23.** 381 **24.** −12

CHAPTER 2

Margin Exercises, Section 2.1, pp. 113–116

1. False **2.** True **3.** Neither **4.** Yes **5.** No **6.** No **7.** 9 **8.** 22 **9.** −5 **10.** 13.2 **11.** −6.5 **12.** −2 **13.** $\frac{31}{8}$

Margin Exercises, Section 2.2, pp. 119–122

1. 15 **2.** $-\frac{7}{4}$ **3.** $-\frac{4}{5}$ **4.** 8 **5.** −18 **6.** 10 **7.** 28 **8.** 7800 **9.** −3

Margin Exercises, Section 2.3, pp. 125–130

1. 5 **2.** 4 **3.** 4 **4.** 39 **5.** $-\frac{3}{2}$ **6.** −4.3 **7.** −3 **8.** 800 **9.** 1 **10.** 2 **11.** 2 **12.** $\frac{17}{2}$ **13.** $\frac{8}{3}$ **14.** −4.3 **15.** 2 **16.** 3 **17.** −2 **18.** $-\frac{1}{2}$

Margin Exercises, Section 2.4, pp. 136–140

1. 35 in., 37 in. **2.** 5 **3.** 228 and 229 **4.** 2875 **5.** Width: 9 ft; length: 12 ft **6.** 30°, 90°, 60°

Margin Exercises, Section 2.5, pp. 145–148

1. 32% **2.** 25% **3.** 225 **4.** 50 **5.** 11.04 **6.** 111,416 mi^2 **7.** \$8400 **8.** \$658

Margin Exercises, Section 2.6, pp. 151–152

1. 2.8 mi **2.** $I = \frac{E}{R}$ **3.** $D = \frac{C}{\pi}$ **4.** $c = 4A - a - b - d$ **5.** $I = \frac{9R}{E}$

Margin Exercises, Section 2.7, pp. 155–160

1. (a) No; (b) no; (c) no; (d) yes; (e) no **2.** (a) Yes; (b) yes; (c) yes; (d) no; (e) yes

3.

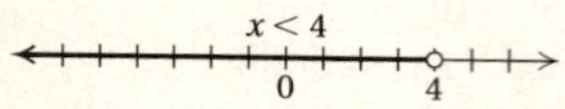

4.

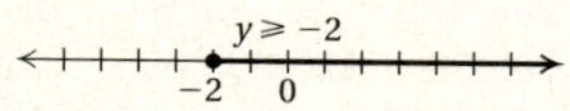

5.

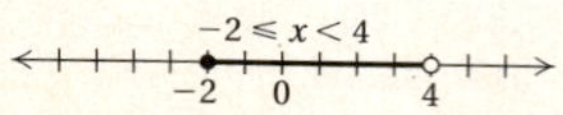

6. $\{x \mid x > 2\}$;

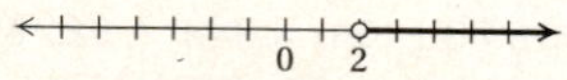

7. $\{x \mid x \leq 3\}$;

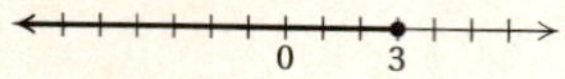

8. $\{x \mid x < -3\}$;

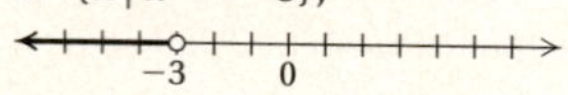

9. $\left\{x \mid x \geq \frac{2}{15}\right\}$ **10.** $\{y \mid y \leq -3\}$

11. $\{x \mid x < 8\}$;

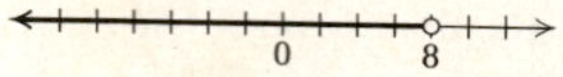

12. $\{y \mid y \geq 32\}$;

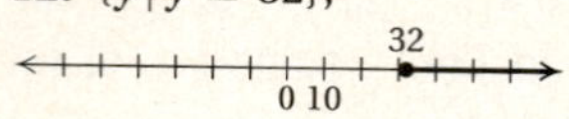

13. $\{x \mid x \geq -6\}$ **14.** $\left\{y \mid y < -\frac{13}{5}\right\}$ **15.** $\left\{x \mid x > -\frac{1}{4}\right\}$ **16.** $\left\{y \mid y \geq \frac{19}{9}\right\}$ **17.** $\left\{y \mid y \geq \frac{19}{9}\right\}$ **18.** $\{x \mid x \geq -2\}$

Margin Exercises, Section 2.8, pp. 165–166

1. $x \leq 8$ **2.** $y > -2$ **3.** $s \leq 180$ **4.** $p \geq \$5800$ **5.** $2x - 32 > 5$ **6.** $\{x \mid x \geq 84\}$ **7.** $\{C \mid C < 1063°\}$

CHAPTER 3

Margin Exercises, Section 3.1, pp. 179–184

1. $5 \cdot 5 \cdot 5 \cdot 5$ **2.** $x \cdot x \cdot x \cdot x \cdot x$ **3.** $3t \cdot 3t$ **4.** $3 \cdot t \cdot t$ **5.** 6 **6.** 1 **7.** 8.4 **8.** 1 **9.** 125 **10.** 3215.36 cm^2 **11.** 119 **12.** 3; -3 **13.** (a) 144; (b) 36; (c) no **14.** 3^{10} **15.** x^{10} **16.** p^{24} **17.** x^5 **18.** a^9b^8 **19.** 4^3 **20.** y^4 **21.** p^9 **22.** a^4b^2 **23.** $\frac{1}{4^3} = \frac{1}{64}$ **24.** $\frac{1}{5^2} = \frac{1}{25}$ **25.** $\frac{1}{2^4} = \frac{1}{16}$ **26.** $\frac{1}{(-2)^3} = -\frac{1}{8}$ **27.** $\frac{4}{p^3}$ **28.** x^2 **29.** 5^2 **30.** $\frac{1}{x^7}$ **31.** $\frac{1}{7^5}$ **32.** b **33.** t^6

Margin Exercises, Section 3.2, pp. 189–194

1. 3^{20} **2.** $\frac{1}{x^{12}}$ **3.** y^{15} **4.** $\frac{1}{x^{32}}$ **5.** $\frac{16x^{20}}{y^{12}}$ **6.** $\frac{25x^{10}}{y^{12}z^6}$ **7.** x^{74} **8.** $\frac{27z^{24}}{y^6x^{15}}$ **9.** $\frac{x^{12}}{25}$ **10.** $\frac{8t^{15}}{w^{12}}$ **11.** $\frac{9}{x^8}$ **12.** 5.17×10^{-4} **13.** 5.23×10^8 **14.** 689,300,000,000 **15.** 0.0000567 **16.** 5.6×10^{-15} **17.** 7.462×10^{-13} **18.** 2.0×10^3 **19.** 5.5×10^2 **20.** $6.\overline{6} \times 10^{-2}$ **21.** 2.3725×10^9 gal

Margin Exercises, Section 3.3, pp. 199–204

1. $4x^2 - 3x + \frac{5}{4}$; $15y^3$; $-7x^3 + 1.1$; answers may vary **2.** -19 **3.** -104 **4.** -13 **5.** 8 **6.** 132 **7.** 360 ft **8.** 7.55 parts per million **9.** $-9x^3 + (-4x^5)$ **10.** $-2y^3 + 3y^7 + (-7y)$ **11.** $3x^2$, $6x$, $\frac{1}{2}$ **12.** $-4y^5$, $7y^2$, $-3y$, -2 **13.** $4x^3$ and $-x^3$ **14.** $4t^4$ and $-7t^4$; $-9t^3$ and $10t^3$ **15.** $5x^2$ and $7x^2$; $3x$ and $-8x$; -10 and 11 **16.** 2, -7, -8.5, 10, -4 **17.** $8x^2$ **18.** $2x^3 + 7$ **19.** $-\frac{1}{4}x^5 + 2x^2$ **20.** $-4x^3$ **21.** $5x^3$ **22.** $25 - 3x^5$ **23.** $6x$ **24.** $4x^3 + 4$ **25.** $-\frac{1}{4}x^3 + 4x^2 + 7$ **26.** $3x^2 + x^3 + 9$ **27.** $6x^7 + 3x^5 - 2x^4 + 4x^3 + 5x^2 + x$ **28.** $7x^5 - 5x^4 + 2x^3 + 4x^2 - 3$ **29.** $14t^7 - 10t^5 + 7t^2 - 14$ **30.** $-2x^2 - 3x + 2$ **31.** $10x^4 - 8x - \frac{1}{2}$ **32.** 4, 2, 1, 0; 4 **33.** x **34.** x^3, x^2, x, x^0 **35.** x^2, x **36.** x^3 **37.** Monomial **38.** None of these **39.** Binomial **40.** Trinomial

Margin Exercises, Section 3.4, pp. 209–212

1. $x^2 + 7x + 3$ **2.** $-4x^5 + 7x^4 + x^3 + 2x^2 + 4$ **3.** $24x^4 + 5x^3 + x^2 + 1$ **4.** $2x^3 + \frac{10}{3}$ **5.** $2x^2 - 3x - 1$ **6.** $8x^3 - 2x^2 - 8x + \frac{5}{2}$ **7.** $-8x^4 + 4x^3 + 12x^2 + 5x - 8$ **8.** $-x^3 + x^2 + 3x + 3$ **9.** $-(12x^4 - 3x^2 + 4x)$; $-12x^4 + 3x^2 - 4x$ **10.** $-(-4x^4 + 3x^2 - 4x)$; $4x^4 - 3x^2 + 4x$ **11.** $-\left(-13x^6 + 2x^4 - 3x^2 + x - \frac{5}{13}\right)$; $13x^6 - 2x^4 + 3x^2 - x + \frac{5}{13}$ **12.** $-(-7y^3 + 2y^2 - y + 3)$; $7y^3 - 2y^2 + y - 3$ **13.** $-4x^3 + 6x - 3$ **14.** $-5x^4 - 3x^2 - 7x + 5$ **15.** $-14x^{10} + \frac{1}{2}x^5 - 5x^3 + x^2 - 3x$ **16.** $2x^3 + 2x + 8$ **17.** $x^2 - 6x - 2$ **18.** $-8x^4 - 5x^3 + 8x^2 - 1$ **19.** $x^3 - x^2 - \frac{4}{3}x - 0.9$ **20.** $2x^3 + 5x^2 - 2x - 5$ **21.** $-x^5 - 2x^3 + 3x^2 - 2x + 2$ **22.** $\frac{7}{2}x^2$ **23.** $\pi x^2 - 2x^2$, or $(\pi - 2)x^2$

Margin Exercises, Section 3.5, pp. 217–220

1. $-15x$ **2.** $-x^2$ **3.** x^2 **4.** $-x^5$ **5.** $12x^7$ **6.** $-8y^{11}$ **7.** $7y^5$ **8.** 0 **9.** $8x^2 + 16x$ **10.** $-15t^3 + 6t^2$ **11.** $5x^6 + 25x^5 - 30x^4 + 40x^3$ **12.** $x^2 + 13x + 40$ **13.** $x^2 + x - 20$ **14.** $5x^2 - 17x - 12$ **15.** $6x^2 - 19x + 15$ **16.** $x^4 + 3x^3 + x^2 + 15x - 20$ **17.** $6y^5 - 20y^3 + 15y^2 + 14y - 35$ **18.** $3x^3 + 13x^2 - 6x + 20$ **19.** $20x^4 - 16x^3 + 32x^2 - 32x - 16$ **20.** $6x^4 - x^3 - 18x^2 - x + 10$

Margin Exercises, Section 3.6, pp. 224–228

1. $x^2 + 7x + 12$ **2.** $x^2 - 2x - 15$ **3.** $2x^2 + 9x + 4$ **4.** $2x^3 - 4x^2 - 3x + 6$ **5.** $12x^5 + 6x^2 + 10x^3 + 5$ **6.** $y^6 - 49$ **7.** $-2x^7 + x^5 + x^3$ **8.** $t^2 + 8t + 15$ **9.** $x^2 - \frac{16}{25}$ **10.** $x^5 + 0.5x^3 - 0.5x^2 - 0.25$ **11.** $8 + 2x^2 - 15x^4$ **12.** $30x^5 - 3x^4 - 6x^3$ **13.** $x^2 - 25$ **14.** $4x^2 - 9$ **15.** $x^2 - 4$ **16.** $x^2 - 49$ **17.** $36 - 16y^2$ **18.** $4x^6 - 1$ **19.** $x^2 + 16x + 64$ **20.** $x^2 - 10x + 25$ **21.** $x^2 + 4x + 4$ **22.** $a^2 - 8a + 16$ **23.** $4x^2 + 20x + 25$ **24.** $16x^4 - 24x^3 + 9x^2$ **25.** $49 + 14y + y^2$ **26.** $9x^4 - 30x^2 + 25$ **27.** $x^2 + 11x + 30$ **28.** $t^2 - 16$ **29.** $-8x^5 + 20x^4 + 40x^2$ **30.** $81x^4 + 18x^2 + 1$ **31.** $4a^2 + 6a - 40$ **32.** $25x^2 + 5x + \frac{1}{4}$ **33.** $4x^2 - 2x + \frac{1}{4}$ **34.** $x^3 - 3x^2 + 6x - 8$

Margin Exercises, Section 3.7, pp. 233–236

1. -7940 **2.** -176 **3.** 433.32 ft^2 **4.** -3, 3, -2, 1, 2 **5.** 3, 7, 1, 1, 0; 7 **6.** $2x^2y + 3xy$ **7.** $5pq - 8$ **8.** $-4x^3 + 2x^2 - 4y + 2$ **9.** $14x^3y + 7x^2y - 3xy - 2y$ **10.** $-5p^2q^4 + 2p^2q^2 + 3p^2q + 6pq^2 + 3q + 5$ **11.** $-8s^4t + 6s^3t^2 + 2s^2t^3 - s^2t^2$ **12.** $-9p^4q + 9p^3q^2 - 4p^2q^3 - 9q^4 + 5$ **13.** $x^5y^5 + 2x^4y^2 + 3x^3y^3 + 6x^2$ **14.** $p^5q - 4p^3q^3 + 3pq^3 + 6q^4$ **15.** $3x^3y + 6x^2y^3 + 2x^3 + 4x^2y^2$ **16.** $2x^2 - 11xy + 15y^2$ **17.** $16x^2 + 40xy + 25y^2$ **18.** $9x^4 - 12x^3y^2 + 4x^2y^4$ **19.** $4x^2y^4 - 9x^2$ **20.** $16y^2 - 9x^2y^4$ **21.** $9y^2 + 24y + 16 - 9x^2$ **22.** $4a^2 - 25b^2 - 10bc - c^2$

Margin Exercises, Section 3.8, pp. 241–244

1. $x^2 + 3x + 2$ **2.** $2x^2 + x - \frac{2}{3}$ **3.** $4x^2 - \frac{3}{2}x + \frac{1}{2}$ **4.** $2x^2y^4 - 3xy^2 + 5y$ **5.** $x - 2$ **6.** $x + 4$ **7.** $x + 4$, R -2, or $x + 4 + \frac{-2}{x + 3}$ **8.** $x^2 + x + 1$

CHAPTER 4

Margin Exercises, Section 4.1, pp. 257–260

1. (a) $12x^2$; (b) $(3x)(4x)$, $(2x)(6x)$; answers may vary **2.** (a) $16x^3$; (b) $(2x)(8x^2)$, $(4x)(4x^2)$; answers may vary **3.** $(8x)(x^3)$, $(4x^2)(2x^2)$, $(2x^3)(4x)$; answers may vary **4.** $(7x)(3x)$, $(-7x)(-3x)$, $(21x)(x)$; answers may vary **5.** $(6x^4)(x)$, $(-2x^3)(-3x^2)$, $(3x^3)(2x^2)$; answers may vary

6. (a) $3x + 6$; (b) $3(x + 2)$ **7.** (a) $2x^3 + 10x^2 + 8x$; (b) $2x(x^2 + 5x + 4)$ **8.** $x(x + 3)$ **9.** $y^2(3y^4 - 5y + 2)$ **10.** $3x^2(3x^2 - 5x + 1)$ **11.** $\frac{1}{4}(3t^3 + 5t^2 + 7t + 1)$ **12.** $7x^3(5x^4 - 7x^3 + 2x^2 - 9)$ **13.** $(x^2 + 3)(x + 7)$ **14.** $(x^2 + 2)(a + b)$ **15.** $(x^2 + 3)(x + 7)$ **16.** $(2t^2 + 3)(4t + 1)$ **17.** $(3m^3 + 2)(m^2 - 5)$ **18.** $(3x^2 - 1)(x - 2)$ **19.** $(2x^2 - 3)(2x - 3)$ **20.** Not factorable using factoring by grouping

Margin Exercises, Section 4.2, pp. 263–266

1. $(x + 4)(x + 3)$ **2.** $(x + 9)(x + 4)$ **3.** $(x - 5)(x - 3)$ **4.** $(t - 5)(t - 4)$ **5.** The positive factor has the larger absolute value. **6.** The negative factor has the larger absolute value. **7.** $x(x + 6)(x - 2)$ **8.** $(y - 6)(y + 2)$ **9.** $(t^2 + 7)(t^2 - 2)$ **10.** $p(p - q - 3q^2)$ **11.** Not factorable **12.** $(x + 4)^2$

Margin Exercises, Section 4.3, pp. 270–272

1. $(2x + 5)(x - 3)$ **2.** $(4x + 1)(3x - 5)$ **3.** $(3x - 4)(x - 5)$ **4.** $2(5x - 4)(2x - 3)$ **5.** $(2x + 1)(3x + 2)$ **6.** $(2a - b)(3a - b)$ **7.** $3(2x + 3y)(x + y)$

Margin Exercises, Section 4.4, p. 276

1. $(2x + 1)(3x + 2)$ **2.** $(4x + 1)(3x - 5)$ **3.** $3(2x + 3)(x + 1)$ **4.** $2(5x - 4)(2x - 3)$

Margin Exercises, Section 4.5, pp. 280–284

1. Yes **2.** No **3.** No **4.** Yes **5.** No **6.** Yes **7.** No **8.** Yes **9.** $(x + 1)^2$ **10.** $(x - 1)^2$ **11.** $(t + 2)^2$ **12.** $(5x - 7)^2$ **13.** $(7 - 4y)^2$ **14.** $3(4m + 5)^2$ **15.** $(p^2 + 9)^2$ **16.** $z^3(2z - 5)^2$ **17.** $(3a + 5b)^2$ **18.** Yes **19.** No **20.** No **21.** No **22.** Yes **23.** Yes **24.** Yes **25.** $(x + 3)(x - 3)$ **26.** $4(4 + t)(4 - t)$ **27.** $(a + 5b)(a - 5b)$ **28.** $x^4(8 + 5x)(8 - 5x)$ **29.** $5(1 + 2t^3)(1 - 2t^3)$ **30.** $(9x^2 + 1)(3x + 1)(3x - 1)$ **31.** $(7p^2 + 5q^3)(7p^2 - 5q^3)$

Margin Exercises, Section 4.6, pp. 290–292

1. $3(m^2 + 1)(m + 1)(m - 1)$ **2.** $(x^3 + 4)^2$ **3.** $2x^2(x + 1)(x + 3)$ **4.** $(3x^2 - 2)(x + 4)$ **5.** $8x(x - 5)(x + 5)$ **6.** $x^2y(x^2y + 2x + 3)$ **7.** $2p^4q^2(5p^2 + 2pq + q^2)$ **8.** $(a - b)(2x + 5 + y^2)$ **9.** $(a + b)(x^2 + y)$ **10.** $(x^2 + y^2)^2$ **11.** $(xy + 1)(xy + 4)$ **12.** $(p^2 + 9q^2)(p + 3q)(p - 3q)$

Margin Exercises, Section 4.7, pp. 298–300

1. $3, -4$ **2.** $7, 3$ **3.** $-\frac{1}{4}, \frac{2}{3}$ **4.** $0, \frac{17}{3}$ **5.** $-2, 3$ **6.** $7, -4$ **7.** 3 **8.** $0, 4$ **9.** $\frac{4}{3}, -\frac{4}{3}$ **10.** $3, -3$

Margin Exercises, Section 4.8, pp. 303–306

1. 5 and −5 **2.** 7 and 8 **3.** −4 and 5 **4.** Length: 5 cm; width: 3 cm **5.** (a) 342; (b) 9 **6.** 22 and 23 **7.** 3 m, 4 m

Chapter 5

Margin Exercises, Section 5.1, pp. 322–326

1. 3 **2.** −8, 3 **3.** None **4.** $\frac{x(2x + 1)}{x(3x - 2)}$ **5.** $\frac{(x + 1)(x + 2)}{(x - 2)(x + 2)}$ **6.** $\frac{-1(x - 8)}{-1(x - y)}$ **7.** 5 **8.** $\frac{x}{4}$ **9.** $\frac{2x + 1}{3x + 2}$ **10.** $\frac{x + 1}{2x + 1}$ **11.** $x + 2$ **12.** $\frac{y + 2}{4}$ **13.** −1 **14.** −1 **15.** −1 **16.** $\frac{a - 2}{a - 3}$ **17.** $\frac{x - 5}{2}$

Margin Exercises, Section 5.2, pp. 331–332

1. $\frac{2}{7}$ **2.** $\frac{2x^3 - 1}{x^2 + 5}$ **3.** $\frac{1}{x - 5}$ **4.** $x^2 - 3$ **5.** $\frac{6}{35}$ **6.** $\frac{x^2}{40}$ **7.** $\frac{(x - 3)(x - 2)}{(x + 5)(x + 5)}$ **8.** $\frac{x - 3}{x + 2}$ **9.** $\frac{(x - 3)(x - 2)}{x + 2}$ **10.** $\frac{y + 1}{y - 1}$

Margin Exercises, Section 5.3, pp. 335–336

1. 144 **2.** 12 **3.** 10 **4.** 120 **5.** $\frac{35}{144}$ **6.** $\frac{1}{4}$ **7.** $\frac{11}{10}$ **8.** $\frac{9}{40}$ **9.** $60x^3y^2$ **10.** $(y + 1)^2(y + 4)$ **11.** $7(t^2 + 16)(t - 2)$ **12.** $3x(x + 1)^2(x - 1)$

Margin Exercises, Section 5.4, pp. 339–342

1. $\frac{7}{9}$ **2.** $\frac{3 + x}{x - 2}$ **3.** $\frac{6x + 4}{x - 1}$ **4.** $\frac{x - 5}{4}$ **5.** $\frac{x - 1}{x - 3}$ **6.** $\frac{10x^2 + 9x}{48}$ **7.** $\frac{9x + 10}{48x^2}$ **8.** $\frac{4x^2 - x + 3}{x(x - 1)(x + 1)^2}$ **9.** $\frac{2x^2 + 16x + 5}{(x + 3)(x + 8)}$ **10.** $\frac{8x + 88}{(x + 16)(x + 1)(x + 8)}$ **11.** $\frac{-2x - 11}{3(x + 4)(x - 4)}$

Margin Exercises, Section 5.5, pp. 347–350

1. $\frac{4}{11}$ **2.** $\frac{5}{y}$ **3.** $\frac{x^2 + 2x + 1}{2x + 1}$ **4.** $\frac{3x - 1}{3}$ **5.** $\frac{4x - 3}{x - 2}$ **6.** $\frac{-x - 7}{15x}$ **7.** $\frac{x^2 - 48}{(x + 7)(x + 8)(x + 6)}$ **8.** $\frac{-8y - 28}{(y + 4)(y - 4)}$ **9.** $\frac{x - 13}{(x + 3)(x - 3)}$ **10.** $\frac{6x^2 - 2x - 2}{3x(x + 1)}$

Margin Exercises, Section 5.6, pp. 355–358

1. $\frac{33}{2}$ **2.** 3 **3.** $\frac{3}{2}$ **4.** $-\frac{1}{8}$ **5.** 1 **6.** 2 **7.** 4

Margin Exercises, Section 5.7, pp. 363–368

1. −3 **2.** 40 km/h, 60 km/h **3.** $\frac{24}{7}$, or $3\frac{3}{7}$ hr **4.** 58 km/L **5.** 0.280 **6.** 124 km/h **7.** 2.4 fish/yd^2 **8.** 100 oz **9.** 42 **10.** 2074

Margin Exercises, Section 5.8, pp. 373–374

1. $M = \frac{fd^2}{km}$ **2.** $b_1 = \frac{2A - hb_2}{h}$ **3.** $f = \frac{pq}{p + q}$

4. $b = \frac{a}{2Q + 1}$

Margin Exercises, Section 5.9, pp. 378–380

1. $\frac{136}{5}$ **2.** $\frac{7x^2}{3(2 - x^2)}$ **3.** $\frac{x}{x - 1}$ **4.** $\frac{136}{5}$ **5.** $\frac{7x^2}{3(2 - x^2)}$

6. $\frac{x}{x - 1}$

CHAPTER 6

Margin Exercises, Section 6.1, pp. 393–396

1. (a) \$200; **(b)** 28% **2. (a)** Approximately \$17,000; **(b)** approximately \$14,000 **3. (a)** The first week; **(b)** the third and fifth weeks

4.–11.

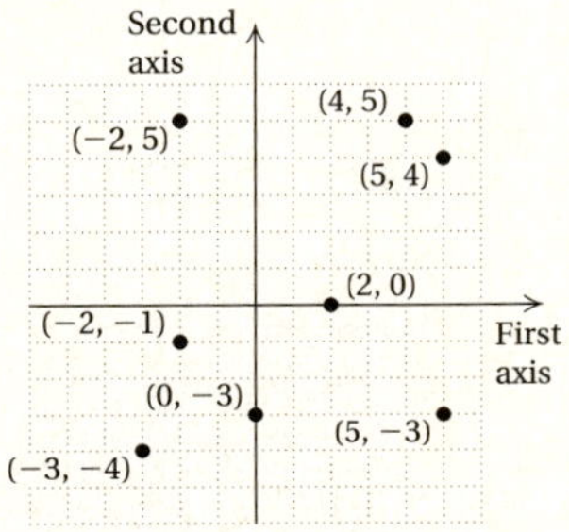

12. Both are negative numbers. **13.** First positive; second negative **14.** I **15.** III **16.** IV **17.** II **18.** A: (−5, 1); B: (−3, 2); C: (0, 4); D: (3, 3); E: (1, 0); F: (0, −3); G: (−5, −4)

Margin Exercises, Section 6.2, pp. 401–406

1. No **2.** Yes

3.

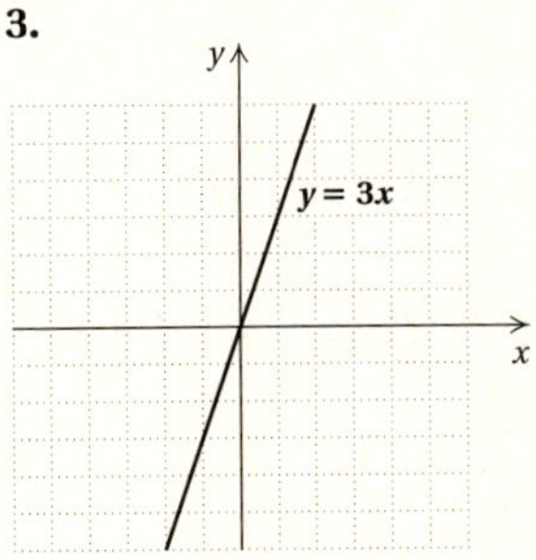

4.

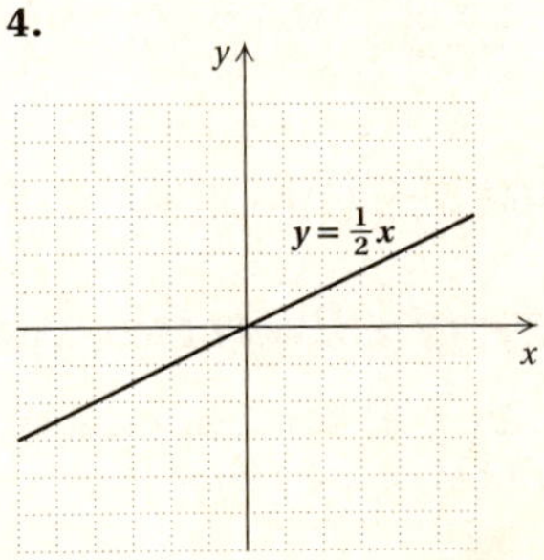

5.

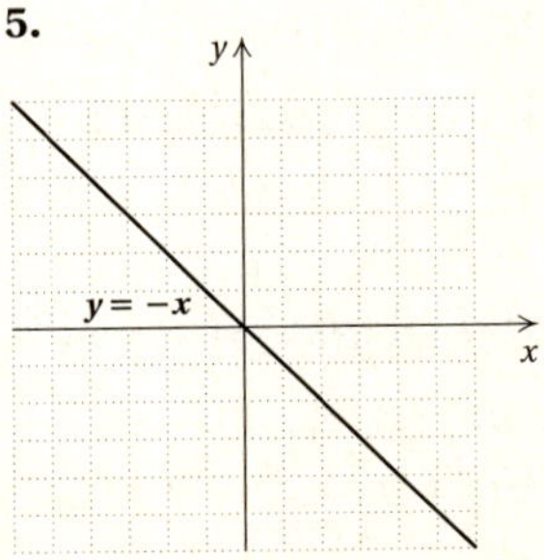

6.

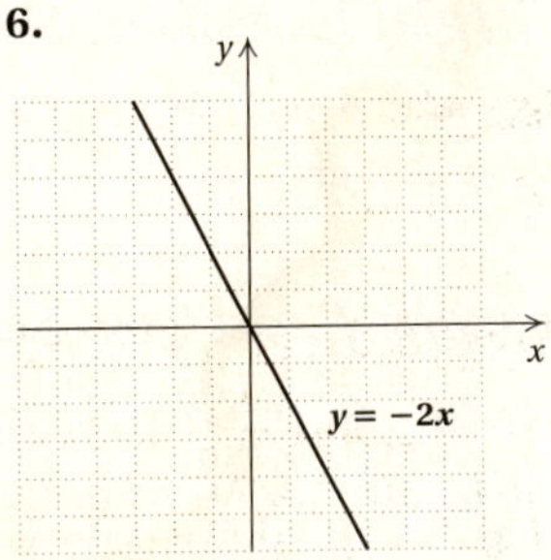

7.

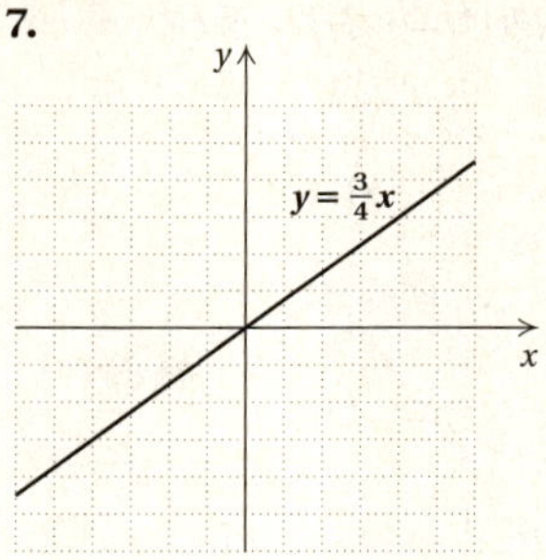

8.

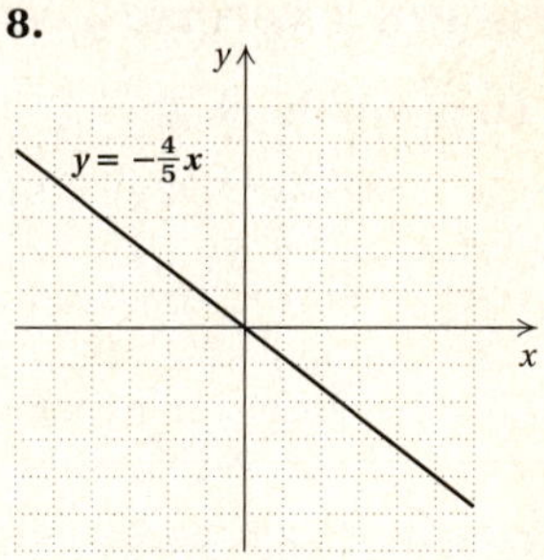

9. $y = x + 3$ looks like $y = x$ moved *up* 3 units.

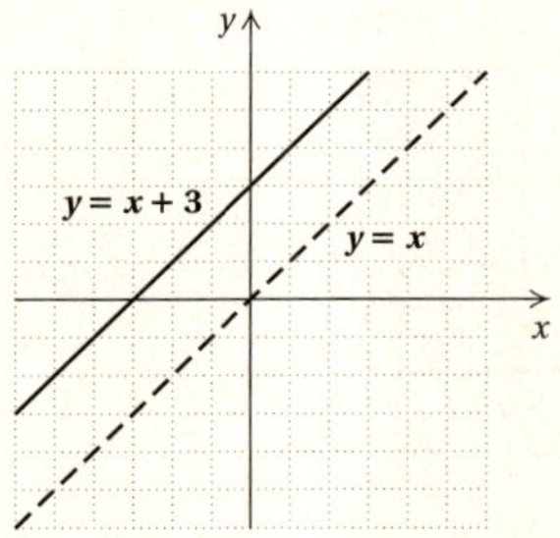

10. $y = x - 1$ looks like $y = x$ moved *down* 1 unit.

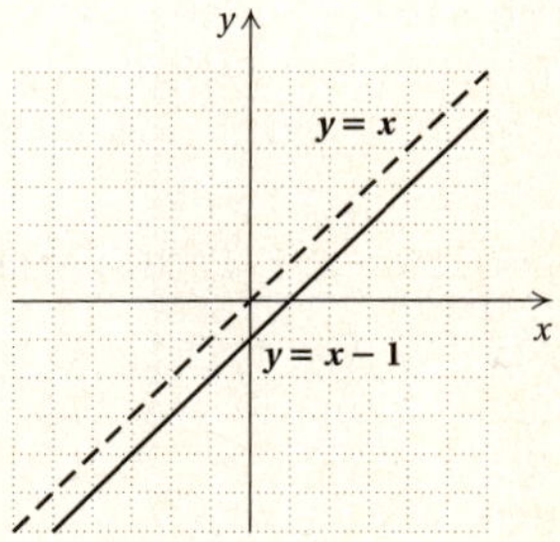

11. $y = 2x + 3$ looks like $y = 2x$ moved *up* 3 units.

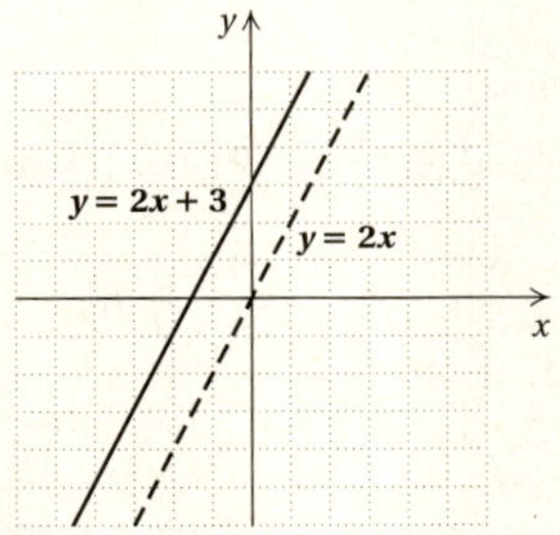

12.

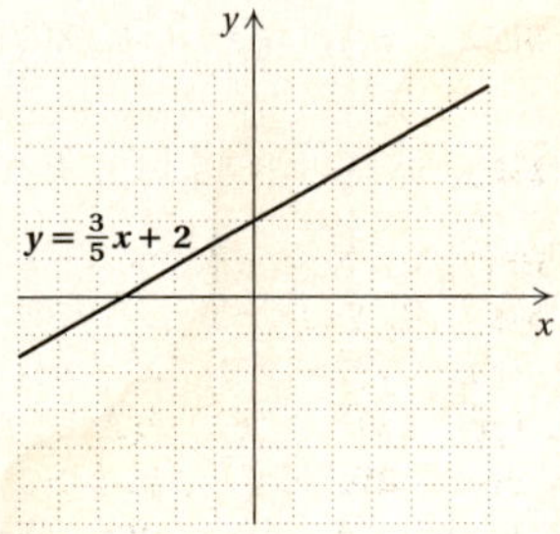

13.

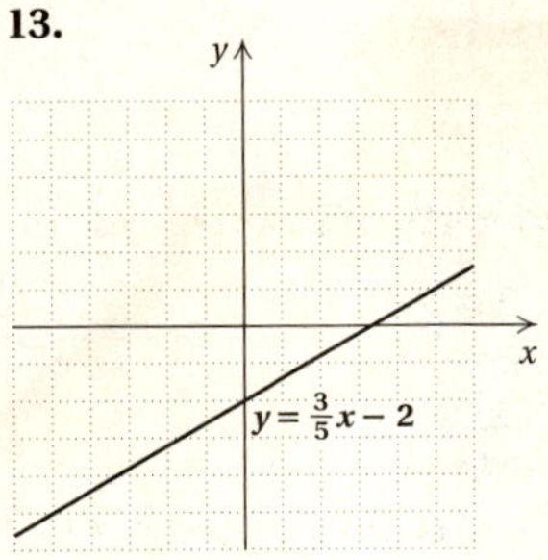

14.

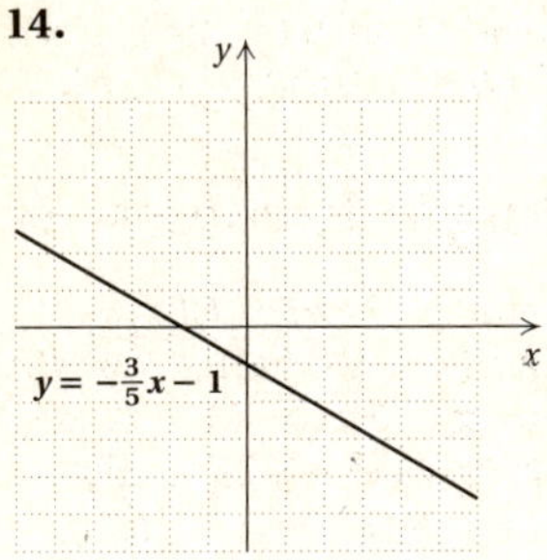

15.

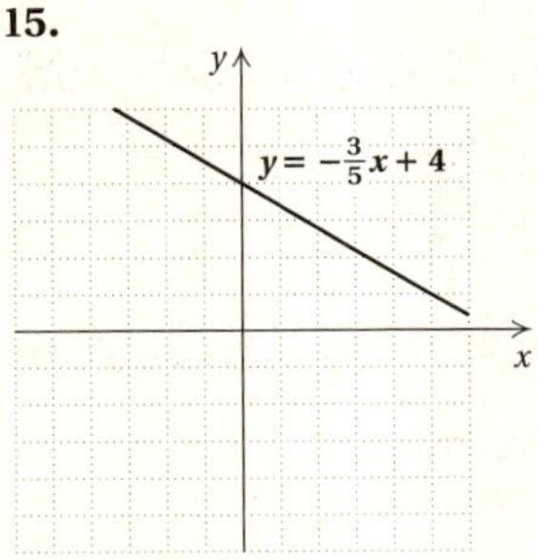

Margin Exercises, Section 6.3, pp. 411–414

1. **(a)** (4, 0); **(b)** (0, 3)

2.

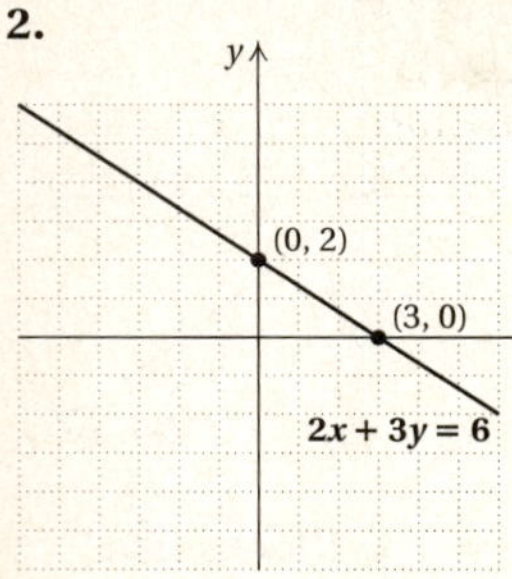

3.

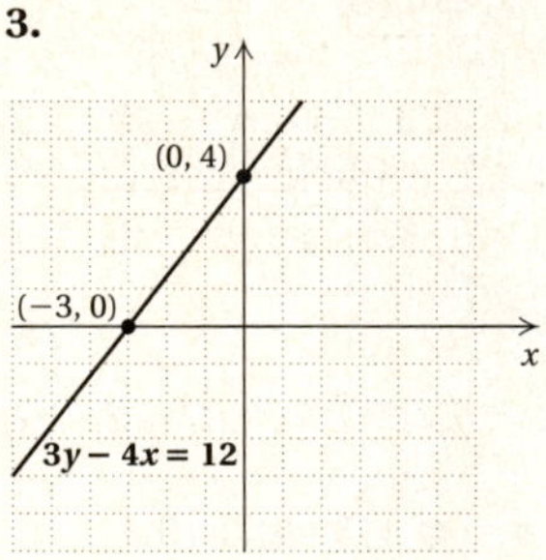

4.

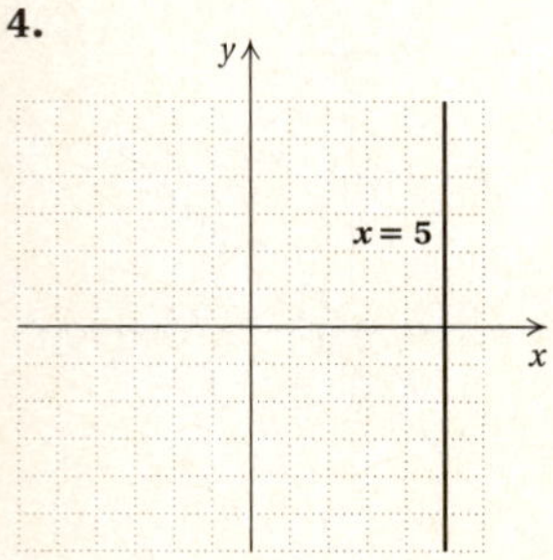

5.

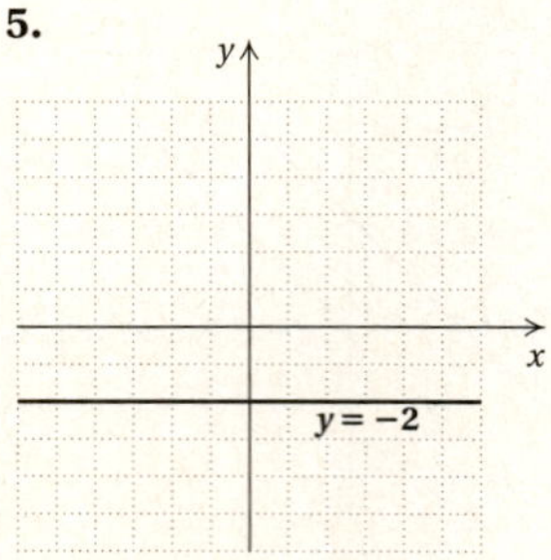

6.

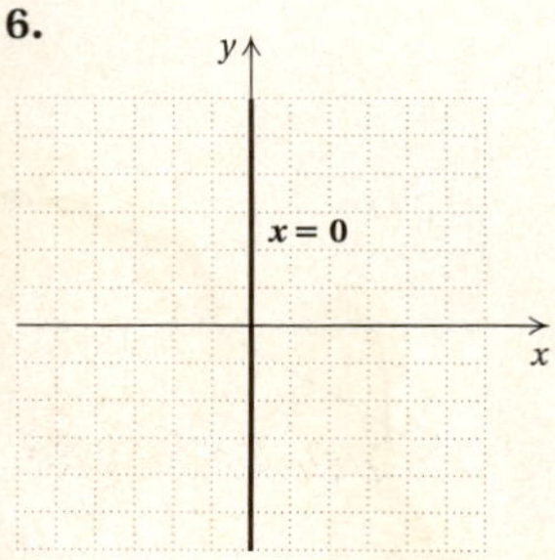

7.

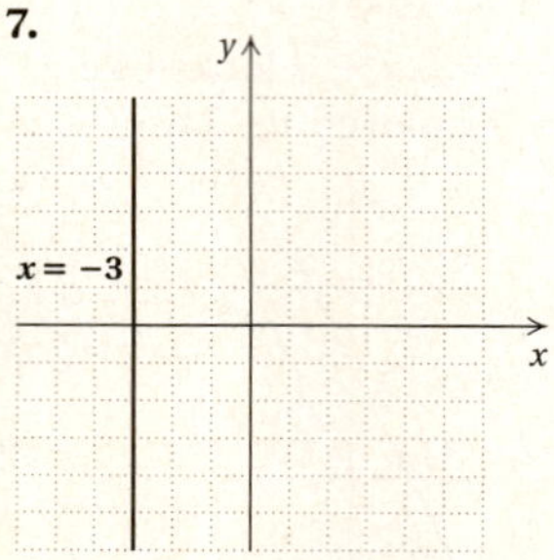

Margin Exercises, Section 6.4, pp. 418–422

1. $\frac{2}{5}$

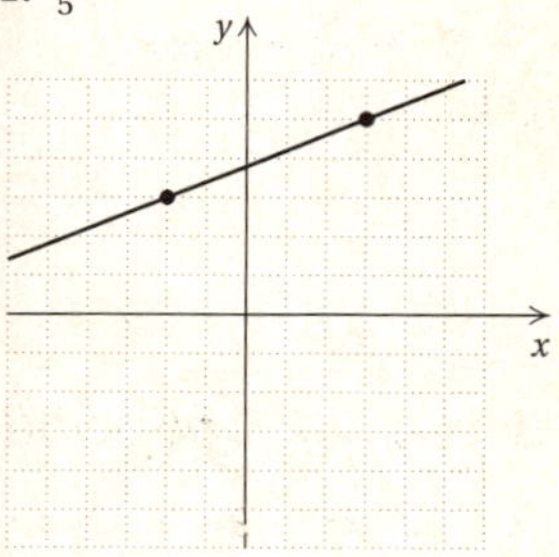

2. $-\frac{5}{3}$

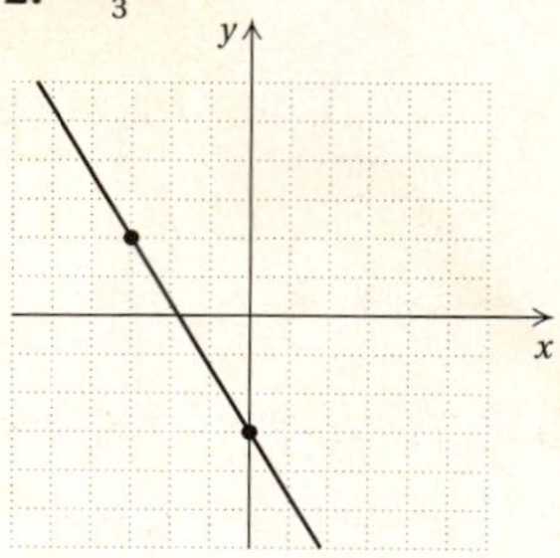

3. -1 **4.** $\frac{5}{4}$ **5.** Undefined **6.** 0
7. 5, (0, 0) **8.** $-\frac{3}{2}$, $(0, -6)$ **9.** $-\frac{3}{4}$, $\left(0, \frac{15}{4}\right)$
10. 2, $\left(0, -\frac{17}{2}\right)$ **11.** $-\frac{7}{5}$, $\left(0, -\frac{22}{5}\right)$ **12.** $y = 3.5x - 23$
13. $y = 5x - 18$ **14.** $y = -3x - 5$ **15.** $y = 6x - 13$
16. $y = -\frac{2}{3}x + \frac{14}{3}$ **17.** $y = x + 2$ **18.** $y = 2x + 4$

Margin Exercises, Section 6.5, p. 426

1. No **2.** Yes **3.** Yes **4.** No

Margin Exercises, Section 6.6, pp. 429–432

1. No **2.** No

3.

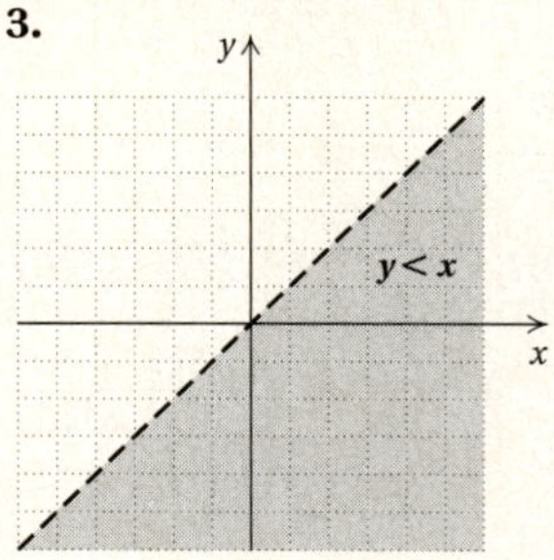

4.

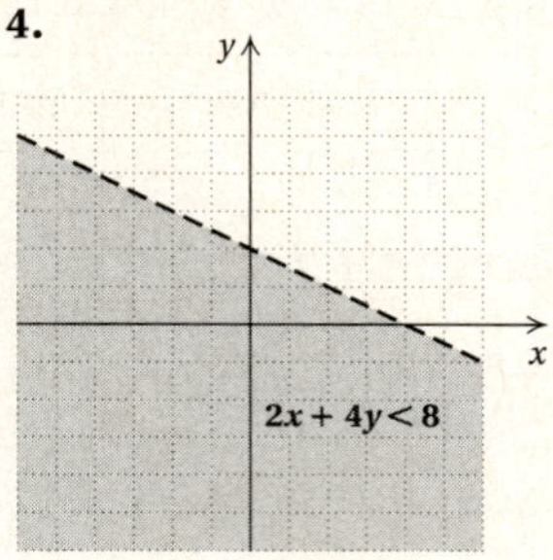

5.

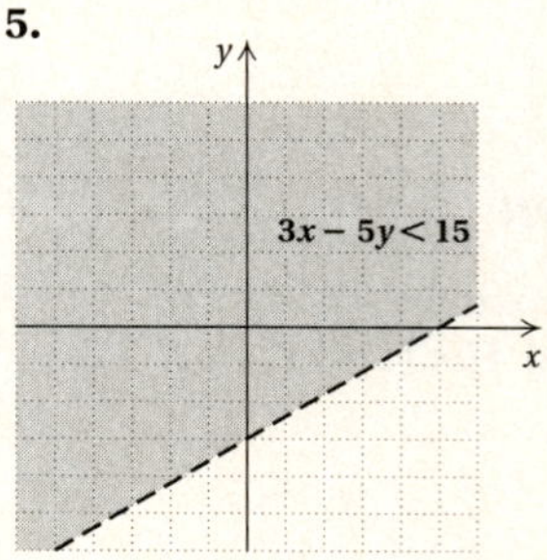

6.

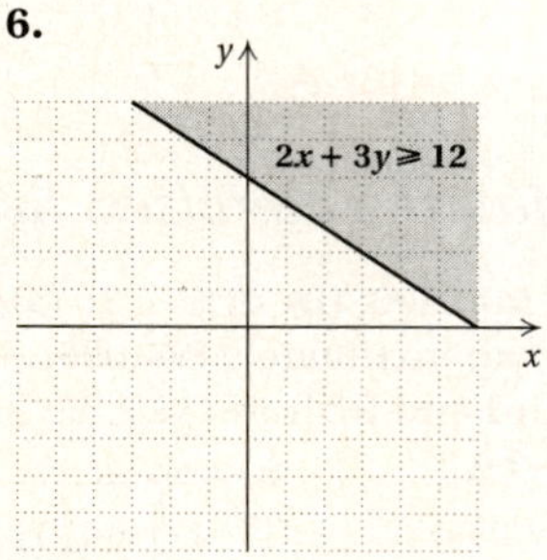

7.

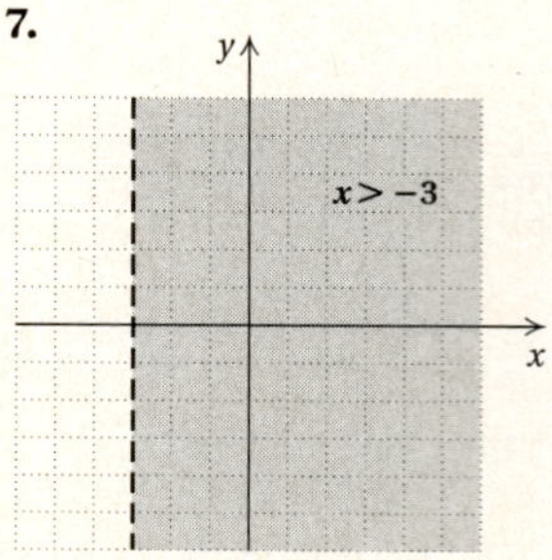

8.

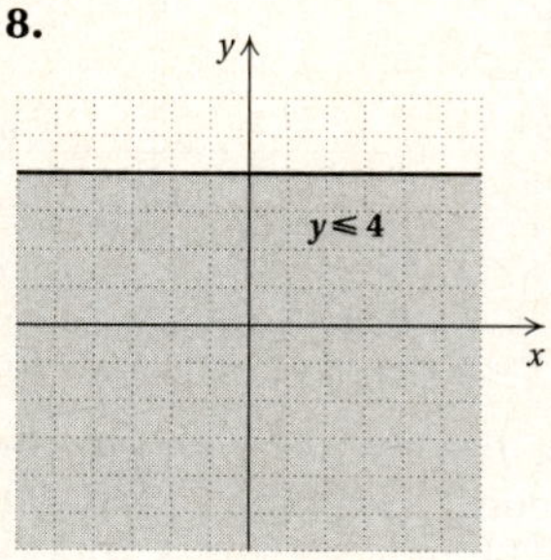

Margin Exercises, Section 6.7, pp. 435–438

1. $y = 7x$ **2.** $y = \frac{5}{8}x$ **3.** \$0.4667; \$0.0194 **4.** 174.24 lb
5. $y = \dfrac{63}{x}$ **6.** $y = \dfrac{900}{x}$ **7.** 8 hr **8.** $7\frac{1}{2}$ hr

CHAPTER 7

Margin Exercises, Section 7.1, pp. 454–456

1. Yes **2.** No **3.** (2, −3) **4.** No solution **5.** Infinite number of solutions

Margin Exercises, Section 7.2, pp. 460–462

1. (3, 2) **2.** (3, −1) **3.** $\left(\frac{24}{5}, -\frac{8}{5}\right)$ **4.** Length: 27.5 cm; width: 10.5 cm

Margin Exercises, Section 7.3, pp. 465–470

1. (3, 2) **2.** (1, −2) **3.** (1, 4) **4.** (−8, 3) **5.** (1, 1)
6. (−2, −1) **7.** $\left(\frac{17}{13}, -\frac{7}{13}\right)$ **8.** No solution **9.** Infinite number of solutions **10.** (1, −1) **11.** 75 miles

Margin Exercises, Section 7.4, pp. 476–481

1. Burger: \$0.99; chicken: \$0.70

2.

	MARVELLA	MALCOLM	
Age now	M	K	→ $M = 26 + K$
Age in 5 yr	$M + 5$	$K + 5$	→ $M + 5 = 2(K + 5)$

Marvella is 47; Malcolm is 21.

3.

ADULTS	CHILDREN	TOTAL	
\$2.10	\$0.75		
x	y	166	→ $x + y = 166$
\$2.10x	\$0.75y	\$293.25	→ $2.10x + 0.75y = 293.25$

125 adults and 41 children

4.

FIRST	SECOND	MIXTURE	
x	y	30 L	→ $x + y = 30$
50%	70%	55%	
50%x	70%y	55% × 30, or 16.5 L	→ $50\%x + 70\%y = 16.5$

22.5 L of 50%, 7.5 L of 70%

5.

A	B	MIXTURE	
\$1.40	\$1.75	\$1.54	
x	y	50	→ $x + y = 50$
1.40x	1.75y	1.54(50), or 77	→ $1.40x + 1.75y = 77$

30 lb of A, 20 lb of B

6. 7 quarters, 13 dimes

Margin Exercises, Section 7.5, pp. 488–490

1. 168 km **2.** 275 km/h **3.** 324 mi **4.** 3 hr

CHAPTER 8

Margin Exercises, Section 8.1, pp. 503–506

1. 6, −6 **2.** 8, −8 **3.** 11, −11 **4.** 12, −12 **5.** 4 **6.** 7 **7.** 10 **8.** 21 **9.** −7 **10.** −13 **11.** 5.477 **12.** −9.899 **13.** **(a)** 20; **(b)** 23 **14.** $45 + x$ **15.** $\dfrac{x}{x+2}$ **16.** $x^2 + 4$ **17.** Yes **18.** No **19.** No **20.** Yes **21.** $|xy|$ **22.** $|xy|$ **23.** $|x - 11|$ **24.** $|x + 4|$ **25.** xy **26.** xy **27.** $x - 11$ **28.** $x + 4$ **29.** $5y$ **30.** $\frac{1}{2}t$

Margin Exercises, Section 8.2, pp. 509–512

1. **(a)** 20; **(b)** 20 **2.** $\sqrt{33}$ **3.** 5 **4.** $\sqrt{x^2 + x}$ **5.** $\sqrt{x^2 - 4}$ **6.** $4\sqrt{2}$ **7.** $x + 7$ **8.** $5x$ **9.** $6m$ **10.** $2\sqrt{23}$ **11.** $x - 10$ **12.** $8t$ **13.** $10a$ **14.** t^2 **15.** t^{10} **16.** h^{23} **17.** $x^3\sqrt{x}$ **18.** $2x^5\sqrt{6x}$ **19.** $3\sqrt{2}$ **20.** 10 **21.** $4x^3y^2$ **22.** $5xy^2\sqrt{2xy}$ **23.** 16.585 **24.** 10.097

Margin Exercises, Section 8.3, pp. 517–520

1. 4 **2.** 5 **3.** $x\sqrt{6x}$ **4.** $\frac{4}{3}$ **5.** $\frac{1}{5}$ **6.** $\frac{6}{x}$ **7.** $\frac{3}{4}$ **8.** $\frac{15}{16}$ **9.** $\frac{7}{y^5}$ **10.** $\frac{\sqrt{15}}{5}$ **11.** $\frac{\sqrt{10}}{4}$ **12.** $\frac{10\sqrt{3}}{3}$ **13.** $\frac{\sqrt{21}}{7}$ **14.** $\frac{\sqrt{5r}}{r}$ **15.** $\frac{8y\sqrt{7}}{7}$ **16.** 0.816 **17.** 0.877

Margin Exercises, Section 8.4, pp. 525–528

1. $12\sqrt{2}$ **2.** $5\sqrt{5}$ **3.** $-12\sqrt{10}$ **4.** $5\sqrt{6}$ **5.** $\sqrt{x+1}$ **6.** $\frac{3}{2}\sqrt{2}$ **7.** $\frac{8\sqrt{15}}{15}$ **8.** $\sqrt{15} + \sqrt{6}$ **9.** $4 + 3\sqrt{5} - 4\sqrt{2} - 3\sqrt{10}$ **10.** $2 - a$ **11.** $25 + 10\sqrt{x} + x$ **12.** 2 **13.** $7 - \sqrt{5}$ **14.** $\sqrt{5} + \sqrt{2}$ **15.** $1 + \sqrt{x}$ **16.** $\frac{21 - 3\sqrt{5}}{22}$ **17.** $\frac{7 + 2\sqrt{10}}{3}$ **18.** $\frac{7 + 7\sqrt{x}}{1 - x}$

Margin Exercises, Section 8.5, pp. 533–536

1. $\frac{64}{3}$ **2.** 2 **3.** $\frac{3}{8}$ **4.** 4 **5.** 1 **6.** 4 **7.** Approximately 313 km **8.** Approximately 16 km **9.** 196 m

Margin Exercises, Section 8.6, pp. 541–542

1. $\sqrt{65} \approx 8.062$ **2.** $\sqrt{75} \approx 8.660$ **3.** $\sqrt{10} \approx 3.162$ **4.** $\sqrt{175} \approx 13.229$ **5.** $\sqrt{325} \approx 18.028$ ft

CHAPTER 9

Margin Exercises, Section 9.1, pp. 555–558

1. $y^2 - 8y = 0$; $a = 1$, $b = -8$, $c = 0$ **2.** $x^2 + 9x - 3 = 0$; $a = 1$, $b = 9$, $c = -3$ **3.** $4x^2 + x + 4 = 0$; $a = 4$, $b = 1$, $c = 4$ **4.** $5x^2 - 21 = 0$; $a = 5$, $b = 0$, $c = -21$ **5.** $0, -\frac{9}{2}$ **6.** $0, \frac{3}{5}$ **7.** $\frac{3}{4}, -2$ **8.** 4, 1 **9.** 13, 5 **10.** **(a)** 14; **(b)** 11

Margin Exercises, Section 9.2, pp. 561–566

1. $\sqrt{10}, -\sqrt{10}$ **2.** 0 **3.** $\frac{\sqrt{6}}{2}, -\frac{\sqrt{6}}{2}$ **4.** 7, −1 **5.** $-4 \pm \sqrt{11}$ **6.** −5, 11 **7.** $1 \pm \sqrt{5}$ **8.** 2, 4 **9.** 2, −10 **10.** $6 \pm \sqrt{13}$ **11.** 5, −2 **12.** $\frac{-3 \pm \sqrt{33}}{4}$ **13.** About 7.9 sec

Margin Exercises, Section 9.3, pp. 570–572

1. $\frac{1}{2}$, -4 **2.** $5, -2$ **3.** $-2 \pm \sqrt{11}$

4. No real-number solutions **5.** $\frac{4 \pm \sqrt{31}}{5}$ **6.** $-0.3, 1.9$

Margin Exercises, Section 9.4, pp. 575–576

1. $L = \frac{r^2}{20}$ **2.** $L = \frac{T^2 g}{4\pi^2}$ **3.** $m = \frac{E}{c^2}$ **4.** $r = \sqrt{\frac{A}{\pi}}$

5. $d = \sqrt{\frac{C}{P} + 1}$ **6.** $n = \frac{1 + \sqrt{1 + 4N}}{2}$

7. $t = \frac{-v + \sqrt{v^2 + 32h}}{16}$

Margin Exercises, Section 9.5, pp. 579–581

1. Length: $\frac{1 + \sqrt{817}}{2} \approx 14.8$ in.;

width: $\frac{-1 + \sqrt{817}}{6} \approx 4.6$ in.

2. 2.3 cm; 3.3 cm **3.** 3 km/h

Margin Exercises, Section 9.6, pp. 589–590

1. $(0, -3)$

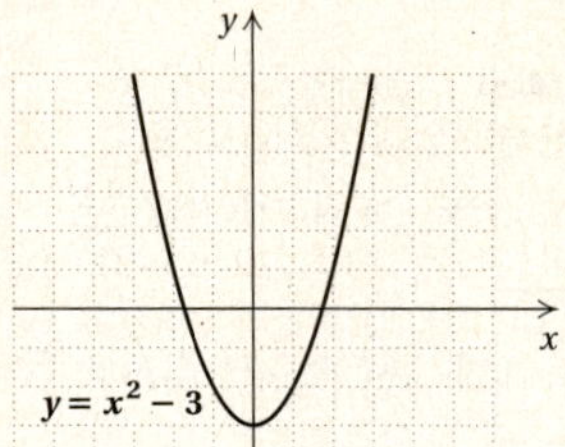

2. $(1, 3)$

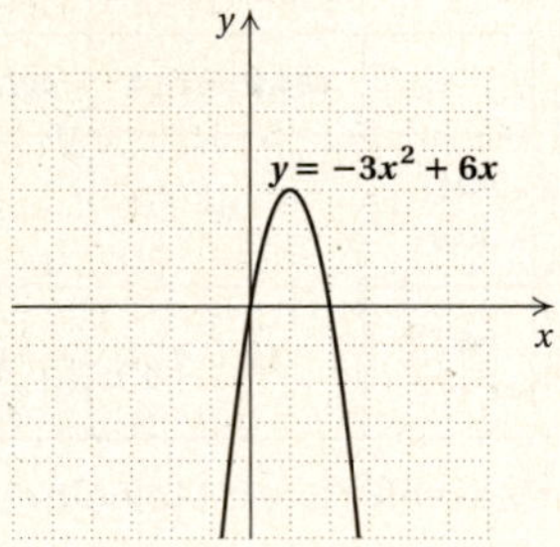

3. $(2, 0)$

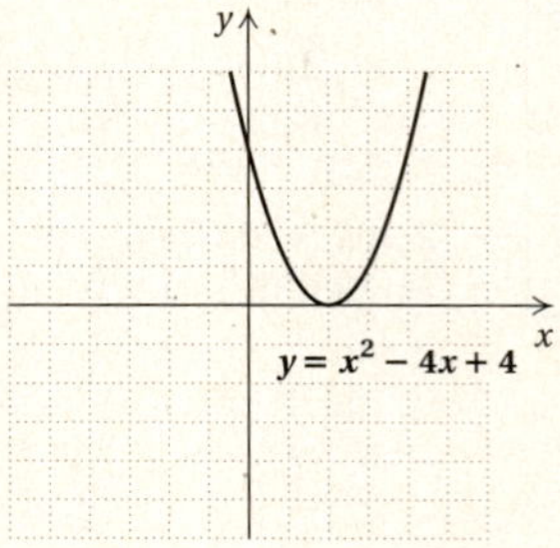

4. $(\sqrt{3}, 0)$; $(-\sqrt{3}, 0)$

5. $(-4, 0)$; $(-2, 0)$

6. $\left(\frac{-2 - \sqrt{6}}{2}, 0\right)$; $\left(\frac{-2 + \sqrt{6}}{2}, 0\right)$

7. None

Exercise Set and Test Answers

Book Diagnostic Pretest, p. xxvii

1. [R.2c] $\frac{8}{15}$ **2.** [R.2c] $\frac{11}{12}$ **3.** [R.3b] 26 **4.** [R.3b] 10.983
5. [1.4a] 1.18 **6.** [1.5a] -11.7 **7.** [1.4b] $-\$9.93$
8. [1.8c] $39a - 84$ **9.** [2.3c] -5 **10.** [2.7e] $\{x \mid x \geq \frac{9}{23}\}$
11. [2.4a] 9 in. and 27 in. **12.** [2.5a] \$1500 **13.** [3.1e] $\frac{x^4}{y}$
14. [3.2b], [3.1d] $-32x^{11}$ **15.** [3.4c] $-x^2 + 3x + 4$
16. [3.6b] $4x^4 - 9$ **17.** [4.5d] $2(x + 9)(x - 9)$
18. [4.3a], [4.4a] $(5x + 1)(x - 3)$ **19.** [4.7b] $-5, 2$
20. [4.8a] 8 m, 17 m **21.** [5.2b] $\frac{x(x - 4)}{2(x + 5)(x - 3)}$
22. [5.4a] $\frac{2 - x}{x(x + 1)(x + 2)}$ **23.** [5.6a] 4
24. [5.7a] 55 mph, 40 mph
25. [6.2b]

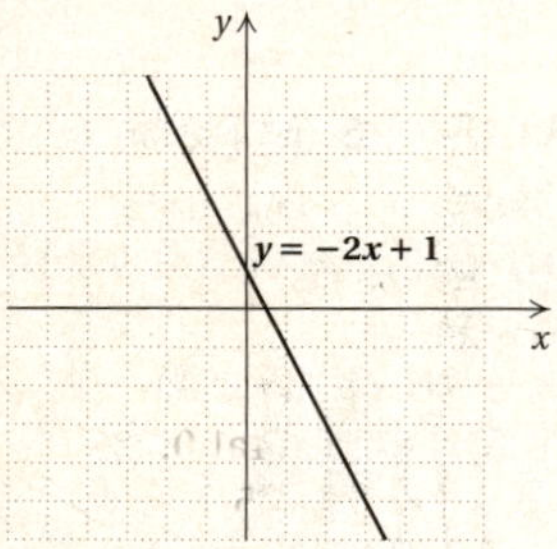

26. [6.4c] $-\frac{2}{3}$; $\left(0, \frac{8}{3}\right)$ **27.** [6.4d] $y = -3x + 11$

28. [6.6b]

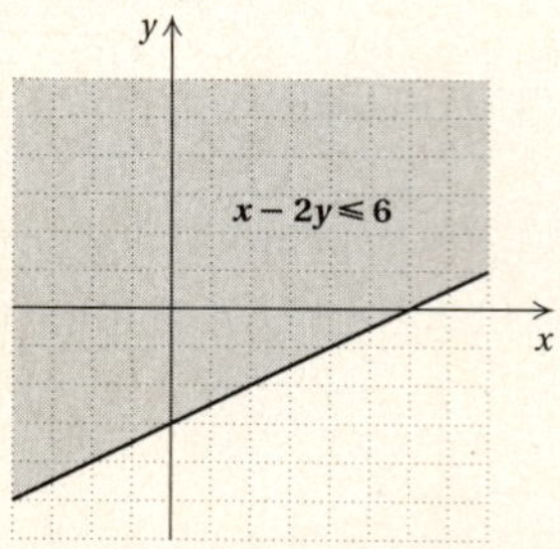

29. [7.2b] $(8, -3)$
30. [7.3b] $\left(\frac{23}{8}, -\frac{3}{16}\right)$
31. [7.4a] 25 liters of each

32. [7.5a] 10 hr **33.** [8.2c] $2xy^2\sqrt{3x}$
34. [8.3a] $\frac{x}{3y}$ **35.** [8.4c] $6 + 3\sqrt{3}$ **36.** [8.5a] $\frac{77}{2}$
37. [9.1c] $-1, \frac{1}{3}$ **38.** [9.3a] No real-number solutions
39. [9.5a] 30 m, 16 m
40. [9.6a]

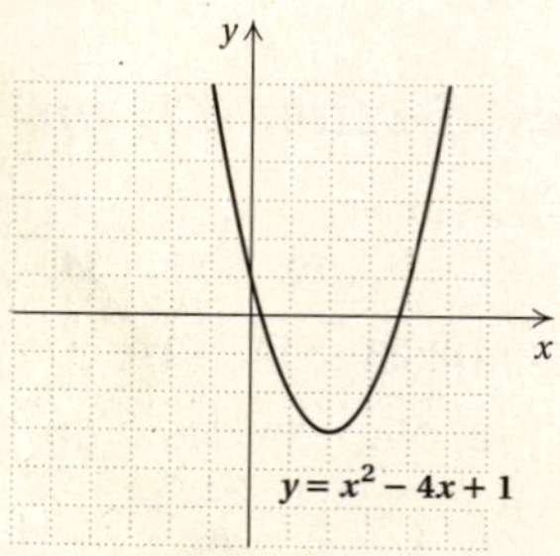

Chapter R

Pretest: Chapter R, p. 2

1. [R.1a] $2 \cdot 2 \cdot 2 \cdot 31$ **2.** [R.1b] 168 **3.** [R.2a] $\frac{10}{15}$
4. [R.2a] $\frac{44}{48}$ **5.** [R.2b] $\frac{23}{64}$ **6.** [R.2b] $\frac{2}{3}$ **7.** [R.2c] $\frac{11}{10}$
8. [R.2c] $\frac{2}{21}$ **9.** [R.2c] $\frac{1}{2}$ **10.** [R.2c] $\frac{5}{21}$ **11.** [R.3a] $\frac{3217}{100}$
12. [R.3a] 0.0789 **13.** [R.3b] 134.0362 **14.** [R.3b] 212.05
15. [R.3b] 350.5824 **16.** [R.3b] 12.4 **17.** [R.3b] $1.\overline{4}$
18. [R.3c] 345.84 **19.** [R.3c] 345.8 **20.** [R.4a] 0.116
21. [R.4b] $\frac{87}{100}$ **22.** [R.4d] 87.5% **23.** [R.5a] 5^4
24. [R.5b] 8 **25.** [R.5b] 1.21 **26.** [R.5c] 18

Exercise Set R.1, p. 7

1. 1, 2, 4, 5, 10, 20 **3.** 1, 2, 3, 4, 6, 8, 9, 12, 18, 24, 36, 72 **5.** $3 \cdot 5$ **7.** $2 \cdot 11$ **9.** $3 \cdot 3$ **11.** $7 \cdot 7$ **13.** $2 \cdot 3 \cdot 3$ **15.** $2 \cdot 2 \cdot 2 \cdot 5$ **17.** $2 \cdot 3 \cdot 3 \cdot 5$ **19.** $2 \cdot 3 \cdot 5 \cdot 7$ **21.** $7 \cdot 13$ **23.** $7 \cdot 17$ **25.** $2 \cdot 2$; 5; 20 **27.** $2 \cdot 2 \cdot 2 \cdot 3$; $2 \cdot 2 \cdot 3 \cdot 3$; 72 **29.** 3; $3 \cdot 5$; 15 **31.** $2 \cdot 3 \cdot 5$; $2 \cdot 2 \cdot 2 \cdot 5$; 120 **33.** 13; 23; 299 **35.** $2 \cdot 3 \cdot 3$; $2 \cdot 3 \cdot 5$; 90 **37.** $2 \cdot 3 \cdot 5$; $2 \cdot 2 \cdot 3 \cdot 3$; 180 **39.** $2 \cdot 2 \cdot 2 \cdot 3$; $2 \cdot 3 \cdot 5$; 120 **41.** 17; 29; 493 **43.** $2 \cdot 2 \cdot 3$; $2 \cdot 2 \cdot 7$; 84 **45.** 2; 3; 5; 30 **47.** $2 \cdot 2 \cdot 2 \cdot 3$; $2 \cdot 2 \cdot 3 \cdot 3$; $2 \cdot 2 \cdot 3$; 72 **49.** 5; $2 \cdot 2 \cdot 3$; $3 \cdot 5$; 60 **51.** $2 \cdot 3$; $2 \cdot 2 \cdot 3$; $2 \cdot 3 \cdot 3$; 36 **53.** **(a)** No; not a multiple of 8; **(b)** no; not a multiple of 8; **(c)** no; not a multiple of 8 or 12; **(d)** yes; it is a multiple of both 8 and 12 and is the smallest such multiple. **55.** 70,200 **57.** Every 420 years

Sidelights: Calculator Corner: Number Patterns, p. 16

1. 9, 1089, 110889, 11108889, 1111088889 **2.** 54, 6534, 665334, 66653334, 6666533334 **3.** 111, 1221, 12321, 123321, 1233321 **4.** 111, 222, 333, 444, 555 **5.** 88, 888, 8888, 88888, 888888 **6.** 9, 98, 987, 9876, 98765 **7.** 48, 408, 4008, 40008, 400008 **8.** 24, 2904, 295704, 29623704, 2962903704 **9.** 1, 121, 12321, 1234321, 123454321 **10.** 81, 9801, 998001, 99980001, 9999800001 **11.** 6006, 60606, 606606, 6066606, 60666606 **12.** 999999, 1999998, 2999997, 3999996, 4999995

Exercise Set R.2, p. 17

1. $\frac{9}{12}$ **3.** $\frac{60}{100}$ **5.** $\frac{104}{160}$ **7.** $\frac{21}{24}$ **9.** $\frac{20}{16}$ **11.** $\frac{2}{3}$ **13.** 4 **15.** $\frac{1}{7}$ **17.** 8 **19.** $\frac{1}{4}$ **21.** 5 **23.** $\frac{17}{21}$ **25.** $\frac{13}{7}$ **27.** $\frac{4}{3}$ **29.** $\frac{1}{12}$ **31.** $\frac{45}{16}$ **33.** $\frac{2}{3}$ **35.** $\frac{7}{6}$ **37.** $\frac{5}{6}$ **39.** $\frac{1}{2}$ **41.** $\frac{13}{24}$ **43.** $\frac{31}{60}$ **45.** $\frac{35}{18}$ **47.** $\frac{10}{3}$ **49.** $\frac{1}{2}$ **51.** $\frac{5}{36}$ **53.** 500 **55.** $\frac{3}{40}$ **57.** $2 \cdot 2 \cdot 7$ **59.** $2 \cdot 2 \cdot 2 \cdot 5 \cdot 5 \cdot 5$ **61.** 48 **63.** $\frac{3}{4}$ **65.** 4 **67.** 1

Exercise Set R.3, p. 25

1. $\frac{53}{10}$ **3.** $\frac{67}{100}$ **5.** $\frac{20{,}007}{10{,}000}$ **7.** $\frac{78{,}898}{10}$ **9.** 0.1 **11.** 0.0001 **13.** 9.999 **15.** 0.4578 **17.** 444.94 **19.** 390.617 **21.** 155.724 **23.** 63.79 **25.** 32.234 **27.** 26.835 **29.** 47.91 **31.** 1.9193 **33.** 13.212 **35.** 0.7993 **37.** 179.5 **39.** 0.1894 **41.** 1.40756 **43.** 3.60558 **45.** 2.3 **47.** 5.2 **49.** 0.023 **51.** 18.75 **53.** 660 **55.** 0.68 **57.** 0.34375 **59.** $1.\overline{18}$ **61.** $0.\overline{5}$ **63.** $2.\overline{1}$ **65.** 745.07; 745.1; 745; 750; 700 **67.** 6780.51; 6780.5; 6781; 6780; 6800 **69.** \$17.99; \$18 **71.** \$346.08; \$346 **73.** \$17 **75.** \$190 **77.** 12.3457; 12.346; 12.35; 12.3; 12

Exercise Set R.4, p. 29

1. 0.63 **3.** 0.941 **5.** 0.01 **7.** 0.0061 **9.** 2.4 **11.** 0.0325 **13.** $\frac{60}{100}$ **15.** $\frac{289}{1000}$ **17.** $\frac{110}{100}$ **19.** $\frac{42}{100{,}000}$ **21.** $\frac{250}{100}$ **23.** $\frac{347}{10{,}000}$ **25.** 100% **27.** 99.6% **29.** 0.47% **31.** 7.2% **33.** 920% **35.** 0.68% **37.** $16.\overline{6}\%$, or $16\frac{2}{3}\%$ **39.** 65% **41.** 29% **43.** 80% **45.** 60% **47.** $66.\overline{6}\%$, or $66\frac{2}{3}\%$ **49.** 175% **51.** 75% **53.** 2.25 **55.** $1.41\overline{6}$ **57.** $0.\overline{90}$ **59.** 32% **61.** 70% **63.** 2700% **65.** 345% **67.** 2.5%

Sidelights: Factors and Sums, p. 34

First row: 48, 90, 432, 63; second row: 7, 18, 36, 14, 12, 6, 21, 11; third row: 9, 2, 2, 10, 8, 10, 21; fourth row: 29, 19, 42

Exercise Set R.5, p. 35

1. 5^4 **3.** 10^3 **5.** 1^3 **7.** 49 **9.** 59,049 **11.** 100 **13.** 1 **15.** 5.29 **17.** 0.008 **19.** 416.16 **21.** $\frac{9}{64}$ **23.** 125 **25.** 1061.208 **27.** 25 **29.** 114 **31.** 33 **33.** 5 **35.** 12 **37.** 324 **39.** 100 **41.** 1000 **43.** 22 **45.** 1 **47.** 4 **49.** 102 **51.** 96 **53.** 24 **55.** 90 **57.** 8 **59.** 1 **61.** 50,000 **63.** $\frac{22}{45}$ **65.** $\frac{19}{66}$ **67.** 31.25% **69.** $2 \cdot 2 \cdot 2 \cdot 2 \cdot 3$ **71.** $\frac{2}{3}$ **73.** 10^2 **75.** 5^2 **77.** $3 = \frac{5+5}{5} + \frac{5}{5}$; $4 = \frac{5+5+5+5}{5}$; $5 = \frac{5(5+5)}{5} - 5$; $6 = \frac{5}{5} + \frac{5 \cdot 5}{5}$; $7 = \frac{5}{5} + \frac{5}{5} + 5$; $8 = 5 + \frac{5+5+5}{5}$; $9 = \frac{5 \cdot 5 - 5}{5} + 5$; $10 = \frac{5 \cdot 5 + 5 \cdot 5}{5}$

Summary and Review: Chapter R, p. 37

1. [R.1a] $2 \cdot 2 \cdot 23$ **2.** [R.1a] $2 \cdot 2 \cdot 2 \cdot 5 \cdot 5 \cdot 7$
3. [R.1b] 416 **4.** [R.1b] 90 **5.** [R.2a] $\frac{12}{30}$ **6.** [R.2a] $\frac{96}{184}$
7. [R.2a] $\frac{40}{64}$ **8.** [R.2a] $\frac{91}{84}$ **9.** [R.2b] $\frac{5}{12}$ **10.** [R.2b] $\frac{51}{91}$
11. [R.2c] $\frac{31}{36}$ **12.** [R.2c] $\frac{1}{4}$ **13.** [R.2c] $\frac{3}{5}$ **14.** [R.2c] $\frac{72}{25}$
15. [R.3a] $\frac{1797}{100}$ **16.** [R.3a] 0.2337 **17.** [R.3b] 2442.905
18. [R.3b] 86.0298 **19.** [R.3b] 9.342 **20.** [R.3b] 133.264
21. [R.3b] 430.8 **22.** [R.3b] 110.483 **23.** [R.3b] 55.6
24. [R.3b] 0.45 **25.** [R.3b] $1.58\overline{3}$ **26.** [R.3c] 34.1
27. [R.4a] 0.047 **28.** [R.4b] $\frac{60}{100}$ **29.** [R.4c] 88.6%
30. [R.4d] 62.5% **31.** [R.4d] 116% **32.** [R.5a] 6^3
33. [R.5b] 1.1236 **34.** [R.5c] 119 **35.** [R.5c] 29
36. [R.5c] 7 **37.** [R.5c] $\frac{103}{17}$ **38.** [R.4d] 0.0000006%

Test: Chapter R, p. 39

1. [R.1a] $2 \cdot 2 \cdot 3 \cdot 5 \cdot 5$ **2.** [R.1b] 120 **3.** [R.2a] $\frac{21}{49}$
4. [R.2a] $\frac{33}{48}$ **5.** [R.2b] $\frac{2}{3}$ **6.** [R.2b] $\frac{37}{61}$ **7.** [R.2c] $\frac{5}{36}$
8. [R.2c] $\frac{11}{40}$ **9.** [R.3a] $\frac{678}{100}$ **10.** [R.3a] 1.895
11. [R.3b] 99.0187 **12.** [R.3b] 1796.58
13. [R.3b] 435.072 **14.** [R.3b] 1.6 **15.** [R.3b] $2.\overline{09}$
16. [R.3c] 234.7 **17.** [R.3c] 234.728 **18.** [R.4a] 0.007
19. [R.4b] $\frac{91}{100}$ **20.** [R.4d] 44% **21.** [R.5b] 625
22. [R.5b] 1.44 **23.** [R.5c] 207 **24.** [R.5c] 20,000
25. [R.2b] $\frac{33}{100}$

CHAPTER 1

Pretest: Chapter 1, p. 42

1. [1.1a] $\frac{5}{16}$ **2.** [1.1b] $78\%x$, or $0.78x$ **3.** [1.1a] 360 ft^2 **4.** [1.3b] 12 **5.** [1.2d] $>$ **6.** [1.2d] $>$ **7.** [1.2d] $>$ **8.** [1.2d] $<$ **9.** [1.2e] 12 **10.** [1.2e] 2.3 **11.** [1.2e] 0 **12.** [1.3b] -5.4 **13.** [1.3b] $\frac{2}{3}$ **14.** [1.3a] -17 **15.** [1.4a] 38.6 **16.** [1.4a] $-\frac{17}{15}$ **17.** [1.3a] -5 **18.** [1.5a] 63 **19.** [1.5a] $-\frac{5}{12}$ **20.** [1.6c] -98 **21.** [1.6a] 8 **22.** [1.4a] 24 **23.** [1.8d] 26 **24.** [1.7c] $9z - 18$ **25.** [1.7c] $-4a - 2b + 10c$ **26.** [1.7d] $4(x - 3)$ **27.** [1.7d] $3(2y - 3z - 6)$ **28.** [1.8b] $-y - 13$ **29.** [1.8c] $y + 18$ **30.** [1.2d] $12 < x$

Exercise Set 1.1, p. 47

1. 23; 28; 41 **3.** 1935 m^2 **5.** 260 mi **7.** 56 **9.** 8 **11.** 1 **13.** 6 **15.** 2 **17.** $b + 7$, or $7 + b$ **19.** $c - 12$ **21.** $4 + q$, or $q + 4$ **23.** $a + b$, or $b + a$ **25.** $y - x$ **27.** $w + x$, or $x + w$ **29.** $n - m$ **31.** $r + s$, or $s + r$ **33.** $2z$ **35.** $3m$ **37.** $89\%x$, or $0.89x$ **39.** $55t$ miles **41.** $2 \cdot 3 \cdot 3 \cdot 3$ **43.** $2 \cdot 2 \cdot 3 \cdot 3 \cdot 3$ **45.** 18 **47.** 60 **49.** $x + 3y$ **51.** $2x - 3$

Exercise Set 1.2, p. 57

1. -1286; 13,804 **3.** $-8{,}000{,}000$ **5.** -3; 128

7.

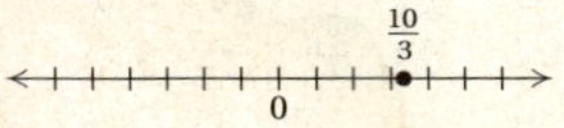

9.

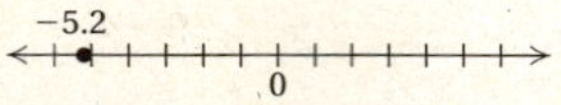

11. -0.875 **13.** $0.8\overline{3}$ **15.** $1.1\overline{6}$ **17.** $0.\overline{6}$ **19.** -0.5 **21.** 0.1 **23.** $>$ **25.** $<$ **27.** $<$ **29.** $<$ **31.** $>$ **33.** $<$ **35.** $>$ **37.** $<$ **39.** $<$ **41.** $<$ **43.** True **45.** False **47.** $x < -6$ **49.** $y \geq -10$ **51.** 3 **53.** 10 **55.** 0 **57.** 24 **59.** $\frac{2}{3}$ **61.** 0 **63.** $-\frac{5}{6}, -\frac{3}{4}, -\frac{2}{3}, \frac{1}{6}, \frac{3}{8}, \frac{1}{2}$

Exercise Set 1.3, p. 63

1. -7 **3.** -6 **5.** 0 **7.** -8 **9.** -7 **11.** -27 **13.** 0 **15.** -42 **17.** 0 **19.** 0 **21.** 3 **23.** -9 **25.** 7 **27.** 0 **29.** 35 **31.** -3.8 **33.** -8.1 **35.** $-\frac{1}{5}$ **37.** $-\frac{8}{9}$ **39.** $-\frac{3}{8}$ **41.** $-\frac{19}{24}$ **43.** $\frac{1}{24}$ **45.** 37 **47.** 50 **49.** -1409 **51.** -24 **53.** 26.9 **55.** -8 **57.** $\frac{13}{8}$ **59.** -43 **61.** $\frac{4}{3}$ **63.** 24 **65.** $\frac{3}{8}$ **67.** 0.57 **69.** 0.529 **71.** 125% **73.** 52% **75.** All positive **77.** Negative

Exercise Set 1.4, p. 69

1. -7 **3.** -4 **5.** -6 **7.** 0 **9.** -4 **11.** -7 **13.** -6 **15.** 0 **17.** 0 **19.** 14 **21.** 11 **23.** -14 **25.** 5 **27.** -7 **29.** -1 **31.** 18 **33.** -10 **35.** -3 **37.** -21 **39.** 5 **41.** -8 **43.** 12 **45.** -23 **47.** -68 **49.** -73 **51.** 116 **53.** 0 **55.** -1 **57.** $\frac{1}{12}$ **59.** $-\frac{17}{12}$ **61.** $\frac{1}{8}$ **63.** 19.9 **65.** -8.6 **67.** -0.01 **69.** -193 **71.** 500 **73.** -2.8 **75.** -3.53 **77.** $-\frac{1}{2}$ **79.** $\frac{6}{7}$ **81.** $-\frac{41}{30}$ **83.** $-\frac{2}{15}$ **85.** 37 **87.** -62 **89.** -139 **91.** 6 **93.** 107 **95.** 219 **97.** \$374 **99.** 7°F **101.** 5832 ft **103.** 125 **105.** 100.5 **107.** 0.583 **109.** $-309{,}882$ **111.** False; $3 - 0 \neq 0 - 3$ **113.** True **115.** True **117.** Up 15 points

Exercise Set 1.5, p. 77

1. -8 **3.** -48 **5.** -24 **7.** -72 **9.** 16 **11.** 42 **13.** -120 **15.** -238 **17.** 1200 **19.** 98 **21.** -72 **23.** -12.4 **25.** 30 **27.** 21.7 **29.** $-\frac{2}{5}$ **31.** $\frac{1}{12}$ **33.** -17.01 **35.** $-\frac{5}{12}$ **37.** 420 **39.** $\frac{2}{7}$ **41.** -60 **43.** 150 **45.** $-\frac{2}{45}$ **47.** 1911 **49.** 50.4 **51.** $\frac{10}{189}$ **53.** -960 **55.** 17.64 **57.** $-\frac{5}{784}$ **59.** 0 **61.** -720 **63.** $-30{,}240$ **65.** 441; -147 **67.** 20; 20 **69.** 72 **71.** 1944 **73.** -17 **75.** **(a)** One must be negative, and one must be positive. **(b)** Either or both must be zero. **(c)** Both must be negative or both must be positive.

Exercise Set 1.6, p. 83

1. -8 **3.** -14 **5.** -3 **7.** 3 **9.** -8 **11.** 2 **13.** -12 **15.** -8 **17.** Undefined **19.** $-\frac{20}{7}$ **21.** $\frac{7}{15}$ **23.** $-\frac{13}{47}$ **25.** $\frac{1}{13}$ **27.** $\frac{1}{4.3}$ **29.** -7.1 **31.** $\dfrac{q}{p}$ **33.** $4y$ **35.** $\dfrac{3b}{2a}$ **37.** $4 \cdot \left(\frac{1}{17}\right)$ **39.** $8 \cdot \left(-\frac{1}{13}\right)$ **41.** $13.9 \cdot \left(-\frac{1}{1.5}\right)$ **43.** $x \cdot y$ **45.** $(3x + 4) \cdot \left(\dfrac{1}{5}\right)$ **47.** $(5a - b)\left(\dfrac{1}{5a + b}\right)$ **49.** $-\frac{9}{8}$ **51.** $\frac{5}{3}$ **53.** $\frac{9}{14}$ **55.** $\frac{9}{64}$ **57.** -2 **59.** $\frac{11}{13}$ **61.** -16.2 **63.** Undefined **65.** $\frac{22}{39}$ **67.** 33 **69.** 87.5% **71.** $-0.\overline{095238}$ **73.** No real numbers **75.** Negative **77.** Positive **79.** Negative

Exercise Set 1.7, p. 93

1. $\dfrac{3y}{5y}$ **3.** $\dfrac{10x}{15x}$ **5.** $-\dfrac{3}{2}$ **7.** $-\dfrac{7}{6}$ **9.** $8 + y$ **11.** nm **13.** $xy + 9$, or $9 + yx$ **15.** $c + ab$, or $ba + c$ **17.** $(a + b) + 2$ **19.** $8(xy)$ **21.** $a + (b + 3)$ **23.** $(3a)b$ **25.** $2 + (b + a)$, $(2 + a) + b$, $(b + 2) + a$; answers may vary **27.** $(5 + w) + v$, $(v + 5) + w$, $(w + v) + 5$; answers may vary **29.** $(3x)y$, $y(x \cdot 3)$, $3(yx)$; answers may vary **31.** $a(7b)$, $b(7a)$, $(7b)a$; answers may vary **33.** $2b + 10$ **35.** $7 + 7t$ **37.** $30x + 12$ **39.** $7x + 28 + 42y$ **41.** 7 **43.** 12 **45.** -12.71 **47.** $7x - 14$ **49.** $-7y + 14$ **51.** $45x + 54y - 72$ **53.** $-4x + 12y + 8z$ **55.** $-3.72x + 9.92y - 3.41$ **57.** $4x$, $3z$ **59.** $7x$, $8y$, $-9z$ **61.** $2(x + 2)$ **63.** $5(6 + y)$ **65.** $7(2x + 3y)$ **67.** $5(x + 2 + 3y)$ **69.** $8(x - 3)$ **71.** $4(8 - y)$ **73.** $2(4x + 5y - 11)$ **75.** $a(x - 1)$ **77.** $a(x - y - z)$ **79.** $6(3x - 2y + 1)$ **81.** $a(3x - 2y + 1)$ **83.** $19a$ **85.** $9a$ **87.** $8x + 9z$ **89.** $7x + 15y^2$ **91.** $-19a + 88$ **93.** $4t + 6y - 4$ **95.** b **97.** $\frac{13}{4}y$ **99.** $8x$ **101.** $5n$ **103.** $-16y$ **105.** $17a - 12b - 1$ **107.** $4x + 2y$ **109.** $7x + y$ **111.** $0.8x + 0.5y$ **113.** $\frac{3}{5}x + \frac{3}{5}y$ **115.** $\frac{89}{48}$ **117.** 144 **119.** $-\frac{5}{24}$ **121.** Not equivalent; $3 \cdot 2 + 5 \neq 3 \cdot 5 + 2$ **123.** Equivalent; commutative law of addition **125.** $q(1 + r + rs + rst)$

Exercise Set 1.8, p. 101

1. $-2x - 7$ **3.** $-5x + 8$ **5.** $-4a + 3b - 7c$ **7.** $-6x + 8y - 5$ **9.** $-3x + 5y + 6$ **11.** $8x + 6y + 43$ **13.** $5x - 3$ **15.** $-3a + 9$ **17.** $5x - 6$ **19.** $-19x + 2y$

21. $9y - 25z$ **23.** $-7x + 10y$ **25.** $37a - 23b + 35c$
27. 7 **29.** -40 **31.** 19 **33.** $12x + 30$ **35.** $3x + 30$
37. $9x - 18$ **39.** $-4x - 64$ **41.** -7 **43.** -7 **45.** -16
47. -334 **49.** 14 **51.** 1880 **53.** 12 **55.** 8 **57.** -86
59. 37 **61.** -1 **63.** -10 **65.** 25 **67.** -7988
69. -3000 **71.** 60 **73.** 1 **75.** 10 **77.** $-\frac{13}{45}$ **79.** $-\frac{23}{18}$
81. -118 **83.** $2 \cdot 2 \cdot 59$ **85.** $\frac{8}{5}$ **87.** 100
89. $6y - (-2x + 3a - c)$ **91.** $6m - (-3n + 5m - 4b)$
93. $-2x - f$

Summary and Review: Chapter 1, p. 107

1. [1.1a] 4 **2.** [1.1b] 19%x, or $0.19x$ **3.** [1.2a] -45, 72
4. [1.2e] 38
5. [1.2b]

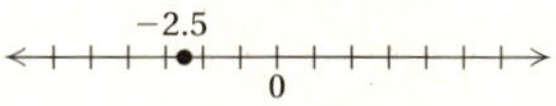

6. [1.2b]

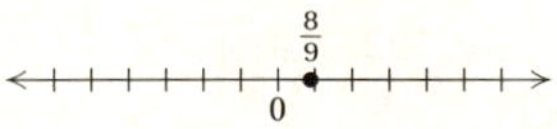

7. [1.2d] $<$ **8.** [1.2d] $>$ **9.** [1.2d] $>$ **10.** [1.2d] $<$
11. [1.3b] -3.8 **12.** [1.3b] $\frac{3}{4}$ **13.** [1.6b] $\frac{8}{3}$ **14.** [1.6b] $-\frac{1}{7}$
15. [1.3b] 34 **16.** [1.3b] 5 **17.** [1.3a] -3 **18.** [1.3a] -4
19. [1.3a] -5 **20.** [1.4a] 4 **21.** [1.4a] $-\frac{7}{5}$
22. [1.4a] -7.9 **23.** [1.5a] 54 **24.** [1.5a] -9.18
25. [1.5a] $-\frac{2}{7}$ **26.** [1.5a] -210 **27.** [1.6a] -7
28. [1.6c] -3 **29.** [1.6c] $\frac{3}{4}$ **30.** [1.8d] 40.4
31. [1.8d] -2 **32.** [1.4b] 8-yd gain **33.** [1.4b] $-\$130$
34. [1.7c] $15x - 35$ **35.** [1.7c] $-8x + 10$
36. [1.7c] $4x + 15$ **37.** [1.7c] $-24 + 48x$
38. [1.7d] $2(x - 7)$ **39.** [1.7d] $6(x - 1)$
40. [1.7d] $5(x + 2)$ **41.** [1.7d] $3(4 - x)$
42. [1.7e] $7a - 3b$ **43.** [1.7e] $-2x + 5y$
44. [1.7e] $5x - y$ **45.** [1.7e] $-a + 8b$
46. [1.8b] $-3a + 9$ **47.** [1.8b] $-2b + 21$ **48.** [1.8c] 6
49. [1.8c] $12y - 34$ **50.** [1.8c] $5x + 24$
51. [1.8c] $-15x + 25$ **52.** [1.2d] True **53.** [1.2d] False
54. [1.2d] $x > -3$ **55.** [R.2c] $\frac{55}{42}$ **56.** [R.5c] $\frac{109}{18}$
57. [R.1a] $2 \cdot 2 \cdot 2 \cdot 3 \cdot 3 \cdot 3 \cdot 3$ **58.** [R.4d] 62.5%
59. [R.4a] 0.0567 **60.** [R.1b] 270 **61.** [1.8d] $-\frac{5}{8}$
62. [1.8d] -2.1

Test: Chapter 1, p. 109

1. [1.1a] 6 **2.** [1.1b] $x - 9$ **3.** [1.1a] 240 ft^2
4. [1.2d] $<$ **5.** [1.2d] $>$ **6.** [1.2d] $>$ **7.** [1.2d] $<$
8. [1.2e] 7 **9.** [1.2e] $\frac{9}{4}$ **10.** [1.2e] 2.7 **11.** [1.3b] $-\frac{2}{3}$
12. [1.3b] 1.4 **13.** [1.3b] 8 **14.** [1.6b] $-\frac{1}{2}$ **15.** [1.6b] $\frac{7}{4}$
16. [1.4a] 7.8 **17.** [1.3a] -8 **18.** [1.3a] $\frac{7}{40}$ **19.** [1.4a] 10
20. [1.4a] -2.5 **21.** [1.4a] $\frac{7}{8}$ **22.** [1.5a] -48
23. [1.5a] $\frac{3}{16}$ **24.** [1.6a] -9 **25.** [1.6c] $\frac{3}{4}$
26. [1.6c] -9.728 **27.** [1.8d] -173 **28.** [1.4b] \$148
29. [1.7c] $18 - 3x$ **30.** [1.7c] $-5y + 5$
31. [1.7d] $2(6 - 11x)$ **32.** [1.7d] $7(x + 3 + 2y)$
33. [1.4a] 12 **34.** [1.8b] $2x + 7$ **35.** [1.8b] $9a - 12b - 7$
36. [1.8c] $68y - 8$ **37.** [1.8d] -4 **38.** [1.8d] 448
39. [1.2d] $-2 \geq x$ **40.** [R.5b] 1.728 **41.** [R.4d] 12.5%
42. [R.1a] $2 \cdot 2 \cdot 2 \cdot 5 \cdot 7$ **43.** [R.1b] 240 **44.** [1.8d] 15
45. [1.8c] $4a$

CHAPTER 2

Pretest: Chapter 2, p. 112

1. [2.2a] -7 **2.** [2.3b] -1 **3.** [2.3a] 2 **4.** [2.1b] 8
5. [2.3c] -5 **6.** [2.3a] $\frac{135}{32}$ **7.** [2.3c] 1
8. [2.7d] $\{x \mid x \geq -6\}$ **9.** [2.7c] $\{y \mid y > -4\}$
10. [2.7e] $\{a \mid a > -1\}$ **11.** [2.7c] $\{x \mid x \geq 3\}$
12. [2.7d] $\{y \mid y < -\frac{9}{4}\}$ **13.** [2.6a] $G = \dfrac{P}{3K}$
14. [2.6a] $a = \dfrac{Ab + b}{3}$ **15.** [2.4a] Width: 34 in.; length: 39 in. **16.** [2.5a] \$460 **17.** [2.4a] 81, 82, 83
18. [2.8b] Numbers less than 17
19. [2.7b]

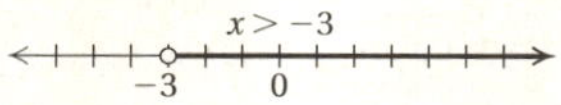

20. [2.7b]

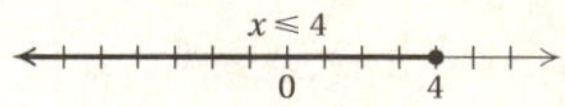

Exercise Set 2.1, p. 117

1. Yes **3.** No **5.** No **7.** Yes **9.** No **11.** No **13.** 4
15. -20 **17.** -14 **19.** -18 **21.** 15 **23.** -14 **25.** 2
27. 20 **29.** -6 **31.** $\frac{7}{3}$ **33.** $-\frac{7}{4}$ **35.** $\frac{41}{24}$ **37.** $-\frac{1}{20}$
39. 5.1 **41.** 12.4 **43.** -5 **45.** $1\frac{5}{6}$ **47.** $-\frac{10}{21}$ **49.** -11
51. $-\frac{5}{12}$ **53.** $-\frac{3}{2}$ **55.** 342.246 **57.** $-\frac{26}{15}$ **59.** -10
61. All real numbers **63.** $-\frac{5}{17}$ **65.** 13, -13

Exercise Set 2.2, p. 123

1. 6 **3.** 9 **5.** 12 **7.** -40 **9.** 1 **11.** -7 **13.** -6
15. 6 **17.** -63 **19.** 36 **21.** -21 **23.** $-\frac{3}{5}$ **25.** $-\frac{3}{2}$
27. $\frac{9}{2}$ **29.** 7 **31.** -7 **33.** 8 **35.** 15.9 **37.** $7x$
39. $8x + 11$ **41.** $x - 4$ **43.** $-10y - 42$ **45.** -8655
47. No solution **49.** No solution **51.** $x = \dfrac{b}{3a}$
53. $x = \dfrac{4b}{a}$

Exercise Set 2.3, p. 131

1. 5 **3.** 8 **5.** 10 **7.** 14 **9.** -8 **11.** -8 **13.** -7
15. 15 **17.** 6 **19.** 4 **21.** 6 **23.** -3 **25.** 1
27. -20 **29.** 6 **31.** 7 **33.** 2 **35.** 5 **37.** 2 **39.** 10
41. 4 **43.** 0 **45.** -1 **47.** $-\frac{4}{3}$ **49.** $\frac{2}{5}$ **51.** -2
53. -4 **55.** $\frac{4}{5}$ **57.** $-\frac{28}{27}$ **59.** 6 **61.** 2 **63.** 6 **65.** 8
67. 1 **69.** 17 **71.** $-\frac{5}{3}$ **73.** -3 **75.** 2 **77.** $\frac{4}{7}$
79. $-\frac{51}{31}$ **81.** 2 **83.** -6.5 **85.** $<$ **87.** -18.7
89. 4.4233464 **91.** $-\frac{7}{2}$ **93.** -2 **95.** 0 **97.** 6
99. $\frac{11}{18}$ **101.** 10

Exercise Set 2.4, p. 141

1. 7543 mi^2 **3.** 481 **5.** 667.5 hr **7.** \$3.30
9. 19 **11.** -68 **13.** 120 **15.** 30 m, 90 m, 360 m
17. 36 and 37 **19.** 56 and 58 **21.** 41, 42, 43
23. 61, 63, 65 **25.** Width: 100 ft; length: 160 ft; area: 16,000 ft^2 **27.** Width: $1\frac{3}{4}$ in.; length: $3\frac{1}{2}$ in.
29. 22.5° **31.** 450.5 mi **33.** 28°, 84°, 68°
35. **(a)** \$10.42; **(b)** 2000 **37.** 120 **39.** 19
41. $4s + 7 = 87$, 20 **43.** 5 half dollars, 10 quarters, 20 dimes, 60 nickels

Exercise Set 2.5, p. 149

1. 20% **3.** 24% **5.** 150 **7.** 2.5 **9.** 546 **11.** 125% **13.** 0.8 **15.** 5% **17.** \$282.20 **19.** 7% **21.** Approximately 86.36% **23.** \$480,000 **25.** \$34.31; \$463.17 **27.** 36 cm^3; 436 cm^3 **29.** \$1.50; \$1.62 **31.** \$32 **33.** 40%; 70%; 95% **35.** \$9.17, not \$9.10 **37.** Approximately 3.7% **39.** They are equal. In A, x is increased to $x + 0.25x = 1.25x$, then decreased to $1.25x - 0.25(1.25x) = 0.9375x$. In B, x is decreased to $x - 0.25x = 0.75x$, then increased to $0.75x + 0.25(0.75x) = 0.9375x$.

Exercise Set 2.6, p. 153

1. $b = \frac{A}{h}$ **3.** $r = \frac{d}{t}$ **5.** $P = \frac{I}{rt}$ **7.** $a = \frac{F}{m}$ **9.** $w = \frac{P - 2l}{2}$ **11.** $r^2 = \frac{A}{\pi}$ **13.** $b = \frac{2A}{h}$ **15.** $m = \frac{E}{c^2}$ **17.** $d = 2Q - c$ **19.** $b = 3A - a - c$ **21.** $y = \frac{C - Ax}{B}$ **23.** $t = \frac{3k}{v}$ **25.** $D^2 = \frac{2.5H}{N}$ **27.** $S = \frac{360A}{\pi r^2}$ **29.** $t = \frac{R - 3.85}{-0.0075}$ **31.** 0.92 **33.** -13.2 **35.** $\frac{1}{6}$ **37.** $b = \frac{2A - ah}{h}$; $h = \frac{2A}{a + b}$ **39.** $a = \frac{Q}{3 + 5c}$ **41.** A quadruples. **43.** A increases by $2h$ units.

Exercise Set 2.7, p. 161

1. **(a)** Yes; **(b)** yes; **(c)** no; **(d)** yes; **(e)** yes **3.** **(a)** No; **(b)** no; **(c)** yes; **(d)** yes; **(e)** no

5.

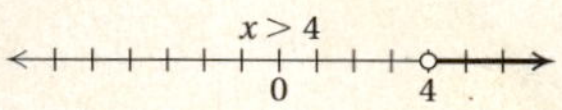

7.

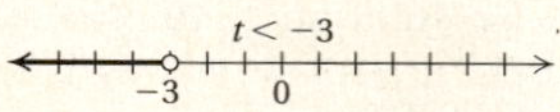

9.

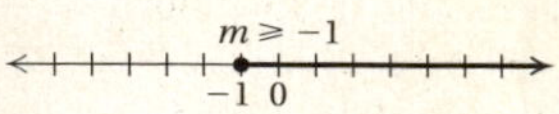

11.

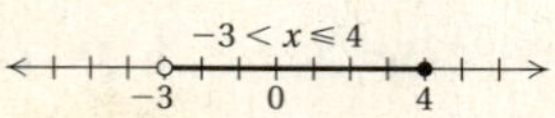

13.

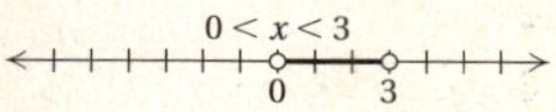

15. $\{x \mid x > -5\}$;

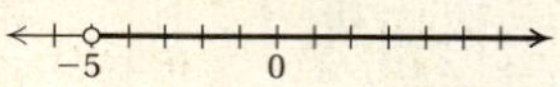

17. $\{x \mid x \le -18\}$;

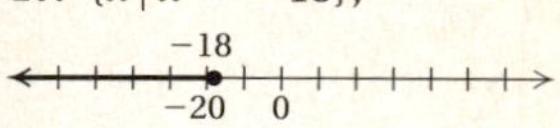

19. $\{y \mid y > -5\}$ **21.** $\{x \mid x > 2\}$ **23.** $\{x \mid x \le -3\}$ **25.** $\{x \mid x < 4\}$ **27.** $\{t \mid t > 14\}$ **29.** $\{y \mid y \le \frac{1}{4}\}$ **31.** $\{x \mid x > \frac{7}{12}\}$

33. $\{x \mid x < 7\}$;

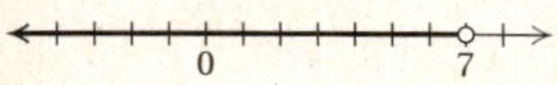

35. $\{x \mid x < 3\}$;

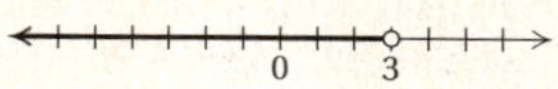

37. $\{y \mid y \ge -\frac{2}{5}\}$ **39.** $\{x \mid x \ge -6\}$ **41.** $\{y \mid y \le 4\}$ **43.** $\{x \mid x > \frac{17}{3}\}$ **45.** $\{y \mid y < -\frac{1}{14}\}$ **47.** $\{x \mid x \le \frac{3}{10}\}$ **49.** $\{x \mid x < 8\}$ **51.** $\{x \mid x \le 6\}$ **53.** $\{x \mid x < -3\}$ **55.** $\{x \mid x > -3\}$ **57.** $\{x \mid x \le 7\}$ **59.** $\{x \mid x > -10\}$ **61.** $\{y \mid y < 2\}$ **63.** $\{y \mid y \ge 3\}$ **65.** $\{y \mid y > -2\}$ **67.** $\{x \mid x > -4\}$ **69.** $\{x \mid x \le 9\}$ **71.** $\{y \mid y \le -3\}$ **73.** $\{y \mid y < 6\}$ **75.** $\{m \mid m \ge 6\}$ **77.** $\{t \mid t < -\frac{5}{3}\}$ **79.** $\{r \mid r > -3\}$ **81.** $\{x \mid x \ge -\frac{57}{34}\}$ **83.** $\{a \mid a < 1\}$ **85.** -74 **87.** $-\frac{5}{8}$ **89.** -38 **91.** -9.4 **93.** True

95.

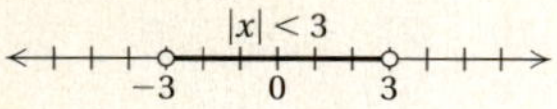

97. All real numbers **99.** $x \ge 6$

Exercise Set 2.8, p. 167

1. $x > 8$ **3.** $y \le -4$ **5.** $n \ge 1300$ **7.** $a \le 500$ **9.** $2 + 3x < 13$ **11.** $\{x \mid x \ge 97\}$ **13.** $\{Y \mid Y \ge 1935\}$ **15.** $\{d \mid d > 25\}$ **17.** $\{x \mid x > 5\}$ **19.** $\{L \mid L < \frac{43}{2}\text{ cm}\}$ **21.** $\{b \mid b > 6\text{ cm}\}$ **23.** $\{x \mid x \ge 21\}$ **25.** $\{t \mid t \le 0.75\text{ hr}\}$ **27.** $\{b \mid b \ge 25\text{ ft}\}$ **29.** 47, 49

Summary and Review: Chapter 2, p. 171

1. [2.1b] -22 **2.** [2.2a] 7 **3.** [2.2a] -192 **4.** [2.1b] 1 **5.** [2.2a] $-\frac{7}{3}$ **6.** [2.1b] 25 **7.** [2.1b] $\frac{1}{2}$ **8.** [2.2a] $-\frac{15}{64}$ **9.** [2.1b] 9.99 **10.** [2.1b] -8 **11.** [2.3b] -5 **12.** [2.3b] $-\frac{1}{3}$ **13.** [2.3a] 4 **14.** [2.3b] 3 **15.** [2.3b] 4 **16.** [2.3b] 16 **17.** [2.3c] 6 **18.** [2.3c] -3 **19.** [2.3c] 12 **20.** [2.3c] 4 **21.** [2.7a] Yes **22.** [2.7a] No **23.** [2.7a] Yes **24.** [2.7c] $\{y \mid y \ge -\frac{1}{2}\}$ **25.** [2.7d] $\{x \mid x \ge 7\}$ **26.** [2.7e] $\{y \mid y > 2\}$ **27.** [2.7e] $\{y \mid y \le -4\}$ **28.** [2.7e] $\{x \mid x < -11\}$ **29.** [2.7d] $\{y \mid y > -7\}$ **30.** [2.7e] $\{x \mid x > -6\}$ **31.** [2.7e] $\{x \mid x > -\frac{9}{11}\}$ **32.** [2.7d] $\{y \mid y \le 7\}$ **33.** [2.7d] $\{x \mid x \ge -\frac{1}{12}\}$

34. [2.7b, e]

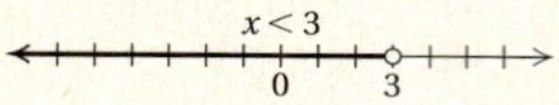

35. [2.7b]

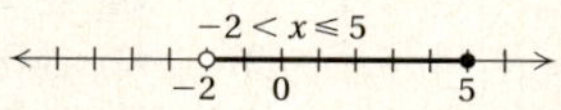

36. [2.7b]

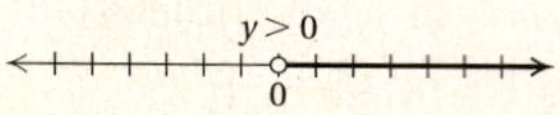

37. [2.6a] $d = \frac{C}{\pi}$ **38.** [2.6a] $B = \frac{3V}{h}$ **39.** [2.6a] $a = 2A - b$ **40.** [2.4a] \$2117 **41.** [2.4a] 27 **42.** [2.4a] 3 m, 5 m **43.** [2.4a] 9 **44.** [2.4a] 57, 59 **45.** [2.4a] Width: 11 cm; length: 17 cm **46.** [2.5a] \$220 **47.** [2.5a] \$26,087 **48.** [2.4a] 35°, 85°, 60° **49.** [2.8b] 86 **50.** [2.8b] $\{w \mid w > 17\text{ cm}\}$ **51.** [R.3a] $1.41\overline{6}$ **52.** [R.3b] 2.3 **53.** [1.3a] -45 **54.** [1.8b] $-43x + 8y$ **55.** [2.5a] \$14,150 **56.** [2.4a] Amazon: 6437 km; Nile: 6671 km **57.** [1.2e], [2.3a] 23, -23 **58.** [1.2e], [2.2a] 20, -20 **59.** [2.6a] $a = \frac{y - 3}{2 - b}$

Test: Chapter 2, p. 173

1. [2.1b] 8 **2.** [2.1b] 26 **3.** [2.2a] -6 **4.** [2.2a] 49 **5.** [2.3b] -12 **6.** [2.3a] 2 **7.** [2.1b] -8 **8.** [2.1b] $-\frac{7}{20}$ **9.** [2.3c] 7 **10.** [2.3c] $\frac{5}{3}$ **11.** [2.3b] 2.5 **12.** [2.7c] $\{x \mid x \le -4\}$ **13.** [2.7e] $\{x \mid x > -13\}$ **14.** [2.7d] $\{x \mid x \le 5\}$ **15.** [2.7d] $\{y \mid y \le -13\}$ **16.** [2.7d] $\{y \mid y \ge 8\}$ **17.** [2.7d] $\{x \mid x \le -\frac{1}{20}\}$ **18.** [2.7e] $\{x \mid x < -6\}$ **19.** [2.7e] $\{x \mid x \le -1\}$

20. [2.7b]

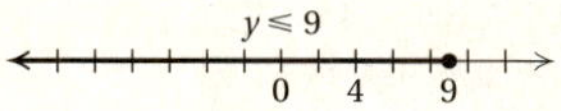

21. [2.7b, e]

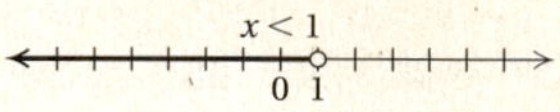

22. [2.7b]

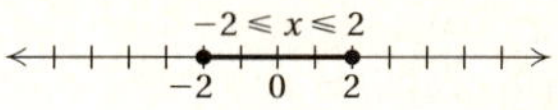

23. [2.4a] Width: 7 cm; length: 11 cm **24.** [2.4a] 6
25. [2.4a] 81, 83, 85 **26.** [2.5a] \$880 **27.** [2.6a] $r = \dfrac{A}{2\pi h}$
28. [2.6a] $l = \dfrac{2w - P}{-2}$ **29.** [2.8b] $\{x \mid x > 6\}$
30. [2.8b] $\{l \mid l \geq 174 \text{ yd}\}$ **31.** [1.3a] $-\frac{2}{9}$ **32.** [1.1a] $\frac{8}{3}$
33. [1.1b] 73%p, or 0.73p **34.** [1.8b] $-18x + 37y$
35. [2.6a] $d = \dfrac{1 - ca}{-c}$, or $\dfrac{ca - 1}{c}$ **36.** [1.2e], [2.3a] 15, −15
37. [2.4a] 60

Cumulative Review: Chapters 1–2, p. 175

1. [1.1a] $\frac{3}{2}$ **2.** [1.1a] $\frac{15}{4}$ **3.** [1.1a] 0 **4.** [1.1b] $2w - 4$
5. [1.2d] > **6.** [1.2d] > **7.** [1.2d] <
8. [1.3b], [1.6b] $-\frac{2}{5}, \frac{5}{2}$ **9.** [1.2e] 3 **10.** [1.2e] $\frac{3}{4}$
11. [1.2e] 0 **12.** [1.3a] −4.4 **13.** [1.4a] $-\frac{5}{2}$ **14.** [1.5a] $\frac{5}{6}$
15. [1.5a] −105 **16.** [1.6a] −9 **17.** [1.6c] −3
18. [1.6c] $\frac{32}{125}$ **19.** [1.7c] $15x + 25y + 10z$
20. [1.7c] $-12x - 8$ **21.** [1.7c] $-12y + 24x$
22. [1.7d] $2(32 + 9x + 12y)$ **23.** [1.7d] $8(2y - 7)$
24. [1.7d] $5(a - 3b + 5)$ **25.** [1.7e] $15b + 22y$
26. [1.7e] $4 + 9y + 6z$ **27.** [1.7e] $1 - 3a - 9d$
28. [1.7e] $-2.6x - 5.2y$ **29.** [1.8b] $-1 + 3x$
30. [1.8b] $-2x - y$ **31.** [1.8b] $-7x + 6$ **32.** [1.8b] $8x$
33. [1.8c] $5x - 13$ **34.** [2.1b] 4.5 **35.** [2.2a] $\frac{4}{25}$
36. [2.1b] 10.9 **37.** [2.1b] $3\frac{5}{6}$ **38.** [2.2a] −48
39. [2.2a] 12 **40.** [2.2a] −6.2 **41.** [2.3a] −3
42. [2.3b] $-\frac{12}{5}$ **43.** [2.3b] 8 **44.** [2.3c] 7 **45.** [2.3b] $-\frac{4}{5}$
46. [2.3b] $-\frac{10}{3}$ **47.** [2.7e] $\{x \mid x < 2\}$ **48.** [2.7e] $\{y \mid y \geq 4\}$
49. [2.7e] $\{y \mid y < -3\}$ **50.** [2.6a] $h = \dfrac{2A}{b + c}$
51. [2.6a] $q = p - 2Q$ **52.** [2.4a] 154 **53.** [2.4a] \$45
54. [2.5a] \$1050 **55.** [2.4a] 50 m, 53 m, 40 m
56. [2.8b] $\{x \mid x \geq 78\}$ **57.** [2.5a] \$24.60
58. [2.5a] \$36,000 **59.** [1.2e], [2.3a] 4, −4
60. [2.3c] All real numbers **61.** [2.3c] No solution
62. [2.3b] 3 **63.** [2.3c] All real numbers
64. [2.6a] $Q = \dfrac{2 - pm}{p}$

CHAPTER 3

Pretest: Chapter 3, p. 178

1. [3.1d] x^2 **2.** [3.1e] $\dfrac{1}{x^7}$ **3.** [3.2b] $\dfrac{16x^4}{y^6}$ **4.** [3.1f] $\dfrac{1}{p^3}$
5. [3.2c] 3.47×10^{-4} **6.** [3.2c] 3,400,000 **7.** [3.3g] 3, 2, 1, 0; 3 **8.** [3.3e] $-3a^3b - 2a^2b^2 + ab^3 + 12b^3 + 9$
9. [3.4a] $11x^2 + 4x - 11$ **10.** [3.4c] $-x^2 - 18x + 27$
11. [3.5b] $15x^4 - 20x^3 + 5x^2$ **12.** [3.6c] $x^2 + 10x + 25$
13. [3.6b] $x^2 - 25$ **14.** [3.6a] $4x^6 + 19x^3 - 30$
15. [3.7f] $4x^2 - 12xy + 9y^2$ **16.** [3.8b] $x^2 + x + 3$, R 8; or $x^2 + x + 3 + \dfrac{8}{x - 2}$

Exercise Set 3.1, p. 185

1. $3 \cdot 3 \cdot 3 \cdot 3$ **3.** $(1.1)(1.1)(1.1)(1.1)(1.1)$ **5.** $(7p)(7p)$
7. $8 \cdot k \cdot k \cdot k$ **9.** 1 **11.** b **13.** 1 **15.** ab **17.** 27
19. 19 **21.** 256 **23.** 93 **25.** 10; 4 **27.** 3629.84 ft^2
29. $\dfrac{1}{3^2} = \dfrac{1}{9}$ **31.** $\dfrac{1}{10^3} = \dfrac{1}{1000}$ **33.** $\dfrac{1}{7^3} = \dfrac{1}{343}$ **35.** $\dfrac{1}{a^3}$
37. y^4 **39.** z^n **41.** 4^{-3} **43.** x^{-3} **45.** 2^7 **47.** 8^{14}
49. x^7 **51.** 9^{38} **53.** $(3y)^{12}$ **55.** $(7y)^{17}$ **57.** 3^3 **59.** $\dfrac{1}{x}$
61. x^{17} **63.** $\dfrac{1}{x^{13}}$ **65.** 1 **67.** 7^3 **69.** 8^6 **71.** y^4
73. $\dfrac{1}{16^6}$ **75.** $\dfrac{1}{m^6}$ **77.** $\dfrac{1}{(8x)^4}$ **79.** 1 **81.** x^2 **83.** x^9
85. $\dfrac{1}{z^4}$ **87.** x^3 **89.** 1 **91.** 25, $\frac{1}{25}$, $\frac{1}{25}$, 25, −25, 25
93. 64%t, or 0.64t **95.** 64 **97.** $\frac{4}{3}$ **99.** No; $(5y)^0 = 1$, but $5y^0 = 5$. **101.** a^{2k} **103.** 2 **105.** No; for example, $(3 + 4)^2 = 49$, but $3^2 + 4^2 = 25$. **107.** < **109.** <
111. Let $x = 1$; then $\dfrac{x + 2}{2} = \dfrac{3}{2}$, but $x = 1$.

Exercise Set 3.2, p. 195

1. 2^6 **3.** $\dfrac{1}{5^6}$ **5.** x^{12} **7.** $16x^6$ **9.** $\dfrac{1}{x^{12}y^{15}}$ **11.** $x^{24}y^8$
13. $\dfrac{9x^6}{y^{16}z^6}$ **15.** $\dfrac{a^8}{b^{12}}$ **17.** $\dfrac{y^6}{4}$ **19.** $\dfrac{8}{y^6}$ **21.** $\dfrac{x^6y^3}{z^3}$
23. $\dfrac{c^2d^6}{a^4b^2}$ **25.** 2.8×10^{10} **27.** 9.07×10^{17}
29. 3.04×10^{-6} **31.** 1.8×10^{-8} **33.** 10^{11}
35. 87,400,000 **37.** 0.00000005704 **39.** 10,000,000
41. 0.00001 **43.** 6×10^9 **45.** 3.38×10^4
47. 8.1477×10^{-13} **49.** 2.5×10^{13} **51.** 5.0×10^{-4}
53. 3.0×10^{-21} **55.** 1.512×10^{10} ft^3; 1.324512×10^{14} ft^3 **57.** 5.681×10^{26} kg
59. $\$2.482 \times 10^9$ **61.** $9(x - 4)$ **63.** $\frac{7}{4}$ **65.** $-\frac{12}{7}$
67. 2.478125×10^{-1} **69.** $\frac{1}{5}$ **71.** 3^{11} **73.** a^n
75. False **77.** False

Exercise Set 3.3, p. 205

1. −18 **3.** 19 **5.** −12 **7.** 2 **9.** 4 **11.** 11
13. 144 ft **15.** 399 **17.** \$258,500 **19.** \$24,000
21. 2, $-3x$, x^2 **23.** $6x^2$ and $-3x^2$ **25.** $2x^4$ and $-3x^4$; $5x$ and $-7x$ **27.** $3x^5$ and $14x^5$; $-7x$ and $-2x$; 8 and −9
29. −3, 6 **31.** 5, 3, 3 **33.** −5, 6, −3, 8, −2 **35.** $-3x$
37. $-8x$ **39.** $11x^3 + 4$ **41.** $x^3 - x$ **43.** $4b^5$
45. $\frac{3}{4}x^5 - 2x - 42$ **47.** x^4 **49.** $\frac{15}{16}x^3 - \frac{7}{6}x^2$
51. $x^5 + 6x^3 + 2x^2 + x + 1$
53. $15y^9 + 7y^8 + 5y^3 - y^2 + y$ **55.** $x^6 + x^4$
57. $13x^3 - 9x + 8$ **59.** $-5x^2 + 9x$ **61.** $12x^4 - 2x + \frac{1}{4}$
63. 1, 0; 1 **65.** 2, 1, 0; 2 **67.** 3, 2, 1, 0; 3
69. 2, 1, 6, 4; 6

71.

TERM	COEFFICIENT	DEGREE OF TERM	DEGREE OF POLYNOMIAL
$-7x^4$	-7	4	
$6x^3$	6	3	
$-3x^2$	-3	2	4
$8x$	8	1	
-2	-2	0	

73. x^2, x **75.** x^3, x^2, x^0 **77.** None missing
79. Trinomial **81.** None of these **83.** Binomial
85. Monomial **87.** 27 **89.** $-\frac{17}{24}$ **91.** -2.6 **93.** $3x^6$
95. 10

Exercise Set 3.4, p. 213

1. $-x + 5$ **3.** $x^2 - 5x - 1$ **5.** $2x^2$
7. $5x^2 + 3x - 30$ **9.** $-2.2x^3 - 0.2x^2 - 3.8x + 23$
11. $12x^2 + 6$ **13.** $-\frac{1}{2}x^4 + \frac{2}{3}x^3 + x^2$
15. $0.01x^5 + x^4 - 0.2x^3 + 0.2x + 0.06$
17. $9x^8 + 8x^7 - 6x^4 + 8x^2 + 4$
19. $1.05x^4 + 0.36x^3 + 14.22x^2 + x + 0.97$ **21.** $-(-5x); 5x$
23. $-(-x^2 + 10x - 2); x^2 - 10x + 2$
25. $-(12x^4 - 3x^3 + 3); -12x^4 + 3x^3 - 3$ **27.** $-3x + 7$
29. $-4x^2 + 3x - 2$ **31.** $4x^4 - 6x^2 - \frac{3}{4}x + 8$
33. $7x - 1$ **35.** $-x^2 - 7x + 5$ **37.** -18
39. $6x^4 + 3x^3 - 4x^2 + 3x - 4$
41. $4.6x^3 + 9.2x^2 - 3.8x - 23$ **43.** $\frac{3}{4}x^3 - \frac{1}{2}x$
45. $0.06x^3 - 0.05x^2 + 0.01x + 1$ **47.** $3x + 6$
49. $11x^4 + 12x^3 - 9x^2 - 8x - 9$ **51.** $x^4 - x^3 + x^2 - x$
53. $5x^2 + 4x$ **55.** $\frac{23}{2}a + 10$ **57.** 6 **59.** $-\frac{7}{22}$ **61.** $\frac{39}{2}$
63. $20 + 5(m - 4) + 4(m - 5) + (m - 5)(m - 4); m^2$
65. $z^2 - 27z + 72$ **67.** $5x^2 - 9x - 1$
69. $4x^3 + 2x^2 + x + 2$

Sidelights: Factors and Sums, p. 220

First row: 90, −432, −63; second row: 7, −18, −36, −14, 12, −6, −21, −11; third row: 9, −2, −2, 10, −8, −8, −8, −10, 21; fourth row: −19, −6

Exercise Set 3.5, p. 221

1. $40x^2$ **3.** x^3 **5.** $32x^8$ **7.** $0.03x^{11}$ **9.** $\frac{1}{15}x^4$ **11.** 0
13. $-24x^{11}$ **15.** $-2x^2 + 10x$ **17.** $-5x^2 + 5x$
19. $x^5 + x^2$ **21.** $6x^3 - 18x^2 + 3x$ **23.** $-6x^4 - 6x^3$
25. $18y^6 + 24y^5$ **27.** $x^2 + 9x + 18$ **29.** $x^2 + 3x - 10$
31. $x^2 - 7x + 12$ **33.** $x^2 - 9$ **35.** $25 - 15x + 2x^2$
37. $4x^2 + 20x + 25$ **39.** $x^2 - \frac{21}{10}x - 1$ **41.** $x^3 - 1$
43. $4x^3 + 14x^2 + 8x + 1$ **45.** $3y^4 - 6y^3 - 7y^2 + 18y - 6$
47. $x^6 + 2x^5 - x^3$ **49.** $-10x^5 - 9x^4 + 7x^3 + 2x^2 - x$
51. $x^4 - x^2 - 2x - 1$ **53.** $6t^4 + t^3 - 16t^2 - 7t + 4$
55. $x^9 - x^5 + 2x^3 - x$ **57.** $x^4 - 1$ **59.** $-\frac{3}{4}$
61. $3(3x - 15y + 5)$ **63.** $78t^2 + 40t$ **65.** $A = \frac{1}{2}b^2 + 2b$
67. 0

Exercise Set 3.6, p. 229

1. $x^3 + x^2 + 3x + 3$ **3.** $x^4 + x^3 + 2x + 2$
5. $y^2 - y - 6$ **7.** $9x^2 + 12x + 4$ **9.** $5x^2 + 4x - 12$
11. $9t^2 - 1$ **13.** $4x^2 - 6x + 2$ **15.** $p^2 - \frac{1}{16}$
17. $x^2 - 0.01$ **19.** $2x^3 + 2x^2 + 6x + 6$
21. $-2x^2 - 11x + 6$ **23.** $a^2 + 14a + 49$
25. $1 - x - 6x^2$ **27.** $x^5 + 3x^3 - x^2 - 3$
29. $3x^6 - 2x^4 - 6x^2 + 4$ **31.** $6x^7 + 18x^5 + 4x^2 + 12$
33. $8x^6 + 65x^3 + 8$ **35.** $4x^3 - 12x^2 + 3x - 9$
37. $4y^6 + 4y^5 + y^4 + y^3$ **39.** $x^2 - 16$ **41.** $4x^2 - 1$
43. $25m^2 - 4$ **45.** $4x^4 - 9$ **47.** $9x^8 - 16$
49. $x^{12} - x^4$ **51.** $x^8 - 9x^2$ **53.** $x^{24} - 9$ **55.** $4y^{16} - 9$
57. $x^2 + 4x + 4$ **59.** $9x^4 + 6x^2 + 1$ **61.** $a^2 - a + \frac{1}{4}$
63. $9 + 6x + x^2$ **65.** $x^4 + 2x^2 + 1$ **67.** $4 - 12x^4 + 9x^8$
69. $25 + 60t^2 + 36t^4$ **71.** $9 - 12x^3 + 4x^6$
73. $4x^3 + 24x^2 - 12x$ **75.** $4x^4 - 2x^2 + \frac{1}{4}$ **77.** $9p^2 - 1$
79. $15t^5 - 3t^4 + 3t^3$ **81.** $36x^8 + 48x^4 + 16$
83. $12x^3 + 8x^2 + 15x + 10$ **85.** $64 - 96x^4 + 36x^8$
87. $t^3 - 1$ **89.** 25; 49 **91.** 56; 16
93. Lamps: 500 watts; air conditioner: 2000 watts; television: 50 watts **95.** $-\frac{41}{7}$ **97.** $8y^3 + 72y^2 + 160y$
99. $-2x^4 + x^3 + 5x^2 - x - 2$ **101.** -7
103. $V = w^3 + 3w^2 + 2w$ **105.** $V = h^3 - 3h^2 + 2h$
107. $10{,}000 - 49 = 9951$ **109.** $5a^2 + 12a - 9$
111. **(a)** $A^2 + AB$; **(b)** $AB + B^2$; **(c)** $A^2 - B^2$;
(d) $(A + B)(A - B) = A^2 - B^2$

Exercise Set 3.7, p. 237

1. -1 **3.** -7 **5.** \$11,236 **7.** \$11,910.16 **9.** 56.52 in^2
11. Coefficients: 1, −2, 3, −5; degrees: 4, 2, 2, 0; 4
13. Coefficients: 17, −3, −7; degrees: 5, 5, 0; 5
15. $-a - 2b$ **17.** $3x^2y - 2xy^2 + x^2$ **19.** $8u^2v - 5uv^2$
21. $20au + 10av$ **23.** $x^2 - 4xy + 3y^2$ **25.** $3r + 7$
27. $-x^2 - 8xy - y^2$ **29.** $2ab - 2$
31. $-2a + 10b - 5c + 8d$ **33.** $6z^2 + 7zu - 3u^2$
35. $a^4b^2 - 7a^2b + 10$
37. $a^4 + a^3 - a^2y - ay + a + y - 1$ **39.** $a^6 - b^2c^2$
41. $y^6x + y^4x + y^4 + 2y^2 + 1$ **43.** $12x^2y^2 + 2xy - 2$
45. $12 - c^2d^2 - c^4d^4$ **47.** $m^3 + m^2n - mn^2 - n^3$
49. $x^9y^9 - x^6y^6 + x^5y^5 - x^2y^2$ **51.** $x^2 + 2xh + h^2$
53. $r^6t^4 - 8r^3t^2 + 16$ **55.** $p^8 + 2m^2n^2p^4 + m^4n^4$
57. $4a^6 - 2a^3b^3 + \frac{1}{4}b^6$ **59.** $3a^3 - 12a^2b + 12ab^2$
61. $4a^2 - b^2$ **63.** $c^4 - d^2$ **65.** $a^2b^2 - c^2d^4$
67. $x^2 + 2xy + y^2 - 9$ **69.** $x^2 - y^2 - 2yz - z^2$
71. $a^2 - b^2 - 2bc - c^2$ **73.** $4xy - 4y^2$ **75.** $2xy + \pi x^2$
77. 33 **79.** $A^3 + 3A^2B + 3AB^2 + B^3$

Exercise Set 3.8, p. 245

1. $3x^4 - \frac{1}{2}x^3 + \frac{1}{8}x^2 - 2$ **3.** $1 - 2u - u^4$ **5.** $5t^2 + 8t - 2$
7. $-4x^4 + 4x^2 + 1$ **9.** $6x^2 - 10x + \frac{3}{2}$ **11.** $9x^2 - \frac{5}{2}x + 1$
13. $6x^2 + 13x + 4$ **15.** $3rs + r - 2s$ **17.** $x + 2$
19. $x - 5 + \dfrac{-50}{x - 5}$ **21.** $x - 2 + \dfrac{-2}{x + 6}$ **23.** $x - 3$
25. $x^4 - x^3 + x^2 - x + 1$ **27.** $2x^2 - 7x + 4$ **29.** $x^3 - 6$
31. $x^3 + 2x^2 + 4x + 8$ **33.** $t^2 + 1$ **35.** 6.8
37. 25,543.75 ft^2 **39.** $\frac{11}{10}$ **41.** $x^2 + 5$
43. $a + 3 + \dfrac{5}{5a^2 - 7a - 2}$ **45.** $2x^2 + x - 3$
47. $a^5 + a^4b + a^3b^2 + a^2b^3 + ab^4 + b^5$ **49.** -5 **51.** 1

Summary and Review: Chapter 3, p. 249

1. [3.1d] $\frac{1}{7^2}$ **2.** [3.1d] y^{11} **3.** [3.1d] $(3x)^{14}$ **4.** [3.1d] t^8

5. [3.1e] 4^3 **6.** [3.1e] $\frac{1}{a^3}$ **7.** [3.1b, e] 1 **8.** [3.2a, b] $9t^8$

9. [3.1d], [3.2a, b] $36x^8$ **10.** [3.2b] $\frac{y^3}{8x^3}$ **11.** [3.1f] t^{-5}

12. [3.1f] $\frac{1}{y^4}$ **13.** [3.2c] 3.28×10^{-5} **14.** [3.2c] 8,300,000
15. [3.2d] 2.09×10^4 **16.** [3.2d] 5.12×10^{-5}
17. [3.2e] 6.205×10^{10} **18.** [3.3a] 10
19. [3.3b] $-4y^5, 7y^2, -3y, -2$ **20.** [3.3h] x^2, x^0
21. [3.3g] 3, 2, 1, 0; 3 **22.** [3.3i] Binomial
23. [3.3i] None of these **24.** [3.3i] Monomial
25. [3.3f] $-2x^2 - 3x + 2$ **26.** [3.3f] $10x^4 - 7x^2 - x - \frac{1}{2}$
27. [3.4a] $x^5 - 2x^4 + 6x^3 + 3x^2 - 9$
28. [3.4a] $2x^5 - 6x^4 + 2x^3 - 2x^2 + 2$
29. [3.4c] $2x^2 - 4x - 6$ **30.** [3.4c] $x^5 - 3x^3 - x^2 + 8$
31. [3.4d], [3.5b] $P = 4w + 8$; $A = w^2 + 4w$
32. [3.6a] $x^2 + \frac{7}{6}x + \frac{1}{3}$ **33.** [3.6c] $49x^2 + 14x + 1$
34. [3.5d] $12x^3 - 23x^2 + 13x - 2$ **35.** [3.6b] $9x^4 - 16$
36. [3.5b] $15x^7 - 40x^6 + 50x^5 + 10x^4$
37. [3.6a] $x^2 - 3x - 28$ **38.** [3.6c] $9y^4 - 12y^3 + 4y^2$
39. [3.6a] $2t^4 - 11t^2 - 21$ **40.** [3.7a] 49
41. [3.7b] Coefficients: 1, −7, 9, −8; degrees: 6, 2, 2, 0; 6
42. [3.7c] $-y + 9w - 5$
43. [3.7c] $m^6 - 2m^2n + 2m^2n^2 + 8n^2m - 6m^3$
44. [3.7d] $-9xy - 2y^2$
45. [3.7e] $11x^3y^2 - 8x^2y - 6x^2 - 6x + 6$
46. [3.7f] $p^3 - q^3$ **47.** [3.7f] $9a^8 - 2a^4b^3 + \frac{1}{9}b^6$

48. [3.8a] $5x^2 - \frac{1}{2}x + 3$

49. [3.8b] $3x^2 - 7x + 4 + \frac{1}{2x + 3}$

50. [1.7d] $25(t - 2 + 4m)$ **51.** [2.3b] $\frac{9}{4}$ **52.** [1.4a] −11.2
53. [2.4a] Width: 125.5 m; length: 144.5 m
54. [3.1d], [3.2a, b], [3.3e] $-28x^8$ **55.** [2.3b], [3.6a] $\frac{94}{13}$

Test: Chapter 3, p. 251

1. [3.1d] $\frac{1}{6^5}$ **2.** [3.1d] x^9 **3.** [3.1d] $(4a)^{11}$ **4.** [3.1e] 3^3

5. [3.1e] $\frac{1}{x^5}$ **6.** [3.1b, e] 1 **7.** [3.2a] x^6

8. [3.2a, b] $-27y^6$ **9.** [3.2a, b] $16a^{12}b^4$ **10.** [3.2b] $\frac{a^3b^3}{c^3}$

11. [3.1d], [3.2a, b] $-216x^{21}$ **12.** [3.1d], [3.2a, b] $-24x^{21}$
13. [3.1d], [3.2a, b] $162x^{10}$ **14.** [3.1d], [3.2a, b] $324x^{10}$

15. [3.1f] $\frac{1}{5^3}$ **16.** [3.1f] y^{-8} **17.** [3.2c] 3.9×10^9

18. [3.2c] 0.00000005 **19.** [3.2d] 1.75×10^{17}
20. [3.2d] 1.296×10^{22}
21. [3.2e] Approximately 2.38×10^2 **22.** [3.3a] −43
23. [3.3d] $\frac{1}{3}$, −1, 7 **24.** [3.3g] 3, 0, 1, 6; 6
25. [3.3i] Binomial **26.** [3.3e] $5a^2 - 6$
27. [3.3e] $\frac{7}{4}y^2 - 4y$ **28.** [3.3f] $x^5 + 2x^3 + 4x^2 - 8x + 3$
29. [3.4a] $4x^5 + x^4 + 2x^3 - 8x^2 + 2x - 7$
30. [3.4a] $5x^4 + 5x^2 + x + 5$
31. [3.4c] $-4x^4 + x^3 - 8x - 3$
32. [3.4c] $-x^5 + 0.7x^3 - 0.8x^2 - 21$
33. [3.5b] $-12x^4 + 9x^3 + 15x^2$ **34.** [3.6c] $x^2 - \frac{2}{3}x + \frac{1}{9}$
35. [3.6b] $9x^2 - 100$ **36.** [3.6a] $3b^2 - 4b - 15$
37. [3.6a] $x^{14} - 4x^8 + 4x^6 - 16$
38. [3.6a] $48 + 34y - 5y^2$
39. [3.5d] $6x^3 - 7x^2 - 11x - 3$
40. [3.6c] $25t^2 + 20t + 4$
41. [3.7c] $-5x^3y - y^3 + xy^3 - x^2y^2 + 19$
42. [3.7e] $8a^2b^2 + 6ab - 4b^3 + 6ab^2 + ab^3$
43. [3.7f] $9x^{10} - 16y^{10}$ **44.** [3.8a] $4x^2 + 3x - 5$
45. [3.8b] $2x^2 - 4x - 2 + \frac{17}{3x + 2}$ **46.** [2.3b] 13
47. [1.7d] $16(4t - 2m + 1)$ **48.** [1.4a] $\frac{23}{20}$
49. [2.4a] 100°, 25°, 55°
50. [3.5b], [3.6a] $V = l^3 - 3l^2 + 2l$ **51.** [2.3b], [3.6a] $-\frac{61}{12}$

Cumulative Review: Chapters 1–3, p. 253

1. [1.1a] $\frac{5}{2}$ **2.** [3.3a] −4 **3.** [3.7a] −14 **4.** [1.2e] 4
5. [1.6b] $\frac{1}{5}$ **6.** [1.3a] $-\frac{11}{60}$ **7.** [1.4a] 4.2 **8.** [1.5a] 7.28
9. [1.6c] $-\frac{5}{12}$ **10.** [3.2d] 2.2×10^{22} **11.** [3.2d] 4×10^{-5}
12. [1.7a] −3 **13.** [1.8b] $-2y - 7$ **14.** [1.8c] $5x + 11$
15. [1.8d] −2 **16.** [3.4a] $2x^5 - 2x^4 + 3x^3 + 2$
17. [3.7d] $3x^2 + xy - 2y^2$ **18.** [3.4c] $x^3 + 5x^2 - x - 7$
19. [3.4c] $-\frac{1}{3}x^2 - \frac{3}{4}x$ **20.** [1.7c] $12x - 15y + 21$
21. [3.5a] $6x^8$ **22.** [3.5b] $2x^5 - 4x^4 + 8x^3 - 10x^2$
23. [3.5d] $3y^4 + 5y^3 - 10y - 12$ **24.** [3.5d] $2p^4 + 3p^3q + 2p^2q^2 - 2p^4q - p^3q^2 - p^2q^3 + pq^3$
25. [3.6a] $6x^2 + 13x + 6$ **26.** [3.6c] $9x^4 + 6x^2 + 1$
27. [3.6b] $t^2 - \frac{1}{4}$ **28.** [3.6b] $4y^4 - 25$
29. [3.6a] $4x^6 + 6x^4 - 6x^2 - 9$ **30.** [3.6c] $t^2 - 4t^3 + 4t^4$
31. [3.7f] $15p^2 - pq - 2q^2$ **32.** [3.8a] $6x^2 + 2x - 3$
33. [3.8b] $3x^2 - 2x - 7$ **34.** [2.1b] −1.2 **35.** [2.2a] −21
36. [2.3a] 9 **37.** [2.2a] $-\frac{20}{3}$ **38.** [2.3b] 2 **39.** [2.1b] $\frac{13}{8}$
40. [2.3c] $-\frac{17}{21}$ **41.** [2.3b] −17 **42.** [2.3b] 2
43. [2.7e] $\{x \mid x < 16\}$ **44.** [2.7e] $\left\{x \mid x \le -\frac{11}{8}\right\}$
45. [2.6a] $h = \frac{A - \pi r^2}{2\pi r}$ **46.** [3.4d] $\pi r^2 - 18$
47. [2.4a] 18 and 19 **48.** [2.4a] 20 ft, 24 ft
49. [2.4a] 10° **50.** [2.4a] −45 **51.** [2.5a] \$3.50
52. [3.1d] y^4 **53.** [3.1e] $\frac{1}{x}$ **54.** [3.2a, b] $-\frac{27x^9}{y^6}$
55. [3.1d, e] x^3 **56.** [3.3d] $\frac{2}{3}$, 4, −6 **57.** [3.3g] 4, 2, 1, 0; 4
58. [3.3i] Binomial **59.** [3.3i] Trinomial
60. [3.4d] $4x - 4$
61. [3.1d], [3.2a, b], [3.4a] $12x^5 - 15x^4 - 27x^3 + 4x^2$
62. [3.4a], [3.6c] $5x^2 - 2x + 10$
63. [3.4a], [3.8b] $4x^2 - 2x + 7$ **64.** [2.3b], [3.6a] $\frac{11}{7}$
65. [2.3b], [3.8b] 1 **66.** [1.2e], [2.3a] −5, 5
67. [2.3b], [3.6a] No solution
68. [2.3b], [3.6a], [3.8b] All real numbers except 5

Chapter 4

Pretest: Chapter 4, p. 256

1. [4.1a] $4(-5x^6)$, $(-2x^3)(10x^3)$, $x^2(-20x^4)$; answers may vary **2.** [4.5b] $2(x+1)^2$ **3.** [4.2a] $(x+4)(x+2)$
4. [4.1b] $4a(2a^4 + a^2 - 5)$
5. [4.3a], [4.4a] $(5x+2)(x-3)$
6. [4.5d] $(9+z^2)(3+z)(3-z)$ **7.** [4.5b] $(y^3-2)^2$
8. [4.1c] $(x^2+4)(3x+2)$ **9.** [4.2a] $(p-6)(p+5)$
10. [4.5d] $(x^2y+8)(x^2y-8)$
11. [4.3a], [4.4a] $(2p-q)(p+4q)$ **12.** [4.7b] 0, 5
13. [4.7a] 4, $\frac{3}{5}$ **14.** [4.7b] $\frac{2}{3}$, -4 **15.** [4.8a] 6, -1
16. [4.8a] Base: 8 cm; height: 11 cm

Exercise Set 4.1, p. 261

1. $(4x^2)(2x)$, $(-8)(-x^3)$, $(2x^2)(4x)$; answers may vary
3. $(-5a^5)(2a)$, $(10a^3)(-a^3)$, $(-2a^2)(5a^4)$; answers may vary
5. $(8x^2)(3x^2)$, $(-8x^2)(-3x^2)$, $(4x^3)(6x)$; answers may vary
7. $x(x-6)$ **9.** $2x(x+3)$ **11.** $x^2(x+6)$
13. $8x^2(x^2-3)$ **15.** $2(x^2+x-4)$
17. $17xy(x^4y^2 + 2x^2y + 3)$ **19.** $x^2(6x^2 - 10x + 3)$
21. $x^2y^2(x^3y^3 + x^2y + xy - 1)$
23. $2x^3(x^4 - x^3 - 32x^2 + 2)$
25. $0.8x(2x^3 - 3x^2 + 4x + 8)$ **27.** $\frac{1}{3}x^3(5x^3 + 4x^2 + x + 1)$
29. $(x^2+2)(x+3)$ **31.** $(x^2+2)(x+3)$
33. $(2x^2+1)(x+3)$ **35.** $(4x^2+3)(2x-3)$
37. $(4x^2+1)(3x-4)$ **39.** $(5x^2-1)(x-1)$
41. $(x^2-3)(x+8)$ **43.** $(2x^2-9)(x-4)$
45. $\{x \mid x > -24\}$ **47.** 27 **49.** $y^2 + 12y + 35$
51. $y^2 - 49$ **53.** $(2x^3+3)(2x^2+3)$
55. $(x^7+1)(x^5+1)$ **57.** Not factorable by grouping

Exercise Set 4.2, p. 267

1. $(x+3)(x+5)$ **3.** $(x+3)(x+4)$ **5.** $(x-3)^2$
7. $(x+2)(x+7)$ **9.** $(b+1)(b+4)$ **11.** $\left(x+\frac{1}{3}\right)^2$
13. $(d-2)(d-5)$ **15.** $(y-1)(y-10)$
17. $(x-6)(x+7)$ **19.** $(x-9)(x+2)$
21. $x(x-8)(x+2)$ **23.** $(y-9)(y+5)$
25. $(x-11)(x+9)$ **27.** $(c^2+8)(c^2-7)$
29. $(a^2+7)(a^2-5)$ **31.** Not factorable
33. Not factorable **35.** $(x+10)^2$ **37.** $(x-25)(x+4)$
39. $(x-24)(x+3)$ **41.** $(x-9)(x-16)$
43. $(a+12)(a-11)$ **45.** $(x-15)(x-8)$
47. $(12+x)(9-x)$, or $-(x+12)(x-9)$
49. $(y-0.4)(y+0.2)$ **51.** $(p+5q)(p-2q)$
53. $(m+4n)(m+n)$ **55.** $(s+3t)(s-5t)$
57. $16x^3 - 48x^2 + 8x$ **59.** $49w^2 + 84w + 36$
61. $16w^2 - 121$ **63.** 15, -15, 27, -27, 51, -51
65. $\left(x+\frac{1}{4}\right)\left(x-\frac{3}{4}\right)$ **67.** $(x+5)\left(x-\frac{5}{7}\right)$
69. $(b^n+5)(b^n+2)$ **71.** $2x^2(4-\pi)$

Exercise Set 4.3, p. 273

1. $(2x+1)(x-4)$ **3.** $(5x+9)(x-2)$
5. $(3x+1)(2x+7)$ **7.** $(3x+1)(x+1)$
9. $(2x-3)(2x+5)$ **11.** $(2x+1)(x-1)$
13. $(3x-2)(3x+8)$ **15.** $(3x+1)(x-2)$
17. $(3x+4)(4x+5)$ **19.** $(7x-1)(2x+3)$
21. $(3x+2)(3x+4)$ **23.** $(3x-7)^2$
25. $(24x-1)(x+2)$ **27.** $(5x-11)(7x+4)$
29. $2(5-x)(2+x)$ **31.** $4(3x-2)(x+3)$
33. $6(5x-9)(x+1)$ **35.** $2(3x+5)(x-1)$
37. $(3x-1)(x-1)$ **39.** $4(3x+2)(x-3)$
41. $(2x+1)(x-1)$ **43.** $(3x+2)(3x-8)$
45. $5(3x+1)(x-2)$ **47.** $x(3x+4)(4x+5)$
49. $x^2(7x-1)(2x+3)$ **51.** $3x(8x-1)(7x-1)$
53. $(5x^2-3)(3x^2-2)$ **55.** $(5t+8)^2$
57. $2x(3x+5)(x-1)$ **59.** Not factorable
61. Not factorable **63.** $(4m+5n)(3m-4n)$
65. $(2a+3b)(3a-5b)$ **67.** $(3a+2b)(3a+4b)$
69. $(5p+2q)(7p+4q)$ **71.** $6(3x-4y)(x+y)$
73. $q = \dfrac{A+7}{p}$ **75.** $y = \dfrac{6-3x}{2}$ **77.** $\{x \mid x > 4\}$
79. $(2x^n+1)(10x^n+3)$ **81.** $(x^{3a}-1)(3x^{3a}+1)$

Exercise Set 4.4, p. 277

1. $(x+7)(x+2)$ **3.** $(x-1)(x-4)$ **5.** $(2x+3)(3x+2)$
7. $(x-4)(3x-4)$ **9.** $(5x+3)(7x-8)$
11. $(2x-3)(2x+3)$ **13.** $(2x^2+5)(x^2+3)$
15. $(2x+1)(x-4)$ **17.** $(3x-5)(x+3)$
19. $(2x+7)(3x+1)$ **21.** $(3x+1)(x+1)$
23. $(2x-3)(2x+5)$ **25.** $(2x-1)(x+1)$
27. $(3x+2)(3x-8)$ **29.** $(3x-1)(x+2)$
31. $(3x-4)(4x-5)$ **33.** $(7x-1)(2x+3)$
35. $(3x+2)(3x+4)$ **37.** $(3x-7)^2$
39. $(24x-1)(x+2)$ **41.** $x^3(5x-11)(7x+4)$
43. $6x(5-x)(2+x)$ **45.** $\{x \mid x \le 8\}$ **47.** $\left\{x \mid x \ge \frac{20}{3}\right\}$
49. $\left\{x \mid x > \frac{26}{7}\right\}$ **51.** $(3x^5-2)^2$ **53.** $(4x^5+1)^2$

Exercise Set 4.5, p. 285

1. Yes **3.** No **5.** No **7.** No **9.** $(x-7)^2$
11. $(x+8)^2$ **13.** $(x-1)^2$ **15.** $(x+2)^2$ **17.** $(q^2-3)^2$
19. $(4y+7)^2$ **21.** $2(x-1)^2$ **23.** $x(x-9)^2$
25. $3(2q-3)^2$ **27.** $(7-3x)^2$ **29.** $5(y^2+1)^2$
31. $(1+2x^2)^2$ **33.** $(2p+3q)^2$ **35.** $(a-3b)^2$
37. $(9a-b)^2$ **39.** $4(3a+4b)^2$ **41.** Yes **43.** No
45. No **47.** Yes **49.** $(y+2)(y-2)$
51. $(p+3)(p-3)$ **53.** $(t+7)(t-7)$
55. $(a+b)(a-b)$ **57.** $(5t+m)(5t-m)$
59. $(10+k)(10-k)$ **61.** $(4a+3)(4a-3)$
63. $(2x+5y)(2x-5y)$ **65.** $2(2x+7)(2x-7)$
67. $x(6+7x)(6-7x)$ **69.** $(7a^2+9)(7a^2-9)$
71. $(a^2+4)(a+2)(a-2)$ **73.** $5(x^2+9)(x+3)(x-3)$
75. $(1+y^4)(1+y^2)(1+y)(1-y)$
77. $(x^6+4)(x^3+2)(x^3-2)$ **79.** $\left(y+\frac{1}{4}\right)\left(y-\frac{1}{4}\right)$
81. $\left(5+\frac{1}{7}x\right)\left(5-\frac{1}{7}x\right)$ **83.** $(4m^2+t^2)(2m+t)(2m-t)$
85. -11 **87.** $-\frac{5}{6}$ **89.** 2 **91.** Not factorable
93. $(x+11)^2$ **95.** $2x(3x+1)^2$
97. $(x^4+2^4)(x^2+2^2)(x+2)(x-2)$ **99.** $3x^3(x+2)(x-2)$
101. $2x\left(3x+\frac{2}{5}\right)\left(3x-\frac{2}{5}\right)$ **103.** $p(0.7+p)(0.7-p)$
105. $(0.8x+1.1)(0.8x-1.1)$ **107.** $x(x+6)$
109. $\left(x+\frac{1}{x}\right)\left(x-\frac{1}{x}\right)$ **111.** $(9+b^{2k})(3-b^k)(3+b^k)$
113. $(3b^n+2)^2$ **115.** $(y+4)^2$ **117.** 9

Exercise Set 4.6, p. 293

1. $3(x + 8)(x - 8)$ **3.** $(a - 5)^2$ **5.** $(2x - 3)(x - 4)$ **7.** $x(x + 12)^2$ **9.** $(x + 2)(x - 2)(x + 3)$ **11.** $3(4x + 1)(4x - 1)$ **13.** $3x(3x - 5)(x + 3)$ **15.** Not factorable **17.** $x(x^2 + 7)(x - 3)$ **19.** $x^3(x - 7)^2$ **21.** $2(2 - x)(5 + x)$, or $-2(x - 2)(x + 5)$ **23.** Not factorable **25.** $4(x^2 + 4)(x + 2)(x - 2)$ **27.** $(1 + y^4)(1 + y^2)(1 + y)(1 - y)$ **29.** $x^3(x - 3)(x - 1)$ **31.** $\frac{1}{9}\left(\frac{1}{3}x^3 - 4\right)^2$ **33.** $m(x^2 + y^2)$ **35.** $9xy(xy - 4)$ **37.** $2\pi r(h + r)$ **39.** $(a + b)(2x + 1)$ **41.** $(x + 1)(x - 1 - y)$ **43.** $(n + p)(n + 2)$ **45.** $(3q + p)(2q - 1)$ **47.** $(2b - a)^2$, or $(a - 2b)^2$ **49.** $(4x + 3y)^2$ **51.** $(7m^2 - 8n)^2$ **53.** $(y^2 + 5z^2)^2$ **55.** $\left(\frac{1}{2}a + \frac{1}{3}b\right)^2$ **57.** $(a + b)(a - 2b)$ **59.** $(m + 20n)(m - 18n)$ **61.** $(mn - 8)(mn + 4)$ **63.** $b^4(ab + 8)(ab - 4)$ **65.** $a^3(a - b)(a + 5b)$ **67.** $\left(a + \frac{1}{5}b\right)\left(a - \frac{1}{5}b\right)$ **69.** $(a^2 - bc)(a^2 + 2bc)$ **71.** $(x - y)(x + y)$ **73.** $(4 + p^2q^2)(2 + pq)(2 - pq)$ **75.** $(1 + 4x^6y^6)(1 + 2x^3y^3)(1 - 2x^3y^3)$ **77.** $(q + 1)(q - 1)(q + 8)$ **79.** $-\frac{14}{11}$ **81.** $X = \dfrac{A + 7}{a + b}$ **83.** $(a + 1)^2(a - 1)^2$ **85.** $(3.5x - 1)^2$ **87.** $(5x + 4)(x + 1.8)$ **89.** $(y + 3)(y - 3)(y - 2)$ **91.** $(a^2 + 1)(a + 4)$ **93.** $(x + 3)(x - 3)(x^2 + 2)$ **95.** $(x + 2)(x - 2)(x - 1)$ **97.** $(y - 1)^3$ **99.** $(y + 4 + x)^2$

Exercise Set 4.7, p. 301

1. $-4, -9$ **3.** $-3, 8$ **5.** $-12, 11$ **7.** $0, -3$ **9.** $0, -18$ **11.** $-\frac{5}{2}, -4$ **13.** $-\frac{1}{5}, 3$ **15.** $4, \frac{1}{4}$ **17.** $0, \frac{2}{3}$ **19.** $0, 18$ **21.** $-\frac{1}{10}, \frac{1}{27}$ **23.** $\frac{1}{3}, -20$ **25.** $0, \frac{2}{3}, \frac{1}{2}$ **27.** $-1, -5$ **29.** $-9, 2$ **31.** $3, 5$ **33.** $0, 8$ **35.** $0, -18$ **37.** $4, -4$ **39.** $-\frac{2}{3}, \frac{2}{3}$ **41.** -3 **43.** 4 **45.** $0, \frac{6}{5}$ **47.** $\frac{5}{3}, -1$ **49.** $\frac{2}{3}, -\frac{1}{4}$ **51.** $\frac{2}{3}, -1$ **53.** $\frac{7}{10}, -\frac{7}{10}$ **55.** $9, -2$ **57.** $\frac{4}{5}, \frac{3}{2}$ **59.** $(a + b)^2$ **61.** -16 **63.** $-\frac{10}{3}$ **65.** $4, -5$ **67.** $9, -3$ **69.** $\frac{1}{8}, -\frac{1}{8}$ **71.** $4, -4$ **73.** **(a)** $x^2 - x - 12 = 0$; **(b)** $x^2 + 7x + 12 = 0$; **(c)** $4x^2 - 4x + 1 = 0$; **(d)** $x^2 - 25 = 0$; **(e)** $40x^3 - 14x^2 + x = 0$

Exercise Set 4.8, p. 307

1. 5 and -5 **3.** 3 and 5 **5.** 14 and 15 **7.** 12 and 14; -12 and -14 **9.** 15 and 17; -15 and -17 **11.** Length: 12 cm; width: 7 cm **13.** 5 **15.** Height: 6 cm; base: 5 cm **17.** 6 km **19.** 5 and 7 **21.** 182 **23.** 12 **25.** 4950 **27.** 25 **29.** Hypotenuse: 17 ft; leg: 15 ft **31.** 5 ft **33.** 37 **35.** 30 cm by 15 cm

Summary and Review: Chapter 4, p. 313

1. [4.1a] $(-10x)(x)$; $(-5x)(2x)$; $(5x)(-2x)$; answers may vary **2.** [4.1a] $(6x)(6x^4)$; $(4x^2)(9x^3)$; $(-2x^4)(-18x)$; answers may vary **3.** [4.5d] $5(1 + 2x^3)(1 - 2x^3)$ **4.** [4.1b] $x(x - 3)$ **5.** [4.5d] $(3x + 2)(3x - 2)$ **6.** [4.2a] $(x + 6)(x - 2)$ **7.** [4.5b] $(x + 7)^2$ **8.** [4.1b] $3x(2x^2 + 4x + 1)$ **9.** [4.1c] $(x^2 + 3)(x + 1)$ **10.** [4.3a], [4.4a] $(3x - 1)(2x - 1)$ **11.** [4.5d] $(x^2 + 9)(x + 3)(x - 3)$ **12.** [4.3a], [4.4a] $3x(3x - 5)(x + 3)$ **13.** [4.5d] $2(x + 5)(x - 5)$ **14.** [4.1c] $(x^3 - 2)(x + 4)$ **15.** [4.5d] $(4x^2 + 1)(2x + 1)(2x - 1)$ **16.** [4.1b] $4x^4(2x^2 - 8x + 1)$ **17.** [4.5b] $3(2x + 5)^2$ **18.** [4.5c] Not factorable **19.** [4.2a] $x(x - 6)(x + 5)$ **20.** [4.5d] $(2x + 5)(2x - 5)$ **21.** [4.5b] $(3x - 5)^2$ **22.** [4.3a], [4.4a] $2(3x + 4)(x - 6)$ **23.** [4.5b] $(x - 3)^2$ **24.** [4.3a], [4.4a] $(2x + 1)(x - 4)$ **25.** [4.5b] $2(3x - 1)^2$ **26.** [4.5d] $3(x + 3)(x - 3)$ **27.** [4.2a] $(x - 5)(x - 3)$ **28.** [4.5b] $(5x - 2)^2$ **29.** [4.5b] $(7b^5 - 2a^4)^2$ **30.** [4.2a] $(xy + 4)(xy - 3)$ **31.** [4.5b] $3(2a + 7b)^2$ **32.** [4.1c] $(m + t)(m + 5)$ **33.** [4.5d] $32(x^2 - 2y^2z^2)(x^2 + 2y^2z^2)$ **34.** [4.7a] $1, -3$ **35.** [4.7b] $-7, 5$ **36.** [4.7b] $-4, 3$ **37.** [4.7b] $\frac{2}{3}, 1$ **38.** [4.7b] $\frac{3}{2}, -4$ **39.** [4.7b] $8, -2$ **40.** [4.8a] 3 and -2 **41.** [4.8a] -18 and -16; 16 and 18 **42.** [4.8a] -19 and -17; 17 and 19 **43.** [4.8a] $\frac{5}{2}$ and -2 **44.** [4.8a] Length: 19 ft; width: 12 ft **45.** [4.8a] Base: 18 m; height: 24 m **46.** [1.6c] $\frac{8}{35}$ **47.** [2.7e] $\left\{x \,|\, x \le \frac{4}{3}\right\}$ **48.** [2.6a] $b = \dfrac{A - a}{2}$ **49.** [3.6d] $4a^2 + 12a + 9$ **50.** [3.6d] $4a^2 - 9$ **51.** [3.6d] $10a^2 - a - 21$ **52.** [4.8a] 2.5 cm **53.** [4.8a] 0, 2 **54.** [4.8a] Length: 12; width: 6 **55.** [4.7b] No solution **56.** [4.7a] $2, -3, \frac{5}{2}$ **57.** [4.6a] a, i; b, k; c, g; d, h; e, j; f, l

Test: Chapter 4, p. 315

1. [4.1a] $(4x)(x^2)$; $(2x^2)(2x)$; $(-2x)(-2x^2)$; answers may vary **2.** [4.2a] $(x - 5)(x - 2)$ **3.** [4.5b] $(x - 5)^2$ **4.** [4.1b] $2y^2(2y^2 - 4y + 3)$ **5.** [4.1c] $(x^2 + 2)(x + 1)$ **6.** [4.1b] $x(x - 5)$ **7.** [4.2a] $x(x + 3)(x - 1)$ **8.** [4.3a], [4.4a] $2(5x - 6)(x + 4)$ **9.** [4.5d] $(2x + 3)(2x - 3)$ **10.** [4.2a] $(x - 4)(x + 3)$ **11.** [4.3a], [4.4a] $3m(2m + 1)(m + 1)$ **12.** [4.5d] $3(w + 5)(w - 5)$ **13.** [4.5b] $5(3x + 2)^2$ **14.** [4.5d] $3(x^2 + 4)(x + 2)(x - 2)$ **15.** [4.5b] $(7x - 6)^2$ **16.** [4.3a], [4.4a] $(5x - 1)(x - 5)$ **17.** [4.1c] $(x^3 - 3)(x + 2)$ **18.** [4.5d] $5(4 + x^2)(2 + x)(2 - x)$ **19.** [4.3a], [4.4a] $(2x + 3)(2x - 5)$ **20.** [4.3a], [4.4a] $3t(2t + 5)(t - 1)$ **21.** [4.2a] $3(m + 2n)(m - 5n)$ **22.** [4.7b] $5, -4$ **23.** [4.7b] $\frac{3}{2}, -5$ **24.** [4.7b] $7, -4$ **25.** [4.8a] $8, -3$ **26.** [4.8a] Length: 10 m; width: 4 m **27.** [1.6c] $-\frac{10}{11}$ **28.** [2.7e] $\left\{x \,|\, x < \frac{19}{3}\right\}$ **29.** [2.6a] $T = \dfrac{I}{PR}$ **30.** [3.6d] $25x^4 - 70x^2 + 49$ **31.** [4.8a] Length: 15; width: 3 **32.** [4.2a] $(a - 4)(a + 8)$ **33.** [4.5d], [4.7a] (c) **34.** [3.6b], [4.5d] (d)

Cumulative Review, Chapters 1–4, p. 317

1. [1.2d] $<$ **2.** [1.2d] $>$ **3.** [1.4a] 0.35 **4.** [1.6c] -1.57 **5.** [1.5a] $-\frac{1}{14}$ **6.** [1.6c] $-\frac{6}{5}$ **7.** [1.8c] $4x + 1$ **8.** [1.8d] -8 **9.** [3.2a, b] $\dfrac{8x^6}{y^3}$ **10.** [3.1d, e] $-\dfrac{1}{6x^3}$ **11.** [3.4a] $x^4 - 3x^3 - 3x^2 - 4$

12. [3.7e] $2x^2y^2 - x^2y - xy$
13. [3.8b] $x^2 + 3x + 2 + \dfrac{3}{x - 1}$ 14. [3.6c] $4t^2 - 12t + 9$
15. [3.6b] $x^4 - 9$ 16. [3.6a] $6x^2 + 4x - 16$
17. [3.5b] $2x^4 + 6x^3 + 8x^2$
18. [3.5d] $4y^3 + 4y^2 + 5y - 4$ 19. [3.6b] $x^2 - \frac{4}{9}$
20. [4.2a] $(x + 4)(x - 2)$ 21. [4.5d] $(2x + 5)(2x - 5)$
22. [4.1c] $(3x - 4)(x^2 + 1)$ 23. [4.5b] $(x - 13)^2$
24. [4.5d] $3(5x + 6y)(5x - 6y)$
25. [4.3a], [4.4a] $(3x + 7)(2x - 9)$
26. [4.2a] $(x^2 - 3)(x^2 + 1)$
27. [4.6a] $2(y - 1)(y + 1)(2y - 3)$
28. [4.3a], [4.4a] $(3p - q)(2p + q)$
29. [4.3a], [4.4a] $2x(5x + 1)(x + 5)$ 30. [4.5b] $x(7x - 3)^2$
31. [4.3a], [4.4a] Not factorable 32. [4.1b] $3x(25x^2 + 9)$
33. [4.5d] $3(x^4 + 4y^4)(x^2 + 2y^2)(x^2 - 2y^2)$
34. [4.2a] $14(x + 2)(x + 1)$
35. [4.6a] $(x^3 + 1)(x + 1)(x - 1)$ 36. [2.3b] 15
37. [2.7e] $\{y \mid y < 6\}$ 38. [4.7a] 15, $-\frac{1}{4}$ 39. [4.7a] 0, -37
40. [4.7b] 5, -5, -1 41. [4.7b] 6, -6 42. [4.7b] $\frac{1}{3}$
43. [4.7b] -10, -7 44. [4.7b] 0, $\frac{3}{2}$ 45. [2.3a] 0.2
46. [4.7b] -4, 5 47. [2.7e] $\{x \mid x \le 20\}$
48. [2.3c] All real numbers 49. [2.6a] $m = \dfrac{y - b}{x}$
50. [2.4a] 50, 52 51. [4.8a] -20 and -18; 18 and 20
52. [4.8a] 6 ft, 3 ft 53. [2.4a] 150 m by 350 m
54. [2.5a] \$6500 55. [4.8a] 17 m
56. [2.4a] 30 m, 60 m, 10 m 57. [2.5a] \$29
58. [4.8a] 18 cm, 16 cm 59. [2.7e], [3.6a] $\left\{x \mid x \ge -\frac{13}{3}\right\}$
60. [2.3b] 22 61. [4.7b] -6, 4
62. [4.6a] $(x - 2)(x + 1)(x - 3)$
63. [4.6a] $(2a + 3b + 3)(2a - 3b - 5)$
64. [4.5a] 25 65. [4.8a] 2 cm

CHAPTER 5

Pretest: Chapter 5, p. 320

1. [5.3c] $(x + 2)(x + 3)^2$ 2. [5.4a] $\dfrac{-b - 1}{b^2 - 4}$, or $\dfrac{b + 1}{4 - b^2}$
3. [5.5a] $\dfrac{1}{y - 2}$ 4. [5.4a] $\dfrac{7a + 6}{a(a + 2)}$ 5. [5.5b] $\dfrac{2x}{x + 1}$
6. [5.1d] $\dfrac{2(x - 3)}{x - 2}$ 7. [5.2b] $\dfrac{x - 3}{x + 3}$ 8. [5.9a] $\dfrac{y + x}{y - x}$
9. [5.6a] -5 10. [5.6a] 0 11. [5.8a] $M = \dfrac{3R}{a - b}$
12. [5.7b] 10.5 hr 13. [5.7a] $\frac{30}{11}$ hr
14. [5.7a] 60 mph, 80 mph

Exercise Set 5.1, p. 327

1. 0 3. 8 5. $-\frac{5}{2}$ 7. 7, -4 9. 5, -5 11. None
13. $\dfrac{(4x)(3x^2)}{(4x)(5y)}$ 15. $\dfrac{2x(x - 1)}{2x(x + 4)}$ 17. $\dfrac{-1(3 - x)}{-1(4 - x)}$
19. $\dfrac{(y + 6)(y - 7)}{(y + 6)(y + 2)}$ 21. $\dfrac{x^2}{4}$ 23. $\dfrac{8p^2q}{3}$ 25. $\dfrac{x - 3}{x}$
27. $\dfrac{m + 1}{2m + 3}$ 29. $\dfrac{a - 3}{a + 2}$ 31. $\dfrac{a - 3}{a - 4}$ 33. $\dfrac{x + 5}{x - 5}$
35. $a + 1$ 37. $\dfrac{x^2 + 1}{x + 1}$ 39. $\dfrac{3}{2}$ 41. $\dfrac{6}{t - 3}$ 43. $\dfrac{t + 2}{2(t - 4)}$
45. $\dfrac{t - 2}{t + 2}$ 47. -1 49. -1 51. -6 53. $-x - 1$
55. $\dfrac{56x}{3}$ 57. $\dfrac{2}{dc^2}$ 59. $\dfrac{x + 2}{x - 2}$ 61. $\dfrac{(a + 3)(a - 3)}{a(a + 4)}$
63. $\dfrac{2a}{a - 2}$ 65. $\dfrac{(t + 2)(t - 2)}{(t + 1)(t - 1)}$ 67. $\dfrac{x + 4}{x + 2}$ 69. $\dfrac{5(a + 6)}{a - 1}$
71. 18 and 20; -18 and -20 73. $(2y^2 + 1)(y - 5)$
75. $(a - 8)^2$ 77. $x + 2y$ 79. $\dfrac{(t - 9)^2(t - 1)}{(t^2 + 9)(t + 1)}$
81. $\dfrac{x - y}{x - 5y}$

Exercise Set 5.2, p. 333

1. $\dfrac{x}{4}$ 3. $\dfrac{1}{x^2 - y^2}$ 5. $\dfrac{x^2 - 4x + 7}{x^2 + 2x - 5}$ 7. $\dfrac{3}{10}$ 9. $\dfrac{1}{4}$
11. $\dfrac{b}{a}$ 13. $\dfrac{(a + 2)(a + 3)}{(a - 3)(a - 1)}$ 15. $\dfrac{(x - 1)^2}{x}$ 17. $\dfrac{1}{2}$
19. $\dfrac{15}{8}$ 21. $\dfrac{15}{4}$ 23. $\dfrac{a - 5}{3(a - 1)}$ 25. $\dfrac{(x + 2)^2}{x}$
27. $\dfrac{3}{2}$ 29. $\dfrac{c + 1}{c - 1}$ 31. $\dfrac{y - 3}{2y - 1}$ 33. $\dfrac{x + 1}{x - 1}$
35. 4 37. $\dfrac{4y^8}{x^6}$ 39. $\dfrac{4x^6}{y^{10}}$ 41. $-\dfrac{1}{b^2}$ 43. $\dfrac{x(x^2 + 1)}{3(x + y - 1)}$

Exercise Set 5.3, p. 337

1. 108 3. 72 5. 126 7. 360 9. 500 11. $\frac{65}{72}$
13. $\frac{29}{120}$ 15. $\frac{23}{180}$ 17. $12x^3$ 19. $18x^2y^2$ 21. $6(y - 3)$
23. $t(t + 2)(t - 2)$ 25. $(x + 2)(x - 2)(x + 3)$
27. $t(t + 2)^2(t - 4)$ 29. $(a + 1)(a - 1)^2$
31. $(m - 3)(m - 2)^2$ 33. $(2 + 3x)(2 - 3x)$
35. $10v(v + 4)(v + 3)$ 37. $18x^3(x - 2)^2(x + 1)$
39. $6x^3(x + 2)^2(x - 2)$ 41. $(x - 3)^2$ 43. $(x + 3)(x - 3)$
45. $(x + 3)^2$ 47. 1440 49. 24 min

Exercise Set 5.4, p. 343

1. 1 3. $\dfrac{6}{3 + x}$ 5. $\dfrac{2x + 3}{x - 5}$ 7. $\dfrac{1}{4}$ 9. $-\dfrac{1}{t}$
11. $\dfrac{-x + 7}{x - 6}$ 13. $y + 3$ 15. $\dfrac{2b - 14}{b^2 - 16}$ 17. $a + b$
19. $\dfrac{5x + 2}{x - 5}$ 21. -1 23. $\dfrac{-x^2 + 9x - 14}{(x - 3)(x + 3)}$ 25. $\dfrac{2x + 5}{x^2}$
27. $\dfrac{41}{24r}$ 29. $\dfrac{4x + 6y}{x^2y^2}$ 31. $\dfrac{4 + 3t}{18t^3}$ 33. $\dfrac{x^2 + 4xy + y^2}{x^2y^2}$
35. $\dfrac{6x}{(x - 2)(x + 2)}$ 37. $\dfrac{11x + 2}{3x(x + 1)}$ 39. $\dfrac{x^2 + 6x}{(x + 4)(x - 4)}$
41. $\dfrac{6}{z + 4}$ 43. $\dfrac{3x - 1}{(x - 1)^2}$ 45. $\dfrac{11a}{10(a - 2)}$
47. $\dfrac{2x^2 + 8x + 16}{x(x + 4)}$ 49. $\dfrac{7a + 6}{(a - 2)(a + 1)(a + 3)}$
51. $\dfrac{2x^2 - 4x + 34}{(x - 5)(x + 3)}$ 53. $\dfrac{3a + 2}{(a + 1)(a - 1)}$
55. $\dfrac{2x + 6y}{(x + y)(x - y)}$ 57. $\dfrac{a^2 + 7a + 1}{(a + 5)(a - 5)}$

59. $\dfrac{5t-12}{(t+3)(t-3)(t-2)}$ **61.** x^2-1 **63.** $\dfrac{1}{8x^{12}y^9}$

65. $\dfrac{1}{x^{12}y^{21}}$ **67.** Perimeter: $\dfrac{16y+28}{15}$; area: $\dfrac{y^2+2y-8}{15}$

69. $\dfrac{(z+6)(2z-3)}{(z+2)(z-2)}$ **71.** $\dfrac{11z^4-22z^2+6}{(z^2+2)(z^2-2)(2z^2-3)}$

Exercise Set 5.5, p. 351

1. $\dfrac{4}{x}$ **3.** 1 **5.** $\dfrac{1}{x-1}$ **7.** $\dfrac{8}{3}$ **9.** $\dfrac{13}{a}$ **11.** $\dfrac{8}{y-1}$

13. $\dfrac{x-2}{x-7}$ **15.** $\dfrac{4}{a^2-25}$ **17.** $\dfrac{2x-4}{x-9}$ **19.** $\dfrac{-9}{2x-3}$

21. $\dfrac{-a-4}{10}$ **23.** $\dfrac{7z-12}{12z}$ **25.** $\dfrac{4x^2-13xt+9t^2}{3x^2t^2}$

27. $\dfrac{2x-40}{(x+5)(x-5)}$ **29.** $\dfrac{3-5t}{2t(t-1)}$ **31.** $\dfrac{2s-st-s^2}{(t+s)(t-s)}$

33. $\dfrac{y-19}{4y}$ **35.** $\dfrac{-2a^2}{(x+a)(x-a)}$ **37.** $\dfrac{9x+12}{(x+3)(x-3)}$

39. $\dfrac{1}{2}$ **41.** $\dfrac{x-3}{(x+3)(x+1)}$ **43.** $\dfrac{18x+5}{x-1}$ **45.** 0

47. $\dfrac{20}{2y-1}$ **49.** $\dfrac{2a-3}{2-a}$ **51.** $\dfrac{z-3}{2z-1}$ **53.** $\dfrac{2}{x+y}$

55. x^5 **57.** $\dfrac{b^{20}}{a^8}$ **59.** $\dfrac{6}{x^3}$ **61.** $\dfrac{1-3x}{(2x-3)(x+1)}$

Exercise Set 5.6, p. 359

1. $\frac{47}{2}$ **3.** -6 **5.** $\frac{24}{7}$ **7.** $-4, -1$ **9.** $4, -4$ **11.** 3
13. $\frac{14}{3}$ **15.** 5 **17.** 5 **19.** $\frac{5}{2}$ **21.** -2 **23.** $\frac{17}{2}$
25. No solution **27.** -5 **29.** $\frac{5}{3}$ **31.** $\frac{1}{2}$
33. No solution **35.** No solution **37.** $\dfrac{1}{a^6b^{15}}$ **39.** $\dfrac{16x^4}{t^8}$
41. $32x^6$ **43.** 7 **45.** No solution **47.** $2, -2$ **49.** 4

Exercise Set 5.7, p. 369

1. $\frac{24}{7}$ **3.** 20 and 15
5. 30 km/h, 70 km/h

	DISTANCE	SPEED	TIME	
Car	150	r	t	→ $150 = r(t)$
Truck	350	$r + 40$	t	→ $350 = (r+40)t$

7. Passenger: 80 mph; freight: 66 mph

	DISTANCE	SPEED	TIME	
Freight	330	$r - 14$	t	→ $330 = (r-14)t$
Passenger	400	r	t	→ $400 = r(t)$

9. 20 mph
11. $6\frac{6}{7}$ hr **13.** $5\frac{1}{7}$ hr **15.** 9 **17.** 2.3 km/h **19.** 66 g
21. 702 km **23.** 1.92 g **25.** 10,000
27. **(a)** 4.8 tons; **(b)** 48 lb **29.** 11 **31.** $9\frac{3}{13}$ days **33.** $\frac{3}{4}$
35. $\dfrac{A}{B}=\dfrac{C}{D}$; $\dfrac{A}{C}=\dfrac{B}{D}$; $\dfrac{D}{B}=\dfrac{C}{A}$; $\dfrac{D}{C}=\dfrac{B}{A}$

37. Ann: 6 hr; Betty: 12 hr **39.** $27\frac{3}{11}$ min

Exercise Set 5.8, p. 375

1. $r=\dfrac{S}{2\pi h}$ **3.** $b=\dfrac{2A}{h}$ **5.** $n=\dfrac{S+360}{180}$

7. $b=\dfrac{3V-kB-4kM}{k}$ **9.** $r=\dfrac{S-a}{S-l}$ **11.** $h=\dfrac{2A}{b_1+b_2}$

13. $B=\dfrac{A}{AQ+1}$ **15.** $p=\dfrac{qf}{q-f}$ **17.** $A=P(1+r)$

19. $R=\dfrac{r_1r_2}{r_1+r_2}$ **21.** $D=\dfrac{BC}{A}$ **23.** $h_2=\dfrac{p(h_1-q)}{q}$

25. $a=\dfrac{b}{K-C}$ **27.** $23x^4+50x^3+23x^2-163x+41$

29. $3(2y^2-1)(5y^2+4)$ **31.** $(y+7)(y-5)$

33. $T=\dfrac{FP}{u+EF}$ **35.** $-40°$

Exercise Set 5.9, p. 381

1. $\dfrac{25}{4}$ **3.** $\dfrac{1}{3}$ **5.** -6 **7.** $\dfrac{1+3x}{1-5x}$ **9.** $\dfrac{2x+1}{x}$ **11.** 8

13. $x-8$ **15.** $\dfrac{y}{y-1}$ **17.** $-\dfrac{1}{a}$ **19.** $\dfrac{ab}{b-a}$

21. $\dfrac{p^2+q^2}{q+p}$ **23.** $4x^4+3x^3+2x-7$ **25.** $(p-5)^2$

27. $50(p^2-2)$ **29.** $\dfrac{(x-1)(3x-2)}{5x-3}$ **31.** $-\dfrac{ac}{bd}$

33. $\dfrac{5x+3}{3x+2}$

Summary and Review: Chapter 5, p. 385

1. [5.1a] 0 **2.** [5.1a] 6 **3.** [5.1a] $6, -6$ **4.** [5.1a] $-6, 5$
5. [5.1a] -2 **6.** [5.1a] 0, 3, 5 **7.** [5.1c] $\dfrac{x-2}{x+1}$

8. [5.1c] $\dfrac{7x+3}{x-3}$ **9.** [5.1c] $\dfrac{y-5}{y+5}$ **10.** [5.1d] $\dfrac{a-6}{5}$

11. [5.1d] $\dfrac{6}{2t-1}$ **12.** [5.2b] $-20t$ **13.** [5.2b] $\dfrac{2x^2-2x}{x+1}$

14. [5.3c] $30x^2y^2$ **15.** [5.3c] $4(a-2)$

16. [5.3c] $(y-2)(y+2)(y+1)$ **17.** [5.4a] $\dfrac{-3x+18}{x+7}$

18. [5.4a] -1 **19.** [5.5a] $\dfrac{4}{x-4}$ **20.** [5.5a] $\dfrac{x+5}{2x}$

21. [5.5a] $\dfrac{2x+3}{x-2}$ **22.** [5.4a] $\dfrac{2a}{a-1}$ **23.** [5.4a] $d+c$

24. [5.5a] $\dfrac{-x^2+x+26}{(x-5)(x+5)(x+1)}$ **25.** [5.5b] $\dfrac{2(x-2)}{x+2}$

26. [5.9a] $\dfrac{z}{1-z}$ **27.** [5.9a] $c-d$ **28.** [5.6a] 8

29. [5.6a] $3, -5$ **30.** [5.7a] $5\frac{1}{7}$ hr **31.** [5.7a] 240 km/h, 280 km/h **32.** [5.7a] -2 **33.** [5.7b] 160

34. [5.7a] 95 mph, 175 mph **35.** [5.8a] $s=\dfrac{rt}{r-t}$

36. [5.8a] $C=\frac{5}{9}(F-32)$, or $C=\dfrac{5F-160}{9}$

37. [4.6a] $(5x^2 - 3)(x + 4)$ **38.** [3.2a, b] $\frac{1}{125x^9y^6}$
39. [3.4c] $-2x^3 + 3x^2 + 12x - 18$
40. [4.8a] Length: 5 cm; width: 3 cm; perimeter: 16 cm
41. [5.1d], [5.2b] $\frac{5(a + 3)^2}{a}$ **42.** [5.5a] $\frac{10a}{(a - b)(b - c)}$
43. **(a)** [5.4a] $\frac{10x + 6}{(x - 3)(x + 3)}$; **(b)** [5.6a] $\frac{1}{10}$;
(c) [5.4a], [5.6a] In part (a), the LCM is used to find an equivalent expression for each rational expression with the LCM as the least common denominator. In part (b), the LCM is used to clear fractions.

Test: Chapter 5, p. 387

1. [5.1a] 0 **2.** [5.1a] -8 **3.** [5.1a] 7, -7 **4.** [5.1a] 1, 2
5. [5.1a] 1 **6.** [5.1a] 0, -3, -5 **7.** [5.1c] $\frac{3x + 7}{x + 3}$
8. [5.1d] $\frac{a + 5}{2}$ **9.** [5.2b] $\frac{(5x + 1)(x + 1)}{3x(x + 2)}$
10. [5.3c] $(y - 3)(y + 3)(y + 7)$ **11.** [5.4a] $\frac{23 - 3x}{x^3}$
12. [5.5a] $\frac{8 - 2t}{t^2 + 1}$ **13.** [5.4a] $\frac{-3}{x - 3}$ **14.** [5.5a] $\frac{2x - 5}{x - 3}$
15. [5.4a] $\frac{8t - 3}{t(t - 1)}$ **16.** [5.5a] $\frac{-x^2 - 7x - 15}{(x + 4)(x - 4)(x + 1)}$
17. [5.5b] $\frac{x^2 + 2x - 7}{(x - 1)^2(x + 1)}$ **18.** [5.9a] $\frac{3y + 1}{y}$ **19.** [5.6a] 12
20. [5.6a] 5, -3 **21.** [5.7a] 4 **22.** [5.7b] 16
23. [5.7a] 45 mph, 65 mph **24.** [5.8a] $t = \frac{g}{M - L}$
25. [4.6a] $(4a + 7)(4a - 7)$ **26.** [3.2a, b] $\frac{y^{12}}{81x^8}$
27. [3.4c] $13x^2 - 29x + 76$
28. [4.8a] 21 and 22; -22 and -21
29. [5.7a] Team A: 4 hr; team B: 10 hr **30.** [5.9a] $\frac{3a + 2}{2a + 1}$

Cumulative Review: Chapters 1–5, p. 389

1. [1.1a] $-\frac{9}{5}$ **2.** [3.1c] 12 **3.** [1.8c] $2x + 6$
4. [3.1d], [3.2a, b] $\frac{27x^7}{4}$ **5.** [3.1e] $\frac{4x^{10}}{3}$ **6.** [5.1c] $\frac{2(t - 3)}{2t - 1}$
7. [5.9a] $\frac{(x + 2)^2}{x^2}$ **8.** [5.1c] $\frac{a + 4}{a - 4}$ **9.** [1.3a] $\frac{17}{42}$
10. [5.4a] $\frac{x^2 + 4xy + y^2}{x^2y^2}$ **11.** [5.4a] $\frac{3z - 2}{z^2 - 1}$
12. [3.4a] $2x^4 + 8x^3 - 2x + 9$ **13.** [1.4a] 2.33
14. [3.7e] $-xy$ **15.** [5.5a] $\frac{1}{x - 3}$
16. [5.5a] $\frac{2x^2 - 14x - 16}{(x + 4)(x - 5)^2}$ **17.** [1.5a] -1.3
18. [3.5b] $6x^4 + 12x^3 - 15x^2$ **19.** [3.6b] $9t^2 - \frac{1}{4}$
20. [3.6c] $4p^2 - 4pq + q^2$ **21.** [3.6a] $3x^2 - 7x - 20$
22. [3.6b] $4x^4 - 1$ **23.** [5.1d] $\frac{2(t - 1)}{t}$
24. [5.1d] $-\frac{2(a + 1)}{a}$ **25.** [3.8b] $3x^2 - 4x + 5$
26. [1.6c] $-\frac{9}{20}$ **27.** [5.2b] $\frac{x - 2}{2x(x - 3)}$ **28.** [5.2b] $-\frac{12}{x}$
29. [4.6a] $(2x + 3)(2x - 3)(x + 3)$
30. [4.2a] $(x + 8)(x - 1)$ **31.** [4.3a], [4.4a] $(3x + 1)(x - 5)$
32. [4.5b] $(4y + 5x)^2$ **33.** [4.2a] $3x(x + 5)(x + 3)$
34. [4.5d] $2(x + 1)(x - 1)$ **35.** [4.5b] $(x - 14)^2$
36. [4.1b, c] $2(y^2 + 3)(2y + 5)$ **37.** [2.3c] -7
38. [4.7a] 0, $-\frac{4}{3}$ **39.** [4.7b] 0, 8 **40.** [4.7b] 4
41. [2.7e] $\{x \mid x \geq -9\}$ **42.** [4.7b] 3, -3 **43.** [2.3b] $-\frac{11}{7}$
44. [5.6a] 3, -3 **45.** [5.6a] $-\frac{1}{11}$ **46.** [4.7a] 0, $\frac{1}{10}$
47. [5.6a] No solution **48.** [5.6a] $\frac{5}{7}$
49. [5.8a] $z = \frac{xy}{x + y}$ **50.** [2.6a] $N = \frac{DT}{3}$
51. [2.4a] 32, 33, 34 **52.** [5.7a] 12 km/h, 10 km/h
53. [5.7a] $\frac{30}{11}$ hr **54.** [2.4a] 34 and 35
55. [4.8a] 16 and 17 **56.** [5.7b] $1\frac{2}{3}$ cups **57.** [4.8a] 7
58. [4.8a] 9 and 11 **59.** [4.7b] 4, $-\frac{2}{3}$
60. [4.7b], [5.6a] 2, -1 **61.** [4.7b], [5.6a], [5.9a] 1, -4
62. [3.1d, e], [4.7b] 5, -5
63. [5.2a], [5.4a] $\frac{-x^2 - x + 6}{2x^2 + x + 5}$ **64.** [3.2d] 5×10^7

CHAPTER 6

Pretest: Chapter 6, p. 392

1. [6.2b]

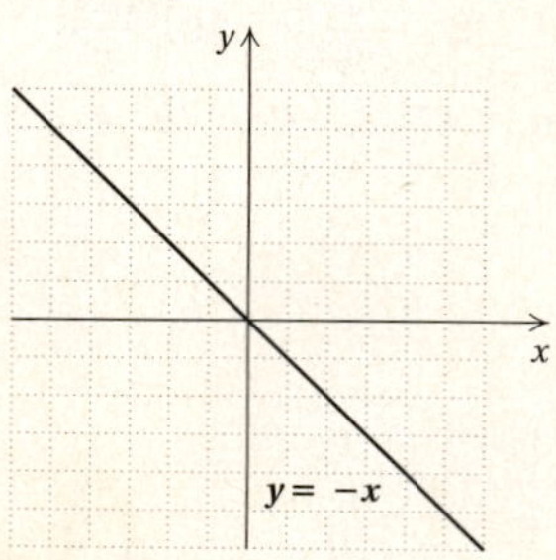

2. [6.3b]

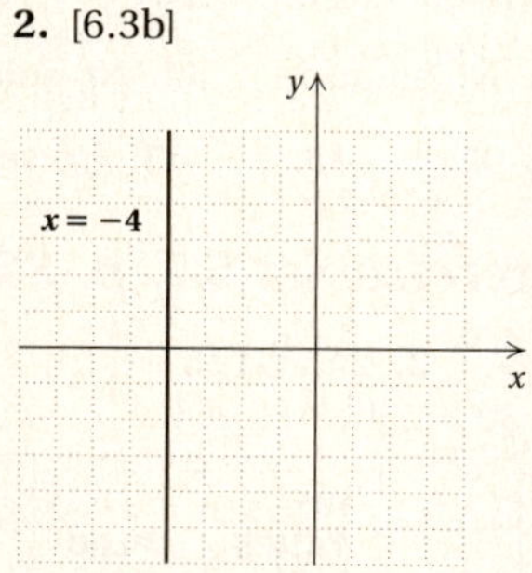

3. [6.3a]

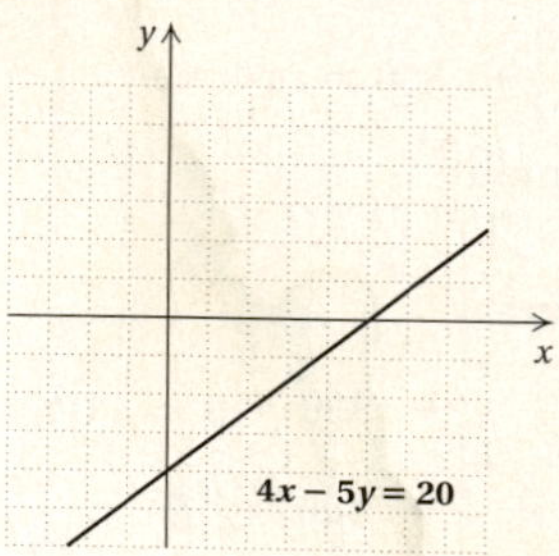

4. [6.2b]

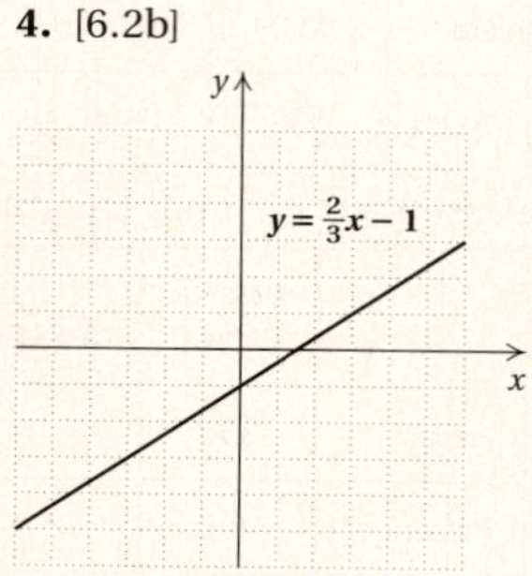

5. [6.1c] III **6.** [6.2a] No **7.** [6.4b] 4 **8.** [6.4b] 0
9. [6.4c] Slope: $\frac{1}{3}$; y-intercept: $\left(0, -\frac{7}{3}\right)$
10. [6.4a] Undefined **11.** [6.4d] $y = x - 4$
12. [6.4d] $y = 4x + 7$ **13.** [6.7a] $y = \frac{5}{2}x$

14. [6.7c] $y = \dfrac{40}{x}$

15. [6.6b]

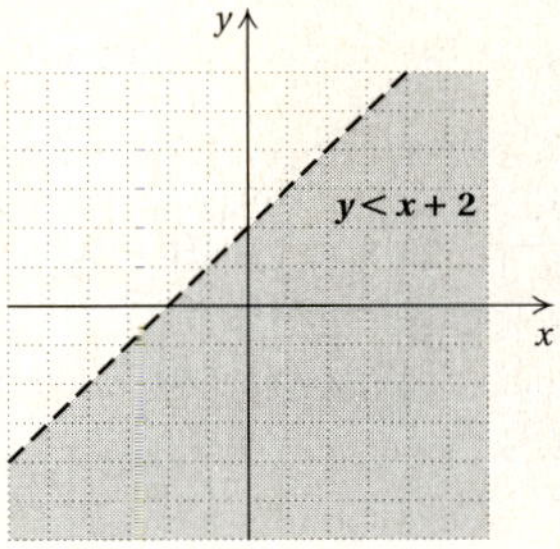

16. [6.6b]

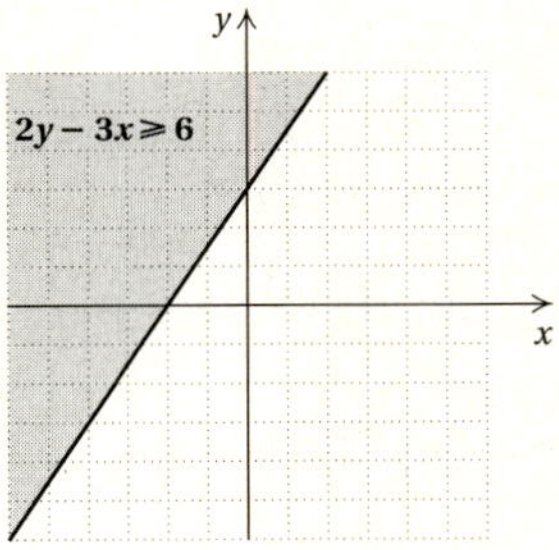

17. [6.5a] Parallel

18. [6.5b] Perpendicular **19.** [6.5b] Perpendicular
20. [6.6a] Yes

Exercise Set 6.1, p. 397

1. 12% **3.** 3.7% **5.** 70.1% **7.** 1743; 360; 270
9. Approximately \$5000 **11.** Approximately \$4000
13. The second week **15.** 590 **17.** 60 **19.** Hiking
21. 1998 **23.** \$17.0 million **25.** \$18.9 million
27.

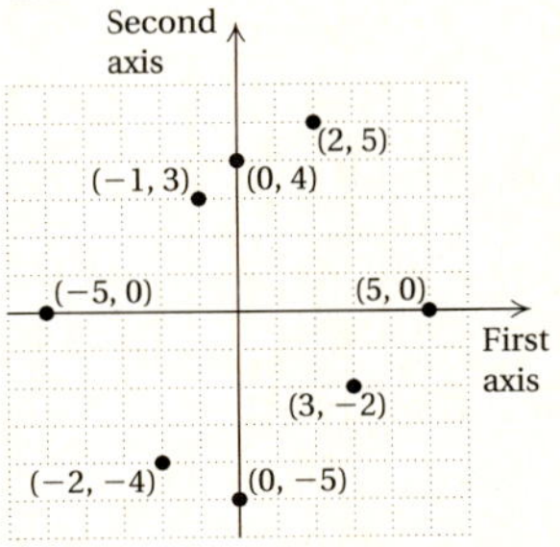

29. II **31.** IV
33. III **35.** I

37. Negative; negative **39.** *A*: (3, 3); *B*: (0, −4); *C*: (−5, 0); *D*: (−1, −1); *E*: (2, 0) **41.** \$5330 **43.** 987
45. 1 **47.** I, IV **49.** I, III **51.** (−1, −5)
53. Answers may vary.

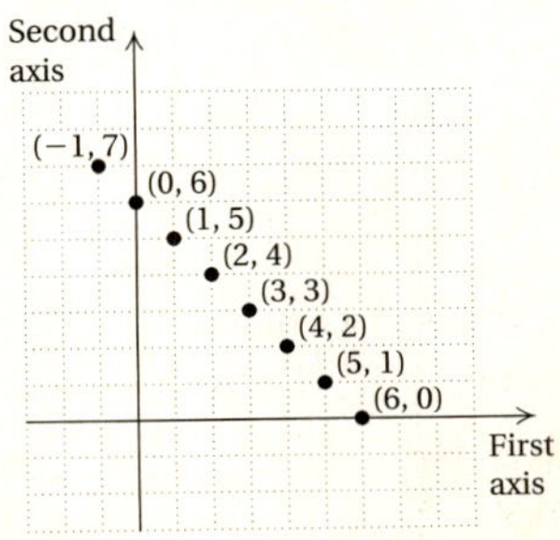

55. 23

Exercise Set 6.2, p. 407

1. Yes **3.** No **5.** No
7.

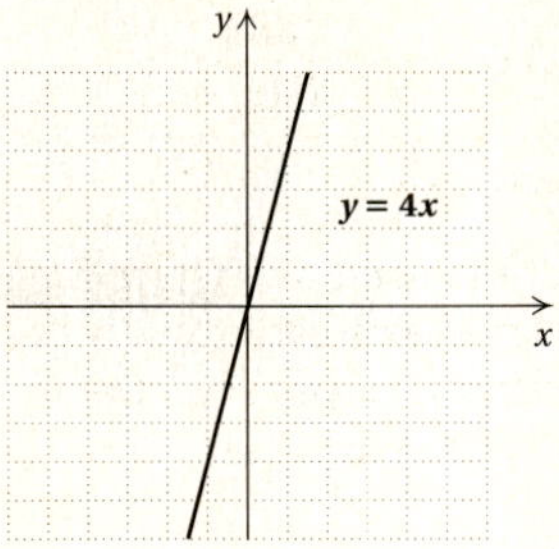

9.

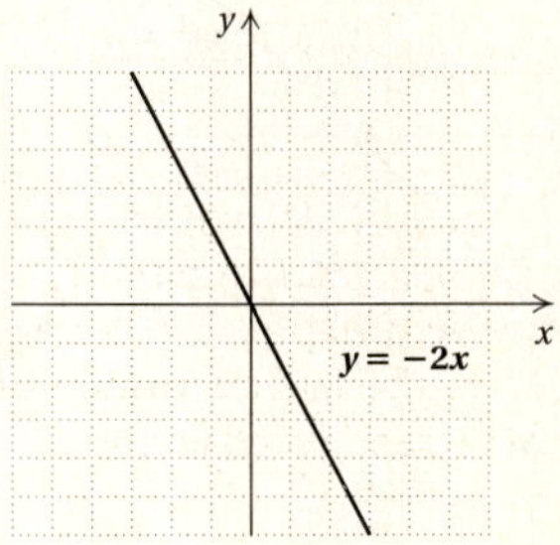

11.

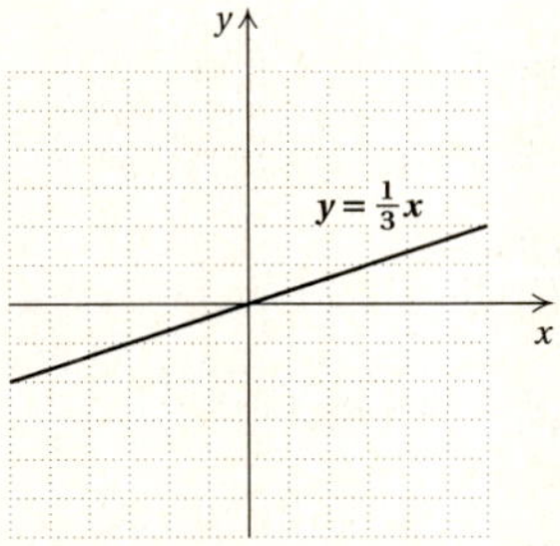

13.

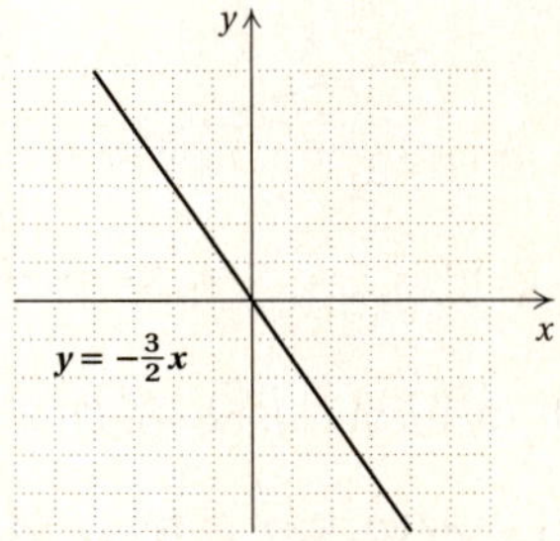

15.

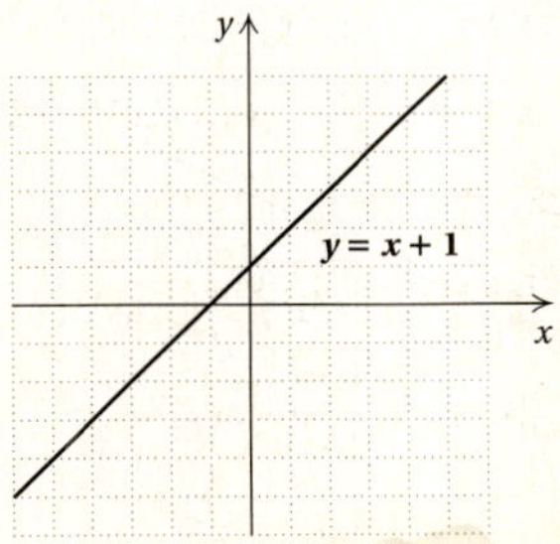

17.

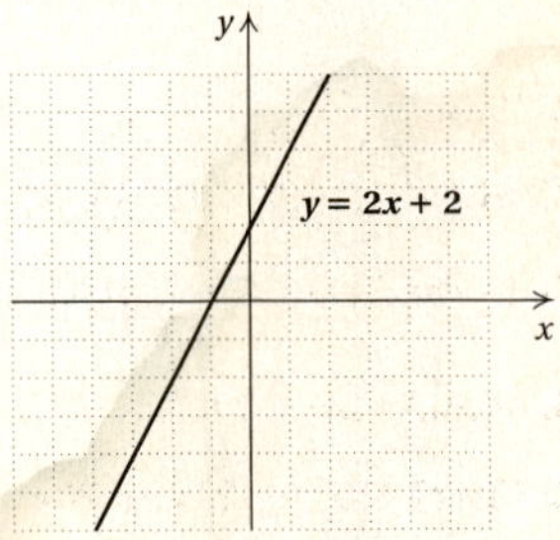

19.

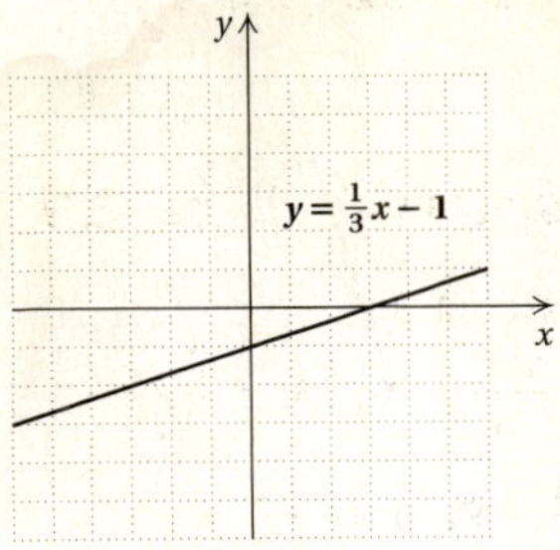

21.

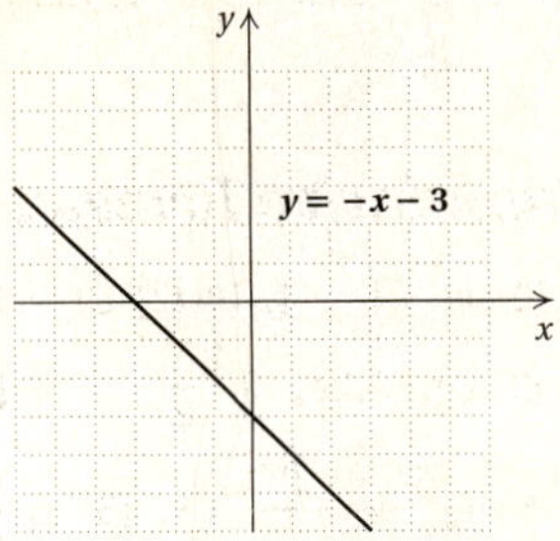

23.

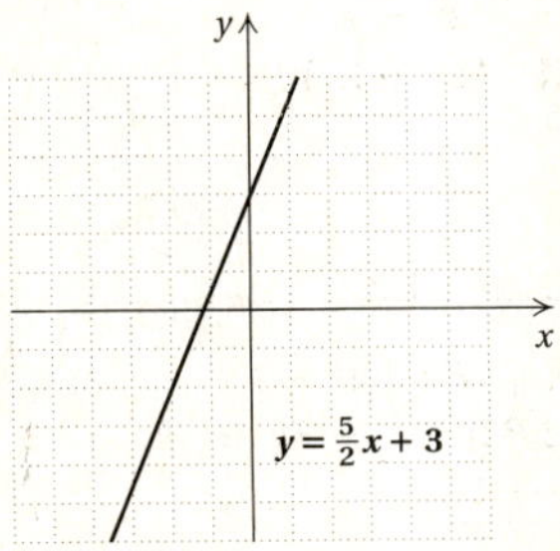

25.

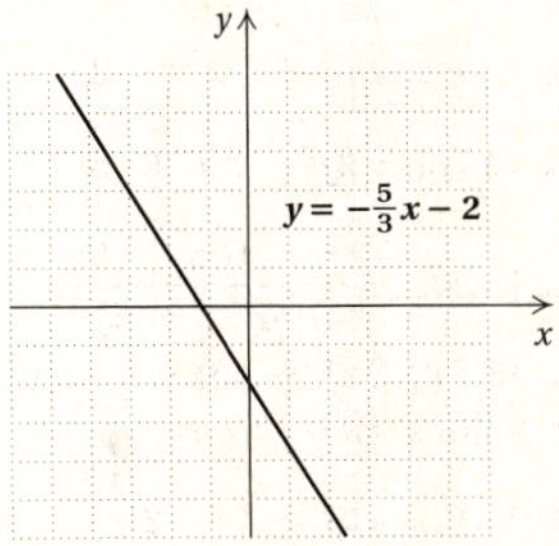

27.

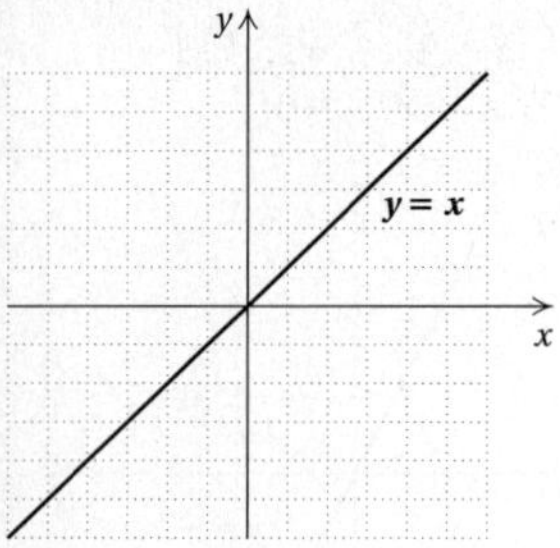

29.

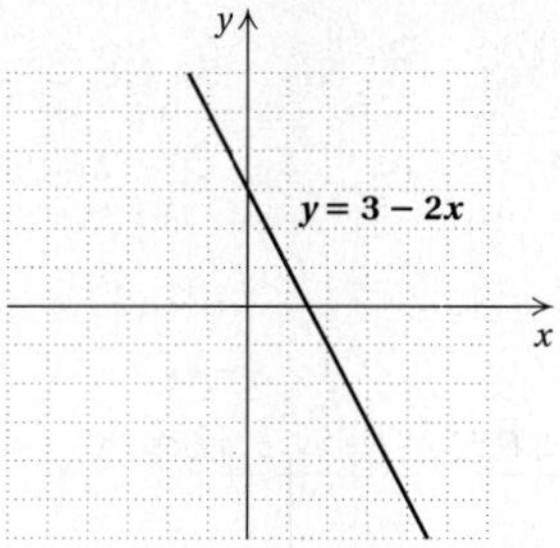

31.

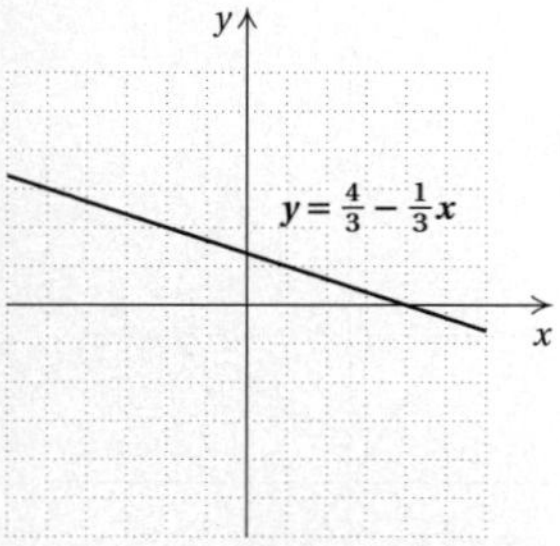

33. 75 km/h
35. $\frac{7}{5}, -\frac{7}{5}$
37. -1

39. (0, 6), (1, 5), (2, 4), (3, 3), (4, 2), (5, 1), (6, 0)

Exercise Set 6.3, p. 415

1.

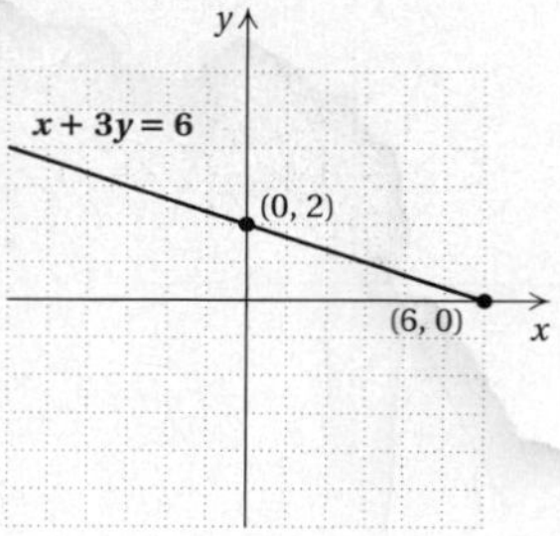

3.

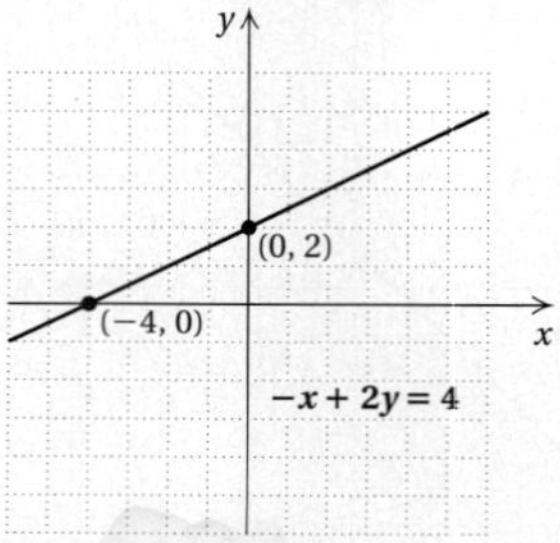

5.

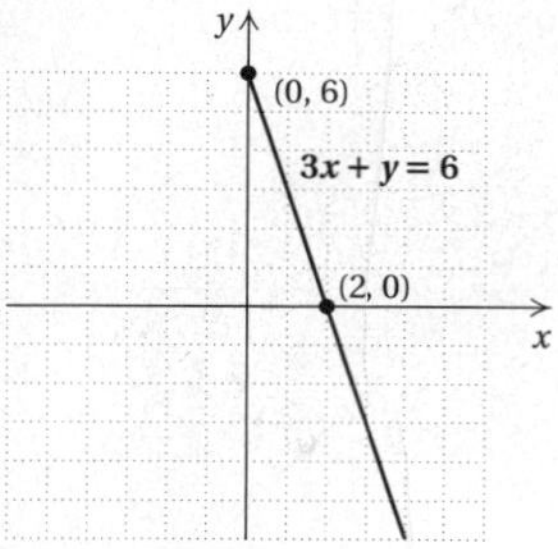

7.

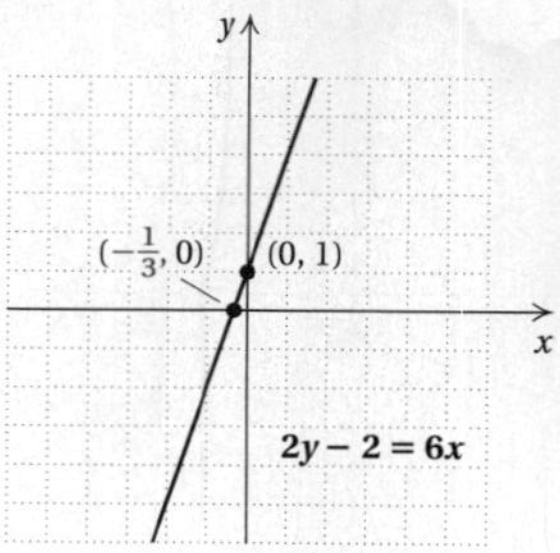

9.

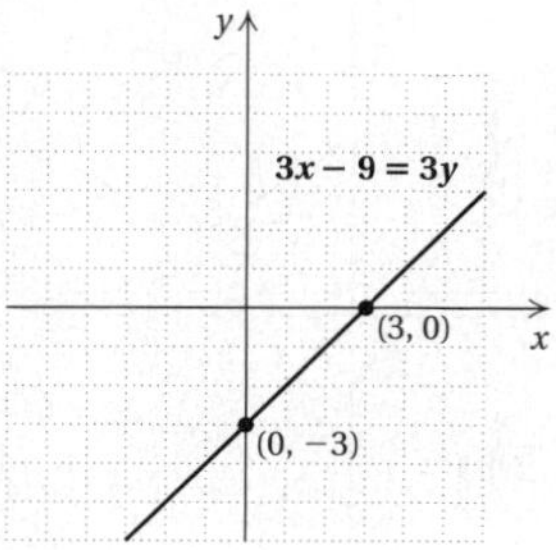

11.

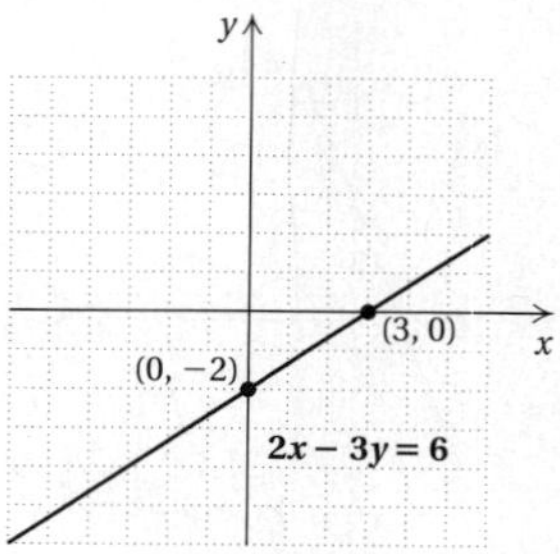

13.

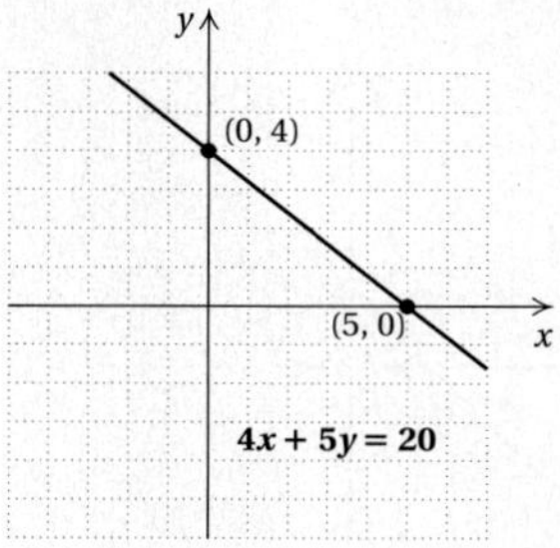

15.

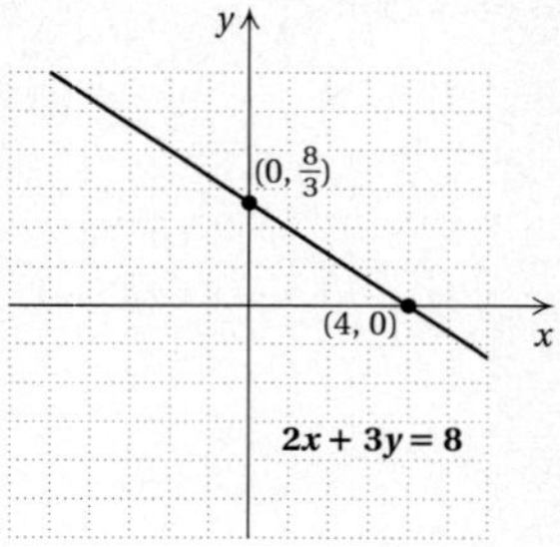

17.

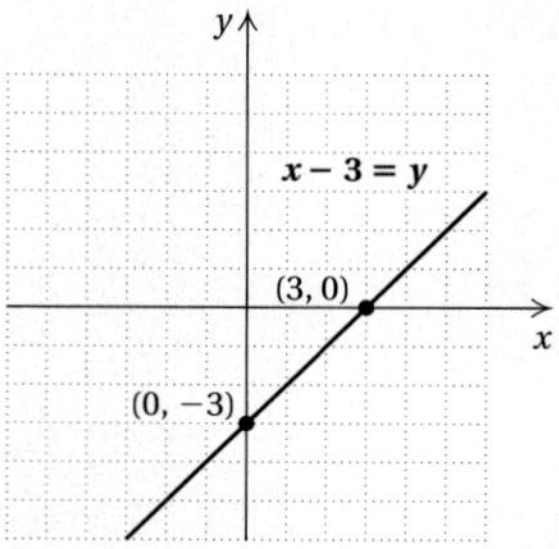

19.

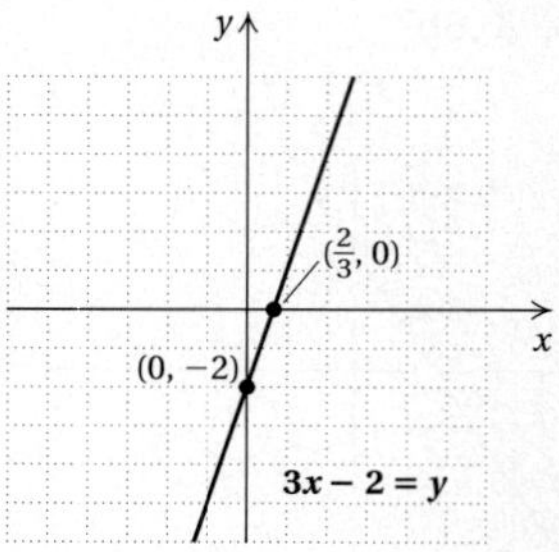

21.

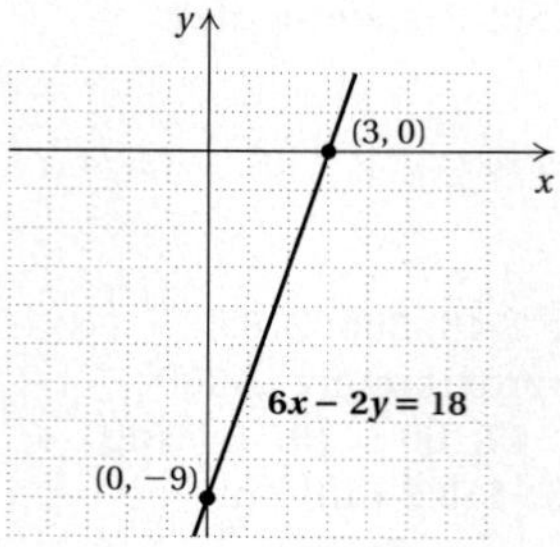

23.

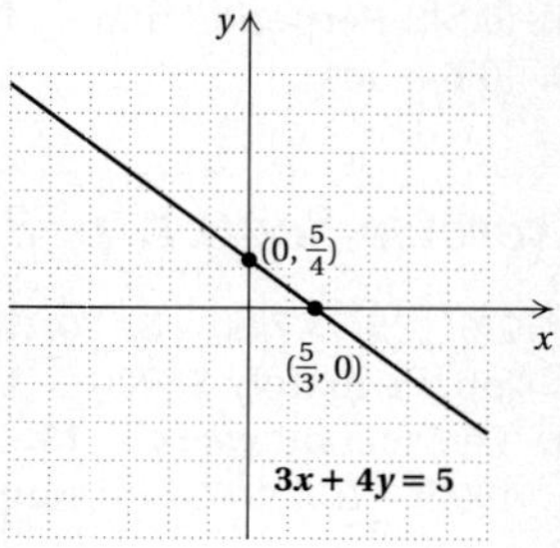

25.

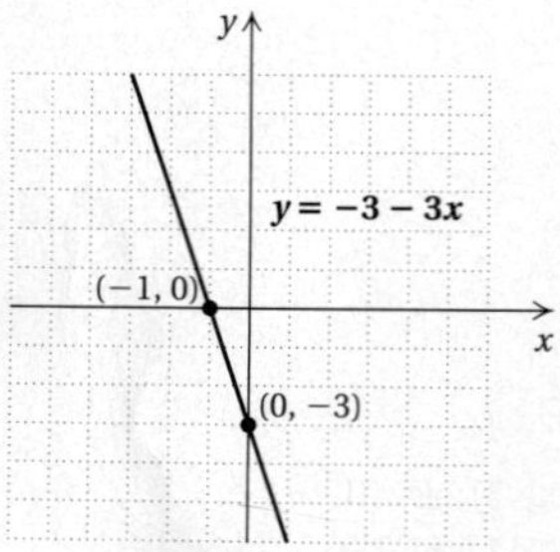

27.

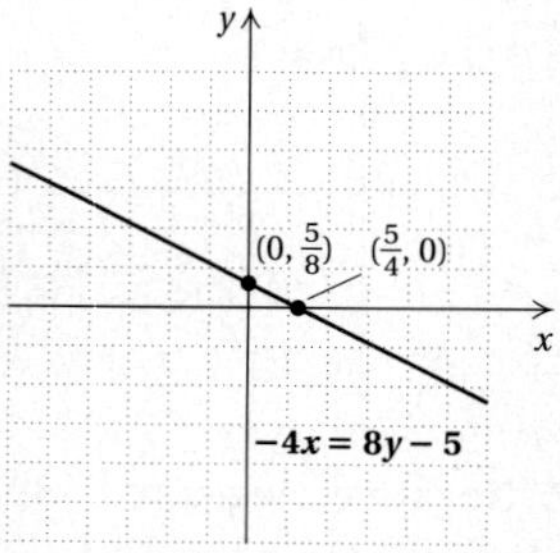

29.

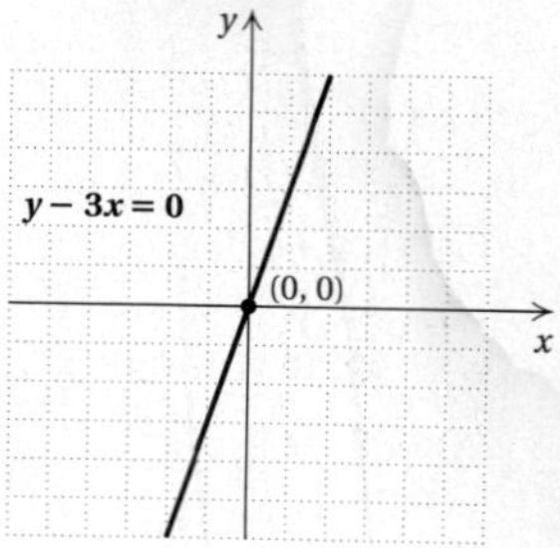

31.

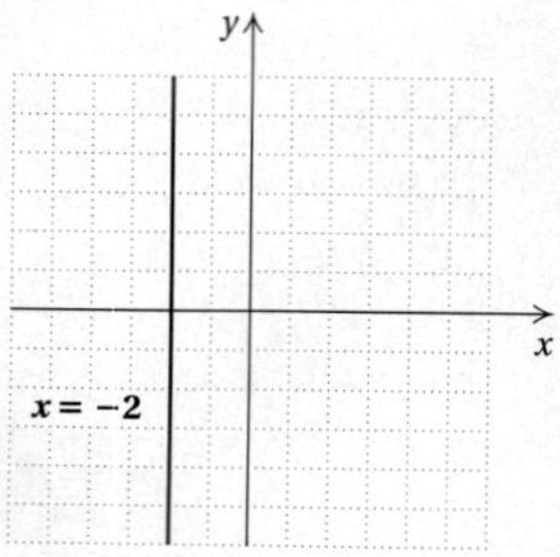

33.

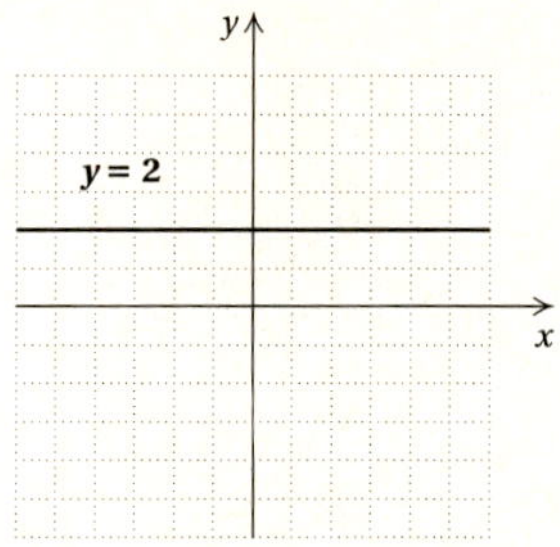

35.

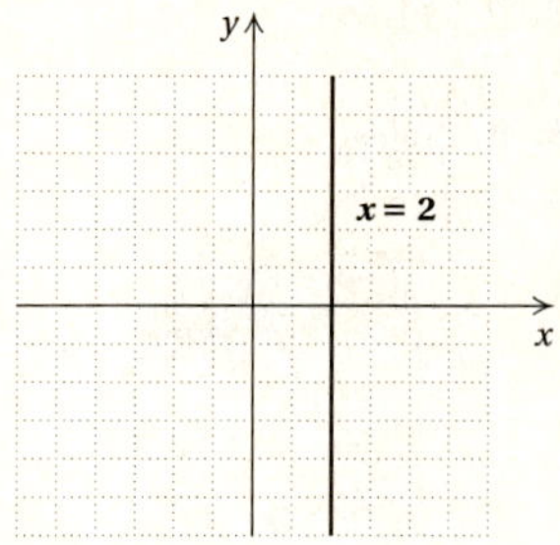

37.

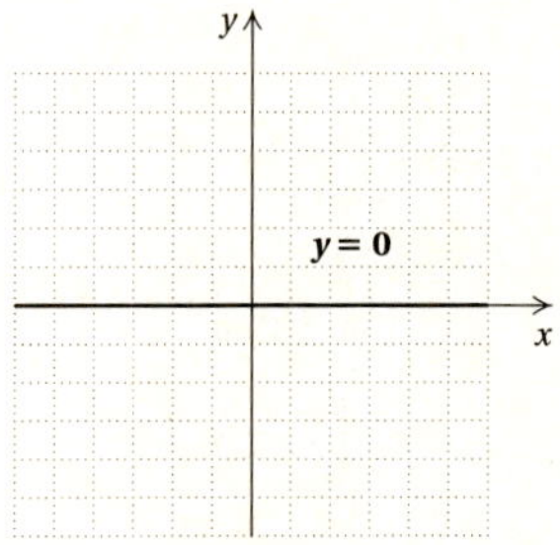

39.

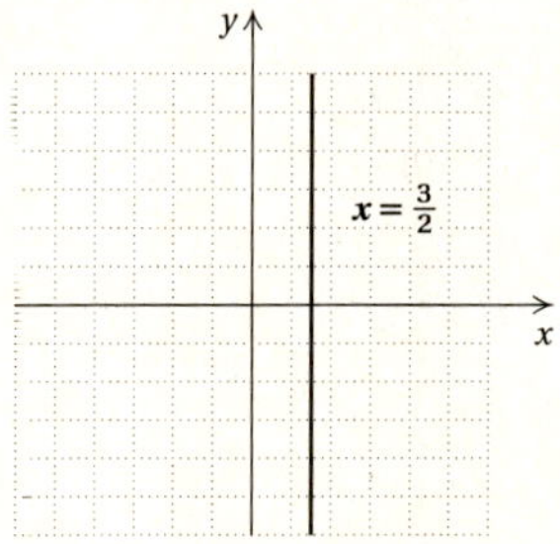

41.

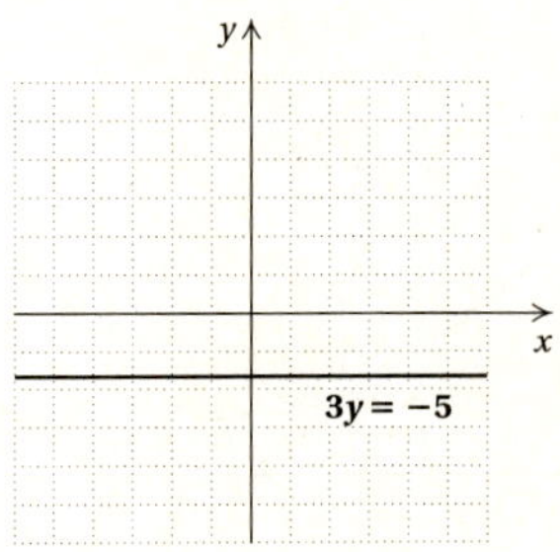

43.

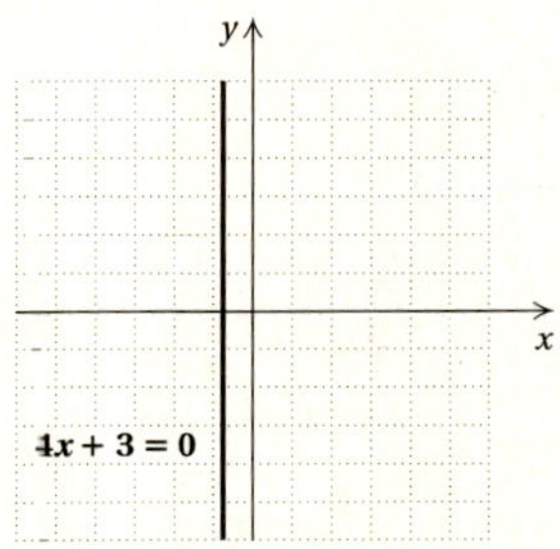

45.

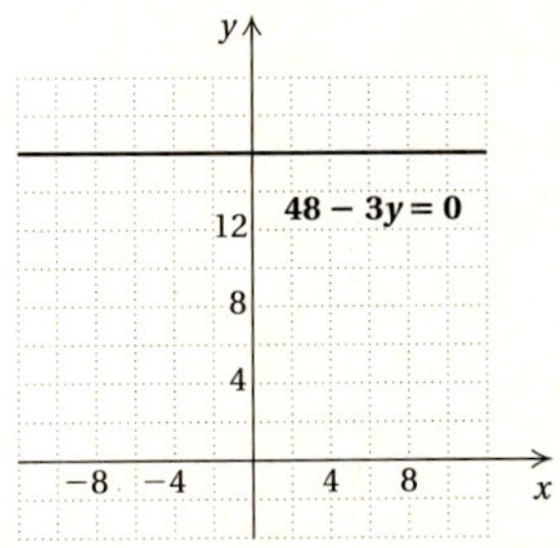

47. $5\frac{5}{11}$ hr

49. −4, 3 51. $-\frac{24}{5}$ 53. $x = 0$ 55. (−3, 6)

Exercise Set 6.4, p. 423

1. 0 **3.** $-\frac{4}{5}$ **5.** $\frac{2}{3}$ **7.** 2 **9.** 0 **11.** Undefined **13.** $-\frac{3}{2}$ **15.** $-\frac{1}{4}$ **17.** 2 **19.** Undefined **21.** 0 **23.** Undefined **25.** 0 **27.** −4, (0, −9) **29.** 1.8, (0, 0) **31.** $-\frac{8}{7}$, (0, −3) **33.** 3, $\left(0, -\frac{5}{3}\right)$ **35.** $-\frac{5}{4}$, (0, 3) **37.** 0, (0, −17) **39.** $y = -2x - 6$ **41.** $y = \frac{3}{4}x + \frac{5}{2}$ **43.** $y = x - 8$ **45.** $y = -3x + 3$ **47.** $y = x + 4$ **49.** $y = -\frac{1}{2}x + 4$ **51.** $y = -\frac{3}{2}x + \frac{13}{2}$ **53.** $y = -4x - 11$ **55.** $\frac{44}{7}$ **57.** −5 **59.** $\frac{68}{109}$ **61.** $y = 3x - 9$ **63.** $y = \frac{3}{2}x - 2$

Exercise Set 6.5, p. 427

1. Yes **3.** No **5.** No **7.** No **9.** Yes **11.** Yes **13.** No **15.** Yes **17.** Yes **19.** Yes **21.** No **23.** In 7 hr **25.** 5 **27.** −6 **29.** $y = 3x + 6$ **31.** $y = -3x + 2$ **33.** $y = \frac{1}{2}x + 1$ **35.** $k = 16$ **37.** A: $y = \frac{4}{3}x - \frac{7}{3}$; B: $y = -\frac{3}{4}x - \frac{1}{4}$

Exercise Set 6.6, p. 433

1. No **3.** Yes

5.

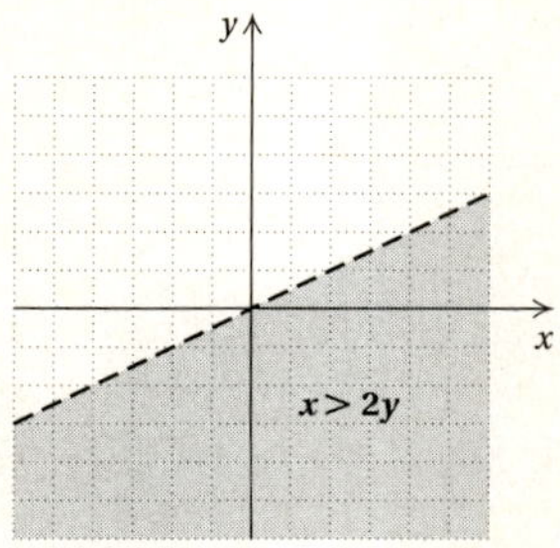

7.

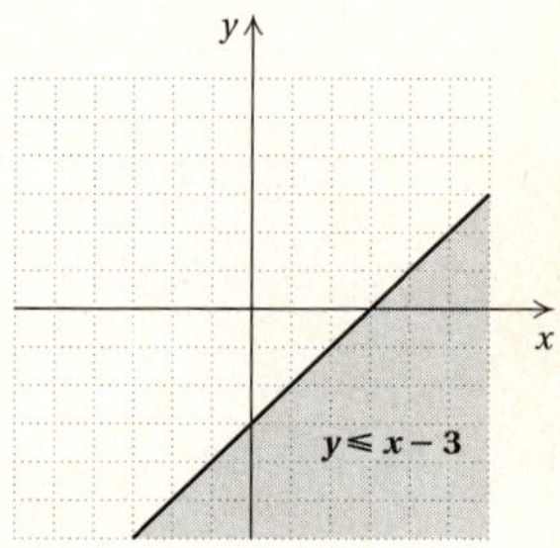

9.

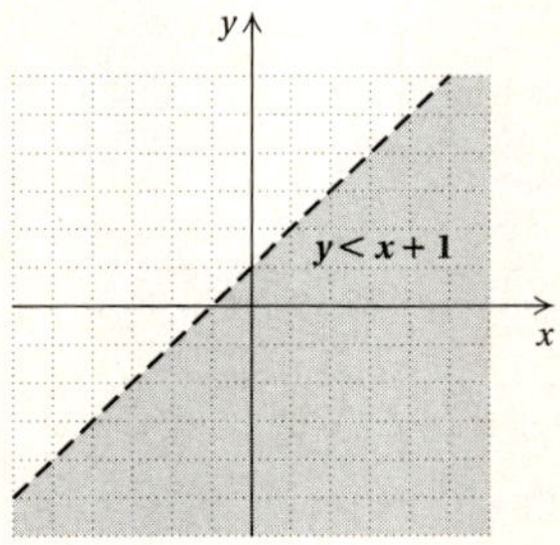

11.

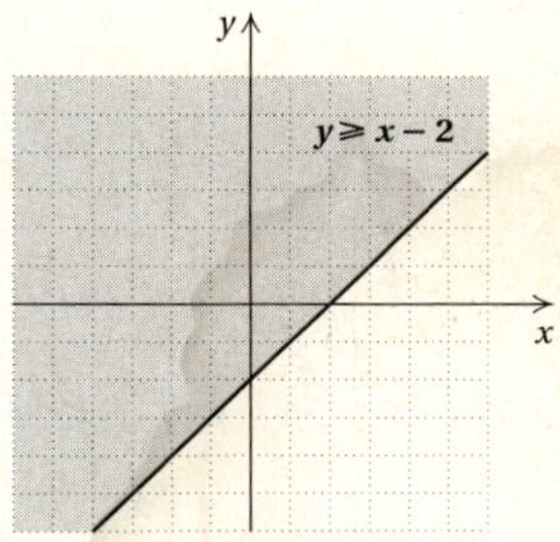

13.

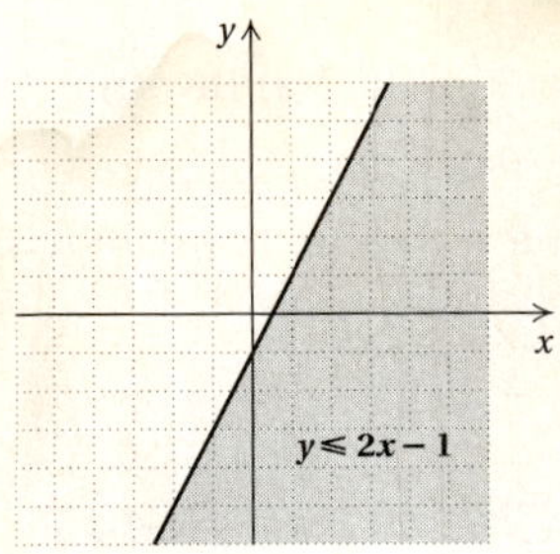

15.

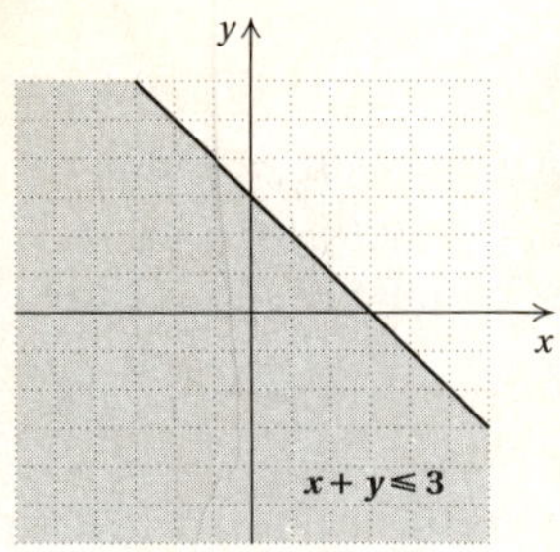

17.

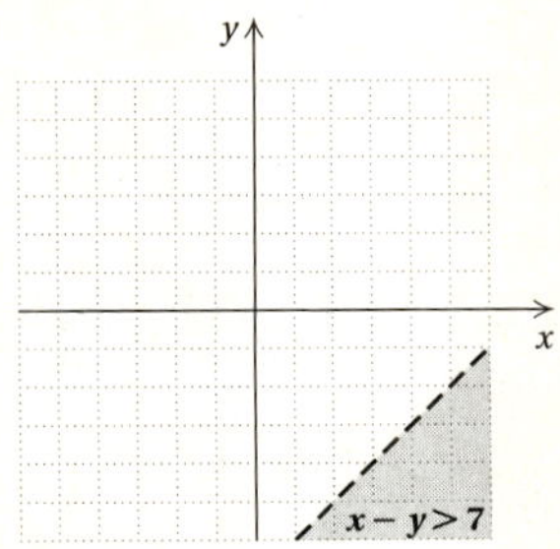

19.

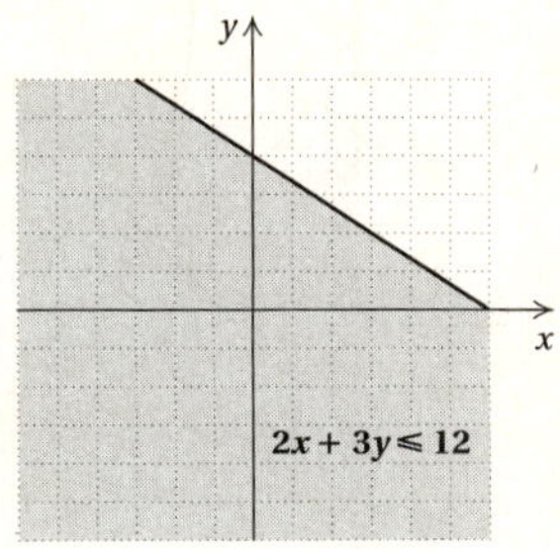

21.

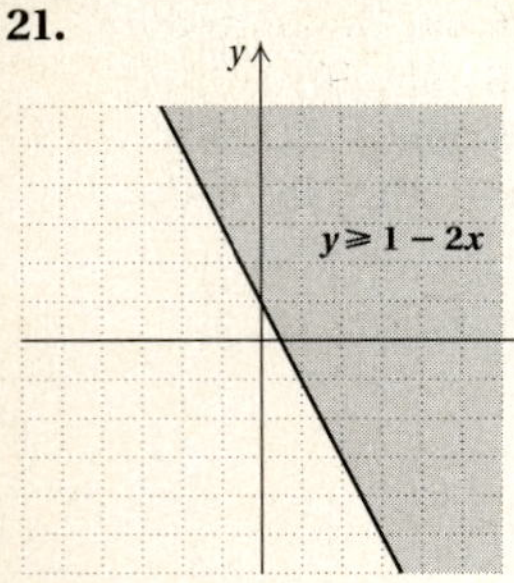

23.

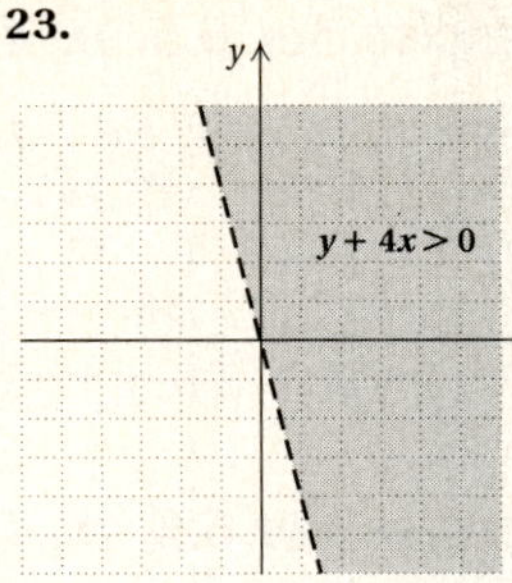

25.

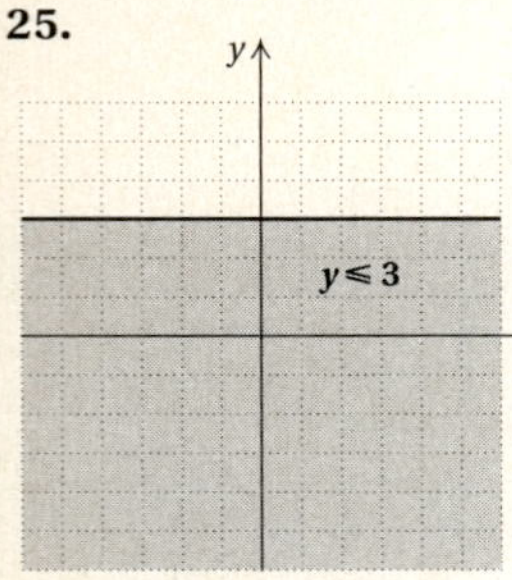

27.

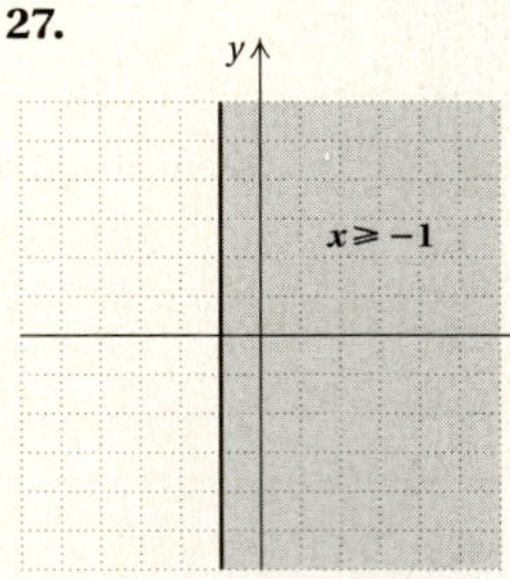

29. $\frac{1}{3}$ **31.** 3 **33.** 4 **35.** $35c + 75a > 1000$

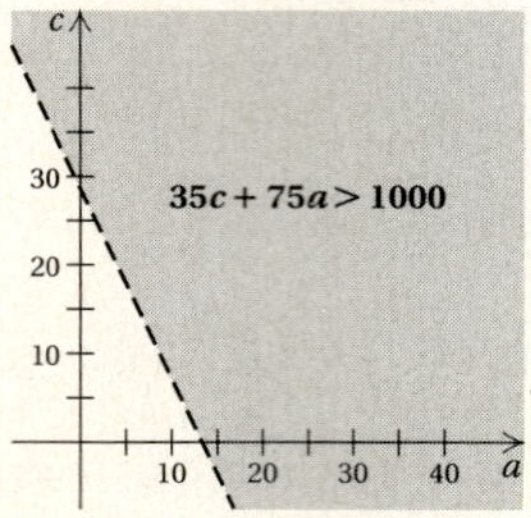

Exercise Set 6.7, p. 439

1. $y = 4x$ **3.** $y = \frac{8}{5}x$ **5.** $y = 3.6x$ **7.** $y = \frac{25}{3}x$
9. \$183.75 **11.** $22\frac{6}{7}$ **13.** $36.\overline{6}$ lb **15.** 3,200,000
17. $y = \frac{75}{x}$ **19.** $y = \frac{80}{x}$ **21.** $y = \frac{1}{x}$ **23.** $y = \frac{2100}{x}$
25. $y = \frac{0.06}{x}$ **27.** **(a)** Direct; **(b)** $69\frac{3}{8}$ **29.** **(a)** Inverse; **(b)** $4\frac{1}{2}$ hr **31.** $3\frac{5}{9}$ amperes **33.** 640 **35.** 8.25 ft
37. $\frac{8}{5}$ **39.** 9, 16 **41.** $\frac{4}{7}, \frac{2}{5}$ **43.** $P = ks$; $k = n$, where n is the number of sides of the polygon **45.** $C = kA$
47. $P^2 = kt$ **49.** $P = kV^3$ **51.** $N = \frac{k}{C}$

Summary and Review: Chapter 6, p. 445

1. [6.1a] **(a)** 9%; **(b)** years 3 and 4
2. [6.1d] $(-5, -1)$
3. [6.1d] $(-2, 5)$
4. [6.1d] $(3, 0)$

5.–7. [6.1b]

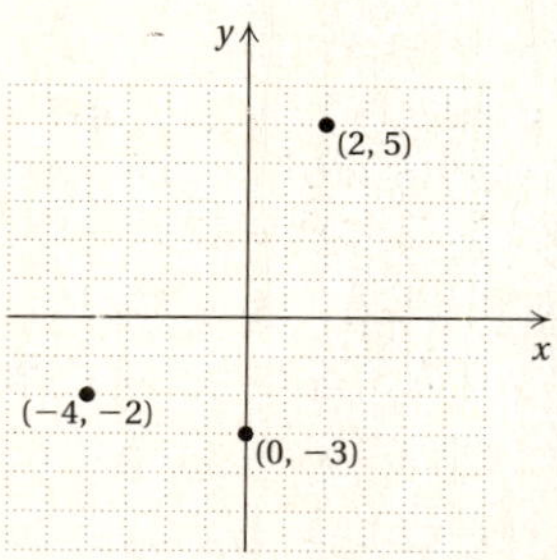

8. [6.1c] IV **9.** [6.1c] III **10.** [6.1c] I **11.** [6.2a] No
12. [6.2a] Yes

13. [6.2b]

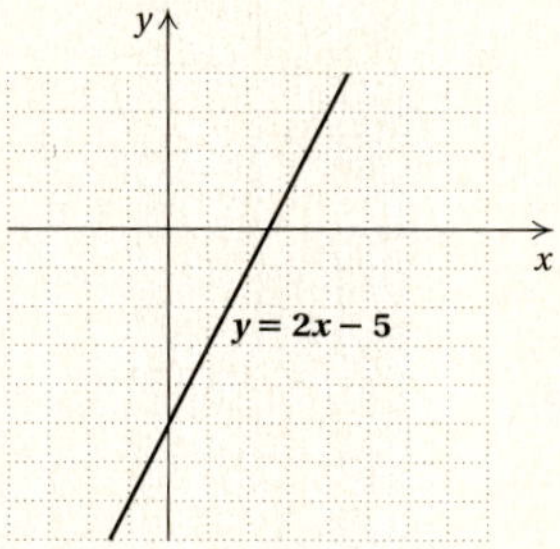

14. [6.2b]

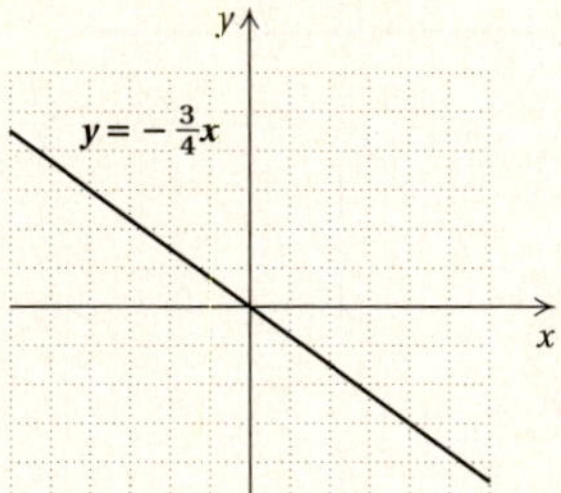

15. [6.2b]

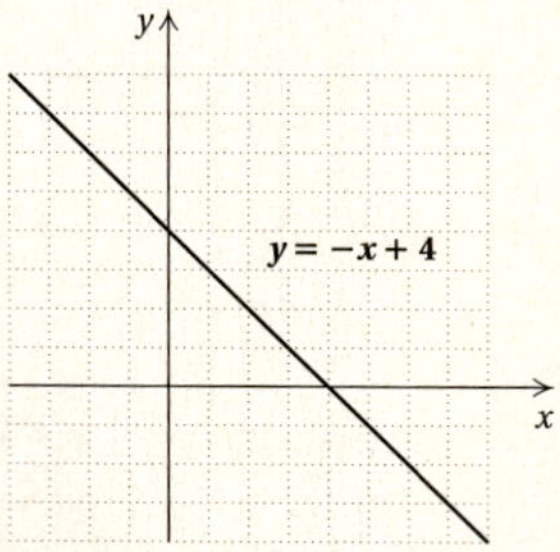

16. [6.2b]

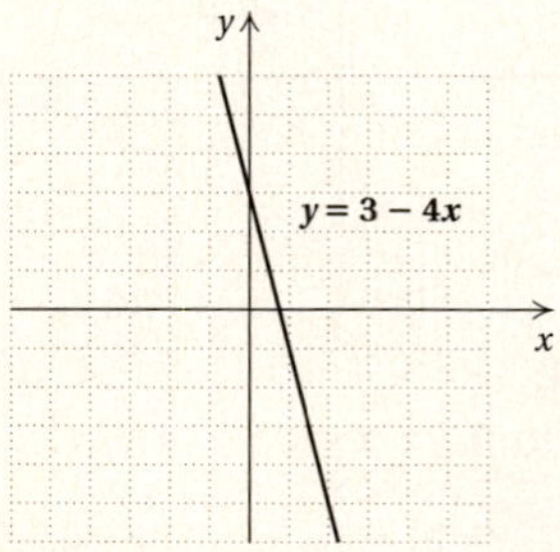

17. [6.3a]

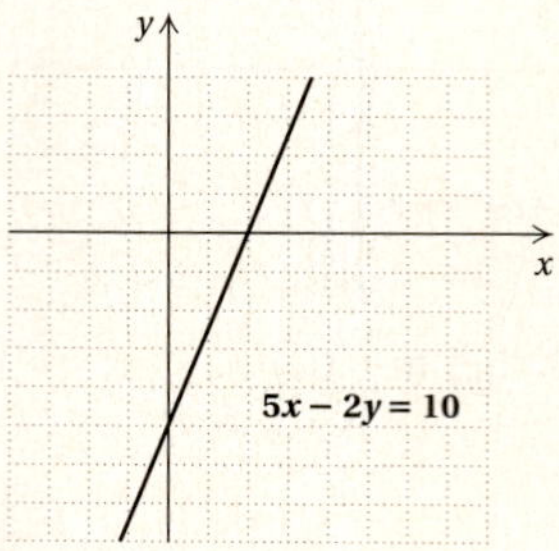

18. [6.3b]

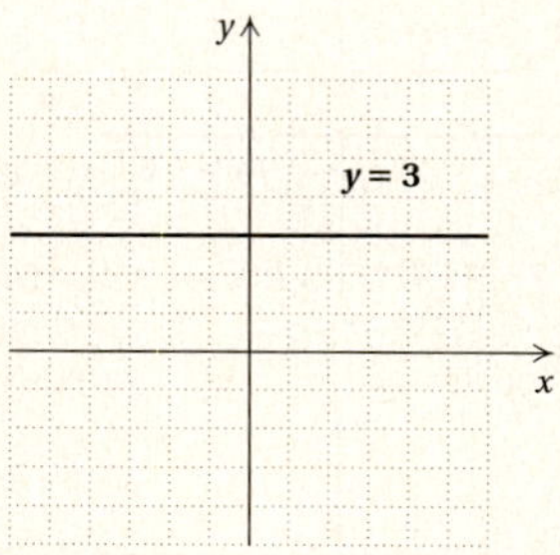

19. [6.3b]

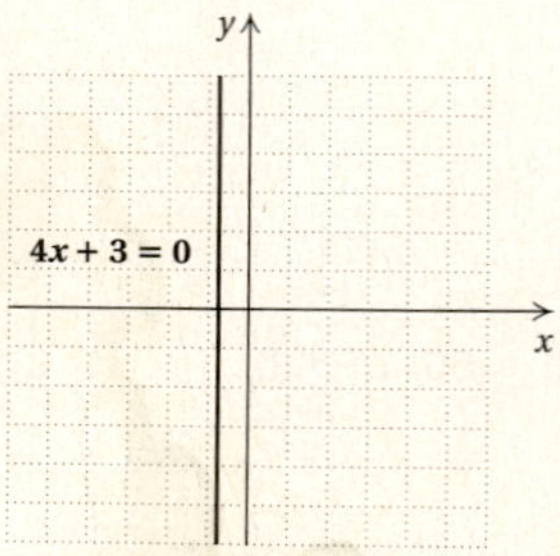

20. [6.3a]

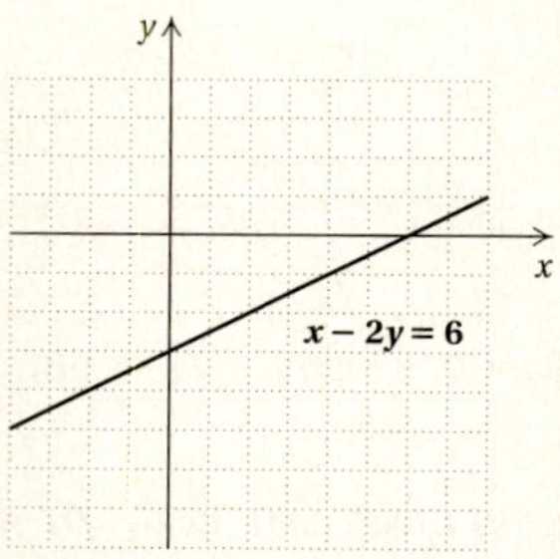

21. [6.4a] $\frac{3}{2}$ **22.** [6.4a] 0 **23.** [6.4a] Undefined
24. [6.4a] 2 **25.** [6.4b] 0 **26.** [6.4b] Undefined
27. [6.4b] $-\frac{4}{3}$ **28.** [6.4c] -9, $(0, 46)$ **29.** [6.4c] -1, $(0, 9)$ **30.** [6.4c] $\frac{3}{5}$, $\left(0, -\frac{4}{5}\right)$ **31.** [6.4d] $y = 3x - 1$
32. [6.4d] $y = \frac{2}{3}x - \frac{11}{3}$ **33.** [6.4d] $y = -2x - 4$
34. [6.4d] $y = x + 2$ **35.** [6.4d] $y = \frac{1}{2}x - 1$
36. [6.5a] Parallel **37.** [6.5b] Perpendicular
38. [6.5a] Parallel **39.** [6.5a, b] Neither **40.** [6.6a] No
41. [6.6a] No **42.** [6.6a] Yes

43. [6.6b]

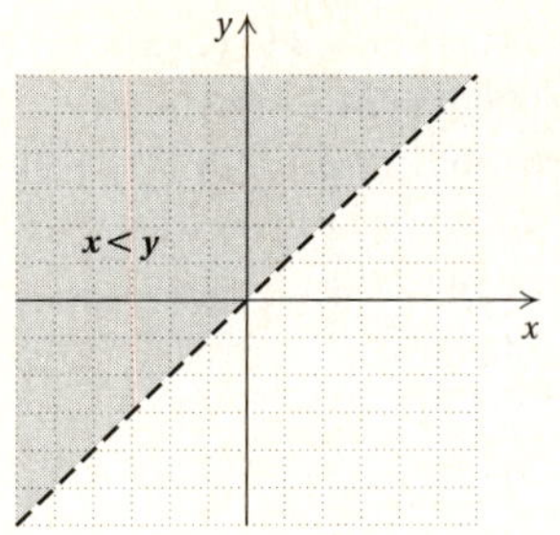

44. [6.6b]

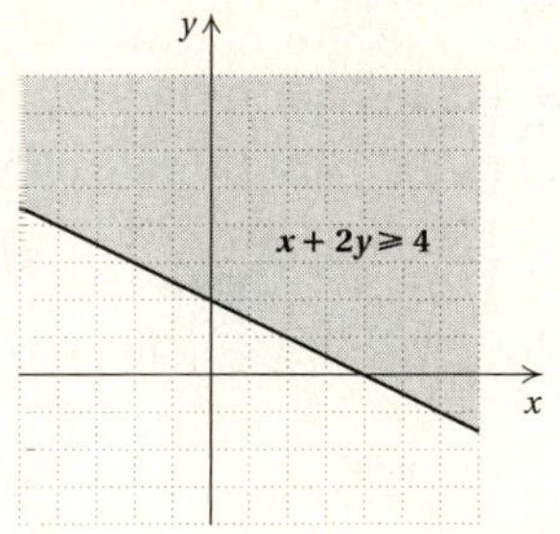

45. [6.6b]

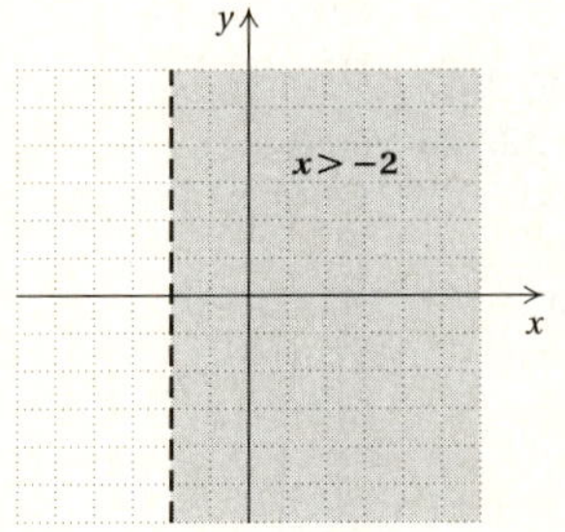

46. [6.7a] $y = 3x$ **47.** [6.7a] $y = \frac{1}{2}x$ **48.** [6.7a] $y = \frac{4}{5}x$

49. [6.7c] $y = \dfrac{30}{x}$ **50.** [6.7c] $y = \dfrac{1}{x}$ **51.** [6.7c] $y = \dfrac{0.65}{x}$

52. [6.7b] \$247.50 **53.** [6.7d] 1 hr **54.** [5.7a] $3\frac{1}{3}$ hr

55. [1.8d] 52 **56.** [5.6a] −4 **57.** [4.7b] 5, −11

Test: Chapter 6, p. 447

1. [6.1a] Bachelor's **2.** [6.1a] 520,000 **3.** [6.1a] 261,800
4. [6.1a] 401,900 **5.** [6.1c] II **6.** [6.1c] III
7. [6.1d] (3, 4) **8.** [6.1d] (0, −4) **9.** [6.2a] Yes

10. [6.2b]

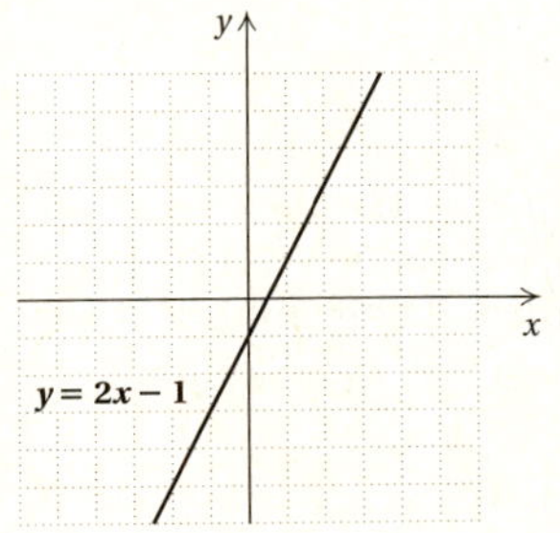

11. [6.3a]

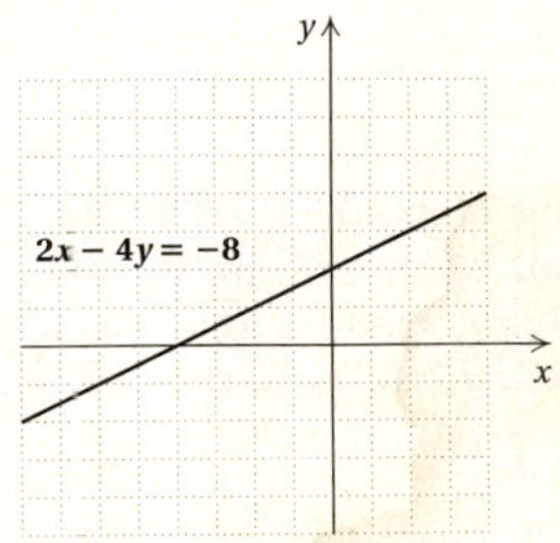

12. [6.3b]

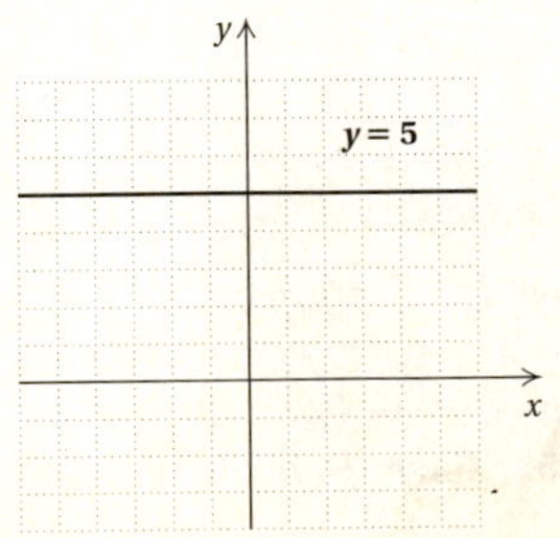

13. [6.2b]

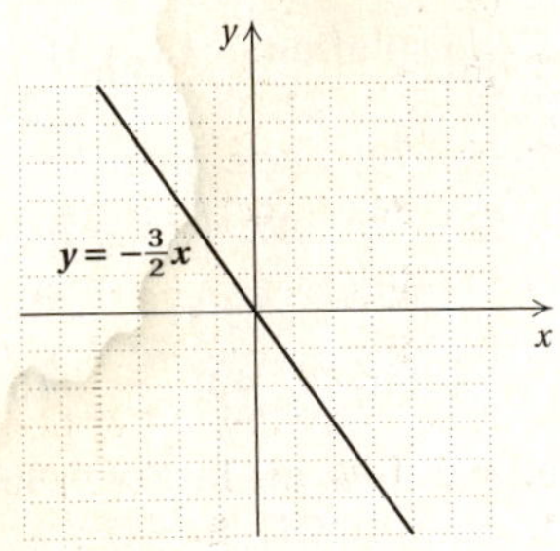

14. [6.3b]

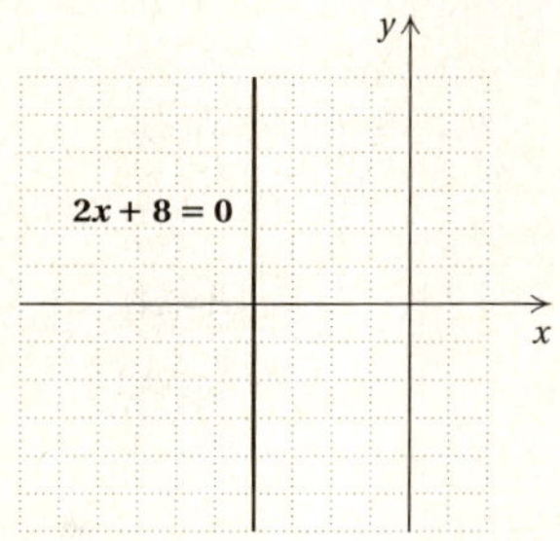

15. [6.4a] Undefined
16. [6.4a] $\frac{7}{12}$
17. [6.4b] 0
18. [6.4b] Undefined
19. [6.4c] 2, $\left(0, -\frac{1}{4}\right)$
20. [6.4c] $\frac{4}{3}$, (0, −2)

21. [6.4d] $y = x + 2$ **22.** [6.4d] $y = -3x - 6$

23. [6.4d] $y = -3x + 4$ **24.** [6.4d] $y = \frac{1}{4}x - 2$

25. [6.5a] Parallel **26.** [6.5a, b] Neither
27. [6.5b] Perpendicular
28. [6.6a] No **29.** [6.6a] Yes
30. [6.6b]

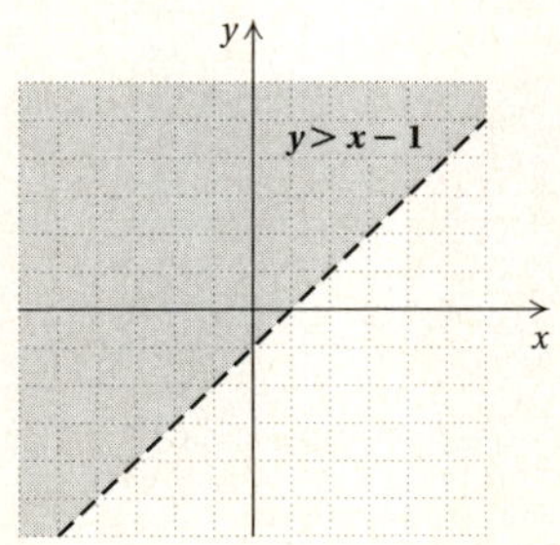

31. [6.6b]

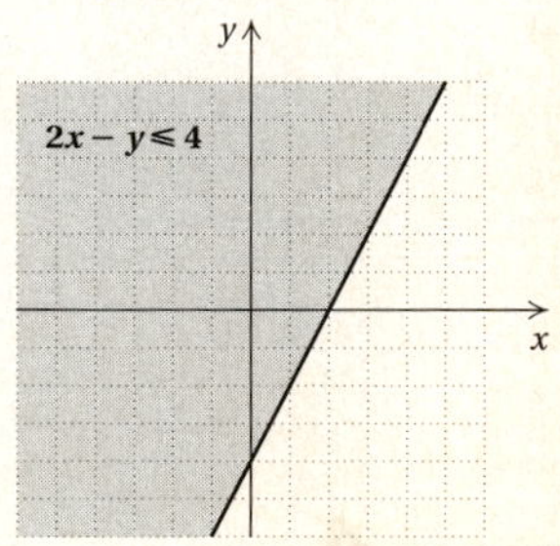

32. [6.7a] $y = 2x$ **33.** [6.7a] $y = 0.5x$ **34.** [6.7c] $y = \dfrac{12}{x}$

35. [6.7c] $y = \dfrac{1}{x}$ **36.** [6.7b] 240 km **37.** [6.7d] $1\frac{1}{5}$ hr

38. [5.7a] Freight: 90 mph; passenger: 105 mph

39. [1.8d] 1 **40.** [4.7b] $\frac{1}{3}$, −5 **41.** [5.6a] No solution

42. [6.1b] Area: 56; perimeter: 30

43. [6.4c, d] $y = \frac{2}{3}x + \frac{11}{3}$

Cumulative Review: Chapters 1–6, p. 449

1. [1.2e] 3.5 **2.** [3.3d] 1, −2, 1, −1 **3.** [3.3g] 3, 2, 1, 0; 3
4. [3.3i] None of these **5.** [3.3e] $2x^3 - 3x^2 - 2$

6. [1.8c] $\dfrac{3}{8}x + 1$ **7.** [3.1e], [3.2a, b] $\dfrac{9}{4x^8}$

8. [5.9a] $\dfrac{8x - 12}{17x}$ **9.** [3.7e] $-2xy^2 - 4x^2y^2 + xy^3$

10. [3.4a] $2x^5 + 6x^4 + 2x^3 - 10x^2 + 3x - 9$

11. [5.1d] $\dfrac{2}{3(y + 2)}$ **12.** [5.2b] 2 **13.** [5.4a] $x + 4$

14. [5.5a] $\dfrac{2x - 6}{(x + 2)(x - 2)}$ **15.** [3.6a] $a^2 - 9$

16. [3.5d] $2x^5 + x^3 - 6x^2 - x + 3$ **17.** [3.6b] $4x^6 - 1$
18. [3.6c] $36x^2 - 60x + 25$ **19.** [3.5b] $12x^4 + 16x^3 + 4x^2$
20. [3.6a] $6x^7 - 12x^5 + 9x^2 - 18$ **21.** [2.3c] 3

22. [4.7b] $\frac{1}{2}$, −4 **23.** [4.7b] −5, 4

24. [2.7e] $\{x \mid x \geq -26\}$ **25.** [4.7a] 0, 4 **26.** [4.7b] 0, 10

27. [4.7b] 20, −20 **28.** [2.6a] $a = \dfrac{t}{x + y}$ **29.** [5.6a] 2

30. [5.6a] No solution **31.** [1.7d] $-2(3 + x + 6y)$

32. [4.2a] $(x - 4)(x - 6)$ **33.** [4.5d] $2(x + 3)(x - 3)$ **34.** [4.1c] $(m^3 - 3)(m + 2)$ **35.** [4.5b] $(4x + 5)^2$ **36.** [4.3a], [4.4a] $(2x + 1)(4x + 3)$ **37.** [4.8a] 4 or -5 **38.** [6.7b] \$78 **39.** [2.5a] \$2500 **40.** [5.7a] 35 mph, 25 mph **41.** [4.8a] 14 ft **42.** [5.7a] 35, 28

43. [6.2b]

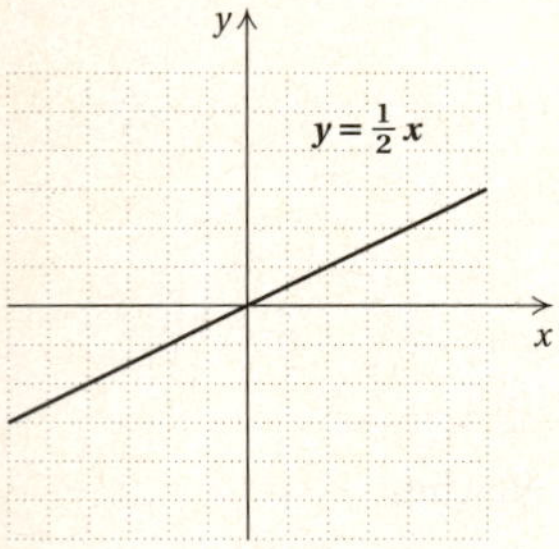

44. [6.3a]

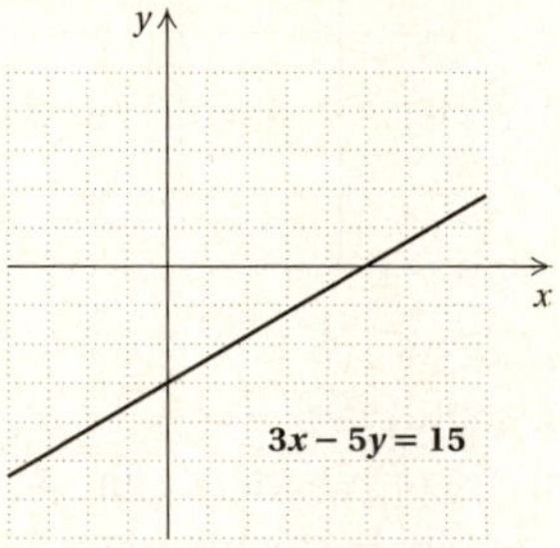

45. [6.3b]

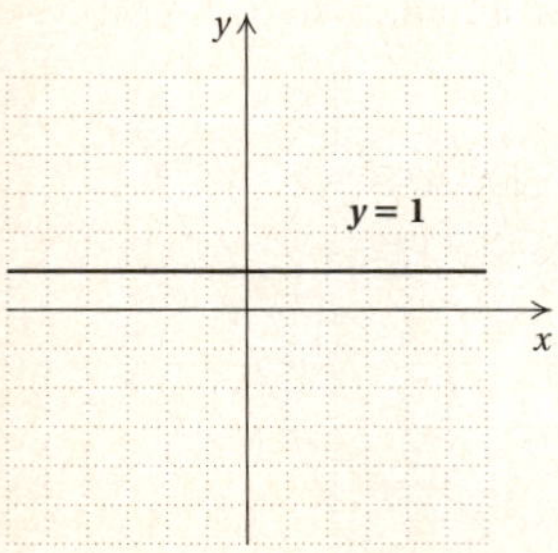

46. [6.6b]

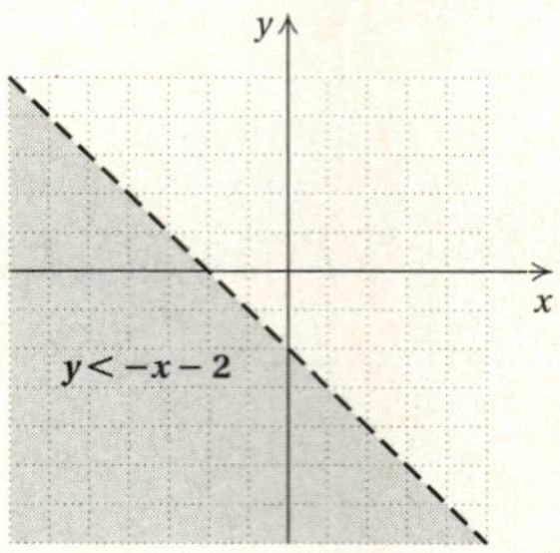

47. [6.6b]

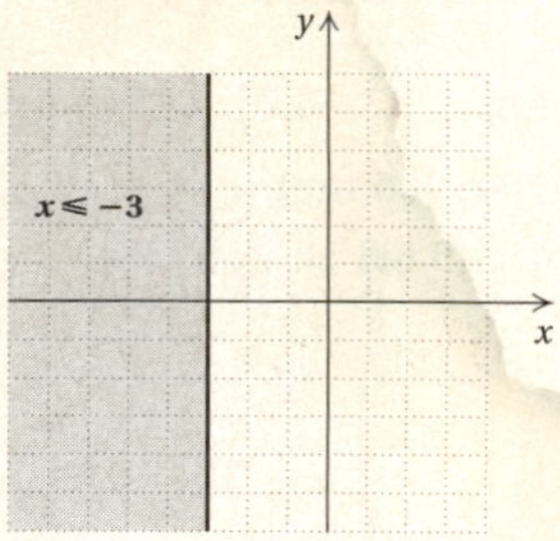

48. [6.7a] $y = \dfrac{2}{3}x$

49. [6.7c] $y = \dfrac{10}{x}$ **50.** [6.4a] Undefined **51.** [6.4a] $-\frac{3}{7}$ **52.** [6.4c] $\frac{4}{3}$, $(0, -2)$ **53.** [6.4d] $y = -4x + 5$ **54.** [6.4d] $y = \frac{1}{6}x - \frac{17}{6}$ **55.** [6.5a, b] Neither **56.** [6.5a] Parallel **57.** [3.4c], [3.6a] 12 **58.** [3.6b, c] $16y^6 - y^4 + 6y^2 - 9$ **59.** [4.5d] $2(a^{16} + 81b^{20})(a^8 + 9b^{10})(a^4 + 3b^5)(a^4 - 3b^5)$ **60.** [4.7a] 4, -7, 12 **61.** [6.4d], [6.5a] $y = \frac{2}{3}x$ **62.** [5.1a], [5.9a] 0, 3, $\frac{5}{2}$

CHAPTER 7

Pretest: Chapter 7, p. 452

1. [7.1a] Yes **2.** [7.1b] No solution **3.** [7.2a] (5, 2) **4.** [7.2b] $(2, -1)$ **5.** [7.3a] $\left(\frac{3}{4}, \frac{1}{2}\right)$ **6.** [7.3b] $(-5, -2)$ **7.** [7.3b] (10, 8) **8.** [7.3c] 50 and 24 **9.** [7.2c] 25° and 65° **10.** [7.5a] 8 hr after the second train leaves

Exercise Set 7.1, p. 457

1. Yes **3.** No **5.** Yes **7.** Yes **9.** Yes **11.** (4, 2) **13.** (4, 3) **15.** $(-3, -3)$ **17.** No solution **19.** (2, 2) **21.** $\left(\frac{1}{2}, 1\right)$ **23.** Infinite number of solutions **25.** $(5, -3)$ **27.** $\dfrac{108}{x^{13}}$ **29.** $\dfrac{2x^2 - 1}{x^2(x + 1)}$ **31.** $\dfrac{9x + 12}{(x - 4)(x + 4)}$ **33.** $A = 2, B = 2$ **35.** $x + 2y = 2,$ $x - y = 8$

Exercise Set 7.2, p. 463

1. (1, 9) **3.** $(2, -4)$ **5.** (4, 3) **7.** $(-2, 1)$ **9.** $(2, -4)$ **11.** $\left(\frac{17}{3}, \frac{16}{3}\right)$ **13.** $\left(\frac{25}{8}, -\frac{11}{4}\right)$ **15.** $(-3, 0)$ **17.** (6, 3) **19.** 16 and 21 **21.** 12 and 40 **23.** 20 and 8 **25.** Length: 380 mi; width: 270 mi **27.** Length: $134\frac{1}{3}$ m; width: $65\frac{2}{3}$ m

29.

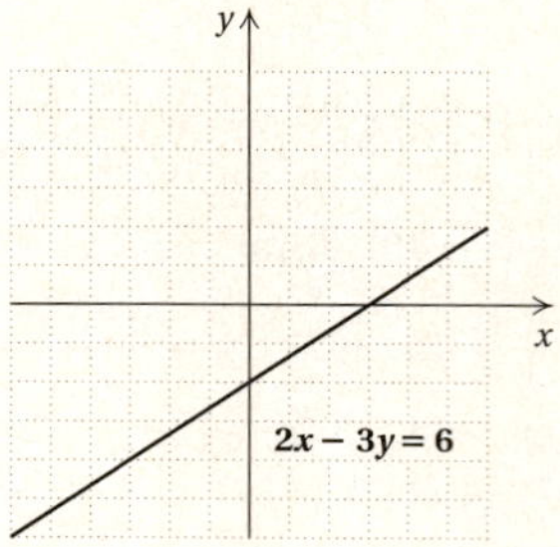

31.

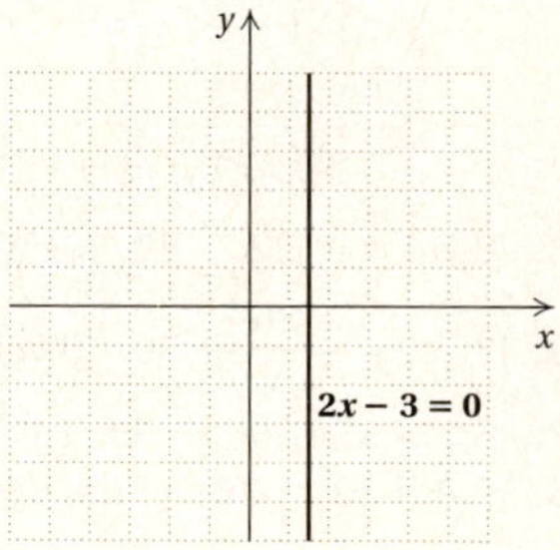

33.

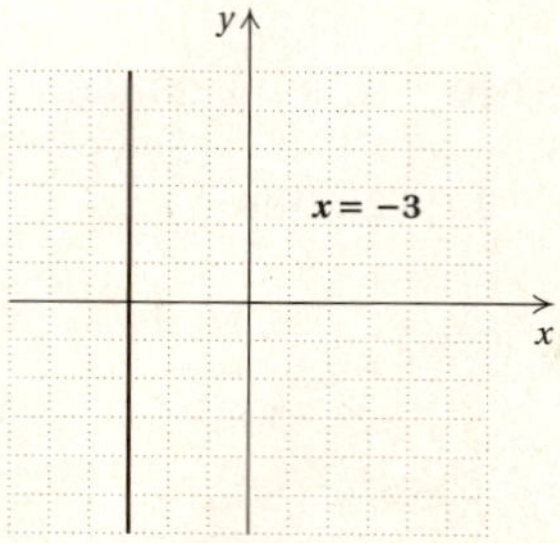

35. (4.3821792, 4.3281211) **37.** $(10, -2)$ **39.** 97 cm, 97 cm, 130 cm **41.** You get $0 = 0$, an equation that is true for any values of x and y. There are infinitely many solutions; the equations have the same graph.

Exercise Set 7.3, p. 471

1. $(6, -1)$ **3.** (3, 5) **5.** (2, 5) **7.** $\left(-\frac{1}{2}, 3\right)$ **9.** $\left(-1, \frac{1}{5}\right)$ **11.** No solution **13.** $(-1, -6)$ **15.** (3, 1) **17.** (8, 3) **19.** (4, 3) **21.** $(1, -1)$ **23.** $(-3, -1)$ **25.** (3, 2) **27.** (50, 18) **29.** Infinite number of solutions **31.** $(2, -1)$ **33.** $\left(\frac{231}{202}, \frac{117}{202}\right)$ **35.** 200 miles **37.** 50° and 130° **39.** 62° and 28° **41.** Hay: 415 hectares; oats: 235 hectares **43.** $\dfrac{a^7}{b^9}$ **45.** $\dfrac{x - 3}{x + 2}$ **47.** $\dfrac{-x^2 - 7x + 23}{(x + 3)(x - 4)}$ **49.** Will is 6; his father is 30. **51.** Base: 10 ft; height: 5 ft **53.** (5, 2) **55.** $(0, -1)$

Exercise Set 7.4, p. 483

1. Huxtables: $\frac{28}{3}$ bags; Simpsons: $\frac{14}{3}$ bags **3.** Kuyatts': 32 years; Marconis': 16 years **5.** Randy is 24; Mandy is 6. **7.** 70 dimes; 33 quarters **9.** Soda: \$0.49; pizza: \$1.50 **11.** 128 cardholders; 75 noncardholders **13.** 17 Upper Box, 12 Lower Reserved

15.

A	B	MIXTURE	
x	y	100 liters	$\rightarrow x + y = 100$
50%	80%	68%	
50%x	80%y	68% × 100, or 68 liters	$\rightarrow 50\%x + 80\%y = 68$

40 L of A; 60 L of B

17. 0.4 gal of Skylite Pink; 0.6 gal of MacIntosh Red

19.

BRAZILIAN	TURKISH	MIXTURE	
\$5	\$8	\$7	
x	y	300	$\rightarrow x + y = 300$
$5x$	$8y$	\$7(300), or \$2100	$\rightarrow 5x + 8y = 2100$

100 lb of Brazilian;
200 lb of Turkish

21. 39 lb of A; 36 lb of B **23.** 12 of A; 4 of B **25.** $6\frac{17}{25}$ large type; $5\frac{8}{25}$ small type **27.** $4\frac{4}{7}$ L **29.** 550 **31.** 43.75 L **33.** \$12,500 at 12%; \$14,500 at 13% **35.** 0.8 oz of flavoring; 19.2 oz of sugar **37.** Bat: \$119.95; ball: \$4.95; glove: \$79.95

Exercise Set 7.5, p. 491

1. 4.5 hr

SPEED	TIME
30	t
46	t

3.

SPEED	TIME	
72	$t + 3$	$\rightarrow d = 72(t + 3)$
120	t	$\rightarrow d = 120t$

$7\frac{1}{2}$ hr after the first train leaves, or $4\frac{1}{2}$ hr after the second train leaves

5.

SPEED	TIME	
$r + 6$	4	$\rightarrow d = (r + 6)4$
$r - 6$	10	$\rightarrow d = (r - 6)10$

14 km/h

7. 384 km **9.** 330 mph **11.** $1\frac{23}{43}$ min after the toddler starts running, or $\frac{23}{43}$ min after the mother starts running **13.** 15 mi **15.** $\frac{x + 2}{x + 3}$ **17.** $\frac{3(x + 4)}{x - 1}$ **19.** $\frac{x^5y^3}{2}$ **21.** Approximately 3603 mi **23.** $5\frac{1}{3}$ mi

Summary and Review: Chapter 7, p. 495

1. [7.1a] No **2.** [7.1a] Yes **3.** [7.1a] Yes **4.** [7.1a] No **5.** [7.1b] (6, −2) **6.** [7.1b] (6, 2) **7.** [7.1b] (0, 5) **8.** [7.1b] No solution **9.** [7.2a] (0, 5) **10.** [7.2b] (−2, 4) **11.** [7.2b] (1, −2) **12.** [7.2a] (−3, 9) **13.** [7.2a] (1, 4) **14.** [7.2a] (3, −1) **15.** [7.3a] (3, 1) **16.** [7.3a] (1, 4) **17.** [7.3a] (5, −3) **18.** [7.3b] (−4, 1) **19.** [7.3b] (−2, 4) **20.** [7.3b] (−2, −6) **21.** [7.3b] (3, 2) **22.** [7.3b] (2, −4) **23.** [7.3b] Infinite number of solutions **24.** [7.2c] 10 and −2 **25.** [7.2c] 12 and 15 **26.** [7.3c] Length: 37.5 cm; width: 10.5 cm **27.** [7.5a] 135 km/h **28.** [7.4a] 297 orchestra, 211 balcony **29.** [7.4a] 40 L of each **30.** [7.4a] Jeff: 27; son: 9 **31.** [7.4a] Asian: 4800 kg; African: 7200 kg **32.** [3.1d] t^8 **33.** [3.1e] $\frac{1}{t^{18}}$ **34.** [5.5a] $\frac{-4x + 3}{(x - 2)(x - 3)(x + 3)}$ **35.** [6.3a]

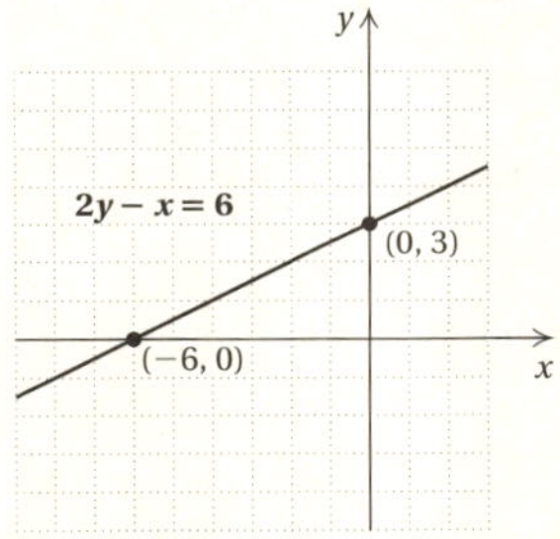

36. [7.1a] $C = 1, D = 3$ **37.** [1.7c], [7.2a] (2, 0) **38.** [7.4a] \$96 **39.** [7.4a] 24

Test: Chapter 7, p. 497

1. [7.1a] No **2.** [7.1b] (2, −1) **3.** [7.2a] (8, −2) **4.** [7.2b] (−1, 3) **5.** [7.2a] No solution **6.** [7.3a] (1, −5) **7.** [7.3b] (12, −6) **8.** [7.3b] (0, 1) **9.** [7.3b] (5, 1) **10.** [7.2c] Length: 2108.5 yd; width: 2024.5 yd **11.** [7.2c] 20 and 8 **12.** [7.5a] 40 km/h **13.** [7.4a] 40 L of A; 20 L of B **14.** [5.5a] $\frac{-x^2 + x + 17}{(x - 4)(x + 4)(x + 1)}$

15. [6.3a]

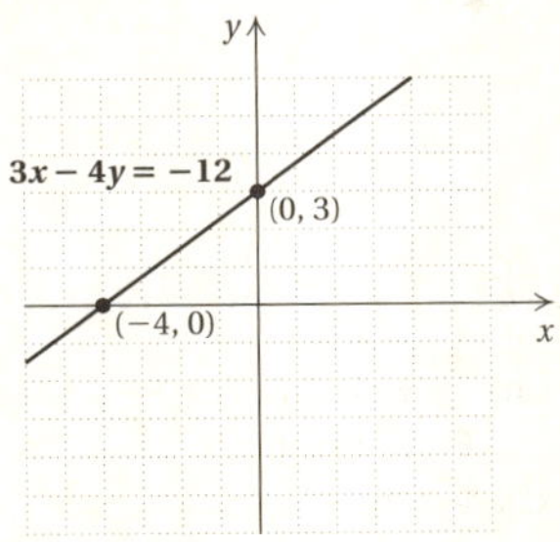

16. [3.1d] $\frac{10x^4}{y^2}$ **17.** [3.1e] $\frac{a^{10}}{b^6}$ **18.** [7.1a] $C = -\frac{19}{2}$; $D = \frac{14}{3}$ **19.** [7.4a] 5

Cumulative Review: Chapters 1–7, p. 499

1. [1.8d] -6.8 **2.** [3.2d] 3.12×10^{-2} **3.** [1.6c] $-\frac{3}{4}$
4. [3.1d, e] $\frac{1}{8}$ **5.** [5.1c] $\frac{x+3}{2x-1}$ **6.** [5.1c] $\frac{t-4}{t+4}$
7. [5.9a] $\frac{x^2(x+1)}{x+4}$ **8.** [3.6a] $2 - 10x^2 + 12x^4$
9. [3.6c] $4a^4b^2 - 20a^3b^3 + 25a^2b^4$ **10.** [3.6b] $9x^4 - 16y^2$
11. [3.5b] $-2x^3 + 4x^4 - 6x^5$ **12.** [3.5d] $8x^3 + 1$
13. [3.6b] $64 - \frac{1}{9}x^2$ **14.** [3.4c] $-y^3 - 2y^2 - 2y + 7$
15. [3.8b] $x^2 - x - 1 + \frac{-2}{2x-1}$ **16.** [5.4a] $\frac{-5x-28}{5x-25}$
17. [5.5a] $\frac{4x-1}{x-2}$ **18.** [5.1d] $\frac{y}{(y-1)^2}$ **19.** [5.2b] $\frac{3(x+1)}{2x}$
20. [4.1b] $3x^2(2x^3 - 12x + 3)$
21. [4.5d] $(4y^2 + 9)(2y + 3)(2y - 3)$
22. [4.3a], [4.4a] $(3x - 2)(x + 4)$ **23.** [4.5b] $(2x^2 - 3y)^2$
24. [4.1b], [4.2a] $3m(m + 5)(m - 3)$
25. [4.6a] $(x + 1)^2(x - 1)$ **26.** [2.3c] -9 **27.** [4.7a] $0, \frac{5}{2}$
28. [2.7e] $\{x \mid x \le 20\}$ **29.** [2.3c] 0.3 **30.** [4.7b] $13, -13$
31. [4.7b] $\frac{5}{3}, 3$ **32.** [5.6a] -1 **33.** [5.6a] No solution
34. [7.2a] $(3, -3)$ **35.** [7.3b] $(-2, 2)$ **36.** [7.3b] Infinite number of solutions **37.** [5.8a] $x = y - 1$
38. [2.4a] 234 **39.** [5.7a] $5\frac{5}{8}$ hr **40.** [4.8a] 6 m, 10 m
41. [5.7b] 204 **42.** [4.8a] 6 cm, 9 cm **43.** [6.7d] 72 ft; 360 **44.** [7.5a] 10 hr **45.** [7.4a] 2.5 L of A; 7.5 L of B
46. [6.7a] $y = 0.2x$ **47.** [6.4a] 0 **48.** [6.4c] $-\frac{2}{3}$, $(0, 2)$
49. [6.4d] $y = -\frac{10}{7}x - \frac{8}{7}$ **50.** [6.4d] $y = 6x - 3$

51. [6.3b]

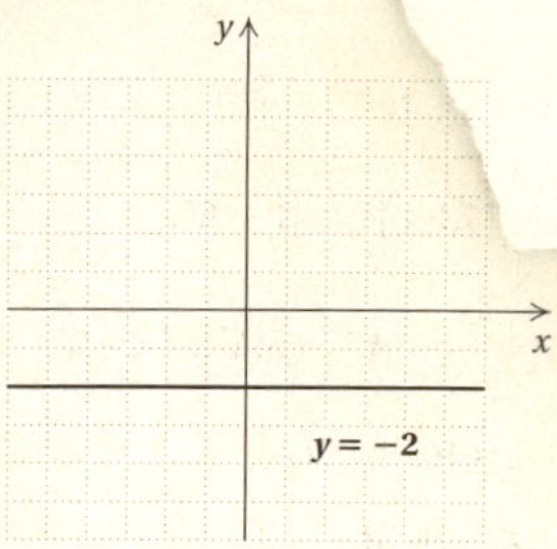

52. [6.3a]

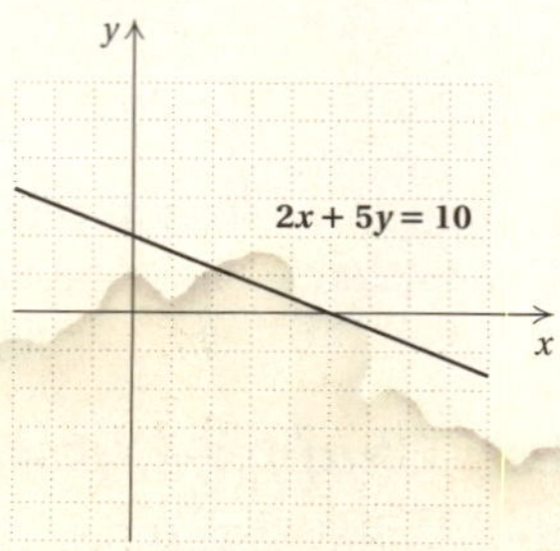

53. [6.6b]

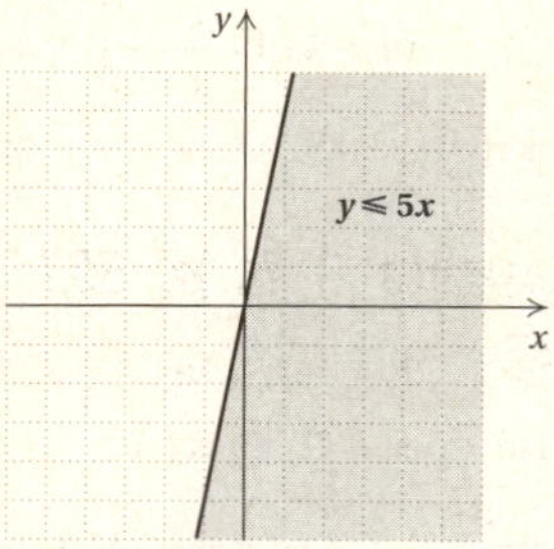

54. [6.6b]

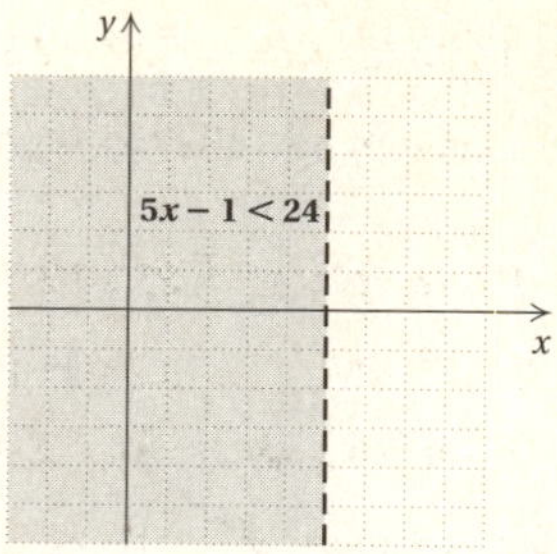

55. [7.1b] No solution **56.** [7.1b] $(1, -1)$
57. [7.1a] $A = -4, B = -\frac{7}{5}$ **58.** [2.7e] No solution
59. [1.8d], [5.2b], [5.5a] 0 **60.** [6.5b] $-\frac{3}{10}$

CHAPTER 8

Pretest: Chapter 8, p. 502

1. [8.1a] $7, -7$ **2.** [8.1d] $3t$ **3.** [8.1e] No **4.** [8.1e] Yes
5. [8.1b] 6.856 **6.** [8.5a] 4 **7.** [8.1f] $2x$ **8.** [8.4a] $12\sqrt{2}$
9. [8.4b] $7 - 4\sqrt{3}$ **10.** [8.4b] 1 **11.** [8.2c] $2\sqrt{15}$
12. [8.4b] $25 - 4\sqrt{6}$ **13.** [8.3a] $\sqrt{5}$ **14.** [8.3b] $2a^2\sqrt{2}$
15. [8.6a] $\sqrt{89} \approx 9.434$ **16.** [8.6b] $\sqrt{193} \approx 13.892$ m
17. [8.3c] $\frac{\sqrt{5x}}{x}$ **18.** [8.4c] $\frac{48 - 8\sqrt{5}}{31}$

Exercise Set 8.1, p. 507

1. $2, -2$ **3.** $3, -3$ **5.** $10, -10$ **7.** $13, -13$ **9.** 2
11. -3 **13.** -6 **15.** -15 **17.** 19 **19.** 2.236
21. 4.123 **23.** 9.539 **25.** **(a)** 13; **(b)** 24 **27.** $a - 4$
29. $t^2 + 1$ **31.** $\frac{3}{x+2}$ **33.** No **35.** Yes **37.** c
39. $3x$ **41.** ab **43.** $34d$ **45.** $x + 3$ **47.** $a - 5$
49. $2a - 5$ **51.** \$5330 **53.** $(x + 1)^2$
55. $\frac{(x+2)(x-2)}{(x+1)(x-1)}$ **57.** 5 **59.** 3.578 **61.** 32.309
63. No solution **65.** $\sqrt{3}$

Exercise Set 8.2, p. 513

1. $2\sqrt{3}$ **3.** $5\sqrt{3}$ **5.** $2\sqrt{5}$ **7.** $10\sqrt{6}$ **9.** $3\sqrt{x}$
11. $4\sqrt{3x}$ **13.** $4\sqrt{a}$ **15.** $8y$ **17.** $x\sqrt{13}$ **19.** $2t\sqrt{2}$
21. $6\sqrt{5}$ **23.** $12\sqrt{2y}$ **25.** $2x\sqrt{7}$ **27.** $\sqrt{2}\,(2x + 1)$
29. $\sqrt{y}(6 + y)$ **31.** x^3 **33.** x^6 **35.** $x^2\sqrt{x}$ **37.** $t^9\sqrt{t}$
39. $(y - 2)^4$ **41.** $2(x + 5)^5$ **43.** $6m\sqrt{m}$ **45.** $2a^2\sqrt{2a}$
47. $2p^8\sqrt{26p}$ **49.** $8x^3y\sqrt{7y}$ **51.** $3\sqrt{6}$ **53.** $3\sqrt{10}$
55. $6\sqrt{7x}$ **57.** $6\sqrt{xy}$ **59.** 13 **61.** $5b\sqrt{3}$ **63.** $2t$
65. $a\sqrt{bc}$ **67.** $2xy\sqrt{2xy}$ **69.** 18 **71.** $\sqrt{10x - 5}$
73. $x + 2$ **75.** $6xy^3\sqrt{3xy}$ **77.** $10x^2y^3\sqrt{5xy}$ **79.** 11.180
81. 18.972 **83.** 26.460 **85.** 11.043 **87.** 20 mph; 54.8 mph **89.** $(-2, 4)$ **91.** $(2, 1)$ **93.** 25,704 ft^2
95. $\sqrt{x - 2}\sqrt{x + 1}$ **97.** $\sqrt{2x + 3}\sqrt{x - 4}$
99. $\sqrt{a + b}\sqrt{a - b}$ **101.** 0.5 **103.** $3a^3$ **105.** $4y\sqrt{3}$
107. $18(x + 1)\sqrt{y(x + 1)}$ **109.** $2x^3\sqrt{5x}$

Exercise Set 8.3, p. 521

1. 3 **3.** 6 **5.** $\sqrt{5}$ **7.** $\frac{1}{5}$ **9.** $\frac{2}{5}$ **11.** 2 **13.** $3y$ **15.** $\frac{4}{7}$
17. $\frac{1}{6}$ **19.** $-\frac{4}{9}$ **21.** $\frac{8}{17}$ **23.** $\frac{13}{14}$ **25.** $\frac{5}{x}$ **27.** $\frac{3a}{25}$
29. $\frac{\sqrt{10}}{5}$ **31.** $\frac{\sqrt{14}}{4}$ **33.** $\frac{\sqrt{3}}{6}$ **35.** $\frac{\sqrt{10}}{6}$ **37.** $\frac{3\sqrt{5}}{5}$
39. $\frac{2\sqrt{6}}{3}$ **41.** $\frac{\sqrt{3x}}{x}$ **43.** $\frac{\sqrt{xy}}{y}$ **45.** $\frac{x\sqrt{5}}{10}$ **47.** $\frac{\sqrt{14}}{2}$
49. $\frac{3\sqrt{2}}{4}$ **51.** $\frac{\sqrt{6}}{2}$ **53.** $\sqrt{2}$ **55.** $\frac{\sqrt{55}}{11}$ **57.** $\frac{\sqrt{21}}{6}$
59. $\frac{\sqrt{6}}{2}$ **61.** 5 **63.** $\frac{\sqrt{3x}}{x}$ **65.** $\frac{4y\sqrt{5}}{5}$ **67.** $\frac{a\sqrt{2a}}{4}$
69. $\frac{\sqrt{42x}}{3x}$ **71.** $\frac{3\sqrt{6}}{8c}$ **73.** $\frac{y\sqrt{xy}}{x}$ **75.** $\frac{3n\sqrt{10}}{8}$

77. 0.577 **79.** 0.935 **81.** 0.289 **83.** 0.707 **85.** 3.801 **87.** 0.850 **89.** (4, 2) **91.** No solution **93.** $\left(-\frac{5}{2}, -\frac{9}{2}\right)$ **95.** 1.57 sec; 3.14 sec; 8.88 sec; 11.10 sec **97.** 1 sec **99.** $\frac{\sqrt{5}}{40}$ **101.** $\frac{\sqrt{5x}}{5x^2}$ **103.** $\frac{\sqrt{3ab}}{b}$ **105.** $\frac{3\sqrt{10}}{100}$ **107.** $\frac{(y-x)}{xy}$

Exercise Set 8.4, p. 529

1. $16\sqrt{3}$ **3.** $4\sqrt{5}$ **5.** $13\sqrt{x}$ **7.** $-9\sqrt{d}$ **9.** $25\sqrt{2}$ **11.** $\sqrt{3}$ **13.** $\sqrt{5}$ **15.** $13\sqrt{2}$ **17.** $3\sqrt{3}$ **19.** $2\sqrt{2}$ **21.** 0 **23.** $(2 + 9x)\sqrt{x}$ **25.** $(3 - 2x)\sqrt{3}$ **27.** $3\sqrt{2x + 2}$ **29.** $(x + 3)\sqrt{x^3 - 1}$ **31.** $(4a^2 + a^2b - 5b)\sqrt{b}$ **33.** $\frac{2\sqrt{3}}{3}$ **35.** $\frac{13\sqrt{2}}{2}$ **37.** $\frac{\sqrt{6}}{6}$ **39.** $\sqrt{15} - \sqrt{3}$ **41.** -62 **43.** 1 **45.** $13 + \sqrt{5}$ **47.** $x - 2\sqrt{xy} + y$ **49.** $-\sqrt{3} - \sqrt{5}$ **51.** $5 - 2\sqrt{6}$ **53.** $\frac{4\sqrt{10} - 4}{9}$ **55.** $5 - 2\sqrt{7}$ **57.** $\frac{12 - 3\sqrt{x}}{16 - x}$ **59.** $\frac{24 + 3\sqrt{x} + 8\sqrt{2} + \sqrt{2x}}{64 - x}$ **61.** $\frac{1}{3}$ hr; the variation constant is the fixed distance. **63.** All **65.** $11\sqrt{3} - 10\sqrt{2}$ **67.** No **69.** True; $(3\sqrt{x + 2})^2 = (3\sqrt{x + 2})(3\sqrt{x + 2}) = (3 \cdot 3)(\sqrt{x + 2} \cdot \sqrt{x + 2}) = 9(x + 2)$.

Exercise Set 8.5, p. 537

1. 36 **3.** 18.49 **5.** 165 **7.** $\frac{621}{2}$ **9.** 5 **11.** 3 **13.** $\frac{17}{4}$ **15.** No solution **17.** No solution **19.** 9 **21.** 12 **23.** 1, 5 **25.** 3 **27.** 5 **29.** No solution **31.** 3 **33.** No solution **35.** 9 **37.** 1 **39.** Approximately 17 km **41.** Approximately 37 m **43.** 180 ft; 500 ft **45.** 12 **47.** $\frac{(x + 7)^2}{x - 7}$ **49.** $\frac{a - 5}{2}$ **51.** 61°, 119° **53.** 2, -2 **55.** $-\frac{57}{16}$ **57.** 13

Exercise Set 8.6, p. 543

1. 17 **3.** $\sqrt{32} \approx 5.657$ **5.** 12 **7.** 4 **9.** 26 **11.** 12 **13.** 2 **15.** $\sqrt{2} \approx 1.414$ **17.** 5 **19.** $\sqrt{18} \approx 4.243$ cm **21.** $\sqrt{26{,}900} \approx 164.012$ yd **23.** About 14,533 ft **25.** $\left(-\frac{3}{2}, -\frac{1}{16}\right)$ **27.** $\left(-\frac{9}{19}, \frac{91}{38}\right)$ **29.** $\sqrt{1525} \approx 39.1$ mi **31.** $12 - 2\sqrt{6} \approx 7.101$ **33.** $\frac{\sqrt{3}}{2} \approx 0.866$

Summary and Review: Chapter 8, p. 547

1. [8.1a] 8, -8 **2.** [8.1a] 20, -20 **3.** [8.1a] 6 **4.** [8.1a] -13 **5.** [8.1b] 1.732 **6.** [8.1b] 9.950 **7.** [8.1b] 10.392 **8.** [8.1b] 17.889 **9.** [8.3d] 0.354 **10.** [8.3d] 0.742 **11.** [8.1d] $x^2 + 4$ **12.** [8.1d] $5ab^3$ **13.** [8.1e] No **14.** [8.1e] Yes **15.** [8.1e] No **16.** [8.1e] No **17.** [8.1e] No **18.** [8.1e] No **19.** [8.1f] m **20.** [8.1f] $x - 4$ **21.** [8.2c] $\sqrt{21}$ **22.** [8.2c] $\sqrt{x^2 - 9}$ **23.** [8.2a] $-4\sqrt{3}$ **24.** [8.2a] $4t\sqrt{2}$ **25.** [8.2a] $x + 8$ **26.** [8.2a] $\sqrt{t - 7}\sqrt{t + 7}$ **27.** [8.2b] x^4 **28.** [8.2b] $m^7\sqrt{m}$ **29.** [8.2c] $2\sqrt{15}$ **30.** [8.2c] $2x\sqrt{10}$ **31.** [8.2c] $5xy\sqrt{2}$ **32.** [8.2c] $10a^2b\sqrt{ab}$ **33.** [8.3b] $\frac{5}{8}$ **34.** [8.3b] $\frac{2}{3}$ **35.** [8.3b] $\frac{7}{t}$ **36.** [8.3c] $\frac{\sqrt{2}}{2}$ **37.** [8.3c] $\frac{\sqrt{2}}{4}$ **38.** [8.3c] $\frac{\sqrt{5y}}{y}$ **39.** [8.3c] $\frac{2\sqrt{3}}{3}$ **40.** [8.3a, c] $\frac{\sqrt{15}}{5}$ **41.** [8.3a, c] $\frac{x\sqrt{30}}{6}$ **42.** [8.4c] $8 - 4\sqrt{3}$ **43.** [8.4a] $13\sqrt{5}$ **44.** [8.4a] $\sqrt{5}$ **45.** [8.4a] $\frac{\sqrt{2}}{2}$ **46.** [8.4b] $7 + 4\sqrt{3}$ **47.** [8.4b] 1 **48.** [8.6a] 20 **49.** [8.6a] $\sqrt{3} \approx 1.732$ **50.** [8.6b] $\sqrt{98} \approx 9.899$ m **51.** [8.5a] 52 **52.** [8.5a] No solution **53.** [8.5b] 9 **54.** [8.5a] 0, 3 **55.** [8.5c] 405 ft **56.** [7.3c] 4300 **57.** [6.7b] $450 **58.** [7.3b] $\left(\frac{22}{17}, -\frac{8}{17}\right)$ **59.** [5.2b] $\frac{(x - 5)(x - 7)}{(x + 7)(x + 5)}$ **60.** [8.1a] 2 **61.** [8.5a] $b = \sqrt{A^2 - a^2}$

Test: Chapter 8, p. 549

1. [8.1a] 9, -9 **2.** [8.1a] 8 **3.** [8.1a] -5 **4.** [8.1b] 10.770 **5.** [8.1b] 9.327 **6.** [8.3d] 1.732 **7.** [8.1d] $4 - y^3$ **8.** [8.1e] Yes **9.** [8.1e] No **10.** [8.1f] a **11.** [8.1f] $6y$ **12.** [8.2c] $\sqrt{30}$ **13.** [8.2c] $\sqrt{x^2 - 64}$ **14.** [8.2a] $3\sqrt{3}$ **15.** [8.2a] $5\sqrt{x - 1}$ **16.** [8.2b] $t^2\sqrt{t}$ **17.** [8.2c] $5\sqrt{2}$ **18.** [8.2c] $3ab^2\sqrt{2}$ **19.** [8.3b] $\frac{3}{2}$ **20.** [8.3b] $\frac{12}{a}$ **21.** [8.3c] $\frac{\sqrt{10}}{5}$ **22.** [8.3c] $\frac{\sqrt{2xy}}{y}$ **23.** [8.3a, c] $\frac{3\sqrt{6}}{8}$ **24.** [8.3a] $\frac{\sqrt{7}}{4y}$ **25.** [8.4a] $-6\sqrt{2}$ **26.** [8.4a] $\frac{6\sqrt{5}}{5}$ **27.** [8.4b] $21 - 8\sqrt{5}$ **28.** [8.4b] 11 **29.** [8.4c] $\frac{40 + 10\sqrt{5}}{11}$ **30.** [8.6a] $\sqrt{80} \approx 8.944$ **31.** [8.5a] 48 **32.** [8.5a] 2, -2 **33.** [8.5b] -3 **34.** [8.5c] About 5000 m **35.** [7.3c] 789.25 yd^2 **36.** [6.7b] $15{,}686\frac{2}{3}$ **37.** [7.3b] (0, 2) **38.** [5.2b] $\frac{x - 7}{x + 6}$ **39.** [8.1a] $\sqrt{5}$ **40.** [8.2b] y^{8n}

Cumulative Review: Chapters 1–8, p. 551

1. [3.3a] -15 **2.** [3.3e] $-4x^3 - \frac{1}{7}x^2 - 2$ **3.** [5.1a] $-\frac{1}{2}$ **4.** [8.1e] No **5.** [8.4b] 1 **6.** [8.1a] -14 **7.** [8.2c] 15 **8.** [8.4b] $3 - 2\sqrt{2}$ **9.** [8.3a, c] $\frac{9\sqrt{10}}{25}$ **10.** [8.4a] $12\sqrt{5}$ **11.** [3.7f] $9x^8 - 4y^{10}$ **12.** [3.6c] $x^4 + 8x^2 + 16$ **13.** [3.6a] $8x^2 - \frac{1}{8}$ **14.** [5.5a] $\frac{4x + 2}{2x - 1}$ **15.** [3.4a, c] $-3x^3 + 8x^2 - 5x$ **16.** [5.1d] $\frac{2(x - 5)}{3(x - 1)}$ **17.** [5.2b] $\frac{(x + 1)(x - 3)}{2(x + 3)}$ **18.** [3.8b] $3x^2 + 4x + 9 + \frac{13}{x - 2}$

19. [8.2a] $\sqrt{2}(x-1)$ **20.** [3.1d] $\frac{1}{x^{12}}$ **21.** [8.3b] $\frac{5}{x^4}$
22. [5.9a] $2(x-1)$ **23.** [4.5d] $3(1+2x^4)(1-2x^4)$
24. [4.1b] $4t(3-t-12t^3)$
25. [4.3a], [4.4a] $2(3x-2)(x-4)$
26. [4.6a] $(2x+1)(2x-1)(x+1)$
27. [4.5b] $(4x^2-7)^2$ **28.** [4.2a] $(x+15)(x-12)$
29. [4.7b] 0, −17 **30.** [2.7e] $\{x \mid x > -6\}$ **31.** [5.6a] $-\frac{12}{5}$
32. [4.7b] −5, 6 **33.** [2.7e] $\left\{x \mid x \leq -\frac{9}{2}\right\}$ **34.** [4.7b] 9, −9
35. [8.5a] 41 **36.** [8.5b] 9 **37.** [2.3b] $\frac{2}{5}$ **38.** [5.6a] $\frac{1}{3}$
39. [7.2a] (4, 1) **40.** [7.3a] (3, −8) **41.** [5.8a] $R = \frac{r^2}{E-r}$
42. [2.6a] $p = \frac{4A}{r+q}$
43. [6.6b]

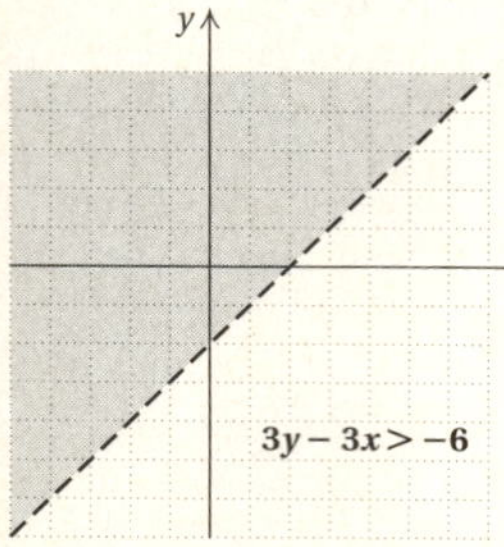

44. [6.3b]

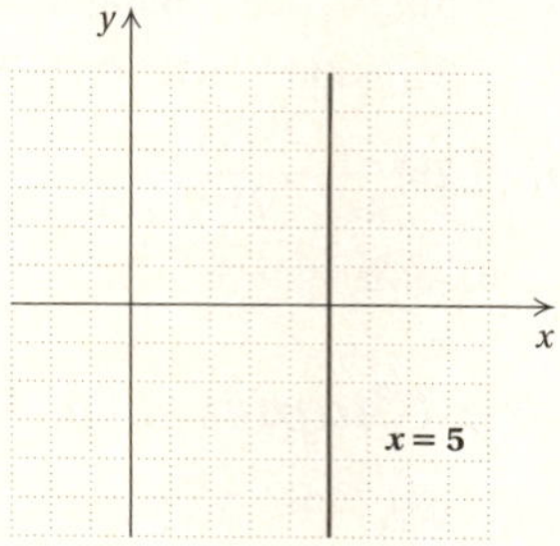

45. [6.3a]

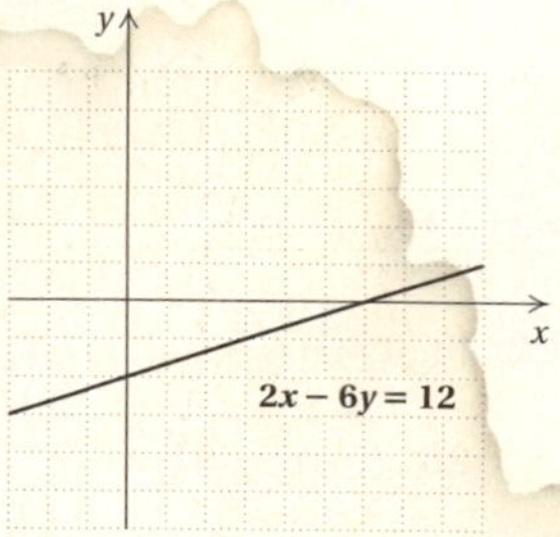

46. [7.1b] (1, 4) **47.** [6.4d] $y = \frac{11}{4}x - \frac{19}{4}$
48. [6.4c] $\frac{5}{3}$, (0, −3) **49.** [6.5b] Perpendicular
50. [6.5a] Parallel **51.** [2.4a] 38°, 76°, 66°
52. [7.4a] Hamburger: \$1.30; milkshake: \$0.90
53. [8.6b] $4\sqrt{3} \approx 6.9$ m **54.** [6.7b] \$7500 **55.** [5.7b] 20
56. [4.8a] 15 m, 12 m **57.** [7.4a] 65 dimes, 50 quarters
58. [5.7b] 357 **59.** [2.5a] \$2600 **60.** [5.7a] 60 mph
61. [1.2d] $<$ **62.** [1.2d, e] $>$ **63.** [7.4a] 300 L
64. [7.3b], [8.5a] (9, 4)

CHAPTER 9

Pretest: Chapter 9, p. 554

1. [9.1c] 3 **2.** [9.2a] $\sqrt{7}, -\sqrt{7}$ **3.** [9.3a] $\frac{-3 \pm \sqrt{21}}{6}$
4. [9.1b] $0, \frac{3}{5}$ **5.** [9.1b] $0, -\frac{5}{3}$ **6.** [9.2b] $-4 \pm \sqrt{5}$
7. [9.2c] $1 \pm \sqrt{6}$ **8.** [9.4a] $n = \frac{p \pm \sqrt{p^2+4A}}{2}$
9. [9.5a] Width: 4 cm; length: 12 cm
10. [9.6b] $\left(\frac{-1+\sqrt{33}}{4}, 0\right), \left(\frac{-1-\sqrt{33}}{4}, 0\right)$
11. [9.5a] 10 km/h
12. [9.6a]

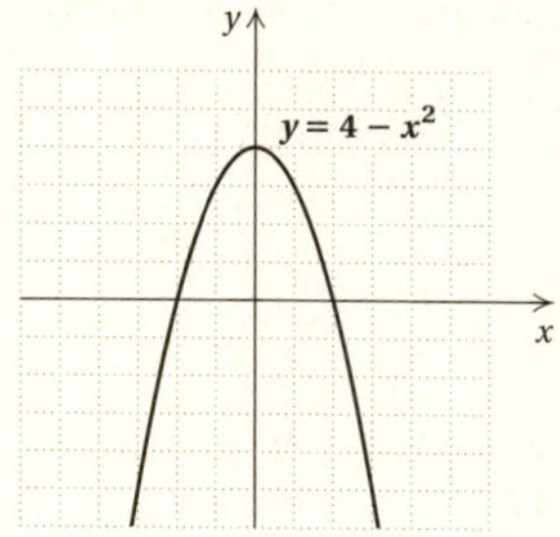

Exercise Set 9.1, p. 559

1. $a=1, b=-3, c=2$ **3.** $7x^2-4x+3=0$; $a=7$, $b=-4$, $c=3$ **5.** $2x^2-3x+5=0$; $a=2$, $b=-3$, $c=5$ **7.** 0, −5 **9.** 0, −2 **11.** $0, \frac{2}{5}$ **13.** 0, −1
15. 0, 3 **17.** $0, \frac{1}{5}$ **19.** $0, \frac{3}{14}$ **21.** $0, \frac{81}{2}$ **23.** −12, 4
25. −5, −1 **27.** −9, 2 **29.** 5, 3 **31.** −5 **33.** 4
35. $-\frac{2}{3}, \frac{1}{2}$ **37.** $4, -\frac{2}{3}$ **39.** $\frac{5}{3}, -1$ **41.** −1, −5
43. −2, 7 **45.** 4, −5 **47.** 4 **49.** 1, −2 **51.** $10, -\frac{2}{5}$
53. 6, −4 **55.** 1 **57.** 5, 2 **59.** No solution **61.** 35
63. 7 **65.** $2\sqrt{5}$ **67.** $\frac{9}{8}$ **69.** $9\sqrt{5}$ **71.** $-\frac{1}{3}, 1$
73. $0, \frac{\sqrt{5}}{5}$ **75.** $-\frac{5}{2}, 1$ **77.** $0, -\frac{b}{a}$ **79.** 81, 1

Exercise Set 9.2, p. 567

1. 11, −11 **3.** $\sqrt{7}, -\sqrt{7}$ **5.** $\frac{\sqrt{15}}{5}, -\frac{\sqrt{15}}{5}$ **7.** $\frac{5}{2}, -\frac{5}{2}$
9. $\frac{7\sqrt{3}}{3}, -\frac{7\sqrt{3}}{3}$ **11.** $\sqrt{3}, -\sqrt{3}$ **13.** $\frac{8}{7}, -\frac{8}{7}$ **15.** −7, 1
17. $-3 \pm \sqrt{21}$ **19.** $-13 \pm 2\sqrt{2}$ **21.** $7 \pm 2\sqrt{3}$
23. $-9 \pm \sqrt{34}$ **25.** $\frac{-3 \pm \sqrt{14}}{2}$ **27.** 11, −5 **29.** 1, −15
31. −2, 8 **33.** −21, −1 **35.** $1 \pm \sqrt{6}$ **37.** $11 \pm \sqrt{19}$
39. $-5 \pm \sqrt{29}$ **41.** $\frac{7 \pm \sqrt{57}}{2}$ **43.** −7, 4
45. $\frac{-3 \pm \sqrt{17}}{4}$ **47.** $\frac{-3 \pm \sqrt{145}}{4}$ **49.** $\frac{-2 \pm \sqrt{7}}{3}$
51. $-\frac{1}{2}, 5$ **53.** $-\frac{5}{2}, \frac{2}{3}$ **55.** About 9.5 sec **57.** About 4.4 sec **59.** $y = \frac{141}{x}$ **61.** $3x\sqrt{2}$ **63.** $3t$ **65.** 12, −12
67. $16\sqrt{2}, -16\sqrt{2}$ **69.** $2\sqrt{c}, -2\sqrt{c}$ **71.** Approximately 49.896, −49.896 **73.** 8, −8 **75.** $\sqrt{11}, -\sqrt{11}$

Exercise Set 9.3, p. 573

1. −3, 7 **3.** 3 **5.** $-\frac{4}{3}, 2$ **7.** $\frac{3}{2}, -\frac{5}{2}$ **9.** −3, 3
11. $1 \pm \sqrt{3}$ **13.** $5 \pm \sqrt{3}$ **15.** $-2 \pm \sqrt{7}$ **17.** $\frac{-4 \pm \sqrt{10}}{3}$
19. $\frac{5 \pm \sqrt{33}}{4}$ **21.** $\frac{1 \pm \sqrt{3}}{2}$ **23.** No real-number solutions **25.** $\frac{5 \pm \sqrt{73}}{6}$ **27.** $\frac{3 \pm \sqrt{29}}{2}$ **29.** $\sqrt{5}, -\sqrt{5}$

31. $-2 \pm \sqrt{3}$ **33.** $\frac{5 \pm \sqrt{37}}{2}$ **35.** $-1.3, 5.3$
37. $-0.2, 6.2$ **39.** $-1.2, 0.2$ **41.** $2.4, 0.3$ **43.** $3x^2\sqrt{3x}$
45. $4\sqrt{5}$ **47.** $2\sqrt{2}$ **49.** $0, 2$ **51.** $\frac{3 \pm \sqrt{5}}{2}$
53. $\frac{-7 \pm \sqrt{61}}{2}$ **55.** $\frac{-2 \pm \sqrt{10}}{2}$

Exercise Set 9.4, p. 577

1. $Q = \frac{P^2}{289}$ **3.** $E = \frac{mv^2}{2g}$ **5.** $r = \frac{1}{2}\sqrt{\frac{S}{\pi}}$
7. $A = \frac{-m + \sqrt{m^2 + 4kP}}{2k}$ **9.** $a = \sqrt{c^2 - b^2}$
11. $t = \frac{\sqrt{s}}{4}$ **13.** $r = \frac{-\pi h + \sqrt{\pi^2 h^2 + \pi A}}{\pi}$
15. $v = 20\sqrt{\frac{F}{A}}$ **17.** $a = \sqrt{c^2 - b^2}$ **19.** $a = \frac{2h\sqrt{3}}{3}$
21. $T = \frac{2 + \sqrt{4 - a(m - n)}}{a}$ **23.** $T = \frac{v^2 \pi m}{8k}$
25. $x = \frac{d\sqrt{3}}{3}$ **27.** $n = \frac{1 + \sqrt{1 + 8N}}{2}$ **29.** $\sqrt{65} \approx 8.062$
31. $\sqrt{41} \approx 6.403$ **33.** $\sqrt{1084} \approx 32.924$ **35.** **(a)** $r = \frac{C}{2\pi}$;
(b) $A = \frac{C^2}{4\pi}$ **37.** $\frac{1}{3a}, 1$

Sidelights: Handling Dimension Symbols, p. 582

1. $15\ \frac{\text{mi}}{\text{hr}}$ **2.** $32\ \frac{\text{km}}{\text{hr}}$ **3.** $3.3\ \frac{\text{m}}{\text{sec}}$ **4.** $19\ \frac{\text{ft}}{\text{min}}$ **5.** 68 ft
6. 28 km/hr **7.** 11 g **8.** 8.6 lb **9.** $4\ \frac{\text{in.-lb}}{\text{sec}}$
10. $24\ \frac{\text{man-hr}}{\text{da}}$ **11.** 12 yd **12.** 220 mi **13.** 16 ft^3
14. $\frac{1}{4}\ \frac{\text{lb}^2}{\text{ft}^2}$ **15.** $\frac{\$970}{\text{day}}$ **16.** $\frac{\$3.20}{\text{hr}}$

Exercise Set 9.5, p. 583

1. Length: 10 cm; width: 7 cm **3.** 10 yd; 24 yd
5. Length: 20 cm; width: 16 cm **7.** Length: 10 m; width: 5 m **9.** 4.6 m; 6.6 m **11.** Length: 5.6 in.; width: 3.6 in. **13.** Length: 6.4 cm; width: 3.2 cm
15. 3 cm **17.** 7 km/h **19.** 8 mph **21.** 4 km/h
23. 36 mph **25.** 1 km/h **27.** $8\sqrt{2}$ **29.** $(2x - 7)\sqrt{x}$
31. $\frac{3\sqrt{2}}{2}$ **33.** 12 and 13; -13 and -12 **35.** $1 + \sqrt{2} \approx$ 2.41 cm **37.** $10\sqrt{2} \approx 14.14$ in.; two 10-in. pizzas

Exercise Set 9.6, p. 591

1. (0, 1)

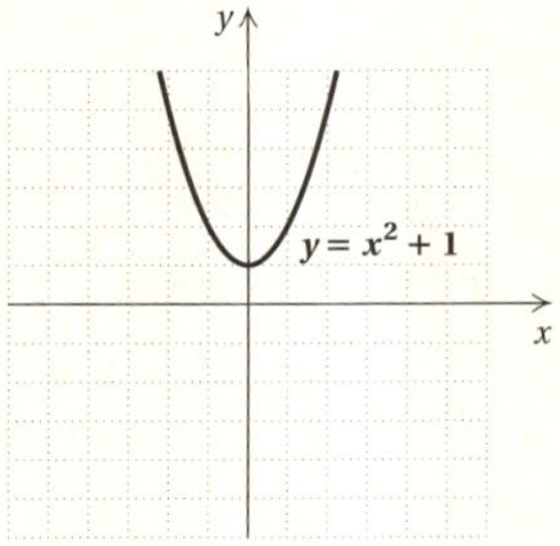

3. (0, 0)

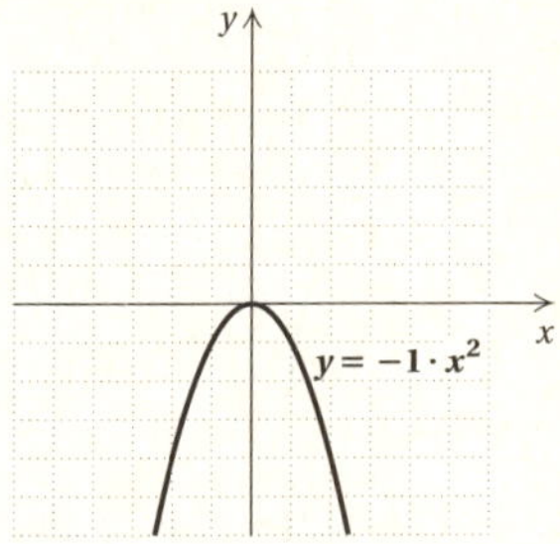

5. (1, 1)

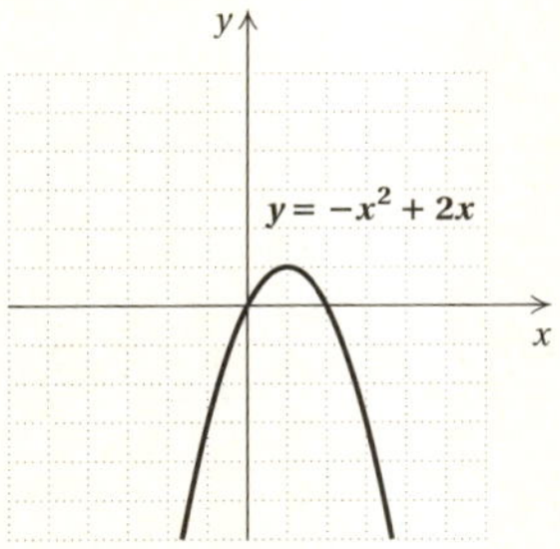

7. $\left(-\frac{1}{2}, \frac{21}{4}\right)$

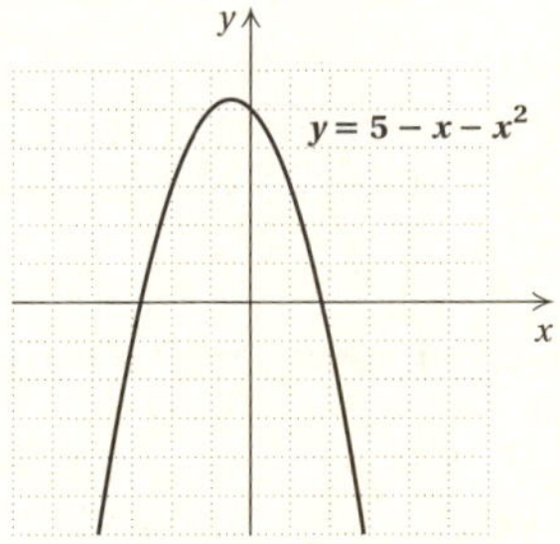

9. (1, 0)

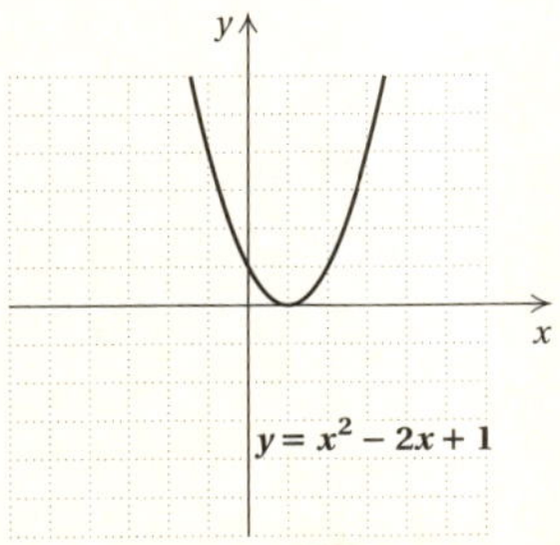

11. (1, 4)

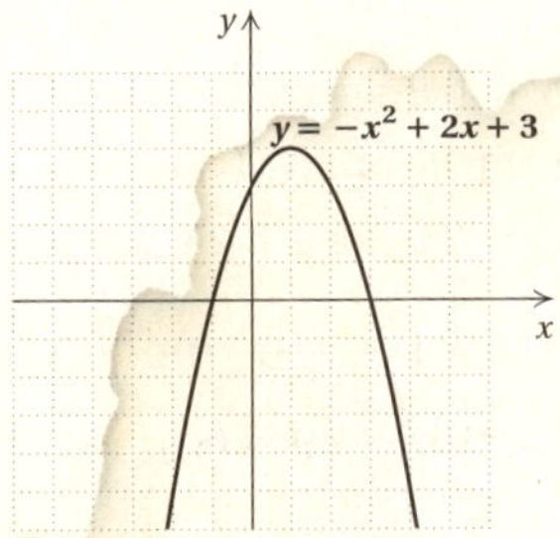

13. (−1, 3)

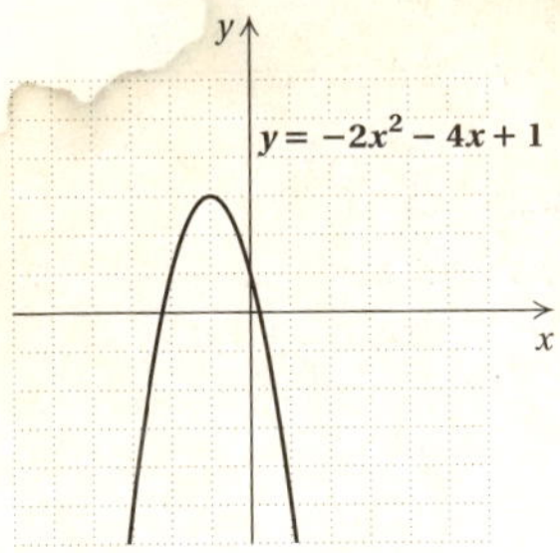

15.

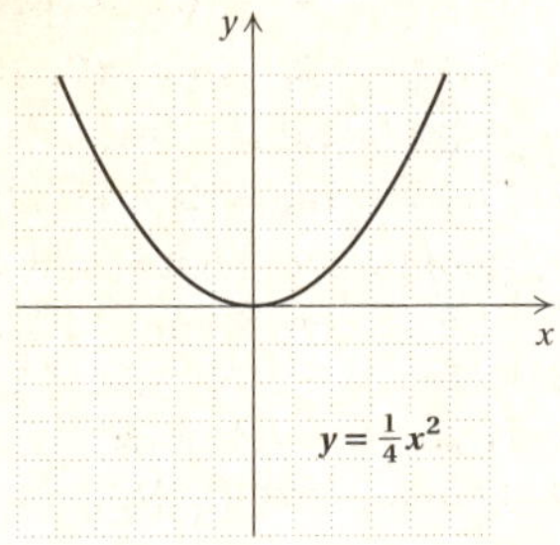

17.

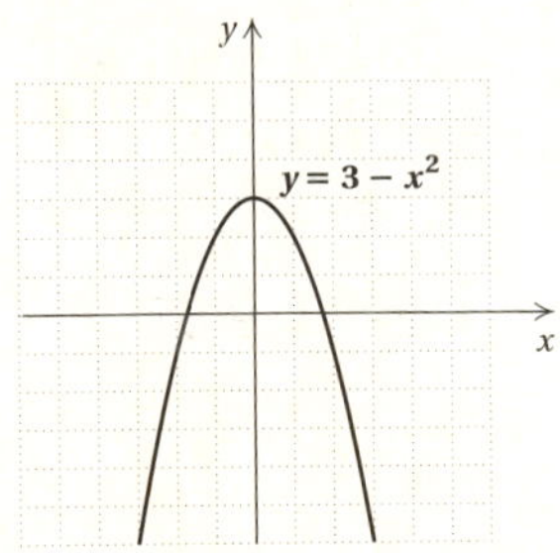

19.

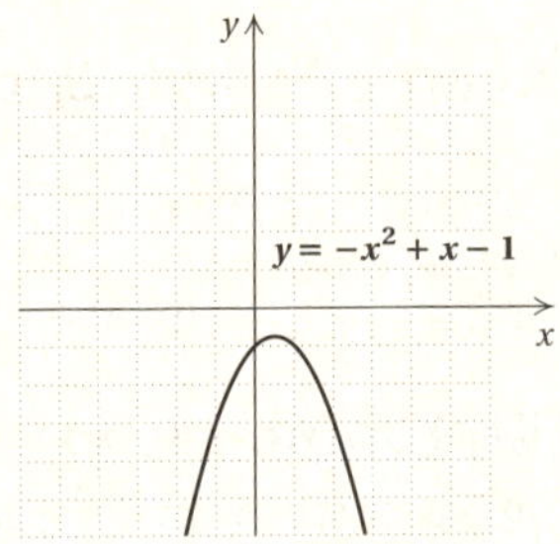

21.

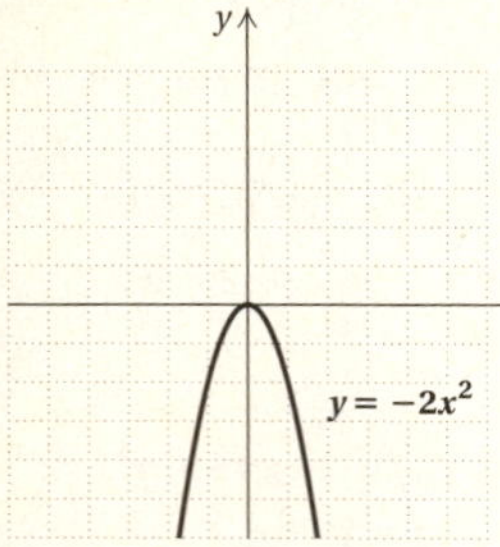

23.

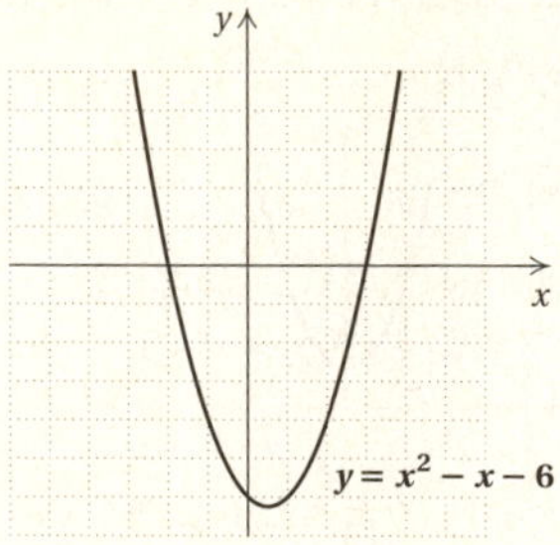

25. $(-\sqrt{2}, 0)$; $(\sqrt{2}, 0)$ **27.** $(-5, 0)$; $(0, 0)$

29. $\left(\dfrac{-1 - \sqrt{33}}{2}, 0\right)$; $\left(\dfrac{-1 + \sqrt{33}}{2}, 0\right)$ **31.** $(3, 0)$

33. $(-2 - \sqrt{5}, 0)$; $(-2 + \sqrt{5}, 0)$ **35.** None

37. $(x + 2)\sqrt{x - 1}$ **39.** $2\sqrt{7}$ **41.** **(a)** After 2 sec; after 4 sec; **(b)** after 3 sec; **(c)** after 6 sec

Summary and Review: Chapter 9, p. 595

1. [9.2a] $\sqrt{3}, -\sqrt{3}$ **2.** [9.1c] $\frac{3}{5}$, 1 **3.** [9.3a] $1 \pm \sqrt{11}$

4. [9.1c] $\frac{1}{3}$, -2 **5.** [9.2b] $-8 \pm \sqrt{13}$ **6.** [9.2a] 0

7. [9.1b] 0, $\frac{7}{5}$ **8.** [9.3a] $\dfrac{1 \pm \sqrt{10}}{3}$ **9.** [9.3a] $-3 \pm 3\sqrt{2}$

10. [9.3a] $\dfrac{2 \pm \sqrt{3}}{2}$ **11.** [9.3a] $\dfrac{3 \pm \sqrt{33}}{2}$

12. [9.3a] No real-number solutions **13.** [9.1b] 0, $\frac{4}{3}$

14. [9.2a] $-2\sqrt{2}, 2\sqrt{2}$ **15.** [9.1c] 3, -5 **16.** [9.1c] 1

17. [9.2c] $\frac{5}{3}$, -1 **18.** [9.2c] $\dfrac{5 \pm \sqrt{17}}{2}$ **19.** [9.3b] 4.6, 0.4

20. [9.3b] -1.9, -0.1 **21.** [9.4a] $T = L(4V^2 - 1)$

22. [9.6a]

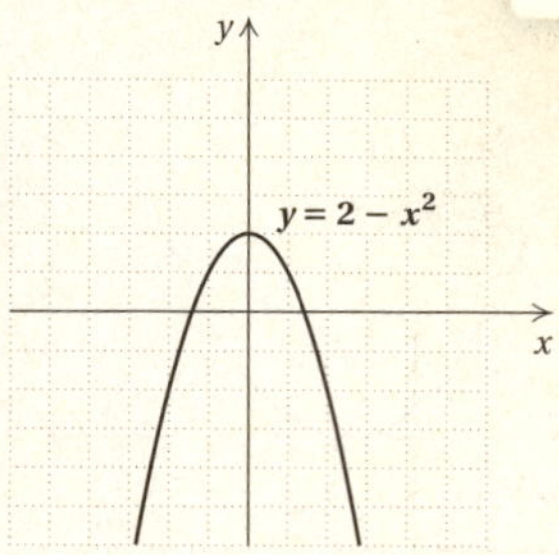

23. [9.6a]

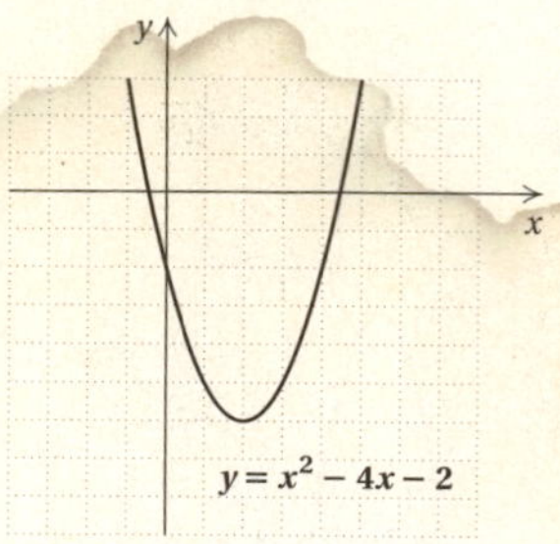

24. [9.6b] $(-\sqrt{2}, 0)$; $(\sqrt{2}, 0)$ **25.** [9.6b] $(2 - \sqrt{6}, 0)$; $(2 + \sqrt{6}, 0)$ **26.** [9.5a] 1.7 m, 4.7 m **27.** [9.5a] Length: 12 m; width: 11 m **28.** [9.5a] 30 km/h **29.** [8.2c] $6\sqrt{a}$

30. [8.2c] $2xy\sqrt{15y}$ **31.** [6.7c] $y = \dfrac{1}{x}$ **32.** [8.6b] $\sqrt{3}$

33. [8.4a] $12\sqrt{11}$ **34.** [8.4a] $4\sqrt{10}$ **35.** [9.5a] 31 and 32; -32 and -31 **36.** [9.2c] 14, -14 **37.** [9.1c], [8.5a] 25

38. [9.5a] $5\sqrt{\pi}$

Test: Chapter 9, p. 597

1. [9.2a] $\sqrt{5}, -\sqrt{5}$ **2.** [9.1b] 0, $-\dfrac{8}{7}$ **3.** [9.1c] -8, 6

4. [9.1c] $-\dfrac{1}{3}$, 2 **5.** [9.2b] $8 \pm \sqrt{13}$ **6.** [9.3a] $\dfrac{1 \pm \sqrt{13}}{2}$

7. [9.3a] $\dfrac{3 \pm \sqrt{37}}{2}$ **8.** [9.3a] $-2 \pm \sqrt{14}$

9. [9.3a] $\dfrac{7 \pm \sqrt{37}}{6}$ **10.** [9.1c] 2, -1 **11.** [9.1c] 2, -4

12. [9.2c] $2 \pm \sqrt{14}$ **13.** [9.3b] 5.7, -1.7

14. [9.4a] $n = \dfrac{-b + \sqrt{b^2 + 4ad}}{2a}$

15. [9.6a]

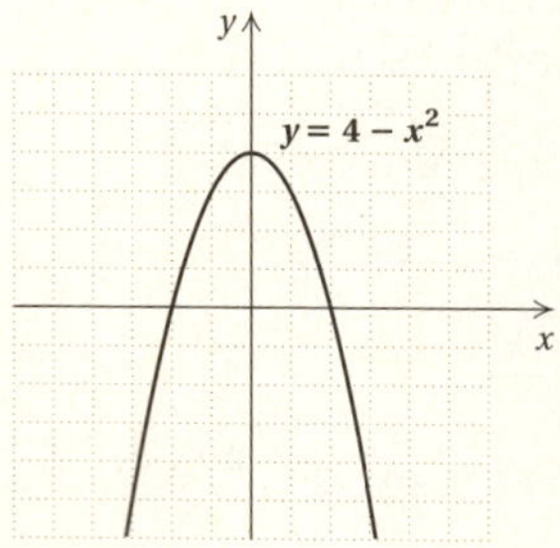

16. [9.6a]

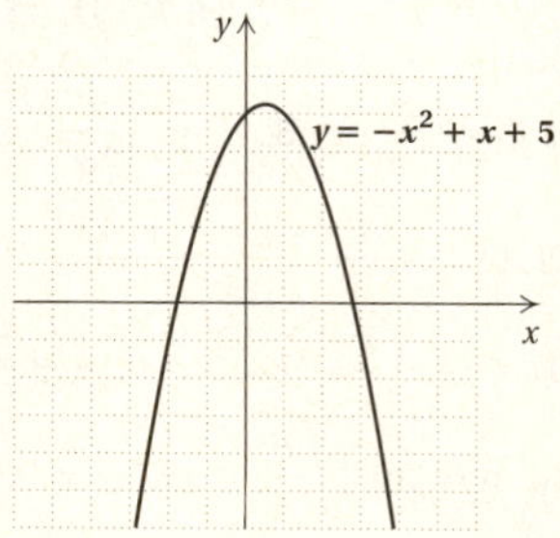

17. [9.6b] $\left(\dfrac{1 - \sqrt{21}}{2}, 0\right)$, $\left(\dfrac{1 + \sqrt{21}}{2}, 0\right)$

18. [9.5a] Length: 6.5 m; width: 2.5 m

19. [9.5a] 24 km/h **20.** [8.4a] $2\sqrt{15}$ **21.** [8.2c] $7xy\sqrt{2x}$

22. [6.7c] $y = \dfrac{4}{x}$ **23.** [8.6b] $\sqrt{5}$ **24.** [9.5a] $5 + 5\sqrt{2}$

25. [7.2b], [9.3a] $1 \pm \sqrt{5}$

Cumulative Review: Chapters 1–9, p. 599

1. [3.1a] $x \cdot x \cdot x$ **2.** [3.1c] 54 **3.** [1.2c] $-0.\overline{27}$

4. [5.3a] 240 **5.** [1.2e] 7 **6.** [1.3a] 9 **7.** [1.4a] 15

8. [1.6c] $-\frac{3}{20}$ **9.** [1.8d] 4 **10.** [1.8b] $-2m - 4$

11. [2.2a] -8 **12.** [2.3b] -12 **13.** [2.3c] 7

14. [4.7b] 3, 5 **15.** [7.2a] (1, 2) **16.** [7.3a] (12, 5)

17. [7.3b] (6, 7) **18.** [4.7b] 3, -2 **19.** [9.3a] $\dfrac{-3 \pm \sqrt{29}}{2}$

20. [8.5a] 2 **21.** [2.7e] $\{x \mid x \geq -1\}$ **22.** [2.3b] 8

23. [2.3b] -3 **24.** [2.7d] $\{x \mid x < -8\}$

25. [2.7e] $\left\{y \mid y \leq \frac{35}{22}\right\}$ **26.** [9.2a] $\sqrt{10}, -\sqrt{10}$

27. [9.2b] $3 \pm \sqrt{6}$ **28.** [5.6a] $\frac{2}{9}$ **29.** [5.6a] -5

30. [5.6a], [9.1b] No solution **31.** [8.5a] 12 **32.** [2.6a] $b = \dfrac{At}{4}$ **33.** [5.8a] $m = \dfrac{tn}{t + n}$ **34.** [9.4a] $A = \pi r^2$

35. [9.4a] $x = \dfrac{b + \sqrt{b^2 + 4ay}}{2a}$ **36.** [3.1d] $\dfrac{1}{x^4}$

37. [3.1e] y^7 **38.** [3.2a, b] $4y^{12}$ **39.** [3.3f] $10x^3 + 3x - 3$

40. [3.4a] $7x^3 - 2x^2 + 4x - 17$ **41.** [3.4c] $8x^2 - 4x - 6$

42. [3.5b] $-8y^4 + 6y^3 - 2y^2$ **43.** [3.5d] $6t^3 - 17t^2 + 16t - 6$ **44.** [3.6b] $t^2 - \frac{1}{16}$ **45.** [3.6c] $9m^2 - 12m + 4$

46. [3.7e] $15x^2y^3 + x^2y^2 + 5xy^2 + 7$ **47.** [3.7f] $x^4 - 0.04y^2$

48. [3.7f] $9p^2 + 24pq^2 + 16q^4$ **49.** [5.1d] $\dfrac{2}{x + 3}$

50. [5.2b] $\dfrac{3a(a - 1)}{2(a + 1)}$ **51.** [5.4a] $\dfrac{27x - 4}{5x(3x - 1)}$

52. [5.5a] $\dfrac{-x^2 + x + 2}{(x + 4)(x - 4)(x - 5)}$ **53.** [4.1b] $4x(2x - 1)$
54. [4.5d] $(5x - 2)(5x + 2)$
55. [4.3a], [4.4a] $(3y + 2)(2y - 3)$ **56.** [4.5b] $(m - 4)^2$
57. [4.1c] $(x^2 - 5)(x - 8)$
58. [4.6a] $3(a^2 + 6)(a + 2)(a - 2)$
59. [4.5d] $(4x^2 + 1)(2x + 1)(2x - 1)$
60. [4.5d] $(7ab + 2)(7ab - 2)$ **61.** [4.5b] $(3x + 5y)^2$
62. [4.1c] $(2a + d)(c - 3b)$
63. [4.3a], [4.4a] $(5x - 2y)(3x + 4y)$ **64.** [5.9a] $-\frac{42}{5}$
65. [8.1a] 7 **66.** [8.1a] -25 **67.** [8.1f] $8x$
68. [8.2c] $\sqrt{a^2 - b^2}$ **69.** [8.2c] $8c^2b\sqrt{3ab}$
70. [8.2a] $5\sqrt{6}$ **71.** [8.2b] $9xy\sqrt{3x}$ **72.** [8.3b] $\dfrac{10}{9}$
73. [8.3b] $\dfrac{8}{x}$ **74.** [8.4a] $12\sqrt{3}$
75. [8.3a, c] $\dfrac{2\sqrt{10}}{5}$ **76.** [8.6a] 40

77. [6.2b]

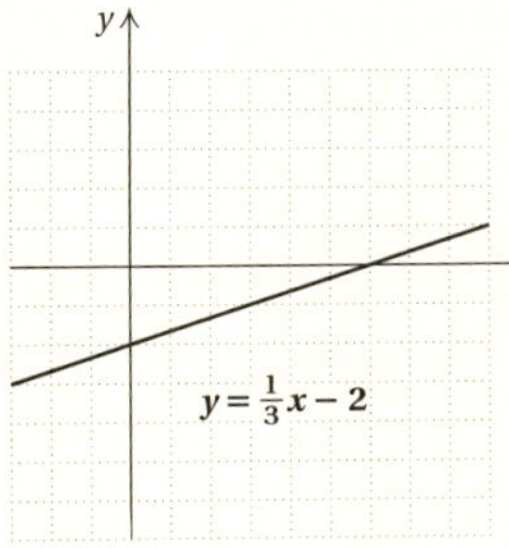

78. [6.3a]

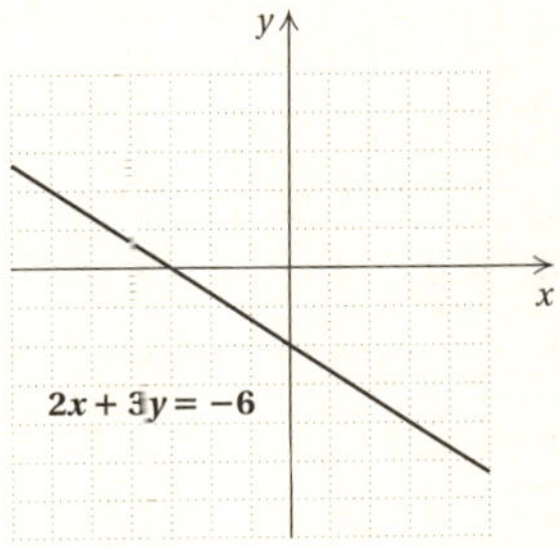

79. [6.3b]

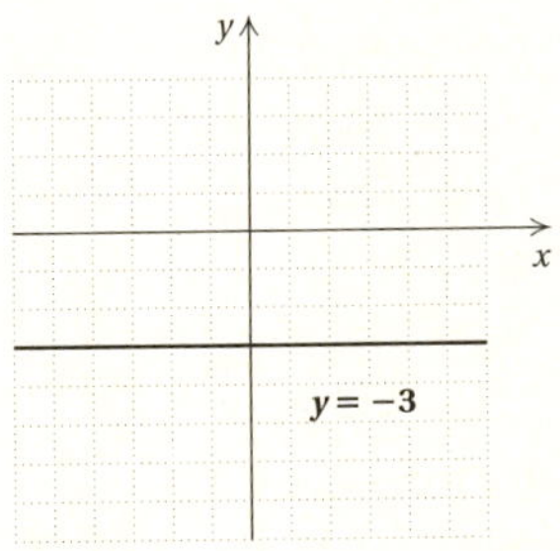

80. [6.6b]

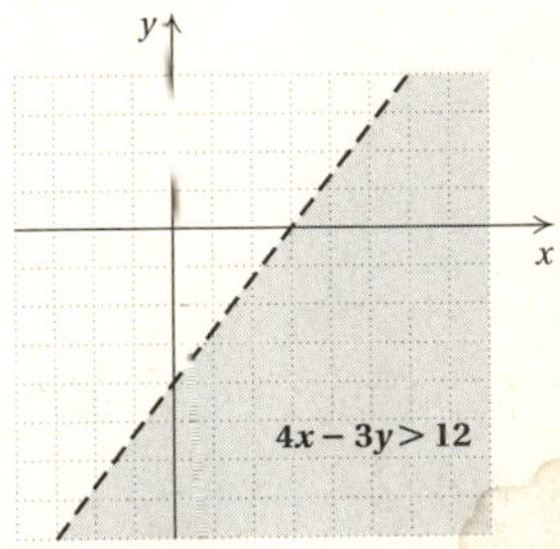

81. [9.6a]

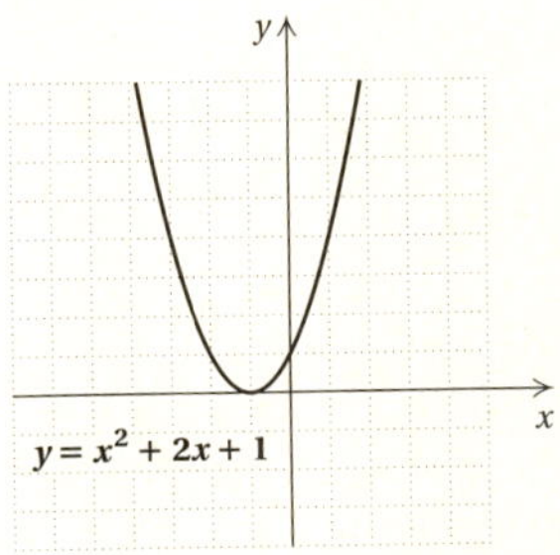

82. [6.6b]

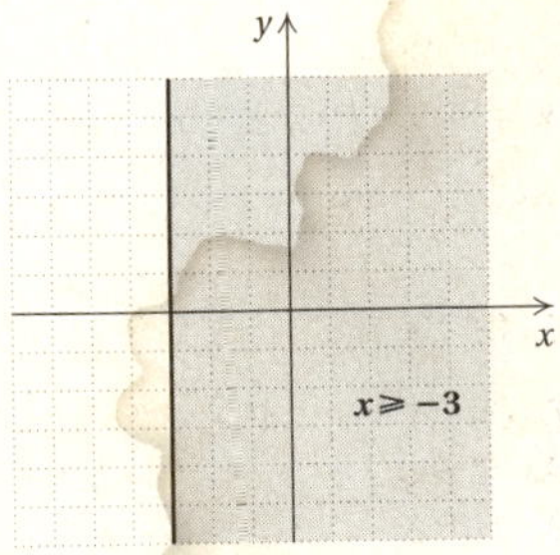

83. [9.2c] $\dfrac{2 \pm \sqrt{6}}{3}$ **84.** [9.3b] -0.2, 1.2 **85.** [2.5a] 25%
86. [2.5a] 60 **87.** [9.5a] 2 km/h **88.** [4.8a] 12 m
89. [2.4a] 6090 **90.** [7.4a] 14 lb of \$1.10; 23 lb of \$0.80
91. [5.7a] $4\frac{4}{9}$ hr **92.** [6.7b] \$451.20; variation constant is the amount earned per hour **93.** [6.5a, b] Neither
94. [6.4c] 2, $(0, -8)$ **95.** [6.4a] 15 **96.** [6.7a] $y = 10x$
97. [6.7c] $y = \dfrac{1000}{x}$ **98.** [1.2e] 12, -12 **99.** [8.1a] $\sqrt{3}$
100. [9.2c] -30, 30 **101.** [8.6a] $\dfrac{\sqrt{6}}{3}$ **102.** [4.5d] Yes
103. [5.1c] No **104.** [3.6c] No **105.** [4.5a], [8.2a] No
106. [8.1f] Yes

FINAL EXAMINATION, P. 603

1. [3.1c] -995 **2.** [5.3a] 48 **3.** [1.2e] 9 **4.** [1.3a] -9.7
5. [1.4a] -5 **6.** [1.5a] $-\frac{6}{7}$ **7.** [1.8b] $19y - 45$
8. [1.8d] 211 **9.** [2.1b] 5.6 **10.** [2.2a] -7
11. [2.3b] -10 **12.** [2.3c] 7 **13.** [4.7b] 6, -4
14. [7.2a] $(4, -3)$ **15.** [7.3b] $(4, 7)$ **16.** [9.1c] $\frac{1}{4}$, $-\frac{1}{2}$
17. [8.5a] 6 **18.** [9.3a] $\dfrac{-3 \pm \sqrt{37}}{2}$ **19.** [2.7e] $\{x \mid x \le \frac{5}{2}\}$
20. [5.8a] $w = \dfrac{1}{A - B}$ **21.** [2.6a] $M = \dfrac{K - 2}{T}$
22. [3.1e] x^{10} **23.** [3.2a] $\dfrac{1}{x^{10}}$ **24.** [3.1d] $\dfrac{1}{x^{12}}$
25. [3.3f] $6y^3 - 3y^2 - y + 9$ **26.** [3.4c] $-2x^2 - 8x + 7$
27. [3.5b] $-6t^6 - 12t^4 - 3t^2$ **28.** [3.5d] $4x^3 - 21x^2 + 13x - 2$ **29.** [3.6b] $x^2 - 64$ **30.** [3.6c] $4m^2 - 28m + 49$
31. [3.7f] $9a^2b^4 + 12ab^2c + 4c^2$ **32.** [3.7f] $9x^4 + 6x^2y - 8y^2$ **33.** [5.1d] $\dfrac{1}{x^2(x + 3)}$ **34.** [5.2b] $\dfrac{3x^4(x - 1)}{4}$
35. [5.4a] $\dfrac{11x - 1}{4x(3x - 1)}$ **36.** [5.5a] $\dfrac{2x + 4}{(x - 3)(x + 1)}$
37. [4.1b] $3x(x^2 - 5)$ **38.** [4.5d] $(4x + 5)(4x - 5)$
39. [4.3a], [4.4a] $(3x - 2)(2x - 3)$ **40.** [4.5b] $(x - 5)^2$
41. [4.1c] $(2x - y)(a + 3b)$
42. [4.5d] $(x^4 + 9y^2)(x^2 - 3y)(x^2 + 3y)$ **43.** [8.2a] $6\sqrt{2}$
44. [8.3a, c] $\dfrac{\sqrt{30}}{5}$ **45.** [8.4a] $13\sqrt{2}$
46. [8.2c] $2a^2b\sqrt{6ab}$

47. [6.3a]

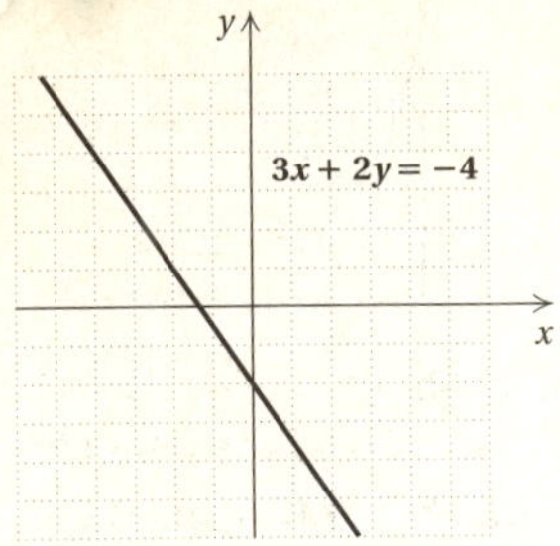

48. [6.3b]

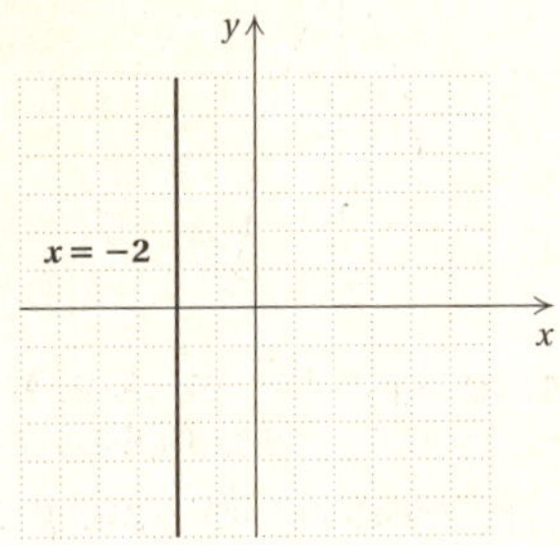

49. [6.6b]

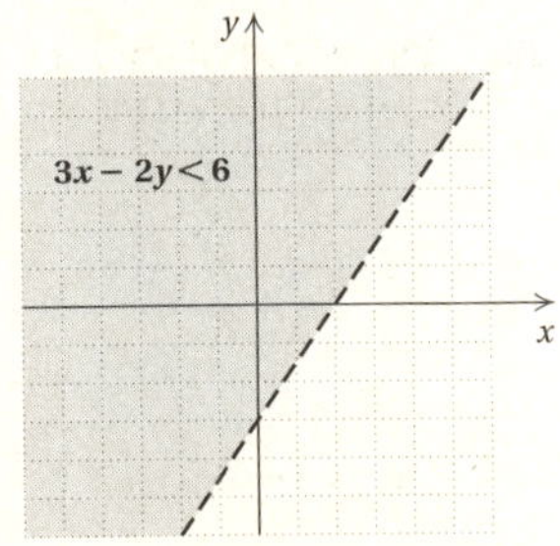

50. [9.6a]

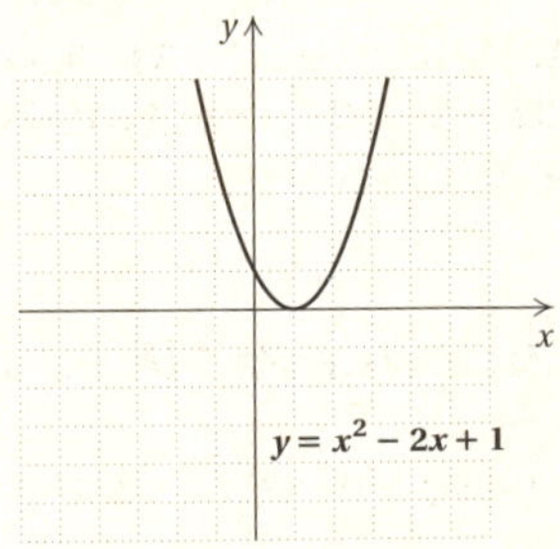

51. [4.8a] 5 and 7; −7 and −5
52. [7.4a] 40 L of A; 20 L of B **53.** [7.5a] 70 km/h
54. [9.5a] Length: 11 m; width: 8 m **55.** [6.4a] $-\frac{4}{3}$
56. [6.5b] Perpendicular **57.** [6.7a] $y = 8x$
58. [6.7c] $y = \dfrac{5000}{x}$ **59.** [2.4a] Square: 15; triangle: 20
60. [9.2c] 144

Appendix: Using a Scientific or Graphing Calculator, p. C–1

1. 6561 **2.** 3375 **3.** 529 **4.** 1024 **5.** 2401
6. 10.4976 **7.** 0.00243 **8.** 81 **9.** 118 **10.** 39 **11.** 71
12. 450 **13.** 100 **14.** 3 **15.** $\frac{13}{24}$ **16.** $\frac{31}{60}$ **17.** $13\frac{1}{12}$, or $\frac{157}{12}$ **18.** $\frac{1}{110}$ **19.** 2 **20.** 2.775 **21.** 1.625 **22.** 0.3125
23. 0.01 **24.** 13 **25.** 250 **26.** 0.042 **27.** 3.317
28. 16.432 **29.** 2.530 **30.** 0.095 **31.** 8 **32.** 16
33. 36 **34.** 153 **35.** −11 **36.** 9 **37.** 114 **38.** Yes
39. No **40.** Yes **41.** No **42.** Yes **43.** −10 **44.** 91
45. 90 **46.** −9 **47.** 30
48. 8.3 09, or 8.3⁰⁹
49. 2.591 −04, or 2.591⁻⁰⁴
50. 4.77 23, or 4.77²³

51. 5.6×10^{16}, or 56,000,000,000,000,000
52. 2×10^{-11}, or 0.00000000002 **53.** 2.6×10^{-4}, or 0.00026 **54.** 3.801×10^{9}, or 3,801,000,000
55. 1.5×10^{-44}

56.
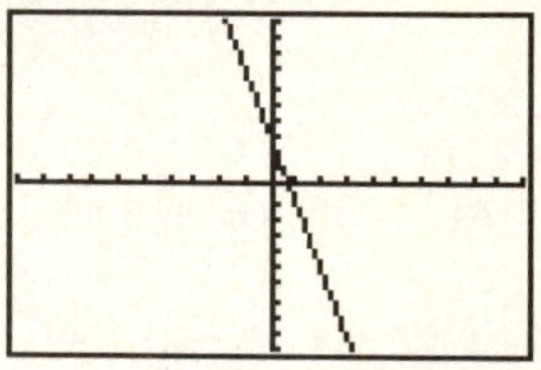

57.
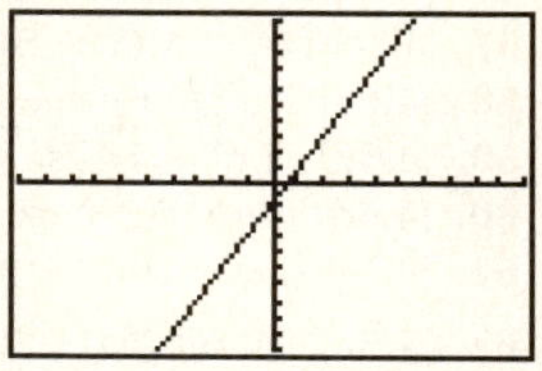

58.
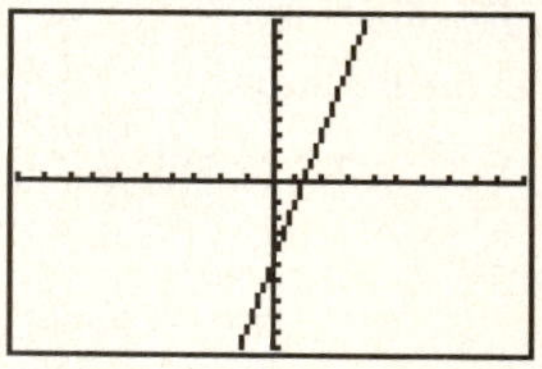

59.
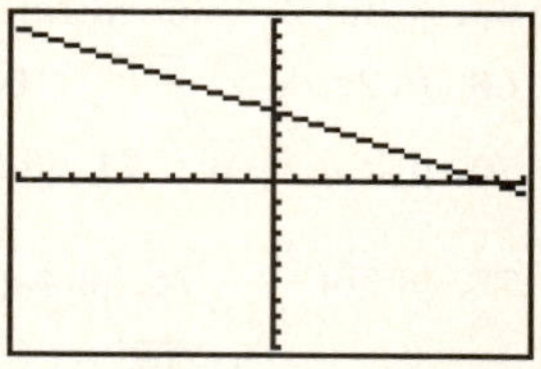

60.
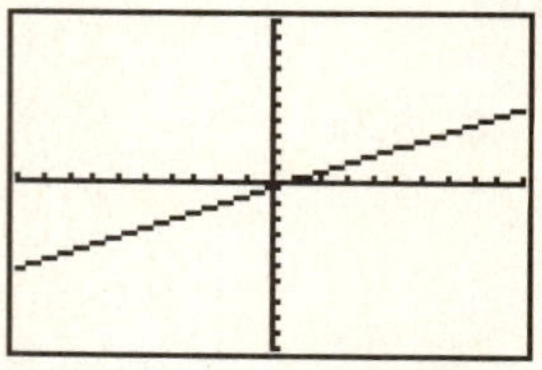

61. (−3.5, −9.5)

62. (7.0, 11.0) **63.** (1.6, 5.7) **64.** (0.9, 0.4)

Index

Q

R